Quantized
Partial Differential Equations

Quantized
Partial Differential Equations

$$(\partial \dot{x}.F^0)_s \nu^k + (\partial \dot{x}_k^p.F^0)_s \nu_p^k \equiv \Xi^0 = 0$$

$$(\partial \dot{x}.F^j)_s \nu^k + (\partial \dot{x}_k^p.F^j)_s \nu_p^k + (\partial \dot{x}_k^0.F^j)_s \nu_0^k$$

$$(\partial \dot{x}_k^{ir}.F^j)_s \nu_{ir}^k + (\partial p^i.F^j)_s \nu_i^{(p)} \equiv \Xi^j = 0,$$

$$(\partial \dot{x}.F^4)_s \nu^k + (\partial \dot{x}_k^p.F^4)_s \nu_p^k + (\partial \theta^0.F^4)_s \nu_0^{(\theta)}$$

$$(\partial \theta^k.F^4)_s \nu_k^{(\theta)} + (\partial \theta^{ir}.F^4)_s \nu_{ir}^{(\theta)} \equiv \Xi^4 = 0$$

$$\leq 3$$

A Prástaro

University of Roma "La Sapienza", Italy

World Scientific

NEW JERSEY · LONDON · SINGAPORE · SHANGHAI · HONG KONG · TAIPEI · CHENNAI

Published by

World Scientific Publishing Co. Pte. Ltd.

5 Toh Tuck Link, Singapore 596224

USA office: Suite 202, 1060 Main Street, River Edge, NJ 07661

UK office: 57 Shelton Street, Covent Garden, London WC2H 9HE

British Library Cataloguing-in-Publication Data
A catalogue record for this book is available from the British Library.

QUANTIZED PARTIAL DIFFERENTIAL EQUATIONS

ISBN 981-238-764-1

Printed in Singapore by World Scientific Printers (S) Pte Ltd

To my family:
Paola, Francesco and Alessandro

PREFACE

A geometric theory of (super) PDE's for the quantum physics is necessarily a theory of noncommutative (super)manifolds. In some previous works we have introduced *quantum (super)manifolds* that appear very useful to globalize the concept of algebras of quantum operators. For such manifolds we have developed also a geometric theory of partial differential equations obtaining theorems of existence of local and global solutions [61–67,69–71]. In some other previous papers we have given also a general proceeding to quantize PDE's [58–63,70]. In particular we have given geometric methods to obtain quantizations as covariant and canonical quantization of PDE's, that are the most useful from the physical point of view. These methods are founded on the geometric properties of (super) PDE's. In this work we further develop our theory of quantized (super) PDE's and quantum PDE's and explicitly prove the relation between them by means of unifying theorems that show as covariant quantizations of (super) PDE's identify quantum (super) PDE's in algebraic sense.

A new direction where Mathematics has accumulated important energies and where we can foresee great results in the next future, is the noncommutative geometry. Really, this is a relatively new area of research in Mathematics, where a central role has been played by A. Connes with his works published starting from fifteen years ago. But we must include, also, in this direction the fundamental works by the Bourbaki's school, the pioneering algebraic-geometric results by the Rome's school (C. Procesi), Moscow's school (Yu. Manin), Polish's school and etc. In some sense A. Connes has built a bridge between geometry and functional analysis, obtaining important results solving interesting geometric problems. These focused, also, attention by theoretical physicists interested to have mathematical tools to utilize in their models for quantum physics.

Furthermore, a new direction of research rised in the last seven years in the framework of noncommutative geometry by us. This new approach develop a noncommutative differential geometry that uses a point of view nearer to the classical differential geometry. In fact, it is formulated on pseudogroups and produces a geometric theory of noncommutative PDE's, extending in a natural way, the great results of the commutative geometric theory of PDE's. This language is more natural to solve problems in the modern quantum field theory, where the central role is played by the geometric structures defining the dynamic field equations. So this new noncommutative geometry offers a new language to solve fundamental problems in physics. As such problems are of central interest for the scientific community, it is expected that, as a push-pull system, all the noncommutative geometry should be more considered by the mathematical community. Let us emphasize that our geometric theory of quantum PDE's is the natural development, in noncommutative sense, of the geometric theory of PDE's. Such a theory, just starting from the pioneering works by the European-American school (E. Cartan, H. Goldschmidt, V. Guillemin, M. Kuraniski, B. Malgrange, D. Spencer, S. Stenberg, W. J. Sweeny,...) is the unique approach that has obtained very general theorems and great results in the theory of PDE's. The actual soldering of the Cartan's theory with the formal theory, (obtained principally by the Moscow's school), from one side and the more recent results (by A.Prástaro) extending algebraic-topological points of views, ((co)bordism groups), to the previous geometric theory of PDE's, from another side, have given to us the opportunity to obtain very general local and global informations on the geometric structure of PDE's and their solutions. These results should also convince analysts that geometric methods are fundamental also in the analysis of PDE's, both commutative or noncommutative. In fact, are just these geometric methods that have given to us the opportunity to solve very crucial open problems in the theory of PDE's. Such are, for example, the existence of global (smooth) solutions for the (quantum) Navier-Stokes equation and the (quantum) (super) Yang-Mills equation. For the last equation a criterion to recognize global solutions with mass-gap were also obtained.

The aim of this work is first to report on such a new theory and them to relate it with a previous formulation of quantized PDE's developed starting from fifteen years ago. This formulation of quantization of PDE's is just founded on the geometric structure of PDE's and gives us a fully covariant description of the covariant and canonical quantizations. (These are the most interesting, from the physical point of view, between all the possible quantizations of PDE's and are shortly referred as *Dirac quantizations* also.) Note that a purely algebraic description of quantum phenomena, i.e., by simply using noncommutative algebras, is not satisfactory as, in this way, we are not more able to recover the meaning of full covariance that, instead, is fundamental for a good description of any natural geometric theory and hence of any physical theory. Our approach to quantum (super)manifolds overcomes this difficulty as uses the noncommutative algebras as a "fundamental algebra of noncommutative numbers", with which to build the models for quantum manifolds. In this way we can naturally extend the classic commutative geometry to a noncommutative context and give also a natural noncommutative meaning of full covariance. We explicitly prove that Dirac covariantly quantized PDE's identify quantum PDE's. So the ring is closed! We can start from a classical theory, i.e., a commutative PDE, canonically quantize it and obtain a quantum PDE. To this last we can apply all our results of the geometric theory of quantum PDE's and obtain a full characterization of our quantum problems.

This work is a collection of three long papers (part. I, part. II, part. III) devoted to the geometric theory of noncommutative manifolds, noncommutative PDE's and quantized PDE's. The category of noncommutative manifolds is that of quantum manifolds as introduced in some our previous works. The aim is to give a formal relation between classical, quantum and quantized PDE's in a unified geometric building. Therefore, this work is divided in three parts. Part I. The first section of this part is devoted to look through some definitions and results in algebraic-topology that are used in all the book. The other three sections are reserved to the description of some suitable categories of noncommutative manifolds called *quantum manifolds* and *quantum supermanifolds* respectively. The starting points for such structures are *quantum algebras* and *quantum superalgebras* respectively. These are particular classes of topological noncommutative algebras, that are useful to locally interpret the concept of quantum fields. The globalization of such structures is made by means of *quantum (super)manifolds* that are built by means of suitable pseudogroups on suitable model spaces. In this context fundamental is the role of the class of differentiability that we call *quantum differentiability*. Then we characterize quantum (super)manifolds by their differential and algebraic-topological properties. The emphasis is put on their (co)homologic and (co)bordism groups. This is made in order to obtain suitable tools to use in the geometric theory of (quantum) PDE's. Part II. Here we develop the geometric theory of quantum (super) PDE's obtaining general theorems that allow us to state existence of local and global solutions. Global properties of solutions are characterized by means of integral (co)bordism groups of quantum (super) PDE's. Furthermore, we apply such results to some important equations of the mathematical physics. Particular attention is devoted to the quantum Navier-Stokes equation and quantum Yang-Mills equation. For such equations we obtain also theorems of existence of global solutions, solving well-known open problems under focus of the mathematical community. Part III. In the first section we consider the geometric theory of PDE's and their characterizations by means of integral (co)bordism groups, following our previous results on this subject. In section two an algebraic theory of PDE's is given also. In section three a theory of spectral measures of PDE's gives us the way to interpret the meaning of quantizations of PDE's on the ground of the mathematical logic, that is just the subject of section four. Furthermore, in section five the specialization on the covariant and canonical quantization of PDE's is made by using purely differential geometric techniques. Here we resume some our previous results on the quantizations of PDE's and give also some further results on the covariant and canonical quantizations of PDE's. In particular, we prove that Dirac covariant quantizations of (super) PDE's identify quantum (super) PDE's. Therefore, for such quantized equations our theory of quantum (super) PDE's can be applied. Many applications are given to equations describing important phenomena of quantum physics also.

Let us emphasize also that the geometric setting to obtain covariant and canonical quantizations of classical PDE's interesting field theory, realizes a bridge between noncommutative PDE's and classic PDE's. This just justifies our use of the word "quantum", according to its original physical meaning for which has been first forget.

Finally two addenda, one devoted to the relations between integral bordism groups for the Navier-Stokes equation and the solution of the fundamental problem on the existence of global smooth solutions of this important PDE (Addendum I), and another that studies the relations between integral bordism groups and constrained variational problems (Addendum II), conclude this book.

Roma, Easter 2003.

Agostino Prástaro

CONTENTS

QUANTIZED PDE's I: NONCOMMUTATIVE MANIFOLDS

Abstract - In this first part we consider *quantum (super)manifolds* as topological spaces locally identified with open sets of some locally convex topological vector spaces built starting from suitable topological algebras, *quantum (super)algebras*. The noncommutative character of such quantum (super)manifolds is given by the underlying noncommutative algebras A. In fact, here A plays the role of "fundamental algebra of numbers", like $\mathbf{K}=\mathbf{R},\mathbf{C}$ does for usual commutative manifolds. Therefore, quantum (super)manifolds are the natural generalization of manifolds, when one substitutes commutative numbers with noncommutative ones. Commutative manifolds are contained into quantum (super)manifolds, as quantum (super)algebras A are required to contain $\mathbf{K}$. This aspect is also reflected by the fact that the class of differentiability Q_w^k for pseudogroup structures defining quantum (super)manifolds contains the usual C^k differentiability for manifolds. In fact, the class of differentiability of such topological manifolds is defined by requireing weak differentiability and Z-linearity of the derivatives, where Z is the centre of the underlying quantum (super)algebras. We give (co)homological characterizations of quantum (super)algebras and quantum (super)manifolds by applying to these noncommutative topological manifolds standard methods of algebraic topology. In particular, we calculate also (co)bordism groups in quantum (super)submanifolds.

1.1 - ALGEBRAIC TOPOLOGY

As in this work we systematically use the language and tools of the algebraic topology, and even if we suppose the reader just introduced in this subject, we consider suitable to start with some fundamental definition and results on the algebraic topology, emphasizing those that will be used in the following. (See also, e.g., refs.[17,77,81–84].)

In all this work we will denote by $\mathbf{K}$ be the field of real numbers $\mathbf{R}$ or complex numbers $\mathbf{C}$. Moreover, we denote by $\mathbf{N}$ the set of natural numbers and by $\mathbf{Z}$ and $\mathbf{Q}$ the fields of rational integers and rational numbers respectively. We will denote by $\mathbf{H}$ and $\mathbf{O}$ respectively the algebras of quaternions and octonions.

Here $\bigcup$ denotes disjoint union and I the segment $I \equiv [0,1] \subset \mathbf{R}$. We will work in the category $\mathcal{T}_{op}$ of topological spaces and continuous maps. For any $X \in Ob(\mathcal{T}_{op})$, we set $X^+ \equiv X/\emptyset = X\bigcup\{*\}$, where $*$ denotes a point. Therefore, X^+ is the *compactified to a point* of the topological space X. Two topological spaces X and Y that are homeomorphic will be denoted by $X \cong Y$. We shall, instead, write $X \simeq Y$ if they are homotopy equivalent, i.e., there exist two maps $f : X \to Y$, $g : Y \to X$, such that $g \circ f \simeq 1_X$ and $f \circ g \simeq 1_Y$, i.e., there exist continuous mappings (homotopies) $F : X \times I \to Y$ and $G : Y \times I \to X$, such that $F|_{X \times \{0\}} = g \circ f$, $F|_{X \times \{1\}} = 1_X$, $G|_{Y \times \{0\}} = f \circ g$, $G|_{Y \times \{1\}} = 1_Y$. The set of all homotopy classes of maps $X \to Y$ is denoted by $[X,Y]$. If $A \subset X$ is a subspace, a homotopy $F : X \times I \to Y$ *relative* to A is such that $F(a,t) = F(a,0)$, $\forall a \in A$, $t \in I$. If $F_0 = f$ and $F_1 = g$, we write $f \simeq_A g$. We say also that F is a *homotopy rel* A. In the following table we resume some fundamental operations in algebraic topology.

TAB.1.1 - Fundamental opertions in algebraic topology

Name	Definition
cone of X	$CX \equiv X \times I/X \times \{1\}$
suspension of X	$SX \equiv X \times I/\{X \times \{0\}\}\bigcup\{X \times \{1\}\}$
wedge product of (X,a) and (Y,b)	$X \vee Y \equiv X \times \{b\}\bigcup/Y \times \{a\}$
smash product of X and Y	$X \wedge Y \equiv X \times Y/(X \vee Y)$

One has the natural equivalence: $[SX, *; Y, y_0] \leftrightarrow [X, x_0; \Omega Y, \omega_0]$, where ΩY is the loop space of Y. This is an isomorphism of groups. One has the following one-to-one correspondence: $[CX, *; Y, y_0] \leftrightarrow [X, x_0; PY, \omega_0]$, where $PY \equiv (Y, y_0)^{(I,0)}$. The map $p : PY \to Y$, $p(\omega) = \omega(1)$ is continuous. One has $p^{-1}(y_0) = \Omega Y$.

Definition 1.1. For any pointed topological space (X, x_0) we define *nth homotopy group* (or *nth Hurewicz group*) $\pi_n(X, x_0) = \pi_0(\Omega^n X, \omega_0)$.

Proposition 1.1. *One has the following isomorphisms:*

$$\pi_n(X, x_0) \cong [S^1, s_0; \Omega^{n-1}, \omega_0] \cong [S^2, s_0; \Omega^{n-2}, \omega_0] \cong \cdots \cong [S^n, s_0; X, x_0].$$

A pointed space (X, x_0) *is called* contractible *if it has the same homotopy type of* $(\{x_0\}, x_0)$. *Then, if* (X, x_0) *is contractible, then* $\pi_n(X, x_0) = 0$, $n \geq 0$.

Example 1.1. (D^n, x) and $(\mathbf{R}^n, x)$ are examples of contractible pointed spaces. Instead S^n is not contractible. In fact, one has: $\pi_r(S^n, s_0) = 0$, for $r < n$, $\pi_r(S^n, s_0) = \mathbf{Z}$, for $r = n$ $(n \geq 1)$. (For $r > n$ the homotopy groups are not known in general.) $\blacksquare$

Definition 1.2. The sequence of pointed sets $(A, a_0) \xrightarrow{f} (B, b_0) \xrightarrow{g} (C, c_0)$ is called *exact* if $\operatorname{im}(f) = g^{-1}(c_0)$. (If above sequence is in the category of groups, and the fixed points are the unities, then above definition coincides with one for exact sequence of group-homomorphisms.) In the category of pointed topological spaces, above sequence is called *exact*, (resp. *coexact*), if for every pointed topological space (W, w_0), the following sequence of pointed sets:

$$[(W, w_0); (A, a_0)] \xrightarrow{\;f_*\;} [(W, w_0); (B, b_0)] \xrightarrow{\;g_*\;} [(W, w_0); (C, c_0)],$$

(resp.

$$[(A, a_0); (W, w_0)] \xleftarrow{\;f^*\;} [(B, b_0); (W, w_0)] \xleftarrow{\;g^*\;} [(C, c_0); (W, w_0)]),$$

is exact.

Definition 1.3. The *nth relative homotopy* set $\pi_n(X, A, x_0)$, $A \subset X$, $x_0 \in A$, is defined by

$$\pi_n(X, A, x_0) = \pi_{n-1}(P(A; x_0, A), \omega_0) = \pi_0(\Omega^{n-1} P(X; x_0, A), \omega_0),$$

where $P(X; x_0, A) \equiv (X, x_0, A)^{(I, 0, 1)}$.

Proposition 1.2. *One has the following exact sequence (exact homotopy sequence of (X, A, x_0)):*

(1.1)
$$\cdots \to \pi_{n+1}(X, A, x_0) \xrightarrow{\partial} \pi_n(A, x_0) \xrightarrow{i_*} \pi_n(X, x_0) \xrightarrow{j_*} \pi_n(X, A, x_0) \xrightarrow{\partial} \cdots$$

$$\pi_1(X, A, x_0) \xrightarrow{\partial} \pi_0(A, x_0) \xrightarrow{i_*} \pi_0(X, x_0)$$

One has the one-to-one correspondence:

$$[CX, X, *; Y, B, y_0] \leftrightarrow [X, x_0; P(Y; y_0, B), \omega_0].$$

Furthermore, one has the following isomorphisms:

$$\pi_n(X, A, x_0) = \pi_{n-1}(P(X; x_0, A), \omega_0) \cong [S^{n-1}, s_0; P(X; x_0, A), \omega_0]$$
$$\cong [CS^{n-1}, S^{n-1}, *; X, A, x_0].$$

Definition 1.4. If $A \subset X$ is a subspace and $i : A \to X$ is the inclusion map, then A is called a retract of X if there is a map $r : X \to A$ with $r \circ i = 1_A$.

Definition 1.5. If $A \subset X$ is a retract and if in addition $i \circ r \simeq 1_X$, then A is called a *deformation retract* of X. If $i \circ r \simeq_A 1_X$, then A is called a *strong deformation retract* of X. The homotopy $H : X \times I \to X$ such that $H_0 = 1_X$, $H_1(X) \subset A$, is called the *deformation* of X into A.

Proposition 1.3. *If $A \subset X$ is a retract of X then we have*

$$\pi_n(X, x_0) \cong \operatorname{im}(i_*) \oplus \ker(r_*) \cong \pi_n(A, x_0) \oplus \ker(r_*).$$

In particular, for $n = 1$, $\ker(r_*)$ is a normal subgroup of $\pi_1(X, x_0)$ and $\alpha \in \pi_1(X, x_0)$ can be written uniquely as $\alpha'.\alpha''$ with $\alpha' \in \operatorname{im}(i_*)$, $\alpha'' \in \ker(r_*)$. $\pi_1(X, x_0)$ is the so-called semi-direct product of $\operatorname{im}(i_*)$ with $\ker(r_*)$. If $A \subset X$ is a deformation retract of X, $x_0 \in A$ and H is homotopy rel x_0 (which will certainly be true for strong deformation retract) then the inclusion $i : (A, x_0) \to (X, x_0)$ is a homotopy equivalence and hence $i_* : \pi_n(A, x_0) \to \pi_n(X, x_0)$ is an isomorphism for all $n \geq 0$. Furthermore, (X, x_0) is contractible iff $\{x_0\}$ is a strong deformation retract of X.

Proof. In fact one has the following split short exact sequence:

$$0 \to \pi_n(A, x_0) \overset{i_*}{\underset{r_*}{\rightleftarrows}} \pi_n(X, x_0) \to \pi_n(X, A.x_0) \to 0.$$

$\square$

Definition 1.6. A pair (X, A) is called 0-*connected* if every path component of X meets A. (X, A) is called n-*connected* iff $\pi_k(X, x) = 0$, for $0 \leq k \leq n$ and all $x \in A$.

Proposition 1.4. *A pair (X, A) is n-connected, $n \geq 0$, iff $i_* : \pi_r(A, x_0) \to \pi_r(X, x_0)$ is a bijection for $r < n$ and a surjection for $r = n$, all $x_0 \in A$. In particular (X, X) is n-connected, all $n \geq 0$.*

Proposition 1.5. *(X, A) is n-connected iff every map $f : (D^n, S^{n-1}) \to (X, A)$ is homotopic rel S^{k-1} to a map into A, $0 \leq k \leq n$ (where $D^0 = *$, $S^{-1} = \emptyset$).*

Definition 1.7. A topological space $(X, \mathcal{T})$ is *nonconnected* if $X = G \bigcup H$, where $G, H \in \mathcal{T}$, otherwise it is *connected*.

Remark 1.1. This definition is not equivalent to that of *arcwise connected*. In fact a space can be connected but not arcwise connected, as can be seen in the following well-known example. $\blacksquare$

Example 1.2. Let us consider the following subsets of $\mathbf{R}^2$: $A \equiv \{(0,y) : \frac{1}{2} \leq y \leq 1\}$, $B \equiv \{(x,y) : y = \sin(\frac{1}{x}), 0 < x \leq 1\}$. As each point of A is accumulation point of B, it follows that $A \bigcup B$ is a connected space, but it is not surely arcwise connected. $\blacksquare$

Definition 1.8. Let $X_0 \subset X$ be a subspace of X, let Y be another topological space and $f : X_0 \to Y$ a continuous map. We set $X \bigcup_f Y \equiv X \bigcup Y/ \sim$, where the equivalence relation $\sim$ is defined by

$$x \sim y \Leftrightarrow \{x = y \text{ if } x \notin X_0, y = f(x), \text{ if } x \in X_0\}.$$

We say that $X \bigcup_f Y$ is obtained by *attaching* (or *gluing*) X to Y by means of the *attaching map* f. Y and $X \setminus X_0$ are canonical subspaces of $X \bigcup_f Y$. This is not true for the attached part X, but can be changed by the canonical continuous map $X \subset X \bigcup Y \to X \bigcup_f Y$. The *mapping cone* is $Y \bigcup_f CX$. In particular, if $X = S^{n-1}$ then $Y \bigcup_f CS^{n-1}$ is called Y *with an n-cell attached*. The projection $q : Y \vee CS^{n-1} \to Y \bigcup_f CS^{n-1}$ restricts to give a map $g : CS^{n-1} \to Y \bigcup_f CS^{n-1}$ with on the interior of CS^{n-1} is a homeomorphism. g is called the *characteristic map of the cell*. Since $CS^{n-1} \simeq D^n$, where D^n is the n-dimensional disk, we may regard g as a map $g : (D^n, S^{n-1}) \to (Y \bigcup_f CS^{n-1}, Y)$. Note that $g|_{S^{n-1}} = f$. More precisely if we have a map $f : \bigvee_\alpha S_\alpha^{n-1} \to Y$ of many $(n-1)$-spheres into Y, then the mapping cone $Y \bigcup_f C(\bigvee_\alpha S^{n-1}) = Y \bigcup_f \bigvee_\alpha (CS_\alpha^{n-1})$ is Y with the n-cells e_α^n attached and its attaching map is $f|_{S_\alpha^{n-1}}$. The characteristic map g_α^n of e_α^n is $g|_{CS_\alpha^{n-1}}$. Attaching 0-cell will mean adding a disjoint point. A *cell space* is a topological space obtained from a finite set of points (i.e., from a finite discrete space) by iterating the procedure of attaching cells of arbitrary dimensions, with the condition that only finitely many cells of each dimension are attached. The initial discrete points may be regarded as 0-dimensional cells. A *cell complex (CW-complex)* is a cell space X where each cell is attached to cells of lower dimension. The subcomplex of all cells of dimension $k \leq n$ is called the *n-skeleton X_n of X*. Then a CW-complex X is the union of the ascending chain of its n-skeleton: $X_0 \subset X_1 \subset \cdots \subset X_n \subset \cdots \subset X$.[1]

---

[1] Note that there is another concept of homotopic maps that just refers to CW-complexes. In fact, we say that two maps $f,g : X \to Y$ are *weakly homotopic* if for any finite CW-complex Z and any map $h : Z \to X$, we have $f \circ h \simeq g \circ h$. We write $f \simeq_w g$. $\simeq_w$ is an equivalence relation

Proposition 1.6. 1) *If $A \subset X$ is a subspace then $X \bigcup_i CA/CA$ is homeomorphic to X/A. Here $i : A \to X$ is the inclusion.*[2]

2) For any map $f : (X, x_0) \to (Y, y_0)$ the sequence

$$(X, x_0) \xrightarrow{f} (Y, y_0) \xrightarrow{j} (Y \bigcup_f CX, *) \xrightarrow{\kappa'} (SX, *) \xrightarrow{Sf} (SY, *) \xrightarrow{Sj} \cdots \to$$

$$(S^n X, *) \xrightarrow{S^n f} (S^n Y, *) \xrightarrow{S^n j} (S^n(Y \bigcup_f CX), *) \xrightarrow{S^n \kappa'} \cdots$$

is coexact. Furthermore there is an action of the group $[SX, *; W, w_0]$ on the set $[Y \bigcup_f CX, *; W, w_0]$ such that the set $f_*^{-1}(x)$, $x \in [Y, y_0; W, w_0]$, are precisely the orbits.

Definition 1.9. Given any map $f : (X, x_0) \to (Y, y_0)$ the *reduced mapping cylinder* $\widetilde{M}_f$ of f is the following space: $\widetilde{M}_f \equiv \{(I \times X/I \times \{x_0\}) \bigvee Y\}/ \sim$, where $\sim$ is obtained from $(I \times X)/I \times \{x_0\}) \bigvee Y$ by identifying $[1, x] \in I \times X/I \times \{x_0\}$ with $f(x) \in Y$, for every $x \in X$.

Proposition 1.7. *One has the canonical projection $q : (I \times X/I \times \{x_0\}) \bigvee Y \to \widetilde{M}_f$. We set $q(t, x) \equiv [t, x]$ and $q(y) \equiv [y]$. One has the following mappings:*

$$\begin{aligned}
i &: (X, x_0) \to (\widetilde{M}_f, *), \; i(x) = [0, x], \\
j &: (Y, y_0) \to (\widetilde{M}_f, *), \; j(y) = [y], \\
r &: (\widetilde{M}_f, *) \to (Y, y_0), \; r([y]) = y, \; r[s, x] = f(x), \\
H &: \widetilde{M}_f \times I \to \widetilde{M}_f, \; H([y], t) = [y], \; H([s, x], t) = [t + s - st, x].
\end{aligned}$$

These satisfy the following relations: $r \circ i = f$, $r \circ j = 1_Y$, $H_0 = 1_{\widetilde{M}_f}$, $H_1 = j \circ r$. r and j are homotopy equivalences. i is a homeomorphism of X onto $i(X) \subset \widetilde{M}_f$. Therefore X can be considered as a subspace of $\widetilde{M}_f$. One has $\widetilde{M}_f/X \cong Y \bigcup_f CX$. One has the following exact sequence (exact homotopy sequence):

(1.2)
$$\cdots \to \pi_{r+1}(\widetilde{M}_f, X, *) \xrightarrow{\partial} \pi_r(X, x_0) \xrightarrow{f_*} \pi_r(Y, y_0) \xrightarrow{\kappa_* \circ j_*} \pi_r(\widetilde{M}_f, X, *) \xrightarrow{\partial} \cdots$$

With this sequence, statements about when f_* is an isomorphism can be converted into statements about the connectivity of the pair $(\widetilde{M}_f, X, *)$.

---

and we write $[X,Y]_w$ or $[X,x_0;Y,y_0]_w$ for the set of all weak homotopy equivalence classes of maps $X \to Y$ or $(X,x_0) \to (Y,y_0)$.

[2] Note that every theorem we state about X/A will hold well for $X^+ \equiv X/\emptyset$.

Definition 1.10. If we are working in $\mathcal{T}_{op}$ rather than in the category $\mathcal{T}_{top,\bullet}$ of pointed topological spaces, then for $f : X \to Y$ we define *unreduced mapping cylinder* the space $M_f \equiv (I \times X) \bigcup Y/ \sim$, with $[1,x] \sim f(x)$, all $x \in X$.

Definition 1.11. A map $f : X \to Y$ is called an n-equivalence, $n \geq 0$, if for all $x_0 \in X$, $f_* : \pi_r(X,x_0) \to \pi_r(Y,y_0)$ is a bijection for $r < n$ and a surjection for $r = n$. f is called a *weak homotopy equivalence* if it is a n-equivalance for all $n \geq 0$. (Every homotopy equivalence in $\mathcal{T}_{op}$ is a weak homotopy equivalence.)

Proposition 1.8. *A map* $f : X \to Y$ *is an n-equivalence, (resp. weak homotopy equivalence), iff for all* $x_0 \in X$, $(\widetilde{M}_f, X)$ *is 0-connected and* $\pi_r(\widetilde{M}_f, X, *) = 0$, *for* $1 \leq r \leq n$, *(resp. for* $1 \leq r$).

Proposition 1.9. *For any pair* (X, A), *if A is 0-connected, then* $\pi_n(X, A, x_0) \cong \pi_n(X, A, x)$, $\forall x_0, x \in A$, $n \geq 1$. *If X is not 0-connected* $\pi_n(X, x_0) \not\cong \pi_n(X, x)$, *in general. 0-connectedness, however, is sufficient in order to have* $\pi_n(X, x_0) \cong \pi_n(X, x)$.

Example 1.3. $\pi_1(S^1, s_0) \cong \mathbf{Z}$. Set $X = S^1 \bigcup \{x_0\}$. Since S^1 is 0-connected, it follows that for every continuous map $f : (S^1, s^0) \to (X, y)$, $y \in X$, $f(S^1)$ must lie in the 0-component of y. Therefore $\pi_1(X, s^0) \cong \pi_1(S^1, s^0) \cong \mathbf{Z}$, but $\pi_1(X, x_0) \cong \pi_1(\{x_0\}, x_0) = 0$. $\blacksquare$

Proposition 1.10. (X, x_0) *can have the same homotopy type as* (Y, y_0) *only if the following conditions are verified:*

(i) $\pi_1(X, x_0) \cong \pi_1(Y, y_0)$.

(ii) $\pi_n(X, x_0)$ *and* $\pi_n(Y, y_0)$ *are isomorphic as modules over* $\mathbf{Z}[\pi_1(X, x_0)]$ *for* $n \geq 2$. *(This is a stronger condition than merely requering that* $\pi_n(X, x_0)$ *and* $\pi_n(Y, y_0)$ *are isomorphic as groups for all* $n \geq 0$.)

Example 1.4. For the unreduced mapping cylinder M_f of a map $f : X \to Y$ the retraction $r : M_f \to Y$ induces an isomorphism $r_* : \pi_q(M_f, x_0) \to \pi_q(Y, y_0)$, $q \geq 0$. Then in the induced exact sequence (1.2) we can replace $\widetilde{M}_f$ with M_f. $\blacksquare$

Example 1.5. If $f : X \to Y$ is a homotopy equivalence in $\mathcal{T}_{op}$, then $f_* : \pi_n(X, x) \to \pi_n(Y, f(x))$ is an isomorphism for all $n \geq 0$, all $x \in X$. $\blacksquare$

Proposition 1.11. (CW-substitute for topological spaces). *1) For any topological space Y one can construct a CW-complex Y' and a weak homotopy equivalence $f : Y' \to Y$. The couple (f, Y') is called a CW-substitute for Y. Any two CW-substitute for Y are homotopy equivalent by an equivalence* *which is unique up to homotopy.*

2) Given a map $h : X \to Y$ of topological spaces, CW-substitutes (X', f), (Y', g) for X and Y respectively, we can find a cellular map $h' : X' \to Y'$ so that $g \circ h' \simeq h \circ f$ and h' is unique up to homotopy.

Example 1.6. A *graph* is a 0-connected 1-dimensional CW-complex.   ∎

Example 1.7. A *tree* is a graph which contains no subcomplexes homeomorphic to S^1. Tree are contractible. Every graph contains a maximal tree. For any graph X with maximal tree T we have $\pi_1(X, x_0) \cong F(T^*)$, the free group generated by the set T^* of all 1-cells not in T.   ∎

Definition 1.12. A map $p : E \to B$ is said to have the *homotopy lifting property* (HLP)) with respect to a space X if for every map $f : X \to E$ and homotopy $G : X \times I \to B$ of $p \circ f : X \to B$ there is a homotopy $\tilde{G} : X \times I \to E$ with $\tilde{G}_0 = f$ and $p \circ \tilde{G} = G$. ($\tilde{G}$ is said to be a *lifting* of G.) Then the following diagram is commutative:

$$
\begin{array}{ccc}
X & \xrightarrow{f} & E \\
{\scriptstyle i_0}\downarrow & \tilde{G}\searrow \quad \downarrow{\scriptstyle p} & \\
X \times I & \xrightarrow[G]{} & B
\end{array}
\quad , \quad i_0(x) = (x, 0), \ (\forall x \in X).
$$

p is called a *fibration* if it has the HLP for all spaces X. p is called a *weak fibration* if it has the HLP for all disks D^n, $n \geq 0$. (Any fibration is also a weak fibration.) If $b_0 \in B$ is the base point, there the space $F = p^{-1}(b_0)$ is called the *fibre* of p. The projection $p_B : B \times F \to B$ is a fibration called the *trivial fibration* over B with fibre F. A *fibre bundle* is a quadruple (B, p, E, F) where p is a map $p : E \to B$ such that B has an open covering $\{U_\alpha\}_{\alpha \in A}$ and for each $\alpha \in A$ there is a homeomorphism $\phi_\alpha : U_\alpha \times F \to p^{-1}(U_\alpha)$ such that $p \circ \phi_\alpha = p_{U_\alpha} : U_\alpha \times F \to U_\alpha$. In other words, locally $p : E \to B$ looks like a trivial fibration. A fibre bundle (B, p, E, F) with F discrete is called a *covering* of B, p a *covering projection* and E *covering space* over B.

Definition 1.13. The coverings $\widetilde{X}$ of X form a subcategory of the category of fibre bundles over X; the total source of this category, if it exists, is called a *universal covering* of X.

Proposition 1.12. *If X is a connected and locally 0-connected space, then a 1-connected covering space is a universal covering. If, moreover, any point $x \in X$ has an open neighbourhood U such that $\pi_1(X, U) = 0$, the universal covering exists and is uniquely determined up to an isomorphism. Thus any* connected manifold has a universal covering. In particular, each connected Lie group has a 1-connected universal covering space which is itself a Lie group such that the covering map is a Lie group homomorphism.

Example 1.8. The group $Spin(m)$ is the universal 1-connected double covering group of $SO(m)$. ∎

Proposition 1.13. *If B is paracompact, for every fibre bundle $(B, p, E.F)$, $p : E \to B$ is a fibration.*

Definition 1.14. A map $p : E \to B$ has a *local cross-section* at a point $x \in B$ if there is a neighborhood U of x in B and a map $s : U \to E$ with $p \circ s = 1_U$.

Proposition 1.14. *1) Any fibre bundle has local cross-section.*

2) Let G be a topological group and $H \subset G$ be a closed subgroup. Then, the map $p : G \to G/H$, $(G/H$ is a group iff H is a normal subgroup of $G)$, has a local cross-section at every point of G/H if p has a local cross-section at the coset H.

Theorem 1.1. *1) For any $(X, x_0), (Y, y_0) \in Ob(\mathcal{T}_{top,\bullet})$ we have an isomorphism*

$$((p_X)_*, (p_Y)_*) : \pi_n(X \times Y, (x_0, y_0)) \cong \pi_n(X, x_0) \times \pi_n(Y, y_0), \ n \geq 0.$$

2) The map $\pi : PX \to X$ is a fibration with fibre ΩX. (PX, ω_0) is contractible for every (X, x_0).

3) Suppose $p : E \to B$ has the HLP with respect to a space $I \times X$, $b_0 \in B' \subset B$ and $E' = p^{-1}(B') \subset E$. Then $P(p) : P(E; e_0, E') \to P(E; b_0, B')$ has the HLP with respect to X. In particular, if p is a fibration (resp. weak fibration) then $P(p)$ is also a fibration (resp. weak fibration).

4) If $p : E \to B$ is a weak fibration, $b_0 \in B' \subset B$ and $E' = p^{-1}(B')$, then $p_n : \pi_n(E, E'; e_0) \to \pi_n(B, B', b_0)$ is a bijection for every $n \geq 1$.

5) One has the following exact homotopy sequence of weak fibration $p : E \to B$:

$$(1.3) \quad \begin{aligned} \cdots \to \pi_{n+1}(B, b_0) \xrightarrow{\partial'} \pi_n(F, e_0) \xrightarrow{i_*} \pi_n(E, e_0) \xrightarrow{p_*} \pi_n(B, b_0) \cdots \\ \cdots \xrightarrow{\partial'} \pi_0(F, e_0) \xrightarrow{i_*} \pi_0(E, e_0) \xrightarrow{p_*} \pi_0(B, b_0). \end{aligned}$$

6) If $p : E \to B$ is a weak fibration with E contractible, then $\partial' : \pi_n(B, b_0) \to \pi_{n-1}(F, e_0)$ is an isomoprhism for $n \geq 1$. In fact, from (1.3) we get the fol-

lowing exact sequences ($n \geq 1$):

$$\begin{array}{ccccccc}
\pi_n(E,e_0) & \overset{p_*}{\to} & \pi_n(B,b_0) & \overset{\partial'}{\to} & \pi_{n-1}(F,e_0) & \to & \pi_{n-1}(E,e_0) \\
\| & & \| & & \| & & \| \\
0 & \overset{p_*}{\to} & \pi_n(B,b_0) & \overset{\partial'}{\to} & \pi_{n-1}(F,e_0) & \to & 0
\end{array}$$

7) If $p : \widetilde{X} \to X$ is a covering of X, as $\pi_n(F,e_0) = 0$, for $n \geq 1$, it follows that $p_* : \pi_n(\widetilde{X},\widetilde{x}_0) \to \pi_n(X,x_0)$ is an isomorphism for all $n > 1$ and a monomorphism for $n = 1$. In fact for $n > 1$ from (1.3) we get the following exact sequence:

$$\pi_n(F,e_0) \equiv 0 \to \pi_n(E,e_0) \overset{p_*}{\to} \pi_n(B,b_0) \to \pi_n(F,e_0) \equiv 0$$

and for $n = 1$ we have

$$\pi_1(F,e_0) \equiv 0 \to \pi_1(E,e_0) \overset{p_*}{\to} \pi_1(B,b_0) \to \pi_0(F,e_0).$$

If $\widetilde{X}$ is 0-connected then the points of F are in correspondence one-to-one with the cosets of $p_*(\pi_1(\widetilde{X},x_0))$ in $\pi_1(X,x_0)$. In fact, from (1.3) we get

$$\pi_1(F,e_0) \equiv 0 \to \pi_1(E,e_0) \overset{p_*}{\to} \pi_1(B,b_0) \to \pi_0(F,e_0) \to \pi_0(E,e_0) = 0.$$

But this exact sequence says that one has the isomorphim

$$\pi_1(B,b_0)/\pi_1(E,e_0) \cong \pi_0(F,e_0).$$

Example 1.9. (*Examples of coverings*). 1) (*Circle*). $S^1 \equiv \{z \in \mathbf{C}; |z| = 1\} \subset \mathbf{C}$. $p : \mathbf{R} \to S^1$, $p(t) = e^{2\pi i t}$, $t \in \mathbf{R}$, is a covering with fibre $\mathbf{Z}$. One has: $\pi_n(S^1,s_0) \cong \pi_n(\mathbf{R},0) = 0$, $n > 1$; $\pi_1(S^1,s_0) \cong \pi_0(F,0) = \pi_0(\mathbf{Z},0) = \mathbf{Z}$. $\blacksquare$

2) (*n-Dimensional torus*). $T^n \equiv S^1 \times \cdots \times S^1$. $p : \mathbf{R}^n \to T^n$, $p(r_1,\cdots,r_n) = (e^{2\pi i r_1},\cdots,e^{2\pi i r_n})$ is a covering with fibre the set of integer lattice points in $\mathbf{R}^n$. Since $\mathbf{R}^n$ is contractible it follows $\pi_k(T^n,*) = 0$, $k \geq 2$. On the other hand $\pi_k(T^n,*) \cong \prod_{1 \leq i \leq n} \pi_k(S^1,s_0) = 0$, $k \geq 2$; $\pi_1(T^n,*) \cong \prod_{1 \leq i \leq n} \mathbf{Z} \cong \mathbf{Z} \oplus \cdots \oplus \mathbf{Z}$. $\blacksquare$

Definition 1.15. An inclusion $i : A \to X$ is said to have the *homotopy extension property* (HEP) with respect to a space Y if for every map $f : X \to Y$ and homotopy $G : A \times I \to Y$ of $f|_A$ there is a homotopy $F :$

$X \times I \to Y$ of f extending G. i is called a *cofibration* if it has the HEP with respect to all spaces Y.[3]

Example 1.10. 1) For any map $f : (X, x_0) \to (Y, y_0)$ the inclusion $j : Y \to Y \bigcup_f CX$ is a cofibration. ∎

2) In particular if X is obtained from A by attaching n-cells,

$$X = A \bigcup_f C(\bigvee_\alpha S_\alpha^{n-1}),$$

then the inclusion $A \to X$ is a cofibration. ∎

3) If (X, A) is a relative CW-complex then the inclusion $i : A \to X$ is a cofibration. ∎

Theorem 1.2. (J.H.C.Whitehead). *A map $f : X \to Y$ between CW-complexes is a homotopy equivalence iff it is a weak homotopy equivalence, i.e., $f_* : \pi_n(X, x_0) \cong \pi_n(Y, y_0)$, $n \geq 0$.*

Remark 1.2. The existence of the map $f : X \to Y$ is important: it is not enough to have $\pi_n(X, x_0) \cong \pi_n(Y, y_0)$, $n \geq 0$, if the isomorphisms are not induced by a map $f : (X, x_0) \to (Y, y_0)$. ∎

Theorem 1.3. (We can construct CW complexes with arbitrary given homotopy groups). 1) *Let G be an abelian group and $n \geq 2$ an integer. There is a complex X with*

$$\pi_r(X, x_0) \cong \begin{cases} G, \, r = n \\ 0, \, r \neq n. \end{cases}$$

2) *Let X be an $(n-1)$-connected CW-complex with $\pi_n(X, x_0) \cong G$ and (Y, y_0) a space with*

$$\pi_r(Y, y_0) \cong \begin{cases} H, \, r = n \\ 0, \, r > n. \end{cases}$$

Let $\phi : G \to H$ be a homomorphism. Then we can find a map $f : (X, x_0) \to (Y, y_0)$ such that the following diagram commutes:

$$\begin{array}{ccc} \pi_n(X, x_0) & \xrightarrow{f_*} & \pi_n(Y, y_0) \\ \| \wr & & \| \wr \\ G & \xrightarrow{\phi} & H \end{array}$$

---

[3] It can be shown that the assumption that $i{:}A{\to}X$ be an inclusion is unnecessary. The other conditions on a cofibration imply that i is a homomorphism of A onto a closed subspace of X.

f is unique up to homotopy.

3) *Any two CW-complexes* X, X', *such that*

$$\pi_r(X, x_0) \cong \pi_r(X', x_0) \cong \begin{cases} G, \; r = n \\ 0, \; r \neq n \end{cases}$$

are homotopy equivalent. Any CW-complex X with such a property is called an Eilenberg-MacLane complex of type (G, n).

Example 1.11. In the following table are reported some examples of Eilenberg-MacLane spaces.

TAB.1.2 - Examples of Eilenberg-MacLane spaces

n	G	$K(G,n)$
1	$\mathbf{Z}$	$K(G,1) \cong S^1$
1	$\mathbf{Z} \times \cdots_n \cdots \times \mathbf{Z}$	$K(G,1) \cong T^n$
1	$\pi_1(M_g^2)$	$K(G,1) \cong M_g^2 \; (\star)$
1	$\mathbf{Z}_m$	$K(G,1) \cong S^\infty / \mathbf{Z}_m \; (\star\star)$
1	$F \equiv$ free group	$K(G,1) \cong S^1 \vee \cdots \vee S^1 \; (\star\star\star)$
2	$\mathbf{Z}$	$K(G,2) \cong \mathbf{CP}^\infty = S^\infty / S^1$

$(\star)$ Surface of genus $g \geq 1$. $(\star\star)$ In particular $K(\mathbf{Z}_2,1) = \mathbf{RP}^\infty = \lim_{N \to \infty} \mathbf{RP}^N$.

$\lim_{N \to \infty} \mathbf{RP}^N$ denotes the "direct limit" or union of the chain $\mathbf{RP}^1 \subset \mathbf{RP}^2 \subset \cdots$.

$(\star\star\star)$ Bouquet of circles.

                                                           ∎

Definition 1.16. Let $\mathcal{A}$ denote the category of abelian groups and homomorphisms. Let $\mathcal{T}_{op}^{2'}$ be the category of topological pairs, where

$$Hom_{\mathcal{T}_{op}^{2'}}((X, A), (Y, B)) = [(X, A); (Y, B)].$$

A *unreduced homology theory* $h_\bullet$ on the category $\mathcal{T}_{op}^{2'}$ is a sequence of functors $h_n : \mathcal{T}_{op}^{2'} \to \mathcal{A}$ for each $n \in \mathbf{Z}$ and natural transformations $\partial_n : h_n \to h_{n-1} \circ R$, $n \in \mathbf{Z}$ (R is the *restriction functor*) satisfying the following two axioms:

i)(*Exactness*): for every pair $(X, A) \in Ob(\mathcal{T}_{op}^{2'})$ the sequence

$$\cdots \to h_{n+1}(X, A) \xrightarrow{\partial_{n+1}} h_n(A,\emptyset) \xrightarrow{h_{n+1}[i]} h_n(X,\emptyset) \xrightarrow{h_n[j]} h_n(X, A) \xrightarrow{\partial_n} \cdots$$

is exact, where $i : (A,\emptyset) \to (X,\emptyset)$ and $j : (X,\emptyset) \to (X, A)$ are the inclusions;

ii) (*Excision*): for every pair $(X, A) \in Ob(\mathcal{T}_{op}^{2'})$ and subset $U \subset A$ with $\overline{U} \subset \overset{\circ}{A}$ the inclusion $j : (X \setminus \overline{U}, A \setminus U) \to (X, A)$ induces an isomorphism $h_n[j]; h_n(X \setminus \overline{U}, A \setminus U) \to h_n(X, A)$, all $n \in \mathbf{Z}$.

Proposition 1.15. 1) *The excision axiom is equivalent to the following statement: For every triad* $(X; A, B)$, $(A, B$ *subspaces of* X *with* $A \bigcup B = X$), *such that* $\overset{\circ}{A} \cup \overset{\circ}{B} = X$ *the inclusion* $j : (A, A \bigcap B) \to (X, B)$ *induces an isomorphism* $j_* : h_n(A, A \cap B) \to h_n(X, B)$, *all* $n \in \mathbf{Z}$. (*In this case we say that the triad* $(X; A, B)$ *is excisive.*)
2) *For every CW-triad* $(X; A, B)$ *the inclusion* $j : (A, A \cap B) \to (X, B)$ *induces an isomorphism* $j_* : h_n(A, A \cap B) \to h_n(X, B)$, *all* $n \in \mathbf{Z}$.
3) *If* $f : (X, A) \to (Y, B)$ *is an homotopy equivalence, then* $f_* : h_n(X, A) \to h_n(Y, B)$ *is an isomorphism, all* $n \in \mathbf{Z}$.
4) *If* $A \subset X$ *is a deformation retract of* X, *then* $h_n(X, A) = 0$, $n \in \mathbf{Z}$.
5) $h_n(X, X) = 0$, $n \in \mathbf{Z}$.
6) *If* $x \in X$ *is any point and* $i : (\{x\},\emptyset) \to (X,\emptyset)$ *is the inclusion, then* i_* *injects* $h_n(\{x\},\emptyset)$ *as a direct summand in* $h_n(X,\emptyset)$ *and in fact:* $h_n(X,\emptyset) \cong h_n(\{x\},\emptyset) \bigoplus h_n(X, \{x\})$.
7) *If* $(X; A, B)$ *is a triad and* $I : (A, B) \to (X, B)$, $J : (X, B) \to (X, A)$ *are the inclusions, then we get an exact sequence:*

$$\cdots \to h_n(A, B) \xrightarrow{I_*} h_n(X, B) \xrightarrow{J_*} h_n(X, A) \xrightarrow{\triangle} h_{n-1}(A, B) \to \cdots$$

where $\triangle$ *is the composite*

$$\begin{array}{ccccc} h_n(X, A) & \xrightarrow{\partial} & h_{n-1}(A,\emptyset) & \xrightarrow{i_*} & h_{n-1}(A, B) \\ \| & & & & \| \\ h_n(X, A) & & \xrightarrow[\triangle]{} & & h_{n-1}(A, B) \end{array}$$

8) *A triad* $(X; A, B)$ *is excisive iff* $(X; B, A)$ *is so.*
9) *If* $A \subset X$ *is a cofibration, then the projection* $p : (X, A) \to (X/A, \{*\})$ *induces an isomorphism* $p_* : h_n(X, A) \to h_n(X/A, \{*\})$ *for all* $n \in \mathbf{Z}$.

10) For any pointed space $(X, x_0) \in Ob(\mathcal{T}_{op,\bullet})$ one has the following iso-morphism given by composition:

$$
\begin{array}{ccc}
h_n(X, \{x_0\}) & \xrightarrow[\cong]{\tilde{\sigma}_n(X,x_0)} & h_{n+1}(SX, \{x\}) \\
\Delta \uparrow \| \wr & & \| \wr \\
h_{n+1}(CX, X) & \xrightarrow[p_*]{} & h_{n+1}(CX/X, \{x\})
\end{array}
$$

11) For every $n \in \mathbf{Z}$ and non-negative integer k:

$$
\tilde{\sigma}_n(S^k, s_0) : h_n(S^k, \{s_0\}) \to h_{n+1}(S^{k+1}, \{s_0\})
$$

is an isomorphism.

12) One has the isomorphism:

$$
h_n(S^k, \{s_0\}) \cong h_{n-k}(S^0, \{s_0\}) \cong h_{n-k}(*, \emptyset), \quad \forall n \in \mathbf{Z}.
$$

The groups $h_n(*, \emptyset)$ are called the *coefficient groups of the homology theory* $h_\bullet$.

13)(Mayer-Vietoris sequence). If $(X; A, B)$ is an excisive triad and $C \subset A \cap B$, then there is an exact sequence (Mayer-Vietoris sequence):

$$
\cdots \xrightarrow{\Delta'} h_n(A \cap B, C) \xrightarrow{\alpha} h_n(A, C) \bigoplus h_n(B, C) \xrightarrow{\beta} h_n(X, C) \xrightarrow{\Delta'} h_{n-1}(A \cap B, C) \to \cdots
$$

where $\alpha(x) = (i_{1*}(x), i_{2*}(x))$, $\beta(x) = i_{3*}(x) - i_{4*}(x)$, Δ' is the composite:

$$
\begin{array}{ccccc}
h_n(X, C) & \xrightarrow{J_{1*}} h_n(X, B) & \xleftarrow[\cong]{j_{1*}} h_n(A, A \cap B) & \xrightarrow{\Delta_1} & h_{n-1}(A \cap B, C) \\
\| & & & & \| \\
h_n(X, C) & & \xrightarrow{\Delta'} & & h_{n-1}(A \cap B, C)
\end{array}
$$

Here $i_1, i_2, i_3, i_4, j_1, J_1$ are the following inclusions:

$$
\begin{array}{ccc}
& (A, C) & \\
\nearrow^{i_1} & & \searrow^{i_3} \\
(A \cap B, C) & & (X, C) \\
\searrow_{i_2} & & \nearrow^{i_4} \\
& (B, C) &
\end{array}
\quad , \qquad
\begin{array}{c}
(A, A \cap B) \xrightarrow{j_1} (X, B) \\[4pt]
(X, C) \xrightarrow{J_1} (X, B).
\end{array}
$$

Furthermore, Δ_1 is the boundary for the triple $(A, A \cap B, C)$.

14) If $(X; A, B)$ is an excisive triad, then the inclusions $i_A : (A; A \cap B) \to (X, A \cap B)$, $i_B : (B; A \cap B) \to (X, A \cap B)$, induce an isomorphism

$$(i_{A*}, i_{B*}) : h_n(A, A \cap B) \bigoplus h_n(B, A \cap B) \underset{\cong}{\to} h_n(X, A \cap B), \ n \in \mathbf{Z}.$$

15) If $X = A \cup B$, $A \cap B = \emptyset$, A and B open sets, then

$$(i_{A*}, i_{B*}) : h_n(A, \emptyset) \bigoplus h_n(B, \emptyset) \underset{\cong}{\to} h_n(X, \emptyset), \ n \in \mathbf{Z}.$$

16) For any two pointed CW-complexes $(X.x_0)$, (Y, y_0), the inclusions $i_X : (X, \{x_0\}) \to (X \vee Y, \{*\})$ and $i_Y : (Y, \{y_0\}) \to (X \vee Y, \{*\})$ induce an isomorphism

$$(i_{X*}, i_{Y*}) : h_n(X, \{x_0\}) \bigoplus h_n(Y, \{y_0\}) \underset{\cong}{\to} h_n(X \vee Y, \{*\}), \ n \in \mathbf{Z}.$$

Example 1.12. Let us consider the torus $T = S^1 \times S^1 = A \cup B$, where A and B are homeomorphic to $S^1 \times I$ and $A \cap B \cong S^1 \cup S^1$. Taking into account that $A \simeq B \simeq S^1$, (hence the projections $p_A : A \to S^1$, $p_B : B \to S^1$ are homotopy equivalences), we get, from the Mayer-Vietoris sequence with $C = \emptyset \subset A \cap B$, the following exact sequence:

$$\cdots \to h_{n+1}(T, \emptyset) \overset{\triangle'}{\to} \quad h_n(A \cap B, \emptyset) \quad \overset{\alpha}{\to} \quad h_n(A, \emptyset) \oplus h_n(B, \emptyset) \quad \overset{\beta}{\to} h_n(T, \emptyset) \overset{\triangle'}{\to} \cdots$$
$$\| \wr \qquad\qquad\qquad\qquad \| \wr$$
$$h_n(S^1, \emptyset) \oplus h_n(S^1, \emptyset) \quad \overset{\alpha'}{\to} \quad h_n(S^1, \emptyset) \oplus h_n(S^1, \emptyset)$$

One can see that the map α' is given by $\alpha'(x, y) = (x + y, x + y)$. It follows that $\ker \alpha' \cong h_n(S^1, \emptyset)$ and $\operatorname{coker} \alpha' \cong h_n(S^1, \emptyset)$. Therefore the above exact sequence yields a collection of short exact sequences:

$$(1.4) \qquad 0 \to h_n(S^1, \emptyset) \to h_n(T, \emptyset) \to h_{n-1}(S^1, \emptyset) \to 0, \ n \in \mathbf{Z}.$$

In order to arrive to an explicit calculation of $h_n(T, \emptyset)$ it is necessary to specify the coefficients groups of $h_\bullet$. For example, in the particular case of the ordinary singular homology with coefficients $\mathbf{Z}$ (see below), we have

$$h_n(S^1, \emptyset) = \begin{cases} \mathbf{Z}, & n = 0, 1 \\ 0, & n \neq 0, 1. \end{cases}$$

Furthermore, taking into account that $h_{n-1}(S^1,\emptyset)$ is free-abelian, the sequence (1.4) splits and we can explicitly calculate the groups $h_n(T,\emptyset)$. In fact, we get:

$$h_n(T,\emptyset) \cong h_{n-1}(S^1,\emptyset) \bigoplus h_n(S^1,\emptyset) \cong \begin{cases} \mathbf{Z}, & n = 0,2 \\ \mathbf{Z} \bigoplus \mathbf{Z}, & n = 1 \\ 0, & n \neq 0,1,2. \end{cases}$$

$\blacksquare$

Definition 1.17. On the category $\mathcal{T}'_{op,\bullet}$ of topological pointed spaces, where

$$Hom_{\mathcal{T}'_{op,\bullet}}((X,x_0);(Y,y_o)) = [(X,x_0),(Y,y_0)],$$

the *suspension functor* S, is dedined by $S(X,x_0) = (SX,*)$, $S[f] = [Sf] = [1_{S^1} \wedge f]$. A *reduced homology theory* $\kappa_\bullet$ on $\mathcal{T}'_{op,\bullet}$ is a collection of functors $\kappa_n : \mathcal{T}'_{op,\bullet} \to \mathcal{A}$ and natural equivalences $\sigma_n : \kappa_n \to \kappa_{n-1} \circ S$, $n \in \mathbf{Z}$, satisfying:

(*Exactness*). For every pointed pair (X,A,x_0) with inclusions $i : (A,x_0) \to (X,x_0)$ and $j : (X,x_0) \to (X \cup CA,*)$, the sequence

$$\kappa_n(A,x_0) \xrightarrow{\kappa_n[i]} \kappa_n(X,x_0) \xrightarrow{\kappa_n[j]} \kappa_n(X \cup CA,*)$$

is exact.

There are two functor axioms which we shall often impose on a reduced homology theory.

(*Wedge axiom*). For every collection $\{(X_\alpha,x_\alpha) : \alpha \in A\}$ of pointed spaces, the inclusions $i_\alpha : X_\alpha \to \bigvee_{\beta \in A} X_\beta$ induce an isomorphism

$$\{i_{\alpha *}\} : \bigoplus_{\alpha \in A} \kappa_\alpha(X_\alpha) \underset{\cong}{\to} \kappa_n(\bigvee_{\alpha \in A} X_\alpha), \quad n \in \mathbf{Z}.$$

(*Weak homotopy equivalence axiom*).[4] If $f : X \to Y$ is a weak homotopy equivalence then $f_* : \kappa_n(X,x_0) \to \kappa_n(Y,f(x_0))$ is an isomorphism for all $n \in \mathbf{Z}$, $x_0 \in X$.

---

[4] Wedge axiom is useful for making deducions about CW-complexes which are not finite, and the weak homotopy equivalence axiom for making deducions about spaces which are not CW-complexes.

Proposition 1.16. 1) *For any single point* $\{*\}$ *we have* $\kappa_n(\{*\}) = 0$, $n \in \mathbf{Z}$.
2) *There are natural transformations* $\hat{\partial}_n$ *and, for each pointed pair* (X, A, x_0), *a long exact sequence:*

$$\cdots \to \kappa_n(A) \xrightarrow{i_*} \kappa_n(X) \xrightarrow{j_*} \kappa_n(X \cup CA) \xrightarrow{\hat{\partial}_n(X,A,x_0)} \kappa_{n-1}(A) \xrightarrow{i_*} \cdots$$

Theorem 1.4. (Correspondence between reduced and unreduced homology theories). 1) *Given an unreduced homology theory* $h_\bullet$, *denote by* $\tilde{h}_\bullet$ *the collection of functors* $\tilde{h}_\bullet : \mathcal{T}'_{op,\bullet} \to \mathcal{A}$, *defined by* $\tilde{h}_n(X, x_0) = h_n(X, \{x_0\})$, $\tilde{h}_n[f] = h_n[f]$, $f : (X, x_0) \to (Y, y_0)$, *and natural transformations* $\tilde{\sigma}_n : \tilde{h}_n \to \tilde{h}_{n+1} \circ S$, *for all* $n \in \mathbf{Z}$. *Then,* $\tilde{h}_\bullet$ *is a reduced homology theory.*
2) *Given a reduced homology theory* $\kappa_\bullet$, *denote by* $\hat{\kappa}_\bullet$ *the collection of functors* $\hat{\kappa}_n : \mathcal{T}'_{op,\bullet} \to \mathcal{A}$ *defined by* $\hat{\kappa}_n(X, A) = \kappa_n(X^+ \cup CA^+)$, $\hat{\kappa}_n[f] = \kappa_n[\hat{f}]$, *where* $\hat{f} : X^+ \cup CA^+ \to Y^+ \cup CB^+$, *is induced by* $f : (X, A) \to (Y, B)$, *and natural transformations* $\hat{\partial}_n : \hat{\kappa}_n \to \hat{\kappa}_{n-1} \circ R$. *Then,* $\hat{\kappa}_\bullet$ *satisfies the exactness axiom. Furthermore, if* $\kappa_\bullet$ *satisfies the WHE axiom then* $\hat{\kappa}_\bullet$ *satisfies the excision axiom and it is hence an unreduced homology theory.*
3) *Furthermore,* $\widehat{\tilde{h}}_\bullet$ *is naturally equivalent to* $h_\bullet$ *and* $\widetilde{\hat{\kappa}}_\bullet$ *to* $\kappa_\bullet$. *Thus there is a one-to-one correspondence between reduced and unreduced theories. It is already clear that to any theorem about unreduced theory there corresponds one about a reduced theory and conversely.*
Proposition 1.17. 1) *Let* (X, x_0) *be a pointed space and let* $\{X_\alpha : \alpha \in A\}$ *be a collection of subspaces ordered by inclusions such that for every compact subset* $C \subset X$ *there is an* X_γ, *with* $C \subset X_\gamma$. *Let* $j_\alpha^\beta : X_\alpha \to X_\beta$ *and* $i_\alpha : X_\alpha \to X$ *be the inclusions. Then the groups* $\pi_n(X_\alpha, x_0)$ *and homomorphisms* $j_{\alpha*}^\beta$ *form a direct system of groups* $(n \geq 2)$ *and*

$$\{i_{\alpha*}\} : \varinjlim \pi_n(X_\alpha, x_0) \to \pi_n(X, x_0)$$

is an isomorphism.
2) *Let* (X, x_0) *be a CW-complex and let* $X^0 \subset \cdots \subset X^n \subset \cdots \subset X$ *be subcomplexes with* $\bigcup_{n \geq 0} X^n = X$, $j_n^m : X^n \to X^m$, $i_n : X^n \to X$ *the inclusions. Then, the groups* $\kappa_q(X^n)$ *and homomorphisms* j_n^m *form a direct system and if* $\kappa_\bullet$ *satisfies the wedge axiom on the category* $\mathcal{W}_\bullet$ *of pointed CW-complexes, then*

$$\{i_{n*}\} : \varinjlim \kappa_q(X^n) \to \kappa_q(X)$$

is an isomorphism for all $q \in \mathbf{Z}$.

3) Let $T_\bullet : \kappa_\bullet \to \kappa'_\bullet$ be a natural transformation of homology theories. If $T_q(S^0) : \kappa_q(S^0) \to \kappa'_q(S^0)$ is an isomorphism for $q < N$ and an epimorphism for $q = N$, then for every finite $(n-1)$-connected CW-complex (X, x_0), $T_q(X) : \kappa_q(X) \to \kappa'_q(X)$ is an isomorphism for $q < n + N$, and an epimorphism for $q = n + N$. If $\kappa_\bullet$, $\kappa'_\bullet$ also satisfy the wedge axiom on the category $\mathcal{W}'_\bullet$ of pointed CW-complexes, with morphisms defined up to homotopy equivalences, X may be infinite. If $\kappa_\bullet$, $\kappa'_\bullet$ also satisfy the WHE axiom, X may be any $(n-1)$-connected space.

Definition 1.18. 1) A *reduced cohomology theory* $\kappa^\bullet$ on $\mathcal{T}'_{op,\bullet}$ is a collection of cofunctors $\kappa^n : \mathcal{T}'_{op,\bullet} \to \mathcal{A}$ and natural equivalences $\sigma^n : \kappa^{n-1} \circ S \to \kappa^n$, $n \in \mathbf{Z}$, satisfying:

(*Exactness*). For every pointed pair (X, A, x_0) with inclusions $i : (A, x_0) \to (X, x_0)$ and $j : (X, x_0) \to (X \cup CA, *)$ the sequence

$$\kappa^n(A, x_0) \xleftarrow{\ \ \kappa^n[i]\ \ } \kappa^n(X, x_0) \xleftarrow{\ \ \kappa^n[j]\ \ } \kappa^n(X \cup CA, *)$$

is exact.

2) *There is a corresponding notion of an unreduced cohomology theory on* $\mathcal{T}'^2_{op}$.

3) (*Wedge axiom*). For every collection $\{(X_\alpha, x_\alpha) : \alpha \in A\}$ of pointed spaces the inclusions $i_\alpha : X_\alpha \to \bigvee_{\beta \in A} X_\beta$ induce an isomorphism

$$\{i_\alpha^*\} : \kappa^q(\bigvee_\alpha X_\alpha) \to \prod_{\alpha \in A} \kappa^q(X_\alpha).$$

Proposition 1.18. 1) Let (X, x_0) be a CW-complex and let $X^0 \subset \cdots \subset X^n \subset \cdots \subset X$ be subcomplexes with $\bigcup_{n \geq 0} X^n = X$, $j_n^m : X^n \to X^m$, $i_n : X^n \to X$ the inclusions, $n \geq m$. Then, $\{\kappa^q(X^n), j_n^{m*}, n, m \in \mathbf{N}\}$ is an inverse system for every $q \in \mathbf{Z}$, and if $\kappa^\bullet$ satisfies the wedge axiom on $\mathcal{W}'_\bullet$, then there is an exact sequence[5]

$$(1.5) \qquad 0 \to \lim^1 \kappa^{q-1}(X^n) \to \kappa^q(X) \xrightarrow{\ \{i_n^*\}\ } \lim^0 \kappa^q(X^n) \to 0.$$

---

[5] Let $\{G_\alpha, j_\alpha^\beta{:}G_\beta{\to}G_\alpha\}_{\alpha,\beta \in A}$ be an inverse system of abelian groups G_α. We denote by $\lim^0 G_\alpha{=}\varprojlim G_\alpha$. Consider the map $\prod_n G_n \xrightarrow{\delta} \prod_n G_n$ defined by $\delta f(u){=}(-1)^n f(u) + (-1)^{n+1} j_n^{n+1}(f(n+1))$, for $f \in \prod_n G_n$. Note that $\ker \delta \subset \prod_n G_n$ is the set of all f such that $j_n^{n+1}(f(n+1)){=}f(un)$, all $n \geq 0$. This is just $\lim^0 G_n{=}\varprojlim G_n$. We define $\lim^1 G_n{=}\mathrm{coker}\,\delta$. Then a short exact sequence, $0{\to}\{A_n\} \xrightarrow{\{\phi_n\}} \{B_n\} \xrightarrow{\{\psi_n\}} \{Cn\}{\to}0$ of inverse systems gives rise to

Therefore, $\{i_n^*\}$ is an isomorphism iff $\lim^1 \kappa^{q-1}(X^n) = 0$.

2) Let $T^\bullet : \kappa^\bullet \to \kappa'^\bullet$ be a natural transformation of cohomology theories satisfying the wedge axiom on $\mathcal{W}'_\bullet$. If $T^q(S^0) : \kappa^q(S^0) \to \kappa'^q(S^0)$ is an isomorphism for $q > N$, epimorphism for $q = N$, then for any n-dimensional CW-complex (X, x_0), $T^q(X) : \kappa^q(X) \to \kappa'^q(X)$ is an isomorphism for $q > n + N$, epimorphism for $q = n + N$.

3) (Extension of reduced (co)homology theory from $\mathcal{W}'_\bullet$ to $\mathcal{T}'_{op,\bullet}$). Suppose that $\kappa_\bullet$ is any reduced homology theory on $\mathcal{W}'_\bullet$ and suppose (X, x_0) is any pointed space; choose a CW-substitute $f : X' \to X$ for X and a base point $x_0' \in X'$ such that $f(x_0') = x_0$. Then define $\kappa_n(X) = \kappa_n(X')$, $n \in \mathbf{Z}$. Then, $\kappa_n(X)$ is such a defined group. If (X, A, x_0) is any pointed pair, then we can choose CW-substitutes $f' : A' \to A$ and $g' : X' \to X$. Let $h : A' \to X'$ be any cellular map such that

$$
\begin{array}{ccc}
A' & \xrightarrow{f'} & A \\
h \downarrow & & \cap \\
X' & \xrightarrow{g'} & X
\end{array}
$$

commute up to homotopy. Then we may replace X' by the mapping cylinder M_h and g' by $g' \circ r$ ($r : M_h \to X'$ the projection). Thus we may as well assume $A' \subset X'$. Now $\hat{g}' : X' \cup CA' \to C \cup CA$ is a weak homotopy equivalence. Hence, from the commutative diagram

$$
\begin{array}{ccccc}
\kappa_n(A) & \xrightarrow{i_*} & \kappa_n(X) & \xrightarrow{j_*} & \kappa_n(X \cup CA, *) \\
\| & & \| & & \| \\
\kappa_n(A') & \xrightarrow{i'_*} & \kappa_n(X') & \xrightarrow{j'_*} & \kappa_n(X' \cup CA', *)
\end{array}
$$

we see that κ_* satisfies the exactness axiom. Hence $\kappa_\bullet$ becomes a reduced homology theory on $\mathcal{T}'_{op,\bullet}$, that satisfies the WHE axiom. Therefore $\hat{\kappa}_\bullet$ is an unreduced homology theory on $\mathcal{T}'_{op}$.

Definition 1.19. An *ordinary homology theory* $\kappa_\bullet$ is a homology theory with $\kappa_n(S^0) = 0$ unless $n = 0$. If $\kappa_0(S^0) = G$ then $\kappa_\bullet$ will be called an ordinary homology theory with *coefficients G*.

---

an exact sequence $0 \to \lim^0 A_n \to \lim^0 B_n \to \lim^0 C_n \to \lim^1 A_n \to \lim^1 B_n \to \lim^1 C_n \to 0$. If $\{\phi_n\}$: $\{G_n\} \to \{G_n'\}$ is a morphism of inverse systems such that each ϕ_n is an isomorphism, then $\lim^0 \phi_n$ and $\lim^1 \phi_n$ are isomorphisms.

Theorem 1.5.   1) *Reduced singular homology $\tilde{H}_\bullet(-;G)$ is an ordinary homology theory on the category $\mathcal{T}'_{op,\bullet}$, with coefficients G.*
2) Any two ordinary homology theories with coefficients G satisfying the wedge and WHE axioms are naturally equivalent.
3) If $H_\bullet(X, A; G)$, (resp. $H^\bullet(X, A; G)$), is the relative singular homology, (resp. cohomology), with coefficents in G, we define the reduced homology and cohomology groups $\tilde{H}_\bullet(-; G)$, $\tilde{H}^\bullet(- : G)$ on $\mathcal{T}'_{op,\bullet}$ by

$$\tilde{H}_n(-; G) = H_n(X, \{x_0\}; G),$$
$$\tilde{H}^n(-; G) = H^n(X, \{x_0\}; G).$$

Since the one point space $\{*\}$ has the free abelian group $S_n(X)$ generated by the singular n-simplexes of $X \equiv \{*\}$ given by $S_n(X) = \mathbf{Z}$, $n \geq 0$, and $d : S_n(X) \to S_{n-1}(X)$ the map $d(m) = m. \sum_{0 \leq j \leq n-1}(-1)^j$, $m \in \mathbf{Z}$, we see that

$$H_n(\{*\}, \emptyset; G) = \begin{cases} G, & n = 0 \\ 0, & n \neq 0. \end{cases}$$

Similarly

$$H^n(\{*\}, \emptyset; G) = \begin{cases} G, & n = 0 \\ 0, & n \neq 0. \end{cases}$$

Therefore, $\tilde{H}_n$, $\tilde{H}^n$ are ordinary homology, respectively cohomlogy, theories.
4) The resulting homology theory $H(G)_\bullet(-)$ associated to the Eilenberg-MacLane spectrum $K(G)^6$ with

$$\pi_n(K(G)) = \begin{cases} G, & n = 0 \\ 0, & n \neq 0 \end{cases}$$

is an ordinary homology theory with coefficients G. So on $\mathcal{T}'_{op,\bullet}$ we have a natural equivalence: $H(G)_\bullet(-) \cong \tilde{H}_\bullet(-; G)$.
5) (Hurewicz isomorphism theorem). If (X, A) is an $(n-1)$-connected pair, $n \geq 2$, in the category $\mathcal{T}'_{op,\bullet}$, and A is 1-connected, then

$$h : \pi_q(X, A, x_0) \to H_q(X, A; \mathbf{Z})$$

is an isomorphism for $q \leq n$ and an epimorphism for $q = n + 1$. h is an epimorphism for $q = n$ even if A only 0-connected.

---

[6]   See Definition 1.20 and Theorem 1.8 below.

6) (Whitehead). *Let $f : X \to Y$ be a map of spaces which are 0-connected.*
(i) *If f is an n-equivalence ($n = \infty$ allowed), then $f_* : \tilde{H}_q(X; \mathbf{Z}) \to \tilde{H}_q(Y; \mathbf{Z})$ is an isomorphism for $q < n$ and an epimorphism for $q = n$.*
(i) *If X, Y are 1-connected and $f_* : \tilde{H}_q(X; \mathbf{Z}) \to \tilde{H}_q(Y; \mathbf{Z})$ is an isomorphism for $q < n$ and an epimorphism for $q = n$, then f is an n-equivalence ($n = \infty$ allowed).*
7) *A map $f : X \to Y$ between 1-connected CW-complexes is a homotopy equivalence iff $f_* : \tilde{H}_n(X; \mathbf{Z}) \to \tilde{H}_n(Y; \mathbf{Z}$ is an isomorphism for all $n \geq 0$.*
8) *If (X, x_0) is a CW-complex with $\dim X \leq n$, then*

$$i_0 : [X, x_0; S^q, s_0] \to \pi_s^q(X) \equiv [E(X), \Sigma^q S^0], \quad (n \geq 2),$$

is a bijection if $q \geq n$ and a surjection if $q = n - 1$.[7] In particular we may give the set (nth cohomotopy set) $\pi^n(X, x_0) \equiv [X, x_0; S^n, s_0]$ a structure of abelian group if $\dim X \leq n$. (So that

$$i_0 : [S^n, s_0; S^q, s_0] \to \pi_s^q(S^n) \cong \pi_s^{q-n}(S^0) = \pi_{n-q}^s(S^0)$$

is an isomorphism.) One has a natural homomorphism·

(1.6)(*stable Hopf-homomorphism*) $\qquad \psi_s : \pi_s^q(X) \to \tilde{H}^q(X; \mathbf{Z}).$

ψ_s is an isomorphism for $q \geq n$ and an epimorphism for $q = n - 1$. Furthermore, we have the unstable Hopf function $\psi : \pi^n(X, x_0) \to \tilde{H}^n(X; \mathbf{Z})$, defined to be $\psi \equiv \psi_s \circ i_0$. ψ is a homomorphism for $\dim X \leq n$.
9) (Hopf isomorphism theorem). *If (X, x_0) is a CW-complex with $\dim X \leq n$ then $\psi : \pi^q(X, x_0) \to \tilde{H}^q(X; \mathbf{Z})$ is an isomorphism for $q \geq n$ and an epimorphism for $q = n - 1$.*
10) *For any CW-complex (X, x_0), $T_i : [X, x_0; Y, y_0] \to \tilde{H}^n(X; G)$ is an isomorphism if $\dim X \leq n$ and an epimorphism if $\dim X = n + 1$. Here Y is $(n - 1)$-connected.*
11) *Suppose $(X^\alpha : \alpha \in A)$ is a directed set of subspaces of the space X ($\alpha \leq \beta \Rightarrow X_\alpha \subset X_\beta$) such that for all compact subsets $C \subset X$ there is an $\alpha \in A$ with $C \subset X^\alpha$. Then the inclusions $i_\alpha : X^\alpha \to X$ induce an isomorphism $\{i_{\alpha*}\} : \varinjlim H_n(X^\alpha; G) \to H_n(X; G)$, for all $n \geq 0$, G.*

---

[7] $E(X)$ is the canonical spectrum associated to the CW-complex X (see Proposition 1.19) and ΣS^0 is the spectrum associated to S^0 (see Proposition 1.23).

Theorem 1.6. *By considering the mapping cylinder M_f of a weak homotopy equivalence $f : X \to Y$ and the resulting exact sequence*

$$\cdots \to h_{n+1}(M_f, X) \to h_n(X, \emptyset) \xrightarrow{f_*} h_n(Y, \emptyset) \to h_n(M_f, X) \to \cdots$$

for an unreduced theory $h_\bullet$, we see that the WHE axiom for $h_\bullet$ is equivalent to either of the following.

(i) If $A \subset X$ is a cofibration and (X, A) is n-connected for all $n \geq 0$, then $h_\bullet(X, A) = 0$.

(ii) If X is n-connected for all $n \geq 0$, then $h_\bullet(\{x_0\}) \to h_\bullet(X)$ is surjective for all $x_0 \in X$.

Example 1.13. $\pi_q^s \equiv \varinjlim \pi_{q+n}(S^n \wedge -)$ satisfies the WHE axiom.   ■

Example 1.14. Suppose $X_1 \subset X_2 \subset \cdots \subset X_n \subset \cdots \subset X$ is a sequence of subspaces of a topological space X such that for each compact subspace $K \subset X$ there is a $k \geq 1$ with $K \subset X_k$. Then we can inductively construct CW-substitutes $f^k : X_k' \to X_k$, $k \geq 1$, such that X_k' is a subcomplex of X_{k+1}' for $k \geq 1$ and $f^{k+1}|_{X_k'} = f^k$. We let $X' \equiv \bigcup_{k \geq 1} X_k'$ with the weak topology and $f' : X' \to X$ be the map with $f'|_{X_k'} = f^k$. Then: (a) $f' : X' \to X$ is a CW-substitute for X; (b) Suppose $\kappa_\bullet$ is a homology theory which satisfies the wedge and weak homotopy equivalence axioms. Then $\{i_{k*}\} : \varinjlim \kappa_q(X_k) \to \kappa_q(X)$ is an isomorphism for $q \in \mathbf{Z}$.   ■

Theorem 1.7. *Suppose that X is a space and $X_0 \subset X_1 \subset X_2 \subset \cdots \subset X_n \subset \cdots \subset X$ is a filtration by subspaces X_n such that $H^\bullet(X_n)$ is finitely generated. Suppose further that $\kappa^\bullet$ is a cohomology theory such that $\kappa^{q-s-1}(S^0)$ is finite whenever $\tilde{H}^s(X_n) \neq 0$. Then one sees, by using the Atiyah-Hirzebruch-Whietehead spectral sequence,[8] that $\kappa^{q-1}(X_n)$ is finite for all n. Then, it follows that*

$$\varprojlim^1 \kappa^{q-1}(X_n) = 0 \Rightarrow \kappa^q(X) \cong \varprojlim^0 \kappa(X_n).$$

Proof. We shall use the following lemmas.

Lemma 1.1. *Let $G_0 \xleftarrow{j_0} G_1 \xleftarrow{j_1} G_2 \leftarrow \cdots \leftarrow G_n \xleftarrow{j_n} G_{n+1} \leftarrow \cdots$ be an inverse system $\{G_n\}$ of abelian groups indexed by the non-negative integers. We say that $\{G_n\}$ satisfies the Mittag-Leffler condition (ML) if for each $n \geq 0$ there exists an $m(n) \geq n$ such that*

$$\mathrm{im}\,[j_n^r : G_r \to G_n] = \mathrm{im}\,[j_n^{m(n)} : G_{m(n)} \to G_n], \quad \forall r \geq m(n).$$

---

[8] See Theorem 1.9.

In other words the image of G_r in G_n becomes stable for sufficiently larger r. Then, if $\{G_n\}$ satisfies the (ML), then $\lim^1 G_n = 0$.

Lemma 1.2. *Suppose* $\{G_n, j_n^m : n, m \in \mathbf{N}\}$ *is an inverse system with each* G_n *finite. Then, the subgroups* $\mathrm{im}\,[j_n^r : G_r \to G_n]$, $r \geq n$, *cannot all be distinct, and in fact (ML) is satisfied.*

Now, the proof of the theorem follows soon. $\qquad\qquad\qquad\qquad\square$

Definition 1.20. A *spectrum* E is a collection $\{(E_n, *) : n \in \mathbf{Z}\}$ of CW-complexes such that SE_n is (or is homeomorphic to) a subcomplex of E_{n+1}, all $n \in \mathbf{Z}$. A *subspectrum* $F \subset E$ consists of subcomplexes $F_n \subset E_n$ such that $SF_n \subset F_{n+1}$. A *cell of dimension* $d' - n'$ in E is a sequence

$$e = \{e_{n'}^{d'}, Se_{n'}^{d'}, S^2 e_{n'}^{d'}, \cdots\},$$

where $e_{n'}^{d'}$ is a cell in $E_{n'}$ that is not the suspension of any cell in $E_{n'-1}$. Thus each cell in each complex E_n is a manifold of exactly one cell of E. We call *cell of dimension* $-\infty$ the subspectrum $* \equiv F \subset E$ such that $F_n = *$ for all n. A spectrum E is called *finite* if it has only finitely many cells. It is called *countable* if it has countably many cells.[9] An Ω-*spectrum* is a spectrum E such that the adjoint $\epsilon' : E_n \to \Omega E_{n+1}$ of the inclusion $\epsilon_n : SE_n \to E_{n+1}$ is always a weak homotopy equivalence.

Proposition 1.19. 1) *If* X *is any CW-complex, then we can define a spectrum* $E(X)$ *by taking*

$$E(X)_n \equiv \begin{cases} *, & n < 0 \\ S^n X, & n \geq 0. \end{cases}$$

2) *Given any collection* $\{E_n, \epsilon_n\}$ *of CW-complexes* $(E_n, *)$ *and cellular maps* $\epsilon.SE_n \to E_{n+1}$ *we can construct a spectrum* $E' \equiv \{E_n'\}$ *and homotopy equivalences* $r_n : E_n' \to E_n$ *such that* $r_{n+1}|_{SE_n'} = \epsilon_n \circ Sr_n$, *i.e., the following diagram is commutative*

$$\begin{array}{ccc} SE_n' & \subset & E_{n+1}' \\ Sr_n \downarrow & & \downarrow r_{n+1} \\ SE_n & \subset & E_{n+1} \end{array}$$

Definition 1.21. A *filtration* of a spectrum E is an increasing sequence $\{E^n : n \in \mathbf{Z}\}$ of subspectra of E whose union is E.

---

[9] E is finite iff there is an integer N such that $E_n = S^{n-N} E_N$ for $n \geq N$ and the complex E_N is finite.

Example 1.15. The *skeletal filtration* $\{E^{(n)}\}$ of E is defined as follows: $E^{(n)}$ is the union of all the cells of E of dimension at most n. ∎

Example 1.16. The *layer filtration* $\{E^\infty\}$ of E is defined as follows: for each cell $e = \{e_n, Se_n, \cdots\}$ of E we can find a finite subspectrum $F \subset E$ of which e is a cell. (For example, let $F_n \subset E_n$ be the subcomplex consisting of e_n and all its faces; then take $F_m = *$, $m < n$, $F_m = S^{m-n}F_n$, $m \geq n$.) Let $l(e)$ be the smallest number of cells in any such F. ($l(e)$ coincides with the number of faces of e_n.) Then we define $E^n = *$, $n \leq 0$, $E^n =$ union of all cells with $l(e) \leq n$, $n > 0$. The terms E^n are called the *layers* of E. One can see that $\{E^n\}$ is a filtration of E. ∎

Definition 1.22. A *function* $f : E \to F$ between spectra is a collection $\{f_n : n \in \mathbf{Z}\}$ of cellular maps $f_n : E_n \to F_n$ such that $f_{n+1}|_{SE_n} = Sf_n$. The inclusion $i : F \hookrightarrow E$ of a subspectrum $F \subset E$ is a function and if $g : E \to G$ is a function then $g|_F = g \circ i$ is also a function.

Definition 1.23. A subspectrum $F \subset E$ is called *cofinal* if for any cell $e_n \subset E_n$ of E there is an m such that $S^m e_n \subset F_{n+m}$.

Remark 1.3. If F is cofinal and $K_n \subset E_n$ is a finite subcomplex, then there is an m such that $S^m K_n \subset F_{n+m}$. ∎

Remark 1.4. Intersection of two cofinal subspectra is cofinal and if $G \subset F \subset E$ are subspectra such that F is cofinal in E on G is cofinal in F, then G is cofinal in E. An arbitrary union of cofinal subspectra is cofinal. ∎

Definition 1.24. Let E and F be spectra. Let S be the set of all pairs (E', f') such that $E' \subset E$ is a cofinal subspectrum and $f' : E' \to F$ is a function. We call *maps* from E to F elements of $Hom(E, F)/\sim$, where $\sim$ is the following equivalence relation:

$$(E', f') \sim (E'', f'') \Leftrightarrow \exists (E''', f''') : \left\{ \begin{array}{l} E''' \subset E' \cap E'', \\ f'|_{E'''} = f''' = f''|_{E'''}, \\ E''' \text{ cofinal.} \end{array} \right\}.$$

(Intuitively maps only need to be defined on each cell.) The *category of spectra* $\mathcal{S}$ is the category where objects are spectra and morphisms are maps.

Remark 1.5. Let E and F be spectra and $f : E \to F$ a function. If $F' \subset F$ is a cofinal subspectrum then there is a cofinal subspectrum $E' \subset E$ with $f(E') \subset F'$. (Composition of maps is now possible!) In the category $\mathcal{S}$ any spectrum is equivalent to any cofinal subspectrum of its. ∎

Proposition 1.20. *If $E \equiv \{E_n\}$ is a spectrum and (X, x_0) is a CW-complex, then we can form a new spectrum $E \wedge X$: we take $(E \wedge X)_n = E_n \wedge X$ with the weak topology.*

Proof. In fact $S(E \wedge X)_n = S(E_n \wedge X) = S^1 \wedge (E_n \wedge X) \cong (S^1 \wedge E_n) \wedge E \subset E_{n+1} \wedge X$. Furthermore, given a map $f : E \to F$ of spectra represented by (E', f') and a map $g : K \to L$ of CW-complexes, we get a map $f \wedge g : E \wedge K \to F \wedge L$ of spectra represented by $(E' \wedge K, f' \wedge g)$, since $E' \wedge K$ is cofinal in $E \wedge K$. $\qquad\square$

Definition 1.25. A *homotopy* between spectra is a map $h : E \wedge I^+ \to F$. There are two maps $i_0 : E \to E \wedge I^+$, $i_1 : E \to E \wedge I^+$, induced by the inclusions of $0, 1$ in I^+. Then, we say two maps of spectra $f_0, f_1 : E \to F$ are *homotopic* if there is a homotopy $h : E \wedge I^+ \to F$ with $h \circ i_0 = f_0$, $h \circ i_1 = f_1$. We shall write $h_0 \equiv h \circ i_0$, $h_1 \equiv h \circ i_1$. In terms of cofinal subspectra we can say that two maps $f_0, f_1 : E \to F$ represented by (E'_0, f'_0), (E'_1, f'_1), respectively, are homotopic if there is a cofinal subspectrum $E'' \subset E'_0 \cap E'_1$ and a function $h'' : E'' \wedge I^+ \to F$ such that $h''_0 = f'_0|_{E''}$, $h''_1 = f'_1|_{E''}$. Homotopy is an equivalence relation, so we may define $[E, F]$ to be the set of equivalence classes of maps $f : E \to F$. Composition passes to homotopy classes. The corresponding category is denoted by $\mathcal{S}'$.

Proposition 1.21. *We have the functor $E : \mathcal{W}_{\bullet} \to \mathcal{S}$, such that $E(X, x_0) = E(X)$, and $E(f) = \{E(f)_n\}$, with $E(f)_n = S^n f : S^n X \to S^n Y$, $n \geq 0$, for any $f : (X, x_0) \to (Y, y_0)$. The functor E embeds $\mathcal{W}_{\bullet}$ into $\mathcal{S}$.*

Proposition 1.22. *To any map $f : E \to F$ between spectra, we can associate another spectrum, the (mapping cone), $F \cup_f CE$.*

Proof. In fact let I be pointed on 0, and set $CE \equiv E \wedge I$. Then the mapping cone of f is the spectrum $(F \cup_f CE)_n = E_n \cup_{f'_n} (E'_n \wedge I)$, where (E', f') represents f. If (E'', f'') is another representative of f, then $\{F_n \cup_{f'_n} (E'_n \wedge I)\}$ and $\{F_n \cup_{f''_n} (E''_n \wedge I)\}$ have a natural cofinal subspectrum $\{F_n \cup_{f'''_n} (E'''_n \wedge I)\}$ and hence are equivalent. $\qquad\square$

Proposition 1.23. *One has a natural, invertible, functor $\Sigma : \mathcal{S} \to \mathcal{S}$ that induces a functor on $\mathcal{S}'$.*

Proof. For any spectrum $E \equiv \{E_n\}$ we can define ΣE to be the spectrum with $\Sigma E_n = E_{n+1}$, $n \in \mathbf{Z}$. For any function $f : E \to F$ we define $\Sigma(f) : \Sigma E \to \Sigma F$ to be the map represented by $(\Sigma E', \Sigma(f'))$. Furthermore, Σ induces a functor on $\mathcal{S}'$ since $f_0 \simeq f_1$ implies $\Sigma(f_0) \simeq \Sigma(f_1)$. We can iterate Σ: $\Sigma^{n+1} = \Sigma \circ \Sigma^n$, $n \geq 1$. Σ has also an inverse defined by

$(\Sigma^{-1}E)_n = E_{n-1}$, $\Sigma^{-1}(f)_n = f_{n-1}$. One has $\Sigma^n \circ \Sigma^m = \Sigma^{n+m}$, for all integers n, m. $\qquad\square$

Remark 1.6. ΣE and $E \wedge S^1$ have the same homotopy type, therefore the suspension is invertible in $\mathcal{S}'$. The higher homotopy groups for topological spaces are very difficult to compute. For spectra that computations are easier. $\qquad\blacksquare$

Proposition 1.24. *We have wedge sums in $\mathcal{S}$: given a collection $\{E^\alpha : \alpha \in A\}$ of spectra, we define $\bigvee_\alpha E^\alpha$ by $(\bigvee_\alpha E^\alpha)_n = \bigvee_\alpha E_n^\alpha$.*

Proof. Since $S(\bigvee_\alpha E_n^\alpha) = \bigvee_\alpha SE_n^\alpha \subset \bigvee_\alpha E_{n+1}^\alpha$, this is a spectrum. $\qquad\square$

Proposition 1.25. *For any collection $\{E^\alpha : \alpha \in A\}$ of spectra, the inclusions $i_\beta : E^\beta \to \bigvee_\alpha E^\alpha$ induce bijections:*

$$\left\{ \begin{array}{l} \{Hom(i_\alpha, 1)\} : Hom_{\mathcal{S}}(\bigvee_\alpha E^\alpha, F) \to \prod_\alpha Hom_{\mathcal{S}}(E^\alpha, F) \\[2mm] \{i_\alpha^*\} : [\bigvee_\alpha E^\alpha, F] \to \prod_\alpha [E^\alpha, F] \end{array} \right\}, \quad \forall F \in Hom(\mathcal{S}).$$

Inclusions of spectra possess the homotopy extension property just as with CW-complexes.

Definition 1.26. (*Homotopy groups of spectra*). We set $\pi_n(E) = [\Sigma^n S^0, E]$, $n \in \mathbf{Z}$.

Proposition 1.26. *One has the isomorphisms of abelian groups:*

$$\left\{ \begin{array}{l} \pi_n(E) \cong \lim_{\to} \pi_{n+k}(E_k, *) \\[2mm] \pi_n(E(X)) \cong \lim_{\to} \pi_{n+k}(S^k X, *) = \pi_n^s(X) \end{array} \right\}, \quad n \in \mathbf{Z}.$$

Note that $\pi_n^s(X)$ may be quite different from $\pi_n(X, x_0)$.

Proposition 1.27. 1) *If $f : E \to F$ is a map of spectra which is a weak homotopy equivalence, then $f_* : [G, E] \to [G, F]$ is a bijection for any spectrum G.*

2) *A map of spectra is a weak homotopy equivalence iff it is a homotopy equivalence.*

Definition 1.27. For any map $f : E \to F$ of spectra we call the sequence

$$E \xrightarrow{f} F \xrightarrow{j} F \cup_f CE$$

a *spectral cofibre sequence*. A *general cofibre sequence*, or simply a *cofibre sequence*, is any sequence

$$G \xrightarrow{g} H \xrightarrow{h} K$$

for which there is a homotopy commutative diagram

$$
\begin{array}{ccccc}
G & \xrightarrow{g} & H & \xrightarrow{h} & K \\
\alpha \downarrow & & \downarrow \beta & & \downarrow \gamma \\
E & \xrightarrow[f]{} & F & \xrightarrow[j]{} & F \cup_f CE
\end{array}
$$

where α, β, γ are homotopy equivalences.

Proposition 1.28. 1) *In the sequence*

$$
E \xrightarrow{f} F \xrightarrow{j} F \cup_f CE \xrightarrow{\kappa'} E \wedge S^1 \xrightarrow{f \wedge 1} F \wedge S^1
$$

each pair of consecutive maps forms a cofibre sequence.

2) *Given a homotopy commutative diagram of spectra and maps, as the following*

$$
\begin{array}{ccccccc}
G & \xrightarrow{g} & H & \xrightarrow{h} & K & \xrightarrow{\kappa} & G \wedge S^1 \\
\alpha \downarrow & & \downarrow \beta & & \downarrow \gamma & & \downarrow \alpha \wedge 1 \\
G' & \xrightarrow{g'} & H' & \xrightarrow{h'} & K' & \xrightarrow[\kappa']{} & G' \wedge S^1
\end{array}
$$

where the rows are cofibre sequences, we can find a map $\gamma : K \to K'$ such that the resulting diagram is homotopy commutative.

3) *If $G \xrightarrow{g} H \xrightarrow{h} K$ is a cofibre sequence, then for any spectrum E the sequences*

$$
[E, G] \xrightarrow{g_*} [E, H] \xrightarrow{h_*} [E, K]
$$

$$
[G, E] \xleftarrow{g^*} [H, E] \xleftarrow{h^*} [K, E]
$$

are exact.

Theorem 1.8. ((Co)homology theories associated with any spectrum). For each $(X, x_0) \in Ob(\mathcal{W}'_\bullet)$ and $n \in \mathbf{Z}$, we have

$$
E_n(X) = \pi_n(E \wedge X) = [\Sigma^n S^0, E \wedge X]
$$

$$
E^n(X) = [E(X), \Sigma^n E] \cong [\Sigma^{-n} S^0 \wedge X, E].
$$

These define (co)homology theories on $\mathcal{W}'_\bullet$ that satisfy the wedge axiom. The coefficient groups of the homology theory $E_\bullet$ are

$$
E_n(S^0) = \pi_n(E \wedge S^0) = \pi_n(E), \quad n \in \mathbf{Z}.
$$

The coefficient groups of the cohomology theory $E^\bullet$ are

$$
E^n(S^0) = [E(S^0), \Sigma^n E] = [S^0, \Sigma^n E] \cong [\Sigma^{-n} S^0, E] = \pi_{-n}(E), \quad n \in \mathbf{Z}.
$$

Furthermore, any map $f : E \to F$ of spectra induces natural transformations $T_\bullet(f) : E_\bullet \to F_\bullet$, $T^\bullet(f) : E^\bullet \to F^\bullet$ of homology and cohomology theories respectively. If f is a homotopy equivalence, then $T_\bullet(f)$ and $T^\bullet(f)$ are natural equivalences. This is the case iff $f_\bullet : \pi_n(E) \to \pi_n(F)$ is an isomorphism for all $n \in \mathbf{Z}$, i.e., $T_\bullet(f)$ is a natural equivalence iff it is an isomorphism on the coefficient groups.

Proof. For $f : (X, x_0) \to (Y, y_0)$ we take $E_n(f) = (1 \wedge f)$ and $E^n(f) = E(f)$.[10] We define $\sigma_n : E_n(X) \to E_{n+1}(SX)$ to be the composite

$$E_n(X) \quad = [\Sigma^n S^0, E \wedge X] \xrightarrow[\cong]{\Sigma} [\Sigma^{n+1} S^0, \Sigma E \wedge X] \to [\Sigma^{n+1} S^0, E \wedge S^1 \wedge X] \cong \quad E_{n+1}(SX)$$
$$\| \qquad\qquad\qquad\qquad\qquad\qquad\qquad\qquad\qquad\qquad\qquad\qquad \|$$
$$E_n(X) \xrightarrow[\sigma_n]{} \qquad\qquad\qquad\qquad\qquad\qquad\qquad E_{n+1}(SX)$$

Then σ_n is a natural equivalence. Furthermore, we define $\sigma^n : E^{n+1}(SX) \to E^n(X)$ to be the composite

$$E^{n+1}(SX) \quad = [E(SX), \Sigma^{n+1} E \wedge X] \xleftarrow[\cong]{i^*} [\Sigma E(X), \Sigma^{n+1} E] \xrightarrow[\cong]{\Sigma^{-1}} [E(X), \Sigma^n E] = \quad E^n(X)$$
$$\| \qquad\qquad\qquad\qquad\qquad\qquad\qquad\qquad\qquad\qquad\qquad\qquad \|$$
$$E^{n+1}(SX) \xrightarrow[\sigma^n]{} \qquad\qquad\qquad\qquad\qquad\qquad\qquad E_n(X)$$

Then σ^n is a natural equivalence too. Let (X, A) be any pointed CW-pair. Since

$$E_n \wedge (X \cup CA) \cong (E_n \wedge X) \cup C(E_n \wedge A), \quad n \in \mathbf{Z},$$

we see that

$$E \wedge A \xrightarrow{1 \wedge i} E \wedge X \xrightarrow{1 \wedge j} E \wedge (X \cup CA)$$

is a cofibre sequence. Therefore,

$$[\Sigma^n S^0, E \wedge A] \xrightarrow{(1 \wedge i)_*} [\Sigma^n S^0, E \wedge X \xrightarrow{(1 \wedge j)_*} [\Sigma^n S^0, E \wedge (X \cup CA)]$$

is exact; but this is just the sequence

$$E_n(A) \xrightarrow{i_*} E_n(X) \xrightarrow{j_*} E_n(X \cup CA).$$

Thus $E_\bullet$ is a homology theory on $\mathcal{W}'_\bullet$. Since $S^n(X \cup CA) \cong S^n X \cup C(S^n A)$, $n \in \mathbf{Z}$, we see that

$$E(A) \xrightarrow{E(i)} E(X) \xrightarrow{E(j)} E(X \cup CA)$$

---

[10] Here we have used the fact that $E(SX)$ is a cofinal subspectrum of $\Sigma E(X)$ and hence the inclusion $i : E(SX) \to \Sigma E(X)$ induces an isomorphism i^*.

is a cofibre sequence. Hence

$$[E(A), \Sigma^n(E)] \xrightarrow{E(i)^*} [E(X), \Sigma^n(E)] \xrightarrow{E(j)^*} [E(X \cup CA), \Sigma^n(E)]$$

is exact; but this is just the sequence

$$E^n(A) \xleftarrow{\;i^*\;} E^n(X) \xleftarrow{\;j^*\;} E^n(X \cup CA).$$

Thus $E^\bullet$ is a cohomology theory on $\mathcal{W}'_\bullet$. Since for any collection $\{X_\alpha : \alpha \in A\}$ of CW-complexes we have $S^n(\bigvee_\alpha X_\alpha) \cong \bigvee_\alpha S^n X_\alpha$ and hence $E(\bigvee_\alpha X_\alpha) \cong \bigvee_\alpha E(X_\alpha)$ we conclude that $\{i_\alpha^*\} : E^n(\bigvee_\alpha X_\alpha) \to \prod_\alpha E^n(X_\alpha)$ is an isomorphism for all $n \in \mathbf{Z}$. In other words $E^\bullet$ satisfies the wedge axiom. One can also prove that $E_\bullet$ satisfies the wedge axiom. $\qquad\square$

Corollary 1.1. *For any spectrum E and any filtration $\{X^n\}$ of a CW-complex X we have an exact sequence*

$$0 \to \overset{1}{\lim} E^{q-1}(X^n) \to E^q(X) \xrightarrow{\{i_n^*\}} \overset{0}{\lim} E^q(X^n) \to 0.$$

Proposition 1.29. *We can extend the cohomology theory $E^\bullet$ to a cohomology theory on the category $\mathcal{S}'$ by simply taking*

$$E^n(F) = [F, \Sigma^n(E)], \quad n \in \mathbf{Z},\ F \in Ob(\mathcal{S}').$$

Furthermore, if $T^\bullet : E^\bullet \to F^\bullet$ is a natural equivalence of cohomology theories on $\mathcal{S}'$, we can show $T^\bullet = T^\bullet(f)$ for some map $f : E \to F$.

Proof. In fact $E^\bullet$ is a cohomology theory in the sense that we have natural equivalences

$$
\begin{array}{ccccc}
E^{n+1}(F \wedge S^1) & \to & E^{n+1}(\Sigma F) & \underset{\cong}{\to} & E^n(F) \\
\| & & & & \| \\
E^{n+1}(F \wedge S^1) & & \underset{\sigma^n}{\to} & & E^n(F)
\end{array}
$$

for all $n \in \mathbf{Z}$, $F \in Ob(\mathcal{S}')$. Furthermore, $E^\bullet$ satisfies the following exactness axiom: For any cofibre sequence $F \xrightarrow{f} G \xrightarrow{g} H$, the sequence

$$E^n(F) \xleftarrow{f^*} E^n(G) \xleftarrow{g^*} E^n(H)$$

is exact. (This axiom is equivalent to the usual one over $\mathcal{W}'_\bullet$.) $\qquad\square$

Proposition 1.30. *A possible extension of the homology theory $E^\bullet$ to a homology theory on the category $\mathcal{S}'$ is the following*

$$E_n(G) = \pi_n(E \wedge G) \equiv [\Sigma^n S^0, E \wedge G].$$

In this case, however, it is not assured that a natural transformation $T_\bullet :$ $E_\bullet \to F_\bullet$ on $\mathcal{S}'$ is of the form $T^\bullet = T^\bullet(f)$ for some map $f : E \to F$.

Proposition 1.31. *If E is an Ω-spectrum, then for every CW-complex (X, x_0) we have a natural isomorphism $E^n(X) \cong [X, x_0; E_n, *]$.*

Example 1.17. (*Example of (co)homology theories associated to spectra*). Let us consider the *sphere spectrum* $S^0 \equiv E(S^0)$. The associated homology theory:

$$S_\bullet^0(X) = \pi_\bullet(S^0 \wedge X) = \{\varinjlim \pi_{n+k}(S^k X, *) \equiv \pi_n^s(X)\},$$

is called *stable homotopy* (of X). Furthermore, the associated cohomology theory:

$$(S^0)^\bullet(X) = \pi_s^\bullet(X) \equiv \{\varinjlim \pi_{n+k}(S^k \wedge X)\},$$

is called *stable cohomotopy* (of X). For any $n \geq 2$ we have the natural map

$$i_0 : \pi_n(X, x_0) \to \varinjlim \pi_{n+k}(S^k X, *) = \pi_n^s(X), \ i_0(x) = \{x\}.$$

We can also define i_0 as follows: any map $f : (S^n, s_0) \to (X, x_0)$ defines a function $E(f) : E(S^n) \to E(X)$. Since $E(S^n)$ is a cofinal subspectrum of ΣS^0, we get a map $\{E(f)\} : \Sigma^n S^0 \to E(X)$, and

$$i_0[f] = [\{E(f)\}] \in [\Sigma^n S^0, E(X)] = \pi_n^s(X).$$

This definition of i_0 applies even for $n = 0$ or 1. i_0 is a homomrphism for $n \geq 1$. The coefficient groups $\pi_n^s(S^0) = \varinjlim \pi_{n+k}(S^k, s_0)$ are called *stable homotopy groups* or *n-stems*, and denoted by π_n^s. These groups are known only through a finite range of $n > 0$. In particular, one has: $\pi_n^s = 0$, $n < 0$, $\pi_0^s \cong \mathbf{Z}$.   ∎

Proposition 1.32. 1) *For every spectrum E we can define a reduced homology theory $E_\bullet(-)$ and a reduced cohomology theory $E^\bullet(-)$ on $\mathcal{T}'_{op,\bullet}$, $\mathcal{T}^{2'}_{op,\bullet}$ and $\mathcal{T}'_{op}$ respectively.*

Proof. In fact, for $X \in Ob(\mathcal{T}'_{op,\bullet})$ we have the following reduced homology theory: $\tilde{E}_\bullet(X) \equiv E_\bullet(X') = \pi_\bullet(E \wedge X')$, where X' is any CW-substitute for

X. Furthermore, for any $(X, A) \in Ob(\mathcal{T}^{2'}_{op,\bullet})$ we have the following reduced homology theory: $E_\bullet(X, A) \equiv \tilde{E}_\bullet(X^+ \cup CA^+)$. Finally for any space $X \in Ob(\mathcal{T}'_{op})$ we have $E_\bullet(X) = E_\bullet(X, \emptyset)$. Furthermore, $E^n(X) = [X, E_n]$. The coefficients of these theories are the groups $E^\bullet(*) \cong E_\bullet(*) = \pi_\bullet(E)$. $\qquad\square$

Theorem 1.9. (Atiyah-Hirzebruch-Whitehead). *Suppose $\{E_n\}$ be a spectrum and X a space. Then, there is a spectral sequence $\{E_r^{\bullet,\bullet}, d_r\}$ with*

$$E_2^{p,q} \cong H^p(X; E^q(*))$$

converging to $E^\bullet(X)$. Furthermore, there is also a spectral sequence $\{E_{\bullet,\bullet}^r, d^r\}$ with

$$E_{p,q}^2 \cong H_p(X; E^q(*))$$

converging to $E_\bullet(X)$.[11] *Here $E_\bullet(-)$ (resp. $E^\bullet(-)$) is the homology (resp. cohomology) associated to the spectrum $\{E_n\}$.*

Theorem 1.10. (Leray-Serre). *If $E_\bullet$ is a homology theory with products satisfying the wedge axiom for CW-complexes and the WHE axiom, then for every fibration $p : E \to B$ orientable with respect to $E_\bullet$,*[12] *and with B 0-connected, there is a spectral sequence $\{E_{p,q}^{'r}, d^r\}$ converging to $E_\bullet(E)$ and having*

$$E_{p.q}^2 \cong H_p(B; E_q(F)).$$

The spectral sequence is natural with respect to a fibre map.

Remark 1.7. The problem of extension of maps and sections of fiber bundles is related to (co)homology theories. In fact we have the following theorems.∎

Theorem 1.11. *Let K be a cell complex and let $L \subset K$ be a subcomplex. Let X be a simply-connected topological space (or at least that is homotopically simple in the sense that $\pi_1(X)$ is abelian and acts trivially on all the groups $\pi_i(X)$, $i > 1$.) A given map $f : L \to X$, can be extended from the subcomplex $L \bigcup K^{i-1}$ to $L \bigcup K^i$, if $\pi_{i-1}(X) = 0$.*

---

[11] Let $\mathcal{U}$ be an abelian category. A *differential object* in $\mathcal{U}$ is a pair (A,d) where $A \in Ob(\mathcal{U})$ and $d \in Hom_\mathcal{U}(A;A)$ such that $d^2 = 0$. let $\mathcal{D}(\mathcal{U})$ be the category of differential objects in $\mathcal{U}$. We call *homology* the additive functor $H:\mathcal{D}(\mathcal{U}) \to \mathcal{U}$, given by $H(A,d)=\ker(d)/\mathrm{im}\,(d)=Z(A)/B(A)$, where $Z(A)$ is the set of *cycles* of A and B is the set of *boundaries* of A. $H(A)$ is the *homology* of (A,d). Then, a *spectral sequence* in the category $\mathcal{U}$ is a sequence of differential objects of $\mathcal{U}$: $\{E_n, d_n\}$, $N=1,2,\cdots$, such that $H(E_n, d_n)=E_{n+1}$, $n=1,2,\cdots$. (See, e.g., ref. [25].)

[12] For the homological definition of orientability see nex Remark 1.8.

Proof. In fact the obstruction to such an extension is determined by an element α_f of the relative cohomology group $H^i(K, L; \pi_{i-1}(X))$. The vanishing of α_f in this group suffices for the map to be extendible. In particular, the extension is assured if $\pi_{i-1}(X) = 0$. (For more details see e.g. ref.[17].)$\square$

Theorem 1.12. *Let* $f, g : K \to X$ *be two maps which coincide on the* $(q-1)$-*skeleton* K^{q-1} *of* K. *On each cell* $\sigma^q \subset K^q$ *the two maps* f *and* g *give rise, via their restrictions, to two maps* $f, g : \sigma^q \to X$ *coinciding on the boundary:* $f|_{\partial\sigma^q} = g|_{\partial\sigma^q}$, *and therefore yielding in combination a map* $S^q \to X$, *determining what is called a "distinguishing element" of* $\pi_q(X)$, *i.e., for each* q-*cell* σ^q *of* K, *we have a difference cochain* $\alpha(\sigma^q, f, g) \in \pi_q(X)$. *Then, the difference cochain may be regarded as belonging to the cohomology group* $H^q(K; \pi_q(X))$.

Theorem 1.13. *If* $X = K(G, n)$ *is a Eilenberg-MacLane space then there is a natural one-to-one correspondence* $[K, X] \leftrightarrow H^n(K; G)$. *In the case* $n = 1$, *the elements of* $H^1(K; G)$ *and* $[K, X]$ *are determined by the homomorphisms* $\pi_1(K) \to G$. *(This theorem remains true even if* G *is non-abelian.)*

Example 1.18. One has a natural one-to-one correspondence $[K^n, S^n] \leftrightarrow H^n(K^n; \mathbf{Z})$, where K^n is an n-dimensional complex. $\qquad\blacksquare$

Theorem 1.14. *Let* $\pi : E \to B$ *be a fibre bundle with base* B *given as a simplicial (or cell) complex and fibre* F. *We shall assume that* B *is simply-connected (or at least that* $\pi_1(B)$ *acts trivially on the groups* $\pi_i(F)$). *We shall assume also that the fibre* F *is simply-connected (or at least homotopically simple). Suppose* $s : B^{q-1} \to E$ *be a croos-section of the fibre bundle above the* $(q-1)$-*skeleton* $B^{q-1} \subset B$. *An obstruction to extending a cross-section may be regarded as an element of* $H^q(B; \pi_{q-1}(F))$. *In particular, if the fibre is the* $(q-1)$-*sphere* S^{q-1}, *then the obstruction* $\alpha \in H^q(B; \pi_{q-1}(F))$ *is an Euler characteristic class of the fibre bundle.*

Proof. Let σ^q be any q-simplex of B. Above the simplex σ^q the fibre bundle is canonically identifiable with the direct product: $\pi^{-1}(\sigma^q) \cong \sigma^q \times F$. As on the boundary $\partial\sigma^q \cong S^{q-1}$ the cross-section $s : \partial\sigma^q \to \partial\sigma^q \times F$ is by assumption, already given. Hence via the projection map onto F we obtain a map $S^{q-1} \to F$, defining an element $\alpha(\sigma^q, s) \in \pi_{q-1}(F)$ for each q-simplex $\sigma^q \subset B^q$. Therefore an *obstruction cocycle* α to the attempted extension of the cross-section s to the q-skeleton B^q, belongs to $H^q(B; \pi_{q-1}(F))$. $\quad\square$

Theorem 1.15. *Let* $\varphi_i : B \to E$, $i = 1, 2$, *be two cross-sections agreeing on* the $(q-1)$-skeleton $B^{q-1} \subset B$. The obstruction to a homotopy between the cross-sections φ_1 and φ_2, $\alpha(\varphi_1, \varphi_2) \in H^q(B; \pi_q(F))$.

Corollary 1.2. *If the fibre is contractible, $(\pi_i(F) = 0$ for all $i)$, then it follows that cross-sections always exist, and moreover that all cross-sections are homotopic.*

Example 1.19. This is the situation for the fiber bundle of positive definite quadratic forms, where the cross-sections are Riemannian metrics. So over a manifold M Riemannian metrics always exist and are homotopic, i.e., any two Riemannian metrics are continuously deformable one into the other. For indefinite metrics of type (p, q) with $p + q = n$, this results is not more valid. In fact in these cases one has $\pi_i(F) = \pi_i(GL(n; \mathbf{R})/O(p, q)) \neq 0$. ∎

Example 1.20. Connections on a fibre bundle $E \to B$, with fibre F, always exist. In fact, such connections can be identified with sections of the fibre bundle of horizontal directions over any point $x \in B$. ∎

Remark 1.8. (*Fundamental homology class of manifold*). Let Λ be a commutative ring. Let M be a fixed n-dimensional manifold, not necessarily compact. Let $K \subset M$ denote a compact subset of M. If $K \subset L \subset M$, one has a natural homomorphism $\rho_K : H_i(M, M \setminus L; \Lambda) \to H_i(M, M \setminus K; \Lambda)$. If $a \in H_i(M, M \setminus L; \Lambda)$, then we call $\rho_K(a)$ the *restriction* of a to K. The groups $H_i(M, M \setminus K; \Lambda)$ are zero for $i > n$. A homology class $a \in H_n(M, M \setminus K; \Lambda)$ is zero iff the restriction $\rho_x(a) \in H_n(M, M \setminus x; \Lambda)$ is zero for each $x \in K$. Let us, now, take $\Lambda = \mathbf{Z}$. Then

$$H_i(M, M \setminus x; \mathbf{Z}) \cong H_i(\mathbf{R}^n, \mathbf{R}^n \setminus \{0\}; \mathbf{Z}) = \begin{cases} 0, & i \neq n \\ \text{infinite cyclic}, & i = n. \end{cases}$$

A *local orientation* μ_x for M at x is a choice of one of two possible generators for $H_n(M, M \setminus x; \mathbf{Z})$. Note that such a μ_x determines local orientations μ_y for all points y in a small neighbourhood of x. In fact, if B is a ball about x, then for each $y \in B$ the isomorphisms

$$H_\bullet(M, M \setminus x; \mathbf{Z}) \xrightarrow{\rho_x} H_\bullet(M, M \setminus B; \mathbf{Z}) \xleftarrow{\rho_y} H_\bullet(M, M \setminus y; \mathbf{Z})$$

determine a local orientation μ_y. An *orientation* for M is a function which assigns to each $x \in M$ a local orientation μ_x which continuously depends on x, i.e., for each x there should exist a compact neighbourhood N and a class $\mu_N \in H_n(M, M \setminus N; \mathbf{Z})$ so that $\rho_y(\mu_N) = \mu_y$ for each $y \in N$. An *oriented manifold* is a manifold M endowed with an orientation. For any oriented manifold M and any compact $K \subset M$, there is one and only one $\mu_K \in H_n(M; M \setminus K; \mathbf{Z})$ which satisfies $\rho_x(\mu_K) = \mu_x$ for each $x \in K$. In particular, if M is compact, then there is one an only one $\mu_M \in H_n(M; \mathbf{Z})$ with the required property. This class $\mu \equiv \mu_M$ is called the *fundamental homology class* of M. As $H_n(M; \mathbf{Z}) \cong \mathbf{Z}^r$, for oriented manifold, with r the number of connected components of M, it follows that $\mu_M = (1, \cdots_r \cdots, 1)$ is the basis of the $\mathbf{Z}$-module $H_n(M; \mathbf{Z})$. For any coefficient domain Λ, the unique homomorphism $\mathbf{Z} \to \Lambda$ gives rise to a class in $H_n(M, M \setminus K; \Lambda)$ that will also be denoted by μ_K. For example, we can take $\Lambda \equiv \mathbf{Z}_2$, so that $\mu_K \in H_n(M, M \setminus K; \mathbf{Z}_2)$. This homology class can be constructed directly for any n-dimensional manifold, without making any assumption of orientability. In particular, if M is a non-orientable compact manifold of dimension n, with r connected components, one has $H_n(M; \mathbf{Z}_2) \cong (\mathbf{Z}_2)^r$. Similar considerations apply to an oriented manifold with boundary. For each compact subset $K \subset M$, there exists a unique class $\mu_K \in H_n(M, (M \setminus K) \cup \partial M; \mathbf{Z})$ with the property that $\rho_x(\mu_K) = \mu_x$ for each $x \in K \cap (M \setminus \partial M)$. In particular, if M is compact, then there is a unique fundamental homology class $\mu_M \in H_n(M, \partial M; \mathbf{Z})$ with the required property. Then, the connecting homomorphism $\partial : H_n(M, \partial M; \mathbf{Z}) \to H_n(\partial M; \mathbf{Z})$ maps μ_M to the fundamental homology class of ∂M. $\blacksquare$

Remark 1.9. (*Stiefel-Whitney characteristic classes and Stiefel-Whitney characteristic numbers*). Given a vector bundle $p : E \to B$, fibre $\mathbf{R}^n$, and bundle group $G = O(n)$, we can form the associated bundle $p_k : E_k \to B$ of orhonormal k-frames with fibre $F_k \equiv V_{n,k}$, the Stiefel manifold of orthonormal k-frames in $\mathbf{R}^n$.[13] As

$$\pi_i(V_{n,k}) = \begin{cases} 0, & i < n - k \\ \mathbf{Z}, & n - k = 2r + 1, \text{ or } k = 1 \\ \mathbf{Z}_2, & n - k = 2r, \end{cases}$$

it follows that for each $k = 1, \cdots, n$ the obstruction to the existence of a cross-section of the fibre bundle $p_k : E_k \to B$ will be an element $\alpha_k \in H^{n-k+1}(B; \pi_{n-k}(V_{n,k}))$. The cohomology class α_k, considered module 2, is

---

[13]   In particular, for $k=n$, $F_k \cong O(n)$, and for $k=1$, $F_1 \cong S^{n-1}$.

called the *kth Stiefel-Whitney class of the vector bundle* $p : E \to B$,[14] and
we write

$$\begin{cases} w_0 = 1, \\ w_q \equiv \alpha_{n-q+1} \ (\text{mod } 2) \ \in H^q(B; \mathbf{Z}_2), \quad q = 1, \cdots, n., \\ w_q = 0, \quad q > n. \end{cases}$$

The polynomial $w(t) \equiv w_0 + w_1 t + \cdots + w_q t^q + \cdots + w_n t^n$ is called the
Stiefel-Whitney polynomial of the vector bundle $p : E \to B$. We call
$w(E) = 1 + w_1 + \cdots + w_n$ the *Stiefel-Whitney class* of $p : E \to B$. If the
manifold M ($\dim M = n$) is orientable, one has $w_1 = 0$. In fact, the natural
mapping $j : BSO(n) \to BO(n)$, which "forgets" the orientation on the oriented n-dimensional planes representing the points in $\hat{G}_{\infty,n} = BSO(n)$,[15]
induces an epimorphism $j^\bullet : H^\bullet(BO(n); \mathbf{Z}_2) \to H^\bullet(BSO(n); \mathbf{Z}_2)$, with
kernel $< w_1 >$, the ideal generated by the first Stiefel-Whitney class $w_1 \in
H^1(BO(n); \mathbf{Z}_2)$. Therefore, for any fiber bundle W_O and W_{SO}, over a
manifold X, with structure groups $O(n)$ and $SO(n)$ respectively, we get
the following commutative diagram:

$$\begin{array}{ccccccccc}
0 & \to & < w_1 > & \to & H^\bullet(BO(n); \mathbf{Z}_2) & \xrightarrow{\pi} & H^\bullet(BSO(n); \mathbf{Z}_2) & \to & 0 \\
& & & & \downarrow \cong & & \downarrow \cong & & \\
0 & \to & < w_1|_X > & \to & Kar^\bullet(W_O; \mathbf{Z}_2) & \xrightarrow{\bar{\pi}} & Kar^\bullet(W_{SO}; \mathbf{Z}_2) & \to & 0 \\
& & & & \cap & & \cap & & \\
& & & & H^\bullet(X; \mathbf{Z}_2) & & H^\bullet(X; \mathbf{Z}_2) & &
\end{array}$$

Therefore W_O admits the reduction to W_{SO} if $\bar{\pi}$ is injective, i.e., if $w_1|_X = 0$.
In particular, X, $\dim X = n$, is orientable iff TX admits the reduction to
$SO(n)$, i.e., its first Stiefel-Whitney class is zero. We get also the following
propositions:

(i) $w_n = \chi(X) \bmod 2$, where $\chi(X)$ is the Euler characteristic of X.

(ii) For the direct product $(E_1 \times E_2, p_1 \times p_2, B_1 \times B_2)$ of vector bundles,
one has $w(t) = \overset{1}{w}(t) \, \overset{2}{w}(t)$, where $\overset{i}{w}(t)$, $i = 1, 2$, are the Stiefel-Whitney
polynomials of the factors.

(iii) Let $E_1 \oplus E_2$ be the Whitney sum of two real vector bundles over
the same base, then $w(E_1 \oplus E_2) = w(E_1)w(E_2)$, i.e., $w_k(E_1 \oplus E_2) =
\sum_{0 \le i \le k} w_i(E_1) w_{k-i}(E_2)$.

---

[14] By the Stiefel-Whitney classes of an n-dimensional manifold M, one means the corresponding classes of TM.

[15] universal classifying space for the group $SO(n)$.

(iv) One has the following isomorphism: $H^\bullet(M; \mathbf{Z}_2) \cong \mathbf{Z}_2[w_1, \cdots, w_n]$, $\dim M = n$.

Let $G_{N,k}$ be the Grassmannian manifold that represents the set of k-dimensional vector spaces of $\mathbf{R}^N$. Then $G_{\infty,k} = BO(k)$ is the universal classifying space for the orthogonal group $O(k)$. Let $f : M \to \mathbf{R}^N$ be an embedding of a k-dimensional manifold M into $\mathbf{R}^N$, (for enough large N). Then we have the following map (*generalized Gauss map*)

$$\tau_M : M \to G_{\infty,k}, \quad x \mapsto T_x M \hookrightarrow T_{f(x)}\mathbf{R}^N \subset T_{f(x)}\mathbf{R}^\infty.$$

This induces the tangent bundle TM from the universal bundle, $V_{\infty,k} \to G_{\infty,k}$, of orthonormal tangent k-frames on M, with respect to the induced metric. So we have the following commutative diagram:

$$\begin{array}{ccc}
\tau_M^* V_{\infty,k} \equiv TM & \longrightarrow & V_{\infty,k} \\
\downarrow & & \downarrow \\
M & \xrightarrow{\quad \tau_M \quad} & G_{\infty,k}
\end{array}$$

Each element $w \in H^s(G_{\infty,k}; \mathbf{Z}_2)$ determines a corresponding mod 2 *characteristic class* $w(M) \equiv \tau_M^* w$, and the *stable* mod 2 *characteristic classes* of M (with respect to the group $O(k)$) are those determined by elements $w \in H^s(BO(k); \mathbf{Z}_2)$ which are pull-backs of elements $\bar{w} \in H^s(BO(k+1); \mathbf{Z}_2)$ via the natural embeddings $\lambda : BO(k) \to BO(k+1)$, induced by the standard embedding $\lambda : O(k) \to O(k+1)$: $w = \lambda^* \bar{w}$. So if $w(M)$ is a stable mod 2 characteristic class of M, then $w(M) = \tau_M^* \lambda^* \bar{w}$, for some $\bar{w} \in H^s(BO(k+1); \mathbf{Z}_2)$.

Let us assume, now, that $M = \partial W$, $\dim W = k+1$, we have: $\bar{w}(W) = \tau_W^* \bar{w}$, and taking into account the inclusion map $i : M \to W$, the restriction to M of the map $\tau_W : W \to BO(k+1)$, satisfies

$$\begin{aligned}
\tau_W|_M &= \tau_W \circ i = \lambda \circ \tau_M \\
&= \tau_M \bigoplus 1 : M \to BO(k+1),
\end{aligned}$$

inducing the Whitney sum $TM \bigoplus_M T_0^0 M$. Therefore $w(M) = i^* \bar{w}(W)$. Now since $M = \partial W$ it follows that, for the fundamental homology class $[M]$, we have $i_*[M] = 0$. Therefore, assuming $w(M) \in H^k(M, \mathbf{Z}_2)$, its evaluation on $[M]$ gives:

$$< w(M), [M] > = < i^* \bar{w}, [M] > = < \bar{w}, i_*[M] > = < \bar{w}, 0 > = 0.$$

Since $[M]$ generate $H_k(M; \mathbf{Z}_2)$[16] it follows that the *Stiefel-Whitney numbers* of M, i.e., the values taken on $[M]$ by its mod 2 stable charactersitic classes of dimension k, are zero. $\blacksquare$

So we have the following theorem.

Theorem 1.16. (Pontrjagin). *If B is a smooth compact $(n+1)$-dimensional manifold with boundary $M \equiv \partial B$, then the Stiefel-Whitney numbers of M are all zero.*

Proof. Here, let us give, also, another direct proof to this important theorem . Let us denote the fundamental homology class of the pair $(B, \partial B)$ by $\mu_B \in H_{n+1}(B, \partial B; \mathbf{Z}_2)$. Then, the natural homomorphism $\partial : H_{n+1}(B, \partial B; \mathbf{Z}_2) \to H_n(\partial B; \mathbf{Z}_2)$ maps μ_B to $\mu_{\partial B}$. For any class $v \in H^n(M; \mathbf{Z}_2)$ one has: $< v, \partial \mu_B > = < \delta v, \mu_B >$, where δ is the natural homomorphism $\delta : H^n(\partial B; \mathbf{Z}_2) \to H^{n+1}(B, \partial B; \mathbf{Z}_2)$. (There is not sign since we are working mod 2.) Consider the tangent bundles $TB|_{\partial B}$ and $T(\partial B) \subset TB|_{\partial B}$. Choosing a Euclidean metric on TB, there is a unique outward normal vector field along ∂B, spanning a trivial line bundle ϵ^1, and it follows that $TB|_{\partial B} \cong T(\partial B) \bigoplus \epsilon^1$. Hence the Stiefel-Whitney classes of $TB|_{\partial B}$ are precisely equal to the Stiefel-Whitney classes w_j of $T(\partial B)$. Using the exact sequence

$$H^n(B; \mathbf{Z}_2) \xrightarrow{i^*} H^n(\partial B; \mathbf{Z}_2) \xrightarrow{\delta} H^{n+1}(B, \partial B; \mathbf{Z}_2)$$

it follows that $\delta(w_1^{r_1} \cdots w_n^{r_n}) = 0$ and therefore

$$< w_1^{r_1} \cdots w_n^{r_n}, \partial \mu_B > = < \delta(w_1^{r_1} \cdots w_n^{r_n}), \mu_B > = 0.$$

As $\partial \mu_B = \mu_{\partial B}$, we can conclude that all Stiefel-Whitney numbers of ∂B are zero. $\square$

Definition 1.28. In the category of closed smooth (resp. oriented) manifolds of dimension n, we can define, by means of bordism properties, an equivalence relation. More precisely, we say that $X_1 \sim X_2$ iff $X_1 \bigcup X_2 = \partial W$, where W is a smooth manifold of dimension $n + 1$. The corresponding set Ω_n (resp. $^+\Omega_n$) of equivalences classes is called the *n-dimensional bordism group* (resp. *oriented n-dimensional bordism group*).

---

[16] If M is a closed and connected manifold of dimension k, admitting a finite triangulation, then $H_k(M; \mathbf{Z}_2) \cong \mathbf{Z}_2$. The fundamental class of M is $[M] = \sum_i v_i^k$, i.e., the sum of all k-simplexes.

Now, the nullity of the Stiefel-Whitney numbers is also a sufficient condition to bording. In fact we have the following.

Theorem 1.17. (Pontrjagin-Thom-Wall). *A closed n-dimensional smooth manifold V, belonging to the category of smooth differentiable manifolds, is bordant in this category, i.e., $V = \partial M$, for some smooth $(n+1)$-dimensional manifold M, iff the Stiefel-Whitney numbers $< w_{i_1} \cdots w_{i_p}, \mu_V >$ are all zero, where $i_1 + \cdots + i_p = n$ is any partition of n and μ_V is the fundamental class of V. Furthermore, the bordism group Ω_n of n-dimensional smooth manifolds is a finite abelian torsion group of the form*

$$\Omega_n \cong \underbrace{\mathbf{Z}_2 \oplus \cdots \oplus \mathbf{Z}_2}_{q},$$

where q is the number of nondyadic partitions of n.[17] *Two smooth closed n-dimensional manifolds belong to the same bordism class iff all their corresponding Stiefel-Whitney numbers are equal. Furthermore, the bordism group $^+\Omega_n$ of closed n-dimensional oriented smooth manifolds is a finitely generated abelian group of the form*

$$^+\Omega_n \cong \mathbf{Z} \bigoplus \cdots \bigoplus \mathbf{Z} \bigoplus \mathbf{Z}_2 \bigoplus \cdots \bigoplus \mathbf{Z}_2,$$

where infinite cyclic summands can occur only if $n \equiv 0 \bmod 4$. Two smooth closed oriented n-dimensional manifolds belong to the same bordism class iff all their corresponding Stiefel-Whitney and Pontrjagin numbers are equal.[18]
Proof. See, e.g., refs.[55,81,87]. □

Definition 1.29. Let a *k-cycle* of M be a couple (N, f), where N is a k-dimensional closed (oriented) manifold and $f : N \to M$ is a differentiable mapping. A *group of cycles (N, f)* of an n-dimensional manifold M is the set of formal sums $\sum_i (N_i, f_i)$, where (N_i, f_i) are cycles of M. The quotient of this group by the cycles equivalent to zero, i.e., the boundaries, gives the *bordism groups $\underline{\Omega}_s(M)$*. We define *relative bordisms $\underline{\Omega}_s(X, Y)$*, for any pair of manifolds (X, Y), $Y \subset X$, where the boundaries are constrained to belong to Y. Similarly we define the *oriented bordism groups $^+\underline{\Omega}_s(M)$* and $^+\underline{\Omega}_s(X, Y)$.

---

[17] A partition $(i_1, \cdots, i_r)$ of n is nondyadic if none of the i_β are of the form $2^s - 1$.

[18] *Pontrjagin numbers* are determined by means of homonymous characteristic classes belonging to $H^\bullet(BG, \mathbf{Z})$, where BG is the classifying space for G-bundles, with $G = S_p(n)$.

Proposition 1.33. 1) *One has* $\underline{\Omega}_s(*) \cong \Omega_s$ *and* $^+\underline{\Omega}_s(*) \cong {}^+\Omega_s$.

2) *For bordisms, the theorem of invariance of homotopy is valid. Furthermore, for any CW-pair* (X, Y), $Y \subset X$, *one has the isomorphisms:* $\underline{\Omega}_s(X, Y) \cong \Omega_s(X/Y)$, $s \geq 0$.

Theorem 1.18. *One has a natural group-homomorphism* $\underline{\Omega}_s(X) \to H_s(X; \mathbf{Z}_2)$. *This is an isomorphism for* $s = 1$. *In general,* $\underline{\Omega}_s(X) \neq H_s(X; \mathbf{Z}_2)$.

Proof. In fact one has the following lemma.

Lemma 1.3. (Quillen).[74] *One has the canonical isomorphism:*

$$\underline{\Omega}_p(X) \cong \bigoplus_{r+s=p} H_r(X; \mathbf{Z}_2) \otimes_{\mathbf{Z}_2} \Omega_s.$$

In particular, as $\Omega_0 = \mathbf{Z}_2$ and $\Omega_1 = 0$, we get $\underline{\Omega}_1(X) \cong H_1(X; \mathbf{Z}_2)$. Note that for contractible manifolds, $H_s(X) = 0$, for $s > 0$, but $\underline{\Omega}_s(X)$ cannot be trivial for any $s > 0$. So, in general, $\underline{\Omega}_s(X) \neq H_s(X; \mathbf{Z}_2)$.

After these results and remarks, the proof of the theorem follows directly. □

Definition 1.30. Let B be a closed differential connected manifold and let $\xi \equiv (p : E \to B, F \equiv \mathbf{R}^n, G)$ be a vector bundle over B with fibre $\mathbf{R}^n$ and structure group $G = O(n)$, $SO(n)$, $U(n)$, $SU(n)$ or $S_p(n)$. Let $\tilde{E} \to B$ be the subbundle of ξ defined by the vectors in the fibers with length ≤ 1. The fiber F' of $\tilde{E}$ is $F' \equiv D^n \subset \mathbf{R}^n$. The boundary $\partial\tilde{E}$ is a fiber bundle with fiber S^{n-1}. The *Thom complex* of the vector bundle ξ is the quotient complex $M(\xi) = \tilde{E}/\partial\tilde{E}$. So $M(\xi)$ is the compactified to a point of E: $M(\xi) \equiv E \cup \{\infty\} \equiv E^+$.

Example 1.21. If $B = BG$, the base space of the universal G-bundle, with fibre $\mathbf{R}^n$, we denote by MG the corresponding Thom complex. In particular, for $G = O(n)$, $SO(n)$, $U(n/2)$, $SU(n/2)$, or $S_p(n/4)$, we denote the corresponding Thom complexes by $MO(n)$, $MSO(n)$, $MU(n/2)$, $MSU(n/2)$ and $MS_p(n/4)$ respectively. In some cases the complexes $MO(s)$, $MSO(s)$ are Eilenberg-MacLane complexes of type $K(G, n)$. The following table resume such cases.

TAB.1.3 - $MO(s)$ and $MSO(s)$ as $K(G,n)$-complexes.

$MO(1)\cong\mathbf{R}\mathbb{P}^\infty\cong K(\mathbf{Z}_2,1)$	$\pi_j=0,\, j>1$
$MSO(1)\cong Me\cong S^1\cong K(\mathbf{Z},1)$	$\pi_j=0,\, j>1$
$MSO(2)\cong\mathbf{C}\mathbb{P}^\infty\cong K(\mathbf{Z},2)$	$\pi_j=0,\, j\neq2$

$\phi(1)=u\in H^n(MG)$ is the fundamental class of $K(G,n)$.

(See Lemma 1.4.)

The Thom complexes $M(\xi)$ are simply connected for $n > 1$. Their homotopy groups are reported in the following table.

TAB.1.4 - Homotopy groups of $M(\xi)$.

$\pi_j(M(\xi))$	Conditions
0	$1\leq j<n$
$\mathbf{Z}_2$	$j=n$, non-orientable fiber bundle
$\mathbf{Z}$	$j=n$, orientable fiber bundle

∎

Theorem 1.19. 1) *A cycle $x \in H_s(M;\mathbf{Z}_2)$, $dim M = n + s$, is realized by means of a closed s-dimensional submanifold $N \subset M$, iff there exists a mapping $f : M \to MO(n)$ such that $f^*u = Dx$, where $u \in H^n(MO(n);\mathbf{Z}_2)$ is a fundamental class and $D : H_s(M;\mathbf{Z}_2) \to H^n(M;\mathbf{Z}_2)$ is the Poincaré duality operator.*

2) Let M be an $(n + s)$-dimensional oriented manifold. A cycle $x \in H_s(M;\mathbf{Z})$ is realized by means of a closed oriented submanifold $N \subset M$ iff there exists a mapping $f : M \to MSO(n)$ such that $f^*u = Dx$. A cycle $x \in H_s(M;\mathbf{Z})$ is realized by means of a closed oriented submanifold $N \subset M$ of trivial normal bundle (i.e., defined by means of a family of nonsingular equations $\psi_1 = 0,\cdots,\psi_k = 0$, in M) iff there exists a mapping $f : M \to Me \cong S^n$ such that $f^*u = Dx$.

3) Similar theorems hold in the cases of realizations of cycles by means of submanifolds with normal bundles endowed with structural groups $U(n/2)$, $SU(n/2)$, $S_p(n/4)$. A mapping $M \to MU(n/2)$, $M \to MSU(n/2)$ and $MS_p(n/4)$ generates such restrictions.

Proof. Let us consider the following definitions and lemmas.

Lemma 1.4. *One has the natural isomorphisms:*

$$\phi : H_i(B; A) \to H_{n+i}(M(\xi); A), \ \phi : H^i(B; A) \to H^{n+i}(M(\xi); A),$$

$$z \mapsto \phi(z) \equiv p^* z \ (\mathrm{mod} \ \partial \tilde{E}), \ i \geq 0, \ n = \dim F,$$

where $A \equiv \mathbf{Z}_2$ *if* $G = O(n)$, $A \equiv \mathbf{Z}$ *if* $G = SO(n)$, $A \equiv \mathbf{Q}$ *if* $G = U(n)$, $S_p(n)$. *More precisely* $\phi = D_{\tilde{E}} \circ D_B$, *where* D_X *are the following duality operators:*

$$D_B : H_q(B) \to H^{m-q}(B), \ \dim B = m,$$

$$D_{\tilde{E}} : H_{m-q}(\tilde{E}) \cong H^{m-q}(B) \to H_{n+m-(m-q)}(\tilde{E}, \partial \tilde{E}) \cong H_{q+n}(M(\xi)), \ q > 0.$$

One has a fundamental class *in the cohomology of Thom of* ξ, *i.e.,* $\phi(1) \in H^n(M(\xi))$. *Furthermore, the following identifications hold:* $M(\xi)/B \equiv \xi$, $M(\xi) \backslash B \cong \{*\}$, *where* B *is identified with a submanifold of* $M(\xi)$ *by means of the zero section.*

Lemma 1.5. *The Stiefel-Whitney class* $w_i \in H^i(B; \mathbf{Z}_2)$ *of a vector bundle* ξ *with base* B *is related to the Thom complex* $M(\xi)$ *by the following relationan:* $w_i = \phi^* s_q^i(\phi(1))$, *where* $\phi : H^q(B; \mathbf{Z}_2) \to H^{n+q}(M(\xi); \mathbf{Z}_2)$ *and* s_q^i *are Steenrod squares* [77,83], *i.e., homomorphisms* $s_q^i : H^n(M(\xi); \mathbf{Z}_2) \to H^{n+i}(M(\xi); \mathbf{Z}_2)$.

Definition 1.31. Let $X \subset Y$ be a smooth submanifold of a smooth manifold Y, of codimension k. Let M be another smooth manifold. Then a smooth map $f : M \to Y$ is said to be *transversally regular* on X if the rank of the map $Df(x) : T_x M \to T_{f(x)} Y / T_{f(x)} X$ is k whenever $f(x) \in X$. In such a case we write $f(M) \pitchfork X$.

The group $O(n)$ contains a subgroup of diagonal matrices $D(n) \subset O(n)$ such that $D(n) \cong \mathbf{Z}_2 \times \cdots \times \mathbf{Z}_2$. Furthermore, one has a canonical mapping between classifying spaces

$$BD(n) \cong \mathbf{RP}_1^\infty \times \cdots \times \mathbf{RP}_n^\infty \xrightarrow{\ i\ } BO(n)$$

and the induced cohomology mapping

$$i^* : H^\bullet(BO(n); \mathbf{Z}_2) \to H^\bullet(BD(n); \mathbf{Z}_2).$$

One can see that i^* is a monomorphism and that $\mathrm{im}\,(i^*)$ is the set of symmetric polynomials in $\chi_1, \cdots, \chi_n$, where $0 \neq \chi_i \in H^1(\mathbf{RP}^\infty; \mathbf{Z}_2)$. Moreover, the Stiefel-Whitney classes are elementary symmetric polynomials:

$i^*(w_q) = \sum_{i_1 < \cdots < i_q} \chi_{i_1} \cdots \chi_{i_q}$. Recall that the mapping $j : BSO(n) \to BO(n)$ induces an epimorphism

$$j^\bullet : H^\bullet(BO(n); \mathbf{Z}_2) \to H^\bullet(BSO(n); \mathbf{Z}_2),$$

with $\ker(j^\bullet)$ generated, as ideal, by the element $w_1 \in H^1(BO(n); \mathbf{Z}_2)$. Now, let $M(\xi)$ be the Thom complex of the universal bundle ξ on $BO(n)$. By means of the zero section we can identify a mapping $f : BO(n) \to M(\xi)$, hence we get also the following morphism

$$f^\bullet : H^\bullet(M(\xi); \mathbf{Z}_2) \to H^\bullet(BO(n); \mathbf{Z}_2),$$

that is a ring monomorphism and $\mathrm{im}\,(f^\bullet)$ is the set of polynomials in w_i divisible by $w_n \in H^n(BO(n); \mathbf{Z}_2)$, where $i^* w_n = \chi_1 \cdots \chi_n$. Furthermore,

$$f^* \phi(1) = w_n,$$
$$f^* \phi(w_i) = s_q^i(w_n) = w_i w_n.$$

In general $f^* \phi(x) = x w_n$. Similar results can be obtained for $H^\bullet(BSO(n); \mathbf{Z}_2)$.

Lemma 1.6. *The bordism groups Ω_s, $^+\Omega_s$, Ω_s^U, Ω_s^{SU}, Ω_s^{Sp}, are canonically isomorphic to the stable homotopy groups. See the following table.*

TAB.1.5 - Bordism groups and stable homotopy groups.

$\Omega_s \cong \pi_{n+s}(MO(n))$
$^+\Omega_s \cong \pi_{n+s}(MSO(n))$
$\Omega_s^U \cong \pi_{n+s}(MU(n/2))$
$\Omega_s^{SU} \cong \pi_{n+s}(MSU(n/2))$
$\Omega_s^{Sp} \cong \pi_{n+s}(MSp(n/4))$

After above results let us prove theorem for $G = O(n)$. Let $N \subset M$ be a s-dimensional closed submanifold of M, $\dim M = n + s$. The normal bundle on N defines the following commutative diagram:

$$
\begin{array}{ccc}
M & \xrightarrow{\ f\ } & MO(n) \\
\cup & & \cup \\
N & \longrightarrow & BO(n)
\end{array}
$$

where the bottom mappping is the classifying map. The complement of the neighbourhood of N in M fully reduces to a point σ^0 by means of the contraction of $\partial\tilde{E}$ in the construction of $M(\xi) = MO(n)$. Then, we get $f^*\phi(1) \equiv f^*u = D[N]$, where $\phi : H^0(BO(n); \mathbf{Z}_2) \to H^n(MO(n); \mathbf{Z}_2)$ (see Lemma 1.4). So if $x \in H_s(M; \mathbf{Z}_2)$ is representable by a submanifold $N \subset M$, then we should have $f^*\phi(1) = Dx$. Conversely, if we assign a mapping $f : M \to MO(n)$ transversally regular along $BO(n) \subset MO(n)$, the reciprocal image of $N \equiv f^{-1}(BO(n))$ is such that $f^*u = D[N]$. $\qquad\square$

Theorem 1.20. *Any cycle $x \in H_n(M; \mathbf{Z})$, $\dim M = n + 1$, can be realized with a closed submanifold. Furthermore, any cycle $x \in H_n(M; \mathbf{Z})$, $n + 1 \le \dim M \le n + 2$, can be realized with an orientable closed submanifold. If $s < n/2$, for any cycle $x \in H_s(M; \mathbf{Z})$, $\dim M = n$, there exists a $\lambda \ne 0$ such that the cycle λx is represented by an s-dimensional submanifold $N \subset M$.*

Theorem 1.21. *Let M be any finite cell complex. For any cycle $x \in H_s(M; \mathbf{Z})$ there exists an $\lambda \ne 0$ such that λx is the image of an s-dimensional manifold N, $\phi : N \to M$, $\phi_*[N] = \lambda x$. One has the natural epimorphism*

$$\underline{\Omega}^{SO}_s(X, Y) \bigotimes \mathbf{Q} \to H_s(X, Y; \mathbf{Q}).$$

Definition 1.32. An *X-structure* on a manifold V is a homotopy class of cross-sections of the bundle of geometric objects with fiber X over V. A *X-manifold* is a manifold V together with an X-structure on V.

Proposition 1.34. *If V is an X-manifold, then so is ∂V.*

Definition 1.33. Given any closed X-manifold V one can define a second X-manifold $-V$ such that $\partial(V \times I) \cong V \bigcup (-V)$. Thus one can define a bordism group for the class of n-dimensional X-manifolds, denoted by $\underline{\Omega}^X_n$ and called the *n-th X-bordism group*.

Spectra are also related to the bordism groups. In fact, one has the following.

Theorem 1.22. (R.Thom). *One has the following isomorphism: $\underline{\Omega}^X_\bullet \cong \pi_\bullet(MX)$, where MX is the spectrum (Thom spectrum) associated to the X-structure.*

Proof. See, e.g., refs.[77,81,83]. $\qquad\square$

Example 1.22. In particular, for $X = BO$, SO, $Spin$, U, SU and S_p, we get results reported in Theorem 1.19 and Lemma 1.6. $\qquad\blacksquare$

Definition 1.34. A *singular X-manifold* in the space Y is a continuous map (*singular simplex*) $f : M \to Y$, where M is a closed X-manifold. Two singular X-manifolds (M, f), (M', f') in Y are called *X-bordant* if there is a pair (W, g) such that W is a compact X-manifold with boundary,

$\partial W \cong M \bigcup (-M')$, and g is a continuous map $g : W \to Y$ such that $g|_M = f$, $g|_{M'} = f'$. The corresponding bordism group, for n-dimensional X-manifolds contained in Y, is denoted by $\underline{\Omega}_n^X(Y)$.

Theorem 1.23. *One has the following isomorphisms:*

$$\underline{\Omega}_n^X(Y) \cong MX_n(Y) \cong \pi_n(MX \wedge Y^+),$$

where $Y^+ \equiv Y \cup \{\infty\} \equiv Y/\emptyset$.

Proof. See, e.g., refs.[77,81,83]. $\qquad\qquad\qquad\qquad\qquad\qquad\qquad\square$

Theorem 1.24. *One has the following isomorphism:*

$$\underline{\Omega}_n^X(Y) \cong E_n(Y^+),$$

where $E_\bullet(Y^+)$ *is the homology induced by the spectrum* MX.

1.2 - QUANTUM ALGEBRAS

Quantum algebras are algebraic-topological structures that aim locally interpret the concept of quantum fields. In fact, when, in Physics, one quantizes a field, locally represented by some **K**-valued functions, one obtains a set of linear operators on some suitable Hilbert space, or locally convex topological vector space. These operators generate some noncommutative algebra. Thus the first step, in order to obtain a geometric, fully covariant theory of quantum phenomena, is to start with a theory about noncommutative algebras that should have enough structure to develop a local theory of noncommutative analysis. This is made by taking the "fundamental algebras of noncommutative numbers" as Fréchet spaces having also an associative **K**-algebra structure, containing **K** as a subalgebra and endowed with **K**-linear morphisms $A \to \mathbf{K}$. We call such structures "quantum algebras". For such structures we study their (co)homology properties, (de Rham, Hochschild, cyclic) and relations with Hopf, Lie and Jordan algebras.

Definition 1.35. A *quantum algebra* is a triplet (A, ϵ, c), where: (i) A is a metrizable, complete, Hausdorff locally convex topological **K**-vector space, (i.e., a Fréchet space), that is also an associative **K**-algebra with unit; (ii) $\epsilon : \mathbf{K} \to Z \subset A$ is a ring homomorphism, where Z is the centre of A : $Z \equiv \{a \in A | ab = ba, \forall b \in A\}$.[19] (The homomorphism ϵ is necessarily

---

[19] In this work we denote the centre of a quantum algebra A by $Z=Z(A)$ instead than A_0 as made in some previous papers on the subject [63,64,69,71]. This is done in order to avoid any confusion with the even part of a quantum superalgebra A, here denoted, instead, by A_0.

injective. Of course A is also an associative Z-algebra with unit.) (iii) $c : A \to \mathbf{K}$ is a $\mathbf{K}$-linear morphism, with $c(e) = 1$, where e is the unit of A. (Furthermore, if we require that A is a $\mathbf{K}$-algebra morphism we say that A is an *augmented quantum $\mathbf{K}$-algebra*).[20] For any $a \in A$ we call $a_C \equiv c(a) \in \mathbf{K}$ the *classic-limit* of a.

Limit cases of quantum algebras are just given by $\mathbf{R}$, $\mathbf{C}$ and $\mathbf{H}$. In fact, we have the following.

Proposition 1.35. *A normed quantum algebra A such that its norm N satisfies the following conditions:*

$$(1.7) \qquad \begin{cases} \text{(i) } N(ab) = N(a)N(b), \\ \text{(ii) } N(a) > 0, \\ \text{(iii) } N(a) = 0 \Rightarrow a = 0, \end{cases} \Bigg\} \ \forall a, b \in A,$$

coincides with one of the following algebras, $\mathbf{R}$, $\mathbf{C}$, $\mathbf{H}$, iff $G(A) = A \setminus \{0\}$, where $G(A)$ is the group of invertible elements of A.

Proof. In fact we can use the following lemma.

Lemma 1.7. *Let k be a field. A normed k-algebra A with unit $e \in A$, such that its norm satisfies conditions (1.7), is a division algebra, i.e., for each nonzero element $a \in A$, there exists its inverse $a^{-1} \in A$, (i.e., $a^{-1}a = aa^{-1} = e$), coincides with one of the following algebras: $\mathbf{R}$, $\mathbf{C}$, $\mathbf{H}$, $\mathbf{O}$.*

Proof of Lemma 1.7. It is standard. (See, e.g., ref.[4].) $\qquad\qquad\square$

Then taking into account also that a quantum algebra is associative, and that, instead, the algebra of octonions is not associative, we conclude the proof. $\qquad\qquad\square$

In the following table we report also some distinguished submanifolds X of

---

[20] For example a super algebra, in the sense of [62], can be considered an augmented quantum algebra. However, in the following, whether no confusion can arise, we shall simply call quantum algebras also augmented quantum algebras.

the algebras $\mathbf{R}$, $\mathbf{C}$, $\mathbf{H}$, $\mathbf{O}$, defined by the condition $X = N^{-1}(1)$.[21]

TAB.1.6 - Unit norm submanifolds of $\mathbf{R}$, $\mathbf{C}$, $\mathbf{H}$, $\mathbf{O}$.

Symbols	Properties
$S^0 \equiv \{1, -1\} \cong \mathbf{Z}_2 \subset \mathbf{R}$	Lie group
$S^1 \cong U(1) \subset \mathbf{C}$	Lie group ([†])
$S^3 \cong SU(2) \subset \mathbf{H}$	Lie group ([†])
$S^7 \cong Sp(2)/Sp(1) \subset \mathbf{O}$	homogeneous space ([†])

([†]) parallelizable manifold,

i.e., with holonomy group the identity element.

Example 1.23. Less trivial examples of quantum algebras can be obtained by a (completed) tensor product $A \equiv B \otimes_{\mathbf{K}} F$, where B is a noncommutative quantum algebra and F is one of the following Fréchet spaces: (a) $F \equiv C^k(\Omega, \mathbf{R})$, space of C^k functions defined on an open subset Ω of $\mathbf{R}^n$; (b) $F \equiv Holom(\Omega; \mathbf{C})$, space of holomorphic functions on an open subset Ω of $\mathbf{C}^n$; (c) $F \equiv \mathbf{C}[[x_1, \cdots, x_n]]$, space of formal power series in n variables $x_1, \cdots, x_n$, with complex coefficients; (d) $F \equiv \mathcal{I}(\mathbf{R}^n; \mathbf{R}^n)$, space of C^∞ functions in $\mathbf{R}^n$ rapidly dcreasing at infinity. One can see that all above spaces, (very important in Mathematical Analysis), are examples of Fréchet spaces, which are not Banach spaces. (For a proof see, e.g., ref. [85].) On the other hand, we can use the algebraic structure of $\mathbf{K}$-algebra on $\mathbf{R}$ and $\mathbf{C}$ to induce a structure of $\mathbf{K}$-algebra on the above spaces F. Furthermore, on such algebras F we can recognize canonical inclusions $\epsilon : \mathbf{K} \to F$ and $\mathbf{K}$-linear mappings $c_F : F \to \mathbf{K}$, identified with an integration $c_F : f \mapsto c_F(f) \equiv \int f\eta$, with respect to a suitable measure on $\mathbf{R}^n$ and $\mathbf{C}^n$. In this way F becomes a commutative quantum algebra. Furthermore, as B and F are Fréchet spaces, it follows that also their completed tensor product has a natural structure of Fréchet space.[22] Furthermore, $A \equiv B \otimes_{\mathbf{K}} F$ has a natural structure of $\mathbf{K}$-algebra with product $(b_1 \otimes f_1)(b_2 \otimes f_2) =$

---

[21]   Note that there are some compact 7-dimensional manifolds that are homeomorphic but not diffeomerphic to S^7. (*Milnor exotic spheres* [49].) Let us emphasize also, that the authomorphism groups of the algebras $\mathbf{R}$, $\mathbf{C}$, $\mathbf{H}$ and $\mathbf{O}$, are the compact semi-simple Lie groups called respectively *orthogonal, unitary, symplectic* and *exceptional groups*.

[22]   If E and F are Fréchet $\mathbf{K}$-spaces, their tensor product $E \otimes_{\mathbf{K}} F$ has a natural structure

$b_1 b_2 \otimes f_1 f_2$. Hence, we recognize also a natural inclusion $\epsilon : \mathbf{K} \to A$, $\lambda \mapsto \lambda \otimes 1 = 1 \otimes \lambda$, and a natural $\mathbf{K}$-linear mapping $c : A \to \mathbf{K}$, given by $c = c_B \otimes c_F$. Therefore, in such cases, we get a structure of noncommutative quantum algebra on A. ∎

In Tab.1.7 we define some useful $\mathbf{K}$-algebras canonically associated to a quantum algebra A.

TAB.1.7 - Derived $\mathbf{K}$-algebras of a quantum algebra A.

quantum extension: $\hat{A} \equiv Hom_{\mathbf{K}}(A;A)$
fullquantum extension: $\widehat{A} \equiv Hom_Z(A;A) \equiv A O A$

Proposition 1.36. 1) $\hat{A}$ and $\widehat{A}$ *have natural structures of Z-modules and quantum algebras.*

2) $\widehat{A}$ *has also a structure of Z-algebra.*

Proof. 1) The structure of Z-module is induced by the structure of Z-module of A, i.e., $(\lambda_2 f_1 + \lambda_2 f_2)(a) = \lambda_1 f_1(a) + \lambda_2 f_2(a)$, $\forall \lambda_1 \in Z$, $f_i \in \widehat{A}$ or $\hat{A}$, $i = 1, 2$, $a \in A$. This means also that $\widehat{A}$, and $\hat{A}$ are $\mathbf{K}$-modules. Furthermore, with respect to the product given by composition we get on both $\widehat{A}$ and $\hat{A}$ a structure of $\mathbf{K}$-algebra. But we can also consider the product induced by the product on A, i.e., $(f_1 f_2)(a) = f_1(a) f_2(a)$. This induces a structure of Z-algebra on $\widehat{A}$, as the maps $A \to A$ are Z-linear. Moreover, $\widehat{A}$ and $\hat{A}$ are also locally convex topological vector spaces with respect to the topology of *simple convergence*, (or *pointwise convergence*), i.e., the topology induced by the (product-) topology on A^A, the locally convex space of all mappings from A to A. (For informations on topological vector spaces see e.g., refs.[7, 38, 78, 85].) Taking into account that $\widehat{A}$ and $\hat{A}$ have canonical structures of Fréchet $\mathbf{K}$-spaces, we can conclude the proof. 2) This proof directly follows from above considerations. □

Remark 1.10. (*Cohomology of quantum algebras*). The product in A is a $\mathbf{K}$-bilinear mapping $\mu : A \times A \to A$. This induces the $\mathbf{K}$-linear mapping $\hat{\mu} : A \bigotimes_{\mathbf{K}} A \to A$, (resp. Z-linear mapping $\hat{\mu} : A \bigotimes_Z A \to A$), by means of

---

of locally convex vector space, (*projective topology*). Furthermore, this admits a canonical completion, denoted by $E \widehat{\otimes} F$, that is also a Fréchet space. (For more informations see, e.g., ref.[78, 85].) In the following for abuse of notation we shall denote the completed tensor product yet by the symbol $\otimes$.

the following commutative diagram, with split exact horizontal lines:

$$(1.8) \quad
\begin{array}{ccccccccc}
 & & 0 & & 0 & & & & \\
 & & \downarrow & & \downarrow & & & & \\
0 & \to & \mathfrak{k} & \to & \mathfrak{a} & & & & \\
 & & \downarrow & & \downarrow & & & & \\
0 & \to & \hat{\Omega}^1(A) & \to & A\otimes_{\mathbf{K}} A & \overset{\hat{\mu}}{\to} & A & \to & 0 \\
 & & \pi \downarrow & & \downarrow & & \| & & \\
0 & \to & \widehat{\Omega}^1(A) & \to & A\otimes_Z A & \underset{\widehat{\mu}}{\to} & A & \to & 0 \\
 & & \downarrow & & \downarrow & & & & \\
 & & 0 & & 0 & & & &
\end{array}$$

The commutative diagram (1.8) is also a diagram of two-sided A-modules considering $A \times A$, $A \otimes_{\mathbf{K}} A$, $A \otimes_Z A$ and A two-sided A-modules in natural a way. Then one has also the following splits: $A \otimes_{\mathbf{K}} A \cong \hat{\Omega}(A) \bigoplus (A \otimes 1)$, $A \otimes_Z A \cong \widehat{\Omega}(A) \bigoplus (A \otimes 1)$. If E is a two-sided A-module let $Der_{\mathbf{K}}(A; E)$, (resp. $Der_Z(A; E)$), denote the two-sided A-module of all the $\mathbf{K}$-derivations, (resp. Z-derivations), $A \to E$. Recall that a *derivation* from A to E is a map $\delta : A \to E$ such that $\delta(a+b) = \delta(a)+\delta(b)$ and $\delta(ab) = (\delta(a))b+a\delta(b)$, $\forall b \in A$. We denote by $Der(A; E)$ the set of all these derivations. We call $\mathbf{K}$-*derivation*, (resp. Z-*derivation*), a derivation such that $\delta|_{\mathbf{K}} = 0$, (resp. $\delta|_Z = 0$). Therefore, a $\mathbf{K}$-derivation, (resp. Z-derivation), is also a $\mathbf{K}$-linear, (resp. Z-linear), map $A \to E$. It follows also that a derivation is a $\mathbf{Z}$-linear map $A \to E$, as $\delta(1) = \delta(1.1) = \delta(1)1 + 1\delta(1) = 2\delta(1)$, hence $\delta(1) = 0$. In particular $Der(A) \equiv Der(A; A)$ is a Lie algebra under the bracket $[\delta_1, \delta_2] = \delta_1\delta_2 - \delta_2\delta_1$. One has the following commutative diagram of two-sided A-modules with exact vertical lines:

$$
\begin{array}{ccccc}
Hom_A(\hat{\Omega}^1(A); E) & \overset{\cong}{} & Der_{\mathbf{K}}(A; E) & \subset & Hom_{\mathbf{K}}(A; E) \\
\uparrow & & \uparrow & & \uparrow \\
Hom_A(\widehat{\Omega}^1(A); E) & \overset{\cong}{} & Der_Z(A; E) & \subset & Hom_Z(A; E) \\
\uparrow & & \uparrow & & \uparrow \\
0 & & 0 & & 0
\end{array}
$$

More precisely, the isomorphisms are given by $f \mapsto f \circ \hat{d}$ and $f \mapsto f \circ \pi \circ \hat{d} \equiv f \circ \widehat{d}$ respectively, with $\hat{d} : A \to \hat{\Omega}^1(A)$ defined by $\hat{d}(a) = 1 \otimes a - a \otimes 1$, $\forall a \in A$. The derivation $\hat{d} : A \to \hat{\Omega}^1(A)$, (resp. $\widehat{d} : A \to \widehat{\Omega}^1(A)$), is called *universal derivation*, (resp. *full universal derivation*), of A. In particular for $E \equiv A$ one has: $Der_{\mathbf{K}}(A) \equiv Der_{\mathbf{K}}(A; A) \cong \hat{\Omega}^1(A)^*$; $Der_Z(A) \equiv Der_Z(A; A) \cong$

$\hat{\Omega}^1(A)^*$.[23] The derivations $\hat{d}$ and $\widehat{d}$ solve the following universal problem. For any derivation $\hat{D} \in Der_{\mathbf{K}}(A; E)$ and $\widehat{D} \in Der_Z(A; E)$, one has a unique two-sided module morphism $i_{\hat{D}} : \hat{\Omega}^1(A) \to E$ and $i_{\widehat{D}} : \widehat{\Omega}^1(A) \to E$, such that $\hat{D} = i_{\hat{D}} \circ \hat{d}$ and $\widehat{D} = i_{\widehat{D}} \circ \widehat{d}$. More precisely one has: $i_{\hat{D}}(a \otimes b) = a\hat{D}(b)$ and $i_{\widehat{D}}(a \otimes b) = a\widehat{D}(b)$. Set:

$\hat{\Omega}^0(A) \equiv A$	$\hat{\Omega}^n(A) \equiv \hat{\Omega}^1(A) \otimes_A \cdots_n \cdots \otimes_A \hat{\Omega}^1(A)$
$\widehat{\Omega}^0(A) \equiv A$	$\widehat{\Omega}^n(A) \equiv \widehat{\Omega}^1(A) \otimes_A \cdots_n \cdots \otimes_A \widehat{\Omega}^1(A)$

One has the following commutative diagram of chain complexes with exact vertical lines:

$$
\begin{array}{ccccccccccccc}
& & 0 & & 0 & & & & & & & & \\
& & \downarrow & & \downarrow & & & & & & & & \\
0 & \to & \mathbf{K} & \to & \hat{\Omega}^0(A) & \xrightarrow{\hat{d}} & \hat{\Omega}^1(A) & \xrightarrow{\hat{d}} & \hat{\Omega}^2(A) & \xrightarrow{\hat{d}} & \cdots & \xrightarrow{\hat{d}} & \hat{\Omega}^n(A) & \xrightarrow{\hat{d}} & \cdots \\
& & \downarrow & & \downarrow & & \downarrow & & \downarrow & & & & \downarrow \\
0 & \to & Z & \to & \widehat{\Omega}^0(A) & \xrightarrow{\widehat{d}} & \widehat{\Omega}^1(A) & \xrightarrow{\widehat{d}} & \widehat{\Omega}^2(A) & \xrightarrow{\widehat{d}} & \cdots & \xrightarrow{\widehat{d}} & \widehat{\Omega}^n(A) & \xrightarrow{\widehat{d}} & \cdots \\
& & \downarrow & & \downarrow & & \downarrow & & \downarrow & & & & \downarrow \\
& & 0 & & 0 & & 0 & & 0 & & & & 0
\end{array}
$$

Note that $\hat{d}$, (resp. $\widehat{d}$), are $\mathbf{K}$-homomorphisms, (resp. Z-homomorphisms), just defined at the zero degree by the following $\hat{d}(a) = 1 \otimes a - a \otimes 1$. (They satisfy the Leibniz identity $\hat{d}(ab) = \hat{d}(a)b + a\hat{d}(b)$.) Similarly for $\widehat{d}$. Then $\hat{d}$, (resp. $\widehat{d}$), extends for $\mathbf{K}$-linearity, (resp. Z-linearity), into a $\mathbf{K}$-homomorphism, (resp. Z-hmomorphism), of degree 1, $\hat{d} : \hat{\Omega}^\bullet(A) \equiv \bigoplus_{n \geq 0} \hat{\Omega}^n(A) \to \hat{\Omega}^\bullet(A)$, (resp. $\widehat{d} : \widehat{\Omega}^\bullet(A) \equiv \bigoplus_{n \geq 0} \widehat{\Omega}^n(A) \to \widehat{\Omega}^\bullet(A)$), such that $\hat{d}^2 = 0$, (resp. $\widehat{d}^2 = 0$), that verifiy the Leibniz identity: $d(\omega.\theta) = d(\omega).\theta + (-1)^n \omega d\theta$ if ω is of degree n and θ of degree p. (Here $d = \hat{d}$ or $d = \widehat{d}$). The elements of the graded differential algebras $\hat{\Omega}^\bullet(A)$ and $\widehat{\Omega}^\bullet(A)$ are called respectively *differential forms* and *full differential forms* of the quantum algebra A. The homomorphisms $A \times A \to \hat{\Omega}^1(A)$, $(a, b) \mapsto a.\hat{d}(b) = a\hat{d}b$, and $A \times A \to \widehat{\Omega}^1(A)$, $(a, b) \mapsto a.\widehat{d}(b) = a\widehat{d}b$, induce the

---

[23] For example if $A \equiv \mathbf{K}[x^1, \dots, x^n] \equiv P$, $Der_{\mathbf{K}}(P)$ is a free P-module of rank n with basis $\{\partial x_1, \dots, \partial x_n\}$ and $Der_{\mathbf{K}}(P)^* \cong \hat{\Omega}^1(P)$ is a free P-module with basis $\{dx^1, \dots, dx^n\}$, dual of $\{\partial x_1, \dots, \partial x_n\}$. Of course, in this case, $Z = P$, $\widehat{\Omega}^1(P)^* = Der_P(P) \subset Hom_P(P; P) = P$.

isomorphisms given in the following table:

$\hat{\Omega}^1(A)\cong A\otimes_{\mathbf{K}}(A/\mathbf{K})$
$\widehat{\Omega}^1(A)\cong A\otimes_Z(A/Z)$
$\hat{\Omega}^n(A)\cong A\otimes_{\mathbf{K}}(A/\mathbf{K})\otimes_{\mathbf{K}}\cdots n\cdots\otimes_{\mathbf{K}}(A/\mathbf{K})$
$\widehat{\Omega}^n(A)\cong A\otimes_Z(A/Z)\otimes_Z\cdots n\cdots\otimes_Z(A/Z)$

In this sense a differential form $\alpha \in \hat{\Omega}^\bullet(A)$, (resp. full differential form $\alpha \in \widehat{\Omega}^\bullet(A)$), of the quantum algebra A can be represented in the form $\alpha = \sum_{n\geq 0}\kappa_n a^0.\hat{d}a^1.\hat{d}a^2\ldots\hat{d}a^n$, $\kappa_n \in \mathbf{K}$, (resp. $\alpha = \sum_{n\geq 0}\kappa_n a^0.\widehat{d}a^1.\widehat{d}a^2\ldots\widehat{d}a^n$, $\kappa_n \in Z$). Furthermore, one has

$$\hat{d}\alpha = \sum_{n\geq 0}\kappa_n 1.\hat{d}a^0.\hat{d}a^1.\hat{d}a^2\ldots\hat{d}a^n, \ \widehat{d}\alpha = \sum_{n\geq 0}\kappa_n 1.\widehat{d}a^0.\widehat{d}a^1.\widehat{d}a^2\ldots\widehat{d}a^n.$$

The products (noted $\sharp$) in $\hat{\Omega}^\bullet(A)$ and $\widehat{\Omega}^\bullet(A)$ are respectively given by the following rules:

$$(\omega.\hat{d}a)\sharp(b.\theta) = \omega\sharp\hat{d}(a.b).\theta - \omega\sharp(a.\hat{d}b).\theta, \ \forall a,b \in A, \ \omega,\theta \in \hat{\Omega}^\bullet(A),$$

$$(\omega.\widehat{d}a)\sharp(b.\theta) = \omega\sharp\widehat{d}(a.b).\theta - \omega\sharp(a.\widehat{d}b).\theta, \ \omega,\theta \in \widehat{\Omega}^\bullet(A).$$

Then $\hat{\Omega}^\bullet(A)$, (resp. $\widehat{\Omega}^\bullet(A)$), is the solution of the following universal problem: for any graded differential $\mathbf{K}$-algebra, (resp. Z-algebra), $\Sigma^\bullet$ and any homomorphism of $\mathbf{K}$-algebras, (resp. Z-algebras), $f : A \to \Sigma^\bullet$, there exists a unique homomorphism of graded $\mathbf{K}$-algebras, (resp. Z-algebras), $f^\bullet : \hat{\Omega}^\bullet(A) \to \Sigma^\bullet$, (resp. $f^\bullet : \widehat{\Omega}^\bullet(A) \to \Sigma^\bullet$), such that $f^\bullet|_{\hat{\Omega}^0(A)=A} = f$, (resp. $f^\bullet|_{\widehat{\Omega}^0(A)=A} = f$). The homology $\hat{H}^\bullet(A)$, (resp. $\widehat{H}^\bullet(A)$), of the chain complex $\{\hat{\Omega}^\bullet(A),\hat{d}\}$, (resp. $\{\widehat{\Omega}^\bullet(A),\widehat{d}\}$), is called the *de Rham cohomology*, (resp. *full de Rham cohomology*), of the quantum algebra A. ∎

Remark 1.11. (*Hochschild (co)homology of quantum algebras*). Let us consider the following graded algebras $T^\bullet(A)$ and $\dot{T}^\bullet(A)$ associated to a quantum algebra A:

$$T^\bullet(A)\equiv\bigoplus_{n\geq 0}T^n(A), \quad T^n(A)\equiv A\otimes_{\mathbf{K}}\cdots n+1\cdots\otimes_{\mathbf{K}}A\cong T^1(A)\otimes_A\cdots n\cdots\otimes_A T^1(A),$$

$$\dot{T}^\bullet(A)\equiv\bigoplus_{n\geq 0}\dot{T}^n(A), \quad \dot{T}^n(A)\equiv A\otimes_Z\cdots n+1\cdots\otimes_Z A\cong\dot{T}^1(A)\otimes_A\cdots n\cdots\otimes_A\dot{T}^1(A).$$

Then $\hat{\Omega}^n(A)$, (resp. $\widehat{\Omega}^n(A)$), is a sub two-sided A-module of $T^n(A)$, (resp. $\dot{T}^n(A)$). More precisely, $\hat{\Omega}^n(A) = \cap_i \ker(\hat{b}_i)$, (resp. $\widehat{\Omega}^n(A) = \cap_i \ker(\widehat{b}_i)$), where $\hat{b}_i$, (resp. $\widehat{b}_i$), are the **K**-homomorphisms, (resp. Z-homomorphisms), $\hat{b}_i : T^n(A) \to T^{n-1}(A)$, (resp. $\widehat{b}_i : \dot{T}^n(A) \to \dot{T}^{n-1}(A)$), defined by $a_0 \otimes a_1 \otimes \ldots \otimes a_n \mapsto a_0 \otimes a_1 \otimes \ldots a_i a_{i+1} \otimes a_{i+2} \ldots \otimes a_n$. The product in $T^\bullet(A)$, as well in $\dot{T}^\bullet(A)$, is given by $(a_0 \otimes a_1 \otimes \ldots \otimes a_n)(b_0 \otimes b_1 \otimes \ldots \otimes b_n) = a_0 \otimes a_1 \otimes \ldots \otimes a_n b_0 \otimes b_1 \otimes \ldots \otimes b_n$. Furthermore, $\hat{\Omega}^\bullet(A)$, (resp. $\widehat{\Omega}^\bullet(A)$), is a subalgebra of $T^\bullet(A)$, (resp. $\dot{T}^\bullet(A)$). For any n one has the following commutative diagrams:

$$
\begin{array}{ccc}
\hat{\Omega}^n(A) & \xrightarrow{\hat{d}} & \hat{\Omega}^{n+1}(A) \\
\gamma \downarrow & & \downarrow \gamma \\
T^n(A) & \xrightarrow{D} & T^{n+1}(A)
\end{array}
\qquad
\begin{array}{ccc}
\widehat{\Omega}^n(A) & \xrightarrow{\widehat{d}} & \widehat{\Omega}^{n+1}(A) \\
\dot{\gamma} \downarrow & & \downarrow \dot{\gamma} \\
\dot{T}^n(A) & \xrightarrow{\dot{D}} & \dot{T}^{n+1}(A)
\end{array}
$$

$$
\begin{array}{ccc}
T^n(A) & \xrightarrow{D} & T^{n+1}(A) \\
\phi \downarrow & & \downarrow \phi \\
\hat{\Omega}^n(A) & \xrightarrow{\hat{d}} & \hat{\Omega}^{n+1}(A)
\end{array}
\qquad
\begin{array}{ccc}
\dot{T}^n(A) & \xrightarrow{\dot{D}} & \dot{T}^{n+1}(A) \\
\dot{\phi} \downarrow & & \downarrow \dot{\phi} \\
\widehat{\Omega}^n(A) & \xrightarrow{\widehat{d}} & \widehat{\Omega}^{n+1}(A)
\end{array}
$$

where the homomorphisms ϕ and $\dot{\phi}$ are respectively given by:

$$(a_0 \otimes a_1 \otimes \ldots \otimes a_n) \mapsto a_0 \hat{d}a_1 . \hat{d}a_2 \ldots \hat{d}a_n; \quad (a_0 \otimes a_1 \otimes \ldots \otimes a_n) \mapsto a_0 \widehat{d}a_1 . \widehat{d}a_2 \ldots \widehat{d}a_n.$$

Note that ϕ, (resp. $\dot{\phi}$), is the inverse of the inclusion γ, (resp. $\dot{\gamma}$). However they are not homomorphisms of algebras. Furthermore, $D(a_0 \otimes a_1 \otimes \ldots \otimes a_n) = 1 \otimes a_0 \otimes a_1 \otimes \ldots \otimes a_n - (a_0 \otimes 1 \otimes a_1 \otimes \ldots \otimes a_n) + \ldots + (-1)^n(a_0 \otimes a_1 \otimes \ldots \otimes a_n \otimes 1)$ and similarly for $\dot{D}$. One has: $D^2 = 0$, $\dot{D}^2 = 0$. Therefore one has also the following commutative diagram with exact vertical lines:

$$
\begin{array}{ccccccccccc}
T^0(A) & \xrightarrow{D} & T^1(A) & \xrightarrow{D} & T^2(A) & \xrightarrow{D} & \ldots & \xrightarrow{D} & T^n(A) & \xrightarrow{D} & \ldots \\
\downarrow & & \downarrow & & \downarrow & & & & \downarrow & & \\
\dot{T}^0(A) & \xrightarrow{\dot{D}} & \dot{T}^1(A) & \xrightarrow{\dot{D}} & \dot{T}^2(A) & \xrightarrow{\dot{D}} & \ldots & \xrightarrow{\dot{D}} & \dot{T}^n(A) & \xrightarrow{\dot{D}} & \ldots \\
\downarrow & & \downarrow & & \downarrow & & & & \downarrow & & \\
0 & & 0 & & 0 & & & & 0 & &
\end{array}
$$

Furthermore, let us denote by $\underline{\hat{\Omega}^n(A)} \subset \hat{\Omega}^n(A)$, (resp. $\underline{\widehat{\Omega}^n(A)} \subset \widehat{\Omega}^n(A)$), the **K**-module, (resp. Z-module), of antisymmetric forms. Note that $\underline{\hat{\Omega}^\bullet(A)} \equiv \bigoplus_{n \geq 0} \underline{\hat{\Omega}^n(A)}$, (resp. $\underline{\widehat{\Omega}^\bullet(A)} \equiv \bigoplus_{n \geq 0} \underline{\widehat{\Omega}^n(A)}$) is not a sub-algebra of $\hat{\Omega}^\bullet(A)$, (resp. $\underline{\widehat{\Omega}^\bullet(A)}$), as the product in $\hat{\Omega}^\bullet(A)$, (resp. $\widehat{\Omega}^\bullet(A)$), does not conserve $\underline{\hat{\Omega}^\bullet(A)}$, (resp. $\underline{\widehat{\Omega}^\bullet(A)}$). One has the following commutative diagrams:

$$
\begin{array}{ccc}
\hat{\Omega}^n(A) & \xrightarrow{\hat{d}} & \hat{\Omega}^{n+1}(A) \\
\cup & & \cup \\
\underline{\hat{\Omega}^n(A)} & \xrightarrow{\hat{d}} & \underline{\hat{\Omega}^{n+1}(A)}
\end{array}
\qquad
\begin{array}{ccc}
\widehat{\Omega}^n(A) & \xrightarrow{\widehat{d}} & \widehat{\Omega}^{n+1}(A) \\
\cup & & \cup \\
\underline{\widehat{\Omega}^n(A)} & \xrightarrow{\widehat{d}} & \underline{\widehat{\Omega}^{n+1}(A)}
\end{array}
$$

Then one has the following commutative diagram with exact vertical lines:

$$
\begin{array}{ccccccccccccc}
 & & 0 & & A & & & & & & & & \\
 & & \downarrow & & \| & & & & & & & & \\
0 & \to & \mathbf{K} & \to & \underline{\hat{\Omega}^0(A)} & \xrightarrow{\hat{d}} & \underline{\hat{\Omega}^1(A)} & \xrightarrow{\hat{d}} & \cdots & \xrightarrow{\hat{d}} & \underline{\hat{\Omega}^n(A)} & \xrightarrow{\hat{d}} & \underline{\hat{\Omega}^{n+1}(A)} & \xrightarrow{\hat{d}} \cdots \\
 & & \downarrow & & \| & & \downarrow & & & & \downarrow & & \downarrow & \\
0 & \to & Z & \to & \underline{\widehat{\Omega}^0(A)} & \xrightarrow{\widehat{d}} & \underline{\widehat{\Omega}^1(A)} & \xrightarrow{\widehat{d}} & \cdots & \xrightarrow{\widehat{d}} & \underline{\widehat{\Omega}^n(A)} & \xrightarrow{\widehat{d}} & \underline{\widehat{\Omega}^{n+1}(A)} & \xrightarrow{\widehat{d}} \cdots \\
 & & \| & & \downarrow & & & & & & \downarrow & & \downarrow & \\
 & & A & & 0 & & & & & & 0 & & 0 &
\end{array}
$$

If A is an augmented quantum algebra then the horizontal lines are exact too. Note also that as a generalization of diagram (1.8) we get the following diagram with splited exact horizontal lines:

$$(1.9)\qquad
\begin{array}{ccccccccc}
 & & 0 & & 0 & & & & \\
 & & \downarrow & & \downarrow & & & & \\
0 & \to & \mathfrak{k}^{n+1} & \to & \mathfrak{a}^n & & & & \\
 & & \downarrow & & \downarrow & & & & \\
0 & \to & \hat{\Omega}^{n+1}(A) & \xrightarrow{\hat{i}_n} & \hat{\Omega}^n(A)\otimes_{\mathbf{K}} A & \xrightarrow{\hat{\mu}_n} & \hat{\Omega}^n(A) & \to & 0 \\
 & & \pi\downarrow & & \downarrow & & \| & & \\
0 & \to & \widehat{\Omega}^{n+1}(A) & \xrightarrow{\widehat{i}_n} & \widehat{\Omega}^n(A)\otimes_{Z} A & \xrightarrow[\widehat{\mu}_n]{} & \widehat{\Omega}^n(A & \to & 0 \\
 & & \downarrow & & \downarrow & & & & \\
 & & 0 & & 0 & & & &
\end{array}
$$

where $\mathfrak{k}^1 = \mathfrak{k}$, $\mathfrak{a}^0 = \mathfrak{k}$, $\hat{i}_n(\omega\hat{d}a) = \omega \otimes a$ and $\widehat{i}_n(\omega\widehat{d}a) = \omega \otimes a$. Therefore, one has the following splittings: $\hat{\Omega}^n(A)\bigotimes_{\mathbf{K}} A \cong \hat{\Omega}^{n+1}(A)\bigoplus(\hat{\Omega}^n(A) \otimes 1)$, $\widehat{\Omega}^n(A)\bigotimes_{Z} A \cong \widehat{\Omega}^{n+1}(A)\bigoplus(\widehat{\Omega}^n(A) \otimes 1)$. Let E be a two-sided A-module. In Tab.1.8 we define some distinguished (co)chain complexes and their homologies that are related to tensor product of quantum algebras.

TAB.1.8 - Hochschild (co)homologies of quantum algebra A.

(Co)ochain Complex	(Co)homology Space
$\{\hat{C}_n^{och}(A;E)\equiv E\otimes_{\mathbf{K}}T_0^n(A),\hat{\partial}_n\}$	$\hat{H}_\bullet^{och}(A;E)$ $\left(\text{Hochschild homology with coeffs.in } E\right)$
$\{\widehat{C}_n^{och}(A;E)\equiv E\otimes_Z\dot{T}_0^n(A),\widehat{\partial}_n\}$	$\widehat{H}_\bullet^{och}(A;E)$ $\left(\text{full Hochschild homology with coeffs. in } E\right)$
$\{\hat{C}_n^{och}(A)\equiv\hat{C}_n^{och}(A;A),\hat{\partial}_n\}$	$\hat{H}_\bullet^{och}(A)$ $\left(\text{Hochschild homology}\right)$
$\{\widehat{C}_n^{och}(A)\equiv\widehat{C}_n^{och}(A;A),\widehat{\partial}_n\}$	$\widehat{H}_\bullet^{och}(A;E)$ $\left(\text{full Hochschild homology}\right)$
$\{\hat{C}_{och}^n(A;E)\equiv Hom_{\mathbf{K}}(\dot{T}_0^n(A),E^*),\hat{\delta}_n\}$	$\hat{H}_{och}^\bullet(A;E)$ $\left(\text{Hochschild cohomology with coeffs.in } E\right)$
$\{\hat{C}_{och}^n(A;E)\equiv Hom_Z(\dot{T}_0^n(A),E^*),\hat{\delta}_n\}$	$\widehat{H}_{och}^\bullet(A;E)$ $\left(\text{full Hochschild cohomology with coeffs.in } E\right)$
$\{\hat{C}_{och}^n(A)\equiv\hat{C}_{och}^n(A;A^*),\hat{\delta}_n\}$	$\hat{H}_{och}^\bullet(A)$ $\left(\text{Hochschild cohomology}\right)$
$\{\widehat{C}_{och}^n(A)\equiv\widehat{C}_{och}^n(A;A^*),\widehat{\delta}_n\}$	$\widehat{H}_{och}^\bullet(A)$ $\left(\text{full Hochschild cohomology}\right)$

$E\equiv$ two-sided A-module.

$\hat{\partial}_n(e\otimes a_1\otimes\cdots\otimes a_n)=ea_0\otimes a_1\otimes\cdots\otimes a_n$

$+\sum_{1\leq j\leq n-1}(-1)^j e\otimes a_1\otimes\cdots\otimes a_j a_{j+1}\otimes\cdots\otimes a_n+(-1)^n a_n e\otimes a_1\otimes\cdots\otimes a_{n-1},$

$e\in E,\ a_i\in A;\ \hat{\delta}^n(\varphi)=\varphi\circ\hat{\partial}_n.\ \widehat{\delta}^n(\varphi)=\varphi\circ\widehat{\partial}_n.$

The Hochschild (co)homologies identify c(ontr)ovariant functors from the category of quantum algebras to the category of $\mathbf{K}$-vector spaces. In the particular case that $A = \mathbf{K}$ one has: $\hat{H}_0^{och}(\mathbf{K}) = \widehat{H}_0^{och}(\mathbf{K}) \cong \mathbf{K}$, $\hat{H}_n^{och}(\mathbf{K}) = \widehat{H}_n^{och}(\mathbf{K}) \cong 0$, for $n > 0$. Important acyclic subcomplexes of Hochschild complexes are ones with chains of the form $a_0 \otimes \cdots \otimes 1 \otimes \cdots \otimes a_n$, i.e., with $a_k = 1$ for some $k > 0$. Let us denote these subcomplexes respectively by $\{\hat{D}_\bullet(A),\hat{\partial}_\bullet\}$ and $\{\widehat{D}_\bullet(A),\widehat{\partial}_\bullet\}$. One has the following isomorphisms:

$\hat{\Omega}^\bullet(A) \cong \hat{C}_\bullet(A)/\hat{D}_\bullet(A)$, $\widehat{\Omega}^\bullet(A) \cong \widehat{C}_\bullet(A)/\widehat{D}_\bullet(A)$. Let us denote yet by $\hat{\partial}_\bullet$ and $\widehat{\partial}_\bullet$ the boundaries on the reduced complexes. Then one has:

$$\hat{\partial}_{n+1}(a_0 da_1 \cdots da_n) = a_0 a_1 da_2 \cdots da_n$$
$$+ \sum_{1 \le j \le n-1} (-1)^j a_0 da_1 \cdots d(a_j a_{j+1}) \cdots da_n$$
$$+ (-1)^n a_n a_0 da_1 \cdots da_{n-1},$$

and similarly for $\widehat{\partial}_{n+1}$. In this way we get the chain complexes $\{\hat{\Omega}^\bullet(A), \hat{\partial}_\bullet\}$ and $\{\widehat{\Omega}^\bullet(A), \widehat{\partial}_\bullet\}$. Then we can use such complexes to compute the Hochschild homologies. Another way to compute such homologies is to compute the homologies of the following chain complexes $E \otimes_{\hat{B}} P_\bullet$, $E \otimes_{\widehat{B}} P_\bullet$, where $\hat{B} \equiv A \otimes_{\mathbf{K}} A^{opp}$, $\widehat{B} \equiv A \otimes_Z A^{opp}$, that are algebras with products $(a_1 \otimes a_2)(a_3 \otimes a_4) = a_1 a_3 \otimes a_4 a_2$, and $P_\bullet$ is a projective resolution of the left $\hat{B}$-module, (resp. $\widehat{B}$-module), E. (A two-sided A-module E can be considered a left-$\hat{B}$-module, (resp. $\widehat{B}$-module), by means of the formula: $(a_1 \otimes a_2)s = a_1 s a_2$). In fact A admits the following projective resolutions of $\hat{B}$-modules, (resp. $\widehat{B}$-modules), (called *bar resolution*, (resp. *full bar resolution*)),

$$
\begin{array}{ccccccccccc}
\cdots \to & T_0^{n+2}(A) & \xrightarrow{\hat{\partial}_{n+1}} & T_0^{n+1}(A) & \xrightarrow{\hat{\partial}_{n+1}} & \cdots & \to & A \otimes_{\mathbf{K}} A & \xrightarrow{\hat{\mu}} & A & \to & 0 \\
& \downarrow & & \downarrow & & & & \downarrow & & \| & & \\
\cdots \to & \dot{T}_0^{n+2}(A) & \xrightarrow{\widehat{\partial}_{n+1}} & \dot{T}_0^{n+1}(A) & \xrightarrow{\widehat{\partial}_{n+1}} & \cdots & \to & A \otimes_Z A & \xrightarrow{\widehat{\mu}} & A & \to & 0 \\
& \downarrow & & \downarrow & & & & \downarrow & & & & \\
& 0 & & 0 & & & & 0 & & & &
\end{array}
$$

By tensoring with respect to E we get the following short exact sequence of complexes: $E \otimes_{\hat{B}} T^\bullet(A) \to E \otimes_{\widehat{A}} \dot{T}^\bullet(A) \to 0$, hence their homologies give the Hochschild homologies with coefficients in E, and the relation: $\hat{H}_\bullet^{och}(A; E) \to \widehat{H}_\bullet^{och}(A; E)$. Note that as $\hat{\partial}_1(a_0 \otimes a_1) = a_0 a_1 - a_1 a_0 \equiv [a_0, a_1] = \widehat{\partial}_1(a_0 \otimes a_1)$, the boundaries $\hat{\partial}_1$ and $\widehat{\partial}_1$ are related to the Lie bracket in A. In the commutative case one has: $\hat{H}_n^{och}(A) = \widehat{H}_n^{och}(A) \equiv H_n^{och}(A)$, $\hat{H}_{och}^n(A) = \widehat{H}_{och}^n(A) \equiv H_{och}^n(A)$ and $\hat{\Omega}^\bullet(A) = \widehat{\Omega}^\bullet(A) \equiv \Omega^\bullet(A)$. Then, if E is a symmetric two-sided A-module it results that $E \otimes_A \Omega^n(A)$ is a direct summand of $H_m^{och}(A; E)$. In the particular case that $A \equiv C^\infty(M)$, where M is a compact manifold, one has the isomorphism: $H_n^{och}(C^\infty(M)) \cong \mathcal{D}_n(M)$, where $\mathcal{D}_n(M) \cong \mathcal{L}(\Omega^n(M); \mathbf{K})$ is the inear *space of de Rham currents of degree n* on M, i.e., the space of continuous linear forms on the space

$\Omega^n(M)$ of differential forms of degree n on M. (This can be referred as *Hochschild-Kostant-Rosenberg-Connes theorem* (HKRC-theorem)[11,27].) $\blacksquare$

Remark 1.12. (*Cyclic Homology of quantum algebras*). Let t denote the generator of the cyclic group $\mathbf{Z}/(n+1)\mathbf{Z} \equiv \mathbf{Z}_{n+1}$. We let t act in $\hat{C}_n^{och}(A)$ by the formula: $t(a_0 \otimes \cdots \otimes a_n) = (-1)^n a_n \otimes a_0 \otimes \cdots \otimes a_{n-1}$. Let $\hat{I}_n^{och}(A)$ denote the following subspace of $\hat{C}_n^{och}(A)$: $\hat{I}_n^{och}(A) \equiv (1-t)\hat{C}_n^{och}(A) \subset \hat{C}_n^{och}(A)$. Then, the Hochschild boundary operator $\hat{\partial}$ maps $\hat{I}_n^{och}(A)$ into $\hat{I}_{n-1}^{och}(A)$, so we have the boundary operator $\bar{\partial} : \hat{C}_n^{och}(A)/\hat{I}_n^{och}(A) \to \hat{C}_{n-1}^{och}(A)/\hat{I}_{n-1}^{och}(A)$. We call *cyclic homology* of A the homology $H_\bullet^{cy}(A)$ of the complex $\{\hat{C}_n^{cy}(A) \equiv \hat{C}_n^{och}(A)/\hat{I}_n^{och}(A), \bar{\partial}_n\}$. The following short exact sequence of complexes:

$$0 \to \{\hat{I}_n^{och}(A), \hat{\partial}_n\} \xrightarrow{i} \{\hat{C}_n^{och}(A), \hat{\partial}_n\} \xrightarrow{\pi} \{\hat{C}_n^{cy}(A), \bar{\partial}_n\} \to 0$$

induces the following long-exact sequence in homology:

$$\cdots \to H_n(\hat{I}_\bullet^{och}(A), \hat{\partial}_\bullet) \xrightarrow{i_*} \hat{H}_n^{och}(A) \xrightarrow{\pi_*} \hat{H}_n^{cy}(A) \xrightarrow{\triangle} H_{n-1}(\hat{I}_\bullet(A), \hat{\partial}_\bullet) \to \cdots$$

where $\triangle$ is the connecting homomorphism. As $\{\hat{I}_\bullet^{och}(A), \hat{\partial}_\bullet\}$ is not acyclic one has: $H_n(\hat{I}_\bullet^{och}(A), \hat{\partial}_\bullet) \cong \hat{H}_n^{cy}(A)$. We get the following long exact sequence:

$$\cdots \to H_{n-1}^{cy}(A) \to \hat{H}_n^{och}(A) \to \hat{H}_n^{cy}(A) \to \hat{H}_{n-2}^{cy}(A) \to \cdots,$$

For duality we get the *cyclic cohomology*. More precisely $\hat{H}_{cy}^\bullet(A)$ of A is the homology of the subcomplex $\{\hat{C}_{cy}^\bullet(A), \hat{\delta}\}$ of the Hochschild complex, where

$$\hat{C}_{cy}^n(A) \equiv \left\{ \begin{array}{l} \varphi \in \hat{C}_{och}^n(A) \equiv Hom_{\mathbf{K}}(T_0^n(A); A^*) \cong Hom_{\mathbf{K}}(T_0^{n+1}(A); \mathbf{K}) \\[4pt] : \varphi(a_{\sigma(0)}, \cdots, a_{\sigma(n)}) = \epsilon(\sigma)\varphi(a_0, \cdots, a_n), \\[4pt] \forall \sigma = \text{cyclic permutation of} \{0, 1, \cdots, n\}. \end{array} \right\}.$$

The relation between Hochschild and cyclic cohomology is given by the following long exact sequence:

(1.10)
$$\cdots \to \hat{H}_{cy}^n(A) \xrightarrow{I} \hat{H}_{och}^n(A) \xrightarrow{B} \hat{H}_{cy}^{n-1}(A) \xrightarrow{S} \hat{H}_{cy}^{n+1}(A) \xrightarrow{I} \hat{H}_{och}^{n+1}(A) \to \cdots$$

where I is the canonical homomorphisms (of degree 0), S is the "periodicity homomorphism" (of degree $+2$) and B is the connecting homomorphism (of degree -1). The periodicity map $S : \hat{H}_{cy}^n(A) \to \hat{H}_{cy}^{n+2}(A)$ defines two directed systems of abelian groups; their inductive limits: $\hat{H}_P^0(A) \equiv \lim_{\to} \hat{H}_{cy}^{2k}(A)$, $\hat{H}_P^1(A) \equiv \lim_{\to} \hat{H}_{cy}^{2k+1}(A)$, form a $\mathbf{Z}_2$-graded group $\hat{H}_P^\bullet(A) \equiv \hat{H}_P^0(A) \oplus \hat{H}_P^1(A)$, called the *periodic cyclic cohomology* of the algebra A. From the exact sequence (1.10) we can, for example, calculate the cyclic cohomology in the particular case where $A \equiv C^\infty(M)$, with M a compact manifold and obtain also the dual counterpart of the HKRC-theorem. In fact, in such a case, one has the following isomorphism relating Hochschild cohomology and de Rham homology [11]:

$$H_{och}^k(C^\infty(M)) \cong Z_k^{dR}(M) \oplus H_{k-2}^{dR}(M) \oplus H_{k-4}^{dR}(M) \oplus \cdots \oplus H_{\#k}^{dR}(M),$$

where $Z_k^{dR}(M)$ is the set of closed k-currents on M, $H_r^{dR}(M)$ is the de Rham homology of degree r, and $\#k = 0$ or 1 according as k is even or odd. In Tab.1.9 we report also some values of the periodic cyclic cohomologies for particular commutative cases:

TAB.1.9 - Periodic cyclic cohomology for particular commutative cases.

$\hat{H}_P^0(\mathbf{K}) \cong \mathbf{K}$	$\hat{H}_P^1(\mathbf{K}) \cong 0$
$\hat{H}_P^0(C^\infty(M)) \cong H_{even}^{dR}(M)$	$\hat{H}_P^1(C^\infty(M)) \cong H_{odd}^{dR}(M)$

M=finite dimensional smooth manifold. $H_\bullet^{dr}(M) \equiv$ de Rham homoloy of M.

We can also talk about *full cyclic homology* $\widehat{H}_\bullet^{cy}(A)$ and *full cyclic cohomology* $\widehat{H}_{cy}^\bullet(A)$, of a quantum algebra A, similarly to what just made for the cyclic homology and cyclic cohomology respectively. ■

In the following we give a relation between quantum algebras, as introduced in Definition 1.35, and Hopf algebras. This is of great interest as possible models for quantum algebras can be the Hopf algebras associated to commutative PDEs, quantized by means of the canonical quantization. (See refs.[59,61–63,70] and also part III of this work.)

Definition 1.36. (*Coalgebras*). A $\mathbf{K}$-*coalgebra* is defined by a triplet $(C, \Delta_C, \epsilon_C)$ where: (a) C is a $\mathbf{K}$-linear space; (b) $\Delta_C \in Hom_{\mathbf{K}-module}(C; C \otimes C)$, (*co-multiplication map*), $\epsilon_C \in Hom_{\mathbf{K}-module}(C; \mathbf{K})$, (*counit map*), such that the following diagrams commute:

$$
\begin{array}{ccc}
C \otimes C \otimes C & \xleftarrow{\ \Delta_C \otimes 1\ } & C \otimes C \\
{\scriptstyle 1 \otimes 1}\big\uparrow & & \big\uparrow{\scriptstyle \Delta_C} \\
C \otimes C & \xleftarrow[\ \Delta_C\ ]{} & C
\end{array}
$$

(*the associative law*) :

$$\textit{(the counitary property)}: \quad \begin{array}{ccccc} \mathbf{K}\otimes C & \xleftarrow{\;\epsilon_C\otimes 1\;} & C\otimes C & \xrightarrow{\;1\otimes\epsilon_C\;} & C\otimes\mathbf{K} \\ \wr\| & & \Delta_C\uparrow & & \|\wr \\ C & = & C & = & C \end{array}$$

A $\mathbf{K}$-*coalgebra morphism* between $\mathbf{K}$-coalgebras (C,Δ_C,ϵ_C), (D,Δ_D,ϵ_D) is a $\mathbf{K}$-linear map $\sigma : C \to D$ satisfying the following conditions: (i)$\Delta_D \circ \sigma = (\sigma\otimes\sigma)\circ\Delta_C$; (ii) $\epsilon_D\circ\sigma = \epsilon_C$. For $\mathbf{K}$-linear spaces M and N we define a $\mathbf{K}$-linear map (*twist map*) $\tau : M\bigotimes N \to N\bigotimes M$, $\tau(x\otimes y) = y\otimes x$, $x\in M$, $y\in N$. We say that a $\mathbf{K}$-coalgebra (C,Δ_C,ϵ_C) is *cocommutative* if $\tau\circ\Delta_C = \Delta_C$. We call $\mathbf{K}$-*coalgebra tensor product* between (C,Δ_C,ϵ_C) and (D,Δ_D,ϵ_D) the following $\mathbf{K}$-coalgebra $(C\bigotimes D,\Delta_{C\otimes D},\epsilon_{C\otimes D})$, where $C\bigotimes D$ is the $\mathbf{K}$-linear space tensor product, $\Delta_{C\otimes D}\equiv(1\otimes\tau\otimes 1)\circ\Delta_C\otimes\Delta_D$, $\epsilon_{C\otimes D}\equiv\epsilon_C\otimes\epsilon_D$. The *dual $\mathbf{K}$-algebra* of a $\mathbf{K}$-coalgebra (C,Δ_C,ϵ_C) is the $\mathbf{K}$-algebra (C^*,μ,η) defined by: (i) $C^*\equiv Hom_{\mathbf{K}-module}(C;\mathbf{K})$; (ii) $\mu : C^*\bigotimes C^*\xrightarrow{\;\rho\;}(C\bigotimes C)^*\xrightarrow{\;\delta^*\;}C^*$, where ρ is the injective $\mathbf{K}$-linear map defined by $\rho(f\otimes g)(x\otimes y) = f(x)g(y)$, $f,g\in C^*$, $x,y\in C$; (iii) $\eta : \mathbf{K}\cong\mathbf{K}^*\xrightarrow{\;\epsilon^*\;}C^*$. Thus we cannot define a $\mathbf{K}$-coalgebra structure on the dual $\mathbf{K}$-linear space C^* of a $\mathbf{K}$-coalgebra! If A is finite dimensional $\mathbf{K}$-linear space, then $\rho : A^*\bigotimes A^*\cong(A\bigotimes A)^*$. So, if (A,μ,ϵ) is a $\mathbf{K}$-algebra, (A^*,Δ,ϵ) is a $\mathbf{K}$-coalgebra, called the dual $\mathbf{K}$-coalgebra of A, where $\Delta\equiv\mu^*\circ\rho^{-1}$, $\epsilon\equiv\mu^*$.

Example 1.24. Let S be a set and denote by $\mathbf{K}S$ the free $\mathbf{K}$-module generated by S. If we define $\mathbf{K}$-linear maps $\Delta : \mathbf{K}S \to \mathbf{K}S\bigotimes\mathbf{K}S$, $\Delta(s) = s\otimes s$, $\epsilon : \mathbf{K}S \to \mathbf{K}$, $\epsilon(s) = 1$, then $(\mathbf{K}S,\Delta,\epsilon)$ becomes a $\mathbf{K}$-coalgebra. The dual $\mathbf{K}$-linear space $(\mathbf{K}S)^*$ of $\mathbf{K}S$ can be identified with the set $(\mathbf{K}S)^*\equiv Map(S;\mathbf{K})$ of all maps from S to $\mathbf{K}$, where the dual $\mathbf{K}$-algebra structure of $\mathbf{K}S$ is given by $(f+g)(s) = f(s)+g(s)$, $(fg)(s) = f(s)g(s)$, $(af)(s) = af(s)$, $f,g\in Map(S;\mathbf{K})$, $s\in S$, $a\in\mathbf{K}$. $\blacksquare$

Example 1.25. If $S\equiv\{c_0,c_1,c_2,\cdots\}$, then $(C\equiv\mathbf{K}S,\Delta,\epsilon)$, with $\Delta(c_n) = \sum_{0\le i\le n}c_i\otimes c_{n-i}$, $\epsilon(c_n) = \delta_{0n}$, ($\delta_{ij}$ is the usual delta of Kronecker), is a $\mathbf{K}$-coalgebra. One has the following $\mathbf{K}$-algebra isomorphism: $C^*\cong\mathbf{K}[[x_1]]$. (For details see e.g., ref.[1].) $\blacksquare$

Example 1.26. If $S\equiv\{s_{ij}\}_{1\le i,j\le n}$, then $(V\equiv\mathbf{K}S,\Delta,\epsilon)$, with $\Delta(s_{ij}) = \sum_{1\le k\le n}s_{ik}\otimes s_{kj}$, $\epsilon(s_{ij}) = \delta_{ij}$, is a $\mathbf{K}$-coalgebra. The dual $\mathbf{K}$-linear space V^* of V is a $\mathbf{K}$-linear space of dimension n^2 isomorphic to the $\mathbf{K}$-algebra $M_n(\mathbf{K})$ of all $n\times n$ square matrices with coefficients in $\mathbf{K}$. In fact, we can represent any element $x\in v^*$ in the form $x = \sum x^{ij}e_{ij}$, where $x^{ij}\in\mathbf{K}$ and

$e_{ij} \in V^*$ defined by $< e_{ij}, s_{kl} > = \delta_{ik}\delta_{jl}$. Therefore, $\{e_{ij}\}_{1 \leq i,j \leq n}$ is a basis for V^* over $\mathbf{K}$. $\blacksquare$

Theorem 1.25. *Given a* $\mathbf{K}$*-linear space* H*, suppose that there are* $\mathbf{K}$*-linear maps* $\mu : H \otimes H \to H$, $\eta : \mathbf{K} \to H$, $\Delta : H \to H \otimes H$, $\epsilon : H \to \mathbf{K}$*, such that* (H, μ, η) *is a* $\mathbf{K}$*-algebra and* (H, Δ, ϵ) *is a* $\mathbf{K}$*-coalgebra. Then, the following conditions are equivalent: (i)* η*,* μ *are* $\mathbf{K}$*-coalgebra morphisms; (ii)* Δ*,* ϵ *are* $\mathbf{K}$*-algebra morphisms; (iii)* $\Delta(fg) = \sum g_{(1)}h_{(1)} \otimes g_{(2)}h_{(2)}$*,* $\Delta(1) = 1$*,* $\epsilon(gh) = \epsilon(g)\epsilon(h)$*,* $\epsilon(1) = 1$*. Note that given a* $\mathbf{K}$*-coalgebra* (C, Δ, ϵ) *we can write* $\Delta(c) = \sum_{1 \leq i \leq n} c_{1i} \otimes c_{2i} \equiv \sum_{(c)} c_{(1)} \otimes c_{(2)}$*,* $c_{1i}, c_{2i} \in C$*. Furthermore, for* $\mathbf{K}$*-linear maps* $f, g : C \to C, \mathbf{K}$*, we write* $(f \otimes g)\Delta(c) = \sum_{(c)} f(c_{(1)}) \otimes g(c_{(2)})$*. Moreover, since the associative law holds, we have* $(\Delta \otimes 1)\Delta(c) = (1 \otimes \Delta)\Delta(c) = \sum_{(c)} c_{(1)} \otimes c_{(2)}$*, and, in general, we define* $\Delta_1 \equiv \Delta$*,* $\Delta_n \equiv (1 \otimes \cdots_{n-1} \cdots \otimes 1 \otimes \Delta)\Delta_{n-1}$*,* $(n > 1)$*, and write* $\Delta_n(c) = \sum_{(c)} c_{(1)} \otimes c_{(2)} \otimes \cdots \otimes c_{(n+1)}$*. The counity property may be expressed by* $c = \sum_{(c)} c_{(1)}\epsilon(c_{(2)}) = \sum_{(c)} \epsilon(c_{(1)})c_{(2)}$*. One has the following equalities:* $\Delta(c) = \sum_{(c)} \epsilon(c_{(2)}) \otimes \Delta(c_{(1)}) = \sum_{(c)} \Delta(c_{(2)}) \otimes \epsilon(c_{(1)})$*.* $\Delta(c) = \sum_{(c)} c_{(1)} \otimes \epsilon(c_{(2)})c_{(3)} = \sum_{(c)} c_{(1)} \otimes \epsilon(c_{(3)})c_{(2)}$*,* $c = \sum_{(c)} \epsilon(c_{(1)})\epsilon(c_{(3)})c_{(2)}$*.*

Definition 1.37. A $\mathbf{K}$-bialgebra is a $\mathbf{K}$-linear space H together with $\mathbf{K}$-linear maps μ, η, Δ, ϵ, that satisfy one of the equivalent conditions of above theorem. A $\mathbf{K}$-linear map $\sigma : H \to \mathbf{K}$ between $\mathbf{K}$-bialgebras that is also a $\mathbf{K}$-algebra morphism and a $\mathbf{K}$-coalgebra morphism, then σ is called a $\mathbf{K}$-bialgebra morphism. If a $\mathbf{K}$-bialgebra H is finite-dimensional as a $\mathbf{K}$-linear space, then a $\mathbf{K}$-bialgebra structure may be defined on its dual $\mathbf{K}$-module H^*, which we call the *dual* $\mathbf{K}$*-bialgebra* of H.

Example 1.27. (*Semigroup* $\mathbf{K}$*-bialgebra*). Let S be a semigroup with an identity element. The $\mathbf{K}$-algebra $\mathbf{K}S$ has a $\mathbf{K}$-coalgebra structure. With respect to these two structures $\mathbf{K}S$ admits a $\mathbf{K}$-bialgebra structure. In particular, if S is a group it is called a *group* $\mathbf{K}$*-algebra*. The dual $\mathbf{K}$-linear space $(\mathbf{K}S)^* \cong Map(S; \mathbf{K})$. If S is a finite set, then $(\mathbf{K}S)^*$ becomes a dual $\mathbf{K}$-bialgebra of $\mathbf{K}S$. The $\mathbf{K}$-coalgebra structure of $(\mathbf{K}S)^*$ is given by $< \Delta f, x \otimes y > = < f, xy >$, $< \epsilon f, xy > = < f, e >$, $f \in Map(S; \mathbf{K})$, $x, y \in S$ and e the identity element of S. $\blacksquare$

Example 1.28. (*Universal enveloping* $\mathbf{K}$*-bialgebra of a* $\mathbf{K}$*-Lie algebra*). Let L be a $\mathbf{K}$-Lie algebra and let $U(L)$ be its universal enveloping $\mathbf{K}$-algebra. Then $U(L)$ has a $\mathbf{K}$-bialgebra structure $(U(L), \mu, \eta, \Delta, \epsilon)$ with $\Delta : U(L) \to U(L \oplus L) \cong U(L) \otimes U(L)$, $\Delta(x) = x + x$, and $\epsilon : U(L) \to \mathbf{K}$ is the $\mathbf{K}$-

algebra morphism induced by the following one: $L \to \{e\}$. ∎

Definition 1.38. (*Hopf algebra*). Given a **K**-coalgebra C and a **K**-algebra A, set $R \equiv Hom_{\mathbf{K}-mod.}(C; A)$. If $f, g \in R$, $f * g \equiv \mu_A \circ (f \otimes g) \circ \Delta_C$ is called to be the *convolution* of f and g. R becomes a **K**-algebra with structure maps $\mu_R(f \otimes g) = f * g$, $\eta_R(\alpha) = \alpha \eta_A \circ \epsilon_C$, $\alpha \in \mathbf{K}$. Furthermore, let H be a **K**-bialgebra and let consider $R \equiv Hom_{\mathbf{K}-mod.}(H; H)$ as a **K**-algebra via convolution. If the identity map 1 of H is a regular element of R with respect to multiplication on R, the inverse S of 1 is called the *antipode* of H. S satisfies the following equivalent conditons: $s * 1 = 1 * s = \eta \circ \epsilon$, $\mu \circ (s \otimes 1) \circ \Delta = \mu \circ (1 \circ s) \circ \Delta = \eta \circ \epsilon$. A **K**-bialgebra with antipode is called a **K**-*Hopf algebra*. A morphism $\sigma : H \to K$ between **K**-Hopf algebras is a **K**-*bialgebra morphism* such that $S_K \circ \sigma = \sigma \circ S_H$, where S_K and S_H are the antipodes of K and H respectively.

Theorem 1.26. *Let S be the antipode of a **K**-Hopf algebra H. One has the following:* (i) $S(gh) = s(h)s(g)$, $g, h \in H$; (ii) $S(1) = 1$, *namely* $S \circ \eta = \eta$; (iii) $\epsilon \circ S = \epsilon$; (iv) $\tau \circ (s \otimes s) \circ \Delta = \Delta \circ S$; (v) *If H is commutative or cocommutative, then* $S^2 = 1$.

Example 1.29. Let $\mathbf{K}G$ be the **K**-bialgebra of a group G. $\mathbf{K}G$ has a natural structure of **K**-Hopf algebra with antipode $S : \mathbf{K}G \to \mathbf{K}G$ defined by $S(x) = x^{-1}$, $x \in G$. ∎

Example 1.30. The universal enveloping **K**-bialgebra $U(L)$ of a **K**-Lie algebra L has a natural structure of **K**-Hopf algebra with antipode $S : U(L) \to U(L)$ the principal anti-authomorphism of $U(L)$. ∎

Example 1.31. Let $\lambda \in \mathbf{C} \setminus \{0\}$. We call *Manin's quantum plane* the following algebra:

$$A_\lambda \equiv \frac{\mathbf{C}[x, y]}{< xy - \lambda yx >}.$$

We call *extended Manin's quantum plane* the following algebra:

$$B_\lambda \equiv \frac{\mathbf{C}[x, x^{-1}, y]}{< xy - \lambda yx, x^{-1}x - 1, xx^{-1} - 1 >}.$$

Then B_λ admits a Hopf algebra structure. ∎

For further informations on Hopf algebras see, e.g., refs.[1,82].

Theorem 1.27. (Quantum algebras and Hopf algebras). 1) *Let A be a quantum algebra. If one defines a **K**-linear morphism $\Delta : A \to A \otimes_{\mathbf{K}} A$ such that* $(\Delta \otimes 1) \circ \Delta = (1 \otimes \Delta) \circ \Delta$, $(c \otimes 1) \circ \Delta = (1 \otimes c) \circ \Delta$, *then* (A, c, Δ)

becomes a **K**-coalgebra, (that we call quantum coalgebra). If further μ and ϵ are **K**-coalgebra morphisms, (that is equivalent Δ and c are **K**-algebra morphisms), i.e., $\Delta \circ \mu = (\mu \otimes \mu) \circ \Delta \otimes \Delta$, $c \circ \mu = c \otimes c$, then $(A, \epsilon, c, \Delta, \mu)$ becomes a **K**-bialgebra, (that we call quantum bialgebra).

2) Let $(A, \epsilon, c, \Delta, \mu)$ be a quantum bialgebra and let us denote by $P(A)$ the set of all *primitive elements* of A, i.e., $P(A) = \{a \in A | \Delta a = a \otimes 1 + 1 \otimes a\}$. Then $P(A)$ is a **K**-linear subspace of A, that has the structure of a **K**-Lie algebra with respect to the bracket $[a, b] \equiv ab - ba$, $\forall a, b \in P(A)$. Furthermore, the **K**-subalgebra $< P(A) >$, generated by $P(A)$, is a **K**-bialgebra of A isomorphic to the universal enveloping **K**-bialgebra $U(P(A))$ of $P(A)$.

3) Let A be a quantum bialgebra and let assume that the identity map 1 of A is a regular element of $R \equiv Hom_{\mathbf{K}-module}(A, A)$ with respect to multiplication defined as the following convolution: $\mu_R(f \otimes g) = f \star g \equiv \mu \circ (f \otimes g) \circ \Delta$. Then A becomes a **K**-Hopf algebra (that we call quantum Hopf algebra). In the following we shall more generally say that a quantum Hopf-algebra is any quantum algebra B that is an extension of another quantum algebra C of the type $C = A \otimes_{\mathbf{K}} H$, where H is an Hopf **K**-algebra and A is a quantum algebra. Therefore one has the following short exact sequence $0 \to A \otimes_{\mathbf{K}} H \to B \to B/C \to 0$. (Note that C is obtained from the Hopf **K**-algebra H by extending the field of scalars from **K** to the quantum algebra A. Of course C is not, in general, an Hopf **K**-algebra and neither B is so.)

4) If A is a quantum Hopf algebra that is commutative or cocommutative, then the antipode s is an involution, i.e., $s^2 = 1$.

Proof. The proofs of 1) and 2) are direct consequences of previous statements and standard results of algebra, (see e.g. refs.[1,82]).

3) Given a **K**-coalgebra C, the set $R \equiv Hom_{\mathbf{K}-module}(C; A)$ has a natural structure of **K**-algebra with structure maps: $\mu_R(f \otimes g) = f \star g \equiv \mu \circ (f \otimes g) \circ \Delta_C$ (convolution product of f and g); $\epsilon_R(\lambda) = \lambda \epsilon \circ c_C$. One has $(f \star g)(a) = f(a_{(1)})g(a_{(2)}), \forall a \in C$, where $\Delta(a) = \sum_{(a)} a_{(1)} \otimes a_{(2)}$. When the identity map 1_A of A is a regular element of R, with $C \equiv A$ and with respect to the multiplication on R, the inverse s of 1_A is called the *antipode* of A. More precisely s satisfies the following equivalent conditions: $\mu \circ (s \otimes 1_A) \circ \Delta = \mu \circ (1_A \otimes s) \circ \Delta = \epsilon \circ c$; $s \star 1 = 1 \star s = \epsilon \circ c$. A bialgebra A with antipode is called a Hopf **K**-algebra. A Hopf algebra A, with antipode

s, satisfies the following properties: (i) s is an anti-algebra morphism; that is $s(1) = 1$ and $s \circ \mu \circ \sigma = \mu \circ s \otimes s$; (ii) s is an anti-coalgebra morphism; that is $\Delta \circ s = \sigma \circ (s \otimes s) \circ \Delta$ and $c \circ s = c$.

4) A **K**-algebra, (resp. **K**-coalgebra), A is said *commutative*, (resp. *cocommutative*), if $\mu = \tau \circ \mu$, (resp. $\Delta = \tau \circ \Delta$), where τ is a linear mapping $\tau : A \otimes_{\mathbf{K}} A \to A \otimes_{\mathbf{K}} A$ (*twist map*) such that $\tau(a \otimes b) = b \otimes a$, $\forall a, b \in A$. Now the proof follows directly from the following lemma.

Lemma 1.8. *Let H be a Hopf **K**-algebra with antipode s. If H is commutative or cocommutative, then $s^2 = id_H$.*

Proof of Lemma 1.8. It is standard. (See e.g. refs.[1,82].) $\qquad\qquad\square$

Remark 1.13. (*Quantum algebra and spectrum*). A *representation* of a quantum algebra A is a **K**-algebra morphism $\pi : A \to Hom_{\mathbf{K}}(E; E) \equiv B_\pi$, where E is a locally convex topological **K**-vector space, such that there exists a **K**-algebra morphism $c_E : B_\pi \to \mathbf{K}$ such that the following diagram

$$
\begin{array}{ccc}
\mathbf{K} & \xrightarrow{\;\epsilon_E\;} & Hom_{\mathbf{K}}(E;E) \\
\downarrow{\scriptstyle \epsilon} & & \| \\
A & \xrightarrow{\;\pi\;} & Hom_{\mathbf{K}}(E;E) \\
\downarrow{\scriptstyle c} & & \downarrow{\scriptstyle c_E} \\
\mathbf{K} & = & \mathbf{K}
\end{array}
$$

is commutative. Here ϵ_E is the canonical mapping given by $\epsilon_E(k) : v \mapsto kv$. B_π is called *model algebra* and $E \equiv E_\pi$ *model space* of the representation π of the quantum algebra A. By using the following commutative diagram

$$
\begin{array}{ccc}
Hom_{\mathbf{K}-algebra}(A; Hom_{\mathbf{K}}(E;E)) & \xrightarrow{\;j\;} & Hom_{\mathbf{K}}(A \otimes_{\mathbf{K}} E; E) \\
\cap & & \| \\
Hom_{\mathbf{K}}(A; Hom_{\mathbf{K}}(E;E)) & \cong & Hom_{\mathbf{K}}(A \otimes_{\mathbf{K}} E; E)
\end{array}
$$

we can identify any representation π of A with an element

$$
j(\pi) \in Hom_{\mathbf{K}}(A \underset{\mathbf{K}}{\otimes} E; E).
$$

We say that a subvector space $F \subset E$ is an *invariant* with respect to a representation π if $j(\pi)(A \otimes_{\mathbf{K}} F) \subset F$. We say that a representation is *irreducible* if it has trivial invariant subobjects only. We call *spectrum* of A the set $Spec(A)$ of all irreducible representations of A. The set $Spec(A)$ is topologized by making the closure $\overline{X}$ of $X \subset Spec(A)$ the following subset of $Spec(A)$: $\overline{X} \equiv \{q \in Spec(A) | q|_{\ker(\pi)} = 0, \forall \pi \in X\}$. The topology obtained in this way is called *Jacobson topology*. One has the following mapping:

$$A \to Map(Spec(A), \mathcal{M}_A) \equiv \mathcal{M}_A{}^{Spec(A)}, \quad a \mapsto \hat{a} : \pi : A \to B_\pi, \quad \pi(a) \in B_\pi,$$

where $\mathcal{M}_A \equiv$ set of all the model algebras of A. We call $\hat{a}$ *Fourier image* (or *Fourier transform*) of $a \in A$. The spectrum is related to ideals of A. In fact, any irreducible representation $\pi \in Spec(A)$ identifies a closed two-sided ideal (*primitive ideal*) $\mathfrak{a}_\pi \subset A$ given by: $\mathfrak{a}_\pi \equiv \ker(\pi)$. Denote by $Prim(A)$ the set of primitive ideals in A. $Prim(A)$ is not suitable to parametrize the space of irreducible representations. In fact, e.g. if A is a simple C^*-algebra, i.e., has no ideals closed under the $*$-operation, $Prim(A)$ consists of a single point. On the other hand, there are simple quantum algebras having many inequivalent irreducible representations (all having kernel 0). A particular important subset of $Spec(A)$ is $Sp(A) \equiv Hom_{\mathbf{K}-algebra}(A; \mathbf{K}) \subset Spec(A)$. (Here $E \equiv \mathbf{K}$.) Of course, on $Sp(A)$ we recognize the Jacobson topology induced by $Spec(A)$. As $\hat{A} \equiv Hom_{\mathbf{K}}(A; A)$, we can consider the natural immersion $A \subset \hat{A}$ as a representation of A. This representation is irreducible iff A has no trivial ideals. In such cases the immersion $A \subset \hat{A}$ belongs to the spectrum $Spec(A)$ of A. Therefore, if A has no trivial ideals, the product of A identifies an element of $Spec(A)$. $\blacksquare$

Remark 1.14. (*Quantum algebras and associated Lie and Jordan algebras*). To a quantum algebra A we can functorially associate a Lie algebra $\mathcal{L}_{ie}(A)$ and a Jordan algebra $\mathcal{J}_{ordan}(A)$. More precisely, the algebras $\mathcal{L}_{ie}(A)$ and $\mathcal{J}_{ordan}(A)$ have the same underlying $\mathbf{K}$-vector space of A, but the products are given respectively by $[a,b]_- \equiv ab - ba$ and $[a,b]_+ \equiv \frac{1}{2}(ab + ba)$. These satisfy the properties reported in the following table:

TAB.1.10 - Relations between (anti)commutators in a quantum algebra.

$[a,b]_- + [b,a]_- = 0, \quad [a,[b,c]_-]_- + [b,[c,a]_-]_- + [c,[a,b]_-]_- = 0$
$[a,b]_+ - [b,a]_+ = 0, \quad [a^2,[b,a]_+]_+ - [[a^2,b]_+,b]_+ = 0$
$\frac{1}{2}[a,b]_- + [a,b]_+ - ab = 0$

We call $\mathcal{L}_{ie}(A)$, (resp. $\mathcal{J}_{ordan}(A)$), the *Lie algebra*, (resp. *Jordan algebra*), associated to A. We call $[a,b]_-$, (resp. $[a,b]_+$) the, *commutator*, (resp. *anticommutator*), of $a,b \in A$. In the following we shall use also the notation: $[b,a]_- \equiv [b,a]$; $\{b,a\} \equiv 2[b,a]_+$. If A is a quantum bialgebra then $\mathcal{L}_{ie}(A)$ contains the set $P(A)$ of all the primitive elements of A as a Lie subalgebra. Furthermore, the $\mathbf{K}$-subalgebra of A generated by $P(A)$ is a quantum subbialgebra of A, isomorphic to the universal enveloping algebra $U(P(A))$

of $P(A)$. (Recall that the universal enveloping algebra $U(L)$ of a Lie **K**-algebra L has a natural structure of **K**-bialgebra. See Example 1.28 and refs.[1,82].) ∎

1.3 - QUANTUM MANIFOLDS

Quantum manifolds are geometric objects that globalize the concept of quantum algebras. This is the unique way to obtain a fully covariant theory of quantum phenomena, that is able to reproduce at quantum level the philosophy of the general relativity. In fact, in the framework of the category of quantum manifolds we are able to obtain a satisfactory geometry of noncommutative manifolds and to develop a geometric theory of noncommutative PDE's. This is, of course, essential in order to give a complete description of quantum phenomena.

Definition 1.39. A *quantum vector space* of dimension $(m_1, \ldots, m_s) \in \mathbf{N}^s$, built on the quantum algebra $A \equiv A_1 \times \ldots \times A_s$, is a locally convex topological **K**-vector space E isomorphic to $A_1^{m_1} \times \ldots A_s^{m_s}$. A *quantum manifold* of dimension $(m_1, \ldots, m_s)$ over a quantum algebra $A \equiv A_1 \times \ldots \times A_s$ of class Q_w^k, $0 \leq k \leq \infty, \omega$, is a locally convex manifold M modelled on E and with a Q_w^k-atlas of local coordinate mappings, i.e., the transiction functions $f : U \subset E \to U' \subset E$ define a pseudogroup of local Q_w^k-homeomorphisms on E, where Q_w^k means *quantum class k*, i.e., weak differentiability C_w^k [38], and derivatives Z-linaires. So for each open coordinate set $U \subset M$ we have a set of $m_1 + \ldots + m_s$ coordinate functions $x^A : U \to A$, (*quantum coordinates*). The *tangent space* T_pM at $p \in M$, is the vector space of the equivalence classes $v \equiv [f]$ of C_w^1 (or equivalently C^1) curves $f : I \to M$, $I \equiv$ open neighborhood of $0 \in \mathbf{R}$, $f(0) = p$; two curves f, f' are equivalent if for each (equivalently, for some) coordinate system μ around p the functions $\mu \circ f$, $\mu \circ f' : I \to A_1^{m_1} \times \ldots \times A_s^{m_s}$ have the same derivative at $0 \in \mathbf{R}$. Then, derived tangent spaces associated to a quantum manifold M can be naturaly defined. In the following table we give some useful definitions of derived tangent spaces of quantum manifolds. We say that a quantum manifold of dimension $(m_1, \ldots, m_s)$ is *classic regular* if it admits a projection $c : M \to M_C$ on a n-dimensional manifold M_C. We will call M_C the *classic limit* of M and in order to emphasize this structure we say that the dimension of M is $(n \downarrow m_1, \ldots, m_s)$.

TAB.1.11 - Derived tangent spaces of a quantum manifold M.

Definition	Name
$(T_pM)^* \equiv Hom_{\mathbf{K}}(T_pM;\mathbf{K})$ (*)	*dual tangent space*
$(T_pM)' \equiv \mathcal{L}_{\mathbf{K}}(T_pM;\mathbf{K})$ (†)	*topological dual tang.space*
$\hat{T}_pM \equiv Hom_{\mathbf{K}}(A;T_pM)$	*quantum tangent space*
$\widehat{T}_pM \equiv AOT_pM \equiv Hom_Z(A;T_pM)$ (**)	*fullquantum tangent space*
$(T_pM)^* \equiv Hom_{\mathbf{K}}(T_pM;A)$ (‡)	*quantum dual tang.space*
$(T_pM)^+ \equiv T_pMOA \equiv Hom_Z(T_pM;A)$	*fullquantum dual tang.space*

(*)T_pM is considered a $\mathbf{K}$-vector space; (**)T_pM is considered a Z-module.

(‡) Do not confuse the symbol "star" with "asterisc".

(†) $\mathcal{L}_{\mathbf{K}}(E;F)=$ space of of continuous $\mathbf{K}$-linear applications $E \to F$; E,F top.v.sps.

Remark 1.15. (*Quantum manifolds and spectrum*). The relation between a quantum manifold M, of dimension $(n \downarrow m_1, \cdots, m_s)$ over the quantum algebra A, and spectrum is stressed by the following canonical mapping $j : M \to Sp(\underline{A}) \subset Spec(\underline{A})$, $p \mapsto j(p) \in Sp(\underline{A})$, $j(p)(f) = c(f(p))$. Let M_C be the classic limit of M, then one has the following commutative and exact diagrams:

$$
\begin{array}{ccccc}
0 & \to & A_C & \to & A \\
 & & \| & & \| \\
0 & \to & C_w^\infty(M_C;A) & \to & C_w^\infty(M;A) \\
 & & \|\wr & & \cup \\
0 & \to & Q_w^\infty(M_C;A) & \to & Q_w^\infty(M;A) \\
 & & \| & & \| \\
0 & \to & \underline{A}_C & \to & \underline{A}
\end{array}
\qquad
\begin{array}{ccccc}
0 & \leftarrow & M_C & \leftarrow & M \\
 & & \downarrow & & \downarrow \\
0 & \leftarrow & Sp(A_C) & \leftarrow & Sp(A) \\
 & & \|\wr & & \downarrow \\
0 & \leftarrow & Sp(\underline{A}_C) & \leftarrow & Sp(\underline{A}) \\
 & & & & \downarrow \\
 & & & & 0
\end{array}
$$

Note that the approach to the noncommutative differential geometry developed by A.Connes [11,12] is related to the (co)homology of the algebra A associated to a quantum manifold, in the sense introduced in this paper. However, quantum manifolds, as introduced in Definition 1.39, appear the most suitable spectral spaces to use in a noncommutative differential geometry that aims to directly extend the geometric theory of PDEs to quantum PDEs, in physical sense. ∎

Remark 1.16. (*Symmetry of the quantum higher derivatives*). Let E and F be two locally convex topological vector spaces and let $X \subset E$ be an open set in E. Assume that $f : X \to F$ is weakly p-times differentiable for some $p \in \mathbf{N}$, $p \geq 1$, (i.e., of class C_w^p). Then, every $x \in X$ and every $k \in \mathbf{N}$, $0 \leq k \leq p$, the k-linear mapping $D^k f(x) : E^k \to F$ is totally symmetric, that is $D^k f(x) \in L_{k,s}(E;F) \equiv$ space of symmetric k-linear mappings $E^k \to F$. In fact, let $x \in X$ and $(h_1,\ldots,h_k) \in E^k$ be fixed. The set $\Omega \equiv \{(t_1,\ldots,t_k) \in \mathbf{R}^k | x + \sum_{1 \leq i \leq k} t_i h_i \in X\}$ is open in $\mathbf{R}^k$. For every fixed $u \in F'$ the real function $\phi : \Omega \to \mathbf{R}$, $\phi(t_1,\ldots,t_k) = u(f(x + \sum_{1 \leq i \leq k} t_i h_i))$ is of class C^p. Then we get $(\partial t_1 \ldots \partial t_k.\phi)(0,\ldots,0) = u(D^k f(x)(h_1,\ldots,h_k))$. Moreover, it is well known that

$$(1.11) \qquad (\partial t_1 \ldots \partial t_k.\phi)(0,\ldots,0) = (\partial t_{\sigma(1)} \ldots \partial t_{\sigma(k)}.\phi)(0,\ldots,0),$$

for any permutation $\sigma \in S_k$. (S_k is the group of permutations of k objects.) Hence we must have: $u(D^k f(x)(h_1,\ldots,h_k)) = u(D^k f(x)(h_{\sigma(1)},\ldots,h_{\sigma(k)}))$. On the other hand as u is an arbitrary functional on F we can take for example one that is not constant on the subset

$$\bigcup_{\sigma \in S_k} D^k f(x)(h_{\sigma(1)},\ldots,h_{\sigma(k)}) \subset F.$$

This implies that in order to satisfy condition (1.11) must necessarily be

$$D^k f(x)(h_1,\ldots,h_k) = D^k f(x)(h_{\sigma(1)},\ldots,h_{\sigma(k)}), \ \forall \sigma \in S_k, \ 0 \leq k \leq p.$$

Finally, taking into account the arbitrariness of the vector $(h_1,\ldots,h_k) \in E^k$, it follows that $D^k f(x) \in L_{k,s}(E;F)$, $0 \leq k \leq p$. ∎

Remark 1.17. (*Algebraic characterization of quantum differentiability*). The derivative $Df(x)$ of a mapping $f : U \subset A_1 \times \cdots \times A_s \to B \equiv B_1^{p_1} \times \cdots \times B_r^{p_r}$ of class Q_w^s, $0 \leq s \leq \infty$, as considered in Definition 1.39 is a differential operator of order zero between the Z-modules $E \equiv A_1^{m_1} \times \cdots \times A_s^{m_s}$ and $F \equiv B_1^{p_1} \times \cdots \times B_r^{p_r}$. Recall that the *set of differential operators of order* $\leq k$ from E to F is defined by $Diff_k(E,F) \equiv \{\triangle \in Hom_{\mathbf{K}}(E;F) | \delta_{a_0,a_1,\cdots,a_k}(\triangle) = 0, \ \forall a_0, a_1, \cdots, a_k \in Z\}$, where $\delta_{a_0,a_1,\cdots,a_k} = \delta_{a_0} \circ \delta_{a_1} \circ \cdots \circ \delta_{a_k}$ with $\delta_a \in Hom_{\mathbf{K}}(Hom_{\mathbf{K}}(E;F); Hom_{\mathbf{K}}(E;F))$, such that $\delta_a(\triangle)(m) = \triangle(am) - a\triangle(m)$, $\forall a \in Z$, $\forall m \in E$. One has the following filtration of $Diff_\bullet(E,F) \equiv \bigcup_{0 \leq k \leq \infty} Diff_k(E,F)$: $Hom_Z(E;F) \cong$

$Diff_0(E,F) \subset Diff_1(E,F) \subset Diff_2(E,F) \subset \cdots \subset Diff_\bullet(E,F)$. Furthermore, one has a two-sided Z-module structure on each $Diff_k(E,F)$ and therefore on $Diff_\bullet(E,F)$: $(a\triangle)(m) = a\triangle(m)$; $(\triangle a)(m) = \triangle(am)$. Of course, at the zero order one has a Z-module structure. Set $Diff_k(F) \equiv Diff_k(Z,F)$. One has the following isomorphism of Z-modules:

$$Diff_k(E,F) \cong Hom_Z(E; Diff_k(F)).$$

Furthermore, $Diff_k(-,F)$ is a representable functor, i.e., one has the following isomorphism of Z-modules:

$$Diff_k(E;F) \cong Hom_Z(\mathcal{I}^k(E);F), \quad \mathcal{I}^k(E) \equiv Z \otimes_{\mathbf{K}} E/\mu_{k+1},$$

where μ_{k+1} is the submodule of $Z \otimes_{\mathbf{K}} E$ generated by all the elements of the form $(\delta^{a_0} \circ \delta^{a_1} \circ \cdots \circ \delta^{a_k})(a \otimes m)$, where δ^a is the following endomorphism of the module $Z \otimes_{\mathbf{K}} E$: $\delta^{a'}(a \otimes m) = a \otimes a'm - a'a \otimes m$. Of course, $\mathcal{I}^0(E) \cong E$. One has the following canonical $\mathbf{K}$-linear map $j_k : E \to \mathcal{I}^k(E)$, given by $j_k(m) = 1 \otimes m \bmod \mu_{k+1}$. j_k is also a differential operator of order $\leq k$, while the module $\mathcal{I}^k(E)$ is generated by elements of the form $j_k(m)$, $m \in E$. Then $\hat{\Omega}^1(Z)$ can be defined by means of the following split short exact sequence:

$$0 \to Z \xrightarrow{i_1} \mathcal{I}^1(Z) \xrightarrow{\pi} \hat{\Omega}^1(Z) \to 0,$$

where $i_1(a) = aj_1(1)$, i.e., $\hat{\Omega}^1(Z) = \mathcal{I}^1(Z)/Z$. Then, the derivation $\hat{d} : Z \to \hat{\Omega}^1(Z)$ is given by composition $\hat{d} = \pi \circ j_1$. Let $Der^n_{\mathbf{K}}(Z;E)$ denotes the Z-module of *n-derivations* from the ring Z to the Z-module E, i.e., $\mathbf{K}$-homomorphisms $f : Z^n \to E$, such that the partial mappings, $f_{a_1,\ldots,a_{i-1},a_{i+1},\ldots,a_n} : Z \to E$, $a \mapsto f(a_1,\ldots,a_{i-1},a,a_{i+1},\ldots,a_n)$, are $\mathbf{K}$-derivations for any integer i, $1 \leq i \leq n$. Then $Der^n_{\mathbf{K}}(Z;-)$ is a representable functor, i.e., one has the following isomorphism of Z-modules:

$$Der^n_{\mathbf{K}}(Z;E) \cong Hom_Z(\dot{T}^n_0(\hat{\Omega}^1(Z));E), \quad Der^{n+1}_{\mathbf{K}}(Z;E) \cong Der_{\mathbf{K}}(Z;Der^n_{\mathbf{K}}(Z;E)).$$

For the corresponding submodules of symmetric and skewsymmetric derivations, one has the following isomorphisms:

$$Der^n_{\mathbf{K},sym}(Z;E) \cong Hom_Z(\dot{S}^n_0(\hat{\Omega}^1(Z));E),$$
$$Der^n_{\mathbf{K},skew}(Z;E) \cong Hom_Z(\dot{\Lambda}^n_0(\hat{\Omega}^1(Z));E).$$

Note that one has the following slpit short exact sequence:

$$0 \to Der_{\mathbf{K}}(\hat{\Omega}(Z); E) \to Hom_{\mathbf{K}}(\mathcal{I}^1(Z); E) \to E \to 0,$$

hence the following split:

$$Hom_{\mathbf{K}}(\mathcal{I}^1(Z); E) \cong Der_{\mathbf{K}}(\hat{\Omega}(Z); E) \bigoplus E.$$

$\blacksquare$

Remark 1.18. T_pM results whether a $\mathbf{K}$-vector space or a topological locally convex $\mathbf{K}$-vector space. On the other hand one has also the following non canonical isomorphism $T_pM \cong A_1^{m_1} \times \cdots \times A_s^{m_s}$, so on T_pM one recognizes also a structure of $(m_1, \ldots, m_s)$-dimensional quantum vector space. T_pM is also an Z-module. Emphasize that, in general, T_pM cannot have a canonical structure of two-sided A-module. Moreover, $\widehat{T}_pM$ and $(T_pM)^+$ are $\mathbf{K}$-modules as well Z-modules. $\widehat{T}_pM$ is also a right $\widehat{A}$-module, with action given by: $v\hat{a} = v \circ \hat{a}$, $v \in \widehat{T}_pM$, $\hat{a} \in \widehat{A}$. $(T_pM)^+$, has, instead, a left $\widehat{A}$-module structure: $\hat{a}\alpha = \hat{a} \circ \alpha$, $\forall \hat{a} \in \widehat{A}$, $\alpha \in (T_pM)^+$. Furthermore, $(T_pM)^+$ has a two-sided A-module structure also given by: $(a\alpha)(v) = a\alpha(v)$, $(\alpha a)(v) = \alpha(v)a$, $\forall v \in T_pM$, $a \in A$. One has the following exact sequence of Z-modules: $\widehat{T}_pM \to T_pM \to 0$ induced by the canonical inclusion $Z \to A$. Note that if M is a quantum manifold of dimension $(m_1, \ldots, m_s)$ over a quantum algebra $A \equiv A_1 \times \ldots \times A_s$, the Z-modules T_pM, $\widehat{T}_pM$ and $(T_pM)^+$, have the same dimension $(m_1, \ldots, m_s)$ but over different quantum algebras. More precisely the situation is reported in the following table.

TAB.1.12 - Local isomorphisms of Z-modules over M (*).

Isomorhism	Dimension	Quantum algebra
$T_pM \cong A_1^{m_1} \times \ldots \times A_s^{m_s}$	$(m_1, \ldots, m_s)$	$A \equiv A_1 \times \ldots \times A_s$
$\widehat{T}_pM \cong \widehat{A}_1^{m_1} \times \ldots \times \widehat{A}_s^{m_s}$	$(m_1, \ldots, m_s)$	$\widehat{A}_1 \times \ldots \times \widehat{A}_s$ (**)
$(T_pM)^+ \cong \widehat{A}_1(A)^{m_1} \times \ldots \times \widehat{A}_s(A)^{m_s}$	$(m_1, \ldots, m_s)$	$\widehat{A}_1(A) \times \ldots \times \widehat{A}_s(A)$ (***)

(*) $\dim M = (m_1, \ldots, m_s)$ over the quantum algebra $A \equiv A_1 \times \ldots \times A_s$.

(**) $\widehat{A}_i \equiv Hom_Z(A; A_i)$. (***) $\widehat{A}_i(A) \equiv Hom_Z(A_i; A)$.

In the following table we report some usefuel definitions of derived tensor bundles associated to a quantum manifold.

TAB.1.13 - Derived tensor bundles of quantum manifold M.

Tensor bundle	Symm.t.b.	Skwesymm.t.b.
$T^r_s M \equiv TM \otimes_{\mathbf{K}} \cdots r \cdots \otimes_{\mathbf{K}} TM \otimes_{\mathbf{K}} T^* M \otimes_{\mathbf{K}} \cdots s \cdots \otimes_{\mathbf{K}} T^* M$	$S^r_s M$	$\Lambda^r_s M$
$\dot{T}^r_s M \equiv TM \otimes_Z \cdots r \cdots \otimes_Z TM \otimes_Z (TM)^+_0 \otimes_Z \cdots s \cdots \otimes_Z (TM)^+_0$	$\dot{S}^r_s M$	$\dot{\Lambda}^r_s M$
$\widehat{T}^r_s M \equiv \widehat{TM} \otimes_Z \cdots r \cdots \otimes_Z \widehat{TM} \otimes_Z (TM)^+ \otimes_Z \cdots s \cdots \otimes_Z (TM)^+$	$\widehat{S}^r_s M$	$\widehat{\Lambda}^r_s M$
$(***) \quad \widetilde{T}^0_s M \equiv (TM)^+ \otimes_A \cdots s \cdots \otimes_A (TM)^+$	$\widetilde{S}^0_s M$	$\widetilde{\Lambda}^0_s M$

$(***)(T_p M)^+$ is considered a two-sided A-module.

Example 1.32. Let $M = A^m$, where A is an extension $0 \to B \to A \to C^0(X) \to 0$ of a noncommutative C^*-algebra B by means of the commutative C^*-algebra $C^0(X)$ of $\mathbf{C}$-valued continuous functions on some topological space X. Then M is a quantum manifold of dimension m over A with global charts. ∎

Example 1.33. Let B be a quantum algebra and let $\mathbf{H}$ be the usual (Hamiltonian) quaternionic algebra. A *quantum B-quaternionic manifold* of dimension n and class Q^k_w, $0 \leq k \leq \infty, \omega$, is a quantum manifold M of dimension n and class Q^k_w over the B-quantum algebra $C \equiv B \otimes_{\mathbf{K}} \mathbf{H}$. Then the quantum coordinates in an open coordinate subset $U \subset M$ are called *B-quaternionic coordinates*, $\{q^k\}_{1 \leq k \leq n}$, $q^k : U \to C$. As a particular case we can take $B = \mathbf{K}$. In this case $C = \mathbf{H}$ and we call such quantum $\mathbf{K}$-quaternionic manifolds simply *quantum quaternionic manifolds*. The *category $\mathcal{C}^B_{\mathbf{H}}$, of quantum B-quaternionic manifolds of class Q^k_w*, is defined by considering as morphisms maps of class Q^k_w, between quantum B-quaternionic manifolds. (For more informations about quantum B-quaternionic manifolds see ref.[68].) ∎

Example 1.34. The *quantum derivative space* $\hat{D}(M, N)$ for quantum manifolds M and N of dimension m and n respectively over a quantum algebra A, is defined by

$$\hat{D}(M, N) \equiv TMOTN \equiv \bigcup_{(p,q) \in M \times N} T_p MOT_q N,$$

where $T_p MOT_q N \equiv Hom_Z(T_q N; T_p M)$ is the space of Z-linear mappings $T_p M \to T_q N$. Let us denote by $Q^s_w(M, N)$ the space of mappings $M \to N$

of class Q_w^s. Then if $f \in Q_w^s(M,N) \Rightarrow Df(p) \in \hat{D}(M,N)$, $\forall p \in M$. If $\pi : W \to M$ is a fiber bundle between quantum manifolds, the *quantum derivative space of sections* of π, $\hat{D}(W)$, is defined by $\hat{D} \equiv d^*\hat{D}(M,W)$, where d is the diagonal mapping $d : M \to M \times M$. *Quantum derivative spaces of higher orders* between quantum manifolds can be defined by the following: $\hat{D}^k(M,N) \equiv \hat{D}(\hat{D}^{k-1}(M,N))$, $\hat{D}^k(W) \equiv \hat{D}(\hat{D}^{k-1}(W))$. One has a natural injection over $M \times N$: $\hat{D}(M,N) \hookrightarrow D(M,N)$, where $D(M,N)$ denotes the derivative space considering M and N as locally convex manifolds [63]. $\hat{D}(M,N)$ results a quantum manifold of dimension $m + n|mn$. In fact, if (x^A) is a quantum coordinate system on M and (y^B) is a quantum coordinate system on N then (x^A, y^B, y_A^B), $1 \leq A \leq m, 1 \leq B \leq n$, where x^A and y^B have entries in A and y_A^B into $\hat{A}$ respectively, is a quantum coordinate system on $\hat{D}(M,N)$. The first *derivative* of a quantum differentiable mapping $f : M \to N$ is a section $Df : M \to \hat{D}(M,N)$ of the fiber bundle $\pi_1 : \hat{D}(M,N) \to M$. The local expression of Df is as follows: $x^A \circ Df = x^A$, $y^B \circ Df = f^B \equiv y^B \circ f$, $y_A^B \circ Df \equiv (\partial x_A f^B)$. The $m \times n$ matrix $(\partial x_A f^B)(p), \forall p \in M$, has entries in $\hat{A}$: $(\partial x_A f^B)(p) \in \hat{A}$. So the local representation of $Df(p)(v)$ is the following: $Df(p)(v) = \big(Df(p)(v)^B\big) \equiv \big(\sum_{1 \leq A \leq m}(\partial x_A f^B)(p)(v^A)\big) \in A^n$, where $v \equiv (v^1, ..., v^m) \in A^m$ is the local representation of $v \in T_pM$. Let $\pi : W \to M$ be a fiber bundle, with $\dim M = m$ and $\dim W = m + n$ over the same quantum algebra A. Let us consider the systems of quantum coordinates reported in the following table:

TAB.1.14 - Quantum coordinates on $\hat{D}(M,W)$.

M	(x^H), $\quad 1 \leq H \leq m$
W	(x^H, y^K), $\quad 1 \leq H \leq m, 1 \leq K \leq n$
$\hat{D}(M,W)$	$(x^H, \bar{x}^{\bar{H}}, y^K, \bar{x}_H^{\bar{H}}, y_H^K)$, $\quad 1 \leq H, \bar{H} \leq m, 1 \leq K \leq n$

Then, $\hat{D}(W)$ is a quantum submanifold of $\hat{D}(M,W)$ characterized by means of the following equations: $x^H = \bar{x}^H$, $x_H^{\bar{H}} = \delta_H^{\bar{H}}$, $1 \leq H, \bar{H} \leq m$. So, (x^H, y^K, y_H^K) are good quantum coordinates on $\hat{D}(W)$. Then $\dim \hat{D}(W) = (m+n, mn)$. The *velocity* of a quantum differentiable curve in M, $f : A \to M$, is the map $\dot{f} \equiv pr_2 \circ Df : A \to \widehat{T}M$, where pr_2 is the canonical projection $pr_2 : \hat{D}(A,M) \cong A \times \widehat{T}M \to \widehat{T}M$. The *differential* of a quantum differentiable map $f : M \to A$ is a map $df \equiv pr_2 \circ Df : M \to (TM)^+$, where pr_2 is, now, the canonical projection $pr_2 : \hat{D}(M, A) \cong A \times (TM)^+ \to (TM)^+$. For a quantum differentiable map $f : A \times M \to N$ we can define its *infinitesimal variation* $\partial f : M \to f_0^* \widehat{T} N$ as $\partial f = D_1 f \circ j$, being j the inclusion $j : M \to A \times M$ given by $p \mapsto (0, p)$. So, if f is a quantum deformation of a quantum differentiable section s of the fiber bundle $\pi : W \to M$, such that $f_0 = s$, then $\partial f : M \to s^* v \widehat{T} W$, where the symbol v denotes "vertical". From the above considerations we get that $\widehat{T} M$ can be identified with the space of vectors tangent to quantum differentiable curves $A \to M$ of M, and $(TM)^+$ can be identified with the space of differentials of quantum differentiable A-valued functions on M. Vice versa, we have that a C_w^∞ vector field $v : M \to TM$, (resp. Q_w^∞ quantum vector field $v : M \to \widehat{T} M$), is called *integrable* if there exists an unique (local) *flow* $\{\phi_t\}_{t \in J \subset \mathbf{R}}$, (resp. $\{\phi_a\}_{a \in B \subset A}$, such that $\partial \phi = v$, where ϕ_t, (resp. ϕ_a), are local diffeomorphisms of M of class C_w^∞, (resp. Q_w^∞), such that $\phi_0 = id_M, \phi_{-t} = \phi_t^{-1}, \phi_{t+t'} = \phi_t \circ \phi_{t'}, \forall t, t' \in J$, (resp. $\phi_0 = id_M, \phi_{-a} = \phi_a^{-1}, \phi_{a+a'} = \phi_a \circ \phi_{a'}, \forall a, a' \in B$). Here, J, (resp. B), is a neighborhood of $0 \in \mathbf{R}$, (resp. $0 \in A$). We have the following criteria for the integrability of vector fields on quantum manifolds.

☐ Let M be a quantum manifold of dimension m, over a Banach quantum algebra (or Von Neumann quantum algebra) A. Let $v \in C_w^\infty(TM)$ be a vector field over M. Then, v is an integrable vector field.

☐ Let A be a quantum algebra and let M be a quantum manifold over A. Every vector field $v \in C_w^\infty(TM)$ is integrable if the following condition is satisfied.

(**HY**) The mapping $\bar{v} : U \subset E \to E$ locally characterizing v, with E a Fréchet space, and U an open subset in E, is such that $D\bar{v}(p)$, $p \in U$, satisfies the generalized Hille-Yosida lemma. So, the linear operators $D\bar{v}(p) \in L(E; E)$ have flows. [(*Generalized Hill-Yosida Lemma*) [94]. Let E be a locally convex, sequentially complete, linear topological space. A linear operator $K \in L(D(E), E)$, with domain $D(E)$ dense in E such that the resolvent $R(r; K) \equiv (r\mathbf{1} - K)^{-1} \in L(E, E)$ exists for $r = 1, 2, \cdots$ is the infinitesimal generator of a uniquely determined equi-continuous semi-group of class (C_0) iff the operators $\{(1 - r^{-1}K)^{-s}\}$ are equi-continuous in $r = 1, 2, \cdots$ and $s = 0, 1, \cdots$.]   ∎

Example 1.35. Let M be a quantum manifold, of dimension m, over a Banach quantum algebra A. Then, a quantum vector field $\zeta \in Q_w^\infty(\widehat{T} M)$ is integrable if it is of Lipshitz type, i.e., ζ is locally represented by a mapping $\bar{v} : U \subset M \to \mathcal{L}_Z(A; A^m)$, where $\mathcal{L}_Z(A; A^m) \equiv \mathcal{L}(A; A^m) \cap Hom_Z(A; A^m)$. (For a proof see ref.[94]). Note that if a quantum vector field $\zeta \in Q_w^\infty(\widehat{T}M)$ is such that the corresponding vector field $\zeta(a) \in C_w^\infty(TM)$, given by $v(a)(p) = v(p)(a)$, $\forall p \in M$, $a \in A$, is integrable, then, we can identify a mapping $\phi_a : [0, \epsilon_a a] \to M$, given by $\phi_p(ta) = p + \int_{[0,t]} D\phi_p(t'a)(a)dt'$, with $D\phi_p(t'a)(a) = \zeta(\psi_p(t'))(a)$, where $\psi : J_a \times U \to M$ is the local flow such that $\partial\psi = \zeta(a)$, $J_a \equiv] - \epsilon_a, \epsilon_a[\subset \mathbf{R}$. (For more details see ref.[63].)

$\square$ Let M and N be quantum manifolds of dimension respectively m and n over a quantum algebra A. For any Q_w^∞-mapping $f : M \to N$ we have the following local representation of the derivative Df: $Df = \partial y_B \circ f \otimes_{\widehat{A}} (\partial x_A f^B)dx^A$, with $(\partial x_A f^B) \in Q_w^\infty(U, \widehat{A})$. For example, if $f \in Q_w^\infty(\widehat{\Lambda}_0^0 M) = Q_w^\infty(M, A)$, we get the following representation for its differential $df : M \to \widehat{\Lambda}_1^0 M \cong J\widehat{D}(\widehat{\Lambda}_0^0 M)$, $df = \sum_{1 \leq A \leq m}(\partial x_A f)dx^A$, $(\partial x_A f) : U \to \widehat{A}$. In particular, if $u : M \to W$ is a section of a quantum fiber bundle $\pi : W \to M$, $\dim W = m + n$, $\dim M = m$, we get:

($\diamond$) $Du = \partial x_B \otimes (\delta_A^B)dx^A + \partial y_B \circ (\partial x_A u^B) \otimes dx^A$, $(\partial x_A u^B) \in Q_w^\infty(U, \widehat{A})$. $\blacksquare$

Example 1.36. Let M be a quantum manifold of dimension m over a quantum algebra A. Let $u : M \to W$ be a section of a quantum fiber bundle $\pi : W \to M$, where the fiber F is a quantum manifold modeled on E^s, i.e., W is a quantum manifold of dimension (m, s) over the quantum algebra $B \equiv A \times E$. Let us assume also that E is an Z-module. One has the following local isomorphism :

$$
\begin{array}{ccc}
Hom_Z(T_pM; T_{u(p)}W) & \cong & \widehat{A}^{m^2} \times \widehat{E}^{ms} \\
\| \wr & & \| \wr \\
Hom_Z(T_pM; T_pM) \times Hom_Z(T_pM; T_qF) & \cong & Hom_Z(T_pM; T_pM) \times Hom_Z(T_pM; E)^s
\end{array}
$$

where $\widehat{E} \equiv Hom_Z(A; E)$ is the *A-fullquantum extension* of E. Now, taking into account the canonical isomorphism :

$$\widehat{T}_qF \bigotimes_{\widehat{A}} (T_pM)^+ \cong Hom_Z(T_pM; T_qF),$$

with $\widehat{T}_qF \equiv Hom_A(A; T_qF)$, and being (x^A, y^B) a fibered quantum coordinate system on W, we see that the derivative of u can be locally written with the same formula ($\diamond$), but with $(\partial x_A u^B) \in Q_w^\infty(U; \widehat{E})$. Furthermore

$Hom_Z(T_pM; T_{u(p)}W)$ is a quantum space of dimension (m^2, ms) over the quantum algebra $C \equiv \widehat{A} \times \widehat{E}$.   ∎

Example 1.37. Let W be a quantum manifold of dimension $m + n$ over A and M a quantum nanifold of dimension m over A. If $\pi : W \to M$ is a fiber bundle, then into $\hat{D}^k(W)$ we can distinguish some further subbundles homonymous to classic case ones. For example the *holonomic k-jet quantum derivative space* :

$$J\hat{D}^k \equiv \left\{ \begin{array}{l} u \in \hat{D}^k(W)\,|\,\exists \quad \text{quantum differentiable sections } s \text{ of } \pi: \\ D^k s(p) = u,\, p = \pi_k(u) \end{array} \right\}.$$

$J\hat{D}^k(W)$ is a quantum manifold of dimension $(m + n, mn, \ldots, m^k n)$ over the quantum algebra $B \equiv A \times \overset{2}{\widehat{A}} \times \overset{}{\widehat{A}} \times \ldots \times \overset{k}{\widehat{A}}$, where $\overset{r}{\widehat{A}} \equiv Hom_Z(\dot{T}_0^r A; A) \equiv (\dot{T}_0^r A)^+$. We say also that $J\hat{D}^k(W)$ is a quantum manifold of dimenssion $(m + n, mn, m^2 n, \cdots, m^k n)$ over B. In fact, a system of quantum coordinates on a k-jet quantum derivative space can be written:

$$\{x^A, y^B, y_A^B, y_{A_1 A_2}^B, \ldots, y_{A_1 \ldots A_k}^B\},\ 1 \le A, A_1, \ldots, A_k \le m,\ 1 \le B \le n,$$

where

$$y_{A_1 \ldots A_r}^B(D^k s(p)) \equiv (\partial x_{A_r} \ldots \partial x_{A_1}.s^B)(p)$$

is a matrix with entries in $\overset{r}{\widehat{A}}$. Locally we can write:

$$D^r s(p) = \partial x_B(p) \otimes_{\widehat{A}} \delta_H^B dx^H(p)$$
$$+ \partial y_B(s(p)) \otimes_{\widehat{A}} (\partial x_H.s^B)(p)dx^H(p) + \partial y_B(s(p)) \otimes_{\widehat{A}} (\partial x_{H_1}\partial x_{H_2}.s^B)(p)dx^{H_1}(p) \bullet_z dx^{H_2}(p)$$
$$+ \cdots + \partial y_B(s(p)) \otimes_{\widehat{A}} (\partial x_{H_1}\cdots\partial x_{H_r}.s^B)(p)dx^{H_1}(p) \bullet_z \cdots \bullet_z dx^{H_r}(p),$$

where $dx^{H_1}(p) \bullet_z \cdots \bullet_z dx^{H_r}(p) \equiv \bar{j}(dx^{H_1}(p) \otimes_Z \cdots \otimes_Z dx^{H_r}(p))|_{\dot{S}_0^r(T_pM)}$, with $\bar{j}$ the canonical Z-homomorphism

$$\bar{j} : T_pM)^+ \otimes_Z \cdots_r \cdots \otimes_Z (T_pM)^+ \to Hom_Z(\dot{T}_0^r(T_pM); \dot{T}_0^r A).$$

Furthermore, $J\hat{D}^k(W)$ is an affine bundle over $J\hat{D}^{k-1}(W)$ with associated vector bundle $\pi_{k,0}^*(\dot{S}_0^k MOvTW)$. So we have the following exact sequence of vector bundles over $J\hat{D}^k(W)$:

$$0 \to \pi^*_{k,0}(\dot{S}_0^k MOvTW) \to vTJ\hat{D}^k(W) \to \pi^*_{k,k-1}vTJ\hat{D}^{k-1} \to 0.$$

The relation with k-jet derivative spaces is given by the following exact commutative diagram:

$$
\begin{array}{ccccccccc}
& & 0 & & 0 & & 0 & & \\
& & \downarrow & & \downarrow & & \downarrow & & \\
0 & \to & \pi^*_{k,0}(\dot{S}^k_0 M O v T W) & \to & v T J \hat{D}^k(W) & \to & \pi^*_{k,k-1} v T J \hat{D}^{k-1}(W) & \to & 0 \\
& & \downarrow & & \downarrow & & \downarrow & & \\
0 & \to & \pi^*_{k,0} Hom_{\mathbf{K}}(S^k_0 M; vTW) & \to & v T J D^k(W) & \to & \pi^*_{k,k-1} v T J D^{k-1}(W) & \to & 0
\end{array}
$$

$\blacksquare$

Example 1.38. Let $\pi : W \to M$ be a fiber bundle such that $\dim W = (m, s)$, over the quantum algebra $B \equiv A \times E$ and $\dim M = m$ over A and such that E is an Z-module. Then $J\hat{D}^k(W)$ is a quantum manifold of dimension $(m, s, ms, m^2 s, \cdots, m^k s)$ over the quantum algebra

$$
C \equiv A \times E \times \overset{1}{\widehat{A}}(E) \times \cdots \times \overset{k}{\widehat{A}}(E),
$$

i.e., $J\hat{D}^k(W)$ is modelled on

$$
A^m \times E^s \times (\overset{1}{\widehat{A}}(E))^{ms} \times \cdots \times (\overset{k}{\widehat{A}}(E))^{m^k s} \equiv A^m \times \overset{k}{\widehat{A}}(E)(s, m),
$$

where

$$
\overset{r}{\widehat{A}}(E) \equiv Hom_Z(\dot{T}^r_0(A); E),
$$
$$
\overset{0}{\widehat{A}}(E) \equiv E.
$$

For example, $J\hat{D}^k(\hat{T}^0_p M)$ is a quantum manifold of dimension of the above type with $s = m^p$ and $E = \dot{T}^p_0(\widehat{A})$. Furthermore, $J\hat{D}^\infty(W)$ is modeled on $A^m \times \overset{\infty}{\widehat{A}}(E)(s, m)$, where $\overset{\infty}{\widehat{A}}(E)(s, m) \equiv \bigoplus_{k \geq 0} \overset{k}{\widehat{A}}(E)(s, m)$. In particular, if $E \equiv A$, $s = n$, we get that $J\hat{D}^k(W)$ has dimension $(m + n, mn, m^2 n, \cdots, m^k n)$ over

$$
D \equiv A \times E^s \times \overset{1}{\widehat{A}} \times \cdots \times \overset{k}{\widehat{A}}
$$

as

$$
\overset{k}{\widehat{A}}(A) = \overset{k}{\widehat{A}}.
$$

We put also

$$\overset{\infty}{\widehat{A}}(E) \equiv \bigoplus_{k\geq 0} \overset{k}{\widehat{A}}(E).$$

Set

$$_k\widehat{A}(E) \equiv \bigoplus_{0\leq i\leq k} \overset{i}{\widehat{A}}(E).$$

Then, one has the natural filtration:

$$E \subset \widehat{A}(E) \subset \ldots \subset {}_k\widehat{A}(E) \subset \cdots \subset \overset{\infty}{\widehat{A}}(E).$$

∎

It is important also characterize quantum submanifolds by means of an implicit quantum function theorem.

Definition 1.40. (*Quantum submanifolds*). Let M be a smooth quantum manifold, i.e., a quantum manifold of class Q_w^∞, of dimension m over a quantum algebra A. A subset $S \subset M$ is called a *quantum submanifold* of M if, for every point $p \in M$, there exists a chart $(W, \chi, A^p \times A^q)$ on M, $p \in W$, with quantum coordinates $\{\xi^C\} \equiv \{x^A, y^B\}$, $\{x^A\} : W \to A^p$, $\{y^B\} : W \to A^q$, $p + q = m$, such that the chart $\chi(S \cap W) = \{(u, v) \in \chi(W) \subset A^p \times A^q : (v^B = 0) \in A^q\}$.

Proposition 1.37. *If $S \subset M$ is a quantum submanifold of M, then there is a natural smooth quantum manifold structure on S.*

Proof. In fact, under the hypotheses of above definition, to every chart (W, χ, A^m) on M there corresponds the chart (U, ϕ, A^p) on S, with $U \equiv S \cap W$, $\phi = \pi \circ \chi$, where π is the natural projection $\pi : A^m \to A^p$. □

Theorem 1.28. (Implicit quantum function theorem). *We can locally define a quantum submanifold S of M as a solution set of a system of equations:*

$$S \equiv \{p \in M : F^\alpha(p) = 0, (F^\alpha) : M \to A^q\}$$

if the jacobian matrix $(\partial\xi_C.F^\alpha)(p)$ has rank q.

Proof. In a chart (W, χ, A^m) on M with quantum coordinates $\{\xi^C\}_{1\leq C\leq m}$, the above equations take the form: $S \equiv \{w \in \chi(W) \subset A^m : \Phi^\alpha(m) = 0\}$, $(\Phi^\alpha) \equiv (F^\alpha \circ \chi^{-1}|_{\chi(W)} : \chi(W) \to A^q)$. Such a set will be a quantum submanifold of M if, in a neighborhood of such a point $w \in$, the coordinates ξ^C can be split into two classes $(\xi^C) = (x^A, y^B)$, in a way that the system $\{\Phi^\alpha = 0\}$ will be solvable in this neighbourhood, with respect to the coordinates (y^B), i.e., if it can be written in the following form: $S = \{y^B = f^B(x^\alpha)\}$. Now, let us consider the following lemmas and definitions.

Lemma 1.9. *Let Λ be a ring and let V be a Λ-module with finite basis, $\dim V = n$. Then one has the Λ-isomorphism: $\Lambda^n(V) \cong \Lambda$.*

Proof of Lemma 1.9. It is standard. (See e.g., ref.[4].) $\qquad\square$

Definition 1.41. The *determinant* of $u \in Hom_\Lambda(V)$ is the element $\det u \in \Lambda$ such that $\Lambda^n(u) : \Lambda^n(V) \to \Lambda^n(V)$, $v_1 \wedge \cdots \wedge v_n \mapsto \det u v_1 \wedge \cdots \wedge v_n$. If $\det u = 1$ we say that u is *unimodular*.

Lemma 1.10. *Let V be a Λ-module of finite dimension. One has the following properties:*

(i) $u, v \in Hom_\Lambda(V) \Rightarrow \det(u \circ v) = (\det u).(\det v)$;

(ii) $\det(id_V) = 1 \in A$.

Lemma 1.11. *Let V be a finite dimensional Λ-module. The following propositions are equivalent:*

(i) *u is bijectif;*

(ii) *u is right invertible;*

(iii) *u is left invertible;*

(iv) *u is surjectif;*

(v) *$\det u$ is invertible in Λ.*

Lemma 1.12. *Let V be a Λ-module of finite dimension. $u \in Hom_\Lambda(V)$. The following conditions are equivalent:*

(i) *u is injective;*

(ii) *$\det u$ is not a zero divisor in Λ.*[24]

Now, if $\Phi : U_1 \times U_2 \subset A^p \times A^q \to A^q$, $\Phi(u_0, v_0) = 0$, $(u_0, v_0) \in U_1 \times U_2$, and the matrix $(\partial y_B.\phi^\alpha)(u, v)$ is continuous and invertible in $\chi(W) = U_1 \times U_2 \subset A^p \times A^q$, there exists a neighbourhood $U' \subset U_1$ of u_0 and a (unique) continuous function $f; U' \to A^q$ such that $v_0 = f(u_0)$ and $\Phi(u, f(u)) = 0$. So, if there is a derivative $(\partial x_A.\Phi^\alpha)(u, v)$ in $U_1 \times U_2$, then the explicit function $(y^B = f^B(x^A))$ is differentiable at U' and we have

$$\left\{ \begin{array}{l} (\partial x_A.f^B)(u) = -(\partial y_S.\Phi^{-1 B})(\partial x_A.\Phi^S)(u, v) \\ v = f(u) \end{array} \right\}.$$

---

[24] If Λ is a field one has the equivalence of Lemma 1.12(ii) and Lemma 1.11(i).

So the theorem is proved. $\qquad\qquad\qquad\qquad\qquad\qquad\qquad\qquad\square$

Remark 1.19. Generalizations of above considerations to quantum manifolds of dimension $(m_1,\cdots,m_s)$ are easily obtained. $\qquad\qquad\qquad\blacksquare$

Let us now characterize the (co)homological structures of quantum manifolds and their (co)bordism properties. This is necessary in order to develop a geometric theory of quantum PDE's.

Remark 1.20. Let M be a quantum manifold of dimension m over a quantum algebra A. One has the following commutative diagram:

$$
\begin{array}{ccccc}
0 & & (\dot\Lambda_0^q M)^+ & \equiv & Hom_Z(\dot\Lambda_0^q M;A)\\[2pt]
\uparrow & & \downarrow & & \downarrow\\[6pt]
(\Lambda_0^q M)^*\otimes_{\mathbf{K}}(A/\mathbf{K}) & & & & \\[2pt]
\uparrow & & & & \\[6pt]
(\Lambda_0^q M)^*\otimes_{\mathbf{K}} A & \longrightarrow & (\Lambda_0^q M)^*\equiv Hom_{\mathbf{K}}(\Lambda_0^q M;A) & \longleftarrow & Hom_Z(\Lambda_0^q M;A)\\[2pt]
\uparrow & & \uparrow & & \\[6pt]
(\Lambda_0^q M)^* & = & (\Lambda_0^q M)^* & & \\[2pt]
\uparrow & & \uparrow & & \\[6pt]
0 & & \Lambda_q^0 M & &
\end{array}
$$

where the first vertical line is exact. $\qquad\qquad\qquad\qquad\qquad\qquad\blacksquare$

Definition 1.42. In the following table we define some useful spaces.

TAB.1.15 - Dstinguished spaces of differential forms on M.

$$\Omega^\bullet(M)\equiv\bigoplus_{q\geq0}\Omega^q(M),\quad \Omega^q(M)\equiv C_w^\infty((\Lambda_0^q M)^*)\cong C_w^\infty\left(\bigcup_{p\in M}A_{q;\mathbf{K}}(T_p M;\mathbf{K})\right)$$
$$\Omega^q(M)\equiv C_w^\infty((\Lambda_0^q M)^*)\cong C_w^\infty\left(\bigcup_{p\in M}A_{q;\mathbf{K}}(T_p M;\mathbf{K})\right)$$

$$\hat\Omega^\bullet(M)\equiv\bigoplus_{q\geq0}\hat\Omega^q(M),\quad \hat\Omega^q(M)\equiv C_w^\infty(Hom_{\mathbf{K}}(\Lambda_0^q M;A))\cong C_w^\infty\left(\bigcup_{p\in M}A_{q;\mathbf{K}}(T_p M;A)\right)$$
$$\hat\Omega^q(M)\equiv C_w^\infty(Hom_{\mathbf{K}}(\Lambda_0^q M;A))\cong C_w^\infty\left(\bigcup_{p\in M}A_{q;\mathbf{K}}(T_p M;A)\right)$$

$$\overset{\bullet}{\Omega}{}^q(M)\equiv C_w^\infty(Hom_{\mathbf{K}}(\Lambda_0^q M;A/\mathbf{K}))\cong C_w^\infty\left(\bigcup_{p\in M}A_{q;\mathbf{K}}(T_p M;A/\mathbf{K})\right)$$

$$\overset{\bullet\bullet}{\Omega}{}(M)\equiv\bigoplus_{q>0}\overset{\bullet}{\Omega}{}^q(M),\quad \overset{\bullet}{\Omega}{}^q(M)\equiv C_w^\infty(Hom_{\mathbf{K}}(\Lambda_0^q M;A/\mathbf{K}))\cong C_w^\infty\left(\bigcup_{p\in M}A_{q;\mathbf{K}}(T_p M;A/\mathbf{K})\right)$$

$$Q_w^\infty\left(\bigcup_{p\in M}A_{q;Z}(T_p M;A)\right)$$

$$\widehat\Omega^\bullet(M)\equiv\bigoplus_{q>0}\widehat\Omega^q(M),\quad \widehat\Omega^q(M)\equiv Q_w^\infty((\dot\Lambda_0^q M)^+)\cong Q_w^\infty\left(\bigcup_{p\in M}A_{q;Z}(T_p M;A)\right)$$

Remark 1.21. The $\mathbf{K}$-spaces $\Omega^q(M)$, $\overset{\bullet}{\Omega}{}^q(M)$, $\widehat\Omega^q(M)$ can be canonically represented in $\hat\Omega^q(M)$ by means of above commutative diagram. One has: $(\dot\Lambda_0^0 M)^+ = \hat\Lambda_0^0 M \equiv M\times A = Hom_{\mathbf{K}}(\Lambda_0^0 M;A)$; $\dot\Lambda_0^0 M \equiv M\times Z$; $\Lambda_0^0 M \equiv M\times\mathbf{K}$. As a consequence one has: $\hat\Omega^0(M) = C_w^\infty(M;A)$, $\dot\Omega^0(M) \equiv C_w^\infty(\dot\Lambda_0^0 M) = C_w^\infty(M;Z)$, $\overset{\bullet}{\Omega}{}^0(M) = C_w^\infty(M;A/\mathbf{K})$, $\widehat\Omega^0(M) = Q_w^\infty(M;A)$. $\blacksquare$

Proposition 1.38. (Exterior product). *One has the following canonical Z-homomorphism (exterior product):*

$$\wedge : \hat{\Omega}^q(M) \times \hat{\Omega}^p(M) \to \hat{\Omega}^{q+p}(M), \quad \wedge : (\omega, \theta) \mapsto \omega \wedge \theta,$$

such that

$$(\omega \wedge \theta)(\zeta_1, \ldots, \zeta_{p+q}) = \sum_{\sigma \in S_{p+q}} \epsilon(\sigma)[\omega(\zeta_{\sigma(1)}, \ldots, \zeta_{\sigma(p)}), \theta(\zeta_{\sigma(p+1)}, \ldots, \zeta_{\sigma(p+q)})]_+,$$

for any $\zeta_1, \ldots, \zeta_{p+q} \in C_w^\infty(TM)$, and $[,]_+$ is the Jordan bracket. Then one has the following formula: $\omega \wedge \theta = (-1)^{pq}\theta \wedge \omega$. The local expression of $\omega \wedge \theta$ in quantum coordinates is the following:

$$\omega \wedge \theta = \sum_{1 \le C_1, \ldots, C_{p+q} \le m} (\omega \wedge \theta)_{C_1 \ldots C_{p+q}} dx^{C_1} \triangle \cdots \triangle dx^{C_{p+q}},$$

where[25]

$$(\omega \wedge \theta)_{C_1 \ldots C_{p+q}} \equiv \omega_{C_1 \ldots C_p} \theta_{C_{p+1} \ldots C_{p+q}} + (-1)^{pq} \theta_{C_1 \ldots C_q} \omega_{C_{q+1} \ldots C_{q+p}},$$

with $\omega_{A_1 \ldots A_p} \theta_{B_1 \ldots B_q} \in C_w^\infty(U, \overset{p+q}{\widehat{A}})$ obtained from $\omega_{A_1 \ldots A_p}$ and $\theta_{B_1 \ldots B_q}$ respectively, by means of the following compositions of homomorphisms:

$$\overset{p}{\widehat{A}} \otimes_Z \overset{q}{\widehat{A}} \to Hom_Z(\dot{T}_0^{p+q}(A); A \otimes_Z A) \xrightarrow{Hom_Z(1,\hat{\mu})} \overset{p+q}{\widehat{A}}.$$

Furthermore,

$$dx^{C_1}(p) \triangle \cdots \triangle dx^{C_r}(p) \equiv \bar{j}(dx^{C_1}(p) \otimes_Z \cdots \otimes_Z dx^{C_r})|_{\dot{\Lambda}_0^r(T_p M)}, \ \forall p \in U$$

where $\bar{j}$ is defined as Example 1.37. Similar exterior products can be defined for differential forms in $\Omega^\bullet(M)$ and $\widehat{\Omega}^\bullet(M)$. Of course in the last case the vector fields ζ, ζ_i belong to $Q_w^\infty(TM)$.

Proposition 1.39. (Exterior differential). *One has the following **K**-homomorphism (exterior differential): $d : \Omega^q(M) \to \Omega^{q+1}(M)$, such that*

$$
\begin{aligned}
d\omega(\zeta_0, \ldots, \zeta_q) = &\frac{1}{q+1} \sum_{\lambda=0}^{q} (-1)^\lambda < \omega(\zeta_0, \ldots, \hat{\zeta}_\lambda, \ldots, \zeta_q) > \\
&+ \frac{1}{q+1} \sum_{\lambda < \mu} (-1)^{\lambda+\mu} \omega([\zeta_\lambda, \zeta_\mu], \zeta_0, \ldots, \hat{\zeta}_\lambda, \ldots, \hat{\zeta}_\mu, \ldots, \zeta_q),
\end{aligned}
$$

(1.12)

---

[25] For abuse of notation we also write $(\omega \wedge \theta)_{C_1 \ldots C_{p+q}} \equiv [\omega_{\sigma(C_1 \ldots C_p}, \theta_{C_{p+1} \ldots C_{p+q})}]_+.$

for all fields $\zeta_0, ..., \zeta_q \in C_w^\infty(TM)$. Here the terms $\hat{\zeta}_\lambda$ and $\hat{\zeta}_\mu$ are omitted. This is a first order differential operator, i.e., one has the following commutative diagram

$$
\begin{array}{ccc}
JD((\Lambda_0^q M)^*) & \longrightarrow & (\Lambda_0^{q+1} M)^* \\
D\alpha \uparrow & & \uparrow d\alpha \\
M & \cong & M
\end{array}
$$

for any $\alpha \in \Omega^q(M)$. Similarly one has Z-homomorphisms $d : \hat{\Omega}^q(M) \to \hat{\Omega}^{q+1}(M)$ and $d : \widehat{\Omega}^q(M) \to \widehat{\Omega}^{q+1}(M)$. This last differential identifies a first order quantum differential operator, i.e., one has the following commutative diagram

$$
\begin{array}{ccc}
J\hat{D}((\dot{\Lambda}_0^q M)^+) & \longrightarrow & (\dot{\Lambda}_0^{q+1} M)^+ \\
D\alpha \uparrow & & \uparrow d\alpha \\
M & \cong & M
\end{array}
$$

for any $\alpha \in \hat{\Omega}^q(M)$. One has the following formula:

$$(1.13) \quad d(\omega \wedge \theta) = (d\omega) \wedge \theta + (-1)^q \omega \wedge (d\theta), \quad \forall \omega \in \Omega^q(M), \quad \theta \in \Omega^p(M).$$

The formula (1.13) holds also for differential forms in $\hat{\Omega}^\bullet(M)$ and $\widehat{\Omega}^\bullet(M)$. If $\alpha \in \Omega^q(M)$ the local expression of $d\alpha$ in quantum coordinates is the following

$$
d\alpha = \sum_{1 \le B, A_1, ..., A_p \le m} (\partial x_B . \alpha_{A_1 ... A_p}) dx^B \triangle dx^{A_1} \triangle \cdots \triangle dx^{A_p},
$$

with $(\partial x_B . \alpha_{A_1 ... A_p}) \in C_w^\infty(U, \mathbf{K})$, where $\alpha_{A_1 ... A_p}$ are the local quantum components of α. A similar expression holds also if $\alpha \in \hat{\Omega}(M)$, (resp. $\alpha \in \widehat{\Omega}(M)$), but with $(\partial x_B . \alpha_{A_1 ... A_p}) \in C_w^\infty(U, \overset{p+1}{\hat{A}})$, (resp. $(\partial x_B . \alpha_{A_1 ... A_p}) \in Q_w^\infty(U, \overset{p+1}{\hat{A}})$). Furthermore, for any quantum smooth mapping $f : M \to M$ between quantum manifolds, one has

$$
(f^* \omega)(\zeta_1, ..., \zeta_q) = \omega(f_* \zeta_1, ..., f_* \zeta_q),
$$
$$
f^*(\omega \wedge \theta) = (f^* \omega) \wedge (f^* \theta),
$$
$$
d(f^* \omega) = f^*(d\omega),
$$

where ω and θ belong to $\Omega^\bullet(M)$, $\hat{\Omega}^\bullet(M)$, or $\widehat{\Omega}^\bullet(M)$.

Proposition 1.40. (Interior product). *For any* $\zeta \in C_w^\infty(TM)$ *one has a* **K**-*homomorphism (interior product):*

$$\zeta\rfloor : \Omega^q(M) \to \Omega^{q-1}(M), \quad \zeta\rfloor : \alpha \mapsto \zeta\rfloor\alpha,$$

$$(\zeta\rfloor\alpha)(\zeta_1,\ldots,\zeta_{q-1}) = \alpha(\zeta,\zeta_1,\ldots,\zeta_{q-1}), \quad \forall \zeta_i \in C_w^\infty(TM).$$

Similar definitions hold for differential forms in $\hat{\Omega}^\bullet(M)$ *and, by restriction to vector fields in* $Q_w^\infty(TM)$, *for differential forms in* $\widehat{\Omega}^\bullet(M)$. *One has the following properties:*

$$\zeta\rfloor(\omega \wedge \theta) = (\zeta\rfloor\omega) \wedge \theta + (-1)^q\omega \wedge (\zeta\rfloor\theta), \quad \forall \zeta \in C_w^\infty(TM),$$

$$\zeta_1(\zeta_2\rfloor\omega) + \zeta_2(\zeta_1\rfloor\omega) = 0, \quad \forall \zeta_1, \zeta_2 \in C_w^\infty(TM).$$

Proposition 1.41. (Lie derivative). *For any* $\zeta \in C_w^\infty(TM)$ *one has a* **K**-*homomorphism (Lie derivative):*

$$\mathcal{L}_\zeta : \Omega^q(M) \to \Omega^q(M), \quad \mathcal{L}_\zeta : \omega \mapsto \mathcal{L}_\zeta\omega,$$

$$(\mathcal{L}_\zeta\omega)(\zeta_1,...,\zeta_q) = \zeta.\omega(\zeta_1,...,\zeta_q) - \sum_{1\leq\lambda\leq q} \omega(\zeta_1,...,[\zeta,\zeta_\lambda],...,\zeta_q),$$

$$\forall \zeta_i \in C_w^\infty(TM), i = 1,...,q.$$

Similar Lie derivatives can be defined for differential forms on $\hat{\Omega}^\bullet(M)$ *and, if* $\zeta, \zeta_i \in Q_w^\infty(TM)$, *for differential forms in* $\widehat{\Omega}^\bullet(M)$ *also obtaining Lie derivatives like Z-homomorphisms. One has the following properties:*

$$\mathcal{L}_\zeta(\omega \wedge \theta) = (\mathcal{L}_\zeta\omega) \wedge \theta + \omega \wedge (\mathcal{L}_\zeta\theta),$$

$$\mathcal{L}_{\zeta_1}(\mathcal{L}_{\zeta_2}\omega) - \mathcal{L}_{\zeta_2}(\mathcal{L}_{\zeta_1}\omega) = \mathcal{L}_{[\zeta_1,\zeta_2]}\omega,$$

$$\mathcal{L}_{\zeta_1}(\zeta_2\rfloor\omega) = [\zeta_1,\zeta_2]\rfloor\omega + \zeta_2\rfloor(\mathcal{L}_{\zeta_1}\omega),$$

$$\mathcal{L}_\zeta\omega = d(\zeta\rfloor\omega) + \zeta\rfloor(d\omega).$$

Remark 1.22. Let M be a quantum manifold of dimension $(n \downarrow m)$. If $\alpha \in \hat{\Omega}^p(M)$, then the pull-back $i^*\alpha$ of α by means of any section $i : M_C \to M$, has the following local representation

$$i^*\alpha = \sum_{1\leq j_1<\cdots<j_p\leq n} \beta_{j_1\ldots j_p} \otimes d\xi^{j_1} \wedge \cdots \wedge d\xi^{j_p},$$

with $\beta_{j_1\ldots j_p} \in C_w^\infty(c(U), A)$, given by

$$\beta_{j_1\ldots j_p} = \sum_{1\leq A_1,\cdots,A_p\leq n} \alpha_{A_1\ldots A_p} \circ i\left((\partial\xi_{[j_1}.i^{A_1}) \otimes \cdots \otimes (\partial\xi_{j_p]}.i^{A_p})\right).$$

Here c is the canonical projection $M \to M_C$ and $(\xi^j)_{1 \le j \le n}$ are coordinates on M_C induced by the quantum coordinates $(x^A)_{1 \le A \le m}$ on $U \subset M$. Furthermore,

$$\alpha_{A_1 \ldots A_p} \in C_w^\infty(U, \overset{p}{\hat{A}})$$

are the local quantum components of α. $\qquad\qquad\qquad\qquad\blacksquare$

Proposition 1.42. (de Rham complexes). *Let M be a quantum manifold of dimension m over a quantum algebra A. One has the following cochain complexes:*

$$
\begin{array}{ccccccccccc}
0 & \to & A & \to & \widehat{\Omega}^0(M) & \overset{d}{\to} & \widehat{\Omega}^1(M) & \overset{d}{\to} & \widehat{\Omega}^2(M) & \overset{d}{\to} & \cdots \\
 & & \| & & \downarrow & & \downarrow & & \downarrow & & \\
0 & \to & A & \to & \hat{\Omega}^0(M) & \overset{d}{\to} & \hat{\Omega}^1(M) & \overset{d}{\to} & \hat{\Omega}^2(M) & \overset{d}{\to} & \cdots \\
 & & \uparrow & & \uparrow & & \uparrow & & \uparrow & & \\
0 & \to & \mathbf{K} & \to & \Omega^0(M) & \overset{d}{\to} & \Omega^1(M) & \overset{d}{\to} & \Omega^2(M) & \overset{d}{\to} & \cdots
\end{array}
$$

The top complex is called the de Rham fullquantum complex *of M. The middle complex is called the* de Rham quantum complex *of M. Finally the bottom complex is called the* de Rham complex *of M.*

One has the following commutative diagram with exact horizontal lines:

$$
\begin{array}{ccccccc}
0 & \to & \Omega^q(M) & \to & \hat{\Omega}^q(M) & \to & \overset{\bullet\,q}{\widehat{\Omega}}(M) & \to & 0 \\
 & & d \downarrow & & d \downarrow & & d \downarrow & & \\
0 & \to & \Omega^{q+1}(M) & \to & \hat{\Omega}^{q+1}(M) & \to & \overset{\bullet\,q+1}{\widehat{\Omega}}(M) & \to & 0
\end{array}
$$

This defines an exterior differential $d : \overset{\bullet\,q}{\widehat{\Omega}}(M) \to \overset{\bullet\,q+1}{\widehat{\Omega}}(M)$.

Definition 1.43. (*de Rham cohomologies*). The *de Rham cohomology*, (resp. *de Rham quantumcohomology*, *de Rham fullquantum cohomology*), of a quantum manifold M is the homology $H^\bullet(M)$, (resp. $\hat{H}^\bullet(M)$, $\widehat{H}^\bullet(M)$), of the de Rham complex, (resp. de Rham quantum complex, de Rham fullquantum complex), on M.

Remark 1.23. In the Tab.1.16 and Tab.1.17 we also define some further useful cochain complexes and the corresponding cohomology spaces. Note that $C_p(M; A)$ is a free two-sided A-module: $C_p(M; A) = \{c = \sum_i c^i u_i | c^i \in A\} \cong A \otimes_{\mathbf{K}} C_p(M; \mathbf{K})$, where u_i are singular p-simplexes $u_i : \triangle^p \subset \mathbf{K}^{p+1} \to M$. For $p < 0$ the group $C_p(M; A)$ is defined to be zero. Let $\partial : C_p(M; A) \to C_{p-1}(M; A)$ be the homomorphism defined by $\partial u =$

$\sum_{0\leq i\leq p}(-1)^i\partial_i u : \Delta^{p-1} \to M$; ∂ is called the *boundary homomorphism*. As $\partial \circ \partial = 0$, one has that $\{C_p(M;A),\partial\}$ is a chain complex. (For 0-chains one has $\partial \equiv 0$.) So $C_p(M;A)$ identifies a covariant functor from the category of quantum manifolds to the category of free two-sided A-modules. $C^p(M;A)$ is a two-sided A-module. The following pairing: $<,>$: $C^p(M;A) \times C_p(M;A) \to A$ can be defined. The corresponding cochain complex $\{C^p(M;A),\delta\}$ is defined by considering the *coboundary* of a p-quantumcochain $\alpha \in C^p(M;A)$ as the p-quantumcochain $\delta\alpha \in C^{p+1}(M;A)$ such that $< \delta\alpha,c > +(-1)^p < \alpha,\partial c >= 0$. Note that in this chapter we take $\mathbf{K} = \mathbf{R}$. In Tab.1.18 we give some useful relations between them. ∎

TAB.1.16 - Cohomologies of quantum manifolds.

Cochain Complex	Chomology Space
$\{\hat{\Omega}^\bullet(M),d\}$	$\hat{H}^\bullet(M)$ $\left(de\ Rham\ quantum\ cohomology\right)$
$\{\widehat{\Omega}^\bullet(M),d\}$	$\widehat{H}^\bullet(M)$ $\left(de\ Rham\ fullquantum\ cohomology\right)$
$\{\Omega^\bullet(M),d\}$	$H^\bullet(M)$ $\left(de\ Rham\ cohomoloy\right)$
$\{\overset{\bullet}{\Omega}{}^\bullet(M),d\}$	$\overset{\bullet}{H}{}^\bullet(M)$ $\left(bullet\ de\ Rham\ quantum\ cohomology\right)$
$\{A\otimes_{\mathbf{K}}\Omega^\bullet(M),d\}$	$H^\bullet(M;A)$ $\left(de\ Rham\ cohomology\ with\ coeffs.\ in\ A\right)$
$\{Z\otimes_{\mathbf{K}}\Omega^\bullet(M),d\}$	$H^\bullet(M;Z)$ $\left(de\ Rham\ cohomology\ with\ coeffs.\ in\ Z\right)$

In the following we will state some important theorem characterizing quantum manifolds. Their proofs can be conduced by using standard results of algebraic (co)homology and algebraic-topology.

Theorem 1.29. 1) (Quantum de Rham theorem). *If M is a quantum manifold of dimension m over A, one has the following canonical exact sequence of two sided A-modules:*

$$0 \to \left[A\bigotimes_{\mathbf{K}} H^p(M) \cong H^p(M;A)\right] \to H^p_s(M;A).$$

In particular, if M is classicregular and $H_p(M_C; \mathbf{K})$ is a finitely generated $\mathbf{K}$-space, then one has the following isomorphism: $H^p(M; A) \cong H_s^p(M; A)$.

2) One has the canonical morphisms of two-sided A-modules : $\widehat{\widehat{H}}{}^p(M) \to \widehat{H}{}^p(M) \to H_s^p(M; A)$.

3) If M is a contractible quantum manifold, then $H_s^p(M; A) \cong H^p(M; A) \cong \widehat{H}{}^p(M) \cong \widehat{\widehat{H}}{}^p(M) = 0$, for $p \neq 0$.

4) If M has s connected components one has: $H_s^0(M; A) \cong H^0(M; A) \cong \widehat{H}{}^0(M) \cong \widehat{\widehat{H}}{}^0(M) \cong A^s$.

5) (Quantum Stokes theorem). $< d\phi, c > +(-1)^{\dim \phi} < \phi, \partial c >= 0$, for any $\phi \in \widehat{\Omega}^{p-1}(M)$ and $c \in C_p(M; A)$.

Definition 1.44. Let M be a quantum manifold. The *cup product* is a linear mapping

$$\cup : C^r(M, \mathbf{K}) \bigotimes_{\mathbf{K}} C^s(M, \mathbf{K}) \to C^{r+s}(M, \mathbf{K})$$

defined as follows:

$$< cc' \equiv c \cup c', [\sigma] >= (-1)^r < c, [\sigma \circ \alpha_r] >< c', [\sigma \circ \beta_s] >\in \mathbf{K}$$

with $\sigma \equiv \triangle^{r+s} \subset \mathbf{K}^{r+s+1} \to M$, where $\alpha_r(t_0, \cdots, t_{r+s}) = (t_0, \cdots, t_r, 0, \cdots, 0)$, $\beta_s(t_0, \cdots, t_{r+s}) = (0, \cdots, 0, t_r, \cdots, t_{r+s})$. The *front r-face* of σ is meant the composition $\sigma \circ \alpha_r : \triangle^r \to M$; the *back s-face* of σ is the composition $\sigma \circ \beta_s : \triangle^s \to M$.

TAB.1.17 - Singular (co)homologies of quantum manifolds.

Cochain Complex	(co)homology Space
$\{C_\bullet(M,\mathbf{K})\,\partial\}$	$H_\bullet(M;\mathbf{K})$ $\left(singular\ homology\right)$
$\{C^\bullet(M,\mathbf{K}),\delta\}$	$H_s^\bullet(M;\mathbf{K})$ $\left(singular\ cohomology\right)$
$\{C_\bullet(M,A) \equiv A \otimes_\mathbf{K} C_\bullet(M;\mathbf{K}),\partial\}$	$H_\bullet(M;A)$ $\left(sing.hom.with\ coeffs.in\ A\right)$
$\{C^\bullet(M,A) \equiv Hom_A(C_\bullet(M;A);A),\delta\}$	$H_s^\bullet(M;A)$ $\left(sing.cohom.with\ coeffs.\ in\ A\right)$

One has the isomorphism: $C^P(X,A) \cong Hom_\mathbf{K}(C_p(M;\mathbf{K});A)$

Proposition 1.43. 1) *The constant cocycle* $1 \in C^0(M)$ *serves as identity element. One has*

$$\delta(cc') = (\delta c)c' + (-1)^r c(\delta c'), \quad c \in C^r(M; \mathbf{K}).$$

2) *One has the corresponding product operation;*

$$\cup : H_s^r(M, \mathbf{K}) \bigotimes_{\mathbf{K}} H_s^s M, \mathbf{K}) \to H_s^{r+s}(M, \mathbf{K}).$$

As one has $ab = (-1)^{rs} ba$, $\forall a \in H_s^r(M, \mathbf{K})$, $b \in H_s^s(M, \mathbf{K})$, *the cohomology* $H_s^\bullet(M, \mathbf{K} \equiv \bigoplus_{r \geq 0} H_s^r(M, \mathbf{K})$ *is a graded commutative ring.*
3) *Let* $X, Y \subset M$ *such that* X *and* Y *are relatively open when considered as subsets of* $X \bigcup Y$. *One has the cup operation:*

$$H_s^r(M, X, \mathbf{K}) \bigotimes_{\mathbf{K}} H_s^s(M, Y, \mathbf{K}) \to H_s^{r+s}(M, X \cup Y, \mathbf{K}).$$

4) *If* M *is classic regular and* $H_p(M_C, \mathbf{K})$ *is a finitely generated* $\mathbf{K}$-*space, then the cup product coincides with the exterior products of the closed differential forms on* M_C.

Proof. The proofs follow from above theorems. $\qquad\square$

Definition 1.45. Let M be a quantum manifold. The *cap product* is a linear mapping

$$\cap : C^r(M, \mathbf{K}) \bigotimes_{\mathbf{K}} C_s(M, \mathbf{K}) \to C_{s-r}(M, \mathbf{K}), \ (r \leq s)$$

defined as follows:

$$< a, b \cap \xi > = < ab, \xi >, \forall a \in C^{s-r}(M, \mathbf{K}).$$

Proposition 1.44. 1) *One has the following rules of the cap product:*

$$(bc) \cap \xi = b \cap (c \cap \xi),$$
$$1 \cap \xi = \xi,$$
$$\partial(b \cap \xi) = (\delta b) \cap \xi + (-1)^{\dim b} b \cap \partial \xi.$$

2) *One has the corresponding product operation;*

$$\cap : H_s^r(M, \mathbf{K}) \bigotimes_{\mathbf{K}} H_s(M, \mathbf{K}) \to H_{s-r}(M, \mathbf{K}).$$

3) If M is classic regular and $H_p(M_C, \mathbf{K})$ is a finitely generated $\mathbf{K}$-space, then the cap product coincides with the contractions of the corresponding differential forms on M_C. In particular, if M_C is a compact and oriented manifold, one has the following isomorphism (Poincaré duality): $H^r_s(M, \mathbf{K}) \cong H_{s-r}(M, \mathbf{K})$, $(r \leq s)$, $a \mapsto a \cap \mu_{M_C}$, where μ_{M_C} is the fundamental class of M_C and $a \in H^r(M_C, \mathbf{K})$.

Proof. The proofs follow from above theorems. $\square$

TAB.1.18 - Relations between (co)homologies of quantum manifolds.

$$\hat{H}^\bullet(M) \cong H^\bullet(M) \oplus \overset{\bullet\bullet}{H}(M)$$

$$H^\bullet(M;A) \cong A \otimes_{\mathbf{K}} H^\bullet(M) \cong A \otimes_{\mathbf{K}} H^\bullet(M_C) \quad (*)$$

$$H^\bullet_s(M;\mathbf{K}) \cong Hom_{\mathbf{K}}(H_\bullet(M;\mathbf{K});\mathbf{K})$$

$$H_\bullet(M;A) \cong A \otimes_{\mathbf{K}} H_\bullet(M;\mathbf{K})$$

$$H^\bullet_s(M;A) \cong Hom_A(H_\bullet(M;A);A) \cong Hom_{\mathbf{K}}(H_\bullet(M;\mathbf{K});A)$$

$$H_\bullet(M;A) \cong A \otimes_{\mathbf{K}} H_\bullet(M;\mathbf{K})$$

The last isomorphism holds if M is classicregular.

$H^p_s(M;A) \cong H^p(M;A)$ iff M is classicregular and

$H_p(M_C;\mathbf{K})$ is a finite generated $\mathbf{K}$-space.

$H^\bullet(M;A) \cong H^\bullet(M;Z) \times H^\bullet(M;A/Z); \ H^0(M;A) \cong A^s$,

s=number of connected components of M.

(Co)homology and (co)bordism theories are strictly related. In fact, one has the following.

Theorem 1.30. ((Co)bordisms in quantum manifolds). *Let us consider the chain complex* $\{C_\bullet(M;A), \partial\}$, *where M is a quantum manifold of dimension m over a quantum algebra A. Set:* $B_\bullet(M;A) \equiv \mathrm{im}\,(\partial)$, $Z_\bullet(M;A) \equiv \ker(\partial)$. *Then, one has the following exact commutative diagram:*

$$
\begin{array}{ccccccccc}
 & & 0 & & 0 & & & & \\
 & & \downarrow & & \downarrow & & & & \\
0 & \to & B_\bullet(M;A) & \to & Z_\bullet(M;A) & \to & H_\bullet(M;A) & \to & 0 \\
 & & \downarrow & & \downarrow & & & & \\
 & & C_\bullet(M;A) & = & C_\bullet(M;A) & & & & \\
 & & \downarrow & & \downarrow & & & & \\
0 & \to & {}^A\underline{\Omega}_{\bullet,s}(M) & \to & Bor_\bullet(M;A) & \to & Cyc_\bullet(M;A) & \to & 0 \\
 & & \downarrow & & \downarrow & & & & \\
 & & 0 & & 0 & & & &
\end{array}
$$

where $H_\bullet(M;A)$ is the singular homology of M with coefficients in A. We call ${}^A\underline{\Omega}_{p,s}(M)$ closed singular p-bordism groups with coefficients in A, (resp. $Bor_p(M;A)$ singular p-bordism groups with coefficients in A, $Cyc_p(M;A)$ singular p-cyclism groups with coefficients in A) in the quantum manifold M, $p \in \{0,1,\ldots,m-1\}$. One has the following properties:

$$b \in [a] \in Bor_\bullet(M;A) \Rightarrow a - b = \partial c, \quad c \in C_\bullet(M;A),$$

$$b \in [a] \in Cyc_\bullet(M;A) \Rightarrow \partial(a - b) = 0,$$

$$b \in [a] \in {}^A\underline{\Omega}_{\bullet,s}(M) \Rightarrow \left\{ \begin{array}{l} \partial a = \partial b = 0 \\ a - b = \partial c, \quad c \in C_\bullet(M;A) \end{array} \right\}.$$

Furthermore, one has the following canonical isomorphism: ${}^A\underline{\Omega}_{\bullet,s}(M) \cong H_\bullet(M;A)$. As $C_\bullet(M;A)$ is a free two-sided projective A-module, one has the unnatural isomorphism: $Bor_\bullet(M;A) \cong {}^A\underline{\Omega}_{\bullet,s}(M) \oplus Cyc_\bullet(M;A)$. For the cochain complex $\{C^\bullet(M;A),\delta\}$ corresponding to $\{C_\bullet(M;A),\partial\}$ we can associate the dual objects of the previous point 1): ${}^A\underline{\Omega}_s^p(M)$ closed singular p-cobordism groups with coefficients in A, $Bor^p(M;A)$ singular p-cobordism groups with coefficients in A and $Cyc^p(M;A)$, singular p-cocyclism groups with coefficients in A, $p \in \{0,1,\ldots,m-1\}$. Then, one has the following canonical isomorphism: ${}^A\underline{\Omega}_s^\bullet(M) \cong H_s^\bullet(M;A)$ and the following unnatural isomorphisms: $Bor^\bullet(M;A) \cong {}^A\underline{\Omega}_s^\bullet(M) \oplus Cyc^\bullet(M;A)$. In the following we shall also call ${}^A\underline{\Omega}_{p,s}(M)$, (resp. ${}^A\underline{\Omega}_s^p(M)$), $p \in \{0,1,\ldots,m-1\}$, singular p-bordism groups, (resp. singular p-cobordism groups), i.e., we will omit the term "closed".

Remark 1.24. (*Quantuminvolutive quantumdistributions*). Let M be a smooth quantum manifold of dimension m over a quantum algebra A. A *quantumdistribution* of dimension n on M is a smooth subbundle $\widehat{\mathbb{D}} \subset \widehat{T}M$ on M, fullquantum extension of some distribution $\mathbb{D} \subset TM$, of dimension n on M: $\widehat{\mathbb{D}} = Hom_Z(A;\mathbb{D}) \equiv A \odot \mathbb{D}$. A (*maximal*) *integral quantum submanifold* of $\widehat{\mathbb{D}}$ is a connected quantum submanifold S of M such that $\widehat{T}_pS \subseteq \widehat{\mathbb{D}}_p$, $\forall p \in S$, (equality implies maximality). Then, one has $\dim S \leq n$, and the quantum submanifold S defined by equations $S \equiv \{p \in M : F^\alpha(p) = 0, \alpha \in I\}$ is integral iff $\zeta.F^\alpha \equiv < \zeta, dF^\alpha > = (\partial x_\beta.F^\alpha) \circ \zeta^\beta = 0$ on S, $\forall \zeta \in Q_w^\infty(\widehat{\mathbb{D}})$, $\alpha \in I$. A quantumdistribution $\widehat{\mathbb{D}} \subset \widehat{T}M$ is *quantuminvolutive* if $[\zeta,\xi]_\bullet \in Q_w^\infty(\widehat{\mathbb{D}})$, $\forall \zeta,\xi \in Q_w^\infty(\widehat{\mathbb{D}})$, or, equivalently, if $[\zeta,\xi] \in Q_w^\infty(\mathbb{D})$, $\forall \zeta,\xi \in Q_w^\infty(\mathbb{D})$. Here $[\zeta,\xi]_\bullet$ is the full quantum vector field identified with the derivation $(\zeta\xi - \xi\zeta)_\bullet : Q_w^\infty(M;A) \to Q_w^\infty(M;A)$. More precisely, $(\zeta\xi - \xi\zeta)_\bullet f = (\partial x_\alpha.f)[(\partial x_\beta.\xi^\alpha)\zeta_\bullet^\beta - (\partial x_\beta.\zeta^\alpha)\xi_\bullet^\beta] \in Q_w^\infty(M;A)$, where the indexes $\bullet$ denote restriction to $Z \subset A$. Note that one has the natural epimorphism $\pi| : \widehat{\mathbb{D}} \to \mathbb{D}$ induced by the following canonical epimorphism $\pi : \widehat{T}M \to TM$. As a consequence, if we get also the epimorphism $\pi|_* : Q_w^\infty(\widehat{\mathbb{D}}) \to Q_w^\infty(\mathbb{D})$. Furthermore, we get that $\pi|_*$ is necessarily a Lie algebra homomorphism, i.e., $[\pi|_*(\zeta), \pi|_*(\xi)] = \pi|_*([\zeta,\xi]_\bullet)$, $\forall \zeta, \xi \in Q_w^\infty(\widehat{\mathbb{D}})$. Therefore, if $\widehat{\mathbb{D}}$ is a quantuminvolutive quantumdistribution it follows that also $\mathbb{D}$ is so. The converse is also true as $\pi|_*$ is an epimorphism of Lie algebras. A n-dimensional quantumdistribution $\widehat{\mathbb{D}} \subset \widehat{T}M$ can be defined as the quantum extension of the common null subbundle $\mathbb{D}$ of some subbundle $\Theta \subset (TM)^+$: $\widehat{\mathbb{D}} = \bigcup_{p\in M} \widehat{\mathbb{D}}_p$, $\widehat{\mathbb{D}}_p \equiv AO\mathbb{D}_p$, $\mathbb{D}_p \equiv \{\zeta \in T_pM :< \zeta, \omega(p) >= 0, \forall \omega \in Q_w^\infty(\Theta)\}$. The distribution $\widehat{\mathbb{D}}$ is quantuminvolutive iff $d\omega(\zeta, \xi) = 0$, $\forall \omega \in Q_w^\infty(\Theta)$ and $\zeta, \xi \in Q_w^\infty(\mathbb{D})$, where d is the exterior quantum differential of ω.

$\square$ (*Quantum Frobenius theorem*). If the quantum distribution $\widehat{\mathbb{D}} \subset \widehat{T}M$ is quantuminvolutive and if its Q_w^∞-vector fields are integrable, then for each point $p \in M$ passes a unique maximal integral quantum manifold of $\widehat{\mathbb{D}}$. In this case a set $S \subset W$ is an integral quantum manifold iff $S = \{x, y : y^B = c^B, B \in I'\}$, where c^B are some constants. In other words, integral quantum submanifolds are *folios* of some chart. (In fact, under our hypotheses we can reproduce analogous standard proof for distributions on ordinary manifolds.)

$\square$ (*Quantumdiffieties*). Quantumdiffieties are diffieties [39,63] in quantumsense. These structures are strictly related to QPDE's. In fact infinite quantumprolongations of QPDE's are quantumdiffieties. More precisely, a *quantumdiffiety* is a pair $\mathcal{M} \equiv (M, \widehat{C}(M))$, where M is a smooth quantum manifold and $\widehat{C}(M) \subset \widehat{T}M$ is an involutive quantumdistribution on M, of dimension n, called *Cartan quantumdistribution*. We call *Cartan quantumdimension* of $\mathcal{M}$ the dimension n of $\widehat{C}(M)$. We call *Cartan quantumfields* local quantumsmooth sections of the bundle $\widehat{C}(M) \to M$. Let $\{x^A, y^i\}_{1 \le A \le n, i \in I}$ be a quantum coordinate system on $U \subset M$, where I is a finite or countable set. Let $(\zeta_1, \ldots, \zeta_m)$ be basis vector fields of $\widehat{C}(M)$ on $U \subset M$. The system of quantum coordinates $\{x^A, y^i\}$ is called *natural* if $\zeta_A = \partial x_A + \sum_{i \in I} \partial y_i A_A^i \equiv \partial_A$, $1 \le A \le m$, where $A_A^i : U \to \widehat{A}$ are Q_w^∞ local functions.

☐ Let $C(M)$ be a distribution on M. $\widehat{C}(M) \subset \widehat{T}M$ is a n-dimensional involutive quantumdistribution, i.e. a quantumdiffiety, iff $\dim \widehat{C}(M) = n$ and $(\partial_B.A_A^i)(Y_\bullet^B \otimes X_\bullet^A) - (\partial_A.A_B^i)(X_\bullet^A \otimes Y_\bullet^B) = 0$, $A, B = 1, ..., n, i, j \in I$. This implies $(\partial_B.A_A^i)|_Z - (\partial_A.A_B^i)|_Z = 0$. (For a proof see ref.[63].)

A n-dimensional smooth quantum submanifold $S \subset M$ is integral iff it has a local representation $S \equiv \{x, y : \quad y^i = s^i(x^1, ..., x^n), i \in I\} \subset M$, where the quantumsmooth function $s^i(x^1, ..., x^n)$ satisfy the contact equations: $\zeta.(y^i - s^i) = 0, \forall \zeta \in Q_w^\infty(\widehat{C}(M)), \forall i \in I$, i.e. iff $(\partial x_A.s^i) - A_A^i = 0$. A quantumdiffiety $\mathcal{F} \equiv (F, \widehat{C}(F))$ is called *quantum subdiffiety* of $\mathcal{M} \equiv (M, \widehat{C}(M))$ if F is a quantum submanifold of M and $\widehat{C}(F) = \widehat{T}F \cap \widehat{C}(M)|_F$. The imbedding $i : \mathcal{F} \to \mathcal{M}$ is a Lie-immersion. If the Cartan quantumdimension of $\mathcal{F}$ and $\mathcal{M}$ coincide, then $\mathcal{F}$ is called a *reduction* of the quantumdiffiety $\mathcal{M}$. In this case any integral quantum submanifold of $\mathcal{F}$ is so for all $\mathcal{M}$, and conversely every integral quantum submanifold of $\mathcal{M}$, lying in F, is an integral quantum submanifold of $\mathcal{F}$. Let $\mathcal{M} \equiv (M, \widehat{C}(M))$ be a quantumdiffiety, then $\mathcal{F} \equiv (F, \widehat{C}(M)|_F)$ will be a reduction of $\mathcal{M}$ iff $\zeta :i_F \to i_F$ for every Cartan field $\zeta \in Q_w^\infty(C(M))$, where $i_F \equiv$ defining two-sided ideal of F, i.e., the set of all functions $f \in Q_w^\infty(M; A)$ such that $f|_F = 0$.

☐ Every quantum vector field $\zeta \in Q_w^\infty(\widehat{T}M)$ can be locally represented as follows: $\zeta = \sum_{1 \le \mu \le n} \partial_\mu \circ \xi^\mu + \sum_{i \in I} \partial_{y_i} \circ \eta^i$, where $\xi^\mu, \eta^i : U \to \widehat{A}$ are local Q_w^∞-functions. Set $\delta y^i \equiv dy^i - \sum_{1 \le \mu \le n} A_\mu^i \circ dx^\mu$, $i \in I$. Then every 1-form $\omega \in Q_w^\infty((TM)^+)$, has the unique representation: $\omega = \sum_{1 \le \mu \le n} \phi_\mu \circ dx^\mu + \sum_{i \in I} \chi_i \circ \delta y^i$, with $\phi_\mu, \chi_i : U \to \widehat{A}$ are local Q_w^∞-functions. One has $< \partial_\mu, \delta y^i > = 0$. The common null subbundle of the 1-form δy^i, $i \in I$, is the restriction of the distribution $C(M)$ on U. Let $f : M \to A$ be a Q_w^∞-function on M. Then we can write $df = \sum_{1 \le \mu \le n}(\partial_\mu.f)dx^\mu + \sum_{i \in I}(\partial_{y_i}.f)\delta y^i$, where we have forgotten the symbol of composition.

☐ Let $\mathcal{M} \equiv (M, \widehat{C}(M))$ be a quantum diffiety of dimension n. Then the exterior derivatives $d\delta y^i$ can be written:

$$d\delta y^i = -(\partial_\nu.A_\mu^i)dx^\nu \triangle dx^\mu - \sum_{1 \le A \le n, j \in I} (\partial_{y_j}.A_A^i)\delta y^i \triangle dx^A, \, i \in I.$$

For any vector field ζ, η we get:

$$
\begin{aligned}
d\delta y^i(\zeta, \eta) =& \frac{1}{2}[(\partial_\nu A^i_\mu)(\eta^\nu \otimes \zeta^\mu) - (\partial_\nu A^i_\mu)(\zeta^\nu \otimes \eta^\mu)] \\
& + \frac{1}{2}[(\partial_{y_j} A^i_A)(\eta^j \otimes \zeta^A) - (\partial_{y_j} A^i_A)(\zeta^j \otimes \eta^A)] \\
=& \frac{1}{2}[(\partial_{y_j} A^i_A)(\eta^j \otimes \zeta^A) - (\partial_{y_j} A^i_A)(\zeta^j \otimes \eta^A)].
\end{aligned}
$$

In particular, $d\delta y^i(\zeta, \eta) = 0$, $\forall i \in I, \zeta, \eta \in Q^\infty_w(C(M))$.   ∎

1.4 - QUANTUM SUPERMANIFOLDS

In this section we give the definitions of quantum superalgebra and quantum supermanifold, and we extend to such structures above results for quantum manifolds. The introduction in a quantum algebra of a $\mathbf{Z}_2$-gradiation allows us to define noncommutative manifolds with a richer structure than quantum manifolds have. These are useful to describe quantum super PDEs, hence quantum supergravity.

Definition 1.46. A *quantum superalgebra* is a quantum algebra (A, ϵ, c) that has also a $\mathbf{Z}_2$-gradiation $A = A_0 \oplus A_1$ such that the following bracket: $[a_1, a_2]_g \equiv a_1 a_2 - (-1)^{|a_1||a_2|} a_2 a_1$, written for homogeneous elements $a_i \in A$, $i = 0, 1$, identifies on A a structure of $\mathbf{Z}_2$-graded Lie algebra, i.e., the following conditions are satisfied: (i) $[a_1, a_2]_g \in A_{|a_1|+|a_2|}$; (ii) $[a_1, a_2]_g = -(-1)^{|a_1||a_2|}[a_2, a_1]_g$; (iii)(*graded Jacobi identity*):

$$
\begin{aligned}
(-1)^{|a_1||a_3|}[[a_1, a_2]_g, a_3]_g + (-1)^{|a_2||a_1|}[[a_2, a_3]_g, a_1]_g \\
+ (-1)^{|a_3||a_2|}[[a_3, a_1]_g, a_2]_g = 0.
\end{aligned}
$$

Here $|\cdots|$ denotes the function $|| : A_i \to \mathbf{Z}_2$ given by $|a| = i$, $\forall a \in A_i$, $i \in \mathbf{Z}_2$. We call $\widetilde{Z} \equiv \{a \in A | [a, b]_g = 0, \forall b \in A\}$ *supercentre* of a quantum superalgebra A. $\widetilde{Z}$ is a $\mathbf{Z}_2$-graded commutative algebra, i.e., a superalgebra: $\widetilde{Z} = \widetilde{Z}_0 \oplus \widetilde{Z}_1$, where $\widetilde{Z}_i = \widetilde{Z} \cap A_i$, $i = 0, 1$. We call *centre* of A the set Z of the elements $a \in A$ that commute with all elements of A. Z results just the even part of $\widetilde{Z}$: $Z = \widetilde{Z}_0$.

Example 1.39. Let us consider the covariant quantization of a classical (i.e., nonquantized) super PDE $E_k \subset J\mathcal{D}^k(W)$. (See Example 3.64 and refs.[62].) Then, the quantum field components $\{\hat{\phi}^I\}$, (*quantum superfields*), satisfy the following quantum commutators: $[\hat{\phi}^I(x), \hat{\phi}^J(y)] \equiv \hat{\phi}^I(x)\hat{\phi}^J(y) - (-1)^{|I||J|}\hat{\phi}^J(y)\hat{\phi}^I(x) = i\hbar G^{IJ}\delta(x, y)\mathbf{1}_{\mathcal{H}}$, where $\mathcal{H}$ is a superHilbert space,

G^{IJ} are the components of the Green superkernel canonically associated to E_k. (For details see ref.[62].) Then, $\{\hat{\phi}^I\}$ generate a quantum algebra A that is $\mathbf{Z}_2$-graded, where above bracket defines a $\mathbf{Z}_2$-graded Lie algebra contained in $L(\mathcal{H})$, hence a quantum superalgebra. Furthermore, the equal-time operators $\{\hat{\phi}^I(x), \dot{\hat{\phi}}^J(y)\}_{x^0=y^0}$, satisfy the following quantum commutation relations: $[\hat{\phi}^I(x), \dot{\hat{\phi}}^J(y)]_{x^0=y^0} = \delta^{IJ}\delta(x,y)_{x^0=y^0}$, $[\hat{\phi}^I(x), \hat{\phi}^J(y)]_{x^0=y^0} = [\dot{\hat{\phi}}^I(x), \dot{\hat{\phi}}^J(y)]_{x^0=y^0} = 0$. Then they define a quantum algebra A_{x^0} that has just a structure of quantum superalgebra. $\blacksquare$

Example 1.40. Let us consider a quantum algebra (B, ϵ, c) and a superalgebra $A = A_0 \oplus A_1$. Then, $C \equiv B \otimes_{\mathbf{K}} A$ is also a quantum algebra with product $\mu_C = \mu_B \otimes \mu_A$: $(b_1 \otimes a_1)(b_2 \otimes a_2) = b_1 b_2 \otimes a_1 a_2$, $b_i \in B$, $a_i \in A$, $i = 1, 2$. The centre $Z(C)$ of C is $Z(C) = Z(B) \otimes_{\mathbf{K}} Z$, where $Z(B)$ is the centre of the quantum algebra B. On C one recognizes a natural $\mathbf{Z}_2$-gradiation: $C = C_0 \oplus C_1 = (B \otimes_{\mathbf{K}} A_0) \bigoplus (B \otimes_{\mathbf{K}} A_1)$. Furthermore, putting: $[c_1, c_2]_g \equiv c_1 c_2 - (-1)^{|c_1||c_2|} c_2 c_1$, we identify on C a structure of quantum superalgebra. More precisely one has the following isomorphism of $\mathbf{K}$-algebras: $\mathcal{L}ieg(C) \cong \mathcal{L}ie(B) \bigotimes_{\mathbf{K}} A$, where $\mathcal{L}ieg(C)$ denotes the $\mathbf{Z}_2$-graded Lie algebra of C and $\mathcal{L}ie(B)$ denotes the structure of Lie algebra on B. In fact one has: $[c_1, c_2]_g \equiv [b_1 \otimes a_1, b_2 \otimes a_2]_g = b_1 b_2 \otimes a_1 a_2 - (-1)^{|a_1||a_2|} b_2 b_1 \otimes a_2 a_1 = b_1 b_2 \otimes a_1 a_2 - (-1)^{|a_1||a_2|+|a_1||a_2|} b_2 b_1 \otimes a_1 a_2 = b_1 b_2 \otimes a_1 a_2 - b_2 b_1 \otimes a_1 a_2 = (b_1 b_2 - b_2 b_1) \otimes a_1 a_2 = [b_1, b_2] \otimes a_1 a_2$. The supercentre $\widetilde{Z}$ of such a quantum superalgebra C is the superalgebra $\widetilde{Z} = Z(B) \otimes_{\mathbf{K}} A$. The centre Z of C is just $\widetilde{Z}(C)_0 = Z(B) \otimes_{\mathbf{K}} A_0$: $Z = \widetilde{Z}(C)_0 = Z(B) \otimes_{\mathbf{K}} A_0$. $\blacksquare$

Remark 1.25. A quantum superalgebra is not necessarily a superalgebra, but a superalgebra A is an example of quantum superalgebra such that its supercentre $\widetilde{Z} = A$, and the centre $Z = A_0$, i.e., Z coincides with the even part of A. (A superalgebra is also an augmented algebra. See note in Definition 1.35.) $\blacksquare$

Definition 1.47. A *quantum vector superspace* of dimension

$$(m_1|n_1, \ldots, m_s|n_s),$$

built on the quantum superalgebra $A \equiv A_1 \times \ldots \times A_s$, $(A_i = (A_i)_0 \times (A_i)_1$, $i = 1, \ldots, s$, are quantum superalgebras), is a locally convex topological $\mathbf{K}$-vector space E isomorphic to $(A_1)_0^{m_1} \times (A_1)_1^{n_1} \times \ldots \times (A_s)_0^{m_s} \times (A_s)_1^{n_s} \equiv A_1^{m_1|n_1} \times \cdots \times A_s^{m_s|n_s} \equiv A^{(m_1|n_1, \cdots, m_s|n_s)}$.

Definition 1.48. A *quantum supermanifold* of dimension $(m_1|n_1, \ldots, m_s|n_s)$ over a quantum superalgebra $A \equiv A_1 \times \ldots \times A_s$ of class Q_w^k, $0 \leq k \leq \infty, \omega$, is a locally convex manifold M modeled on E and with a Q_w^k-atlas of local coordinate mappings, i.e., the transiction functions $f : U \subset E \to U' \subset E$ define a pseudogroup of local Q_w^k-homeomorphisms on E, where Q_w^k means C_w^k, i.e., weak differentiability, and derivatives Z-linaires, where $Z = Z(A) = Z(A_1) \times \ldots \times Z(A_s)$ is the centre of A. So for each open coordinate set $U \subset M$ we have a set of $m_1 + n_1 + \ldots + m_s + n_s$ coordinate functions $x^A : U \to A$, (*quantum coordinates*), such that $m_1 + \ldots + m_s$ take values into $A_0 = (A_1)_0 \times \ldots \times (A_s)_0$ and $n_1 + \ldots + n_s$ take values into $A_1 = (A_1)_1 \times \ldots \times (A_s)_1$.

Definition 1.49. A quantum supermanifold of dimension $(m_1|n_1, \ldots, m_s|n_s)$ is called *classic superegular* if it admits a projection $\pi_{SC} : M \to M_{SC}$ on a $m|n$-dimensional supermanifold M_{SC} [62]. We will call M_{SC} the *super-classic limit* of M and in order to emphasize this structure we say that the dimension of M is $(m|n \downarrow m_1|n_1, \ldots, m_s|n_s)$.

Definition 1.50. A quantum supermanifold of dimension $(m_1|n_1, \ldots, m_s|n_s)$ is called *classic regular* if it admits a projection $\pi_C : M \to M_C$ on a n-dimensional manifold M_C. We will call M_C the *classic limit* of M and in order to emphasize this structure we say that the dimension of M is $(n \downarrow m_1|n_1, \ldots, m_s|n_s)$.

Remark 1.26. As the odd part A_1 of a quantum superalgebra $A = A_0 \oplus A_1$ is not, in general, a quantum algebra, it follows that a quantum super-space of dimension $(m|n)$ over A is not a quantum space of dimension (m, n) over A. Hence a quantum supermanifold of dimension $(m|n)$ over A is not a quantum manifold of dimension (m, n) over A. Note also that a quantum supermanifold of dimension $(m|n)$ over A cannot be properly considered a quantum manifold of dimension $m + n$ over A, as open sets of $A_0^m \times A_1^n$ are not open sets of A^{m+n}. However, a quantum superalgebra, forgetting its $\mathbf{Z}_2$-gradiation, remains a quantum algebra. Therefore we can consider a quantum supermanifold as it should be a quantum manifold all the times when the $\mathbf{Z}_2$-gradiation of the underlying superalgebra can be forgotten. In this sense we can generalize all our previous results for quantum manifolds to quantum supermanifolds all the times that they do not depend on the dimension. For example, for the tangent space T_pM of a quantum supermanifold M hold all the definitions and results given for quantum manifold. (See Tab.1.11, Tab.1.12 and Tab.1.13.) Taking into account the $\mathbf{Z}_2$-gradiation of the underlying quantum superalgebra A we can also emphasize some further isomorphism. For example, $\widehat{T_pM}$ and $(T_pM)^+$ have natural $\mathbf{Z}_2$-gradiations and the local components of their vectors have natural $\mathbf{Z}_2$-gradiation too. See Tab.1.19. Furthermore, for quantum (jet) derivative spaces in the category of quantum supermanifolds hold definitions and results just stated for quantum manifolds, so we will adopt for these spaces the same notation given for quantum manifolds. $\blacksquare$

TAB.1.19 - $\mathbf{Z}_2$-gradiation of $\widehat{T_pM}$ and $(T_pM)^+$.

$$\widehat{T_pM}\cong(\widehat{T_pM})_0\oplus(\widehat{T_pM})_1\equiv Hom_Z(A_0;T_pM)\oplus Hom_Z(A_1;T_pM)$$

$$\cong(\widehat{A_0}(A))^{m|n}\times(\widehat{A_1}(A))^{n|m}\subset(\widehat{A})^{m+n}$$

$$\zeta\in\widehat{T_pM},\ \zeta=\partial x_A\zeta^A,\ \zeta^A\in\widehat{A},\ 1\leq A\leq m+n$$

$$\zeta\in(\widehat{T_pM})_0,\ \zeta=\partial x_A\zeta^A,\ \zeta^A\in Hom_Z(A_0;A_i),\ 1\leq A\leq m+n,\ i\in\mathbf{Z}_2,\quad |\zeta^A|=i$$

$$\zeta\in(\widehat{T_pM})_1,\ \zeta=\partial x_A\zeta^A,\ \zeta^A\in Hom_Z(A_1;A_i),\ 1\leq A\leq m+n,\ i\in\mathbf{Z}_2,\quad |\zeta^A|=i+1$$

$$(T_pM)^+\cong(T_pM)_0^+\oplus(T_pM)_1^+\equiv Hom_Z(T_pM;A_0)\oplus Hom_Z(T_pM;A_1)$$

$$\cong(\widehat{A^0}(A))^{m|n}\times(\widehat{A^1}(A))^{n|m}\subset(\widehat{A})^{m+n}$$

$$\alpha\in(T_pM)^+,\ \alpha=\alpha_A dx^A,\ \alpha_A\in\widehat{A},\ 1\leq A\leq m+n$$

$$\alpha\in(T_pM)_0^+,\ \alpha=\alpha_A dx^A,\ \alpha_A\in Hom_Z(A_i;A_0),\ 1\leq A\leq m+n,\ i\in\mathbf{Z}_2,\quad |\alpha_A|=i$$

$$\alpha\in(T_pM)_1^+,\ \alpha=\alpha_A dx^A,\ \alpha_A\in Hom_Z(A_i;A_1),\ 1\leq A\leq m+n,\ i\in\mathbf{Z}_2,\quad |\alpha_A|=i+1$$

$\dim_A(T_pM)=(m|n)$. $T_pM\cong Hom_Z(Z;T_pM)\Rightarrow|\zeta|=0,\ \forall\zeta\in T_pM$.

$\widehat{A_i}(A)\equiv Hom_Z(A_i;A),\ \widehat{A^i}(A)\equiv Hom_Z(A;A_i),\ i\in\mathbf{Z}_2$.

Example 1.41. Supermanifolds, in the sense of ref.[62], can be considered quantum supermanifolds. In these cases the underlying quantum algebra A is a superalgebra, hence an augmented quantum algebra, i.e., $c:A\to\mathbf{K}$ is a $\mathbf{K}$-algebra morphism. $\blacksquare$

Example 1.42. Let $M=A_0^m\times A_1^n$, where A is an extension $0\to B\to A\to C^0(X)\to 0$ of a noncommutative C^*-superalgebra B by means of the commutative C^*-algebra $C^0(X)$ of $\mathbf{C}$-valued continuous functions on some topological space X. Then M is a quantum supermanifold of dimension $(m|n)$ over A with global charts. $\blacksquare$

Example 1.43. Let B be a quantum superalgebra and let $\mathbf{H}$ be the usual (Hamiltonian) quaternionic algebra. A *quantum B-quaternionic superman-*

ifold of dimension $(m|n)$ and class Q_w^k, $0 \leq k \leq \infty, \omega$, is a quantum supermanifold M of dimension $(m|n)$ and class Q_w^k over the B-quantum superalgebra $C \equiv B \otimes_{\mathbf{K}} \mathbf{H}$. Then the quantum coordinates, in an open coordinate subset $U \subset M$, are called B-*quaternionic coordinates*, $\{q^k\}_{1 \leq k \leq n}$, $q^k : U \to C$. As a particular case we can take $B = \mathbf{K}$. In this case $C = \mathbf{H}$ and we call such quantum $\mathbf{K}$-quaternionic supermanifolds simply *quantum quaternionic manifolds*. The *category* $\mathcal{C}_{\mathbf{H}}^B$, *of quantum B-quaternionic supermanifolds* of class Q_w^k, is defined by considering as morphisms maps of class Q_w^k, between quantum B-quaternionic supermanifolds. (See ref.[68] for information about quantum B-quaternionic manifolds.) ∎

Example 1.44. The *quantum derivative space* $\hat{D}(M, N)$ for quantum supermanifolds M and N, of dimension $(m|n)$ and $(r|s)$ respectively, is defined by $\hat{D}(M, N) \equiv TMOTN \equiv \bigcup_{(p,q) \in M \times N} T_pMOT_qN$, where $T_pMOT_qN \equiv Hom_Z(T_qN; T_pM)$ is the space of Z-linear mappings $T_pM \to T_qN$, assumed that they are Z-modules with Z the centre of the superalgebra underlying M. Let us denote by $Q_w^s(M, N)$ the space of mappings $M \to N$ of class Q_w^s. Then if $f \in Q_w^s(M, N) \Rightarrow Df(p) \in \hat{D}(M, N), \forall p \in M$. If $\pi : W \to M$ is a fiber bundle between quantum supermanifolds, the *quantum derivative space for sections* of π, $\hat{D}(W)$, is defined by $\hat{D} \equiv d^* \hat{D}(M, W)$, where d is the diagonal mapping $d : M \to M \times M$. *Quantum derivative spaces of higher orders* between quantum supermanifolds can be defined putting $\hat{D}^k(M, N) \equiv \hat{D}(\hat{D}^{k-1}(M, N)), \hat{D}^k(W) \equiv \hat{D}(\hat{D}^{k-1}(W))$. One has a natural injection over $M \times N$: $\hat{D}(M, N) \hookrightarrow D(M, N)$, where $D(M, N)$ denotes the derivative space simply considering M and N as locally convex manifolds [63]. $\hat{D}(M, N)$ results a quantum supermanifold of finite dimension too. The first *derivative* of a quantum differentiable mapping $f : M \to N$ is a section $Df : M \to \hat{D}(M, N)$ of the fiber bundle $\pi_1 : \hat{D}(M, N) \to M$. The local expression of Df is given by a matrix $(\partial x_A f^B)(p)$, $\forall p \in M$, with entries in $Hom_Z(A; B) \equiv \hat{A}(B)$, where A and B are the quantum superalgebras underlying respectively M and N. More precisely $(\partial x_A f^B)(p) \in Hom_Z(A_{|A|}; B_{|B|}) \subset \hat{A}(B)$. Therefore $|(\partial x_A f^B)(p)| = |A| + |B|$. ∎

Example 1.45. Let $\pi : W \to M$ be a fiber bundle, in the category of quantum supermanifolds, such that $\dim W = (m|n, r|s)$, over the quantum superalgebra $B \equiv A \times E$ and $\dim M = (m|n)$ over A and such that E is a $Z(A) \equiv Z$-module. Then the k-jet-derivative space $J\hat{D}^k(W)$ is a quantum supermanifold modeled on the quantum superalgebra

$$B_k \equiv \prod_{0 \leq s \leq k} \left(\prod_{i_1 + \cdots + i_s \in \mathbf{Z}_2, i_r \in \mathbf{Z}_2} \overset{s}{\widehat{A}}_{i_1 \cdots i_s}(E) \right),$$

where

$$\overset{s}{\widehat{A}}_{i_1 \cdots i_s}(E) \equiv Hom_Z(A_{i_1} \otimes_Z \cdots \otimes_Z A_{i_s}; E),$$

$$\overset{0}{\widehat{A}}(E) \equiv A \times E,$$

$$\overset{1}{\widehat{A}}_i(E) \equiv \widehat{A}_0(E) \times \widehat{A}_1(E) \equiv Hom_Z(A_0; E) \times Hom_Z(A_1; E).$$

Each $\overset{s}{\widehat{A}}_{i_1 \cdots i_s}(E)$ is a quantum superalgebra with $\mathbf{Z}_2$-gradiation induced by E. Hence

$$\overset{s}{\widehat{A}}_{i_1 \cdots i_s}(E)_q \equiv Hom_Z(A_{i_1} \otimes_Z \cdots \otimes_Z A_{i_s}; E_p),$$

$$i_r, p, q \in \mathbf{Z}_2, \ q \equiv i_1 + \cdots + i_s + p.$$

In fact, one has the local isomorphism reported in Tab.1.20.

TAB.1.20 - Local isomorphism of $J\hat{D}^k(W)$.

$J\hat{D}^k(W) \equiv \prod_{0 \leq s \leq k} \left(\prod_{i_1 + \cdots + i_s \in \mathbf{Z}_2, i_r \in \mathbf{Z}_2} (\overset{s}{\widehat{A}}_{i_1 \cdots i_s}(E))^{m^{s_1} n^{s_2} r(s_2) \mid m^{s_1} n^{s_2} s(s_2)} \right)$
$r(s_2) = \{r \ (s_2 = even); \ s \ (s_2 = odd)$
$s(s_2) = \{s \ (s_2 = odd); \ r \ (s_2 = even)$
$s_1 = \text{number of } i_r = 0; \ s_2 = \text{number of } i_r = 1$

Note that, forgetting the $\mathbf{Z}_2$-gradiation on A and E respectively, one has the following isomorphism of quantum algebras:

$$B_k \cong A \times E \times \widehat{A}(E) \times \overset{2}{\widehat{A}}(E) \times \cdots \overset{k}{\widehat{A}}(E),$$

with

$$\overset{s}{\widehat{A}}(E) \equiv Hom_Z(\dot{T}^s_0(A); E).$$

In the particular case that $n = s = 0$, i.e., in the category of quantum manifolds, we get:

$$J\hat{D}^k(W)|_U \equiv A^m \times E^r \times (\widehat{A}(E))^{mr} \times (\overset{2}{\widehat{A}}(E))^{m^2 r} \times \cdots (\overset{k}{\widehat{A}}(E))^{m^k r}.$$

(Note that we write also $\overset{s}{\widehat{A}}(A) = \overset{s}{\widehat{A}}$.) Then, if $(x^A, y^B)_{1 \le A \le m+n, 1 \le B \le r+s}$ are fibered quantum coordinates on the quantum supermanifold W over M, then the induced fibered coordinates on $J\hat{D}^k(W)$ over M are

$$(x^A, y^B, y^B_A, \cdots, y^B_{A_1 \cdots A_k}),$$

with the following gradiations: $|x^A| = |A|$, $|y^B| = |B|$, $|y^B_{A_1 \cdots A_s}| = |B| + |A_1| + \cdots |A_s|$. Note that there is not symmetry in the indexes A_i. $J\hat{D}^k(W)$ is an affine bundle over $J\hat{D}^{k-1}(W)$ with associated vector bundle

$$\pi^*_{k,0}(\dot{S}^k_0 MOvTW) \equiv \pi^*_{k,0}(Hom_Z(\dot{S}^k_0 MO; TW)).$$

So we have the following exact sequence of vector bundles over $J\hat{D}^k(W)$:

$$0 \to \pi^*_{k,0}(\dot{S}^k_0 MOvTW) \to vTJ\hat{D}^k(W) \to \pi^*_{k,k-1}vTJ\hat{D}^{k-1} \to 0.$$

The relation with k-jet derivative spaces, in the category of manifolds modeled on locally convex topological vector spaces [63], is given by the following exact commutative diagram:

$$
\begin{array}{ccccccccc}
& & 0 & & 0 & & 0 & & \\
& & \downarrow & & \downarrow & & \downarrow & & \\
0 & \to & \pi^*_{k,0}(\dot{S}^k_0 MOvTW) & \to & vTJ\hat{D}^k(W) & \to & \pi^*_{k,k-1}vTJ\hat{D}^{k-1}(W) & \to & 0 \\
& & \downarrow & & \downarrow & & \downarrow & & \\
0 & \to & \pi^*_{k,0}Hom_{\mathbf{K}}(S^k_0 M; vTW) & \to & vTJD^k(W) & \to & \pi^*_{k,k-1}vTJD^{k-1}(W) & \to & 0
\end{array}
$$

$\blacksquare$

Remark 1.27. We can obtain the (co)homological and (co)bordism characterizations of quantum supermanifolds, first by forgetting the $\mathbf{Z}_2$-gradiations of the underlying quantum superalgebras, and directly by using the corresponding previously considered characterizations of quantum manifolds. (See Remark 1.20, Definition 1.42, Tab.1.15 and Remark 1.21.) Then, taking into account the $\mathbf{Z}_2$-gradiations, we get further structures that we shall consider below.

$\blacksquare$

Proposition 1.45. 1) *One has the canonical Z-homomorphism ((graded) exterior product) reported in Tab.1.21.*

TAB.1.21 - (Graded) exterior product.

$$\wedge:\widehat{\Omega}^q(M)\times\widehat{\Omega}^p(M)\to\widehat{\Omega}^{q+p}(M), \quad \wedge:(\omega,\theta)\mapsto\omega\wedge\theta$$

$$(\omega\wedge\theta)(\zeta_1,\ldots,\zeta_{p+q})=\sum_{\sigma\in S_{p+q}}\epsilon(\sigma)[\omega(\zeta_{\sigma(1)},\ldots,\zeta_{\sigma(p)}),\theta(\zeta_{\sigma(p+1)},\ldots,\zeta_{\sigma(p+q)})]+$$

$$\zeta_1,\ldots,\zeta_{p+q}\in Q_w^\infty(TM)$$

Here $\epsilon(\sigma)$ is the signature of the permutation $\sigma\in S_{p+q}$, and $[,]_+$ is the (graded) Jordan bracket defined by $[a,b]_+\equiv\frac{1}{2}(ab+(-1)^{(|a||b|)}ba)$, where $(|a||b|)=|a||b|$ for graded exterior product and $(|a||b|)=0$ for exterior product, (i.e., forgetting the $\mathbf{Z}_2$-gradiation).

2) One has $(\omega\wedge\theta)(\zeta_1,\cdots,\zeta_{p+q})=\epsilon(\sigma)(\omega\wedge\theta)(\zeta_{\sigma(1)},\cdots,\zeta_{\sigma(p+q)})$. Then one has the following formula: $\omega\wedge\theta=(-1)^{pq+(|\omega||\theta|)}\theta\wedge\omega$.

3) The local expression of $\omega\wedge\theta$ in quantum coordinates is given by Tab.1.22.[26]

TAB.1.22 - Local expression of $\omega\wedge\theta$.

$$\omega\wedge\theta=\sum_{1\leq C_1,\ldots,C_{p+q}\leq m}(\omega\wedge\theta)_{C_1\ldots C_{p+q}}dx^{C_1}\triangle\cdots\triangle dx^{C_{p+q}}$$

$$(\omega\wedge\theta)_{C_1\ldots C_{p+q}}=\omega_{C_1\ldots C_p}\theta_{C_{p+1}\ldots C_{p+q}}+(-1)^{pq+(|\omega||\theta|)}\theta_{C_1\ldots C_q}\omega_{C_{q+1}\ldots C_{q+p}}$$

with $\omega_{A_1\ldots A_p}\theta_{B_1\ldots B_q}\in Q_w^\infty(U,\overset{p+q}{\widehat{A}})$ obtained from $\omega_{A_1\ldots A_p}$ and $\theta_{B_1\ldots B_q}$ respectively, by means of the following compositions of homomorphisms:

$$\overset{p}{\widehat{A}}\otimes_Z\overset{q}{\widehat{A}}\to Hom_Z(\dot{T}_0^{p+q}(A);A\otimes_Z A)\xrightarrow{Hom_Z(1,\hat{\mu})}\overset{p+q}{\widehat{A}}.$$

Furthermore,

$$dx^{C_1}(p)\triangle\cdots\triangle dx^{C_r}(p)\equiv\bar{j}(dx^{C_1}(p)\otimes_Z\cdots\otimes_Z dx^{C_r})|_{\Lambda_0^r(T_pM)}, \forall p\in U,$$

where $\bar{j}$ is the canonical homomorphism

$$Hom_Z(T_pM;A)\otimes_Z\cdots\otimes_Z Hom_Z(T_pM;A)\to Hom_Z(T_pM\otimes_Z\cdots\otimes_Z T_pM;A\otimes_Z\cdots\otimes_Z A).$$

---

[26] Note that, for abuse of notation, we shall also write $(\omega\wedge\theta)_{C_1\ldots C_{p+q}}=[\omega_{C_1\ldots C_p},\theta_{C_{p+1}\ldots C_{p+q}}]_+$. In the particular case of quantum manifolds, i.e. when $|\omega|=|\theta|=0$, we get that the graded exterior product coincides with the exterior product.

Note that in the commutative case $A \equiv \mathbf{K}$, the formula in Tab.1.22 reduces to the well known form: $\omega \wedge \theta = \sum_{1 \le C_1 < \ldots < C_{p+q} \le m} (\omega \wedge \theta)_{C_1 \ldots C_{p+q}} dx^{C_1} \wedge \cdots \wedge dx^{C_{p+q}}$, where $(\omega \wedge \theta)_{C_1 \ldots C_{p+q}}$ are skewsymmetric in the indexes.

Remark 1.28. We denote also by the same symbol $\widehat{\Omega}^\bullet(M)$, (resp. $\Omega^\bullet(M)$, resp. $\hat{\Omega}^\bullet(M)$), the corresponding algebra, with respect to the (graded) exterior product, of fullquantum differential forms, (resp. differential forms, resp. quantum differential forms), on the quantum supermanifold M (of dimension $(m|n)$ over A). $\blacksquare$

Remark 1.29. Note that neither the exterior product $\alpha \wedge \beta$ of two fullquantum differential forms α and β, that locally should be respectively written in the following form (*reduced form*):

$$\left\{ \begin{aligned} \alpha &= \sum_{1 \le i_1 < \cdots < i_p \le m} \alpha_{i_1 \cdots i_p} dx^{i_1} \wedge \cdots \wedge dx^{i_p} \\ \beta &= \sum_{1 \le j_1 < \cdots < j_q \le m} \beta_{j_1 \cdots j_q} dx^{j_1} \wedge \cdots \wedge dx^{j_q} \end{aligned} \right\}, \quad \alpha_{i_1 \cdots i_p}(p), \quad \beta_{j_1 \cdots j_q}(p) \in \widehat{A}, \quad \forall p \in U \subset M,$$

can, in general, be locally represented with $\widehat{A}$-valued components, i.e., in the reduced form too,

$$\alpha \wedge \beta = \sum_{1 \le j_1 < \cdots < j_{p+q} \le m} (\alpha \wedge \beta)_{j_1 \cdots j_{p+q}} dx^{j_1} \wedge \cdots \wedge dx^{j_{p+q}},$$

with $(\alpha \wedge \beta)_{j_1 \cdots j_{p+q}}(p) \in \widehat{A}$, $\forall p \in U \subset M$. This is immediately seen, e.g., for two 1-forms $\alpha = \alpha_j dx^j$, $\beta = \beta_k dx^k$, in the category of quantum manifolds. In fact one has: $(\alpha \wedge \beta)(\zeta_1 \wedge \zeta_2) = \frac{1}{2}\{\alpha_j \beta_k dx^j \otimes dx^k - \beta_k \alpha_j dx^k \otimes dx^j\}(\zeta_1 \wedge \zeta_2) = \frac{1}{2}(\alpha_j \beta_k - \beta_j \alpha_k) dx^j \otimes dx^k(\zeta_1 \wedge \zeta_2)$. This means that $\alpha \wedge \beta = \sum_{1 \le j,k \le m} \frac{1}{2}(\alpha_j \beta_k - \beta_j \alpha_k) dx^j \triangle dx^k$, where $(\alpha_j \beta_k - \beta_j \alpha_k)(p) \in \overset{2}{\widehat{A}}$, $\forall p \in U \subset M$. Of course in the commutative case $A = \mathbf{K}$ we can also write $\alpha \wedge \beta$ in the reduced form as $\alpha_j \beta_k = \beta_k \alpha_j$. Therefore, the noncommutativity of the algebra A is the obstruction that the exterior product of reduced fullquantum forms is a reduced fullquantum form too. In fact, in the particular case where $\alpha = dx^r = \delta^r_j dx^j$, $\beta = dx^s = \delta^s_k dx^k$, we get $(\alpha \wedge \beta) = \delta^r_j \delta^s_k (\frac{1}{2}(dx^j \otimes dx^k - dx^k \otimes dx^j)) = dx^r \wedge dx^s$. So, condition necessary and sufficient that the exterior product of two fullquantum differential forms of reduced form is of reduced form too is that at least the components of one of these forms are Z-valued. From the above considerations it follows also that the local expression in reduced form of a full quantum differential form is not, in general, invariant with respect to changments of quantum coordinates, In fact, for example, one has:

$$d\bar{x}^{i_1}\wedge d\bar{x}^{i_2} = \sum_{1\leq j_1,j_2\leq m}\frac{1}{2}\left[\sum_{\sigma\in S_2}\epsilon(\sigma)(\partial x_{j_1}.\bar{x}^{i_{\sigma(1)}})\otimes(\partial x_{j_2}.\bar{x}^{i_{\sigma(2)}})\right]dx^{j_1}\triangle dx^{j_2}.$$

In the particular case that $A\equiv \mathbf{K}$, one has, instead, that the reduced form is an invariant expression. In fact, one has:

$$d\bar{x}^{i_1}\wedge d\bar{x}^{i_2}=\sum\nolimits_{1\leq j_1,j_2\leq m}(\partial x_{j_1}.\bar{x}^{i_1})\otimes(\partial x_{j_2}.\bar{x}^{i_2})dx^{j_1}\wedge dx^{j_2}$$
$$=\sum\nolimits_{1\leq j_1<j_2\leq m}\left[(\partial x_{j_1}.\bar{x}^{i_1})(\partial x_{j_2}.\bar{x}^{i_2})-(\partial x_{j_2}.\bar{x}^{i_1})(\partial x_{j_1}.\bar{x}^{i_2})\right]dx^{j_1}\wedge dx^{j_2}.$$

$\blacksquare$

Proposition 1.46. 1) *One has the canonical Z-homomorphism $d:\widehat{\Omega}^q(M)\to\widehat{\Omega}^{q+1}(M)$, (exterior differential), given in Tab.1.23.*

TAB.1.23 - Exterior differential.

$$d:\widehat{\Omega}^q(M)\to\widehat{\Omega}^{q+1}(M)$$
$$d\omega(\zeta_0,...,\zeta_q)=\frac{1}{q+1}\sum_{\lambda=0}^{q}(-1)^\lambda<\omega(\zeta_0,...,\hat{\zeta}_\lambda,...,\zeta_q),\zeta_\lambda>$$
$$+\frac{1}{q+1}\sum_{\lambda<\mu}(-1)^{\lambda+\mu}\omega([\zeta_\lambda,\zeta_\mu],\zeta_0,...,\hat{\zeta}_\lambda,...,\hat{\zeta}_\mu,...,\zeta_q)$$
$$\zeta_0,...,\zeta_q\in Q_w^\infty(TM)$$

Here the terms $\hat{\zeta}_\lambda$ and $\hat{\zeta}_\mu$ are omitted. This is a first order differential operator, i.e., one has the following commutative diagram:

$$
\begin{array}{ccc}
J\hat{D}((\dot{\Lambda}_0^q M)^+) & \longrightarrow & (\dot{\Lambda}_0^{q+1}M)^+ \\
D\alpha\uparrow & & \uparrow d\alpha \\
M & \cong & M
\end{array}
$$

for any $\alpha\in\widehat{\Omega}^q(M)$. One has the following formula: $d(\omega\wedge\theta)=(d\omega)\wedge\theta+(-1)^q\omega\wedge(d\theta)$, $\forall\omega\in\widehat{\Omega}^q(M)$, $\theta\in\widehat{\Omega}^p(M)$.

2) If $\alpha\in\widehat{\Omega}^q(M)$ the local expression of $d\alpha$, in quantum coordinates, is the following

$$(1.14)\qquad d\alpha = \sum_{1\leq B,A_1,\cdots,A_p\leq m}(\partial x_B.\alpha_{A_1\cdots A_p})dx^B\triangle dx^{A_1}\triangle\cdots\triangle dx^{A_p},$$

with $(\partial x_B . \alpha_{A_1 \cdots A_p}) \in Q_w^\infty(U, \overset{p+1}{\widehat{A}})$.

3) *For any quantum smooth mapping $f : M \to M$ between quantum manifolds, one has:*

$$(f^*\omega)(\zeta_1, \ldots, \zeta_q) = \omega(f_*\zeta_1, \ldots, f_*\zeta_q), \quad f^*(\omega\wedge\theta) = (f^*\omega)\wedge(f^*\theta), \quad d(f^*\omega) = f^*(d\omega),$$

where ω and θ belong to $\widehat{\Omega}^\bullet(M)$.

Proposition 1.47. *For any $\zeta \in Q_w^\infty(TM)$ one has a canonical Z-homomorphism (interior product) reported in Tab.1.24, with its properties.*

TAB.1.24 - Interior product and its properties.

$\zeta\rfloor : \widehat{\Omega}^q(M) \to \widehat{\Omega}^{q-1}(M), \quad \zeta\rfloor : \alpha \mapsto \zeta\rfloor\alpha$
$(\zeta\rfloor\alpha)(\zeta_1, \ldots, \zeta_{q-1}) = \alpha(\zeta, \zeta_1, \ldots, \zeta_{q-1}), \quad \forall \zeta_i \in Q_w^\infty(TM)$
$\zeta\rfloor(\omega\wedge\theta) = (\zeta\rfloor\omega)\wedge\theta + (-1)^q \omega\wedge(\zeta\rfloor\theta), \quad \forall \zeta \in Q_w^\infty(TM)$
$\zeta_1(\zeta_2\rfloor\omega) + \zeta_2(\zeta_1\rfloor\omega) = 0, \quad \forall \zeta_1, \zeta_2 \in Q_w^\infty(TM)$

Proposition 1.48. *For any $\zeta \in Q_w^\infty(TM)$ one has a canonical Z-homomorphism (Lie derivative) given in Tab.1.25.*

TAB.1.25 - Lie derivative and its properties.

$\mathcal{L}_\zeta : \widehat{\Omega}^q(M) \to \widehat{\Omega}^q(M), \quad \mathcal{L}_\zeta : \omega \mapsto \mathcal{L}_\zeta\omega$
$(\mathcal{L}_\zeta\omega)(\zeta_1, \ldots, \zeta_q) = \zeta.\omega(\zeta_1, \ldots, \zeta_q) - \sum_{1 \le \lambda \le q} \omega(\zeta_1, \ldots, [\zeta, \zeta_\lambda], \ldots, \zeta_q), \quad \forall \zeta_i \in Q_w^\infty(TM), i = 1, \ldots, q.$
$\mathcal{L}_\zeta(\omega\wedge\theta) = (\mathcal{L}_\zeta\omega)\wedge\theta + \omega\wedge(\mathcal{L}_\zeta\theta); \quad \mathcal{L}_{\zeta_1}(\mathcal{L}_{\zeta_2}\omega) - \mathcal{L}_{\zeta_2}(\mathcal{L}_{\zeta_1}\omega) = \mathcal{L}_{[\zeta_1, \zeta_2]}\omega$
$\mathcal{L}_{\zeta_1}(\zeta_2\rfloor\omega) = [\zeta_1, \zeta_2]\rfloor\omega + \zeta_2\rfloor(\mathcal{L}_{\zeta_1}\omega); \quad \mathcal{L}_\zeta\omega = d(\zeta\rfloor\omega) + \zeta\rfloor(d\omega)$

Remark 1.30. Let M be a quantum supermanifold of dimension $(\bar{n} \downarrow m|n)$. If $\alpha \in \widehat{\Omega}^p(M)$, then the pull-back $i^*\alpha$ of α by means of any section $i : M_C \to M$, has the following local representation

$$(1.15) \qquad i^*\alpha = \sum_{1 \le j_1 < \cdots < j_p \le n} \beta_{j_1 \ldots j_p} \otimes d\xi^{j_1} \wedge \cdots \wedge d\xi^{j_p},$$

with $\beta_{j_1 \ldots j_p} \in Q_w^\infty(c(U), A)$, given by

$$\beta_{j_1 \ldots j_p} = \sum_{1 \le A_1, \cdots, A_p \le \bar{n}} \alpha_{A_1 \ldots A_p} \circ i\left((\partial\xi_{[j_1} . i^{A_1}) \otimes \cdots \otimes (\partial\xi_{j_p]} . i^{A_p})\right).$$

Here c is the canonical projection $M \to M_C$ and $(\xi^j)_{1 \le j \le \bar{n}}$ are coordinates on M_C induced by the quantum coordinates $(x^A)_{1 \le A \le m+n}$ on $U \subset M$. Furthermore, $\alpha_{A_1 \dots A_p} \in Q_w^\infty(U, \overset{p}{\widehat{A}})$ are the local quantum components of α. Note that $\beta_{j_1 \dots j_p}$ are skewsymmetric in the indexes $(j_1, \cdots, j_p)$. The same result can be obtained if $i : N \to M$ is the embedding of a n-dimensional manifold N into M, with (ξ^k) coordinates on N. In particular, if $\alpha \in \widehat{\Omega}^p(M)$, $\beta \in \widehat{\Omega}^q(M)$, we get

$$
\left\{
\begin{array}{l}
i^*(\alpha \wedge \beta) = \displaystyle\sum_{1 \le j_1 < \cdots < j_{p+q} \le \bar{n}} (i^*(\alpha \wedge \beta))_{j_1 \cdots j_{p+q}} d\xi^{j_1} \wedge \cdots \wedge d\xi^{j_{p+q}} \\[2ex]
(i^*(\alpha \wedge \beta))_{j_1 \cdots j_{p+q}} \equiv \bar{\alpha}_{[j_1 \cdots j_p} \bar{\beta}_{j_{p+1} \cdots j_{p+q}]} = \bar{\alpha}_{[j_1 \cdots j_p} \bar{\beta}_{j_{p+1} \cdots j_{p+q}]} \\[2ex]
+(-1)^{pq} \bar{\beta}_{[j_1 \cdots j_p} \bar{\alpha}_{j_{p+1} \cdots j_{p+q}]} \\[2ex]
\bar{\alpha}_{j_1 \cdots j_p} \equiv \alpha_{C_1 \cdots C_p} \circ i((\partial \xi_{j_1} . i^{C_1}) \otimes \cdots \otimes (\partial \xi_{j_p} . i^{C_p})) \in A \\[2ex]
\bar{\beta}_{j_1 \cdots j_q} \equiv \alpha_{C_1 \cdots C_q} \circ i((\partial \xi_{j_1} . i^{C_1}) \otimes \cdots \otimes (\partial \xi_{j_p} . i^{C_q})) \in A
\end{array}
\right\} .
$$

Let us emphasize that $(i^*(\alpha \wedge \beta))_{j_1 \cdots j_{p+q}}$ is skewsymmetric in the permutation of all the indexes. ∎

Remark 1.31. Let M be a quantum supermanifold of dimension $(m|n)$ over a quantum superalgebra A. All the considerations on the de Rham complexes and de Rham cohomologies given for quantum manifolds remain valid for quantum supermanifolds also. (See Proposition 1.42, Definition 1.43, Remark 1.23,Tab.1.16, Tab.1.17, Tab.1.18, Theorem 1.29, Definition 1.44, Proposition 1.43, Definition 1.45, Proposition 1.44 and Theorem 1.30.) ∎

Let us, now, emphasize the effects of the $\mathbf{Z}_2$-gradiation on the quantun superalgebra underlying a quantum supermanifold.

Theorem 1.31. *Let M be a quantum supermanifold. Then we recognize a bigradiation in the corresponding (co)homology spaces and (co)bordism groups. In Tab.1.26 we report just important bigraded (co)homology spaces.*

TAB.1.26 - Singular bigraded (co)homologies of quantum supermanifolds.

Bigr.Cochain Complex	Bigr.(Co)hom. Space
$\{C_{\bullet\mid\bullet}(M,A),\partial\}$	$H_{\bullet\mid\bullet}(M;A)$ $\left(\textit{bigraded singular homology with coefficients in } A\right)$
$\{C^{\bullet\mid\bullet}(M,A),\delta\}$	$H_s^{\bullet\mid\bullet}(M;A)$ $\left(\textit{bigraded singular cohomology with coefficients in } A\right)$

$H_{\bullet}(M,\mathbf{K})$ is a $\mathbf{K}$-module.

One has the isomorphism: $C^p(M,A)\cong Hom_{\mathbf{K}}(C_p(M;\mathbf{K});A)$

$C_{p\mid q}(M,A)\equiv(A_0\otimes_{\mathbf{K}}C_p(M;\mathbf{K})\oplus(A_1\otimes_{\mathbf{K}}C_q(M;\mathbf{K}))$

$C^{p\mid q}(M,A)\equiv Hom_{\mathbf{K}}(C_p(M;\mathbf{K});A_0)\oplus Hom_{\mathbf{K}}(C_q(M;\mathbf{K});A_1)$.

One has the following exact commutative diagram:

$$
\begin{array}{ccccccc}
& & 0 & & 0 & & \\
& & \downarrow & & \downarrow & & \\
0 & \to & B_{\bullet\mid\bullet}(M;A) & \to & Z_{\bullet\mid\bullet}(M;A) & \to & H_{\bullet\mid\bullet}(M;A) \to 0 \\
& & \downarrow & & \downarrow & & \\
& & C_{\bullet\mid\bullet}(M;A) & = & C_{\bullet\mid\bullet}(M;A) & & \\
& & \downarrow & & \downarrow & & \\
0 \to {}^A\underline{\Omega}_{\bullet\mid\bullet,s}(M) & \to & Bor_{\bullet\mid\bullet}(M;A) & \to & Cyc_{\bullet\mid\bullet}(M;A) & \to & 0 \\
& & \downarrow & & \downarrow & & \\
& & 0 & & 0 & &
\end{array}
$$

where

$$H_{\bullet\mid\bullet}(M;A) \cong (A_0\otimes_{\mathbf{K}}H_{\bullet}(M;\mathbf{K}))\bigoplus(A_1\otimes_{\mathbf{K}}H_{\bullet}(M;\mathbf{K}))$$

is the bigraded singular homology of M with coefficients in A. We call ${}^A\underline{\Omega}_{p\mid q,s}(M)$ *closed singular $(p\mid q)$-bordism groups with coefficients in A, (resp. $Bor_{p\mid q}(M;A)$ singular $(p\mid q)$-bordism groups with coefficients in A, $Cyc_{p\mid q}(M;A)$ singular $(p\mid q)$-cyclism groups with coefficients in A) in the quantum supermanifold M, $p\in\{0,1,\ldots,m-1\}$, $q\in\{0,1,\ldots,n-1\}$. One has the following properties:*

$$
(1.16)\qquad
\left\{
\begin{array}{l}
b\in[a]\in Bor_{\bullet\mid\bullet}(M;A)\Rightarrow a-b=\partial c,\quad c\in C_{\bullet\mid\bullet}(M;A)\\[2ex]
b\in[a]\in Cyc_{\bullet\mid\bullet}(M;A)\Rightarrow\partial(a-b)=0\\[2ex]
b\in[a]\in{}^A\underline{\Omega}_{\bullet\mid\bullet,s}(M)\Rightarrow
\left\{
\begin{array}{l}
\partial a=\partial b=0\\[1ex]
a-b=\partial c,\quad c\in C_{\bullet\mid\bullet}(M;A)
\end{array}
\right\}
\end{array}
\right\}.
$$

Furthermore, one has the following canonical isomorphism:

$$^A\underline{\Omega}_{\bullet|\bullet,s}(M) \cong H_{\bullet|\bullet}(M;A).$$

So $Bor_{\bullet|\bullet}(M;A)$ is the extension of $Cyc_{\bullet|\bullet}(M;A)$ by means of $^A\underline{\Omega}_{\bullet|\bullet,s}(M)$. For the cochain complex $\{C^{\bullet|\bullet}(M;A),\delta\}$ corresponding to $\{C_{\bullet|\bullet}(M;A),\partial\}$ we can associate the dual objects of the previous point 1): $^A\underline{\Omega}_s^{p|q}(M)$ closed singular $(p|q)$-cobordism groups with coefficients in A, $Bor^{p|q}(M;A)$ singular $(p|q)$-cobordism groups with coefficients in A and $Cyc^{p|q}(M;A)$, singular $(p|q)$-cocyclism groups with coefficients in A, $p \in \{0,1,\dots,m-1\}$, $q \in \{0,1,\dots,n-1\}$. Then, one has the following isomorphisms:

$$^A\underline{\Omega}_s^{p|q}(M) \cong H_s^{p|q}(M;A) \cong Hom_{\mathbf{K}}(H_p(M;\mathbf{K});A_0) \bigoplus Hom_{\mathbf{K}}(H_q(M;\mathbf{K});A_1)$$

$$\cong Hom_{\mathbf{K}}(\underline{\Omega}_{p,s}(M);A_0) \bigoplus Hom_{\mathbf{K}}(\underline{\Omega}_{q,s}(M);A_1)$$

$$\cong {}^{A_0}\underline{\Omega}_s^p(M) \bigoplus {}^{A_1}\underline{\Omega}_s^q(M).$$

Furthermore, $Bor^{\bullet|\bullet}(M;A)$ is the extension of $^A\underline{\Omega}_s^{\bullet|\bullet}(M)$ by means of $Cyc^{\bullet|\bullet}(M;A)$.

Proof. The proof follows directly from standard techniques of homological algebra and algebraic topology. $\qquad\square$

The graded extension of Theorem 1.29 is the following.

Theorem 1.32. (Superquantum de Rham theorem). *Let M be a quantum supermanifold of dimension $(m|n)$ over the quantum superalgebra A. We call superquantum de Rham cohomology with coefficients in A the following bigraded spaces: $H^{p|q}(M;A) \equiv (A_0 \otimes_{\mathbf{K}} H^p(M)) \bigoplus (A_1 \otimes_{\mathbf{K}} H^q(M))$. This is induced by the bigraded de Rham complex $\{\Omega^{\bullet|\bullet}(M),d^{\bullet|\bullet}\}$, where $\Omega^{p|q}(M) = (A_0 \otimes_{\mathbf{K}} \Omega^p(M)) \bigoplus (A_1 \otimes_{\mathbf{K}} \Omega^q(M))$. Then, one has the following canonical exact sequence of Z-modules:*

$$0 \to H^{p|q}(M;A) \to {}^A\underline{\Omega}_s^{p|q}(M).$$

In particular, if M is a classic regular quantum supermanifold and $H_p(M_C;\mathbf{K})$, $H_q(M_C;\mathbf{K})$ are finitely generated $\mathbf{K}$-spaces, then one has the following canonical isomorphism: $H^{p|q}(M;A) \cong {}^A\underline{\Omega}_s^{p|q}(M;A)$.

Proof. It is a direct consequence of above results and Theorem 1.29 too. $\square$

QUANTIZED PDE's II: NONCOMMUTATIVE PDE's

Abstract - Here we give a geometric theory of PDE's in the category of quantum (super)manifolds. This theory is the natural extension of the geometric theory of PDE's in the category of commutative (super)manifolds. Emphasis is put on some new algebraic topological techniques that allow us to calculate the integral (co)bordism groups of quantum (super)PDE's, hence to characterize global properties of solutions of quantum (super)PDE's. Many applications to important equations of quantum field theory are considered also. (On this subject see also refs.[63–72].)

2.1 - QUANTUM PDE's

In this section we consider quantum PDE's as the natural extension in the category of quantum PDE's of the usual geometric definition of PDE's in the category of commutative manifolds. (For fundamental informations on the classic, i.e., commutative, geometric theory of PDEs see, e.g., refs.[8,9,19,21,37,39,41,53,54,58,62–66].) For such equations we formulate general theorems of existence of local and global solutions. More precisely we get theorems for formal integrability, complete integrability and Cauchy problems solubility. Moreover, we get also general theorems that allow us to calculate the integral (co)bordism groups of such equations. These allow us to obtain theorems of global existence of solutions. Furthermore, we characterize all the conservation laws of quantum PDE's by means of suitable noncommutative Hopf algebras, *quantum Hopf algebras*. The methods, developed to calculate the integral bordism groups of quantum PDE's, allow us to explicitly determain also such Hopf algebras.

Definition 2.1. A *quantum PDE* (QPDE) of order k on the fibre bundle $\pi : W \to M$, defined in the category of quantum manifolds, is a subfibrebundle $\hat{E}_k \subset J\hat{\mathcal{D}}^k(W)$ of the k-jet-quantum derivative space $J\hat{\mathcal{D}}^k(W)$ over M. A QPDE $\hat{E}_k$ is *quantum regular* if the r-quantum prolongations $\hat{E}_{k+r} \equiv J\hat{\mathcal{D}}^r(\hat{E}_k) \cap J\hat{\mathcal{D}}^{k+r}(W)$ are subbundles of $\pi_{k+r,k+r-1} : J\hat{\mathcal{D}}^{k+r}(W) \to$

$J\hat{D}^{k+r-1}(W)$, $\forall r \geq 0$. Furthermore, we say that $\hat{E}_k$ is *formally quantumintegrable* if $\hat{E}_k$ is quantum regular and if the mappings $\hat{E}_{k+r+1} \to \hat{E}_{k+r}$, $\forall r \geq 0$, and $\pi_{k,0} : \hat{E}_k \to W$ are surjective. In the following we shall consider QPDEs on a fiber bundle $\pi : W \to M$, where M is a quantum manifold of dimension m on the quantum algebra A and W is a quantum manifold of dimension (m,s) on the quantum algebra $B \equiv A \times E$, where E is also an Z-algebra. The *quantum symbol* $\dot{g}_{k+r}$ of $\hat{E}_{k+r}$ is a family of Z-modules over $\hat{E}_k$ characterized by means of the following short exact sequence of Z-modules: $0 \to \pi^*_{k+r}\dot{g}_{k+r} \to vT\hat{E}_{k+r} \to \pi^*_{k+r,k+r-1}vT\hat{E}_{k+r-1}$. Then one has the following complex of Z-modules over $\hat{E}_k$ (δ-*quantum complex*):

$$0 \to \dot{g}_m \xrightarrow{\delta} TMO\dot{g}_{m-1} \xrightarrow{\delta} \dot{\Lambda}_0^2 MO\dot{g}_{m-2} \xrightarrow{\delta}$$
$$\cdots \xrightarrow{\delta} \dot{\Lambda}_0^{m-k} MO\dot{g}_k \xrightarrow{\delta} \delta(\dot{\Lambda}_0^{m-k} MO\dot{g}_k) \to 0.$$

We call *Spencer quantumcohomology* of $\hat{E}_k$ the homology of such complex. We denote by $\{H_q^{m-j,j}\}_{q \in \hat{E}_k}$ the homology at $(\dot{\Lambda}_0^j MO\dot{g}_{m-j})_q$. We say that $\hat{E}_k$ is *r-quantumacyclic* if $H_q^{m,j} = 0$, $m \geq k$, $0 \leq j \leq r$, $\forall q \in \hat{E}_k$. We say that $\hat{E}_k$ is *quantuminvolutive* if $H_q^{m,j} = 0$, $m \geq k$, $j \geq 0$. We say that $\hat{E}_k$ is δ-*regular* if there exists an integer $\kappa_0 \geq \kappa$, such that $\dot{g}_{\kappa_0}$ is quantum involutive or 2-quantumacyclic.

First to go on the integrability properties of QPDE's let us first recall some important definitions and results on Noetherian rings and Noetherian spaces.

Definition 2.2. Let R be a nonzero associative noncommutative ring with unit 1. Let M_R denote a right R-module. A chain of $n+1$ subsets $A_0 \supset A_1 \supset \cdots \supset A_n$ is said to have *length n*. If $M_R \neq 0$ and M has exactly two submodules, namely M and 0, then M is a *simple* or *irreducible* module. A *semisimple* module is a module which is a direct sum of simple modules; and if the simple modules are pairwise isomorphic, the module is called *isotypic*. If a module M_R has the property that each descending chain $M = M_0 \supset M_1 \supset M_2 \supset \cdots$ must terminate after a finite number of steps, is called to satisfy the *descending chain condition* (d.c.c.) and then M is called an *Artinian module*. The dual concept, using ascending chains of submodules of M, is the *ascending chain condition* (a.c.c.). Modules satisfying the a.c.c. are called *Noetherian*.

Theorem 2.1. 1) If $N \triangleleft M$ then M is Noetherian or Artinian iff both N and M/N have that property.

2) (Jordan-Holder). *A module M_R satisfies both the a.c.c. and the d.c.c iff there is an upper bound, n say, on the lengths of chains of submodules of M. If that is so, then every chain of submodules of M can be refined to one of length s, say $M = M_0 \supset M_1 \supset \cdots \supset M_n = 0$. The factors M_i/M_{i+1} are simple.*

3) *The following conditions on a semisimple module M_R are equivalent:*

(i) *M_R satisfies the a.c.c.;*

(ii) *M_R satisfies the d.c.c.;*

(iii) *M_R has finite length.*

4) *The following conditions on a module M_R are equivalent:*

(i) *M_R is Noetherian;*

(ii) *each submodule of M_R is finitely generated;*

(iii) *every non-empty set of submodules of M_R has has a maximal number.*

Definition 2.3. Considering a ring R like a (left or right) R-module, then we can define *left (right) Noetherian* and *left (right) Artinian rings.*

Proposition 2.1. *The following conditions on a ring R are equivalent:*

(i) *R is right Noetheirian;*

(ii) *R satisfies the a.c.c on right ideals;*

(iii) *every right ideal of R is finitely generated;*

(iv) *every non-empty set of right ideals of R has a maximal member.*

Definition 2.4. A ring R is *simple* if R has precisely two ideals, 0 and R. This is equivalent to asserting that 0 is the unique maximal ideal of R.

Theorem 2.2. *The following conditions on a ring R are equivalent:*

(i) *R is a simple right Artinian ring;*

(ii) *$R \cong M_n(D)$, for some n and some (uniquely determined) division ring D; ($M_n(D)$ denotes the ring of all $n \times n$ matrices over D);*

(iii) *$R \cong End(M_S)$, where M is a semisimple isotypic S-module of length n over some ring S.*

Theorem 2.3. (Hopkin). *If R is right Artinian then R is right Noetherian. (The converse is false.)*

Example 2.1. (*Hilbert basis theorem*). If A is a Noetherian ring, the polynomial ring $A[\chi_1, \cdots, \chi_n]$ in n indeterminates is also a Noetherian ring.∎

Proposition 2.2. A graded ring $A \equiv \bigoplus_{0 \le r \le \infty} A_r$, ($A_r A_s \subseteq A_{r+s}$, $r, s \ge 0$), is Noetherian iff A_0 is Noetherian and A is finitely generated as an A_0-algebra.

Proof. In fact, in a graded ring A, A_0 is a subring of A and each A_r is an A_0-module. Then, from above results it follows directly the proposition.$\square$

Example 2.2. $A = \kappa[\chi_1, \cdots, \chi_n] = \bigoplus_{0 \leq r \leq \infty} A_r$, where A_r is the set of all homogeneous polynomials of degree r and κ is a field, is a Noetherian ring.■

Example 2.3. If A is a ring and $\mathfrak{a}$ is an ideal of A, one has the canonically associated graded ring $\bigoplus_{0 \leq r \leq \infty} \mathfrak{a}^r$, with $\mathfrak{a}^0 = A$. If A is Noetherian, then $\mathfrak{a}$ is finitely generated, say $\mathfrak{a} = < a_1, \cdots, a_p >$ and we have $\bigoplus_{0 \leq r \leq \infty} \mathfrak{a}^r = A[a_1, \cdots, a_p]$ as a quotient of the polynomial ring $A[\chi_1, \cdots, \chi_p]$. Taking into account the inclusion $\mathfrak{a}^r \supseteq \mathfrak{a}^{r+1}$, we get also the graded ring $\bigoplus_{0 \leq r \leq \infty} \mathfrak{a}^r/\mathfrak{a}^{r+1}$ which is generated over $\mathfrak{a}^0/\mathfrak{a}^1 \cong A/\mathfrak{a}$ by elements of $\mathfrak{a}^1/\mathfrak{a}^2 = \mathfrak{a}/\mathfrak{a}^2$. In particular, if $\mathfrak{a}$ is generated by $(a_1, \cdots, a_p)$ and $(\bar{a}_1, \cdots, \bar{a}_p)$ denote the image of $(a_1, \cdots, a_p)$ in $\mathfrak{a}^1/\mathfrak{a}^2$, one has

$$\bigoplus_{0 \leq r \leq \infty} \mathfrak{a}^r/\mathfrak{a}^{r+1} = (A/\mathfrak{a})[\bar{a}_1, \cdots, \bar{a}_p]$$

that is a quotient of the polynomial ring $(A/\mathfrak{a})[\chi_1, \cdots, \chi_p]$. If $a \in \mathfrak{a}^r/\mathfrak{a}^{r+1}$ and $b \in \mathfrak{a}^s/\mathfrak{a}^{s+1}$, we may define the product $(a + \mathfrak{a}^{r+1})(b + \mathfrak{a}^{s+1}) = ab + \mathfrak{a}^{r+s+1}$. Furthermore, if A is a Noetherian ring, the above graded ring is Noetherian too.   ■

Definition 2.5. A non-empty topological space X is said to be *irreducible* if every pair of non-empty open sets in X intersects (thus X is as far as possible from being Hausdorff).

Proposition 2.3. *The following propositions are equivalent:*

(i) *X is an irreducible topological space;*

(ii) *X is not the union of two proper closed subsets;*

(iii) *If F_i ($1 \leq i \leq n$) are closed subsets which cover X, then $X = F_i$ for some i;*

(iv) *Every non-empty open set is dense in X;*

(v) *Every open set in X is connected.*

Example 2.4. Let X be an infinite set, and topologize X by taking the closed subsets to be X itself and all finite subsets of X. Then, X is irreducible.   ■

Example 2.5. Any irreducible algebraic variety, with the Zariski topology, is an irreducible space.   ■

Definition 2.6. A subset Y of a space X is *irreducible* if Y is irreducible in the induced topology.

Proposition 2.4. 1) If $(F_i)_{1\leq i\leq n}$ is a finite closed covering of a space X, and if Y is an irreducible subset of X, then $Y \subseteq F_i$ for some i.

2) If X is irreducible, every non-empty open subset of X is irreducible.

3) Let $(U_i)_{1\leq i\leq n}$ be a finite space covering of a space X, the U_i being non-empty. Then X is irreducible iff each U_i is irreducible and meets each U_j.

4) If Y is a subset of X, then Y is irreducible iff $\overline{Y}$ is irreducible.

5) The image of an irreducible set under a continuous map is irreducible.

6) X has maximal irreducible subsets, they are all closed and they cover X (the *Zorn's lemma*).

7) If $x \in X$, then $\{x\}$ is irreducible and therefore so is $\overline{\{x\}}$.

Definition 2.7. The maximal irreducible subsets of X are called the *irreducible components* of x.

Remark 2.1. Irreducibility is in some ways analogous to, but stronger than, connectedness. ∎

Definition 2.8. If V is an irreducible subset of X and $V = \overline{\{x\}}$ for some $x \in X$, then x is a *generic point* of V. If $y \in \overline{\{x\}}$, y is a *specialization* of x. The closed set $\overline{\{x\}}$ is the *locus* of x. A subset Y of a space X is *locally closed* if Y is the intersection of an open set and a closed set in X, or equivalently if Y is open in its closure $\overline{Y}$, or equivalently again if every $y \in Y$ has an open neighbourhood U_y in X such that $Y \cap U_y$ is closed in U_y.

Definition 2.9. A topological space X is *Noetherian* iff the closed subsets of X satisfy the descending chain condition.

Theorem 2.4. 1) *The following conditions are equivalent:*

(i) *X is Noetherian.*

(ii) *The open sets in X satisfy the ascending chain conditions.*

(iii) *Every open subset of X is quasi-compact (i.e., compact but not necessarily Hausdorff).*

(iv) *Every subset of X is quasi-compact.*

2) *A Noetherian space is quasi-compact.*

3) *Every subset of a Noetherian space (with the induced topology) is Noetherian.*

4) *Let X be a topological space and let $(X_i)_{1\leq i\leq n}$ be a finite covering of X. If X_i are Noetherian, then so is X.*

5) *If X is Noetherian, the number of irreducible components of X is finite.*

Definition 2.10. The *dimension* of a topological space X is the integer $n \le \infty$ such that in X there exist chains of distinct irreducible closed sets $X_0 \subset X_1 \subset \cdots \subset X_n$ and there do not exist chains with more terms.

Now, we are ready to state the following two theorems that are fundamental in order to characterize the integrability properties of QPDEs.

Theorem 2.5. Let $\hat{E}_k \subset J\hat{D}^k(W)$ be a quantum regular QPDE, on the fiber bundle $\pi : W \to M$. Then there exists a filtration $\{X_q\}$ of $\hat{E}_k$:

$$\emptyset = X_{-1} \subset X_0 \subset X_1 \subset \cdots \subset X_N = \hat{E}_k$$

such that the corresponding spectral sequence $\{E^r_{p,q}, d^r\}$ (homological formal spectral sequence of $\hat{E}_k$), converges to $H_\bullet(\hat{E}_k)$. Furthermore, to the above filtration of $\hat{E}_k$ there also corresponds a spectral sequence $\{E^{p,q}_r, d_r\}$ (cohomological formal spectral sequence of $\hat{E}_k$), converging to $H^\bullet(\hat{E}_k)$. Finally, if $\hat{E}_k$ is formally quantum integrable, then one has $H_\bullet(\hat{E}_k) \cong H_\bullet(\hat{E}_\infty)$, $H^\bullet(\hat{E}_k) \cong H^\bullet(\hat{E}_\infty)$.

Proof. For a detailed proof see ref.[67]. Here let us only stress that the proof is based on the fact that $\hat{E}_k$ is locally a Noetherian space, then the following filtration

$$\hat{E}_k \equiv Y_0 \supset \pi_{k+1,k}((\hat{E}_k)_{+1}) \equiv Y_1 \supset \cdots \supset \pi_{k+q,k}((\hat{E}_k)_{+q}) \equiv Y_q \supset \cdots$$

must be stationary for large enough q. $\qquad\qquad\square$

Theorem 2.6. (δ-Poincaré lemma for quantum PDE's)[67]. Let $\hat{E}_k \subset J\hat{D}^k(W)$ be a quantum regular QPDE. If Z is a Noetherian **K**-algebra, then $\hat{E}_k$ is a δ-regular QPDE.

Theorem 2.7. (Criterion of formal quantum integrability)[67]. Let $\hat{E}_k \subset J\hat{D}^k(W)$ be a quantum regular, δ-regular QPDE. Then if $\hat{g}_{k+r+1}$ is a bundle of Z-modules over $\hat{E}_k$, and $\hat{E}_{k+r+1} \to \hat{E}_{k+r}$ is surjective for $0 \le r \le m$, then $\hat{E}_k$ is formally quantumintegrable.

An *initial condition* for QPDE $\hat{E}_k \subset J\hat{D}^k(W)$ is a point $q \in \hat{E}_k$. A *solution* of $\hat{E}_k$ passing for the initial condition q is a m-dimensional (over A) quantum manifold $N \subset \hat{E}_k$ such that $q \in N$ and such that N can be represented in a neighboroud of any of its points $q' \in N$, except for a nonwhere dense subset $\Sigma(N) \subset N$ of dimension $\le m-1$, as image of the k-derivative $D^k s$ of some Q^k_w-section s of $\pi : W \to M$. We call $\Sigma(N)$ the set of *singular points* (of *Thom-Bordman type*) of N. If $\Sigma(N) = \emptyset$ we say that N is a *regular solution* of $\hat{E}_k \subset J\hat{D}^k(W)$. Furthermore, let us denote by $\hat{J}^k_m(W)$ the k-jet of

m-dimensional (over A) quantum manifolds contained into W. One has the natural embeddings $\hat{E}_k \subset J\hat{\mathcal{D}}^k(W) \subset \hat{J}_m^k(W)$. Then, with respect to the embedding $\hat{E}_k \subset \hat{J}_m^k(W)$ we can consider solutions of $\hat{E}_k$ as m-dimensional (over A) quantum manifolds $V \subset \hat{E}_k$ such that V can be representable in the neighborhood of any of its points $q' \in V$, except for a nonwhere dense subset $\Sigma(V) \subset V$, of dimension $\leq m - 1$, as $N^{(k)}$, where $N^{(k)}$ is the k-quantum prolongation of a m-dimensional (over A) quantum manifold $N \subset W$. In the case that $\Sigma(V) = \emptyset$, we say that V is a *regular solution* of $\hat{E}_k \subset \hat{J}_m^k(W)$. Of course, solutions V of $\hat{E}_k \subset \hat{J}_m^k(W)$, even if regular ones, are not, in general diffeomorphic to their projections $\pi_k(V) \subset M$, hence are not representable by means of sections of $\pi : W \to M$.

Therefore, above theorems allow us to obtain existence theorems of local solutions. (See nota in Definition 2.12.)

Remark 2.2. (*Integral quantum planes and classic limits of solutions*). We call *regular quantumplane* at the point $q \in \hat{J}_m^k(W)$ the linear subspace in $T_q \hat{J}_m^k(W)$ of the form, $T_q N^{(k)}$, for some Q_w^∞ quantum submanifold $N \subset W$, of dimension m over A, passing for $a \in W$, with $a = \pi_{k,0}(q)$. Any point $q \in \hat{J}_m^k(W)$ identifies a unique regular quantumplane L_q at $q' \equiv \pi_{k,k-1}(q) \in \hat{J}_m^{k-1}(W)$: $L_q \subset T_{q'} \hat{J}_m^{k-1}(W)$. Set: $I_k(W) \equiv \bigcup_{u \in \hat{J}_m^k(W)} I_k(W)_u$, $I_k(W)_u \equiv$ Grassmannian of integral quantumplanes at q. An *integral quantumplane* at a point $u \in \hat{J}_m^k(W)$ is defined to be a m-dimensional quantumspace, subspace of $(\mathbb{E}_k)_u$, tangent to some integral quantum manifold of the Cartan distribution $\mathbb{E}_k \subset T\hat{J}_m^k(W))$. Let $I(\hat{E}_{k+s})$ be the fiber bundle of Grassmannian m-dimensional integral quantumplanes of the Cartan quantumdistribution $\mathbb{E}(\hat{E}_{k+s})$ on $\hat{E}_{k+s}$ being $\hat{E}_k \subset \hat{J}_m^k(W)$ a QPDE. If $\hat{E}_{k+s} = \hat{J}_m^{k+s}(W)$ one has $I(\hat{J}_m^{k+s}(W)) = I_{k+s}(W)$. Furthermore, note that here and in the following for p-dimensional integral quantum manifold V, $0 \leq p \leq m$, with boundary ∂V, (or eventually with $\partial V = \emptyset$), we mean an element $V \in C_p(\hat{E}_{k+h}, A)$, $h \geq 0$, such that $\widehat{T}V \subset \widehat{\mathbb{E}}_{k+h}$. So, if $V = \sum_i a^i u_i$, $a_i \in A$, one has $\partial V = \sum_i (-1)^i a^i \partial_i u$. Then, the canonical homomorphism $c : A \to \mathbf{K}$ induces the canonical mapping $V \to V_C = \sum_i \lambda^i u_i$, with $\lambda^i \equiv c(a^i) \in \mathbf{K}$, and one has $\partial V_C = \sum_i (-1)^i \lambda^i \partial_i u$, where $V_C \in C_p(\hat{E}_{k+h}; \mathbf{K})$ is the classic limit of V. Finally, note that by means of the monomorphism $\epsilon : \mathbf{K} \to A$, we can induce an injection $\epsilon : V_C \to V$, such that $c \circ \epsilon = id_V$, hence ϵ is a section of $V \to V_C$. In other words V canonically contains its classic limit V_C.

Theorem 2.8. *Let W be a quantum manifold (resp. $\pi : W \to M$ a fiber bundle) like above specified and let $\hat{E}_k \subset \hat{J}_n^k(W)$ (resp. $\hat{E}_k \subset J\hat{D}_n^k(W)$) be a QPDE. One has the following.*

1) Solutions V of $\hat{E}_k$ are branched coverings of some n-dimensional quantum submanifold $N \subset W$, (resp. coverings of M) with branche points $p \subset \pi_k(\Sigma(V))$. Set $V' \equiv V \setminus \Sigma(V)$ and $N' \equiv N \setminus \pi_k(\Sigma(V))$, (resp. $M' \equiv M \setminus \pi_k(\Sigma(V))$). Then $\pi_k : V' \to N'$ (resp. $\pi_k : V' \to M'$) is a covering map.

2) In particular, if $\Sigma(V) = \emptyset$, then $\pi_k : V \to N$ (resp. $\pi_k : V \to M$) is a covering map. Furthermore, if $\pi_1(N) = 0$, (resp. $\pi_1(M) = 0$), i.e., N (resp. M) is simply connected, then V is globally diffeomorphic to N, (resp. M).

3) If the base space N (resp. M) is connected, then the number of sheets of the covering $V' \to N'$ (resp $V' \to M'$) is independent of the choice of the point $a \in N$ (resp. $p \in M$).

4) In the case of QPDE's $\hat{E}_k \subset \hat{J}_n^k(W)$, if the manifolds V and N are closed, then the branched coverings (resp. coverings) are with a finite number of sheets. A similar proposition holds for the quantum manifolds V and M in the case of solutions of QPDE's $\hat{E}_k \subset J\hat{D}^k(W)$.

Proof. We shall use some standard results about coverings. For their proofs see, e.g., ref.[17]. Let us work in the category of topological spaces.

Lemma 2.1. (Existence and uniqueness of path liftings). *If $p : \tilde{M} \to M$ is a covering of M, then for every path $\gamma : [a,b] \to M$ and every $q_0 \in \tilde{M}$ over $\gamma(a)$, there is one and only one lifting $\tilde{\gamma}$ of γ starting at q_0, i.e., $\tilde{\gamma}(a) = q_0$, $p \circ \tilde{\gamma} = \gamma$.*

Lemma 2.2. (Liftings of homotopies over covering spaces). *Let $p : \tilde{M} \to M$ be a covering of M. Let B another topological space, $h : B \times [0,1] \to M$ a continuous map and let $h_0 : B \to \tilde{M}$ be a continuous lifting of h_0, i.e., $p \circ \tilde{h}_0 = h \circ i_0$, where i_0 is the canonical inclusion $i_0 : B \times \{0\} \to B \times [0,1]$. The map $\tilde{h} : B \times [0,1] \to \tilde{M}$, $\tilde{h}(b,t) = \tilde{h}_0(t)$, starting at $\tilde{h}_0(b)$, is continuous. The following diagram is commutative:*

$$
\begin{array}{ccc}
B \times \{0\} & \xrightarrow{\ \tilde{h}_0\ } & \tilde{M} \\
\downarrow & & \| \\
B \times [0,1] & \xrightarrow{\ \tilde{h}\ } & \tilde{M} \\
\| & & \downarrow{\scriptstyle p} \\
B \times [0,1] & \xrightarrow[\ h\ ]{} & \tilde{M}
\end{array}
$$

Lemma 2.3. (Monodromy). Let $p : \tilde{M} \to M$ be a covering and $\alpha, \beta :$ $[0,1] \to M$ two paths in M which are homotopic with fixed end points, i.e., there is a homotopy $h : [0,1] \times [0,1] \to M$ with $h_0 = \alpha$, $h_t(0) = \alpha(0)$, $h_1 = \beta$, $h_t(1) = \alpha(1)$, $\forall t \in [0,1]$. If $\tilde{\alpha}, \tilde{\beta}$ are liftings of α and β respectively, starting at same point q_0, then they end at the same point as well: $\tilde{\alpha}(1) = \tilde{\beta}(1)$.

Definition 2.11. The *monodromy* (or *discrete holonomy*) of a covering $p :$ $\tilde{M} \to M$ is the homomorphism

$$\sigma : \pi_1(M, p_0) \to S_{p^{-1}(p_0)}, \quad [\gamma] \mapsto \sigma(x_1, x_2, \cdots),$$

where $S_{p^{-1}(p_0)}$ is the group of permutations of the fibre $p^{-1}(p_0) \equiv (x_1, x_2, \cdots)$. $\sigma(\pi_1(M, p_0)) \subset S_{p^{-1}(p_0)}$ is called *monodromy group* (or *discrete holonomy group*).

Example 2.6. Let us consider the universal covering $p : \mathbf{R} \to S^1$, $t \mapsto e^{2\pi i t}$. Then $p^{-1}(0) = \{n = 0, \pm 1, \pm 2, \cdots\}$. $\pi_1(S^1) = \mathbf{Z}$. Let a be the generator of $\mathbf{Z}$. Then $\sigma(a) : n \mapsto n+1$, hence the monodromy $\sigma(a)$ is the translation. ∎

Lemma 2.4. (Behaviour of the groups of homotopy relative to covering spaces). Let $p : (\tilde{M}, q_0) \to (M, p_0)$ be a covering in the category of pointed topological spaces. Then, the induced group homomorphism $p_* :$ $\pi_k(\tilde{M}, q_0) \to \pi_k(M, p_0)$ is an isomorphism for all $n > 1$ and a monomorphism for $n = 1$. Furthermore, if $\tilde{M}$ is 0-connected and a locally trivial covering space with type fibre F, then the points of F are in a one-to-one correspondence with the cosets of the characteristic group $G(\tilde{M}, q_0)$ in $\pi_1(M, p_0)$, where $G(\tilde{M}, q_0) \equiv p_*(\pi_1(\tilde{M}, q_0)) \subset \pi_1(M, p_0)$.

☐ A loop α in (M, p_0) lifts to a loop $\tilde{\alpha}$ in $(\tilde{M}, p_0)$ iff $[\alpha] \in G(\tilde{M}, q_0)$.

☐ If the characteristic subgroups of the covering spaces $(\tilde{M}_1, q_1)$, $(\tilde{M}_2, q_2)$, are contained in one another, the covering space with the smaller group canonically covers the other space; $G(\tilde{M}_1, q_1) \subset G(\tilde{M}_2, q_2) \subset \pi_1(M, p_0)$ implies the following commutative diagram of coverings:

$$
\begin{array}{ccc}
(\tilde{M}_1, q_1) & \longrightarrow & (\tilde{M}_2, q_2) \\
\downarrow & & \downarrow \\
(M, p_0) & = & (M, p_0)
\end{array}
$$

☐ (Liftability criterion of maps on covering spaces). Let $p : (\tilde{M}, q_0) \to (M, p_0)$ be a covering mapping. Let $f : (B, b_0) \to (M, p_0)$ be a continuous map. In order that f should be liftable, i.e., there exists a unique map

$\tilde{f} : (B, b_0) \to (\tilde{M}, q_0)$ such that $p \circ \tilde{f} = f$, it is necessary that $f_*(\pi_1(B, b_0)) \subset G(\tilde{M}, q_0)$.

Lemma 2.5. *Let* $p : \tilde{M} \to M$ *be a covering. If M is connected, then the number of sheets of the covering is independent of the choice of point* $p \in M$.

Lemma 2.6. *Let V and M be closed quantum manifolds of the same dimension n. Let $p : V \to M$ be a map of class Q_w^1 such that $\ker Dp(q) = 0$, $\forall q \in V$. Then, the map p is a covering with a finite number of leaves.*

Proof of Lemma 2.6. For the continuity it follows that around each point $q \in V$ there is a neighbourhood U_q such that the restriction of p to U_q is a diffeomorphism. Hence, taking into account the compactness of V, each point $x \in M$ can have only finitely many preimages. Let $y_1, \cdots, y_m$ be the distinct preimages under p of any particular $x \in M$, and let $V_1, \cdots, V_m$ be respectively pairwise non-intersecting neighbourhoods of these preimages, on each of which p is a diffeomorphism. Then, for some sufficiently small neighbourhood U_x of x we shall have $p^{-1}(U_x) \subset V_1 \bigcup \cdots \bigcup V_m$. So the proof is complete. $\qquad\square$

Lemma 2.7. *Let N and M be closed quantum manifolds of the same dimension n. Let $p : N \to M$ be a map of class Q_w^1 such that $\ker Dp(q) = 0$, $\forall q \in N \setminus \Sigma(N)$, $\ker Dp(q) \neq 0$, $\forall q \in \Sigma(N)$, where $\Sigma(N)$ is of dimension at most $n-2$ (so that the set $p(\Sigma(N))$ does not separate M into disjoint parts). Then, the map $p : N' \equiv N \setminus \Sigma(N) \to M' \equiv M \setminus p(\Sigma(N))$ is a finite-sheet covering map. The map p is called a covering map branched along $f(\sigma(N))$ and the points of the set $p(\sigma(N)) \subset M$ are called the* branch points of the *(branched) covering p.*

Proof of Lemma 2.7. Let ϵ be a positive real number enough small. Let U_ϵ be the open ϵ-neighbourhood of $p(\sigma(N)) \subset M$. Let us consider $M_\epsilon \equiv M \setminus U_\epsilon$, $N_\epsilon \equiv N \setminus p^{-1}(U_\epsilon)$. Each manifold $_\epsilon$ and M_ϵ is connected and compact. Then, we can use above lemma for the map $p : N_\epsilon \to M_\epsilon$. By taking $\epsilon \to 0$, we get the proof. $\qquad\square$

After above lemmas the proof of the theorem follows directly. $\qquad\square$

Definition 2.12. 1) We say that a quantum PDE $\hat{E}_k \subset J\hat{D}^k(\bar{W})$, or $\hat{E}_k \subset \hat{J}_n^k(\bar{W})$, is *completely quantum integrable* if for any initial condition $q \in \hat{E}_k$ passes a solution.[27]

---

[27] In analytic quantum case the formal quantum integrability implies the complete quantum

2) We say that a QPDE $\hat{E}_k \subset J\hat{D}^k(\bar{W})$, or $\hat{E}_k \subset \hat{J}_n^k(\bar{W})$, is *universally quantum regular*, or *universally formally quantum integrable*, if there exists a formally quantum integrable QPDE $(\hat{E}_k)_f \subset \hat{E}_k$ such that $((\hat{E}_k)_f)_{+\infty} \cong (\hat{E}_k)_{+\infty}$, where the index $+\infty$ denotes ∞-quantum prolongation. In such a case all the regular (quantum) solutions of $\hat{E}_k$ coincide with the ones of $(\hat{E}_k)_f$.[28]

Theorem 2.9. (Cauchy problem for QPDE's and characteristic quantum numbers). *The Cauchy problem for (singular) quantumsolutions of QPDE can be formulated in the following way. A Cauchy data for the system $\hat{E}_k \subset \hat{J}_m^k(W)$ is defined by a $(m-1)$-dimensional integral classicregular quantum manifold $\sigma \subset \hat{E}_{k+r}$ of the Cartan quantumdistribution $\mathbb{E}_{k+r}$, beside a section h of the bundle $I(\hat{E}_{k+r})|\sigma \to \sigma$, such that $h(\bar{q}) \supset T_{\bar{q}}\sigma$, for all $\bar{q} \in \sigma$. Here $I(\hat{E}_{k+r})$ is the Grassmannian of integral quantum planes on $\hat{E}_{k+r}$. A quantumsolution of the Cauchy problem with initial data, (σ, h), is defined to be an m-dimensional integral quantum manifold V, with boundary, such that $V \subset \hat{E}_{k+r}$, $\partial V = \sigma$ and $T_{\bar{q}}V = h(\bar{q}), \forall \bar{q} \in \sigma$. With each quantumcohomology class $\omega \in H^{m-1}(I(\hat{E}_{k+r}); A)$ and each Cauchy data (σ, h) we associate the characteristic quantum number $\chi_\omega(\sigma, h) =< h^*\omega, Z_\sigma >$, where $Z_\sigma \in H_{m-1}(\sigma; A)$ is a fundamental cycle of σ. For solvability of the Cauchy problem (σ, h) for QPDE's it is necessary that all the characteristic quantum numbers $\chi_\omega(\sigma, h)$ be equal to zero.*

Proof. We shall use the following lemmas.

Lemma 2.8. *Each m-dimensional integral quantum manifold $V \subset \hat{J}_m^k(W)$ determines a tangential map $i_V : V \to I_k(W)$ given by $i_V(q) = T_qV \in I_k(W)_q$. Then, any integral quantum manifold $V \subset \hat{J}_m^k(W)$ determines a map $H^i(I_k(W); A) \to H^i(V; A), \omega \mapsto i_V^*\omega$, where $i_V^*\omega$ is the characteristic quantumclass on V corresponding to $\omega \in H^i(I_k(W); A)$.*

Then we have the following important lemmas that extend previous one for manifolds and supermanifolds [61,62].

Lemma 2.9. 1) *Let $\hat{E}_k \subset \hat{J}_m^k(W)$ be a QPDE. Each cohomology quantumclass $\omega \in H^i(I(\hat{E}_{k+s}); A)$ defines a characteristic quantumclass $\omega_V \equiv i_V^*\omega$ on any singular solution $V \subset \hat{E}_{k+s}$ of $\hat{E}_k$. $H^\bullet(I(\hat{E}_{k+s}); A)$ is an algebra over $H^\bullet(\hat{E}_{k+s}; A)$.*

---

integrability. (See ref.[67].)

[28] This situation is very frequently met in the QPDEs of the Mathematical Physics. (See e.g., the next section 2.2 and ref.[69] where the quantum Navier-Stokes equation is considered.)

2) *Let $\hat{E}_k$ be formally quantumintegrable QPDE then the fibre bundles $\pi_{k+s,k+s-1} : \hat{E}_{k+s} \to \hat{E}_{k+s-1}$, $s \geq 1$, are affine subbundles in $\pi_{k+s,k+s-1} : \hat{J}_m^{k+s}(W) \to \hat{J}_m^{k+s-1}(W)$ and, hence $H^\bullet(\hat{E}_{k+s}; A) \cong H^\bullet(\hat{E}_k; A)$. If V is a non-singular integral quantum manifold, then all its quantumcharecteristics classes are zero in dimension ≥ 1.*

Proof of Lemma 2.9. See ref.[67]. $\qquad\qquad\qquad\qquad\qquad\qquad\qquad$ □

Lemma 2.10. *Lie quantum transformations, that are local diffeomorphisms*

$$\phi : \hat{J}_m^k(W) \to \hat{J}_m^k(W),$$

of class Q_w^∞, which preserve the Cartan quantumdistributions, $T(\phi)(\mathbb{E}_k)_u = (\mathbb{E}_k)_{\phi(u)}$, $\forall\, u \in \hat{J}_m^k(W)$, preserve also the cohomology space $H^i(I_k(W); A)$ (if they are not just isomorphic). Characteristic Lie quantumtransformations, that are Lie quantumtransformations that transform each integral quantum manifold in itself, preserve the space of characteristic quantum classes of any integral quantum manifold. In particular, any characteristic Lie quantumtransformation that preserves also the classes of $H^i(I_k(W); A)$ preserves characteristic quantumclasses of any integral quantum manifold. As quantum manifolds involved are classicregular, we have the following canonical isomorphisms: $H^{m-1}(\sigma; A) \cong A \bigotimes_{\mathbf{K}} H^{m-1}(\sigma_C)$; $H_{m-1}(\sigma; A) \cong A \bigotimes_{\mathbf{K}} H_{m-1}(\sigma_C; \mathbf{K})$, where σ_C is the classic-limit of σ. (In this way has sense to talk of fundamental cycle of σ.) Hence, the characteristic quantum numbers can be written as follows: ($\spadesuit$) $\chi_\omega(\sigma, h) =< h^*\omega, Z_\sigma >=< a^i \otimes \beta_i, 1 \otimes Z_{\sigma_C} >= a^i < \beta_i, Z_C >$, where $a^i \in A, \beta_i \in H^{m-1}(\sigma_C)$, and Z_{σ_C} is a fundamental cycle of σ_C. On the other and, if $V \subset \hat{E}_{k+r}$ is an integral quantum manifold such that $\partial V = \sigma$, we get also $\partial V_C = \sigma_C$. Then, taking into account that for ordinary manifolds, a necessary condition for solvability of Cauchy problem is that all characteristic numbers $< \beta_i, Z_C >$ must be zero, from ($\spadesuit$) we get that, in order to solve the Cauchy problem for QPDE's, all the characteristic quantum numbers $\chi_\omega(\sigma, h)$ must be zero too. $\qquad\qquad\qquad\qquad\qquad\qquad\qquad\qquad$ □

Theorem 2.10. 1) *Let $\hat{E}_k \subset \hat{J}_m^k(W)$ be a k-order QPDE on the fiber bundle $\pi : W \to M$, $\dim M = m$. Let $(N_i, h_i), i = 0, 1$, be a couple of Cauchy data for the system $\hat{E}_k$, defined on $\hat{E}_{k+r}$. Then one has a canonical mapping: $\phi_{(N,h)} : H^{m-1}(\hat{I}_{k+r}(W); A) \to A$, where $N \equiv N_0 \times 0 \bigcup N_1 \times 1$, and h is the natural extension of $h_i, i = 0, 1$, to N. We call $H^{m-1}(\hat{I}_{k+r}(W); A)$ the $(m-1)$-Leray-Serre quantumcohomology of $\hat{J}_m^{k+r}(W)$. Then, in order* to exists an integral bord V bording N_0 and N_1 it is necessary that the mapping $\phi_{(N,h)}$ should be zero.

2) Furthermore, in order the topology do not change passing from N_0 to N_1 it is necessary that all the quantumcharacteristic classes $i_V^*\omega$ should be zero, $\forall\,\omega\in H^i(\hat{I}_{k+r}(W);A), i\geq 1$.

Proof. 1) In fact, $\phi_{(N,h)}(\omega),\forall\,\omega\in H^{m-1}(\hat{I}_{k+r}(W);A)$, is the characteristic quantum number as defined in the proof of Theorem 2.9:

$$\phi_{(N,h)}(\omega) = a^i < \beta_i, Z_{N_C} >$$
$$= a^i\Big[\int_{Z_{N_{0C}}}\beta_i - \int_{Z_{N_{1C}}}\beta_i\Big]$$
$$= a^i\Big[\int_{V_C}d\beta_i\Big] = 0,$$

as $d\beta_i = 0$. So we can conclude that $\phi_{(N,h)}$ is a zero mapping on the A-module $H^{m-1}(\hat{I}_{k+r}(W);A)$.

2) In fact, the topology does not change passing from N_0 to N_1 iff the integral quantum manifold V that bords N_0 to N_1 is a cylinder. But this requires that all the quantumcharacteristic classes $i_V^*\omega$ are zero, in dimension $i\geq 1$. $\qquad\square$

Now, in order to study the structure of global solutions it is necessary to consider the integral bordism groups of QPDEs. In refs.[63,67,69] we extended to QPDEs our previous results on the determination of integral bordism groups of PDEs [63–65]. Let us recall some definitions and results. (For their proofs see ref.[67].)

A smooth integral (classicregular) quantum manifold $V\subset \hat{J}^k_m(W)$ will be called *admissible* if the following properties hold. **(A)** The set $\Sigma(V)\subset V$ of singular points of V has no open subsets and no *frozen singularities*, i.e., if V is a compact closed integral classicregular quantum manifold of dimension p, $0\leq p\leq m-1$, over a quantum algebra A, for V passes at least one smooth quantum integral manifold Y, of dimension $(p+1)$, over the quantum algebra A, such that $\Sigma(Y)$ has no open subsets. Furthermore, we assume that the set of singular points $\Sigma(V)$ can be solved by means of integral deformations. Moreover, we say that V is *integral admissible*, with respect to a QPDE $\hat{E}_k\subset\hat{J}^k_m(W)$, if V is admissible in above sense, is contained into $\hat{E}_k$, and the $(p+1)$-dimensional integral quantum manifold Y above considered is contained into $\hat{E}_k$ too. (In the following we shall consider only regular integral admissible quantum manifolds, i.e., integral admissible quantum manifolds, $N \subset \hat{E}_k \subset \hat{J}^k_m(W)$, diffeomorphic to their projections into W, by means of the mapping $\pi_{k,0} : \hat{E}_k \to W$.) Let $\hat{E}_k \subset \hat{J}^k_m(W)$ be a QPDE. Let $N_1, N_2 \subset \hat{E}_k$ be two integral admissible smooth quantum manifolds of dimension p, $0 \leq p \leq m - 1$. Then, we say that N_1 and N_2 are *integral bordant*, $N_1 \sim_{\hat{E}_k} N_2$, with respect to $\hat{E}_k$, if there exists a quantum integral manifold $V \subset \hat{E}_k$, such that V is a piecewise admissible integral quantum manifold and $\partial V = N_1 \bigcup N_2$. Note that V is globally smooth if on V there exists an integrable nonsingular vector field, transverse to $N_1 \bigcup N_2$, (e.g., interior normal on N_1 and exterior normal on N_2). Furthermore, if we go to the infinity quantum prolongation $\hat{E}_\infty$ of E_k, i.e., $V \subset \hat{E}_\infty$, then V is diffeomorphic to a global smooth quantum submanifold of W. Furthermore, we say that N_1 and N_2 are *quantum integral bordant* if above conditions are weakned by requiring that $V \subset \hat{J}^k_m(W)$. The integral bordism, (resp. quantum integral bordism), is an equivalence relation and we denote by $\Omega_p^{\hat{E}_k}$, (resp. $\Omega_p(\hat{E}_k)$), the corresponding set of equivalence classes $[N]_{\hat{E}_k}$, (resp. $[N]_{\overline{\hat{E}_k}}$). The operation of taking disjoint union $\bigcup$ defines a sum $+$ on $\Omega_p^{\hat{E}_k}$, (resp. $\Omega_p(\hat{E}_k)$), such that $\Omega_p^{\hat{E}_k}$, (resp. $\Omega_p(\hat{E}_k)$), becomes an abelian group. The class $[\emptyset]_{\hat{E}_k}$, (resp. $[\emptyset]_{\overline{\hat{E}_k}}$), defines the zero element. Furthermore, for $k = 0$ we put $\Omega_p^{\hat{E}_0} = \Omega_p(\hat{E}_0) = {}^A\underline{\Omega}_{p,s}(W)$. We call $\Omega_p^{\hat{E}_k}$, (resp. $\Omega_p(\hat{E}_k)$), $p \in \{0, 1, \dots, m - 1\}$, the *integral p-bordism group*, (resp. *quantum p-bordism group*), of $\hat{E}_k$. The relation between integral bordism groups, quantum bordism groups and singular bordism groups is given by means of the following commutative diagram:

$$
\begin{array}{ccccccc}
 & & & & 0 & & \\
 & & & & \downarrow & & \\
 & & & & K_p(\hat{E}_k) & & \\
 & & & & \downarrow & & \\
0 \to & K_p^{\hat{E}_k} & \to & \Omega_p^{\hat{E}_k} & \xrightarrow{\hat{i}_p} \Omega_p(\hat{E}_k) & \to & 0 \\
 & & & 0 \to & \downarrow \quad \searrow \hat{j}_p & & \downarrow \\
 & & & & \widehat{\Omega}_p(\hat{E}_k) \to & {}^A\underline{\Omega}_{p,s}(W) & \\
 & & & & \downarrow & & \downarrow \\
 & & & & 0 & A \bigotimes_{\mathbf{K}} H_p(W;\mathbf{K}) & \\
 & & & & & \downarrow & \\
 & & & & & 0 &
\end{array}
$$

where

$$K_p(\hat{E}_k) \equiv \left\{ [N]_{\overline{\hat{E}_k}} \in \Omega_p(\hat{E}_k) \,\middle|\, \begin{array}{l} \exists V \subset W,\ (p+1)-\text{quantum manifold over } A, \\[2mm] \partial V = \pi_{k,0}(N) \cong N \end{array} \right\}$$

$$K_p^{\hat{E}_k} \equiv \left\{ [N]_{\hat{E}_k} \in \Omega_p^{\hat{E}_k} \,\middle|\, \begin{array}{l} \exists V \subset \hat{J}_m^k(W),\ (p+1)-\text{admissible integral quantum manifold over } A, \\[2mm] \partial V = N \end{array} \right\}.$$

Therefore, we have the following.

Theorem 2.11. 1) *The quantum p-bordism groups $\Omega_p(\hat{E}_k)$ is an extension of a subgroup of $^A\underline{\Omega}_{p,s}(W) \cong A \bigotimes_{\mathbf{K}} H_p(W;\mathbf{K})$, and the integral p-bordism group $\Omega_p^{\hat{E}_k}$ is an extension of the quantum p-bordism group.*

2) Let us assume that admissible integral quantum manifolds bording admissible integral quantum manifolds of dimension $0 \le p \le m-1$, contained into $\hat{E}_k$, are smooth regular, for a k-order QPDE $\hat{E}_k \subset \hat{J}_m^k(W)$. Then one has the following: (i) If $\hat{E}_k \subset \hat{J}_m^k(W)$ is a formally quantum integrable QPDE, then one has the following isomorphisms: $\Omega_p^{\hat{E}_k} \cong \Omega_p(\hat{E}_k) \cong \Omega_p^{\hat{E}_{k+h}} \cong \Omega_p^{\hat{E}_\infty} \cong \Omega_p(\hat{E}_{k+h}) \cong \Omega_p(\hat{E}_\infty),\ p \in \{0,\dots,m-1\}$. (ii) Furthermore, if W is p-connected, $p \in \{0,\dots,m-1\}$, one has the following isomorphisms: $\Omega_p(\hat{E}_k) \cong {}^A\underline{\Omega}_{p,s}(W),\ p \in \{0,\dots,m-1\}$.

Above considerations can be generalized to include more sophisticated quantum solutions of a quantum PDE $\hat{E}_k$. More precisely one has the following.

Theorem 2.12. *Let $\hat{E}_k \subset \hat{J}_m^k(W)$ be a QPDE and let B be a quantum algebra. Let us consider the following chain complex (bar quantum chain complex of $\hat{E}_k$): $\{\bar{C}_\bullet(\hat{E}_k;B), \bar{\partial}\}$, where $\bar{C}_p(\hat{E}_k;B)$ is the free two-sided B-module of formal linear combinations with coefficients in B, $\sum \lambda_i c_i$, where c_i is a singular p-chain $f : \triangle^p \to \hat{E}_k$ that extends on a neighborhood $U \subset \mathbf{R}^{p+1}$, such that f on U is differentiable and $Tf(\triangle^p) \subset \hat{\mathbb{E}}_k$. Then, one has the following exact commutative diagram:*

$$
\begin{array}{ccccccccc}
& & 0 & & 0 & & \\
& & \downarrow & & \downarrow & & \\
0 & \to & \bar{B}_\bullet(\hat{E}_k;B) & \to & \bar{Z}_\bullet(\hat{E}_k;B) & \to & \bar{H}_\bullet(\hat{E}_k;B) & \to & 0 \\
& & \downarrow & & \downarrow & & \\
& & \bar{C}_\bullet(\hat{E}_k;B) & = & \bar{C}_\bullet(\hat{E}_k;B) & & \\
& & \downarrow & & \downarrow & & \\
0 & \to & {}^B\Omega_{\bullet,s}^{\hat{E}_k} & \to & \bar{B}or_\bullet(\hat{E}_k;B) & \to & \bar{C}yc_\bullet(\hat{E}_k;B) & \to & 0 \\
& & \downarrow & & \downarrow & & \\
& & 0 & & 0 & &
\end{array}
$$

where the meaning of the symbols and their names are directly understood by the similar commutative for quantum manifolds given in Theorem 1.30. In particular, we call (closed) integral singular p-bordism groups, with coefficients in B, of $\hat{E}_k$ the groups $^B\Omega^{\hat{E}_k}_{p,s}$, $p \in \{0, 1, \ldots, m-1\}$. (If $B = \mathbf{K}$ we omit the apex B). In the following table we report some relations between objects considered in above commutative diagram:

TAB.2.1 - Relations between (closed) integral singular bordism groups of $\hat{E}_k$.

$$^B\Omega^{\hat{E}_k}_{\bullet,s} \cong \bar{H}_\bullet(\hat{E}_k; B) \cong \Omega^{\hat{E}_k}_{\bullet,s} \bigotimes_{\mathbf{K}} B$$

$$\bar{B}or_\bullet(\hat{E}_k; B) \cong {}^B\Omega^{\hat{E}_k}_{\bullet,s} \bigoplus \bar{C}yc_\bullet(\hat{E}_k; B)$$

$$\Omega^{\hat{E}_k}_{\bullet,s} \equiv {}^{\mathbf{K}}\Omega^{\hat{E}_k}_{\bullet,s} \cong \bar{H}_\bullet(\hat{E}_k; \mathbf{K})$$

Proof. The proof follows directly from standard techniques of homological algebra and algebraic topology. □

Definition 2.13. We call *quantum singular p-bordism groups*, with coefficients in B, of $\hat{E}_k$ the relative closed integral singular bordism groups of $\hat{J}^k_m(W)$ with respect to $\hat{E}_k$, and we will denote them by $^B\Omega_{p,s}(\hat{E}_k)$, $p \in \{0, 1, \ldots, m-1\}$.[29]

Theorem 2.13. *By means of the cochain complex $\{\bar{C}^\bullet(\hat{E}_k; B), \bar{\delta}\}$, corresponding to $\{\bar{C}_\bullet(\hat{E}_k; B), \bar{\partial}\}$, we obtain objects dual of ones considered in Theorem 2.12). In particular, one has the (canonical) isomorphisms reported in the following table.*

TAB.2.2 - Relations between (closed) integral singular cobordism groups of $\hat{E}_k$.

$$(^B\Omega^{\hat{E}_k}_s)^\bullet \cong \bar{H}^\bullet(\hat{E}_k; B) \cong (\Omega^{\hat{E}_k}_s)^\bullet \bigotimes_{\mathbf{K}} B \cong Hom_B({}^B\Omega^{\hat{E}_k}_{\bullet,s}; B)$$

$$\bar{B}or^\bullet(\hat{E}_k; B) \cong (^B\Omega^{\hat{E}_k}_s)^\bullet \bigoplus \bar{C}yc^\bullet(\hat{E}_k; B)$$

Proof. The proof follows directly from above results. □

Definition 2.14. Then we call (*closed*) *integral singular p-cobordism groups*, with coefficients in B, of $\hat{E}_k$:

$$^B\Omega^{p,(\hat{E}_k)}_s \equiv Hom_B({}^B\Omega_{p,s}\hat{E}_k; B), \ p \in \{0, 1, \ldots, m-1\}.$$

In the following table we define some characteristic numbers that naturally extend to quantum PDEs similar definitions made for PDEs [61,63]. (See

---

[29] Do not confuse $^B\Omega_{p,s}(\hat{E}_k)$ with $^B\underline{\Omega}_{p,s}(\hat{E}_k)$!

also Definition 3.20 and ref.[67].)

TAB.2.3 - Quantum characteristic numbers for a quantum PDE $\hat{E}_k$.

Name	Definition
Quantum Stiefel-Whitney char.numbers	$\hat{w}[N]=b\otimes w[N_C]\in B$ (†)
Quantum Euler char.numbers	$\hat{\chi}[N]=b\otimes\chi[N_C]\in B$ (‡)
Quantum sing.integral char.numbers	$\hat{i}[N]\in B$
Quantum sing.quantum char.numbers	$\hat{q}[N]\in B$

(†) $w[N_C]$=Stiefel-Whitney char.numbers of N_C;(‡) $\chi[N_C]$=Euler char.number of N_C.

Then one has the following properties.

Theorem 2.14. 1) *The equivalence classes* $[N]^s_{\hat{E}_k}$ *in the groups* $^B\Omega^{\hat{E}_k}_{\bullet,s}$ *are characterized by means of the quantum characteristic numbers defined in above table. More precisely one has:*

$$N' \in [N]^s_{\hat{E}_k} \Leftrightarrow \left\{\hat{i}[N'] = \hat{i}[N]\right\} \Rightarrow \{\hat{w}[N'] = \hat{w}[N], \hat{\chi}[N'] = \hat{\chi}[N]\}.$$

2) *The equivalence classes* $[N]^s_{\overline{\hat{E}_k}}$ *in the groups* $^B\Omega_{\bullet,s}(\hat{E}_k)$ *are characterized by means of the quantum characteristic numbers defined in Tab.2.3. More precisely one has:*

$$N' \in [N]^s_{\overline{\hat{E}_k}} \Leftrightarrow \{\hat{q}[N'] = \hat{q}[N]\} \Rightarrow \{\hat{w}[N'] = \hat{w}[N], \hat{\chi}[N'] = \hat{\chi}[N]\}.$$

3) *One has the unnatural splits reported in Tab.2.4.*

TAB.2.4 - Splits of singular bordism groups of QPDE $\hat{E}_k$.

$\Omega_{p,s}(\hat{E}_k)\cong\Omega^{\jmath^k_m(W)}_{p,s}\times\overline{\Omega^{\hat{E}_k}_{p-1,s}}$	$\Omega^{\jmath^k_m(W)}_{p,s}\cong\Omega^{\hat{E}_k}_{p,s}\times\overline{\Omega_{p,s}(\hat{E}_k)}$
$\Omega^{\jmath^k_m(W)}_{p,s}\equiv\text{im }(b_p)\cong\ker(c_p)$	$\overline{\Omega^{\hat{E}_k}_{p-1,s}}\equiv\text{im }(c_p)\cong\text{coim }(c_p)\cong\text{coker }(b_p)$
$\Omega^{\hat{E}_k}_{p,s}\equiv\text{im }(a_p)\cong\ker(b_p)$	$\overline{\Omega_{p,s}(\hat{E}_k)}\equiv\text{im }(b_p)\cong\text{coim }(b_p)\cong\text{coker }(a_p)$

by means of the following exact sequences of vector spaces:

$$\Omega^{\hat{E}_k}_{m-1,s} \xrightarrow{a_{m-1}} \Omega^{\jmath^k_m(W)}_{m-1,s} \xrightarrow{b_{m-1}} \Omega_{m-1,s}(\hat{E}_k) \xrightarrow{c_{m-1}} \Omega^{\hat{E}_k}_{m-2,s} \xrightarrow{a_{m-2}}$$

$$\cdots \xrightarrow{a_0} \Omega^{\hat{E}_k}_{0,s} \xrightarrow{b_0} \Omega^{\jmath^k_m(W)}_{0,s} \xrightarrow{c_0} \Omega_{0,s}(\hat{E}_k) \to 0.$$

Proof. The proofs follow directly by means of standard techniques of homological algebra and algebraic topology. $\qquad\square$

Theorem 2.15. *(Topology transitions as quantum tunnelling effects). Let $\hat{E}_k \subset \hat{J}^k_m(W)$ be a k-order QPDE on the fiber bundle $\pi : W \to M$, $\dim M = m$. Let Y_0 and Y_1 be Cauchy hypersurfaces data of $\hat{E}_k$. Let us assume that there are closed domains $N_i \subset Y_i, i = 0, 1$, such that there exists a (singular) bordism $L \in \Omega(\hat{E}_k)$ between N_0 and $N_1 : \partial L = N_0 \times 0 \bigcup N_1 \times 1$. Then there exists an admissible Morse function $f : L_C \to [a, b]$, where L_C is the classic limit of L, such that $(N_0)_C = f^{-1}(a), (N_1)_C = f^{-1}(b)$, such that:*

(A)(Simple quantum tunnelling effect). If f has a critical point q of index k, then there exists a k-cell $e^k \subset L_C - (N_1)_C$ and a $(m-k)$-cell $e_*^{(m-k)} \subset L_C - (N_0)_C$ such that: (i) $e^k \cap (N_0)_C = \partial e^k$; (ii) $e_*^{(m-k)} \cap (N_1)_C = \partial e_*^{(m-k)}$; (iii) There is a deformation retraction of L_C onto $(N_0)_C \cup e^k$; (iv) There is a deformation retraction of L_C onto $(N_1)_C \cup e_*^{(m-k)}$; (v) $e_*^{(m-k)} \cap e^k = q$; $e_*^{(m-k)} \pitchfork e^k$. (Here $\pitchfork$ denotes transverse.)

(B)(Multi quantum tunnelling effect). If f is of type $(\nu_0, ..., \nu_m)$, such that f has just one critical value c, $a < c < b$, then there are disjoint k-cells $e^k_i \subset L_C - (N_1)_C$, and disjoint $(m-k)$-cells $(e_*)^{(m-k)}_i \subset L_C - (N_0)_C$, $1 \le i \le \nu_k, k = 0, ..., m$, such that: (i) $e^k_i \cap (N_0)_C = \partial e^k_i$; (ii) $(e_*)^{(m-k)}_i \cap (N_1)_C = \partial(e_*)^{(m-k)}_i$; (iii) There is a deformation retraction of L_C onto $(N_0)_C \cup \{\bigcup_{i,k} e^k_i\}$; (iv) There is a deformation retraction of L_C onto $(N_1)_C \cup \{\bigcup_{i,k}(e_*)^{(m-k)}_i\}$; (v) $(e_*)^{(m-k)}_i \cap e^k_i = q_i \in L_C$; $(e_*)^{(m-k)}_i \pitchfork e^k_i$. In the cases (A) and (B), L changes topology by passing from N_0 to N_1.

(C)(No topology transition). If f has no critical point then $L_C \cong (N_0)_C \times I$, where $I \equiv [0, 1]$. In this case L does not change topology by passing from N_0 to N_1.

Proof. The proof can be conduced on L_C by using Theorem 23 in ref.[61]. Then taking into account that $c : L \to L_C$ is a vector fiber bundle the conclusions on L get out from the fact that $H^\bullet(L; A) \cong A \bigotimes_K H^\bullet(L_C)$. $\square$

The following theorem gives a costructive method to obtain integral bords, for fixed boundary $X = X_1 \bigcup X_2$, also for non-linear QPDE's.

Theorem 2.16. *(Relation between integral bordism and quantumpropagators). Let us assume that $\hat{E}_k \equiv \ker_f K \subset J\hat{D}^k(W)$ is a QPDE, where K is a k-order differential operator in the category of quantum manifolds. Then, the quantumpropagator $\hat{G}[s|i]$ identifies a classicregular integral bord $L \subset \Omega(\hat{E}_k)$, such that if $\hat{G}[s|i]$ satisfies the boundary conditions $X_1 \bigcup X_2$,* then $\partial L = X_1 \bigcup X_2$.

Proof. The proof can be conduced directly translating in the present quantum context one given for commutative manifolds [61] or supermanifolds [62]. (See also Section 3.5. In ref.[67] we have given also a constructive geometric method to build distribuive Green kernels for integral manifolds bording initial and final Cauchy manifolds in nonlinear QPDE's.) $\quad\square$

Remark 2.3. The spaces of conservation laws of QPDEs, identified with a spectral terme $\hat{E}_1^{0,q}$, $0 \leq q \leq m - 1$, (similarly to the classic case), can be related to the integral bordism groups by means of suitable noncommutative Hopf algebras. This has been made in refs.[67]. $\quad\blacksquare$

2.2 - THE QUANTUM NAVIER-STOKES EQUATION

Here we will apply above geometrical picture to the quantum Navier-Stokes equation. This is the analogous of the usual Navier-Stokes equation in this noncommutative context. Let us introduce some preliminary geometric objects in order to clear the framework where the quantum Navier-Stokes equation is implemented.

Definition 2.15. The *quantum Galilean space-time*, with respect to a quantum algebra A, is an affine space $(M, \mathbf{M}, \alpha)$, where the space of free vectors is given by $\mathbf{M} \equiv A \otimes_{\mathbf{R}} \mathbf{M}_C$, where $\mathbf{M}_C$ is the space of free vectors of the Galilean space-time $(M_C, \mathbf{M}_C, \alpha_C)$ [56]. (In other words M is obtained from M_C by extending the ring of scalars from $\mathbf{R}$ to A.)

Then the quantum Galilean space-time M is a quantum manifold of dimension 4 and class Q_w^ω. It is easy to see that the two-sided A-module $\mathbf{M} \equiv A \otimes_{\mathbf{R}} \mathbf{M}_C$ is of dimension 4. In fact, for any basis $\{e_\alpha\}_{0 \leq \alpha \leq 3}$ of $\mathbf{M}_C$ and point $\hat{O} \in M$ one has a *quantum affine frame* $\left(\hat{O}, \{\hat{e}_\alpha \equiv 1 \otimes e_\alpha\}_{0 \leq \alpha \leq 3}\right)$ on M. This induces a global chart $\phi : M \cong A^4$, given by $\phi^\alpha(p) = (p - \hat{O})^\alpha \equiv x^\alpha(p)$, where $p - \hat{O} = \sum_{0 \leq \alpha \leq 3}(p - \hat{O})^\alpha \hat{e}_\alpha$. Furthermore, a change of chart is given by the following affine transformation $\bar{x}^\alpha = A_\beta^\alpha x^\beta + y^\alpha$, where $(A_\beta^\alpha) \in GL(4; A)$. Note, also, that M has not a canonical structure of classicregular quantum manifold, even if one has global embeddings $i : M_C \to M$. In fact, if M is pointed at $\hat{O} \in M$ and M_C at $O \in M_C$, then one has an epimorphism $c_{\hat{O},O} : M \to M_C$, given by $c_{\hat{O},O}(p) \equiv p_C$, where $p_C - O = \underline{c}(p - \hat{O})$, with $\underline{c}$ the canonical epimorphism $\underline{c} : A \otimes_{\mathbf{R}} \mathbf{M}_C \to \mathbf{M}_C$ given by composition $A \otimes_{\mathbf{R}} \mathbf{M}_C \to \mathbf{R} \otimes_{\mathbf{R}} \mathbf{M}_C \cong \mathbf{M}_C$. Of course the epimorphism $c_{\hat{O},O}$ is not canonical as it depends on the points $\hat{O}$ and O. Finally, it is easy to see that the epimorphisms $c_{\hat{O},O} : M \to M_C$ have global sections $i : M_C \to M$.

Proposition 2.5. *On the quantum Galilean space-time we canonically recognize a quantum vertical metric.*

Proof. In fact on each tangent space $T_p M \cong \mathbf{M} = A \otimes_{\mathbf{R}} \mathbf{M}_C$ one recognizes a *vertical subspace* $v T_p M \cong A \otimes_{\mathbf{R}} \mathbf{S}$. Here $\mathbf{S}$ is the canonical 3-dimensional subaspace of $\mathbf{M}_C$ of space-like free vectors. (For more details see ref.[56].) Then the quantum vertical metric $\widehat{g}$ on M is given by a mapping $\widehat{g} : M \to Hom_Z(v\dot{T}_0^2 M; A)$ such that $\widehat{g}(p)(\sum_K a^K \otimes v_K, \sum_J b^J \otimes u_J) = \widehat{\mu}(a^K, b^J)\bar{g}(v_K, u_J) \in A$, where $\bar{g}$ is the metric on $\mathbf{M}_C$ and $\widehat{\mu}$ is the product in A. By using the canonical split $v\dot{T}_0^2 M \cong v\dot{S}_0^2 M \bigoplus v\dot{\Lambda}_0^2 M$, we can also split $\widehat{g}$ in its symmetric part $\widehat{g}_{(s)}$ and skewsymmetric part $\widehat{g}_{(a)}$, i.e. $\widehat{g} = \widehat{g}_{(s)} + \widehat{g}_{(a)}$. In the following, for abuse of notation we shall denote $g_{(s)}$ yet by g. $\qquad\square$

Remark 2.4. Any embedding $i : M_C \to M$ of the Galilean space-time into the quantum Galilean space-time M, can be interpreted as a *non-relativistic quantum frame*. The representation of any geometric object $s : M \to W$ defined on M, by means of a non-relativistic quantum frame, is the pull-back $i^* s$ of s by means of i. For any global embedding $i : M_C \to M$ one has $i^* \widehat{g} : M_C \to A \otimes_{\mathbf{R}} v S_2^0 M_C$. Furthermore we get $\bar{g} = (c \otimes 1) \circ i^* \widehat{g} : M_C \to v S_2^0 M_C$, where $c \otimes 1 : A \otimes_{\mathbf{R}} v S_2^0 M_C \to \mathbf{R} \otimes_{\mathbf{R}} v S_2^0 M_C \cong v S_2^0 M_C$ is the canonical projection. Moreover, we can also admit that on M_C a classical frame is also defined, that is a flow ϕ with time-like velocity, in the sense of the Galilean space-time. $\qquad\blacksquare$

Definition 2.16. The *quantum Galilean group*, $\hat{\mathbb{G}}$, with respect to a quantum algebra A, is the symmetry group of the quantum Galilean space-time, (with respect to A).

Proposition 2.6. $\hat{\mathbb{G}}$ *has the following structure:* $\hat{\mathbb{G}} \cong A^4 \times L\mathbb{G}$, *where* $L\mathbb{G}$ *is the linear Galilean group* [56]. *The corresponding quantum Lie algebra is* $\mathfrak{g} = A^4 \oplus \mathfrak{lg}(4)$, *where* $\mathfrak{lg}(4)$ *is the Lie algebra of the linear Galilean group.*

Definition 2.17. A *quantum motion* is a map $m : A \to M$, such that its velocity $\dot{m}(a)(b) \in T_p M$ is time-like, i.e., for any $a, b \in A$, $\dot{m}(a)(b) = c^I \otimes v_I$, with $c^I \in A$ and v_I time-like vectors of the Galilean space-time M_C. (Hence $\dot{m}(a) \notin v\widehat{T}_p M$, $p = m(a) \in M$.) Let us denote by $J\hat{D}(M) \subset \widehat{TM}$ the space of the velocities of quantum motions. We call *adapted quantum coordinates* to M ones $(x^\alpha) = (x^0, x^k)_{1 \le k \le 3}$, such that the 0-lines are quantum motions.

Proposition 2.7. 1) *One has a canonical quantum 1-form* $\widehat{\sigma} : M \to (TM)^+$, *such that* $\widehat{\sigma}(p) : T_pM \to A$ *is the* Z-*homomorphism given by composition:* $T_pM \cong \mathbf{M} \equiv A \otimes_{\mathbf{R}} \mathbf{M}_C \xrightarrow{1 \otimes \underline{\tau}} A \otimes_{\mathbf{R}} \mathbf{R} \cong A$, *where* $\underline{\tau}$ *is the derivative of the affine universal time function* $\tau : M_C \to \mathbf{R}$. *(An origin of times has been fixed.)*

2) *With adapted coordinates we have the following representation of the velocity* $\dot{m}$ *of a quantum motion* $m : A \to M$: $\dot{m} = \partial x_0 \circ m + \partial x_k \circ m \dot{m}^k$.

Proof. 1) The proof directly follows from the definition.

2) An adapted quantum coordinate system is one where $dx^0 = \widehat{\sigma}$. Then, the velocity $\dot{m} : A \to \widehat{TM}$ of a quantum motion $m : A \to M$ is characterized by the following property $< \dot{m}(a), \widehat{\sigma}(m(a)) > = 1_A$. Therefore, taking into account that $\dot{m}(a) = \partial x_\alpha(m(a))(\dot{m}^\alpha(a))$, $\forall a \in A$, we get $< (\widehat{\sigma} = dx^0)(m(a)), \dot{m}(a) > = m^0(a) = a$, $\forall a \in A$. This means that $m^0 = 1_A \in \widehat{A}$. Note also that on the Galilean quantum space-time does not exist an unversal quantum time $M \to A$. In other words the canonical 1-form $\widehat{\sigma} : M \to (TM)^+$ does not admit a canonical potential, i.e. quantum 0-form $\widehat{\tau} : M \to A$, such that $\widehat{\sigma} = d\widehat{\tau}$. $\qquad\qquad\square$

Definition 2.18. A *full quantum non-relativistic frame* is a couple $\Phi \equiv (\phi, \widehat{\tau}_\phi)$, where: (i) $\phi : A \times M \to M$ is a flow into M such that its velocity $\dot{\phi} : M \to \widehat{TM} \cong M \times \widehat{\mathbf{M}}$, is time-like; (ii) $\widehat{\tau}_\phi : M \to A$ is an affine map such that $d\widehat{\tau}_\phi = \widehat{\sigma}$, *(quantum proper time)*. We say that ϕ is *inertial* if $\dot{\phi} = const.$ and that it is *rigid* if it identifies a 1-parameter subgroup of the quantum Galilean group $\widehat{G}$. A full quantum (non-relativistic) frame identifies a split of M: $j_\phi : M \cong S_\phi \times A$, where S_ϕ is the set of flow lines of ϕ, *(proper space)*. More precisely one has $j_\phi(p) = (\phi_p, \widehat{\tau}_\phi(p))$. If ϕ is inertial then S_ϕ is an affine space with space of free vectors $\mathbf{S}_\phi \cong A \otimes_{\mathbf{R}} \mathbf{S}$. We say that a system of quantum coordinates $\{x^\alpha\}$ on M is adapted to ϕ if it is dapted to M and $\partial x_0 = \dot{\phi}$. If $i : M_C \to M$ is a non-relativistic quantum frame with a classic frame $\phi_C : \mathbf{R} \times M_C \to M_C$ defined on M_C, then we say that a full quantum non-relativistic frame $\Phi \equiv (\phi, \widehat{\tau}_\phi)$ is ϕ_C-*compatible* if the flow induced on $i(M_C)$ by ϕ, restricted to $\mathbf{R}$, coincides with the flow induced by ϕ_C.

Proposition 2.8. 1) *On the quantum Galilean space-time,* M, *one has a canonical connection, induced by the affine structure of* M, *that is a* Z-*homomorphism* $G : TTM \to vTTM$, *over* TM. *This induces the canonical connections* $\widehat{G} : T\widehat{TM} \to v T\widehat{TM}$ *and* $\widehat{G} : \widehat{TTM} \to v\widehat{TTM}$, *over* $\widehat{TM}$, *such* *that the following diagram is commutative:*

$$
\begin{array}{ccccc}
\widetilde{TTM} & \equiv & Hom_Z(A;\widetilde{TTM}) & \cong & A^4\times(\widehat{A})^4\times(\widehat{A})^4\times(\overset{2}{\widehat{A}})^4 \\[4pt]
\widehat{G}\downarrow & & \downarrow Hom_Z(1_A;\widehat{G}) & & \downarrow \\[4pt]
\widetilde{vTTM} & \equiv & Hom_Z(A;\widetilde{vTTM}) & \cong & A^4\times(\widehat{A})^4\times\{0\}\times(\overset{2}{\widehat{A}})^4
\end{array}
$$

2) If $\zeta : M \to \widehat{TM}$ is a quantum vector field on M with the following representation in quantum adapted coordinates $\zeta = \partial x_\alpha \zeta^\alpha$, where $\zeta^\alpha : U \subset M \to \widehat{A}$ are the local components of ζ, then the absolute differential $\nabla\zeta$, induced by the canonical connection $\widehat{G}$, is given by

$$\nabla\zeta = Hom_Z(1_{TM};\widehat{G})\circ D\zeta : M \to Hom_Z(TM;\zeta^*v\widehat{TTM}) \cong Hom_Z(TM;\widehat{TM}).$$

Its representation in coordinates is given by: $\nabla\zeta = \partial x_\gamma \otimes_{\widehat{A}} [\widehat{G}^\gamma_{\alpha\omega}\zeta^\omega + (\partial x_\alpha.\zeta^\gamma)]dx^\alpha$, where $\widehat{G}^\gamma_{\alpha\omega} : M \to \overset{2}{\widehat{A}}$ are the connection coefficients of the quantum Galilean connection $\widehat{G}$.

Proof. Emphasize that we have used the isomorphisms

$$Hom_Z(T_pM;T_{\zeta(p)}\widehat{TM}) \cong T_{\zeta(p)}\widehat{TM} \otimes_{\widehat{A}} (T_pM)^+ \cong (\widehat{A})^{16} \times (\overset{2}{\widehat{A}})^{16}.$$

Hence we can write

$$D\zeta(p) = \partial x_\alpha(\zeta(p)) \otimes_{\widehat{A}} \delta^\alpha_\beta dx^\beta + \partial\dot{x}_\alpha(\zeta(p)) \otimes_{\widehat{A}} (\partial x_\beta.\zeta^\alpha)(p)dx^\beta(p),$$

where $\delta^\alpha_\beta \in \widehat{A}$ and $(\partial x_\alpha.\zeta^\gamma)(p) \in \overset{2}{\widehat{A}} \cong Hom_Z(A;\widehat{A})$. Therefore, we get:

$$
\begin{aligned}
(\widehat{G} \otimes_{\widehat{A}} 1) \circ D\zeta(p) = &\, \partial\dot{x}_\gamma(\zeta(p))\widehat{G}^\gamma(\partial x_\alpha(\zeta(p))) \otimes_{\widehat{A}} \delta^\alpha_\beta dx^\beta \\
&+ \partial\dot{x}_\gamma(\zeta(p)) \otimes_{\widehat{A}} (\partial x_\beta.\zeta^\alpha)(p)dx^\beta(p).
\end{aligned}
$$

Then by putting $\widehat{G}^\gamma(\partial x_\beta(\zeta(p))) = \widehat{G}^\gamma_\beta \circ (\partial x_\omega\zeta^\omega)(p) \equiv \widehat{G}^\gamma_{\beta\omega}\zeta^\omega$, we get above formula. Let us denote by $\overset{\vee}{\widehat{G}}$ the quantum connection induced by $\widehat{G}$ on the dual bundle $(TM)^+$ of $\widehat{TM}$. Then, for the connection coefficients $\overset{\vee}{\widehat{G}}{}^\gamma_{\alpha\beta}$ one has: $\overset{\vee}{\widehat{G}}{}^\gamma_{\alpha\beta} = -\widehat{G}^\gamma_{\alpha\beta}$. Furthermore, we have also $\widehat{G}^0_{\alpha\beta} = 0$, (as $\nabla\widehat{\sigma} = 0$). Finally, if the quantum coordinates are adapted to an inertial quantum frame, then $\widehat{G}^k_{00} = \widehat{G}^k_{i0} = \widehat{G}^k_{0i} = 0$. $\square$

Remark 2.5. The quantum canonical vertical metric $\widehat{g}$ has, in adapted quantum coordinates, the following representation: $\widehat{g} = \widehat{g}_{ij} dx^i \otimes dx^j$, where $\widehat{g}_{ij} : M \to \overset{2}{\widehat{A}}$. In fact one has the following: $\widehat{g}(p) \in Hom_Z(v\dot{S}_0^2(T_pM); A) \cong (\overset{2}{\widehat{A}})^9$. Of course, in general, $\widehat{g}_{ij} \neq \widehat{g}_{ji}$. On the other hand, for any embedding $i : M_C \to M$, one has $i^*\widehat{g} = g_{ij} dx^i \otimes dx^j : M \to A \otimes_{\mathbf{R}} vS_2^0(M_C)$, where $g_{ij} : U_C \subset M_C \to A$ are local symmetric functions in the indeces $i, , j$. Note that $(\widehat{g}_{ij})$ is a 3×3 invertible matrix with entries into $\overset{2}{\widehat{A}}$. In fact, if we take an adapted frame $(1 \otimes \theta^k)_{1 \leq k \leq 3}$ of $A \otimes_{\mathbf{R}} S$ we can represent $\widehat{g}$ in the 3×3 invertible matrix $(\widehat{g}_{ij}) = (g_{ij})$. So we can identify the corresponding controvariant quantum metric $\widehat{g} : M \to v\widehat{S_0^2M} \equiv Hom_Z(A; vS_0^2M)$ that satisfies the condition $\widehat{g} \circ \widehat{g} = 1_A$, i.e., for any $p \in M$, one has the following commutative diagram:

$$
\begin{array}{ccc}
vS_0^2(T_pM) & = & vS_0^2(T_pM) \\[4pt]
\uparrow {\scriptstyle \widehat{g}(p)} & & \downarrow {\scriptstyle \underline{g}(p)} \\[4pt]
A & \xrightarrow{\ \ 1_A\ \ } & A
\end{array}
$$

Therefore, one has $\widehat{g}^{ij}\widehat{g}_{js} = \delta_s^i$, where $\widehat{g}^{ij} : M \to \widehat{\dot{T}_0^2(A)} \equiv Hom_Z(A; \dot{T}_0^2(A))$ are the components of $\widehat{g}$ in the adapted quantum coordinates: $\widehat{g} = \partial x_i \otimes \partial x_j \widehat{g}^{ij}$. ∎

Proposition 2.9. *Similarly to the classic case, from the relation $\nabla \widehat{g} = 0$ we get that the space-like connection coefficients are related to the metric by a Levi-Civita type relation:*

$$
\hat{G}_{ik}^j = \frac{1}{2}[(\partial x_k . \widehat{g}_{is}) + (\partial x_i . \widehat{g}_{sk}) - (\partial x_i . \widehat{g}_{ks})]\widehat{g}^{sj}.
$$

Note that the products like $(\partial x_k . \widehat{g}_{is})\widehat{g}^{sj}$ are made by composition:

$$
A \xrightarrow{\ \widehat{g}^{sj}\ } A \otimes_Z A \xrightarrow{\ (\partial x_k . \widehat{g}_{is})\ } \widehat{A}.
$$

Here we have used the canonical isomorphism $\overset{2}{\widehat{A}} \cong Hom_Z(A; \widehat{A})$.

Theorem 2.17. *Let us, now, consider the following fiber bundle: $\pi : W \equiv J\hat{D}(M) \times_M (\dot{\Lambda}_0^0 M)^+ \to M$. A section $s = (v, p)$ of this fiber bundle over M represents a field of velocities and isobaric pressure. With respect to*

     Quantized Partial Differential Equations

adapted quantum coordinates, on M we can induce the quantum coordinates as reported in Tab.2.5.

TAB.2.5 - Quantum coordinates and model spaces on quantum manifolds

Quantum Manifolds	Quantum Coordinates	Model Spaces
M	(x^α), $0\leq\alpha\leq3$	A^4
W	$(x^\alpha,\dot{x}^i,p)$, $1\leq i\leq3$	$A^5\times(\widehat{A})^3$
$J\hat{D}(W)$	$(x^\alpha,\dot{x}^i,p,\dot{x}^i_\beta,p_\beta)$, $0\leq\beta\leq3$	$A^5\times(\widehat{A})^7\times(\overset{2}{\widehat{A}})^{12}$
$J\hat{D}^2(W)$	$(x^\alpha,\dot{x}^i,p,\dot{x}^i_\beta,p_\beta,\dot{x}^i_{\alpha\beta},p_{\alpha\beta})$, $0\leq\alpha,\beta\leq3$	$A^5\times(\widehat{A})^7\times(\overset{2}{\widehat{A}})^{28}\times(\overset{3}{\widehat{A}})^{48}$

Then, the *quantum Navier-Stokes equation* (QNS), i.e., the Navier-Stokes equation written in the category of quantum manifolds, is a second order equation, that, in adapted coordinates, is represented in Tab.2.6.

TAB.2.6 - Quantum Navier-Stokes equation (QNS)

(A)	$F^0\equiv\hat{G}^j_{jk}\dot{x}^k+\dot{x}^i_s\delta^s_i=0$
(B)	$F^j\equiv[\dot{x}^j_0+\hat{G}^j_{ik}\dot{x}^i\dot{x}^k+\dot{x}^j_k\dot{x}^k]\rho+p_i\widehat{g}^{ji}$ $-2\{\hat{G}^j_{ir}\widehat{g}^{pr}[\hat{G}^i_{ps}\dot{x}^s+\dot{x}^i_p]+\hat{G}^i_{is}\widehat{g}^{pj}[\hat{G}^s_{pr}\dot{x}^r+\dot{x}^s_p]$ $+(\partial x_i.\widehat{g}^{pj})[\hat{G}^i_{ps}\dot{x}^s+\dot{x}^i_p]+\widehat{g}^{pj}[(\partial x_i.\hat{G}^i_{ps})\dot{x}^s+\hat{G}^i_{ps}\dot{x}^s_i+\dot{x}^i_{ip}]\}\chi+(\partial x_k.f)\widehat{g}^{kj}\rho=0$

$F^0\colon J\hat{D}(W)\to\widehat{A}$, $F^j\colon J\hat{D}^2(W)\to\overset{2}{\widehat{A}}$; $(1\leq i,j,k,p,r,s\leq3)$.

Let us assume that Z is a Noetherian **K**-algebra. Then the quantum Navier-Stokes equation (QNS) is not a formally quantumintegrable QPDE of second order on the fiber bundle $\pi:W\to M$. Into (QNS) we can distinguish an important sub-QPDE: $(\widehat{QNS})\subset J\hat{D}^2(W)\subset\hat{J}^2_4(W)$ that is formally quantumintegrable. The local expression of $(\widehat{QNS})$ is given in Tab.2.7

TAB.2.7 - Formally quantumintegrable sub-Navier-Stokes equation $(\widehat{QNS})$

(A)	$F^0\equiv\hat{G}^j_{jk}\dot{x}^k+\dot{x}^i_s\delta^s_i=0$
(B)	$F^0_\alpha\equiv(\partial x_\alpha.\hat{G}^j_{jk})\dot{x}^k+\hat{G}^j_{jk}\dot{x}^k_\alpha+\dot{x}^i_{s\alpha}\delta^s_i+\hat{G}^j_{jk}\dot{x}^k+\dot{x}^i_s\delta^s_i=0$
(C)	$F^j\equiv[\dot{x}^j_0+\hat{G}^j_{ik}\dot{x}^i\dot{x}^k+\dot{x}^j_k\dot{x}^k]\rho+p_i\widehat{g}^{ji}$ $-2\{\hat{G}^j_{ir}\widehat{g}^{pr}[\hat{G}^i_{ps}\dot{x}^s+\dot{x}^i_p]+\hat{G}^i_{is}\widehat{g}^{pj}[\hat{G}^s_{pr}\dot{x}^r+\dot{x}^s_p]$ $+(\partial x_i.\widehat{g}^{pj})[\hat{G}^i_{ps}\dot{x}^s+\dot{x}^i_p]+\widehat{g}^{pj}[(\partial x_i.\hat{G}^i_{ps})\dot{x}^s+\hat{G}^i_{ps}\dot{x}^s_i+\dot{x}^i_{ip}]\}\chi+(\partial x_k.f)\widehat{g}^{kj}\rho=0$

$F^0\colon J\hat{D}(W)\to\widehat{A}$, F^j, $F^0_\alpha\colon J\hat{D}^2(W)\to\overset{2}{\widehat{A}}$, $(1\leq i,j,k.p,r,s\leq3)$, $(0\leq\alpha\leq3)$.

Proof. We shall consider the following Definition.

Definition 2.19. The *acceleration* $\ddot{m}$ of a quantum motion m is defined by

$$\ddot{m} = \nabla_{\dot{m}}\dot{m} : A \to Hom_Z(A; v\widehat{TM}) \cong Hom_Z(\dot{T}_0^2(A); vTM) \cong A^4 \times (\widehat{A})^3.$$

Therefore, we call $Hom_Z(A; v\widehat{TM})$ the *space of accelerations* of quantum motions.

This means that a quantum body force B takes its values just into the space of accelerations of quantum motions and, in quantum adapted coordinates, it is identified by three functions $\widehat{A}$-valued. As a consequence, we get also that a *quantum stress* is a field of geometric objects $P : M \to Hom_Z(A; v\widehat{T_0^2 M}) \cong A^4 \times (Hom_Z(A; \dot{T}_0^2(A)))^9$. Therefore, with respect to quantum adapted coordinates we get the following representations: $B = \partial x_k B^k$, with $B^k : M \to \widehat{A}$, and $P = \partial x_i \otimes \partial x_j P^{ij}$, with $P^{ij} : M \to Hom_Z(A; \dot{T}_0^2(A))$. Here we have used the following compositions: $A \otimes_Z A \xrightarrow{B^k(p)} A \xrightarrow{\partial x_k(p)} vT_pM$, $A \otimes_Z A \xrightarrow{P^{ij}(p)} \dot{T}_0^2(A) \xrightarrow{\partial x_i(p) \otimes \partial x_j(p)} vT_pM \otimes_Z vT_pM$, for any $p \in M$. We use, now, the following quantum rheological constitutive equation: $P = -pg + \wp$, with $\wp = 2\mathfrak{e}\chi$, where χ is the *quantum viscosity*, here considered as a constant belonging to $\widehat{A}$. $\mathfrak{e}$ is the *full quantum infinitesimal deformation*, given by $\mathfrak{e} = j \circ < v\nabla v_\wedge, \widehat{g} > : M \to Hom_Z(A; v\dot{T}_0^2 M)$, with $v\nabla$ the vertical absolute differential and $v_\wedge = \partial x_k v^k$ the space-component, with respect to an inertial full quantum frame, of the velocity of the quantum flow. j is the canonical homomorphism $v\widehat{TM} \otimes_Z v\widehat{TM} \to Hom_Z(A; v\dot{T}_0^2 M) \cong Hom_Z(A \otimes_Z A; v\dot{T}_0^2 M)$. The action of χ on $\mathfrak{e}$ is induced by the right $\widehat{A}$-module structure of $Hom_Z(A; v\dot{T}_0^2 M)$. The expression of $\mathfrak{e}$ in quantum adapted coordinates is the following: $\mathfrak{e} = \partial x_j \otimes \partial x_k e^{jk}$, with $e^{jk} = \widehat{g}^{ij}[\hat{G}_{is}^k v^s + (\partial x_i . v^k)] : M \to Hom_Z(A \otimes_Z A; A \otimes_Z A)$. Finally, the divergence of P gives, in components, $P_{/j}^{ij} = -p_{/j}\widehat{g}^{ij} + \wp_{/j}^{ij} : M \to \widehat{A}$, with $\wp_{/j}^{ij} = 2\{\hat{G}_{jr}^i \widehat{g}^{pr}[\hat{G}_{ps}^j v^s + (\partial x_p . v^j)] + \hat{G}_{js}^j \widehat{g}^{pi}[\hat{G}_{pr}^s v^r + (\partial x_p . v^s)] + (\partial x_j . \widehat{g}^{pi})[\hat{G}_{ps}^j v^s + (\partial x_p . v^j)] + \widehat{g}^{pi}[(\partial x_j . \hat{G}_{ps}^j)v^s + \hat{G}_{ps}^j(\partial x_j . v^s) + (\partial x_j \partial x_p . v^j)]\}\chi.$ Then the quantum equation of continuity can be written $\mathrm{div}\, v = 0$, and the quantum equation of motion is $(\nabla_v v)\rho = \mathrm{div}\, P + B\rho = 0$. Here ρ is the mass density, considered a constant belonging to $\widehat{A}$. We can take only the symmetric part $\mathfrak{e}_{(s)}$ of $\mathfrak{e}$, as we have the canonical split $\mathfrak{e} = \mathfrak{e}_{(s)} + \mathfrak{e}_{(a)}$, induced by the split $v\dot{T}_0^2 M \cong v\dot{S}_0^2 M \oplus v\dot{\Lambda}_0^2 M$. In this case P is also symmetric. It is important to emphasize the following lemma.

Lemma 2.11. *The fiber bundle $\pi : W \equiv J\hat{D}(M) \times_M (\dot{\Lambda}_0^0 M)^+ \to M$ admits global sections and all the sections are homotopic.*

Proof of Lemma 2.11. In fact, the fiber bundle $\pi : W \equiv J\hat{D}(M) \times_M (\dot{\Lambda}_0^0 M)^+ \to M$ is contractile, hence we can use Corollary 1.2. $\qquad\square$

Now, we can formally reformulate the quantum Navier-Stokes equation (QNS) on the quantum Galilean space-time M as a subquantum manifold of the second order jet-derivative space $J\hat{D}^2(W)$, given by equations reported in Tab.2.6. Note that we have assumed that a full quantum non-relativistic inertial frame $\Phi \equiv (\phi, \hat{\tau}_\phi)$ has been assumed and the quantum coordinates (x^α) on M are adapted to ϕ. Furthermore, with respect to this frame the body force B is conservative: $B = -v\mathrm{grad}\, f = -\partial x_k(\partial x_j.f)\hat{g}^{jk}$, where the quantum potential f is a function $f : M \to \hat{A}$. Furthermore, the vertical gradient is defined by: $v\mathrm{grad}\, f = <\widehat{\hat{g}}, vdf>$, where $\widehat{\hat{g}} : M \to Hom_Z(A; v\widehat{\dot{S}_0^2 M})$ is the canonical quantum extension of $\hat{g}$, given by $\widehat{\hat{g}}(p)(a) = \hat{g}(p)(a)$, $\forall p \in M$ and $a \in A$. Therefore, in adapted quantum coordinates, we get that $\widehat{\hat{g}}$ is represented by functions $\widehat{\hat{g}}^{ij} : M \to Hom_Z(\dot{T}_0^2(A); \dot{T}_0^2(A)) \cong Hom_Z(A; \widehat{\dot{T}_0^2(A)})$, given by $\widehat{\hat{g}}^{ij}(p)(a) = \hat{g}^{ij}(p)$. Furthermore, the vertical differential $vdf = (\partial x_k.f)dx^k$ is performed by using the composition: $vT_pM \xrightarrow{dx^k(p)} A \xrightarrow{(\partial x_k.f)(p)} \hat{A}$, $\forall p \in M$. In this way we get $B^j = -(\partial x_k.f)\hat{g}^{kj} : M \to \overset{2}{\hat{A}}$. (QNS) is not a formally quantumintegrable QPDE as, similarly to the classic case, we have not the surjectivity of the mapping $(QNS)_{+1} \to (QNS)$, where $(QNS)_{+1}$ is the first prolongation of (QNS). On the other hand, we can see that the sub-quantum equation $\widehat{(QNS)}$ is formally quantumintegrable. Let us, now, go in some details, (even if these can be easily reproduced from a similar proof for the classic case [65].) Let us consider the following isomorphisms induced by adapted quantum coordinates: $J\hat{D}^2(W) \cong A^5 \times (\hat{A})^7 \times (\overset{2}{\hat{A}})^{28} \times (\overset{3}{\hat{A}})^{48}$, $J\hat{D}^3(W) \cong A^5 \times (\hat{A})^7 \times (\overset{2}{\hat{A}})^{28} \times (\overset{3}{\hat{A}})^{112} \times (\overset{4}{\hat{A}})^{192}$. Therefore we have also $(QNS) \cong A^5 \times (\hat{A})^6 \times (\overset{2}{\hat{A}})^{25} \times (\overset{3}{\hat{A}})^{48}$, $(QNS)_{+1} \cong A^5 \times (\hat{A})^6 \times (\overset{2}{\hat{A}})^{21} \times (\overset{3}{\hat{A}})^{100} \times (\overset{4}{\hat{A}})^{192}$. Furthermore, the first prolongation $(\dot{g}_2)_{+1} \subset Hom_Z(\dot{S}_0^3 M; vTW)$ of the symbol $\dot{g}_2 \subset Hom_Z(\dot{S}_0^2 M; vTW)$ of (QNS) is characterized by the equations: $\widehat{g}^{pj}\dot{X}^i_{ip\alpha} = 0$. As $Hom_Z(\dot{S}^3_0(T_pM); vT_qW) \cong (\overset{3}{\widehat{A}})^{64} \times (\overset{4}{\widehat{A}})^{192}$, it

follows that $(\dot{g}_2)_{+1} \cong (\overset{3}{\widehat{A}})^{52} \times (\overset{4}{\widehat{A}})^{192}$. Therefore we get, with respect to

the quantum algebra $B^{(1)} \cong A \times \widehat{A} \times \overset{2}{\widehat{A}} \times \overset{3}{\widehat{A}} \times \overset{4}{\widehat{A}}$, one has $\dim(QNS)_{+1} =$
$(5,6,21,100,192) \neq \dim(QNS) + \dim((\dot{g}_2)_{+1}) = (5,6,25,100,192)$. Hence
(QNS) cannot be formally quantumintegrable. Let us, instead, consider
equation $(\widehat{QNS})$. One has the following local isomorphisms: $(\widehat{QNS}) \cong$
$A^5 \times (\widehat{A})^6 \times (\overset{2}{\widehat{A}})^{21} \times (\overset{3}{\widehat{A}})^{48}$, $(\widehat{QNS})_{+1} \cong A^5 \times (\widehat{A})^6 \times (\overset{2}{\widehat{A}})^{21} \times (\overset{3}{\widehat{A}})^{84} \times (\overset{4}{\widehat{A}})^{192}$,

$(\widehat{\dot{g}_2})_{+1} \cong (\overset{3}{\widehat{A}})^{36} \times (\overset{4}{\widehat{A}})^{192}$, where $\widehat{\dot{g}_2}$ is the symbol of $(\widehat{QNS})$. Therefore we get
$\dim(\widehat{QNS}) + \dim(\widehat{\dot{g}_2})_{+1} = (5,6,21,84,192) = \dim(\widehat{QNS})_{+1}$. Let us, now,
show that this property can be extended to any prolongation $(\widehat{QNS})_{+r}$,
$\forall r \geq 1$. In fact, in Tab.2.8 we report local isomorphisms for $J\hat{D}^{2+r}(W)$,
$((\widehat{QNS}))_{+r}$ and $(\widehat{\dot{g}_2})_{+r}$.

TAB.2.8 - Local isomorphisms for prolonged quantum manifolds.

$$J\hat{D}^{2+r}(W) \cong A^5 \times \prod_{1 < s < 2+r} (\widehat{A})^{[4^s + 4^{s-1} \times 3]} \times (\overset{3+r}{\widehat{A}})^{[4^{2+r} \times 3]}$$

$$(\widehat{QNS})_{+r} \cong A^5 \times \prod_{1 < s < 2+r} (\widehat{A})^{[4^s + 4^{s-1} \times 3] - [4^{s-1} + 4^{s-2} \times 3]} \times (\overset{3+r}{\widehat{A}})^{[4^{2+r} \times 3]}$$

$$(\widehat{\dot{g}_2})_{+r} \cong (\overset{2+r}{\widehat{A}})^{[4^{2+r} - 4^{r+1} - 4^r \times 3]} \times (\overset{3+r}{\widehat{A}})^{[4^{2+r} \times 3]}$$

$(r \geq 1)$.

Then it is easy to see that $\dim(\widehat{QNS})_{+r} = \dim(\widehat{QNS})_{+(r-1)}) + \dim(\dot{g}_2)_{+r}$.
Therefore, we can conclude that the mapping $(\widehat{QNS})_{+(r+1)} \rightarrow (\widehat{QNS})_{+r}$
is surjective $\forall r \geq 1$. So taking into account that $(\widehat{QNS})$ is quantum regular, that Z is a Noetherian **K**-algebra, hence that $(\widehat{QNS})$ is δ-regular and
that $(\widehat{\dot{g}_2})_{+r}$ is a bundle of Z-modules over $(\widehat{QNS})$, we can conclude, by using
Theorem 2.6 and Theorem 2.7 that $(\widehat{QNS})$ is formally quantumintegrable.$\square$

Theorem 2.18. (Theorem of existence of regular solutions for (QNS)). *For
any point $q \in (\widehat{QNS}) \subset (QNS)$ we can construct a regular solution of
(QNS).*

Proof. In fact $(\widehat{QNS})$ is a completely quantumintegrable QPDE, i.e., for
any point $q \in (\widehat{QNS}) \subset J\hat{D}^2(W)$ there exists a neighborhood U of $p =$
$\pi_2(q) \in M$ such that there exists a section s of $\pi : W|_U \rightarrow U$ such that

$D^2 s(U) \subset (\widehat{QNS})$ and $D^2 s(p) = q$. The complete integrability of $(\widehat{QNS})$ follows from the fact that it is an algebraic formally quantumintegrable QPDE, so we can use the theorem of existence of regular solutions for quantumanalytic formally quantumintegrable QPDEs. (See also ref.[67].) Furthermore, as $(\widehat{QNS})$ is contained into (QNS) the theorem is proved.$\square$

Remark 2.6. Note that even if the quantum Navier-Stokes equation is not formally quantum integrable, it is universally quantum regular in the sense of Definion 2.12. In fact $((\widehat{QNS}))_{+\infty} = (QNS)_{+\infty}$. $\blacksquare$

Theorem 2.19. (Cauchy problem for (QNS)). *Any (regular) space-like integral quantum submanifold $N \subset (\widehat{QNS}) \subset (QNS)$ of dimension 3 that satisfies the initial conditions and the transversality conditions, with respect to an integrable time-like vector field $\zeta \in \mathfrak{s}(\hat{C}_2(QNS))$, generates a (regular) solution of (QNS) if N has no frozen singularities. Here $\hat{C}_2(\widehat{QNS})$ is the following ideal $\hat{C}_2(\widehat{QNS}) \equiv < \omega^I, \omega^I_\alpha, \epsilon^0 \equiv dF^0, \epsilon^0_\alpha \equiv dF^0_\alpha, \epsilon^j \equiv dF^j >$, with $\omega^I \equiv dy^I - y^I_\alpha dx^\alpha$, $\omega^I_\alpha \equiv dy^I_\alpha - y^I_{\alpha\beta} dx^\alpha$ the quantum contact differential 1-forms on $\hat{J}^2_4(W)$. More precisely, let us denote $\hat{C}_2(\widehat{QNS}) = < \Omega_I >$. Then, the initial and transversal conditions are the following: $\zeta \notin \mathfrak{d}(TN)$, $\psi^*\Omega_I = 0$, $\psi^*(\zeta \rfloor \Omega_I) = 0$, where $\psi : \Xi \to (\widehat{QNS})$ is the mapping that represents the manifold N. Here $\mathfrak{s}(\mathcal{I})$ denotes the space of infinitesimal symmetries of an ideal $\mathcal{I}$ of quantum differential forms and $\mathfrak{d}(TN)$ denotes the $\Omega^0(N)$-module of vector fields on N, with $\Omega^0(N)$ the space of smooth zero-forms on N. Furthermore, N is considered space-like with respect to the full quantum inertial frame $\Phi \equiv (\phi, \hat{\tau}_\phi)$, i.e., with respect to the induced fiber bundle structure $\hat{\tau}_\phi : M \to A$, that allows us to write $M = \bigcup_{a \in A} M_a$. Therefore, N is space-like if, for some $a \in A$, one has $N \subset (\widehat{QNS})_a$, where $(\widehat{QNS})_a \equiv (\widehat{QNS}) \bigcap J\hat{D}^2(W)_a$, with $J\hat{D}^2(W)_a$ the fiber of $J\hat{D}^2(W)$ over $a \in A$, in the following fiber structure over A: $J\hat{D}^2(W) \xrightarrow{\pi_{2,0}} W \xrightarrow{\pi} M \xrightarrow{\hat{\tau}_\phi} A$.*

Proof. As the flow of a vector field ζ with above properties transforms, into a suitable neighbourood of the zero, a space-like integral manifold into another spacelike integral manifold, it is enough, now, to prove that $\mathfrak{s}(\hat{C}_2(\widehat{QNS}))$ is not zero for integrable time-like vector fields. On the other hand, as the vector field ∂x_0 is a symmetry for the QPDEs $(\widehat{QNS})$ and (QNS), it follows that $\mathfrak{s}(\hat{C}_2(\widehat{QNS})) \neq 0$ and it contains integrable time-like vector fields. $\square$

Above theorems prove that the class of admissible integral quantum manifolds, (with respect to the definition given in previous section), of dimension 3 is not empty in the (QNS) equation. Now, in order to characterize the structure of the global solutions of the quantum Navier-Stokes equation we shall calculate its p-bordism groups, $0 \le p \le 3$. One has the following.

Theorem 2.20. *The integral bordism groups of (QNS) are resumed in Tab.2.9.*

Proof. The proof can be directly gained by considering the sub-equation $(\widehat{QNS})$. For this, in fact, we can apply Theorem 2.11(2)(ii). However, the proof can be also directly obtained by considering all the equation (QNS). In fact it is enough to reproduce in this context the proof given in last paper quoted in refs.[65] for calculate the integral bordism groups of the classic Navier-Stokes equation. $\qquad\square$

TAB.2.9 - Integral bordism groups in (QNS)

p	$\Omega_p^{(QNS)}$	$\Omega_p^{(QNS)+\infty}$	$\Omega_{p,s}^{(QNS)}$	$\Omega_{p,s}^{(QNS)+\infty}$
0	A	A	A	A
1,2,3	0	0	0	0

Corollary 2.1. (Tunnel effects in the quantum Navier-Stokes equation). *In the set of p-dimensional, $2 \le p \le 4$, integral manifolds of (NS) one has ones that change their sectional topology.*

Remark 2.7. As $\Omega_3^{(QNS)+\infty} = 0$, we get, as a by-product, the existence of quantum smooth global solutions of the quantum Navier-Stokes equation.$\blacksquare$

Remark 2.8. (*Observed quantum Navier-Stokes equation*). Let us represent the quantum Navier-Stokes equation $(QNS) \subset J\hat{D}^2(W)$ by means of a quantum frame represented by means of an embedding $i : M_C \to M$. As the (quantum) continuity equation is represented by a $\hat{A}$-valued scalar field on M, its observation gives simply the composition by means of i. More complex is, instead, the situation for the quantum motion equation, that is represented by a field $s \equiv (\nabla_v v)\rho - \operatorname{div} P - B\rho : M \to Hom_Z(A; \widehat{vTM})$. We can represent such a field as a $\hat{A}$-valued vertical differential 1-form i^*s on M_C: $i^*s : M_C \to \hat{A} \otimes_{\mathbf{K}} (vTM_C)^*$. In fact one has the following commutative diagram:

$$
\begin{array}{ccc}
M & \xrightarrow{\;s\;} & Hom_Z(A;\widehat{vTM}) \cong Hom_Z(A;(vTM)^+) \cong Hom_Z(A \bigotimes_Z vTM;A) \\
 & & \downarrow \\
i^*s \downarrow & & Hom_Z(A \bigotimes_{\mathbf{K}} vTM;A) \\
 & & \downarrow i_* \\
\widehat{A} \bigotimes_{\mathbf{K}} (vTM_C)^* \quad\cong & & Hom_Z(A \bigotimes_{\mathbf{K}} vTM_C;A)
\end{array}
$$

If s has the following local representation $s = \partial x_k s^k$, where $s^k : M \to \overset{2}{\widehat{A}}$, then one has the following local representation on M_C: $i^*s = (i^*s)_p d\xi^p$, with $(i^*s)_p \equiv (\partial \xi_p.i^j) g_{kj} s^k : M_C \to \widehat{A}$, where (ξ^α) are coordinates on M_C induced by (x^α) on M. By summarasing, the observed (QNS) can be written as follows:

$$
(QNS)[i] : \left\{
\begin{array}{l}
0 = F^0 \circ i : M_C \to \widehat{A} \\[2mm]
0 = (\partial \xi_p.s^j) g_{kj} F^k : M_C \to \widehat{A}
\end{array}
\right\}_{1 \leq p \leq 3}
$$

Of course, one has that $(QNS)[i]$ is not formally quantumintegrable and that the observed equation $\widehat{(QNS)}[i]$ is instead formally quantumintegrable. Therefore, for $(QNS)[i]$ hold conclusions similar to previous ones for (QNS).∎

2.3 - QUANTUM SUPER PDE's

In this section we define quantum super PDEs (QSPDEs) and we extend to such equations our previously formulated geometric theory of quantum PDEs (QPDEs). (See also refs.[71].) In particular, we adapt to QSPDE's our previous theorem to recognize formal integrability for QPDE's, obtaining an analogous theorem for QSPDE's. Similarly we obtain a theorem to solve the Cauchy problems for QSPDE's. Furthermore, we characterize integral (co)bordism groups of QSPDE's and relate them to the Reinhart integral bordism groups of commutative manifolds. Finally, we give a theorem that allows us to determine all the conservation laws of QSPDE's by means of suitable quantum Hopf superalgebras.

Definition 2.20. A *quantum super PDE* (QSPDE) of order k on the fibre bundle $\pi : W \to M$, defined in the category of quantum supermanifolds, is a subfibrebundle $\widehat{E}_k \subset J\widehat{D}^k(W)$ of the jet-quantum derivative space $J\widehat{D}^k(W)$ over M.

Remark 2.9. Note that, all the intrinsic definitions and results, given in the previous section for QPDEs, continue to hold for QSPDEs, when we forget the $\mathbf{Z}_2$-gradiation of the quantum superalgebras. On the other hand, all the regular solutions $V \subset \hat{E}_k$ of a QSPDE $\hat{E}_k \subset J\hat{D}^k(W)$, where $\pi : W \to M$ is a fiber bundle in the category of quantum supermanifolds, have a natural structure of quantum supermanifold induced from the quantum supermanifold M of the basis as $V \cong M$. In the following we shall consider QSPDEs on a fiber bundle $\pi : W \to M$, where M is a quantum supermanifold of dimension $(m|n)$ on the quantum superalgebra A and W is a quantum supermanifold of dimension $(m|n,r|s)$ on the quantum superalgebra $B \equiv A \times E$, where E is also a Z-module, with $Z = Z(A)$ the centre of A. Let us emphasize that the quantum symbol $(\dot{g}_k)_q$, $q \in \hat{E}_k$, of a QSPDE $\hat{E}_k \subset J\hat{D}^k(W)$ is a Z-submodule of $Hom_Z(\dot{S}_0^k(T_pM); vT_{\bar{q}}W)$, $p = \pi_k(q) \in M$, $\bar{q} = \pi_{k,0}(q) \in W$. One has the following local isomorphism:

$$Hom_Z(\dot{S}_0^k(T_pM); vT_{\bar{q}}W)$$

$$\cong \prod_{i_1+\cdots+i_k \in \mathbf{Z}_2, i_r \in \mathbf{Z}_2} (\widehat{A}\,_{i_1\cdots i_k}^{\,k}(E))^{m^{k_1}n^{k_2}r(k_2)|m^{k_1}n^{k_2}s(k_2)}.$$

Here k_1 and k_2 are respectively the numbers of $i_r = 0$ and $i_r = 1$. Therefore, $(\dot{g}_k)_q$, can be locally characterized by a matrix $(\zeta_{A_1\cdots A_k}^B(p))$ with entries into the quantum superalgebras $\widehat{A}\,_{i_1\cdots i_k}^{\,k}(E)$. On the other hand as $(\dot{g}_k)_q \in vT_q\hat{E}_k$, for any $\zeta(q) \in (\dot{g}_k)_q$, we can also write $\zeta(q) = \partial y_B^{A_1\cdots A_k}(q)(\zeta_{A_1\cdots A_k}^B(q))$ such that

$$(\partial y_B^{A_1\cdots A_k}.F^I)(q)(\zeta_{A_1\cdots A_k}^B(q)) = 0,$$

where $\partial y_B^{A_1\cdots A_k}(q) \in Hom_Z(B_k; vT_qJ\hat{D}^k(W))$, with B_k the quantum superalgebra underlying $J\hat{D}^k(W)$ and $F^I : J\hat{D}^k(W) \to B_k$ are the functions locally characterizing $\hat{E}_k$. One has the following complex of Z-modules over $\hat{E}_k$ (*δ-quantum complex*):

$$0 \to \dot{g}_m \xrightarrow{\delta} TMO\dot{g}_{m-1} \xrightarrow{\delta} \dot{\Lambda}_0^2MO\dot{g}_{m-2} \xrightarrow{\delta}$$

$$\cdots \xrightarrow{\delta} \dot{\Lambda}_0^{m-k}MO\dot{g}_k \xrightarrow{\delta} \delta(\dot{\Lambda}_0^{m-k}MO\dot{g}_k) \to 0.$$

The definitions of *Spencer quantumcohomology, r-quantumacyclicity, quantuminvolutiveness* and *δ-regularity* for a QSPDE do not change with respect to ones given for QPDEs. Furthermore, Theorem 2.5, Theorem 2.6 and Theorem 2.7 on the formal properties of QPDE's, continue to hold also for QSPDE's. (Their proofs can be copied by the proofs of similar theorems

given in ref.[66] for QPDEs. In fact the proofs given there are intrinsic.)
So we have the following.   ∎

Theorem 2.21. (Formal properties and spectral sequences of QSPDE's).
Let $\hat{E}_k \subset J\hat{D}^k(W)$ be a quantum regular QSPDE on the fiber bundle
$\pi : W \to M$. Then, there exists a filtration $\{X_q\}$ of $\hat{E}_k$: $\emptyset = X_{-1} \subset X_0 \subset X_1 \subset \cdots \subset X_N = \hat{E}_k$ such that the corresponding spectral sequence $\{E^r_{p,q}, d^r\}$, (homological formal spectral sequence of $\hat{E}_k$), converges to $H_\bullet(\hat{E}_k)$. Therefore, to the above filtration there also corresponds a spectral sequence $\{E^{p,q}_r, d^r\}$, (cohomological formal spectral sequence of $\hat{E}_k$), such that it converges to $H^\bullet(\hat{E}_k)$. Finally, if $\hat{E}_k$ is formally quantum integrable, then one has: $H_\bullet(\hat{E}_k) \cong H_\bullet(\hat{E}_\infty)$, $H^\bullet(\hat{E}_k) \cong H^\bullet(\hat{E}_\infty)$.

Theorem 2.22. (δ-Poincaré lemma for QSPDE's). Let $\hat{E}_k \subset J\hat{D}^k(W)$ be a quantum regular QSPDE. If Z is a Noetherian **K**-algebra, then $\hat{E}_k$ is a δ-regular QSPDE.

Theorem 2.23. (Criterion of formal quantum integrability for QSPDE's).
Let $\hat{E}_k \subset J\hat{D}^k(W)$ be a quantum regular, δ-regular QSPDE. Then if $\dot{g}_{k+r+1}$ is a bundle of Z-modules over $\hat{E}_k$, and $\hat{E}_{k+r+1} \to \hat{E}_{k+r}$ is surjective for $0 \leq r \leq h$, for suitable h, then $\hat{E}_k$ is a formally quantumintegrable QSPDE.

Definition 2.21. An *initial condition* for QSPDE $\hat{E}_k \subset J\hat{D}^k(W)$ is a point $q \in \hat{E}_k$. A *solution* of $\hat{E}_k$ passing for the initial condition q is a quantum supermanifold $N \subset \hat{E}_k$ such that $q \in N$, $\dim_A N = (m|n)$, and such that N can be represented in a neighboroud of any of its points $q' \in N$, except for a nonwhere dense subset $\Sigma(N) \subset N$ of dimension $\leq (m-1|n-1)$, as image of the k-derivative $D^k s$ of some Q^k_w-section s of $\pi : W \to M$. We call $\Sigma(N)$ the set of *singular points* (of Thom-Bordman type) of N. If $\Sigma(N) = \emptyset$ we say that N is a *regular solution* of $\hat{E}_k \subset J\hat{D}^k(W)$.

Definition 2.22. 1) We denote by $\hat{J}^k_{m|n}(W)$ the *k-jet space for quantum supermanifolds* of dimension $(m|n)$ (over A) contained in the quantum supermanifold W.

2) We call *QSPDE*, of order k, for quantum supermanifolds, contained into W and of dimension $(m|n)$ over A, a quantum supermanifold $\hat{E}_k \subset \hat{J}^k_{m|n}(W)$.

3) We say that the QSPDE $\hat{E}_k \subset \hat{J}^k_{m|n}(W)$ is *completely quantum superintegrable* if for any point $q \in \hat{E}_k$ passes a quantum supermanifold, $V \subset \hat{E}_k$, of dimension $(m|n)$ over A, that is the k-quantum prolongation, $V = N^{(k)}$, of a $(m|n)$-dimensional (over A) quantum supermanifold $N \subset W$.

Remark 2.10. Note that, by using the natural embedding $J\hat{D}^k(W) \subset \hat{J}^k_{m|n}(W)$, we can consider QSPDEs $\hat{E}_k \subset J\hat{D}^k(W)$ like QSPDEs $\hat{E}_k \subset \hat{J}^k_{m|n}(W)$, hence we can consider solutions of $\hat{E}_k$ as $(m|n)$-dimensional (over A) quantum supermanifolds $V \subset \hat{E}_k$ such that V can be representable in the neighborhood of any of its points $q' \in V$, except for a nonwhere dense subset $\Sigma(V) \subset V$, of dimension $\leq (m-1|n-1)$, as $N^{(k)}$, where $N^{(k)}$ is the k-quantum prolongation of a $(m|n)$-dimensional (over A) quantum supermanifold $N \subset W$. In the case that $\Sigma(V) = \emptyset$, we say that V is a *regular solution* of $\hat{E}_k \subset \hat{J}^k_{m|n}(W)$. Of course, solutions V of $\hat{E}_k \subset \hat{J}^k_{m|n}(W)$, even if regular ones, are not, in general diffeomorphic to their projections $\pi_k(V) \subset M$, hence are not representable by means of sections of $\pi : W \to M$. $\blacksquare$

Remark 2.11. We call *regular quantum superplane* at the point $q \in \hat{J}^k_{m|n}(W)$ the linear subspace in $T_q \hat{J}^k_{m|n}(W)$ of the form, $T_q N^{(k)}$, for some Q^∞_w quantum supermanifold $N \subset W$, of dimension $(m|n)$ over A, passing for $a \in W$, with $a = \pi_{k,0}(q)$. Any point $q \in \hat{J}^k_{m|n}(W)$ identifies a unique regular quantum superplane L_q at $q' \equiv \pi_{k,k-1}(q) \in \hat{J}^{k-1}_{m|n}(W)$: $L_q \subset T_{q'} \hat{J}^{k-1}_{m|n}(W)$. Set: $I_k(W) \equiv \bigcup_{u \in \hat{J}^k_{m|n}(W)} I_k(W)_u$, $I_k(W)_u \equiv$ Grassmannian of integral quantum superplanes at q. An *integral quantum superplane* at a point $u \in \hat{J}^k_{m|n}(W)$ is defined to be a $(m|n)$-dimensional quantum superspace, subspace of $(\mathbb{E}_k)_u$, tangent to some integral quantum supermanifold of the Cartan distribution $\mathbb{E}_k \subset T\hat{J}^k_{m|n}(W))$. Let $I(\hat{E}_{k+s})$ be the fiber bundle of Grassmannian $(m|n)$-dimensional integral quantum superplanes of the Cartan quantumdistribution $\mathbb{E}(\hat{E}_{k+s})$ on $\hat{E}_{k+s}$ being $\hat{E}_k \subset \hat{J}^k_{m|n}(W)$ a QSPDE. If $\hat{E}_{k+s} = \hat{J}^{k+s}_{m|n}(W)$ one has $I(\hat{J}^{k+s}_{m|n}(W)) = I_{k+s}(W)$. Note that here and in the following for $(p|q)$-dimensional integral quantum supermanifold V, $0 \leq p \leq m$, $0 \leq q \leq n$, with boundary ∂V, (or eventually with $\partial V = \emptyset$), we mean an element $V \in C_{p,q}(\hat{E}_{k+h}, A)$, $h \geq 0$, such that $\widehat{T}V \subset \widehat{\mathbb{E}}_{k+h}$. So, if $V = \sum_i a^i u_i + \sum_j b^j v_i$, $a_i \in A_0$, $b_j \in A_1$, one has $\partial V = \sum_i (-1)^i a^i \partial_i u + \sum_j (-1)^j b^j \partial_j v$. Then, the canonical homomorphism $c : A \to \mathbf{K}$ induces the canonical mapping $V \to V_C = \sum_i \lambda^i u_i + \sum_j \mu^j v_j$, with $\lambda^i \equiv c(a^i) \in \mathbf{K}$, $\mu^j \equiv c(b^j) \in \mathbf{K}$, and one has $\partial V_C = \sum_i (-1)^i \lambda^i \partial_i u + \sum_j (-1)^j \mu^j \partial_j v$, where $V_C \in C_{p+q}(\hat{E}_{k+h}; \mathbf{K})$ is the classic limit of V. Finally, note that by means of the monomorphism $\epsilon : \mathbf{K} \to A$, we can induce an injection $\epsilon : V_C \to V$, such that $c \circ \epsilon = id_V$, hence ϵ is a section of $V \to V_C$. In other words V canonically contains its classic limit V_C. (But there is not, in general, a canonical superclassic limit V_{SC} of V.) $\blacksquare$

Theorem 2.24. *Let W be a quantum supermanifold (resp. $\pi : W \to M$ a fiber bundle) like above specified and let $\hat{E}_k \subset \hat{J}^k_{m|n}(W)$ (resp. $\hat{E}_k \subset J\hat{D}^k(W)$) be a QSPDE. One has the following.*

1) Solutions V of $\hat{E}_k$ are branched coverings of some $(m|n)$-dimensional quantum supermanifold $N \subset W$, (resp. coverings of M) with branche points $p \in \pi_k(\Sigma(V))$. Set $V' \equiv V \setminus \Sigma(V)$ and $N' \equiv N \setminus \pi_k(\Sigma(V))$, (resp. $M' \equiv M \setminus \pi_k(\Sigma(V)))$. Then $\pi_k : V' \to N'$ (resp. $\pi_k : V' \to M'$) is a covering map.

2) In particular, if $\Sigma(V) = \emptyset$, then $\pi_k : V \to N$ (resp. $\pi_k : V \to M$) is a covering map. Furthermore, if $\pi_1(N) = 0$, (resp. $\pi_1(M) = 0$), i.e., N (resp. M) is simply connected, then V is globally diffeomorphic to N, (resp. M).

3) If the base space N (resp. M) is connected, then the number of sheets of the covering $V' \to N'$ (resp $V' \to M'$) is independent of the choice of the point $a \in N$ (resp. $p \in M$).

4) In the case of QPDE's $\hat{E}_k \subset \hat{J}^k_{m|n}(W)$, if the quantum supermanifolds V and N are closed, then the branched coverings (resp. coverings) are with a finite number of sheets. A similar proposition hold for the quantum supermanifolds V and M in the case of solutions of QSPDE's $\hat{E}_k \subset J\hat{D}^k(W)$.

Proof. The proof is similar to one for Theorem 2.8. $\square$

For QSPDEs in the category of quantum analytic supermanifolds, above theorem allows us to obtain existence theorems of local solutions of class Q^ω_w, similarly to what just known in the (super)classic case [61,62,64]. More precisely, one has the following.

Theorem 2.25. *Let $\pi : W \to M$ be a fiber bundle in the category of quantum supermanifolds of class Q^ω_w, where W is a quantum supermanifold of dimension $(m|n, s|r)$ over the quantum superalgebra $B \equiv A \times E$ and M is a quantum supermanifold of dimension $(m|n)$ over A. Furthermore, E is also a $Z \equiv Z(A)$-module, with Z a Noetherian $\mathbf{K}$-algebra. Let $\hat{E}_k \subset J\hat{D}^k(W)$ be a QSPDE such that the following conditions are satisfied: (i) $\hat{E}_k$ is quantum regular; (ii) The symbol $\hat{g}_{k+r+1}$ of $\hat{E}_{k+r+1}$ is a bundle of Z-modules over $\hat{E}_k$, $r \geq 0$; (iii) $\hat{E}_{k+r+1} \to \hat{E}_{k+r}$ is surjective with $0 \leq r \leq l$, for suitable l. Then, for any point $q \in \hat{E}_k$, passes a quantum supermanifold $V \subset \hat{E}_k$ of dimension $(m|n)$, over A, diffeomorphic to its projection on M, (i.e., V is a regular Q^ω_w-solution of $\hat{E}_k \subset J\hat{D}^k(W)$).*

Proof. In fact, under our hypotheses we can apply Theorem 2.21, hence $\hat{E}_k$ is formally quantum integrable. As a consequence, for any point $q \in \hat{E}_k$, we can find a point $q' \in \hat{E}_{k+1}$, such that $\pi_{k+1,k}(q') = q$, that identifies the tangent space at q, $T_q N^{(k)}$, of a regular supermanifold $N^{(k)} \subset \hat{E}_k$, of dimension $(m|n)$ over A, obtained as the k-quantum prolongation of a $(m|n)$-dimensional (over A) quantum supermanifold $N \subset W$, that is image of some Q_w^ω-section $s : M \to W$, defined in a neighbourhood of $p \equiv \pi_k(q) \in M$. $\qquad\qquad\square$

Corollary 2.2. *Let $\hat{E}_k \subset J\hat{D}^k(W)$ be a formally quantum integrable QSPDE in the category of quantum supermanifolds of class Q_w^ω. Then the QSPDE $\hat{E}_k \subset \hat{J}_{m|n}^k(W)$ is completely quantum superintegrable.*

Theorem 2.26. (Cauchy problem for QSPDE's). *Any (regular) integral quantum supermanifold $N \subset \hat{E}_k$, of dimension $(m-1|n-1)$, that satisfies the initial conditions and the transversality conditions, with respect to an integrable vector field ζ belonging to the infinitesimal symmetry algebra of $\hat{E}_k \subset \hat{J}_{m|n}^k(W)$, generates a (regular) solution of $\hat{E}_k$.*

Proof. In fact the flow of a vector field ζ, with the above properties, transforms, into a suitable neighbourhood of the zero, regular integral manifolds of $\hat{E}_k$ into other regular integral manifolds of $\hat{E}_k$. Note that such solutions are, in general, of class Q_w^∞, even if the integral quantum supermanifold N is of class Q_w^ω. $\qquad\qquad\square$

Remark 2.12. Now, in order to study the structure of global solutions it is necessary to consider the integral bordism groups of QSPDEs embedded in the spaces $\hat{J}_{m|n}^k(W)$. For a smooth integral (classicregular) quantum supermanifold $V \subset \hat{J}_{m|n}^k(W)$ we can naturally extend the definitions of "admissibility", "frozen singularities", integrable admissibility" and so on just made for bordisms in QPDEs. Let us denote by $\Omega_{p|q}^{\hat{E}_k}$, (resp. $\Omega_{p|q}(\hat{E}_k)$), the integral bordism groups, (resp. quantum bordism groups), for $(p|q)$-dimensional quantum supermanifolds of a QSPDE $\hat{E}_k \subset \hat{J}_{m|n}^k(W)$. We call $\Omega_{p|q}^{\hat{E}_k}$, (resp. $\Omega_{p|q}(\hat{E}_k)$), $p \in \{0, 1, \ldots, m-1\}$, $q \in \{0, 1, \ldots, n-1\}$, the *integral $(p|q)$-bordism group*, (resp. *quantum $(p|q)$-bordism group*), of $\hat{E}_k$. The relation between integral bordism groups, quantum bordism groups and singular bordism groups is given by means of a commutative diagram similar to one for quantum manifolds, and that allows us to state the following.∎

Theorem 2.27. 1) *The quantum $(p|q)$-bordism groups $\Omega_{p|q}(\hat{E}_k)$ is an ex-*

tension of a subgroup of ${}^A\underline{\Omega}_{p|q,s}(W) \cong H_{p|q}(W;A)$, and the integral $(p|q)$-bordism group $\Omega_{p|q}^{\hat{E}_k}$ is an extension of the quantum $(p|q)$-bordism group.

2) Let us assume that admissible integral quantum supermanifolds bording admissible integral quantum supermanifolds of dimension $(p|q)$, $0 \le p \le m-1$, $0 \le q \le n-1$, contained into $\hat{E}_k$, are smooth regular, for a k-order QSPDE $\hat{E}_k \subset \hat{J}^k_{m|n}(W)$. Then one has the following: (i) If $\hat{E}_k \subset \hat{J}^k_{m|n}(W)$ is a formally quantum integrable QSPDE, then one has the following isomorphisms: $\Omega_{p|q}^{\hat{E}_k} \cong \Omega_{p|q}(\hat{E}_k) \cong \Omega_{p|q}^{\hat{E}_{k+h}} \cong \Omega_{p|q}^{\hat{E}_\infty} \cong \Omega_{p|q}(\hat{E}_{k+h}) \cong \Omega_{p|q}(\hat{E}_\infty)$, with $p \in \{0,\ldots,m-1\}$ and $q \in \{0,\ldots,n-1\}$. (ii) Furthermore, if W is $p+q$-connected, $p \in \{0,\ldots,m-1\}$, $q \in \{0,\ldots,n-1\}$, one has the following isomorphisms: $\Omega_{p|q}(\hat{E}_k) \cong {}^A\underline{\Omega}_{p|q,s}(W)$, with $p \in \{0,\ldots,m-1\}$ and $q \in \{0,\ldots,n-1\}$.

Above considerations can be generalized to include more sophisticated quantum solutions of QSPDEs. More precisely one has the following.

Theorem 2.28. *Let $\hat{E}_k \subset \hat{J}^k_{m|n}(W)$ be a QSPDE and let B be a quantum superalgebra. Let us consider the following chain complex (bigraded bar quantum chain complex of $\hat{E}_k$): $\{\bar{C}_{\bullet|\bullet}(\hat{E}_k;B), \bar{\partial}\}$, induced by the $\mathbf{Z}_2$-gradiation of B on the corresponding bar quantum chain complex of $\hat{E}_k$, i.e., $\{\bar{C}_\bullet(\hat{E}_k;B), \bar{\partial}\}$. (See Section 1.) Then, one has the following exact commutative diagram:*

$$
\begin{array}{ccccccccc}
& & 0 & & 0 & & & & \\
& & \downarrow & & \downarrow & & & & \\
0 & \to & \bar{B}_{\bullet|\bullet}(\hat{E}_k;B) & \to & \bar{Z}_{\bullet|\bullet}(\hat{E}_k;B) & \to & \bar{H}_{\bullet|\bullet}(\hat{E}_k;B) & \to & 0 \\
& & \downarrow & & \downarrow & & & & \\
& & \bar{C}_{\bullet|\bullet}(\hat{E}_k;B) & = & \bar{C}_{\bullet|\bullet}(\hat{E}_k;B) & & & & \\
& & \downarrow & & \downarrow & & & & \\
0 & \to & {}^B\Omega_{\bullet|\bullet,s}^{\hat{E}_k} & \to & \bar{B}or_{\bullet|\bullet}(\hat{E}_k;B) & \to & \bar{C}yc_{\bullet|\bullet}(\hat{E}_k;B) & \to & 0 \\
& & \downarrow & & \downarrow & & & & \\
& & 0 & & 0 & & & &
\end{array}
$$

where the meaning of the symbols and their names are directly understood by the similar commutative diagrams previously given in Theorem 1.30. In particular, we call (closed) integral singular $(p|q)$-bordism groups, with coefficients in B, of $\hat{E}_k$ the groups ${}^B\Omega_{p|q,s}^{\hat{E}_k}$, $p \in \{0,1,\ldots,m-1\}$, $q \in \{0,1,\ldots,n-1\}$. (If $B = \mathbf{K}$ we omit the apex B). In Tab.2.10 we report *some relations between objects considered in above commutative diagram.*

TAB.2.10 - Relations between bigraded (closed) integral singular bordism groups of $\hat{E}_k$.

${}^B\Omega^{\hat{E}_k}_{p\|q,s} \cong \bar{H}_{q\|q}(\hat{E}_k;B) \cong (\Omega^{\hat{E}_k}_{p,s}\otimes_{\mathbf{K}} B_0) \bigoplus (\Omega^{\hat{E}_k}_{q,s}\otimes_{\mathbf{K}} B_1)$
$\Omega^{\hat{E}_k}_{\bullet\|\bullet,s} \equiv {}^{\mathbf{K}}\Omega^{\hat{E}_k}_{\bullet\|\bullet,s} \cong \bar{H}_{\bullet\|\bullet}(\hat{E}_k;\mathbf{K})$

Definition 2.23. We call *quantum singular $(p\|q)$-bordism groups*, with coefficients in B, of $\hat{E}_k$ the relative closed integral singular bordism groups of $\hat{J}^k_{m\|n}(W)$ with respect to $\hat{E}_k$, and we will denote them by ${}^B\Omega_{p\|q,s}(\hat{E}_k)$, $p \in \{0,1,\ldots,m-1\}$, $q \in \{0,1,\ldots,n-1\}$.

Theorem 2.29. *By means of the bigraded cochain complex $\{\bar{C}^{\bullet\|\bullet}(\hat{E}_k;B),\bar{\delta}\}$, dual of the bigraded chain complex $\{\bar{C}_{\bullet\|\bullet}(\hat{E}_k;B),\bar{\partial}\}$, we obtain objects dual of ones above considered. In particular, one has the (canonical) isomorphisms reported in Tab.2.11.*

TAB.2.11 - Relations between bigraded (closed) integral singular cobordism groups of $\hat{E}_k$.

$({}^B\Omega^{\hat{E}_k}_s)^{p\|q} \cong \bar{H}^{p\|q}(\hat{E}_k;B) \cong Hom_{\mathbf{K}}({}^{\mathbf{K}}\Omega^{\hat{E}_k}_{p,s};B_0) \bigoplus Hom_{\mathbf{K}}({}^{\mathbf{K}}\Omega^{\hat{E}_k}_{q,s};B_1)$
$\bar{Bor}^{\bullet\|\bullet}(\hat{E}_k;B) \cong ({}^B\Omega^{\hat{E}_k}_s)^{\bullet\|\bullet} \bigoplus \bar{Cyc}^{\bullet\|\bullet}(\hat{E}_k;B)$

Definition 2.24. Then we call *(closed) integral singular $(p\|q)$-cobordism groups*, with coefficients in B, of $\hat{E}_k$:

$${}^B\Omega^{p\|q,}_s(\hat{E}_k), \ p \in \{0,1,\ldots,m-1\}, \ q \in \{0,1,\ldots,n-1\}.$$

Remark 2.13. The quantum characteristic numbers, just defined for QPDE's in Definition 2.14, are now $\mathbf{Z}_2$-graded. These characterize the actual bigraded integral and quantum bordism groups. ∎

Example 2.7. The methods here developed allow us, for example, classify singular solutions of a QSPDE $\hat{E}_k$, with B given by the following quantum superalgebra: $B = A \otimes_{\mathbf{Z}} \mathbf{Z}_t \cong A/tA$, $t \in \mathbf{N}$, where A is a quantum superalgebra. Such solutions can be considered the natural extension, in the category of quantum supermanifolds, of the well known $\mathbf{Z}_t$-manifolds, introduced in algebraic topology by D. Sullivan [51]. (These last are classified by the bordism groups $\Omega_p \otimes_{\mathbf{Z}} \mathbf{Z}_t$ [10].) ∎

Theorem 2.30. (Relation between integral bordism groups in QSPDE's and Reinhart integral bordism groups of commutative manifolds). *Let* $N_0, N_1 \subset \hat{E}_k \subset \hat{J}^k_{m|n}(W)$ *be two integral bording, admissible integral manifolds of* $\hat{E}_k$ *of dimension* $p|q$, *i.e.,* $N_0 \bigcup N_1 = \partial V$, $V \subset \hat{E}_k$, *integral quantum supermanifold of dimension* $(p+1|q+1)$. *Then, we get that also* $(N_0)_C \bigcup (N_1)_C = \partial V_C$ *iff* $(N_0)_C$ *and* $(N_1)_C$ *have the same Stiefel-Whitney and Euler characteristic numbers.*

Proof. Note that integral bordisms in QSPDEs are defined also by considering the integrable characteristic fullquantum vector fields defined on the integral quantum supermanifolds of dimension $(p+1|q+1)$ that cobord admissible integral quantum supermanifolds of dimension $(p|q)$. If $\partial V = N$, we say that N has *positive orientation, (negative orientation),* into V if with respect to the characteristic vector field $\zeta : V \to TV$, $\zeta|_N$ is directed outside (resp. inside) V. Then, considering the canonical classic limit V_C of a quantum integral supermanifold V, and the canonical embedding $i : V_C \to V$, we can see that if $\zeta : V \to \widehat{TV}$ is such a characteristic vector field, then it is defined also a vector field $\zeta_C : V_C \to TV_C$ such that the following diagram is commutative

$$
\begin{array}{ccc}
V & \xrightarrow{\ \zeta\ } & \widehat{TV} \\
i \uparrow & & \uparrow i_* \\
V_C & \xrightarrow[\ \zeta_C\ ]{} & TV_C
\end{array}
$$

where i_* is the embedding induced by i and the epimorphism $c : A \to \mathbf{K}$. Namely, $i_* \equiv Hom(c; T(i)) : Hom_{\mathbf{K}}(\mathbf{K}; TV_C) \to Hom_Z(A; TV|_{V_C})$. If $\phi : A \times V \to V$ is the quantum flow associated to ζ and $\phi_C : \mathbf{K} \times V_C \to V_C$ is the flow associated to ζ_C, then one has also the following commutative diagram:

$$
\begin{array}{ccc}
A \times V & \xrightarrow{\ \phi\ } & V \\
\epsilon \times i \uparrow & & \uparrow i \\
\mathbf{K} \times V_C & \xrightarrow[\ \phi_C\ ]{} & V_C
\end{array}
$$

So, if $\partial V = N$ one has $\partial V_C = N_C$. Let us denote by $\Omega_p^{\hat{E}_k}{}_c$ the integral p-bordism group of the classic limits of integral quantum supermanifolds of $\hat{E}_k$. Now, one has the following canonical short exact sequence: $\Omega_p^{\hat{E}_k} \to \Omega_p^{\hat{E}_k}{}_c \to 0$ induced by the epimorphisms $N \to N_C$, namely $[N]_{\hat{E}_k} \mapsto [N_C]_{\hat{E}_k}$. Then, by denoting $\Omega_p^\uparrow$ the Reinhart p-bordism groups and Ω_p the p-bordism

group for smooth finite dimensional manifolds respectively [65,81], one has
the following exact commutative diagram

$$
\begin{array}{ccccccccc}
\Omega_p^{\hat{E}_k} & \to & \Omega_{\underset{c}{p}}^{\hat{E}_k} & \to & 0 & & & & \\
 & & \downarrow & & & & & & \\
0 & \to & K_p^\uparrow & \to & \Omega_p^\uparrow & \to & \Omega_p & \to & 0
\end{array}
$$

that gives a relation between integral bordism groups in QSPDEs and Reinhart p-bordism groups. This has as consequence that if $N_0 \bigcup N_1 = \partial V$,
then we have also $(N_0)_C \bigcup (N_1)_C = \partial V_C$ iff $(N_0)_C$ and $(N_1)_C$ have the
same Stiefel-Whitney and Euler characteristic numbers. $\qquad\square$

Remark 2.14. The spaces of conservation laws of QSPDEs, similarly to
QPDEs, identify quantum Hopf algebras that now are $\mathbf{Z}_2$-graded with a
natural structure of quantum superalgebra induced. (For informations on
noncommutative Hopf algebras see e.g. refs.[67]. Quantum Hopf algebras
are generalizations of such algebras [82].) $\qquad\blacksquare$

Definition 2.25. A *quantum Hopf superalgebra D* is a quantum Hopf algebra, in the sense of definition given in ref.[67], (see Theorem 1.1 there),
when the extension is obtained by means of a quatum superalgebra. More
precisely, D is an extension

$$0 \to C \equiv A \otimes_{\mathbf{K}} H \to D \to D/C \to 0,$$

where H is an Hopf $\mathbf{K}$-algebra and A is a quantum superalgebra.

The following theorem gives an important structure for the spaces of quantum conservation laws of QSPDEs and generalizes Theorem 5.9 in ref.[67].
(See also ref.[71].)

Theorem 2.31. *The space of conservation laws of a QSPDE $\hat{E}_k \subset \hat{J}^k_{m|n}(W)$
has a natural structure of quantum Hopf superalgebra. We call such a space
the quantum Hopf superalgebra of $\hat{E}_k$.*

Proof. Let us consider the following lemma.

Lemma 2.12. *Let $\hat{E}_k \subset \hat{J}^k_{m|n}(W)$ be a QSPDE. Then there exists a double
bigraded spectral sequence $\{\hat{E}_{r|s}^{\bullet|\bullet,\bullet|\bullet}(\hat{E}_\infty), d_{r|s}\}$ canonically associated to
$\hat{E}_k$, (Cartan quantum superspectral sequence), that converges to $\hat{H}^{\bullet|\bullet}(\hat{E}_\infty)$.
Furthermore, if $\hat{E}_k$ is path-connected and formally quantum integrable, one
has $\hat{H}^{\bullet|\bullet}(\hat{E}_k) \cong \hat{H}^{\bullet|\bullet}(\hat{E}_\infty)$.*

Proof of Lemma 2.12. Let us consider the bigraded complex $\{\widehat{\Omega}^{\bullet|\bullet}(\hat{E}_\infty), d_{\bullet|\bullet}\}$, where $\widehat{\Omega}^{p|q}(\hat{E}_\infty) \equiv (A_0 \otimes_{\mathbf{K}} \widehat{\Omega}^p(\hat{E}_\infty)) \bigoplus (A_1 \otimes_{\mathbf{K}} \widehat{\Omega}^q(\hat{E}_\infty))$. One has the isomorphism:

$$\{\widehat{\Omega}^{\bullet|\bullet}(\hat{E}_\infty), d_{\bullet|\bullet}\} \cong \{A_0 \otimes_{\mathbf{K}} \widehat{\Omega}^\bullet(\hat{E}_\infty), d_\bullet\} \bigoplus \{A_1 \otimes_{\mathbf{K}} \widehat{\Omega}^\bullet(\hat{E}_\infty), d_\bullet\}.$$

Therefore, we can identify a spectral sequence

$$(2.1) \qquad \hat{E}_{r|s}^{u|v,p|q}(\hat{E}_\infty) = (A_0 \otimes_{\mathbf{K}} \hat{E}_r^{u,p}(\hat{E}_\infty)) \bigoplus (A_1 \otimes_{\mathbf{K}} \hat{E}_s^{v,q}(\hat{E}_\infty)),$$

where $\hat{E}_r^{\bullet,\bullet}(\hat{E}_\infty)$ is the Cartan quantum spectral sequence associated to $\hat{E}_k$, (see ref.[67]). Then, the rest of the proof is a direct consequence of (2.1) and Theorem 3.6 in ref.[67]. $\qquad\square$

Now, a $(p|q)$-conservation quantum superlaw of a QSPDE $\hat{E}_k \subset \hat{J}^k_{m|n}(W)$ is any function $f : \Omega_{p|q}^{\hat{E}_\infty} \to B \equiv \widehat{\dot{A}}(E) \equiv \bigoplus_{r \geq 0} Hom_Z(\dot{T}_0^r(A); E)$. On the other hand, some conservation quantum superlaw can be also identified by means of differential forms on the infinity prolongation $\hat{E}_\infty$ of $\hat{E}_k$. Thus, we can also consider the spectral term $\hat{E}_{1|1}^{0|0,p|q}$, $0|0 \leq p|q \leq m-1|n-1$ as the "space of $(p|q)$-conservation quantum superlaws" of $\hat{E}_k$. Set $\hat{\mathbf{H}}_{p|q}(\hat{E}_\infty) \equiv \mathrm{Map}(\Omega_{p|q}^{\hat{E}_\infty}; B)$. One has a canonical mapping: $j : \hat{E}_{1|1}^{0|0,p|q} \to \hat{\mathbf{H}}_{p|q}(\hat{E}_\infty)$. Let us denote $j(\hat{E}_{1|1}^{0|0,p|q}) \equiv \hat{E}^{0,p|q}$ and $< \hat{E}^{0,p|q} > \subset \hat{\mathbf{H}}_{p|q}(\hat{E}_\infty)$ the quantum Hopf superalgebra generated by $\hat{E}^{0,p|q}$. If $\hat{E}^{0,p|q} = 0$ we put $< \hat{E}^{0,p|q} > = \hat{\mathbf{H}}_{p|q}(\hat{E}_\infty)$. Emphasize that $\hat{\mathbf{H}}_{p|q}(\hat{E}_\infty)$ is just a quantum Hopf superalgebra in the just specified sense. (The proof is similar to one given in ref.[67] for the quantum Hopf algebra of a QPDE.) $\qquad\square$

Example 2.8. (*Quantum super d'Alembert equation*). Let $A = A_0 \oplus A_1$ be a quantum superalgebra with $Z = Z(A)$ Noetherian and $\mathbf{K} = \mathbf{R}$. Let us consider the following trivial fiber bundle $\pi : W \equiv A^3 \to A^2 \equiv M$ with quantum coordinates $(x, y, u) \mapsto (x, y)$. Then, the quantum super d'Alembert equation is the QSPDE $\widehat{(d'A)} \subset J\hat{D}^2(W) \subset \hat{J}_2^2(W)$, defined by means of the following $\hat{A}$-valued Q_w^∞-function $F \equiv uu_{xy} - u_x u_y : J\hat{D}^2(W) \to \overset{2}{\hat{A}} \subset \overset{\infty}{\hat{A}}$. By forgetting the $\mathbf{Z}_2$-gradiation of A, we get the same situation just considered in ref.[67]. So, $\widehat{(d'A)} \subset J\hat{D}^2(W)$ is just a formally quantum integrable QSPDE of dimension $(3, 2, 2)$ over the quantum algebra $B \equiv A \times \overset{1}{\hat{A}} \times \overset{2}{\hat{A}}$, in the open quantum submanifold $u \neq 0$, and also completely quantum integrable there. On the other hand, by considering that M has a natural structure of quantum supermanifold of dimension $(2|2)$ over A, it follows also that for any initial condition, i.e., any point $q \in \widehat{(d'A)} \setminus u^{-1}(0)$, passes a quantum supermanifold of dimension $(2|2)$ over A, solution of $\widehat{(d'A)}$. Therefore, also the QSPDE $\widehat{(d'A)} \subset \hat{J}^2_{2|2}(W)$ is completely quantum super-integrable in the quantum supermanifold $u \neq 0$. We can also state that $\widehat{(d'A)}$ is completely quantum superintegrable, as it is algebraic in the open set $\widehat{(d'A)} \setminus u^{-1}(0)$. Furthermore, by applying Theorem 2.25 we get that the integral $(1|1)$-bordism group of $\widehat{(d'A)}$ is trivial. In fact, we have: $\Omega_{1|1}^{\widehat{(d'A)}} \cong$

$$^A\underline{\Omega}_{1|1}(W) \cong (Z \otimes_{\mathbf{K}} H_1(W; \mathbf{K})) \bigoplus (A_1 \otimes_{\mathbf{K}} H_1(W; \mathbf{K})) = 0 \cong \Omega_{1|1}^{\widehat{(d'A)}\infty}.$$ This is enough to state the existence of global smooth solutions of the QSPDE $\widehat{(d'A)}$ for any admissible compact closed smooth Cauchy data. Furthermore, as a consequence of the actual structure of the integral bordism group $\Omega_{1|1}^{\widehat{(d'A)}\infty}$, we get also that the quantum Hopf superalgebra of $\widehat{(d'A)}$ is

$$\hat{\mathbf{H}}_{1|1}((\widehat{d'A})_{+\infty}) \cong \hat{\overset{\infty}{A}} \equiv \bigoplus_{r \geq 0} Hom_Z(\dot{T}^r_0(A); A). \qquad \blacksquare$$

Example 2.9. (*Quantum super Navier-Stokes equation*). In the previous section we have considered the Navier-Stokes equation for quantum fluids as a QPDE $(NS) \subset J\hat{D}^2(W)$, where $\pi : W \to M$ is a suitable fiber bundle over the quantum Galilean space-time M. (See also ref.[69].) Now, we can extend such considerations to quantum superfluids, i.e., quantum fluids in the category of quantum supermanifolds, by obtaining a QSPDE that formally appears the same as written in ref.[69], but with an underlying quantum superalgebra A. Then by using results in ref.[69] we can prove that when A has Noetherian centre $Z = Z(A)$, (NS) contains a formally quantum integrable QSPDE $\widehat{NS} \subset J\hat{D}^2(W)$, that is also completely quantum superintegrable. Then, by using Theorem 2.25 one can prove that the integral $(3|3)$-bordism group $\Omega_{3|3}^{\widehat{(NS)}} = 0 = \Omega_{3|3}^{\widehat{(NS)}+\infty}$. This is enough to state the existence of global smooth solutions of the QSPDE (NS) for any admissible smooth boundary condition contained int $\widehat{(NS)}$. Furthermore, we get also that the quantum Hopf superalgebra of $\widehat{(NS)}$ is

$$\hat{\mathbf{H}}_{3|3}((\widehat{NS})_{+\infty}) \cong \hat{\overset{\infty}{A}} \equiv \bigoplus_{r \geq 0} Hom_Z(\dot{T}^r_0(A); A). \qquad \blacksquare$$

2.4 - THE QUANTUM SUPER YANG-MILLS EQUATIONS

In this section we shall apply our previous geometric theory of QSPDEs to study very important partial differential equations of quantum physics that describe the superunification of the fundamental forces of the nature with the gravity. Here the situation is completely different to what could happear at macroscopic level, where gravity should not influence the interactions between microscopic particles, as their masses are too small to produce sensible gravitational effects. From the quantum point of view, instead, quantum gravity, has a very strong influence on the interactions between fundamental particles. The quantum space-time is a sort of foam of quantum black holes, that are like laboratories producing all the particles. (For a modern exposition of the physical principles of quantum gravity, see, e.g. refs.[96–100] and references quoted there. In particular, for general informations on quantum field theory see e.g., references quoted in [96].) In this section we aim just to show that our mathematical theory of QSPDEs gives a suitable mathematical framework to describe such microscopic worlds, and to solve basic open problems in quantum field theory.

Definition 2.26. A *quantum Lie supergroup* of dimension $(d|N)$ over a quantum superalgebra A is a group G that has also a structure of quantum supermanifold of dimension $(d|N)$ over A, compatible with the structure of group.

TAB.2.12 - Fullquantum Lie superalgebra.

$$\widehat{C}^C_{AB} = -(-1)^{|A||B|}\widehat{C}^C_{BA}$$

$$(-1)^{|A||C|}\widehat{C}^E_{AD}\widehat{C}^D_{BC} + (-1)^{|B||A|}\widehat{C}^E_{BD}\widehat{C}^D_{CA} + (-1)^{|C||B|}\widehat{C}^E_{CD}\widehat{C}^D_{AB} = 0$$

Remark 2.15. If M is a quantum supermanifold of dimension $(p|q)$ over the quantum superalgebra $A = A_0 \oplus A_1$, we get that any vector field $\zeta :$ $M \to TM$ identifies a $\mathbf{K}$-derivation on the algebra $C^\infty_w(M, \mathbf{K})$ at $p \in M$. Furthermore has sense to talk of the product of such derivations $\zeta\xi$ by means of the formula

$$\zeta\xi.f = (\partial x_A \partial x_B.f)\xi^B \otimes_{\mathbf{K}} \zeta^A + (\partial x_B.f)(\partial x_A.\xi^B)\zeta^A,$$

where above formula is meant by composition, (for any $p \in U$):

$$\mathbf{K} \otimes_{\mathbf{K}} \mathbf{K} \xrightarrow{\xi^B(p)\otimes\zeta^A(p)} A \otimes_{\mathbf{K}} A \xrightarrow{(\partial x_A\partial x_B.f)(p)} \mathbf{K},$$

$$\mathbf{K} \xrightarrow{\zeta^A(p)} A \xrightarrow{(\partial x_A.\xi^B)(p)} A \xrightarrow{(\partial x_B.f)(p)} \mathbf{K}.$$

Note that above compositions identify numbers of $\mathbf{K}$. In fact, $\mathbf{K} \otimes_{\mathbf{K}} \mathbf{K} \cong \mathbf{K}$ and $Hom_{\mathbf{K}}(\mathbf{K}; \mathbf{K}) \cong \mathbf{K}$. So we can define the following Lie bracket on $C_w^\infty(TM)$: $[\zeta, \xi] \equiv \zeta\xi - \xi\zeta$. In this way $C_w^\infty(TM)$ becomes a Lie algebra that we denote by $\mathcal{L}(M)$, and call it the *quantum Lie algebra of vector fields* C_w^∞ on M. Furthermore, a fullquantum vector field $\zeta \in Q_w^\infty(\widehat{TM})$, locally written $\zeta = \partial x_A \zeta^A$, with $\zeta^A \in Q_w^\infty(U, \widehat{A})$, identifies a Z-derivation at $p \in M$ on $Q_w^\infty(M, A)$, if $\zeta(p) \in Hom_Z(A; T_pM)$ is restricted to $Z \subset A$. In fact, for any $f \in Q_w^\infty(M, A)$ and $p \in M$, $(\partial x_A.f)(p)\zeta^A(p)$ is given by composition:

$$Z \xrightarrow{\zeta^A(p)} A \xrightarrow{(\partial x_A.f)(p)} A.$$

As $(\partial x_A.f)(p)\zeta^A(p) \in Hom_Z(Z; A) \cong A$ we get that $(\partial x_A.f)\zeta^A$ identifies an A-valued function of class Q_w^∞ on M. In this way we can also consider the product of two fullquantum vector fields $\zeta, \xi \in Q_w^\infty(\widehat{TM})$ as a Z-derivation on $Q_w^\infty(M, A)$. Hence, we can consider the bracket

$$[\zeta, \xi] \equiv \zeta\xi - (-1)^{|\xi||\zeta|}\xi\zeta = \partial x_A[\zeta, \xi]^A,$$

where

$$[\zeta, \xi]^A = \left((\partial x_B.\zeta^A)\xi_\bullet^B - (-1)^{|\xi^A||\zeta^B|}(\partial x_B.\xi^A)\zeta_\bullet^B \right) \in Q_w^\infty(M; \widehat{A}),$$

where $\xi_\bullet^B$, (resp. $\zeta_\bullet^B$), denote the restriction of ξ^B, (resp. ζ^B), to Z. Then $Q_w^\infty(\widehat{TM})$ becomes a quantum superalgebra that we call *the fullquantum Lie superalgebra* of M and we denote by $\widehat{\mathcal{L}}(M)$. Finally, if G is a quantum Lie supergroup of dimension $(m|n)$, over quantum superalgebra A, one has the subalgebras $\mathcal{L}_G(G) \subset \mathcal{L}(G)$, $\widehat{\mathcal{L}}_G(G) \subset \widehat{\mathcal{L}}(G)$ respectively of the vector fields G-invariants and fullquantum vector fields G-invariants on G. Then on the tangent space T_eG, at the unit $e \in G$, one has a structure of Lie quantum algebra, (resp. Lie quantum superalgebra), denoted by $\mathfrak{g}$, (resp. $\widehat{\mathfrak{g}}$). One has $\widehat{\mathfrak{g}} = Hom_Z(A; \mathfrak{g})$. We get the following short exact sequence of Z-modules: $\widehat{\mathfrak{g}} \to \mathfrak{g} \to 0$. If $\{Z_K\}_{1 \leq K \leq m+n}$ is a system of generators for the right $\widehat{A}$-module $\widehat{\mathfrak{g}}$, then we get

$$(2.2) \qquad\qquad [Z_K, Z_H] = Z_I \widehat{C}_{KH}^J, \quad \widehat{C}_{KH}^J \in \widehat{A}.$$

The $\widehat{C}_{KH}^J$ are called *fullquantum structure constants* of G. They satisfy the conditions reported in Tab.2.12. Note that any element $\zeta \in \mathfrak{g}$ can be represented in the form $\zeta = \sum_K Z_K(\zeta_\bullet^K)$, if $\hat{\zeta} = Z_K\zeta^K$ is a vector of $\hat{\mathfrak{g}}$ that projects on $\zeta \in \mathfrak{g}$. Therefore, the projection of relations (2.2) on $\mathfrak{g}$ gives:

$$[Z_K, Z_H]_\bullet = Z_I \hat{C}^J_{KH}|_\bullet, \quad \hat{C}^J_{KH}|_\bullet \equiv \hat{C}^J_{KH}|_Z \in A.$$

We set $C^J_{KH} \equiv \hat{C}^J_{KH}|_\bullet \in A$ and we call them *quantum structure constants* of G.   ∎

Definition 2.27. A *quantum super Yang-Mills* PDE is the dynamic equation of a gauge theory, built in the category of quantum supermanifolds, hence with structure group a quantum Lie supergroup.

Theorem 2.32. *A quantum super Yang-Mills PDE, over a quantum superspacetime M, can be considered a QSPDE $(\widehat{YM}) \subset J\hat{D}^\infty(\bar{C})$, on the fiber bundle $\bar{\pi} : \bar{C} \equiv Hom_Z(TM; \mathfrak{g}) \to M$, where $\mathfrak{g}$ is the quantum Lie superalgebra of the structure group and Z is the centre of the quantum superalgebra A underlying M.*

Proof. We shall use the following fundamental lemma.

TAB.2.13 - Local quantum coordinates on $\pi{:}C{\to}M$ and local isomorphisms.

$M \cong A^{4\|N_1} \subset A^m$	$(x^\alpha, \theta^a)_{0\leq\alpha\leq3;1\leq a\leq N_1} \equiv (X^A)_{1\leq A\leq(4+N_1)} \equiv m$ $(x^\alpha,\theta^a){:}U{\subset}M{\to}A_0{\times}A_1, \quad \|x^\alpha\|=0, \quad \|\theta^a\|=1$
$G \cong A^{d\|N_2} \subset A^n$	$(y^i, \psi^b)_{1\leq i\leq d;1\leq b\leq N_2} \equiv (Y^B)_{1\leq B\leq(d+N_2)} \equiv n$ $(y^i,\psi^b){:}U{\subset}G{\to}A_0{\times}A_1, \quad \|y^i\|=0, \quad \|\psi^b\|=1$
$P \cong A^{4+d\|N_1+N_2} \subset A^{m+n}$	$(x^\alpha,\theta^a,y^i,\psi^b) \equiv (X^A,Y^I)_{1\leq A\leq m,1\leq I\leq n} \equiv (V^I)_{1\leq I\leq m+n}$ $x^\alpha,y^i{:}U{\subset}P{\to}A_0, \; \theta^a,\psi^b{:}U{\subset}P{\to}A_1$

$$C \cong A^{4+d\|N_1+N_2} \times (\widehat{A}{}^1_{\,0}(A))^{(4+d)d\|(4+d)N_2} \times (\widehat{A}{}^1_{\,1}(A))^{(N_1+N_2)N_2\|(N_1+N_2)d}$$

$$\subset A^{m+n} \times (\widehat{A}{}^1)^{(m+n)n}$$

$$(X^A, Y^I, \dot{Y}^I_A, \dot{Y}^I_J)_{1\leq A\leq m,1\leq I,J\leq n},$$

$$|X^A|=|A|, \quad |Y^I|=|I|, \quad |\dot{Y}^I_A|=|I|+|A|, \quad |\dot{Y}^I_J|=|I|+|J|$$

Lemma 2.13. *Let $A = A_0 \oplus A_1$ be a quantum superalgebra. Let $\pi_P : P \to M$ be a principal fiber bundle, where M is a $(4|N_1)$-dimensional quantum supermanifold over A, (quantum superspacetime); the structure group of P is a quantum Lie supergroup G of dimension $(d|N_2)$; P is* hence a quantum supermanifold of dimension $(4 + d|N_1 + N_2)$. Let us put quantum coordinates on P as reported in Tab.2.13. A *quantum pseudoconnection*, that is a principal connection $_1\mu$ of class Q_w^∞ on the trivial fiber bundle $P \times \mathfrak{g} \to P$, where $\mathfrak{g}$ is the quantum Lie superalgebra corresponding to G, is a section of the following fiber bundle over P: $\pi_E : \underline{E} \equiv Hom_Z(TP;\mathfrak{g}) \to P$. A *quantum configuration* is a section c of class Q_w^∞ of $\pi : C \equiv Hom_Z(TP;\mathfrak{g}) \to M$, with $\pi \equiv \pi_P \circ \pi_E$, such that $c = _1\mu \circ s$, where $s : M \to P$ is a section of class Q_w^∞ of π_P. Then, a *quantum super Yang-Mills equation* $(\widehat{YM}) \subset J\hat{D}^\infty(C)$ is the variational QSPDE for quantum configurations, corresponding to a fixed quantum Lagrangian density $\mathcal{L}$ on C: $\mathcal{L} = LdX^1\triangle \cdots \triangle dX^m \in Q_w^\infty(Hom_Z(\dot{\Lambda}_0^m(J\hat{D}^\infty(C)); A))$. $L(q) \in \overset{m}{\hat{A}}, \forall q \in J\hat{D}^\infty(C))$. Then the local expression of $(\widehat{YM})$ is given by the equations reported in Tab.2.14, where $\bar{R}_{A_1 A_2}^K$ and $\bar{\mu}_K^A$ are the curvature and pseudoconnection components induced on M by means of any section $s : M \to P$ of class Q_w^∞. The *blow-up structure* $\pi^*C(P) \subset \underline{E}$, implies that there are acceptable only quantum pseudoconnections that can be written in the following form $_1\mu = (\partial X_\alpha + Z_K\bar{\mu}_\alpha^K)\otimes dX^\alpha$, where $C(P) \cong J\hat{D}(P)/G$ is the fiber bundle over M of the principal quantum connections on P. Recall, (see Proposition 1.45), that

$$[\bar{\mu}_{A_1}^I, \bar{\mu}_{A_2}^J]_+ \equiv (\bar{\mu}^I \wedge \bar{\mu}^J)_{A_1 A_2} = \bar{\mu}_{A_1}^I \bar{\mu}_{A_2}^J - (-1)^{|I||J|}\bar{\mu}_{A_1}^J \bar{\mu}_{A_2}^I,$$
$$[\bar{\mu}_H^I, \bar{R}_{A_1 A_2}^J]_+ \equiv (\bar{\mu}^I \wedge \bar{R}^J)_{H A_1 A_2}$$
$$= \bar{\mu}_H^I \bar{R}_{A_1 A_2}^J + (-1)^{|I||J|}\bar{R}_{H A_1}^J \bar{\mu}_{A_2}^I.$$

If we use the relations in Tab.2.12 of the full quantum structure constants, we can also write the fields $\bar{R}_{A_1 A_2}^K$ in the following equivalent way:

$$\bar{R}_{A_1 A_2}^K = (\partial X_{A_1} \cdot \bar{\mu}_{A_1}^K) + \hat{C}_{IJ}^K \bar{\mu}_{A_2}^I \bar{\mu}_{A_1}^J.$$

Furthermore, $[i]$ is the length of multi-index i: $[i] = i_1 + \cdots + i_m$. Here $\{\partial_B\}$ denotes the local basis for Cartan fields on $J\hat{D}^\infty(C)$ and $\partial_i \equiv (\partial_1)^{i_1} \cdots (\partial_m)^{i_m}$, where $\partial_\mu = \partial X_\mu + \sum_{k,A} \bar{\mu}_{k\mu}^A \partial\bar{\mu}_A^k$.

TAB.2.14 - Local expression of $(\widehat{YM}) \subset J\hat{D}^{\infty}(C)$ and Bianchi identity $(B) \subset J\hat{D}^2(C)$.

(*Field equations*) $E_K^A \equiv \frac{\delta L}{\delta \bar{\mu}_K^A} \equiv \sum_{i \in I} (-1)^{[i]} \partial_i \left(\frac{\partial L}{\partial \bar{\mu}_{Ki}^A} \right) = 0$	$((\widehat{YM}))$
(*Fields*) $F_{A_1 A_2}^K \equiv \bar{R}_{A_1 A_2}^K - \left[(\partial X_{A_2} \cdot \bar{\mu}_{A_1}^K) + \frac{1}{2} \widehat{C}_{IJ}^K [\bar{\mu}_{A_2}^I, \bar{\mu}_{A_1}^J]_+ \right] = 0$	
(*Bianchi identities*) $B_{HA_1 A_2}^K \equiv (\partial X_H \cdot \bar{R}_{A_1 A_2}^K) + \frac{1}{2} \widehat{C}_{IJ}^K [\bar{\mu}_H^I, \bar{R}_{A_1 A_2}^J]_+ = 0$	(B)

$$F_{A_1 A_2}^K : \Omega_1 \subset J\hat{D}(C) \to \overset{2}{\widehat{A}}; \quad B_{HA_1 A_2}^K : \Omega_2 \subset J\hat{D}^2(C) \to \overset{3}{\widehat{A}}; \quad E_K^A : \Omega_k \subset J\hat{D}^{\infty}(C) \to \overset{\infty}{\widehat{A}}.$$

Proof of Lemma 2.13. We shall use some preliminary definitions and results.
Remark 2.16. (*Nonholonomic quantum jet-derivative spaces*). Let $\pi : W \to M$ be a fiber bundle in the category of quantum supermanifolds, where $\dim_A M = (m|n)$ and $\dim_B W = (m|n, r|s)$, with $B \equiv A \times E$ a quantum superalgebra with E also a Z-module, where Z is the centre of A. The *k-semi-holonomic quantum jet-derivative space* on $\pi : W \to M$ is the quantum supermanifold $\bar{J}\hat{D}^k(W) \subset J\hat{D}(\bar{J}\hat{D}^{k-1}(W))$ defined by means of the following exact commutative diagram:

$$
\begin{array}{ccccccc}
0 & \to & \bar{J}\hat{D}^k(W) & \to & J\hat{D}(\bar{J}\hat{D}^{k-1}(W)) & \overset{J\hat{D}(\bar{\pi}_{k-1,k-2})}{\underset{\bar{f}_{k-1}}{\rightrightarrows}} & J\hat{D}(\bar{J}\hat{D}^{k-2}(W)) \\
& & & & \downarrow & & \| \\
0 & \to & \bar{J}\hat{D}^{k-1}(W) & & \to & & J\hat{D}(\bar{J}\hat{D}^{k-2}(W)) \\
& & & & \downarrow & & \\
& & & & 0 & &
\end{array}
$$

We put $\bar{J}\hat{D}(W) = J\hat{D}(W)$. $\bar{\pi}_{k,k-1} : \bar{J}\hat{D}^k(W) \to \bar{J}\hat{D}^{k-1}(W)$ is an affine fiber bundle with associated vector bundle

$$\bar{\pi}_{k-1,0}^*(\dot{T}_0^k MO_Z vTW)$$
$$\equiv \bar{\pi}_{k-1,0}^* Hom_Z(\dot{T}_0^k M; vTW).$$

The *k-sesquiholonomic quantum jet-derivetive space* $\check{J}\hat{D}^k(W) \subset \bar{J}\hat{D}^k(W)$ over $\pi : W \to M$ is defined by means of the following exact commutative diagram:

$$
\begin{array}{ccccccc}
0 & \to & \check{J}\hat{D}^k(W) & \to & J\hat{D}(J\hat{D}^{k-1}(W)) & \overset{J\hat{D}(\pi_{k-1,k-2})}{\underset{f_{k-1}}{\rightrightarrows}} & J\hat{D}(J\hat{D}^{k-2}(W)) \\
& & & & \downarrow & & \| \\
0 & \to & \bar{J}\hat{D}^{k-1}(W) & & \to & & J\hat{D}(J\hat{D}^{k-2}(W)) \\
& & & & \downarrow & & \\
& & & & 0 & &
\end{array}
$$

One has $\check{J}\hat{D}^2(W) = \bar{J}\hat{D}^2(W)$. $\check{\pi}_{k,k-1} : \check{J}\hat{D}^k(W) \to \check{J}\hat{D}^{k-1}(W)$ is an affine bundle with associated vector bundle

$$\pi_{k-1,0}{}^*(TMO_Z(\dot{S}_0^{k-1}MO_ZvTW))$$
$$\equiv \pi_{k-1,0}{}^*Hom_Z(TM; Hom_Z(\dot{S}_0^{k-1}M; vTW)).$$

This definition is similar to one for the (super)classic case [62,73]. However, as the quantum coordinates $\{x^A, y^B, y_A^B, y_{AC}^B, \cdots, y_{A_1\cdots A_k}^B\}$ on the k-holonomic spaces $J\hat{D}^k(W)$ have not symmetry in the indexes $A_1, \ldots, A_p$, $1 \le p \le k$, we cannot distinguish more such non-holonomic spaces simply on the ground of the quantum coordinates as made in the (super)classic case [62,73]. If $\pi : W \to M$ is a fiber bundle over M with algebraic structure on the fibres, then the fiber bundles $\bar{\pi}_k : \bar{J}\hat{D}^k(W) \to M$, (resp. $\check{\pi}_k : \check{J}\hat{D}^k(W) \to M$, resp. $\pi_k : J\hat{D}^k(W) \to M$), have the same algebraic structures on the fibers over M. A k-holonomic quantum connection, (resp. k-sesquiholonomic quantum connection, resp. k semiholonomic quantum connection), on $\pi : W \to M$ is a subbundle $\hat{C}_k \subset J\hat{D}^k(W)$, (resp. $\check{C}_k \subset \check{J}\hat{D}^k(W)$, resp. $\bar{C}_k \subset \bar{J}\hat{D}^k(W)$), identified by means of a section $\hat{\rceil} : J\hat{D}^{k-1}(W) \to J\hat{D}^k(W)$, (resp. $\check{\rceil} : \check{J}\hat{D}^{k-1}(W) \to \check{J}\hat{D}^k(W)$, resp. $\bar{\rceil} : \bar{J}\hat{D}^{k-1}(W) \to \bar{J}\hat{D}^k(W)$), of $\pi_{k,k-1}$, (resp. $\check{\pi}_{k,k-1}$, resp. $\bar{\pi}_{k,k-1}$), i.e., $\hat{C}_k$ is a k-(holonomic) QSPDE: $\bar{C}_k \equiv \bar{\rceil}(\bar{J}\hat{D}^{k-1}(W)) \subset \bar{J}\hat{D}^k(W)$, (resp. $\check{C}_k$ is a k-(sesquiholonomic) QSPDE: $\check{C}_k \equiv \check{\rceil}(\check{J}\hat{D}^{k-1}(W)) \subset \check{J}\hat{D}^k(W)$, resp. $\bar{C}_k$ is a k-semiholonomic QSPDE: $\check{C}_k \equiv \check{\rceil}(\check{J}\hat{D}^{k-1}(W)) \subset \check{J}\hat{D}^k(W)$). Similarly to the superclassic case [62] one can prove that the first quantum prolongation $\rceil^{(1)} : J\hat{D}^k(W) \to J\hat{D}(J\hat{D}^k(W))$ of a k-holonomic quantum connection $\rceil : J\hat{D}^{k-1}(W) \to J\hat{D}^k(W)$ has values into the $(k+1)$-sesquiholonomic quantum jet-derivative space $\check{J}\hat{D}^{k+1}(W)$: $\rceil^{(1)} : J\hat{D}^k(W) \to \check{J}\hat{D}^{k+1}(W)$. Note that the second quantum prolongation $\rceil^{(2)}$ of $\rceil$ does not take values into $\check{J}\hat{D}^{k+2}(W)$ but into a different subspace of $\bar{J}\hat{D}^{k+2}(W)$. So we can write $\rceil^{(2)} : J\hat{D}^{k+1}(W) \to \bar{J}\hat{D}^{k+2}(W)$. Furthermore, $\bar{C}_k$ identifies (and it is identifed by) an affine morphism $\bar{\lceil} : \bar{J}\hat{D}^k(W) \to \bar{\pi}_{k-1,0}^*(\dot{T}_0^k MO_ZvTW)$ over $\bar{J}\hat{D}^{k-1}(W)$, whose fibre derivative is $vD\bar{\lceil} = 1$. One has $\bar{\lceil} \circ \bar{\rceil} = 0$, $\bar{\lceil} = id_{\bar{J}\hat{D}^k(W)} \circ \bar{\rceil} \circ \bar{\pi}_{k,k-1}$. Let $\{x^A, y^B, \cdots, y_{A_1\cdots A_{k-1}}^B, \bar{y}_{A_1\cdots A_k}^B\}$ be the naturally associated fibered quantum coordinates on $\bar{\pi}_{k-1,0}^*(\dot{T}_0^k MO_ZvTW)$.

Then, the following local expression for $\bar{\lceil}$ is reported in Tab.2.15.

TAB.2.15 - Local expression for $\bar{\lceil}$.

$$x^A \circ \bar{\lceil} = x^A, \cdots, y^B_{A_1 \cdots A_{k-1}} \circ \bar{\lceil} = y^B_{A_1 \cdots A_{k-1}}$$

$$\bar{y}^B_{A_1 \cdots A_k} \circ \bar{\lceil} = y^B_{A_1 \cdots A_k} + \bar{\lceil}^B_{A_1 \cdots A_k}$$

$$-\bar{\lceil}^B_{A_1 \cdots A_k} = y^B_{A_1 \cdots A_k} \circ \bar{\lceil}$$

$(\bar{\lceil}^B_{A_1 \cdots A_k}$ are $\overset{k}{\hat{A}}(E)$-valued local functions on $\bar{J}\hat{D}^{k-1}(W)$.) Hence the local expression of $\bar{C}_k$ is the following: $y^B_{A_1 \cdots A_k} + \bar{\lceil}^B_{A_1 \cdots A_k} = 0$. If $\pi : W \to M$ is endowed with an algebraic structure on the fibers then, the functor $\bar{J}\hat{D}^k(-)$ extends this algebraic structure to the fiber bundle $\bar{\pi}_k : \bar{J}\hat{D}^k(W) \to M$. Then, if one recognizes the same algebraic structure on the fibers of a k-semiholonomic quantum connection $\bar{C}_k = \bar{\lceil}(\bar{J}\hat{D}^{k-1}(W)) \subset \bar{J}\hat{D}^k(W)$, we say that $\bar{C}_k$ is an algebraic k-semiholonomic quantum connection on $\pi : W \to M$. This is equivalent to say that $\bar{\lceil}(\bar{J}\hat{D}^{k-1}(W)) \to \bar{J}\hat{D}^k(W)$ is a morphism over M with respect to this algebraic structure.   ■

Example 2.10. 1) If $\pi : W \to M$ is a vector bundle, a linear k-semiholonomic quantum connection $\bar{C}_k \equiv \bar{\lceil}(\bar{J}\hat{D}^{k-1}(W)) \subset \bar{J}\hat{D}^k(W)$ can be locally written as follows:

$$(2.3) \qquad \lceil^B_{A_1 \cdots A_k} = y^D \lceil^B_{A_1 \cdots A_k D} + \cdots + y^D_{C_1 \cdots C_{k-1}} \lceil^{B \quad\quad C_1 \cdots C_{k-1}}_{A_1 \cdots A_k D}$$

where $\lceil^{B \quad\quad C_1 \cdots C_p}_{A_1 \cdots A_k D}$ are $Hom_Z(\dot{T}^k_0(A) \otimes_Z Hom_Z(\dot{T}^p_0(A); E); E)$-valued local functions on M. (Note that $\dot{T}^0_0(A) = Z$.)   ■

2) If $\pi : E \to M$ is an affine fiber bundle, (in the category of quantum supermanifolds), an affine k-semiholonomic quantum connection $\bar{C}_k = \bar{\lceil}(\bar{J}\hat{D}^{k-1}(E)) \subset \bar{J}\hat{D}^k(E)$ can be locally characterized by:

$$(2.4) \quad \lceil^B_{A_1 \cdots A_k} = \tilde{\lceil}^B_{A_1 \cdots A_k} + y^D \lceil^B_{A_1 \cdots A_k D} + \cdots + y^D_{C_1 \cdots C_{k-1}} \lceil^{B \quad\quad C_1 \cdots C_{k-1}}_{A_1 \cdots A_k D}$$

where $\lceil^{B \quad\quad C_1 \cdots C_p}_{A_1 \cdots A_k D}$ are as in above example and $\tilde{\lceil}^B_{A_1 \cdots A_k}$ are $\overset{k}{\hat{A}}(E)$-valued local functions on M.   ■

If $\bar{C}_k$ is a k-semiholonomic quantum connection on $\pi : W \to M$, then we define *k-semiholonomic covariant map of a section* $u : M \to \bar{J}\hat{D}^k(W)$ the following section: $\overline{\nabla}u \equiv \bar{\lceil} \circ u : M \to \dot{T}^k_0 MO_Z s^* vTW$, with $s \equiv \bar{\pi}_{k-1,0} \circ u :$

$M \to W$. Similar definitions and results hold for k-sesquiholonomic and k-holonomic quantum connections. In particular, if $\bar{\rceil}$ is such that there exists a (holonomic) k-quantum connection $\rceil : J\hat{D}^{k-1}(W) \to J\hat{D}^k(W)$ such that the following diagram is commutative

$$
\begin{array}{ccc}
J\hat{D}^{k-1}(W) & \overset{\bar{\rceil}}{\to} & J\hat{D}^k(W) \\
\cup & & \cup \\
J\hat{D}^{k-1}(W) & \overset{\rceil}{\to} & J\hat{D}^k(W)
\end{array}
$$

then we say that $\bar{\rceil}$ is the *blow up* of $\rceil$, and for any section $s : M \to W$ we have defined the k-covariant derivative $\overset{k}{\overline{\nabla}} s \equiv \overline{\nabla} D^k s : M \to \dot{S}_0^k MO_Z s^* vTW$. The local expressions are reported in Tab.2.16.

TAB.2.16 - Local expression for $\overset{k}{\overline{\nabla}} s$.

$$\overline{\nabla} u \equiv \partial y_B \otimes (\overline{\nabla} u)^B_{A_1 \cdots A_k} dx^{A_1} \otimes \cdots \otimes dx^{A_k}$$

$$\bar{y}^B_{A_1 \cdots A_k} \circ \overline{\nabla} u \equiv (\overline{\nabla} u)^B_{A_1 \cdots A_k} = u^B_{A_1 \cdots A_k} + \lceil^B_{A_1 \cdots A_k} \circ \bar{\pi}_{k,k-1} \circ u$$

$$\overset{k}{\overline{\nabla}} s \equiv \partial y_B \otimes (\overset{k}{\overline{\nabla}} s)^B_{A_1 \cdots A_k} dx^{A_1} \otimes \cdots \otimes dx^{A_k}$$

$$\bar{y}^B_{A_1 \cdots A_k} \circ \overset{k}{\overline{\nabla}} \equiv (\overset{k}{\overline{\nabla}} s)^B_{A_1 \cdots A_k} = (\partial x_{A_1} \cdots \partial x_{A_k} \cdot s^B) + \lceil^B_{A_1 \cdots A_k} \circ D^{k-1} s$$

The *semiholonomic quantum s-prolongation* of a k-semiholonomic quantum connection $\bar{\rceil}$ on W is given by the following: $\bar{\rceil}^{(s)} \equiv J\hat{D}(\bar{\rceil}^{(s-1)})|_{\bar{J}\hat{D}^{k+s-1}(W)}$, where $\bar{\rceil}^{(1)} = J\hat{D}(\bar{\rceil})|_{\bar{J}\hat{D}^k(W)} : \bar{J}\hat{D}^k(W) \to J\hat{D}(\bar{J}\hat{D}^k(W))$. Furthermore, $\bar{\rceil}^{(s)} : \bar{J}\hat{D}^{k+s-1}(W) \to \bar{J}\hat{D}^{k+s}(W)$. The s-semiholonomic quantum prolongation of $\bar{C}_k$ is $\bar{C}_{k+s} = \bar{\rceil}^{(s)}(\bar{C}_{k+s-1})$. Furthermore if $\pi : W \equiv E \to M$ is a vector bundle, there is an isomorphism of vector bundles over $\bar{J}\hat{D}^{k-1}(E)$: $\bar{J}\hat{D}^k(E) \cong \dot{T}_0^k MO_Z E \times_M \bar{J}\hat{D}^{k-1}(E)$ induced by means of the following splitted exact sequences of vector bundles over M:

$$0 \to \dot{T}_0^k MO_Z E \underset{\bar{\rceil}}{\rightleftharpoons} \bar{J}\hat{D}^k(E) \underset{\bar{\rceil}}{\rightleftharpoons} \bar{J}\hat{D}^{k-1}(E) \to 0.$$

A fullquantum $(p+1)$-differential form $\mu : M \to (\dot{\Lambda}_0^{p+1} M)^+$ on M identifies a quantum connection on the fiber bundle $\pi : (\dot{\Lambda}_0^p M)^+ \to M$. In fact we recognize the following subbundle $C_1 \equiv j(\pi^* \widetilde{C}_1) \subset J\hat{D}((\dot{\Lambda}_0^p M)^+)$, where

$\widetilde{C}_1 = \mu(M) \subset (\dot{\Lambda}_0^{p+1} M)^+$, and j is the homomorphism $j = \bar{j} \circ \delta \circ i : \pi^* \widetilde{C}_1 \to J\hat{D}((\dot{\Lambda}_0^p M)^+)$ over $(\dot{\Lambda}_0^p M)^+$ given by means of the following commutative diagram:

$$
\begin{array}{ccccccc}
\pi^* \widetilde{C}_1 & \xrightarrow{i} & \pi^*(\dot{\Lambda}_0^{p+1} M)^+ & \xrightarrow{\delta} & \pi^*\left(TM O_Z(\dot{\Lambda}_0^p M)^+\right) & \xrightarrow{\bar{j}} & J\hat{D}\left((\dot{\Lambda}_0^p M)^+\right) \\
\downarrow & & \downarrow & & \downarrow & & \downarrow \\
(\dot{\Lambda}_0^p M)^+ & = & (\dot{\Lambda}_0^p M)^+ & = & (\dot{\Lambda}_0^p M)^+ & = & (\dot{\Lambda}_0^p M)^+
\end{array}
$$

Therefore, if $\mu_{A_1 \cdots A_{p+1}} : U \subset M \to \overset{p+1}{\widehat{A}}$ are the local components, in quantum coordinates, of a $(p+1)$-full quantum differential form μ on M, then C_1 is defined by the coefficients: $-\lceil_{C_1 \cdots C_p D} \equiv y_{C_1 \cdots C_p D} \circ \rceil = \mu_{A_1 \cdots A_p D} \circ \pi$. As a by-product we get also that any fullquantum p-differential form $\beta : M \to (\dot{\Lambda}_0^p M)^+$ stands for a quantum connection on the fiber bundle $\pi : (\dot{\Lambda}_p^0 M)^+ \to M$. In fact, we can take $\mu = d\beta$. Furthermore, the exterior differential of quantum p-differential forms $d : J\hat{D}((\dot{\Lambda}_0^p M)^+) \to (\dot{\Lambda}_0^{p+1} M)^+$, can be considered the skewsymmetrized restriction of the covariant derivative of the canonical quantum connection on $(\dot{\Lambda}_0^p M)^+ \to M$ generated by the 0-section $0 : M \to (\dot{\Lambda}_0^p M)^+$:

(2.5)
$$
\left\{
\begin{array}{l}
d\beta = A(\nabla\beta) = \sum_{1 \le A_1, \cdots, A_p \le m} \left[\lceil_{A_1 \cdots A_p C} \circ D\beta + (\partial x_C . \beta_{A_1 \cdots A_p}) \right] dx^C \triangle dx^{A_1} \triangle \cdots \triangle dx^{A_p} \\
\quad = \sum_{1 \le C, A_1, \cdots, A_{p+1} \le m} (\partial x_C . \beta_{A_1 \cdots A_p}) dx^C \triangle dx^{A_1} \triangle \cdots \triangle dx^{A_p}
\end{array}
\right\}
$$

as $\lceil_{A_1 \cdots A_p C} = 0$. If $\pi : E \to M$ is a vector bundle, (in the category of quantum supermanifolds), and $\bar{\rceil}$ is a semiholonomic quantum connection on π, (that is also a holonomic quantum connection), then $\bar{\rceil}$ defines a *covariant exterior differential*: $_\rceil d : J\hat{D}(Hom_Z(\dot{\Lambda}_0^\bullet M; E)) \to Hom_Z(\dot{\Lambda}_0^\bullet M; E)$ defined by means of the following commutative diagram:

$$
\begin{array}{ccc}
J\hat{D}(Hom_Z(\dot{\Lambda}_0^\bullet M; E)) & \xrightarrow{\ _\rceil d\ } & Hom_Z(\dot{\Lambda}_0^\bullet M; E) \\
\bar{\rceil}^j \| \wr & & \uparrow pr_2 \\
S \oplus Hom_Z(\dot{\Lambda}_0^\bullet M; E) & = & S \oplus Hom_Z(\dot{\Lambda}_0^\bullet M; E),
\end{array}
$$

where $S = \oplus_p S^p$ is a graded vector bundle over M. Then if the expression in quantum coordinates of a E-valued graded quantum differential form β is $\beta = \sum_p \partial y_B \otimes \beta^B{}_{A_1 \cdots A_p} dx^{A_1} \triangle \cdots \triangle dx^{A_p}$, we get:

(2.6)
$$
_\rceil d\beta = \sum_{A, A_r} \partial y_B \otimes \left[(\partial x_A . \beta^B{}_{A_1 \cdots A_p}) + \sum_{C_r, D} \lceil^{B C_1 \cdots C_p}_{A A_1 \cdots A_p D} \beta^D{}_{C_1 \cdots C_p} \right] dx^A \triangle dx^{A_1} \triangle \cdots \triangle dx^{A_p}.
$$

A *quantum k-pseudoconnection* on E is a k-semiholonomic quantum connection $\bar{C}_k = \bar{\daleth}(\bar{J}\hat{D}^{k-1}(E)) \subset \bar{J}\hat{D}^k(E))$ on the fiber bundle $\pi : E \to M$ such that $\bar{C}_k$ is freely generated by $\underline{\widetilde{C}}_k \subset \bar{J}\hat{D}^k(E)$ and $\bar{J}\hat{D}^{k-1}(E)$, that is $\bar{C}_k = \widetilde{\bar{C}}_k \times \bar{J}\hat{D}^{k-1}(E)$, where $\widetilde{\bar{C}}_k = \daleth\mu(M) \subset Hom_Z(\dot{\Lambda}_0^k M; E)$ for some E-valued k-differential form $\daleth\mu$ on M. The embedding $\underline{\widetilde{C}}_k \subset \bar{J}\hat{D}^k(E)$ is obtained by means of the following exact commutative diagram:

$$
\begin{array}{ccccccccc}
0 & \to & \dot{T}_0^k M O_Z E & \to & \bar{J}\hat{D}^k(E) & \to & \bar{J}\hat{D}^{k-1}(E) & \to & 0 \\
& & \uparrow & & \| & & & & \\
0 & \to & \dot{\Lambda}_0^k M O_Z E & \to & \bar{J}\hat{D}^k(E) & & & & \\
& & \uparrow & & & & & & \\
& & 0 & & & & & &
\end{array}
$$

Then a quantum **Z**-*graded pseudoconnection* on E is a subbundle $\bar{C}$ of the vector bundle $\oplus_{k\geq 0}\bar{J}\hat{D}^k(E)$ over M that can be written as $\bar{C} = \oplus_{k\geq 0}\bar{C}_k$, where $\bar{C}_k$ is a quantum k-pseudoconnection on E. Hence $\bar{C}$ is identified with a E-valued quantum **Z**-graded differential form $\daleth\mu : M \to Hom_Z(\dot{\Lambda}_0^\bullet M; E)$. Then we define *quantum curvature* $\daleth R = \daleth d\daleth\mu : M \to Hom_Z(\dot{\Lambda}_0^\bullet M; E)$, that is a E-valued quantum **Z**-graded differential form on M. One has the following identity: (*Bianchi identity*) $\daleth d\daleth R = \daleth d\daleth d\daleth\mu = 0$. $\qquad\blacksquare$

Remark 2.17. (*Quantum connections on quantum principal fiber bundles*). In this appendix $(P, M, \pi; G)$ is a G-principal fibre bundle in the category of quanatum supermanifolds. Any quantum principal connection $\kappa : M \to C(P) \equiv J\hat{D}(P)/G$ ($\equiv$ fibre bundle of quantum principal connections) on $(P, M, \pi; G)$ is characterized by a section $\mathbf{k}$ of $\pi^*C(P) \to P$ such that the following diagram is commutative:

$$
\begin{array}{ccccc}
Hom_Z(TP;\mathfrak{g}) & \overset{j}{\leftarrow} & \pi^*C(P) & \to & C(P) \\
\omega_\mathbf{k} \quad \uparrow & & \downarrow\uparrow\,\underline{\mathbf{k}} & & \downarrow\uparrow\,\mathbf{k} \\
P & = & P & \underset{\pi}{\to} & M
\end{array}
$$

More precisely one has $\underline{\mathbf{k}} = (id_P, \check{\mathbf{k}})$, where $\check{\mathbf{k}} = \mathbf{k} \circ \pi$, and $\omega_\mathbf{k}$ is the Ehresmann connection associated to $\mathbf{k}$. j is the canonical embedding over P given by $j(p, \mathbf{k}(x)) = \omega_\mathbf{k}(p)$, with $x = \pi(p)$. Set $\overline{\pi^*C(P)} \equiv j(\pi^*C(P)) \subset Hom_Z(TP;\mathfrak{g})$. Let $\pi_E : E \to P$ be a vector fiber bundle over P. Let C denote the fiber bundle $\dot{\Lambda}^\bullet P O_Z E$ over M with projection π_C and let $\underline{E}$ denote the fiber bundle $\dot{\Lambda}^\bullet P O_Z E$ over P with projection $\pi_{\underline{E}}$. In other words, the following diagram is commutative:

$$
\begin{array}{ccc}
C \equiv Hom_Z(\dot{\Lambda}^\bullet P; E) & = & \underline{E} \\
\pi_C \downarrow & & \downarrow \pi_{\underline{E}} \\
M & \xleftarrow{\;\;\pi\;\;} & P
\end{array}
$$

One has a canonical epimorphism of fibre bundles over $Hom_Z(\dot{\Lambda}^\bullet P; E)$:

$$
\mathbf{n} : \pi^*_{k,0} J\hat{D}^k(\underline{E}) \to J\hat{D}^k(C), \; \mathbf{n}(D^k s(x)),
$$

$$
D^k{}_\rceil\mu(s(x))) = D^k({}_\rceil\mu \circ s)(x) = D^k c(x),
$$

where $\pi_{k,0} : J\hat{D}^k(P) \to P$ is the canonical epimorphism.

A *k-order quantum gauge system with blow-up structure* is a k-order gauge continuum system $G(M) \equiv (\mathcal{B}, \hat{E}_k)$, [57], in the category of quantum supermanifolds, such that: (i) $\mathcal{B}$ is a superbundle of geometric objects $\mathcal{B} \equiv (P, C; \mathbf{B})$, (see also refs. [57,63]),[30] being P a principal fiber bundle $(P, M, \pi; G)$, where G is a quantum Lie supergroup called *quantum gauge structure group* and M is a quantum supermanifold called *base quantum supermanifold*. C is a fiber bundle over M (called *configuration bundle*) $\pi_C : C \to M$, with $C \equiv Hom_Z(\dot{\Lambda}^\bullet P; E)$. $\mathbf{B}$ is the natural functor $\mathbf{B} : \mathcal{P} \to \mathcal{C}$, where $\mathcal{P}$, (resp. $\mathcal{C}$), is the category whose objects are open subbundles of P, (resp. C), and whose morphisms are the local fibre bundle endomorphisms between these objects. The *blow up structure* is given by an embedding morphism $j : \pi^*C(P) \to \underline{E}$ over P given by means of the

---

[30] Recall that a *structure of superbundle of geometric objects* is given by means of two fiber bundles $C \xrightarrow{\pi_C} M \xrightarrow{\pi_B} B$ over the same base M and a covariant functor $\mathbb{B}{:}\mathcal{C}(B) \to \mathcal{C}(C)$, where $\mathcal{C}(B)$, (resp. $\mathcal{C}(C)$), is the category whose objects are open subbundles of B, (resp. C), and whose morphisms are the local fibre bundle authomorphisms between those objects such that: (i) if $B|_U \in Ob(\mathcal{C}(B))$, then $\mathbb{B}(B|_U){=}\pi_C^{-1}(U) \in Ob(\mathcal{C}(C))$; (ii) if $f \in Hom(\mathcal{C}(B))$ with $f \equiv (f_B, f_M){:}B|_U \to B|_{U'}$, then $\mathbb{B}(f) \in Hom(\mathcal{C}(C))$ and satisfies: (a) $\pi_C \circ \mathbb{B}(f){=}f_M \circ \pi_C$; (b) if $B|_U \in Ob(\mathcal{C}(B))$, $\bar{U} \subset U$, then $\mathbb{B}(f)|_{\pi_C^{-1}(\bar{U})}(\bar{U}){=}\mathbb{B}(f|_{B|_{\bar{U}}})$. C is called the *total bundle* and B the *base bundle*. A section of π_C is called a *superfield of geometric objects*. (See also refs.[56,57].) Note that in order that $(H, W, \mathbb{B})$ be a superbundle of geometric objects it is enough that $(\pi_H{:}H \to J\mathcal{D}^p(W), \mathbb{H})$ be a *natural bundle*, (or a bundle of *geometric objects* in the sense of [50,54]). In fact, in such a case we have $\mathbb{B} \equiv \mathbb{H} \circ J\mathcal{D}^p(-)$. Furthermore, if $(W, M, \overline{\mathbb{B}})$ is also a natural bundle, then $(H, M, \widetilde{\mathbb{B}})$ is also a natural bundle with $\widetilde{\mathbb{B}} \equiv \mathbb{H} \circ J\mathcal{D}^p(-) \circ \overline{\mathbb{B}}$.

following commutative diagram with exact lines:

$$
\begin{array}{ccc}
\underline{E} & = & Hom_Z(\Lambda^\bullet P;E) \\
j \uparrow & & \uparrow \\
0 \quad \to \quad \pi^*C(P) & \to & Hom_Z(TP;E) \\
& & \uparrow \\
& & 0
\end{array}
$$

$\hat{E}_k$ is a k-order formally quantumintegrable QSPDE on C, $\hat{E}_k \subset J\hat{D}^k(C)$, given by: $\hat{E}_k = \mathbf{n}(\pi^*_{k,0}\underline{\hat{E}}_k)$, where $\underline{\hat{E}}_k$ is a k-order formally quantuminte-grable QSPDE on $\pi_E : \underline{E} \to P$, $\underline{\hat{E}}_k \subset J\hat{D}^k(\underline{E})$, such that $\mathbf{b}^k(\underline{\hat{E}}_k) \subset \overline{C(P)}^k$, being $\mathbf{b}^k : J\hat{D}^k(\underline{E}) \to J\hat{D}^k(TPO_ZE)$ the canonical epimorphism over P induced by the canonical epimorphism $b : \underline{E} \to TPO_ZE$ over P. $\blacksquare$

Remark 2.18. The set $\underline{Sol}(\hat{E}_k)$ of regular Q^∞_w-solutions of $\hat{E}_k \subset J\hat{D}^k(C)$, is characterized by the sections $c \in Q^\infty_w(C)$ with $c = {}_1\mu \circ s$, where $s = \pi_{\underline{E}} \circ c$, ${}_1\mu = \mathbf{Z}$-graded quantum pseudoconnection on P, such that $D^k{}_1\mu(s(x)) \in \underline{E}_k$, $s \in Q^\infty_w(P)$ and $b \circ {}_1\mu : P \to TPO_ZE$ is the corresponding quantum pseudoconnection that coincides with an Ehresmann quantum connection on P. $\blacksquare$

Let us emphasize, now, that $J\hat{D}^k(C)$, $k \geq 1$, is a quantum supermani-fold modeled on the quantum superalgebra B_k, that, forgetting the $\mathbf{Z}_2$-gradiation on A, is just the quantum algebra $B_k = A \times \hat{A} \times \overset{2}{\widehat{A}} \times \cdots \times \overset{k+1}{\widehat{A}}$. Therefore, $(\widehat{YM})$ is a quantum supermanifold contained into $J\hat{D}^\infty(C) = \lim_{\leftarrow k} J\hat{D}^k(C)$. In particular, $J\hat{D}^2(C)$ is a quantum supermanifold modeled on the quantum superalgebra

$$
B_2 \equiv A \times \Big(\prod_{i \in \mathbf{Z}_2} \overset{1}{\widehat{A}}_i(A) \Big) \times \Big(\prod_{i,j \in \mathbf{Z}_2} \overset{2}{\widehat{A}}_{ij}(A) \Big) \times \Big(\prod_{i,j,r \in \mathbf{Z}_2} \overset{3}{\widehat{A}}_{ijr}(A) \Big)
$$

$$
= A \times \overset{1}{\widehat{A}}_0(A) \times \overset{1}{\widehat{A}}_1(A) \times \overset{2}{\widehat{A}}_{00}(A) \times \overset{2}{\widehat{A}}_{10}(A) \times \overset{2}{\widehat{A}}_{01}(A) \times \overset{2}{\widehat{A}}_{11}(A)
$$

$$
\times \overset{3}{\widehat{A}}_{000}(A) \times \overset{3}{\widehat{A}}_{100}(A) \times \overset{3}{\widehat{A}}_{010}(A) \times \overset{3}{\widehat{A}}_{110}(A) \times \overset{3}{\widehat{A}}_{001}(A) \times \overset{3}{\widehat{A}}_{101}(A) \times \overset{3}{\widehat{A}}_{011}(A) \times \overset{3}{\widehat{A}}_{111}(A)
$$

of dimension $(a,b,c,d,e,f,g,h,i,l,m,o,p,q)$, where the symbols are reported in Tab.2.17

TAB.2.17 - Expressions for symbols ($a,b,c,d,e,f,g,h,i,l,m,o,p,q$).

$a = \alpha\vert\gamma$	$b = 8d(1+d)\vert N_2\beta$	$c = N_2\delta\vert d\delta$
$d = 4d\beta\vert 4N_2\beta$	$e = N_1 N_2 \beta\vert 4d\delta$	$f = 4N_2\delta\vert 4d\delta$
$g = N_1 d\delta\vert N_1(N_2\gamma+4\alpha)$	$h = 16d\alpha\vert 16N_2\alpha$	$i = 4N_1 N_2\alpha\vert 4N_1$
$l = N_1\alpha\vert 4N_1 d\alpha$	$m = N_1^2 d\alpha\vert N_1^2 N_2\alpha$	$n = 16N_2\gamma\vert 16d\gamma$
$o = 4N_1 d\gamma\vert 4N_1 N_2\gamma$	$p = 4N_1 d\gamma\vert 4N_1 N_2\gamma$	$q = N_1^2 N_2\gamma\vert N_1^2 d\gamma$

$\alpha \equiv 4+d, \quad \beta \equiv 8+d, \quad \gamma \equiv N_1+N_2, \quad \delta \equiv 2N_1+N_2.$

Therefore the Bianchi identity identifies a QSPDE $(B) \subset J\hat{D}^2(C)$ of second order of dimension: $\dim_{B_2}(B) = (a, b, c, d, e, f, g, \bar{h}, \bar{i}, \bar{l}, \bar{m}, \bar{o}, \bar{p}, \bar{q})$, with the symbols specified in Tab.2.17 and in Tab.2.18

TAB.2.18 - Expressions for symbols ($\bar{h},\bar{i},\bar{l},\bar{m},\bar{o},\bar{p},\bar{q}$).

$\bar{h} = 16d\alpha\vert 16N_2\alpha$	$\bar{i} = 4N_1 N_2\alpha\vert 4N_1\alpha d$	$\bar{l} = N_1\alpha\vert 4N_1 d\alpha$
$\bar{m} = N_1^2 d\alpha\vert N_1^2 N_2\alpha$	$\bar{n} = 16N_2\gamma\vert 16d\gamma$	$\bar{o} = 4N_1 d\gamma\vert 4N_1 N_2\gamma$
$\bar{p} = 4N_1 d\gamma\vert 4N_1 N_2\gamma$	$\bar{q} = N_1^2 N_2\gamma\vert N_1^2 d\gamma$	

Note that one has the local isomorphism over P: $\underline{E} \equiv Hom_Z(TP;\mathfrak{g}) \cong (\widehat{A})^{\alpha d\vert\alpha N_2} \times (\widehat{A})^{\gamma N_2\vert\gamma d}$. Therefore, we can locally represent a quantum pseudoconnection by means of fullquantum differential 1-forms on P: $\lceil\mu^K = \mu_A^K dV^A$, with $\mu_A^K \in Q_w^\infty(U;\widehat{A})$. One has the following structure of μ_A^K, taking into account the $\mathbf{Z}_2$-gradiation of A: $\mu_A^K \in Q_w^\infty(U; Hom_Z(A_{\vert A\vert}; A_{\vert K\vert}))$. Furthermore, one has a natural Z-bilinear morphism

$$\underline{E} \times \underline{E} \xrightarrow{\;[,]\;} Hom_Z(\dot{\Lambda}_0^2 M;\mathfrak{g}),$$

over P defined by means of the following formula:

$$[\phi,\psi]^C = \sum_{A,B} \widehat{C}_{AB}^C(\phi^A \wedge \psi^B) = \sum_{A,B,H_1,H_2} \widehat{C}_{AB}^C(\phi^A \wedge \psi^B)_{H_1 H_2} dV^{H_1} \triangle dV^{H_2},$$

where $\psi^B = \psi_H^B dV^H$, $\phi^A = \phi_H^A dV^H$, and

$$(\phi^A \wedge \psi^B)_{H_1 H_2} \equiv \phi_{H_1}^A \psi_{H_2}^B - (-1)^{\vert A\vert\vert B\vert}\psi_{H_1}^B \phi_{H_2}^A$$

is the graded exterior products of the fullquantum differential forms ψ^B and ϕ^A. $\widehat{C}^C_{AB} \in \widehat{A}$ are the fullquantum structure constants of G. They satisfy the relations reported in Tab.2.12. Above bracket can be extended in a natural way on the fiber bundle $Hom_Z(\dot{\Lambda}^\bullet(M); \mathfrak{g})$, over P, yet denoted $\underline{E}$, where the fibres are **K**-algebras, with a $\mathbf{Z}_2$-gradiation induced by $\mathfrak{g}$. Then, this fiber bundle results a fiber bundle of quantum superalgebras over P. The same structure is then recognized on $Q_w^\infty(\underline{E})$. In fact one has that the relations reported in Tab.2.19 are satisfied for any $\phi, \psi, \rho \in Q_w^\infty(\underline{E})$.

TAB.2.19 - The quantum superalgebra $Q_w^\infty(\underline{E})$.

$$[\phi,\psi] \in Q_w^\infty(\underline{E})_{|\phi|+|\psi|}$$

$$[\phi,\psi] = -(-1)^{|\phi||\psi|}[\psi,\phi]$$

$$(-1)^{|\phi||\rho|}[[\phi,\psi],\rho] + (-1)^{|\rho||\psi|}[[\rho,\phi],\psi] + (-1)^{|\phi||\psi|}[[\psi,\rho],\phi] = 0$$

The curvature $_1R = {}_1d_1\mu = d_1\mu + \frac{1}{2}[_1\mu, _1\mu]$, and the corresponding Bianchi identities, have the local representations given in Tab.2.20.

TAB.2.20 - Bianchi identity.

$$(\text{Curvature})\ R^K_{A_1A_2} = (\partial V_{A_1} \cdot \mu^K_{A_2}) + \frac{1}{2}\widehat{C}^K_{IJ}[\mu^I_{A_1}, \mu^J_{A_2}]+$$

$$(\text{Bianchi identities})\ (\partial V_B \cdot R^K_{A_1A_2}) + \frac{1}{2}\widehat{C}^K_{IJ}[\mu^I_B, R^J_{A_1A_2}]_+ = 0$$

$$R^K_{A_1A_2} \in Q_w^\infty(U,\overset{2}{\widehat{A}}),\ U \subset P.$$

Taking into account the $\mathbf{Z}_2$-gradiation of the quantum superalgebra A we can see that the homogeneous components $R^K_{A_1A_2}$ belong to some sub-spaces of $\overset{2}{\widehat{A}}$ dpending on the $\mathbf{Z}_2$-gradiation of A. More precisely $R^K_{A_1A_2} \in Q_w^\infty(U, Hom_Z(A_{|A_1|} \otimes_Z A_{|A_2|}; A_{|K|}))$. The $\mathbf{Z}_2$-gradiations of the fullquantum forms μ^K and R^K are respectively $|\mu^K| = |K|$ and $R^K = |K|$. Now, for any section $s : M \to P$ of class Q_w^∞ a quantum pseudoconnection field $_1\mu$ and its corresponding quantum curvature field $_1R$ identify a $\mathfrak{g}$-valued quantum differential 1-form and 2-form respectively on the quantum superspace-time M:

$$(2.7) \qquad \left\{ \begin{array}{ll} \bar{\mu} \equiv s^*{}_1\mu : M \to Hom_Z(TM; \mathfrak{g}), & \bar{\mu}^K = \mu^K_A dX^A \\[2mm] \bar{R} \equiv s^*{}_1R : M \to Hom_Z(\dot{\Lambda}^2_0 M; \mathfrak{g}), & \bar{R}^K = \bar{R}^K_{AB} dX^A \triangle dX^B \end{array} \right\},$$

with $\bar\mu_A^K \in Q_w^\infty(U;\widehat A)$, $\bar R_{AB}^K \in Q_w^\infty(U,\overset{2}{\widehat A})$, $U \subset M$. Equations in Tab.2.20, evalued on the quantum superspace-time M by means of s, give the two last equations of the $\widehat{(YM)}$ in Tab.2.14. The first is the Euler-Lagrange equation corresponding to a quantum Lagrangian L. (See refs.[62,67] for variational calculus on supermanifolds and quantum manifolds.) The blow-up structure is the natural constraint for principal connections.

We are ready, now, to conclude the proof of our theorem, by emphasizing that sections $\bar\mu : M \to Hom_Z(TM;\mathfrak{g})$ represent connections of our gauge theory evalued on the quantum superspacetime M and that $\bar R : M \to Hom_Z(\dot\Lambda_0^2 M;\mathfrak{g})$ are the corresponding curvature 2-forms just evalued on M. Therefore the Yang-Mills QSPDE can be considered defined in the quantum jet-derivative space $J\hat D^\infty(\bar C)$, where $\bar C$ is the fiber bundle over M defined by $\bar C \equiv Hom_Z(TM;\mathfrak{g})$. Finally note that as the fiber bundles $Hom_Z(T;\mathfrak{g}) \to M$ and $Hom_Z(\dot\Lambda_0^2 M;\mathfrak{g}) \to M$ have contractibe fibers, it follows that sections $\bar\mu$ and $\bar R$ are globally defined on M. (See Corollary 1.2). $\qquad\square$

Corollary 2.3. *The evaluation of the $\widehat{(YM)}$ at the macroscopic level is obtained by means of the pull-back of any pseudoconnection and the corresponding curvature on the 4-dimensional globally hyperbolic space-time N embedded into M by means of some map $i : N \to M$. The corresponding quantum super Yang-Mills equation $\widehat{(YM)}[i] \subset J\hat D^\infty(i^*C)$ can be locally written in the form reported in Tab.2.21.*

TAB.2.21 - Local expression of $\widehat{(YM)}[i]\subset J\hat D^\infty(i^*C)$ and Bianchi identity $(B)[i]\subset J\hat D^2(i^*C)$.

(Field equations) $\frac{\delta L}{\delta\bar\mu_K^A} \equiv \sum_{i\in I}(-1)^{[i]}\partial_i\left(\frac{\partial L}{\partial\bar\mu_{Ki}^A}\right)=0$	$(\widehat{(YM)}[i])$
(Fields) $\tilde R_{\beta\gamma}^K - \left[(\partial\xi_{[\beta}\cdot\tilde\mu_{\gamma]}^K)+\frac{1}{2}\widehat C_{IJ}^K\tilde\mu_{[\gamma}^I\tilde\mu_{\beta]}^J\right]=0$ *(Bianchi identities)* $(\partial\xi_{[\alpha}\cdot\tilde R_{\beta\gamma]}^K)+\frac{1}{2}\widehat C_{IJ}^K\tilde\mu_{[\alpha}^I\tilde R_{\beta\gamma]}^J=0$	$((B)[i])$

where $\tilde R_{\alpha\beta}^K$ and $\tilde\mu_\alpha^K$ are the components of curvature and pseudoconnection respectively induced by means of the embedding $i : N \to M$, and (ξ^α) are coordinates induced on N.

Proof. In fact one has:

$$(2.8)\qquad \left\{\begin{array}{l} \tilde\mu\equiv i^*\bar\mu{:}M_C\to\mathfrak{g}\otimes_{\mathbf K}(TM_C)^*, \quad \tilde\mu=Z_K\tilde\mu_\gamma^K\otimes d\xi^\gamma \\[2mm] \tilde R\equiv i^*\bar R{:}M_C\to\mathfrak{g}\otimes_{\mathbf K}\Lambda_2^0 M_C, \quad \tilde R=\sum_{\gamma<\delta}Z_K\tilde R_{\gamma\delta}^K\otimes d\xi^\gamma\wedge d\xi^\delta \end{array}\right\},$$

with $\tilde{\mu}_\gamma^K \in Q_w^\infty(\tilde{U}; \widehat{A})$, $\tilde{U} \subset i(M_C) \subset M$. Then equations in Tab.2.21 are the pullback of equations in Tab.2.14 by means of i. $\qquad\square$

Remark 2.19. A *quantum supersymmetry* of $(\widehat{YM})$ is a (local) diffeomorphism $f : P \to P$ of class Q_w^∞ that does not come from the action of the group structure G on P and such that the natural diffeomorphism $\mathbf{B}(f)$ induced on $\bar{C}$, lifted on $J\hat{D}^\infty(\bar{C})$ preserves $(\widehat{YM}) \subset J\hat{D}^\infty(\bar{C})$. Any quantum supersymmetry of $(\widehat{YM})$ transforms any solution c of $(\widehat{YM})$ into a new solution c'. (For informations on the geometric structure of the classical Yang-Mills equation, see e.g. refs.[13,57,58,73].) $\qquad\blacksquare$

Theorem 2.33. *Let $(\widehat{YM}) \subset J\hat{D}^\infty(\bar{C})$ be a formally quantum integrable QSPDE, (or at least an universal quantum regular QSPDE), with nontrivial quantum supersymmetry group, such that $A = A_0 \oplus A_1$ is a quantum superalgebra with Noetherian center Z. Then one has:*

1)(Existence of local (smooth) solutions). Considering $(\widehat{YM})$ in the category of quantum supermanifolds of class Q_w^ω, for any initial condition $q \in (\widehat{YM})$, passes a local solution. Hence $(\widehat{YM})$ is completely quantum superintegrable.

2)(Existence of local (smooth) solutions for Cauchy problems). For any admissible integral quantum (smooth) supermanifold $N \subset (\widehat{YM})$ of dimension $(3|N_1 - 1)$ passes a local solution that is a quantum supermanifold of dimension $(4|N_1)$.

3)(Existence of global (smooth) solutions for any (smooth) boundary condition). Let us assume that the quantum space-time M is $(p+q)$-connected with $p \in \{0, 1, 2, 3\}$, $q \in \{0, \cdots, N_1 - 1\}$. Then, for any admissible integral quantum (smooth) supermanifold $N \subset (\widehat{YM})$ of dimension $(3|N_1 - 1)$, with prefixed evolution of the boundary ∂N into $(\widehat{YM})$, passes a (smooth) global solution that is a global quantum supermanifold of dimension $(4|N_1)$.

Proof. 1) and 2) are direct consequences of Theorem 2.22, Theorem 2.23 and Theorem 2.26 respectively.

3) As $(\widehat{YM})$ is formally quantum integrable and the configuration bundle is $(p + q)$-connected, $0 \le p + q \le 2 + N_1$, one has from Theorem 2.27 that its integral bordism groups $\Omega_{p|q}^{(\widehat{YM})}$ are given by $\Omega_{p|q}^{(\widehat{YM})} \cong (A_0 \otimes_{\mathbf{K}} H_p(C; \mathbf{K})) \bigoplus (A_1 \otimes_{\mathbf{K}} H_q(C; \mathbf{K}))$, for $0|0 \le p|q \le 3|N_1 - 1$. In particular, $\Omega_{3|N_1-1}^{(\widehat{YM})} \cong (A_0 \otimes_{\mathbf{K}} H_3(C; \mathbf{K})) \bigoplus (A_1 \otimes_{\mathbf{K}} H_{N_1-1}(C; \mathbf{K})) = 0$. Hence, as a by-product of this result we get the existence of global (smooth) solutions

for any (smooth) boundary condition. $\qquad\square$

Definition 2.28. A *quantum relativistic frame* is a quantum frame [67], $\hat{\psi} \equiv (i : N \to M, \tau : N \to \mathbf{R})$, where i is an embedding of the 4-dimensional globally hyperbolic space-time N into M, and τ, (*proper-time*), is a differential numerical function of constant rank 1, that identifies a foliation on N by means of the flow associated to its time-like gradient: $\dot{\psi} \equiv \mathrm{grad}\,(\tau) : N \to TN$. In coordinates (x^α), adapted to such a function τ, one has: $x^0 = \tau$, $\dot{\psi} = \partial x_0$. Then $i(N) \subset M$ is a globally hyperbolic 4-dimensional manifold endowed with a global foliation of space-like codimension 3.

Corollary 2.4. *If we consider* $(\widehat{YM})$ *with respect to a quantum relativistic frame, above theorem gives global solutions with respect to the proper time of the quantum relativistic frame.*

Remark 2.20. Let us emphasize that the variational equation obtained by a first order quantum Lagrangian density of reduced form $\mathcal{L} = L dX^1 \wedge \cdots \wedge dX^m$, with $L(q) \in \widehat{A}$, $\forall q \in J\hat{D}(\bar{C})$, are second order QSPDE. In fact, in such a case one has that $E_K^A \equiv (\partial \bar{\mu}_K^A . L) - (\partial_B . (\partial \bar{\mu}_K^{BA} . L)) = 0$ is a $\overset{3}{\widehat{A}}$-valued function on $J\hat{D}^2(\bar{C})$. In fact, E_K^A are $Hom_Z(\widehat{A}; \widehat{A})$-valued functions on $J\hat{D}^2(\bar{C})$. On the other hand one has the following canonical Z-homomorphism: $Hom_Z(\widehat{A}; \widehat{A}) \cong Hom_Z(\widehat{A} \otimes_Z A; A) \overset{j}{\to} Hom_Z(A \otimes_Z A \otimes_Z A; A)$, where j is defined by $j(\alpha) = (\bar{\mu} \otimes 1_A) \circ \alpha$, for any $\alpha \in Hom_Z(\widehat{A} \otimes_Z A; A)$. Here $\bar{\mu}$ is the Z-homomorphism $A \otimes_Z A \to \widehat{A}$ induced by the multiplication $\hat{\mu}$, i.e. $\bar{\mu}(a \otimes b)(c) = abc$. $\qquad\blacksquare$

In order to go into some important applications of Theorem 2.32, let us consider in some details quantum (pseudo) Riemannian (super) manifolds.

Definition 2.29. (*Quantum (pseudo) Riemannian manifolds*). Let M be a quantum manifold of dimension m on the quantum algebra A. We call *fullquantum metric* $\hat{g}$ on M a Q_w^∞-section of $Hom_Z(TM \otimes_Z TM; A) \cong (\widehat{TM})^+ \otimes_{\widehat{A}} (TM)^+$ over M. We call such quantum manifolds, *quantum (pseudo) Riemannian manifolds*.

Lemma 2.14. *Let M be a quantum (pseudo) Riemannian manifold of dimension m, over a quantum algebra A. Then, one has the canonical Z-isomorphism* $\widehat{TM} \cong (TM)^+$.

Proof of Lemma 2.14. Taking into account the following (canonical) isomorphism $Hom_Z(T_p M \otimes_Z T_p M; A) \cong Hom_Z(T_p M; (T_p M)^+) \cong (\overset{2}{\widehat{A}})^{m^2}$,

$\forall p \in M$, we get the following local representation of $\widehat{g}$: $\widehat{g} = \sum_{\alpha,\beta} \widehat{g}_{\alpha\beta} dx^{\alpha} \otimes dx^{\beta} = \sum_{\alpha,\beta} \partial \dot{x}^{\alpha} \otimes \widehat{g}_{\alpha\beta} dx^{\alpha}$, where $\widehat{g}_{\alpha\beta}(p) \in \widehat{A})^{m^2}$ and $(x^{\alpha}, \dot{x}_{\alpha})$ are quantum coordinates induced on $(TM)^+$. Note that to the quantum coordinates $\dot{x}_{\alpha} : \Omega \subset (TM)^+ \to \widehat{A}$ correspond the fullquantum vectors $\partial \dot{x}^{\alpha}(p) \in Hom_Z(\widehat{A}; (T_pM)^+)$. Then by using the canonical A-monomorphism $\underline{j} : A \to \widehat{A}$, identified by $1 \mapsto 1_A$, we get the following homomorphism

$$Hom_Z(\widehat{A}; (T_pM)^+) \to Hom_Z(A; (T : pM)^+).$$

Hence $\{\partial \dot{x}^{\alpha}(p)\}$ identify vectors of the left $\widehat{A}$-module

$$Hom_Z(A; (T_pM)^+) \equiv (\widehat{T_pM})^+$$

that, for abuse of notation we also denote by $\{\partial \dot{x}^{\alpha}(p)\}$. As locally

$$Hom_Z(A; (T_pM)^+) \cong Hom_Z(A; \widehat{A})^m,$$

we see that a vector $v \in Hom_Z(A; (T_pM)^+)$, can be locally represented as $v = \partial \dot{x}^{\alpha}(p)v_{\alpha}$, $v_{\alpha} \in Hom_Z(A; \widehat{A})$. Furthermore, by considering the canonical short exact sequence $\widehat{A} \xrightarrow{\bar{j}} A \to 0$, (obtained by restriction on $Z \subset A$, i.e., $\bar{j}(\alpha) = \alpha|_Z \in Hom_Z(Z; A) \cong A$, that is equivalent to say $\bar{j}(\alpha) = \alpha(1)$), we get the following short exact sequence

$$0 \to Hom_Z(A; (T_pM)^+ \to Hom_Z(\widehat{A}; (T_pM)^+).$$

In this way we can identify $Hom_Z(A; (T_pM)^+)$ with a quantum subspace of $Hom_Z(\widehat{A}; (T_pM)^+)$. In fact, the vectors $v \in Hom_Z(\widehat{A}; (T_pM)^+)$ are the vectors $\zeta = \partial \dot{x}_{\alpha}(p)\zeta^{\alpha} \in Hom_Z(\widehat{A}; (T_pM)^+)$ such that $\zeta^{\alpha} \in Hom_Z(A; \widehat{A}) \subset Hom_Z(\widehat{A}; \widehat{A})$. Note also that $\bar{j}$ is a split on the left of the following short exact sequence of Z-modules: $0 \to A \overset{1}{\underset{\bar{j}}{\rightleftarrows}} \widehat{A} \to \widehat{A}/A \to 0$. Now, by using the canonical isomorphism

$$Hom_Z(T_pM \otimes_Z T_pM; A) \cong Hom_Z(T_pM; Hom_Z(T_pM; A)),$$

we see that a fullquantum metric $\widehat{g}$ on M induces a Z-homomorphism $\widehat{g}'(p) : T_pM \to (T_pM)^+$, for all $p \in M$. In quantum coordinates one has that if $u = \partial x_i(p)(u^i) \in T_pM$, $u^i \in A$, then $\widehat{g}'(p)(u) = \widehat{g}_{\alpha\beta}(p)(u^{\alpha})dx^{\beta}(p) \equiv$

$u_\beta(p)dx^\beta(p)$, with $u_\beta(p) \equiv \widehat{g}_{\alpha\beta}(p)(u^\alpha) \in \widehat{A}$, i.e., $u_\beta(p)(a) = \widehat{g}_{\alpha\beta}(p)(u^\alpha \otimes a)$, $\forall a \in A$. Then, we say that (u_β) are the *covariant components* of u, where (u^i) are its *controvariant components*. By applying the covariant functor $Hom_Z(A; -)$ to the morphism $\widehat{g}'(p)$, we get that a fullquantum metric identifies also a Z-homomorphism $\widehat{g}'(p)_* \equiv Hom_Z(1_A; \widehat{g}'(p))$: $\widehat{T_pM} \to (\widehat{T_pM})^+ \equiv Hom_Z(A; (T_pM)^+)$, given by $\widehat{g}'(p)_*(\zeta) = \widehat{g}'(p) \circ \zeta$, for any $\zeta \in Q_w^\infty(\widehat{T_pM})$. In quantum coordinates one has $\widehat{g}'(p)_*(\zeta(a))(v) = \widehat{g}(p)_{\alpha\beta}(\zeta^\alpha(a) \otimes v^\beta)$, $\forall a \in A$, and $v \in T_pM$. We say that $\widehat{g}$ is *nondegenerate* if $\widehat{g}'(p)$ is injective, i.e., one has the following short exact sequence:

$$0 \to T_pM \xrightarrow{\widehat{g}'(p)} (T_pM)^+, \ \forall p \in M.$$

Note that this does not, in general, mean that $\widehat{g}(p)(u, u) \neq 0$, $\forall 0 \neq u \in T_pM$, $\forall p \in M$. But, instead, it means that $\widehat{g}(p)(u, v)$ cannot be zero for any $v \in T_pM$, if $0 \neq u \in T_pM$. Furthermore, as $Hom_Z(A; -)$ is a left exact functor, if $\widehat{g}$ is nondegenerate, it follows that it is also exact the following short exact sequence:

$$0 \to \widehat{T_pM} \xrightarrow{\widehat{g}'(p)_*} (\widehat{T_pM})^+, \ \forall p \in M.$$

In quantum coordinates we can write $\widehat{g}'(p)(\zeta)(v) = \widehat{g}(p)_{\alpha\beta}(\zeta^\alpha|_Z \otimes v^\beta)$, $\forall v \in T_pM$, $\zeta \in \widehat{T_pM}$. This is equivalent to say that $\widehat{g}'(p)_*$, induces a Z-isomorphism $\widehat{T_pM} \cong (T_pM)^+$, $\forall p \in M$. Therefore, if on a quantum manifold of dimension m, over a quantum algebra A, it is defined a nondegenerate fullquantum metric $\widehat{g}$, we get a canonical Z-isomorphism $\widehat{TM} \cong (TM)^+$ over M. $\qquad\square$

Remark 2.21. For abuse of notation, we will denote the isomorphism $\widehat{TM} \cong (TM)^+$ yet by $\widehat{g}'$. So, by considering the canonical isomorphism

$$Hom_Z(\widehat{T_pM}; (T_pM)^+) \cong Hom_Z(\widehat{A}; (T_p)^+) \otimes_{Hom_Z(\widehat{A};\widehat{A})} Hom_Z(\widehat{T_pM}; \widehat{A}),$$

we see that $\widehat{g}'(p)$ has the following local representation:

$$\widehat{g}'(p) = \sum_{\alpha,\beta} \partial\dot{x}^\beta(p) \otimes \widehat{g}_{\alpha\beta}(p)d\dot{x}^\alpha(p), \ \widehat{g}_{\alpha\beta}(p) \in Hom_Z(\widehat{A}; \widehat{A}).$$

For the inverse isomorphism $\bar{g}'(p) : (T_pM)^+ \cong \widehat{T_pM}$, by considering the canonical isomorphism

$$Hom_Z((T_pM)^+; \widehat{T_pM}) \cong Hom_Z(\widehat{A}; \widehat{T_pM}) \otimes_{Hom_Z(\widehat{A};\widehat{A})} Hom_Z((T_pM)^+; \widehat{A}),$$

we get the following local representation: $\bar{g}'(p) = \sum_{\alpha,\beta} \partial\dot{x}_\beta(p) \otimes \widehat{g}^{\alpha\beta}(p) d\dot{x}_\alpha(p)$, $\widehat{g}^{\alpha\beta}(p) \in Hom_Z(\widehat{A}; \widehat{A})$. Therefore, $\forall p \in M$, the following conditions are verified:

(2.9)
$$\left\{ \begin{aligned} \bar{g}'(p) \circ \widehat{g}'(p) &= 1_{\widehat{T_pM}} \in Hom_Z(\widehat{T_pM}; \widehat{T_pM}) \cong Hom_Z(\widehat{A}; \widehat{T_pM}) \\ &\qquad\qquad \otimes_{Hom_Z(\widehat{A};\widehat{A})} Hom_Z(T_pM; \widehat{A}) \\ \widehat{g}'(p) \circ \bar{g}'(p) &= 1_{(T_pM)^+} \in Hom_Z((T_pM)^+; (T_pM)^+) \cong Hom_Z(\widehat{A}; T_pM) \\ &\qquad\qquad \otimes_{Hom_Z(\widehat{A};\widehat{A})} Hom_Z((T_pM)^+; \widehat{A}) \end{aligned} \right\}.$$

As a consequence, we get that equations (2.9), in quantum coordinates, give the the following conditions:

(2.10)
$$\left\{ \begin{array}{l} \widehat{g}_{\alpha\beta}(p)\widehat{g}^{\beta\gamma}(p)=\delta^\gamma_\alpha \in Hom_Z(\widehat{A};\widehat{A}) \\[2em] \widehat{g}^{\alpha\beta}(p)\widehat{g}_{\beta\gamma}(p)=\delta^\alpha_\gamma \in Hom_Z(\widehat{A};\widehat{A}) \end{array} \quad \begin{array}{ccccc} \widehat{A} & \xrightarrow{\widehat{g}^{\beta\gamma}(p)} & \widehat{A} & \xrightarrow{\widehat{g}_{\alpha\beta}(p)} & \widehat{A} \\ \| & & & & \| \\ \widehat{A} & \xrightarrow{\delta^\gamma_\alpha} & & & \widehat{A} \\[1.5em] \widehat{A} & \xrightarrow{\widehat{g}_{\beta\gamma}(p)} & \widehat{A} & \xrightarrow{\widehat{g}^{\alpha\beta}(p)} & \widehat{A} \\ \| & & & & \| \\ \widehat{A} & \xrightarrow{\delta^\alpha_\gamma} & & & \widehat{A} \end{array} \right\}.$$

Note that the matrix $(\widehat{g}_{\alpha\beta})$ is the inverse of $(\widehat{g}^{\alpha\beta})$, and vice versa. It is important to recognize that on a quantum (pseudo) Riemannian manifold, $(M, \widehat{g})$, one has a canonical Z-homomorphism $(T_pM)^+ \to T_pM$, $\forall p \in M$, given by composition $(T_pM)^+ \overset{\bar{g}'(p)}{\cong} \widehat{T_pM} \to T_pM$. For abuse of notation we shall denote such homomorphism by $\bar{g}'(p)$ too. This morphism is a split on the left of the following short exact sequence:

$$0 \to T_pM \underset{\bar{g}'(p)}{\overset{\widehat{g}'(p)}{\rightleftarrows}} (T_pM)^+ \to (T_pM)^+/T_pM \to 0, \ \forall p \in M.$$

By considering the following canonical isomorphism

$$Hom_Z(T_pM; (T_pM)^+) \cong Hom_Z(A; (T_pM)^+) \otimes_{\widehat{A}} (T_pM)^+,$$

we can locally represent the mapping $\widehat{g}'(p) : T_pM \to (T_pM)^+$ as follows:

$$\left\{ \begin{aligned} \widehat{g}'(p) &= \partial\dot{x}^\beta(p) \otimes \widehat{g}_{\alpha\beta}(p) dx^\alpha(p), \\ \widehat{g}_{\alpha\beta}(p) &\in Hom_Z(A; \widehat{A}) \cong \overset{2}{\widehat{A}}. \end{aligned} \right\}$$

Similarly, by considering the canonical isomorphism

$$Hom_Z((T_pM)^+; T_pM) \cong \widehat{T_pM} \otimes_{\widehat{A}} (T_pM)^{++},$$

one has the following local representation of $\bar{g}'(p) : (T_pM)^+ \to T_pM$:

$$\left\{ \begin{array}{l} \bar{g}'(p) = \partial x_\beta(p) \otimes \widehat{g}^{\alpha\beta}(p) d\dot{x}_\alpha(p), \\ \widehat{g}^{\alpha\beta}(p) \in Hom_Z(\widehat{A}; A). \end{array} \right\}$$

Taking into account the previous split one has also:

$$\sum_\beta \widehat{g}^{\beta\gamma}(p)\widehat{g}_{\alpha\beta}(p) = \delta^\gamma_\alpha \in \widehat{A},$$

where the product is given by composition, i.e., one has the following commutative diagram:

$$
\begin{array}{ccccc}
A & \xrightarrow{\widehat{g}_{\alpha\beta}(p)} & \widehat{A} & \xrightarrow{\widehat{g}^{\beta\gamma}(p)} & A \\
\| & & & & \| \\
A & & \xrightarrow{\delta^\gamma_\alpha} & & A
\end{array}
$$

        ■

Definition 2.30. We define *dual fullquantum metric* on the quantum manifold M a section $\widehat{g}$ (of class Q_w^∞) of the fiber bundle $Hom_Z((TM)^+ \otimes_Z (TM)^+; \widehat{A})$ over M.

Remark 2.22. By considering the following canonical isomorphisms:

$$Hom_Z((T_pM)^+ \otimes_Z (T_pM)^+; \widehat{A}) \cong Hom_Z((T_pM)^+; Hom_Z((TM)^+; \widehat{A}))$$

$$\cong Hom_Z(\widehat{A}; Hom_Z((TM)^+ \otimes_Z (TM)^+; \widehat{A})) \otimes_{Hom_Z(\widehat{A};A)} Hom_Z((T_p)^+; \widehat{A})$$

we see that a dual fullquantum metric $\widehat{g}$ can be locally written as follows:

$$\widehat{g} = \partial \ddot{x}_\beta(p) \otimes \widehat{g}^{\alpha\beta}(p) d\dot{x}_\alpha(p), \; \widehat{g}^{\alpha\beta}(p) \in Hom_Z(\widehat{A} \otimes_Z \widehat{A}; \widehat{A})$$

where $\ddot{x}^\alpha(p) : Hom_Z((T_pM)^+; \widehat{A}) \to Hom_Z(\widehat{A}; A)$ are local quantum coordinates induced by x^α on M.   ■

One has the following.

Lemma 2.15. *Let $(M, \widehat{g})$ be a quantum (pseudo) Riemannian manifold of dimension m over A.*

1) One has the following canonical isomorphism of fiber bundles of Z-modules over M:

$$Hom_Z((TM)^+ \otimes_Z (TM)^+; \widehat{A}) \cong Hom_Z(\widehat{TM} \otimes_Z \widehat{TM}; \widehat{A})$$

$$\cong Hom_Z(\widehat{TM}; Hom_Z(\widehat{TM}; \widehat{A}))$$

$$\cong Hom_Z(\widehat{A}; Hom_Z(\widehat{TM}; \widehat{A})) \otimes_{Hom_Z(\widehat{A}; \widehat{A})} Hom_Z(\widehat{TM}; \widehat{A})$$

2) One has the following short exact sequence of fiber bundles of Z-modules over M:

$$Hom_Z(\widehat{TM} \otimes_Z \widehat{TM}; \widehat{A}) \to Hom_Z(TM \otimes_Z TM; A) \to 0$$

3) A dual fullquantum metric $\widehat{g}$ identifies a section $\widehat{g}_{ext} : M \to Hom_Z(\widehat{TM} \otimes_Z \widehat{TM}; \widehat{A})$ and a section $\widehat{g}_{rest} : M \to Hom_Z(TM \otimes_Z TM; A)$ such that the following diagram is commutative:

$$\begin{array}{ccccc}
Hom_Z((TM)^+ \otimes_Z (TM)^+; \widehat{A}) & \cong & Hom_Z(\widehat{TM} \otimes_Z \widehat{TM}; \widehat{A}) & \to & Hom_Z(TM \otimes_Z TM; A) \\
\widehat{g} \uparrow & & \uparrow \widehat{g}_{ext} & & \uparrow \widehat{g}_{rest} \\
M & = & M & = & M
\end{array}$$

So on a quantum (pseudo) Riemannian manifold $(M, \widehat{g})$, to each dual fullquantum metric $\widehat{g}$, we can associate a fullquantum metric $\widehat{g}_{rest}$. In particular, if $\widehat{g}_{rest} = \widehat{g}$ we say that $\widehat{g}$ is the *dual fullquantum metric* of $\widehat{g}$. Furthermore, the dual fullquantum metric of $\widehat{g}$ is uniquely identified on a quantum (pseudo) Riemannian manifold.

Proof of Lemma 2.15. 1) In fact in such a case one has the canonical isomorphism $(T_pM)^+ \cong \widehat{T_pM}$, $\forall p \in M$.

2) The nondegenerate fullquantum metric $\widehat{g}$ induces the following short exact sequence $0 \to T_pM \to \widehat{T_pM}$. Furthermore, taking into account the canonical epimorphism $\widehat{A} \to A$, we get the result.

3) It follows directly from the above results. More precisely, the uniqueness of the dual it follows from the epimorphism $\widehat{TM} \cong (TM)^+ \to TM$ and the following monomorphism $A \to \widehat{A}$, that allow us to write the following canonical short exact sequence: $0 \to Hom_Z(TM \otimes_Z TM; A) \to Hom_Z(\widehat{TM} \otimes_Z \widehat{TM}; \widehat{A})$ of fiber bundles of Z-modules over M. $\qquad\square$

Remark 2.23. We can, also, define the dual metric of a quantum metric, as an object $\bar{g}(p) \in Hom_Z(A; T_pM \otimes_Z T_pM) \equiv \widehat{T_pM \otimes_Z T_pM}$, such that the following conditions are verified: $\widehat{g}(p) \circ \bar{g}(p) = 1_A$; $\bar{g}(p) \circ \widehat{g}(p) = 1_{T_pM}$. In quantum coordinates we can write: $\bar{g}(p) = \partial x_\alpha \otimes \partial x_\beta \widehat{g}^{\alpha\beta}$, with $\widehat{g}^{\alpha\beta}(p) \in Hom_Z(A : A \otimes_Z A)$, such that the following conditions are verified:

$$(2.11) \quad \left\{ \begin{array}{c} \widehat{g}_{\alpha\beta}(p)\widehat{g}^{\alpha\beta}(p)=1_A \qquad \begin{array}{ccccc} & A & \xrightarrow{\widehat{g}^{\alpha\beta}(p)} & A\otimes_Z A & \xrightarrow{\widehat{g}_{\alpha\beta}(p)} & A \\ & \| & & & & \| \\ & A & \xrightarrow{\quad 1_A \quad} & & & A \end{array} \\[3em] \widehat{g}^{\alpha\beta}\widehat{g}_{\alpha\beta}=1_{A\otimes_Z A} \qquad \begin{array}{ccccc} & A\otimes_Z A & \xrightarrow{\widehat{g}_{\alpha\beta}(p)} & \widehat{A} & \xrightarrow{\widehat{g}^{\alpha\beta}(p)} & A\otimes_Z A \\ & \| & & & & \| \\ & A\otimes_Z A & \xrightarrow{\quad 1_{A\otimes_Z A} \quad} & & & A\otimes_Z A \end{array} \end{array} \right\}.$$

Note that in the commutative case, $A = \mathbf{K}$, one has $\widehat{g}_{\alpha\beta}\widehat{g}^{\alpha\beta} = \widehat{g}^{\alpha\beta}\widehat{g}_{\alpha\beta} = \mathrm{tr}\,(\widehat{g})$. Therefore conditions (2.11) mean that the dual metric is normalized with respect to the trace. This definition of dual metric, (def2), is not, in general, equivalent to the previous one, (def1). In fact

$$Hom_Z(A; A \otimes_Z A) \ncong Hom_Z(\widehat{A}; A).$$

This difference disappears when $A = \mathbf{K}$. However, on a quantum ((pseudo) Riemannian) manifold there is a relation between them.   ∎

Lemma 2.16. *On a quantum ((pseudo) Rimennian) manifold $(M,\widehat{g})$ of dimension m on A, a dual fullquantum metric as given in (def2) can be represented into one of type (def1) and it is uniquely identified by $\widehat{g}$.*

Proof of Lemma 2.16. In fact, on a quantum (pseudo) Riemannian manifold one has a canonical Z-monomorphism $\widehat{g}'(p) : T_pM \to (T_pM)^+$. Then, one has the following commutative diagram with exact lines:

$$\begin{array}{ccc} & 0 & \\ & \downarrow & \\ Hom_Z(A;T_pM\otimes_Z T_pM) & \xrightarrow{j_*} & Hom_Z((T_pM)^+\otimes_Z(T_pM)^+;\widehat{A}) \\ \| & & \uparrow \\ Hom_Z(A;T_pM\otimes_Z T_pM) & \xrightarrow{\quad} & Hom_Z(T_pM\otimes_Z T_pM;A) \\ & & \uparrow \\ \downarrow & & Hom_Z(T_pM\otimes_Z T_pM;A\otimes_Z A) \\ & & \uparrow \\ Hom_Z(A;(T_pM)^+\otimes_Z(T_pM)^+) & \xrightarrow{\quad} & (T_pM)^+\otimes_Z(T_pM)^+ \xrightarrow{\quad} 0 \end{array}$$

So $\bar{g}(p) \in Hom_Z(A; T_pM \otimes_Z T_pM)$ can be identified with an object belonging to $Hom_Z(T_pM \otimes_Z T_pM; A)$, $\forall p \in M$. (If M is simply a quantum manifold endowed with a fullquantum metric, the Z-homomorphism $\widehat{g}'(p)$ is not more injective, but we get the same a representation $Hom_Z(A; T_pM \otimes_Z T_pM) \to Hom_Z((T_pM)^+ \otimes_Z (T_pM)^+; \widehat{A})$.) Moreover, as locally, one has also the following commutative diagram:

$$
\begin{array}{ccc}
Hom_Z(A; T_pM \otimes_Z T_pM) & \to & Hom_Z(T_pM \otimes_Z T_pM; A) \\
\| \wr & & \| \wr \\
Hom_Z(A; A \otimes_Z A) & & Hom_Z(A \otimes_Z A; A) \\
\downarrow & & \| \wr \\
Hom_Z(A; \widehat{A}) & \cong & \overset{2}{\widehat{A}}
\end{array}
$$

we see that the matrix $(\widehat{g}^{\alpha\beta}(p))$, with entries into $Hom_Z(A; A \otimes_Z A)$, can be also represented with a matrix with entries into $\overset{2}{\widehat{A}}$. Furthermore, note that j_* can be explicitly given by $j_*(\zeta)(\alpha \otimes \beta)(a) = (\alpha \otimes \beta)(\zeta(a))$, $\forall \zeta \in Hom_Z(A; T_pM \otimes_Z T_pM)$, $\alpha \otimes \beta \in (T_pM)^+ \otimes_Z (T_pM)^+$. This is a monomorphism. Another monomorphism is the following: $i_* : Hom_Z(T_pM \otimes_Z T_pM; A) \to Hom_Z((T_pM)^+ \otimes_Z (T_pM)^+; A)$. Therefore one has the following exact diagrams:

$$
\begin{array}{ccccc}
0 & \to & Hom_Z(T_pM \otimes_Z T_pM; A) & \overset{i_*}{\to} & Hom_Z((T_pM)^+ \otimes_Z (T_pM)^+; A) \\
 & & & & \uparrow j_* \\
 & & & & Hom_Z(A; T_pM \otimes_Z T_pM) \\
 & & & & \uparrow \\
 & & & & 0
\end{array}
$$

So, we say that $\bar{g} : M \to Hom_Z(A; TM \otimes_Z TM)$ is the *dual metric* of $\widehat{g} : M \to Hom_Z(TM \otimes_Z TM; A)$ if $i_* \circ \widehat{g} = j_* \circ \bar{g}$. Then, under these conditions $\bar{g}$ is uniquely identified by $\widehat{g}$ and vice versa. In fact one has the following relations: $\widehat{g} = i_*^{-1} \circ j_* \circ \bar{g}$ and $\bar{g} = j_*^{-1} \circ i_* \circ \widehat{g}$. In local quantum coordinates we get: $\widehat{g} = \partial \dot{x}^\beta \otimes \widehat{g}_{\alpha\beta} dx^\alpha$, $\bar{g} = \partial \dot{x}_{\alpha\beta} \widehat{g}^{\alpha\beta}$, with $\widehat{g}_{\alpha\beta}(p) \in Hom_Z(A \otimes_Z A; A)$, $\widehat{g}^{\alpha\beta}(p) \in Hom_Z(A; A \otimes_Z A)$, such that conditions (2.11) are verified. $\quad\square$

Remark 2.24. By considering the canonical isomorphism

$$
Hom_Z(\dot{T}^2_0(T_pM); A) \cong Hom_Z(\dot{S}^2_0(T_pM); A) \bigoplus Hom_Z(\dot{\Lambda}^2_0(T_pM); A),
$$

we see that a fullquantum metric $\widehat{g}$ splits into a symmetric and skewsymmetric parts: $\widehat{g} = \widehat{g}_{(s)} + \widehat{g}_{(a)}$. In quantum coordinates one has

$$
\widehat{g} = \sum_{\alpha,\beta} \widehat{g}_{\alpha\beta} dx^\alpha \otimes dx^\beta = \left(\sum_{\alpha,\beta} \widehat{g}_{\alpha\beta} dx^\alpha \bullet dx^\beta \right) + \left(\sum_{\alpha,\beta} \widehat{g}_{\alpha\beta} dx^\alpha \triangle dx^\beta \right),
$$

with $\widehat{g}_{\alpha\beta} \in \overset{2}{\widehat{A}}$. Therefore, we cannot, in general, distinguish between $\widehat{g}$, $\widehat{g}_{(s)}$, and $\widehat{g}_{(a)}$, simply by looking to their quantum components. (Of course if $A = \mathbf{K}$, the situation is completely different.) $\blacksquare$

Remark 2.25. We can generalize above considerations on quantum (pseudo) Riemannian manifolds, to quantum (pseudo) Riemannian supermanifolds also. More precisely a *quantum (pseudo) Riemannian supermanifold* $(M, \widehat{\mathfrak{g}})$ is a quantum supermanifold M of dimension $(m|n)$ over a quantum superalgebra A, endowed with a Q_w^∞ section $\widehat{g} : M \to Hom_Z(TM \otimes_Z TM; A)$ such that the induced homomorphisms $T_pM \to (T_pM)^+$, $\forall p \in M$, are injective. In quantum coordinates $\widehat{g}(p)$ is represented by a matrix $\widehat{g}_{\alpha\beta}(p) \in \overset{2}{\widehat{A}}_{00}(A) \times \overset{2}{\widehat{A}}_{10}(A) \times \overset{2}{\widehat{A}}_{01}(A) \times \overset{2}{\widehat{A}}_{11}(A)$. The corresponding dual quantum metric gives $\widehat{g}^{\alpha\beta}(p) \in \overset{2}{\widehat{A}}{}^{00}(A) \times \overset{2}{\widehat{A}}{}^{10}(A) \times \overset{2}{\widehat{A}}{}^{01}(A) \times \overset{2}{\widehat{A}}{}^{11}(A)$, with $\overset{2}{\widehat{A}}{}^{ij}(A) \equiv Hom_Z(A; A_i \otimes_Z A_j)$, $i, j \in \mathbf{Z}_2$, such that $\widehat{g}_{\gamma\beta}(p)\widehat{g}^{\alpha\beta}(p) = \delta_\gamma^\alpha \in \widehat{A}$, $\widehat{g}^{\alpha\beta}(p)\widehat{g}_{\gamma\beta}(p) = \delta_\gamma^\alpha \in Hom_Z(A \otimes_Z A; A \otimes_Z A)$.

Remark 2.26. (*Quantum dual curvature and quantum Lagrangian for free quantum (super) Yang-Mills equation*). Let us work on a quantum (pseudo) Riemannian manifold $(M, \widehat{g})$ of dimension m on A. We shall first prove that to any quantum curvature $\bar{R} : M \to Hom_Z(\dot{\Lambda}^2(TM); \mathfrak{g})$ there corresponds a *dual quantum curvature* $\underline{R} : M \to Hom_Z(\dot{\Lambda}^2(TM)^+; \mathfrak{g}^+)$, identified by means of the fullquantum metric $\widehat{g}$ on M and the scalar product $\underline{g}$ on $\mathfrak{g}$, i.e., by using the Z-homomorphisms $\bar{g}'(p) : (T_pM)^+ \to T_pM$, $\forall p \in M$, and $\underline{g}' : \mathfrak{g} \to \mathfrak{g}^+$. Note that by using the following isomorphisms, (the first canonical and the last induced by local quantum coordinates):

$$Hom_Z(\dot{\Lambda}^2(TM); \mathfrak{g}) \cong \widehat{\mathfrak{g}} \bigotimes_{\widehat{A}} \left(\dot{\Lambda}_0^2(T_pM)\right)^+ \cong \widehat{A}^d \bigotimes_{\widehat{A}} (\overset{2}{\widehat{A}})^{m^2} \cong (\overset{2}{\widehat{A}})^{dm^2},$$

we can locally write $\bar{R}$ in the following way: $\bar{R} = Z_K \otimes \bar{R}_{AB}^K dx^A \triangle dx^B$, $\bar{R}_{AB}^K(p) \in \overset{2}{\widehat{A}} \cong Hom_Z(A; \widehat{A})$, $\forall p \in U \subset M$. Similarly, by using the following isomorphisms:

$$Hom_Z(\dot{\Lambda}^2(TM)^+; \mathfrak{g}^+) \cong Hom_Z(\widehat{A}; \mathfrak{g}) \bigotimes_{Hom_Z(\widehat{A};\widehat{A})} Hom_Z(\dot{\Lambda}_0^2(T_pM)^+; \widehat{A})$$

$$\cong Hom_Z(\widehat{A};\widehat{A})^d \bigotimes_{Hom_Z(\widehat{A};\widehat{A})} Hom_Z(\widehat{A} \otimes_Z \widehat{A};\widehat{A})^{m^2}$$

$$\cong Hom_Z(\widehat{A} \otimes_Z \widehat{A};\widehat{A})^{dm^2}$$

we can locally write $\underline{R}$ in the following way: $\underline{R} = \widehat{Z}^H \otimes \bar{R}_H^{AB} d\dot{x}^A \triangle d\dot{x}^B$, $\bar{R}_H^{AB}(p) \in Hom_Z(\widehat{A} \otimes_Z \widehat{A}; \widehat{A})$, $\forall p \in U \subset M$, where $(\widehat{Z}^H)$ is the basis on $Hom_Z(\widehat{A}; \mathfrak{g}^+)$ induced by the basis (Z_K) on $\widehat{\mathfrak{g}}$, and $\dot{x}_B : (TM)^+ \to \widehat{A}$ are vertical quantum coordinates on the fiber bundle $(TM)^+ \to M$. In fact, one has the following homomorphisms, (the first induced by the quantum metrics, the second is a canonical isomorphism and the last induced by local quantum coordinates):

$$Hom_Z((T_pM)\otimes_Z(T_pM);\mathfrak{g}) \xrightarrow{j} Hom_Z\left((T_pM)^+\otimes_Z(T_pM)^+;\mathfrak{g}^+\right)$$

$$\cong Hom_Z(\widehat{A};\mathfrak{g}^+) \bigotimes_{Hom_Z(\widehat{A};\widehat{A})} Hom_Z\left((T_pM)^+\otimes_Z(T_pM)^+;\widehat{A}\right)$$

$$\cong Hom_Z(\widehat{A};\widehat{A})^d \bigotimes_{Hom_Z(\widehat{A};\widehat{A})} Hom_Z(\widehat{A}\otimes_Z\widehat{A};\widehat{A})^{m^2}$$

$$\cong Hom_Z(\widehat{A}\otimes_Z\widehat{A};\widehat{A})^{dm^2}$$

if $\Xi \in Q_w^\infty\left(Hom_Z((TM)\otimes_Z(TM);\mathfrak{g})\right)$, we can locally write

$$j_*(\Xi) \subset Hom_Z((TM)^+ \otimes_Z (TM)'; \mathfrak{g}^+)$$

in the following way:

$$\left\{ \begin{array}{l} j_*(\Xi) = \widehat{Z}^H \otimes j_*(\Xi)_H^{AB} d\dot{x}^A \otimes d\dot{x}^B \\[2mm] \bar{j}_*(\Xi)_H^{AB}(p) = \underline{g}_{KH}(p)\Xi_{A_1A_2}^K(p)\widehat{g}^{AA_1}(p)\otimes\widehat{g}^{BA_2}(p) \in Hom_Z(\widehat{A}\otimes_Z\widehat{A};\widehat{A}) \end{array} \right\}, \quad \forall p \in U \subset M,$$

where the product is given by composition:

$$\widehat{A} \otimes_Z \widehat{A} \xrightarrow{\widehat{g}^{AA_1}(p)\otimes\widehat{g}^{BA_2}(p)} A \otimes_Z A \xrightarrow{\Xi_{A_1A_2}^K(p)} A \xrightarrow{\underline{g}_{KH}} \widehat{A}$$

and $\Xi = Z_K \otimes \Xi_{A_1A_2}^K dx^{A_1} \triangle dx^{A_2} \in Q_w^\infty\left(Hom_Z(TM \otimes_Z TM; \mathfrak{g})\right)$. As a consequence we get that the local quantum components of the dual metric are related to the original ones by the following formula:

$$\bar{R}_H^{AB}(p) = \underline{g}_{KH}\bar{R}_{A_1A_2}^K(p)\widehat{g}^{AA_1}(p) \otimes \widehat{g}^{BA_2}(p), \quad \forall p \in U \subset M.$$

Now, by considering the canonical isomorphisms:

$$
1)\left\{
\begin{aligned}
Hom_Z(T_pM\otimes_Z T_pM;\mathfrak{g}) &\cong \widehat{\mathfrak{g}}\bigotimes_{\widehat{A}} Hom_Z(T_pM\otimes_Z T_pM;A)\\
&\cong \widehat{\mathfrak{g}}\bigotimes_{\widehat{A}} Hom_Z(T_pM;Hom_Z(T_pM)^+)\\
&\cong \widehat{\mathfrak{g}}\bigotimes_{\widehat{A}}\left[Hom_Z(A;(T_pM)^+)\otimes_{\widehat{A}}(T_pM)^+\right]
\end{aligned}
\right\}
$$

$$
2)\left\{
\begin{aligned}
&Hom_Z\big((T_pM)^+\otimes_Z(T_pM)^+;\mathfrak{g}^+\big)\\
&\cong Hom_Z(\widehat{A};\mathfrak{g})\bigotimes_{Hom_Z(\widehat{A};A)} Hom_Z((T_pM)^+\otimes_Z(T_pM)^+;\widehat{A})\\
&\cong Hom_Z(\widehat{A};\mathfrak{g}^+)\bigotimes_{Hom_Z(\widehat{A};A)} Hom_Z((T_pM)^+;Hom_Z((T_pM)^+;\widehat{A}))\\
&\cong Hom_Z(\widehat{A};\mathfrak{g}^+)\\
&\bigotimes_{Hom_Z(\widehat{A};A)}\left[Hom_Z(\widehat{A};Hom_Z((T_pM)^+;\widehat{A}))\otimes_{Hom_Z(\widehat{A};A)} Hom_Z((T_pM)^+;\widehat{A})\right]
\end{aligned}
\right\}
$$

we get the following pairing:

$$
<,>:Hom_Z\big((T_pM)^+\otimes_Z(T_pM)^+;\mathfrak{g}^+\big)\times Hom_Z(T_pM\otimes_Z T_pM;\mathfrak{g})\to Hom_Z(A;Hom_Z(\widehat{A};\widehat{A})),
$$

$$
<\zeta\otimes\xi\otimes\beta,\bar\zeta\otimes\bar\xi\otimes\bar\beta>=<\zeta,\bar\zeta>\otimes_{Hom_Z(\widehat{A};A)}\bar{j}(<\xi,\bar\xi>)\otimes_{\widehat{A}}<\beta,\bar\beta>,
$$

where: (i) $<\zeta,\bar\zeta>\in Hom_Z(\widehat{A};\widehat{A})$ is defined by composition: $<\zeta,\bar\zeta>(\alpha)=\zeta(\alpha)\circ\bar\zeta\circ\alpha$, $\forall\alpha\in\widehat{A}$; (ii) $<\xi,\bar\xi>\in Hom_Z(\widehat{A};Hom_Z(A;\widehat{A}))\cong Hom_Z(\widehat{A}\otimes_Z A;\widehat{A})\overset{\bar{j}}{\cong} Hom_Z(A;Hom_Z(\widehat{A};\widehat{A}))$ is defined by composition: $<\xi,\bar\xi>(\alpha)=\xi(\alpha)\circ\bar\xi\circ\alpha$, $\forall\alpha\in\widehat{A}$; (iii) $<\beta,\bar\beta>=\beta(\bar\beta)\in\widehat{A}$. As a consequence, we get a canonical pairing between a fullquantum curvature $\bar{R}$ and its dual $\underline{R}$ that in local quantum coordinates can be written as follows:

$$
<\underline{R}(p),\bar{R}(p)>=\sum_{H,i,j}\bar{R}^{ij}_H(p)\bar{R}^H_{ij}(p)\in Hom_Z(A;Hom_Z(\widehat{A};\widehat{A})),\quad \forall p\in U\subset M,
$$

where the product is given by composition:

$$
\begin{array}{ccc}
A\ \xrightarrow{\ \bar{R}^H_{ij}(p)\ } & \widehat{A}\ \xrightarrow{\ \bar{R}^{ij}_H(p)\ } & Hom_Z(\widehat{A};\widehat{A})\\
\| & & \|\\
A\ \xrightarrow{\ \bar{R}^{ij}_H(p)\bar{R}^H_{ij}(p)\ } & & Hom_Z(\widehat{A};\widehat{A})
\end{array}
$$

Now, by considering the canonical Z-homomorphism $\bar\mu:A\otimes_Z A\to\widehat{A}$ induced by the product $\widehat\mu:A\otimes_Z A\to A$, i.e., $\bar\mu(a\otimes b)(x)=abx\in A$,

$\forall x \in A$, we can see that $< \underline{R}(p), \bar{R}(p) > \in \overset{4}{\widehat{A}} \equiv Hom_Z(\dot{T}_0^4(A); A)$. In fact, one has the following canonical Z-isomorphisms:

$$Hom_Z(A; Hom_Z(\widehat{A}; \widehat{A})) \cong Hom_Z(A; Hom_Z(\widehat{A} \otimes_Z A; A))$$
$$\cong Hom_Z(A \otimes_Z \widehat{A} \otimes_Z A; A).$$

Therefore, by composing with the following Z-homomorphism:

$$\bar{\mu}_* : Hom_Z(A \otimes_Z \widehat{A} \otimes_Z A; A) \to Hom_Z(A \otimes_Z A \otimes_Z A \otimes_Z A; A) \equiv \overset{4}{\widehat{A}},$$

we can conclude that $< \underline{R}(p), \bar{R}(p) >$ can be identified with an element of $\overset{4}{\widehat{A}}$. This is equivalent to say that $\mathcal{L} = L dx^1 \triangle \cdots \triangle dx^4 \in Q_w^\infty(\widehat{\Omega}^4(J\hat{D}(C)))$, where (x^α) are quantum coordinates on M and $L(q) = \frac{1}{2} < \underline{R}(p), \bar{R}(p) > \in \overset{4}{\widehat{A}}$, $\forall q \in M$, $\pi_1(q) = p \in M$. So we get the following result. ∎

Lemma 2.17. *Let $(M, \widehat{g})$ be a quantum (pseudo) Riemannian manifold of dimension 4 on A. Let $\pi : C \to M$ be a fiber bundle over M defined as in Theorem 2.32, with $\mathfrak{g}$ endowed with a nondegerate fullquantum metric. Then, it is canonically defined a quantum Lagrangian function of class Q_w^∞, $L : J\hat{D}(C) \to \overset{4}{\widehat{A}}$, or a quantum Lagrangian density $\mathcal{L} = L dx^0 \triangle dx^1 \triangle dx^2 \triangle dx^3 \in Q_w^\infty(\widehat{\Omega}^4(J\hat{D}(C)))$, on the fiber bundle π, with $L \circ Dc = \frac{1}{2}\bar{R}_K^{AB}\bar{R}_{AB}^K$, $\forall c \in Q_w^\infty(C)$.*

Proof of Lemma 2.17. It follows directly from above considerations. Note that the quantum Euler-Lagrange form $\mathcal{E}[L]$ is given by

$$\mathcal{E}[L] = \frac{\delta L}{\delta \bar{\mu}_A^K} \delta \bar{\mu}_K^A \triangle dX^0 \triangle dX^1 \triangle dX^2 \triangle dX^3,$$

where $\delta \bar{\mu}_K^A = d\bar{\mu}_K^A - \sum_B \bar{\mu}_{KB}^A dX^B$, therefore

$$\delta \bar{\mu}_K^A \triangle dX^0 \triangle dX^1 \triangle dX^2 \triangle dX^3 \in Hom_Z(\dot{\Lambda}_0^5 J\hat{D}(\bar{C}); \widehat{A} \otimes_Z \dot{T}_0^4 A)$$

and $\frac{\delta L}{\delta \bar{\mu}_A^K}(q) \in Hom_Z(\widehat{A}; \overset{4}{\widehat{A}}) \cong Hom_Z(\widehat{A} \otimes_Z \dot{T}_0^4 A; A)$, for any $q \in J\hat{D}(\bar{C})$. Taking into account the canonical Z-homomorphism

$$Hom_Z(\widehat{A} \otimes_Z \dot{T}_0^4 A; A) \to Hom_Z(\dot{T}_0^6 A; A) \equiv \overset{6}{\widehat{A}},$$

we get that $\frac{\delta L}{\delta \bar{\mu}_A^K}$ identify local $\overset{6}{\widehat{A}}$-valued functions $E_A^K \equiv \frac{\delta L}{\delta \bar{\mu}_A^K} \circ \pi_{5,2}$: $J\hat{D}^5(\bar{C}) \to \overset{6}{\widehat{A}}$, that allow us to recognize $(\widehat{YM}) \subset J\hat{D}^5(\bar{C})$, since locally one has the isomorphism $J\hat{D}^5(\bar{C}) \cong A^m \times \prod_{1 \le s \le 6}(\widehat{A})^{m^s n}$. (Similar results hold working in the category of quantum supermanifods.) $\qquad\square$

Remark 2.27. Note also that we can define *dual quantum curvature* of $\bar{R}$ a Q_w^∞-section $\underline{R} : M \to Hom_Z(\mathfrak{g}; \dot{\Lambda}_0^2(M))$. In fact, on a quantum (pseudo) Riemannian manifold $(M, \hat{g})$, one has the following commutative diagram, for any $p \in M$:

$$
\begin{array}{ccc}
Hom_Z(\mathfrak{g}; \dot{\Lambda}^2(T_pM)) & \xrightarrow{\ j_* \ } & Hom_Z(\dot{\Lambda}^2(T_p)^+; \mathfrak{g}^+) \\
\downarrow & & \uparrow \\
Hom_Z(\mathfrak{g}; \dot{\Lambda}^2(T_pM)^+) & & Hom_Z(\dot{\Lambda}^2(T_p); \mathfrak{g}^+) \\
\downarrow & & \| \wr \\
Hom_Z(\mathfrak{g}; Hom_Z(\dot{\Lambda}^2(T_p); A \otimes_Z A)) & & Hom_Z(\dot{\Lambda}^2(T_pM) \otimes_Z \mathfrak{g}; A) \\
\downarrow & & \| \wr \\
Hom_Z(\mathfrak{g} \otimes_Z \dot{\Lambda}^2(T_p); A \otimes_Z A) & \longrightarrow & Hom_Z(\mathfrak{g} \otimes_Z \dot{\Lambda}^2(T_pM); A)
\end{array}
$$

More precisely, $j_*(\zeta)(\alpha \otimes \beta)(u) = (\alpha \otimes \beta)(\zeta(u))$, $\forall \zeta \in Hom_Z(\mathfrak{g}; \dot{\Lambda}^2(T_pM))$, $\alpha \otimes \beta \in \dot{\Lambda}^2(T_pM)^+$, $u \in \mathfrak{g}$. Therefore, for any $p \in M$, $\underline{R}(p)$ can be represented as an object blonging to $Hom_Z(\dot{\Lambda}^2(T_pM); \mathfrak{g}^+)$. Note, that j_* is a monomorphism. Furthermore, one has, also, the following monomorphism: $i_* : Hom_Z(\dot{\Lambda}^2(T_pM); \mathfrak{g}) \to Hom_Z(\dot{\Lambda}^2(T_pM)^+; \mathfrak{g}^+)$. So we get the following exact diagrams, for any $p \in M$:

$$
\begin{array}{ccccc}
0 & \longrightarrow & Hom_Z(\dot{\Lambda}^2(T_pM); \mathfrak{g}) & \xrightarrow{\ i_+ \ } & Hom_Z(\dot{\Lambda}^2(T_pM)^+; \mathfrak{g}^+) \\
& & & & \uparrow {\scriptstyle j_*} \\
& & & & Hom_Z(\mathfrak{g}; \dot{\Lambda}^2(T_pM)) \\
& & & & \uparrow \\
& & & & 0
\end{array}
$$

We say that $\underline{R} : M \to Hom_Z(\mathfrak{g}; \dot{\Lambda}_0^2(M))$ is the *dual curvature* of the curvature $\bar{R} : M \to Hom_Z(\dot{\Lambda}_0^2(M); \mathfrak{g})$ if the following condition is verified: $j_* \circ \underline{R} = i_* \circ \bar{R}$. Then, under these conditions, $\underline{R}$ is uniquely identified by $\bar{R}$ and vice versa. In fact one has: $\bar{R} = i_*^{-1} \circ j_* \circ \underline{R}$ and $\underline{R} = j_*^{-1} \circ i_* \circ \bar{R}$. In local quantum coordinates we get: $\bar{R} = Z_K \otimes R_{\alpha\beta}^K dx^\alpha \otimes dx^\beta$, $\underline{R} = \partial \dot{x}_{\alpha\beta} R_H^{\alpha\beta} \otimes Z^H$, with $R_{\alpha\beta}^K(p) \in Hom_Z(A \otimes_Z A; A)$, and $R_K^{\alpha\beta}(p) \in Hom_Z(A; A \otimes_Z A)$, where the relation between $R_{\alpha\beta}^K(p)$ and $R_K^{\alpha\beta}(p)$ is given by: $R_H^{\alpha\beta}(p) =$

$\widehat{g}^{\alpha\delta}(p)R^{K}_{\delta\gamma}(p)\widehat{g}^{\gamma\beta}(p)\underline{g}_{HK}$, where the product is given by composition:

$$
\begin{array}{ccc}
A & \xrightarrow{\quad R^{\alpha\beta}_{H} \quad} & A\otimes_{Z}A \\
\| & & \| \\
A \xrightarrow[\underline{g}_{HK}]{} A \xrightarrow[\widehat{g}^{\gamma\beta}(p)]{} A\otimes_{Z}A \xrightarrow[R^{K}_{\delta\gamma}(p)]{} A \xrightarrow[\widehat{g}^{\alpha\delta}(p)]{} A\otimes_{Z}A
\end{array}
$$

where $\underline{g}_{HK} \in \widehat{A}$ denotes, for abuse of notation, the metric $\underline{g}_{HK} \in Hom_{Z}(A;\widehat{A})$ composed with the natural epimorphism $Hom_{Z}(A;\widehat{A}) \to \widehat{A}$. $\quad\blacksquare$

Remark 2.28. All above considerations on the dual curvature can be generalized on quantum (pseudo) Riemannian supermanifolds. So we get the following results. $\quad\blacksquare$

Lemma 2.18. *On a 4-dimensional quantum (pseudo) Riemannian (super) manifold, endowed with a fiber bundle $\bar{\pi} : \bar{C} \to M$ as defined in Theorem 2.32, we recognize the following quantum Lagrangian function $L : J\hat{D}(C) \to \widehat{A}$, $L \circ Dc = \frac{1}{2}R^{H}_{\alpha\beta}R^{\alpha\beta}_{H}$, $\forall c \in Q^{\infty}_{w}(C)$. Hence one has the following first order Lagrangian density of reduced form: $\mathcal{L} = L dx^{0} \wedge dx^{1} \wedge dx^{2} \wedge dx^{3}$. Thus the corresponding Euler-Lagrange equation is a second order QPDE, (resp. QSPDE).*

Proof of Lemma 2.18. If M is a quantum manifold, one has the following commutative diagram of fiber bundles of Z-modules over M:

$$
\begin{array}{ccc}
Hom_{Z}(\mathfrak{g};\dot{\Lambda}^{2}_{0}M) \times Hom_{Z}(\dot{\Lambda}^{2}_{0}M;\mathfrak{g}) & \xrightarrow{\ <,>\ } & M \times \widehat{A} \\
\| \wr j & & \| \\
(\widehat{\dot{\Lambda}^{2}_{0}M} \otimes_{A} \widehat{\mathfrak{g}^{+}}) \times (\widehat{\mathfrak{g}} \otimes_{A} \widehat{(\dot{\Lambda}^{2}_{0}M)^{+}}) & \longrightarrow & M \times \widehat{A}
\end{array}
$$

Then, $L \circ Dc = \frac{1}{2} < \bar{R}, \underline{R} >$. Moreover, if M is a quantum supermanifold, the isomorphism j, in above commutative diagram, does not more exist, but the pairing $<,>$ is defined as the trace of the Z-homomorphism $\bar{R} \circ \underline{R} : \mathfrak{g} \to \mathfrak{g}$. In quantum coordinates we get the same formula. Note that the quantum Euler-Lagrange form $\mathcal{E}[L]$ can be locally written as follows:

$$
\mathcal{E}[L] = \frac{\delta L}{\delta\bar{\mu}^{K}_{A}} \delta\bar{\mu}^{A}_{K} \triangle dX^{0} \triangle dX^{1} \triangle dX^{2} \triangle dX^{3},
$$

where $\delta\bar{\mu}^{A}_{K} = d\bar{\mu}^{A}_{K} - \sum_{B} \bar{\mu}^{A}_{KB} dX^{B}$, therefore

$$
\delta\bar{\mu}^{A}_{K} \triangle dX^{0} \triangle dX^{1} \triangle dX^{2} \triangle dX^{3} \in Hom_{Z}(\dot{\Lambda}^{5}_{0}J\hat{D}(\bar{C}); \widehat{A} \otimes_{Z} A)
$$

and $\frac{\delta L}{\delta \bar{\mu}_A^K}(q) \in Hom_Z(\widehat{A}; \widehat{A}) \cong Hom_Z(\widehat{A} \otimes_Z A; A)$, for any $q \in J\widehat{D}(\bar{C})$. On the other hand, taking into account the canonical Z-homomorphism

$$Hom_Z(\widehat{A} \otimes_Z A; A) \to Hom_Z(\dot{T}_0^3 A; A) \equiv \overset{3}{\widehat{A}},$$

we get that $\frac{\delta L}{\delta \bar{\mu}_A^K}$ identify local $\overset{3}{\widehat{A}}$-valued functions $E_A^K \equiv \frac{\delta L}{\delta \bar{\mu}_A^K} \circ \pi_{5,2}$: $J\widehat{D}^2(\bar{C}) \to \overset{3}{\widehat{A}}$, that allow us to recognize $(\widehat{YM}) \subset J\widehat{D}^2(\bar{C})$, since locally one has the isomorphism $J\widehat{D}^2(\bar{C}) \cong A^m \times \prod_{1 \leq s \leq 3}(\widehat{A})^{m^s n}$. (Similar results hold working in the category of quantum supermanifods.) $\quad\square$

Now we are ready to go in the announced applications of Theorem 2.32.

TAB.2.22 - $N=2$ Super Poincaré algebra.

$$[J_{\alpha\beta}, J_{\gamma\delta}] = \eta_{\beta\gamma} J_{\alpha\delta} + \eta_{\alpha\delta} J_{\beta\gamma} - \eta_{\alpha\gamma} J_{\beta\delta} - \eta_{\beta\delta} J_{\alpha\gamma}$$

$$[P_\alpha, P_\beta] = -8e^2 J_{\alpha\beta}, \qquad [J_{\alpha\beta}, P_\gamma] = \eta_{\beta\gamma} P_\beta - \eta_{\alpha\gamma} P_\beta$$

$$[J_{\alpha\beta}, Q_{\gamma i}] = (\sigma_{\alpha\beta})_\gamma^{\mu j} Q_{\mu j}, \qquad [Q_{\beta i}, Q_{\mu j}] = (C\gamma^\alpha)_{\beta\mu} \delta_{ij} P_\alpha + C_{\beta\mu} \epsilon_{ij} \overline{Z}$$

Example 2.11. (*Quantum supergravity $N = 2$*). This model realizes at quantum level Einstein's dream to unifying electromagnetism and gravity by considering two real (one complex) gravitino $\psi_\mu^{\alpha i}$, the photon and the graviton. In this model divergence disappears, i.e., it does not necessitate renormalization. For basic informations on classical supergravity see eg. refs.[62,73,86–90]. Let us recast, here, in our noncommutative geometrical setting an example of supergravity previously considered in the literature in a purely (super)commutative (geometric) framework. We shall start with the *quantum $N = 2$ super Poincaré group* over a quantum superalgebra $A = Z \oplus A_1$,[31] that is a quantum Lie supergroup G having as quantum Lie superalgebra $\mathfrak{g}$ one identified by the following infinitesimal generators: $\{Z_K\}_{1 \leq K \leq 19} \equiv \{J_{\alpha\beta}, P_\alpha, \overline{Z}, Q_{\beta i}\}_{0 \leq \alpha, \beta \leq 3; 1 \leq a \leq 2}$, such that $J_{\alpha\beta} = -J_{\beta\alpha}, P_\alpha, \overline{Z} \in Hom_Z(A_0; \mathfrak{g})$, $Q_{\beta i} \in Hom_Z(A_1; \mathfrak{g})$, and such that nonzero $\mathbf{Z}_2$-graded brackets are reported in Tab.2.22. Here $C_{\alpha\beta}$ is the antisymmetric charge conjugation matrix, $\sigma_{\beta\mu} = \frac{1}{4}[\gamma_\beta, \gamma_\mu]$, with γ^μ the Dirac

---

[31] Recall that the *quantum Poincaré group*, with respect to a quantum algebra A, is the symmetry group of the quantum Minkowski space-time over A. Its quantum Lie algebra is $\mathfrak{so}(3,1) \times A^4$. (For details see ref.[67].)

matrices. $\overline{Z}$ commutes with all the other ones. Then, with reference of above notation, one has $\dim G = (d|N_2) = (11|8)$, and we will consider the following principal bundle in the category of quantum supermanifolds: P is a quantum supermanifold of dimension $(15|8)$; M is a quantum supermanifold of dimension $(4|N_1) = (4|0)$. Therefore M is a 4-dimensional commutative manifold that we assume also be globally hyperbolic. Then a pseudoconnection can be written by means of the following fullquantum differential 1-forms on P: $_\rceil\mu^K = \mu_H^K dY^H$, $(\mu_H^K) = (\frac{1}{2}\omega_H^{\alpha\beta}, \theta_H^\mu, A_H, \psi_H^{aj})$. With respect to a section $s : M \to P$ we get: $(s^*_\rceil\mu)^K = \bar{\mu}_\gamma^K dX^\gamma$, $(\bar{\mu}_\gamma^K) = (\frac{1}{2}\bar{\omega}_\gamma^{\alpha\beta}, \bar{\theta}_\gamma^\mu, \bar{A}_\gamma, \bar{\psi}_\gamma^{aj})$, where $\bar{\omega}_\gamma^{\alpha\beta}$ is the usual Levi-Civita connection, $\bar{\theta}_\gamma^\mu$ is the vierbein, $\bar{A}_\gamma$ is the electromagnetic field and $\bar{\psi}_\gamma^{aj}$ is the usual spin $\frac{3}{2}$ field. The blow-up structure: $\pi^*\hat{C}(P) \hookrightarrow \underline{E}$ implies that are acceptable only quantum pseudoconnections that can be written in the form

$$_\rceil\mu = (\partial x_\alpha + Z_K \mu_\alpha^K) \otimes dx^\alpha,$$

where $\mu_\alpha^K \in Q_w^\infty(\pi(U), \widehat{A})$, $U \subset P$. ($\hat{C}(P) \cong J\hat{D}(P)/G$ is the fiber bundle, over M, of principal quantum connections on the G-principal fiber bundle $\pi : P \to M$) The corresponding curvature can be written in the form:

$$_\rceil R_{\beta\alpha}^K = (\partial x_\beta \mu_\alpha^K) + C_{IJ}^K[\mu_\beta^I, \mu_\alpha^J]_+.$$

The local expression of the *dynamic equation* $\hat{E}_2 \subset J\hat{D}^2(C)$ is resumed in Tab.2.23, where

$$L : J\hat{D}(\underline{E}) \to \widehat{A}$$

is a quantum Lagrangian function of first order. (Its Lagrangian density is of reduced form.) In Tab. 2.24 is reported the corresponding equation $\hat{E}_2[i] \subset J\hat{D}^2(i^*\bar{C})$ on a macroscopic shell, identified by a (local) embedding $i : N \to M$, (or better, with respect to a quantum relativistic frame). Possible Lagrangian densities are polinomial in the curvature, (see example below), hence we can assume that they give formally quantum integrable QSPDEs. Let us, now, calculate the quantum and integral bordism groups of $\hat{E}_2$ and its fullquantum p-Hopf superalgebras. Let us assume that $\hat{E}_2$ is formally quantumintegrable. Then the quantum and integral bordism groups in $\hat{E}_2$ are reported in Tab.2.25. There are also reported the fullquantum p-Hopf algebras $\hat{\mathbf{H}}_p((\hat{E}_2)_{+\infty})$, $0 \leq p \leq 3$, of $\hat{E}_2$. Note that we have used the fact that $\hat{C}(P) \to M$ is a contractible fiber bundle of dimension $(4|0, 44|32)$ over the quantum superalgebra

$$A \times \widehat{A} = (A_0 \times A_1) \times \widehat{A}\,{}^1_0(A) \times \widehat{A}\,{}^1_1(A).$$

The same quantum and integral bordism groups can be recognized also for $\hat{E}_2[i]$. In fact, as $\hat{E}_2$ is a formally quantum integrable QSPDE on the fiber bundle $\bar{\pi} : \bar{C} \to M$, we can apply Theorem 2.27 to obtain the quantum and integral bordism groups in $\hat{E}_2$. Of course the same conclusions hold also for $\hat{E}_2[i]$ as this equation is formally quantumintegrable too. As a by-product of above Table 2.25 we get that 1-dimensional admissible integral closed quantum submanifolds contained into $\hat{E}_2$, (resp. $\hat{E}_2[i]$), can propagate and interact between them by means of 2-dimensional admissible integral quantum manifolds contained into $\hat{J}^2_4(\bar{C})$, (resp. $\hat{J}^2_4(i^*\bar{C})$), or by means of 2-dimensional admissible integral quantum manifolds contained into $\hat{E}_2$, (resp. $\hat{E}_2[i]$), in such a way to generate (quantum) tunnel effects. Finally, as a consequence of the triviality of the 3-dimensional integral bordism grooups: $\Omega_3^{\hat{E}_2} \cong \Omega_3^{\hat{E}_\infty} \cong 0$, we get the existence of global quantum (smooth) solutions of such equations. As a by-product we get that the quantum supergravity equation $\hat{E}_2 \subset J\hat{D}^2(\bar{C})$ admits global solutions having a change of sectional topology (*quantum tunnel effects*). In general these solutions are not globally representable as second derivative of sections of the fiber bundle $\bar{C} \to M$.[32] In fact a bounadry value problem for $\hat{E}_k$ can be directly implemented in the manifold $\hat{E}_k \subset J\hat{D}^2(\bar{C}) \subset \hat{J}^2_4(\bar{C})$ by requering that a 3-dimensional compact space-like (for some $t = t_0$), admissible integral manifold $B \subset \hat{E}_k$ propagates in $\hat{E}_k$ in such a way that the boundary ∂B describes a fixed 3-dimensional time-like integral manifold $Y \subset \hat{E}_k$. (We shall require that the boundary ∂B of B is orientable.) Y is not, in general, a closed (smooth) manifold. However, we can solder Y with two other compact 3-dimensional integral manifolds X_i, $i = 1, 2$, in such a way that the result is a closed 3-dimensional (smooth) integral

---

[32] Such solutions well interpret the meaning of "quantum geometrodynamics" as first conjectured by J.A.Wheeler [91]. (Compare also with the more recent approach on the "topological quantum field theory" by Atiayah. M. and Witten, E. (See, e.g. refs.[2] and works quoted there by these authors.))

manifold $Z \subset \hat{E}_k$.

TAB.2.23 - Dynamic Equation $\hat{E}_2 \subset J\hat{D}^2(\bar{C})$ and Bianchi identity.

Fields equations $(\hat{E}_2)$	$(\partial\omega_{ab}^{\gamma}.L)-\partial_\mu(\partial\omega_{ab}^{\gamma\mu}.L)=0 \ \left(\textit{curvature equation}\right)$ $(\partial\theta_{\alpha}^{\gamma}.L)-\partial_\mu(\partial\theta_{\alpha}^{\gamma\mu}.L)=0 \ \left(\textit{torsion equation}\right)$ $(\partial\psi_{\beta i}^{\gamma}.L)-\partial_\mu(\partial\psi_{\beta i}^{\gamma\mu}.L)=0 \ \left(\textit{gravitino equation}\right)$ $(\partial A^{\gamma}.L)-\partial_\mu(\partial A^{\gamma\mu}.L)=0 \ \left(\textit{Maxwell's equation}\right)$
Bianchi identity (B)	$(\partial x_\gamma.R_{\beta\alpha}^{ab})+2[\omega_{e\gamma}^{a},R_{\beta\alpha}^{eb}]_+=0$ $(\partial x_\gamma.R_{\beta\omega}^{\alpha})+[\omega_{\gamma}^{\alpha b},R_{\beta\omega b}]_++(C\gamma^\alpha)_{\delta\mu}[\psi_{j\gamma}^{\delta},\rho_{\beta\omega}^{\mu j}]_+=0$ $(\partial x_\gamma.\rho_{\omega\alpha}^{\beta i})+(\sigma_{ab})_{\delta j}^{\beta i}[\omega_{\gamma}^{ab},\rho_{\omega\alpha}^{\delta j}]_+=0$ $(\partial x_\gamma F_{\beta\alpha})+C_{\delta\mu}\epsilon_{ij}[\psi_{\gamma}^{\delta i},\rho_{\beta\alpha}^{\mu j}]_+=0$
Fields	$R_{\mu\nu}^{ab}=(\partial x_\mu.\omega_{\nu}^{ab})+2[\omega_{e\mu}^{a},\omega_{\nu}^{eb}]_+ \ \left(\textit{curvature}\right)$ $R_{\mu\nu}^{\alpha}=(\partial x_\mu.\theta_{\nu}^{\alpha})+[\omega_{\beta\mu}^{\alpha},\theta_{\nu}^{\beta}]_++(C\gamma^\alpha)_{\beta\delta}[\psi_{j\mu}^{\beta},\psi_{\nu}^{\delta j}]_+ \ \left(\textit{torsion}\right)$ $\rho_{\mu\nu}^{\beta i}=(\partial x_\mu.\psi_{\nu}^{\beta i})+(\sigma_{ab})_{\gamma j}^{\beta i}[\omega_{\mu}^{ab},\psi_{\nu}^{\gamma j}]_+ \ \left(\textit{gravitino}\right)$ $F_{\mu\nu}=(\partial x_\mu.A_\nu)+C_{\beta\gamma}\epsilon_{ij}[\psi_{\mu}^{\beta i},\psi_{\nu}^{\gamma j}]_+ \ \left(\textit{electromagnetic field}\right)$

TAB.2.24 - Dynamic Equation on macroscopic shell: $\hat{E}_2[i] \subset J\hat{D}^2(i^*\bar{C})$ and Bianchi identity.

Fields equations $(\hat{E}_2[i])$	$(\partial\omega_{ab}^{\gamma}.L)-\partial_\mu(\partial\omega_{ab}^{\gamma\mu}.L)=0 \ \left(\textit{curvature equation}\right)$ $(\partial\theta_{\alpha}^{\gamma}.L)-\partial_\mu(\partial\theta_{\alpha}^{\gamma\mu}.L)=0 \ \left(\textit{torsion equation}\right)$ $(\partial\psi_{\beta i}^{\gamma}.L)-\partial_\mu(\partial\psi_{\beta i}^{\gamma\mu}.L)=0 \ \left(\textit{gravitino equation}\right)$ $(\partial A^{\gamma}.L)-\partial_\mu(\partial A^{\gamma\mu}.L)=0 \ \left(\textit{Maxwell's equation}\right)$
Bianchi identity $(B[i])$	$(\partial x_{[\gamma}.R_{\beta\alpha]}^{ab})+2\omega_{e[\gamma}^{a}R_{\beta\alpha]}^{eb}=0$ $(\partial x_{[\gamma}.R_{\beta\omega]}^{\alpha})+\omega_{[\gamma}^{\alpha b}R_{\beta\omega]b}+\frac{1}{2}(C\gamma^\alpha)_{\delta\mu}\psi_{j[\gamma}^{\delta}\rho_{\beta\omega]}^{\mu j}=0$ $(\partial x_{[\gamma}.\rho_{\omega\alpha]}^{\beta i})+\frac{1}{2}(\sigma_{ab})_{\delta j}^{\beta i}\omega_{[\gamma}^{ab}\rho_{\omega\alpha]}^{\delta j}=0$ $(\partial x_{[\gamma}F_{\beta\alpha]})+\frac{1}{2}C_{\delta\mu}\epsilon_{ij}\psi_{[\gamma}^{\delta i}\rho_{\beta\alpha]}^{\mu j}=0$
Fields	$R_{\mu\nu}^{ab}=(\partial x_{[\mu}.\omega_{\nu]}^{ab})+2\omega_{e[\mu}^{a}\omega_{\nu]}^{eb} \ \left(\textit{curvature}\right)$ $R_{\mu\nu}^{\alpha}=(\partial x_{[\mu}.\theta_{\nu]}^{\alpha})+\omega_{\beta[\mu}^{\alpha}\theta_{\nu]}^{\beta}+\frac{1}{2}(C\gamma^\alpha)_{\beta\delta}\psi_{j[\mu}^{\beta}\psi_{\nu]}^{\delta j} \ \left(\textit{torsion}\right)$ $\rho_{\mu\nu}^{\beta i}=(\partial x_{[\mu}.\psi_{\nu]}^{\beta i})+\frac{1}{2}(\sigma_{ab})_{\gamma j}^{\beta i}\omega_{[\mu}^{ab}\psi_{\nu]}^{\gamma j} \ \left(\textit{gravitino}\right)$ $F_{\mu\nu}=(\partial x_{[\mu}.A_{\nu]})+\frac{1}{2}C_{\beta\gamma}\epsilon_{ij}\psi_{[\mu}^{\beta i}\psi_{\nu]}^{\gamma j} \ \left(\textit{electromagnetic field}\right)$

More precisely, we can take $X_1 = B$ so that $\tilde{Z} \equiv X_1 \bigcup_{\partial B} Y$ is a 3-dimensional compact integral manifold such that $\partial\tilde{Z} \equiv C$ is a 2-dimensional space-like integral manifold. We can assume that C is an orientable manifold. Then, from Tab. 2.25 it follows that $\partial X_2 = C$, for some space-like compact 3-dimensional integral manifold $X_2 \subset \hat{E}_k$. Set $Z \equiv \tilde{Z} \bigcup_C X_2$. Therefore, one has $Z = X_1 \bigcup_{\partial B} Y \bigcup_C X_2$. Then from Tab. 2.25 it follows also that there exists a 4-dimensional integral (smooth) manifold $V \subset \hat{E}_k$ such that $\partial V = Z$. Hence the integral manifold V is a solution of our boundary value problem between the times t_0 and t_1, where t_0 and t_1 are the times corresponding to the boundaries where are soldered X_i, $i = 1, 2$ to Y. Now, this process can be extended for any $t_2 > t_1$. So we are able to find (smooth) solutions for any $t > t_0$, hence (smooth) solutions for any $t > t_0$, therefore, global (smooth) solutions. Remark that in order to assure the smoothness of the global solution so built it is enough to develop such construction in the infinity prolongation $(\hat{E}_k)_{+\infty}$ of $\hat{E}_k$. Finally note that in the set of solutions of $\hat{E}_k$ there are ones that have change of sectional topology. In fact the 3-dimensional integral bordism groups are trivial: $\Omega_3^{\hat{E}_k} = 0 = \Omega_3^{(\hat{E}_k)_{+\infty}}$. Now, if the 3-dimensional space-like compact domain B describes a region where is present a "black hole", a solution like one described in the previous point 1) represents an *evaporating black hole*. The point where there is the singularity of the characteristic flow of the solution is the *explosive end* [97] of the evaporation process with production of new particles and radiation described by the outgoing solution. In fact, in order to obtain such solutions we must have a Cauchy integral data with a geometric black hole B embedded in a compact 3-dimensional integral manifold N, $B \subset N$, such that its boundary ∂N propagates with a fixed flow. In this geometric context a black hole is a 3-dimensional compact space-like integral manifold $B \subset \hat{E}_k$ like a disk, where the curvature of the boundary ∂B is very high, (of physical dimension of 1-fermi, i.e. the range of the strong interactions). So it is a "non-naked singularity", or in the physical language of Hawking it is a "singularity with its event horizon". (For informations on this subject see e.g., refs.[24,97], references quoted there and in [98].) Then a solution, with quantum tunnel effect of such boundary problem, can describe a vaporation process of such black hole. Our theorem of the integral bordism groups of above point 1) assures the existence of such solutions and a way to build them. It is important to note that all the integral characteristic numbers are conserved through an evaporating black-hole. In fact, recall (see Theorem 5.7 in ref.[67] and previous results) that a necessary and sufficient condition for two integral admissible, closed, compact, quantum manifolds of the same dimension, $N_0, N_1 \subset \hat{E}_k$, contained into a QSPDE $\hat{E}_k \subset \hat{J}_n^k(W)$, and with orientable classic limit $(N_i)_C$, $i = 0, 1$, be the boundary of an admissible integral quantum manifold V of $\hat{E}_k$, $\partial V = N_0 \bigcup N_1$, is that all the integral characteristic numbers $\hat{i}[N_0]$ of N_0 should equal to ones of N_1: $\hat{i}[N_0] = \hat{i}[N_1]$. In our case N_i, $i = 0, 1$, are just compact closed orientable commutative manifolds of dimension q, $0 \le q \le 3$. So, as a by-product of the above quoted theorem, we get that through a quantum tunnel effect all the "admissible conservation laws" are preserved. In this model of $N = 2$ extended quantum supergravity the possible particles involved in an evaporation process of quantum black hole are gravitons, photons and gravitinos (spin $\frac{3}{2}$). By adopting other N-extended quantum supergravity models, we can also consider gravitons, N gravitinos, $\frac{1}{2}N(N-1)$ vectors, hence an evaporization process of quantum black holes should interest production of gravitinos, gravitons, quarks, photons, electrons, etc... all the possible particles. Therefore, N-extended quantum supergravity appears a good theoretical model for a quantum superunified high energy physics in interaction with gravity. $\blacksquare$

We are ready, now, to consider an application of above general setting that completely solves an important mathematical problem in quantum gauge theories.

TAB.2.25 - Quantum and integral bordism groups in $\hat{E}_2$ and $\hat{E}_2[i]$.

p	$\Omega_p(X)$	Ω_p^X	$\hat{\mathbf{H}}_p(X)_{+\infty})$
0	A	A	B^A
$1 \le p \le 3$	0	0	B

$B \equiv \widehat{A}^{\infty}; X \equiv \hat{E}_2, \hat{E}_2[i]$.

Theorem 2.34. (Free quantum super Yang-Mills equation). *The free quantum super Yang-Mills equation admits (smooth) global solutions, for any (smooth) boundary conditions. The invertibility of the corresponding Hamiltonians are necessary and sufficient conditions that such solutions have mass-gap, i.e., their energy content should be different from zero.*

Proof. Let us assume that M is the *quantum super Minkowsky spacetime*, (with respect to a quantum superalgebra A). This is a superextension of the Minkowsky spacetime. Therefore, M is an affine space with structure $(M, \mathbf{M}, \alpha)$, where the space of free vectors $\mathbf{M}$ is given by $\mathbf{M} \equiv A \otimes_{\mathbf{R}} \mathbf{M}_C$, where $\mathbf{M}_C$ is the space of free vectors of the Minkowsky space-time $(M_C, \mathbf{M}_C, \alpha_C)$ and A is just a quantum superalgebra. Then, M results a quantum manifold of dimension 4 over A, but also a quantum supermanifold of dimension $(4|4)$ over A. Let us denote by $(X^\alpha)_{0 \leq \alpha \leq 3}$ the quantum coordinates $X^\alpha : M \to A$, and let us denote by $(x^\alpha, \theta^a)_{0 \leq \alpha \leq 3, 1 \leq a \leq 4}$ the corresponding quantum supercoordinates, i.e., $x^\alpha : M \to A_0$, $\theta^a : M \to A_1$. There are global embeddings $i : M_C \to M$, but there is not a canonical projection $M \to M_C$. Hence, M is not classic regular in a canonical way, (even if there are surjections $M \to A^4 \to \mathbf{R}^4$. Furthermore, on M is defined a canonical *fullquantum* metric, that is a section $\widehat{g} : M \to Hom_Z(\dot{T}_0^2 M; A)$, given by $\widehat{g}(p)(a^k \otimes v_k) \otimes (b^j \otimes u_j)) = a^k b^j \bar{g}(v_k, u_j)$, where $\bar{g}$ is the canonical Minkowsky metric on $\mathbf{M}_C$. (We have used the canonical isomorphism $T_p M \cong \mathbf{M} \cong A \otimes_{\mathbf{R}} \mathbf{M}_C$.) The fullquantum metric $\widehat{g}$ identifies a canonical injective mapping $\widehat{g}'$ of $T_p M$ into the fullquantum dual $(T_p M)^+$, i.e., one has the following exact sequence of Z-modules:

$$0 \to T_p M \xrightarrow{\widehat{g}'} (T_p M)^+,$$

for any $p \in M$. In fact, if $0 \neq v = a^j \otimes v_j \in T_p M$, it follows that $\widehat{g}'(v) = \widehat{g}_v = 0$ iff $\widehat{g}_v(u) = \widehat{g}_v(b^i \otimes u_i) = a^j b^i \bar{g}(v_j, u_i) = 0$, $\forall u \in T_p M$. In particular, it should be $0 = a^j b^i \bar{g}(v_j, u_i)$ with $b^i \in \mathbf{K}$ and $u_i \neq 0$. But this is possible iff whether $a^j = 0$ or $v_j = 0$, i.e., iff $v = 0$, that contradicts our hypothesis. Hence, $\ker('\widehat{g}) = \{0\} \subset T_p M$. As a consequence, the fullquantum metric $\widehat{g}$ is nondegenerate. Note that $\widehat{g}(u, u)$ can be zero even if $u \neq 0$. For example if $u = a \otimes v \in T_p M \cong A \otimes_{\mathbf{R}} \mathbf{M}_C$, with $v^2 = \bar{g}(v, v) \neq 0$, and $a \in A$ is nilpotent with $a^2 = 0$, we get $\widehat{g}(u, u) = a^2 v^2 = 0$. Note that on any tangent space $T_p M \cong \mathbf{M} \equiv A \otimes_{\mathbf{R}} \mathbf{M}_C$ we can take a basis $\{1 \otimes e_\alpha\}_{0 \leq \alpha \leq 3}$, where $\{e_\alpha\}$ is a basis on $\mathbf{M}_C$. In such a basis $\widehat{g}(p)$ is represented by the symmetric nondegenarate matrix $\widehat{g}_{\alpha\beta} = \bar{g}_{\alpha\beta} \in \mathbf{R} \subset A$. With respect to this fact we can state that if $(\bar{g}_{\alpha\beta})$ has signature $(+ - - -)$, then also $(\widehat{g}_{\alpha\beta})$ has the same signature. For any embedding $i : M_C \to M$ one has the following commutative diagram:

$$
\begin{array}{ccccc}
A \otimes_{\mathbf{R}} S_2^0(M_C) & \longleftarrow & A \otimes_{\mathbf{R}} T_2^0(M_C) & \longleftarrow & (\dot{T}_0^2 M)^+ \\
{\scriptstyle i^* \widehat{g}_{(s)}} \uparrow & & \uparrow {\scriptstyle i^* \widehat{g}} & & \uparrow {\scriptstyle \widehat{g}} \\
M_C & = & M_C & \xrightarrow{i} & M
\end{array}
$$

where we have considered the canonical split $T_2^0(M_C) \cong S_2^0(M_C) \oplus \Lambda_2^0(M_C)$. The *quantum (special) super Poincaré group* (with respect to A) is the symmetry group $\hat{\mathbf{P}}$ of the quantum oriented super Minkowsky spacetime M. Hence, $\hat{\mathbf{P}} = A^4 \times SO(3,1)$ and it is a quantum Lie supergroup of dimension $(8|4)$. Hence $\hat{\mathbf{P}}$ is a central extension of $\mathbf{P}$ by means of the subspace $\mathfrak{b} \subset A^4$ defined by means of the following exact commutative diagram:

$$
\begin{array}{ccccccccc}
0 & \to & \mathfrak{b} & \to & \hat{\mathbf{P}} & \to & \mathbf{P} & \to & 1 \\
 & & \| & & \| & & \| & & \| \\
0 & \to & \mathfrak{b} & \to & SO(3,1) \times A^4 & \to & SO(3,1) \times \mathbf{R}^4 & \to & 1
\end{array}
$$

such that on $\mathfrak{b}$ is defined a bilinear map $\rho : \mathfrak{b} \times \mathfrak{b} \to \mathfrak{p}$ which satisfies the following requirements:

(i)(symmetry) $\rho(u,v) = \rho(v,u)$, $u, v \in \mathfrak{b}$;

(ii) $Ad(x)(\rho(u,v)) = [x, \rho(u,v)] = \rho(\tau(x)(u), v) + \rho(u, \tau(x)(v))$, $x \in \mathfrak{p}$, $u, v \in \mathfrak{b}$, with $\tau : \mathfrak{p} \to L(\mathfrak{b})$, a Lie algebra representation which gives to $\mathfrak{b}$ the structure of $\mathfrak{p}$-module;

(iii) $\tau(\rho(u,v))(w) + \tau(\rho(v,w))(u) + \tau(\rho(w,u))(v) = 0$, $u, v, w \in \mathfrak{b}$.

In fact, under such hypotheses, the tangent space

$$
T_e\hat{\mathbf{P}} \equiv \hat{\mathfrak{p}} = A^4 \oplus \mathfrak{so}(3,1) \cong \hat{\mathfrak{p}}_0 \oplus \hat{\mathfrak{p}}_1 = (\mathfrak{so}(3,1) \oplus A_0^4) \bigoplus A_1^4
$$

is a quantum algebra $\mathbf{Z}_2$-graded and endowed with a structure of $\mathbf{Z}_2$-graded Lie algebra, hence it is a quantum superalgebra (see Definition 1.46). (This situation is similar to one just considered in ref.[73] for classic, i.e., non-quantized, supergravity.) Then, a *free quantum super Yang-Mills equation* is a quantum super Yang-Mills QSPDE of second order $(\widehat{YM}) \subset J\hat{D}^2(\bar{C})$, with respect to a G-principal fibre bundle $\pi : P \to M$, where M is just the quantum super Minkowsky spacetime with respect to a quantum superalgebra A, and such that the field equations come from the following first order Lagrangian function:

$$
\begin{cases}
L : J\hat{D}(\bar{C}) \to \hat{A}, \\
L \circ D\bar{c} = \dfrac{1}{2} \bar{R}_{\alpha\beta}^K \bar{R}_K^{\alpha\beta}, \\
\forall \bar{c} \in Q_w^\infty(\bar{C}),
\end{cases}
$$

(i.e., the corresponding Lagrangian density is of reduced form). Here the rising and lowering of indexes is obtained by means of the fullquantum metrics $\widehat{g}$ on M and $\underline{g}$ on $\mathfrak{g}$ respectively.[33] Therefore in such a case the local expression of $(\widehat{YM})$ results given by the equations reported in Tab.2.26. Note that the quantum super Yang-Mills equation is now

$$E_K^A \equiv (\partial \bar{\mu}_K^A . L) - (\partial_B (\partial \bar{\mu}_K^{AB} . L)) = 0.$$

This is the local expression of the quantum Euler-Lagrange equation with

$$L = \frac{1}{2} \bar{R}_{\alpha\beta}^S \bar{R}_S^{\alpha\beta},$$

where

$$
\begin{aligned}
\bar{R}_{\alpha\beta}^S =& \bar{\mu}_{\beta\alpha}^S + \frac{1}{2}\widehat{C}_{IJ}^S [\bar{\mu}_\alpha^I, \bar{\mu}_\beta^J]_+ \\
=& \bar{\mu}_{\beta\alpha}^S + \frac{1}{2}\widehat{C}_{IJ}^S (\bar{\mu}_\alpha^I \bar{\mu}_\beta^J - (-1)^{|I||J|}\bar{\mu}_\alpha^J \bar{\mu}_\beta^I) \\
=& \bar{\mu}_{\beta\alpha}^S + \frac{1}{2}\widehat{C}_{IJ}^S \bar{\mu}_\alpha^I \bar{\mu}_\beta^J - (-1)^{|I||J|}\frac{1}{2}\widehat{C}_{IJ}^S \bar{\mu}_\alpha^J \bar{\mu}_\beta^I) \\
=& \bar{\mu}_{\beta\alpha}^S + \frac{1}{2}\widehat{C}_{IJ}^S \bar{\mu}_\alpha^I \bar{\mu}_\beta^J - (-1)^{|I||J|}\frac{1}{2}\widehat{C}_{JI}^S \bar{\mu}_\alpha^I \bar{\mu}_\beta^J) \\
=& \bar{\mu}_{\beta\alpha}^S + \frac{1}{2}\widehat{C}_{IJ}^S \bar{\mu}_\alpha^I \bar{\mu}_\beta^J + \frac{1}{2}\widehat{C}_{IJ}^S \bar{\mu}_\alpha^I \bar{\mu}_\beta^J) \\
=& \bar{\mu}_{\beta\alpha}^S + \widehat{C}_{IJ}^S \bar{\mu}_\alpha^I \bar{\mu}_\beta^J.
\end{aligned}
$$

Furthermore one has:

$$
\begin{aligned}
\bar{R}_{\alpha\beta}^S =& g^{\alpha\delta} \bar{R}_{\delta\gamma}^L g^{\gamma\beta} \underline{g}_{SL} \\
=& g^{\alpha\delta} [\bar{\mu}_{\gamma\delta}^L + \widehat{C}_{IJ}^L \bar{\mu}_\delta^{\bar{I}} \bar{\mu}_\gamma^{\bar{J}}] g^{\gamma\beta} \underline{g}_{SL}.
\end{aligned}
$$

Let us calculate the two terms, in the quantum Euler-Lagrange equations,

---

[33] For example, if $G \equiv \widehat{P}$, $\underline{g}{:}\widehat{\mathfrak{p}} \times \widehat{\mathfrak{p}} \to A$ is induced by the Killing metric on $\mathfrak{so}(3,1)$ and the product on A^4.

by means of quantum curvature forms:

$$
\begin{aligned}
(\partial\bar\mu_K^A.L) =& \frac{1}{2}(\partial\bar\mu_K^A.\bar R_{\alpha\beta}^S)\bar R_S^{\alpha\beta} + \frac{1}{2}\bar R_{\alpha\beta}^S(\partial\bar\mu_K^A.\bar R_S^{\alpha\beta}) \\
=& \frac{1}{2}\widehat C_{IJ}^S(\delta_\alpha^A\delta_K^I\bar\mu_\beta^J + \bar\mu_\alpha^I\delta_\beta^A\delta_K^J)\bar R_S^{\alpha\beta} \\
& + \frac{1}{2}\bar R_{\alpha\beta}^S[g^{\alpha\delta}\widehat C_{\bar I J}^L(\delta_\delta^A\delta_K^{\bar I}\bar\mu_\gamma^{\bar J} + \bar\mu_\delta^{\bar I}\delta_\gamma^A\delta_K^{\bar J})]g^{\gamma\beta}\underline g_{SL} \\
=& \frac{1}{2}\widehat C_{KJ}^S\bar\mu_\beta^J\bar R_S^{A\beta} + \frac{1}{2}\widehat C_{IK}^S\bar\mu_\alpha^I\bar R_S^{\alpha A} \\
& + \frac{1}{2}\bar R_{\alpha\beta}^S g^{\alpha A}\widehat C_{KJ}^L\bar\mu_\gamma^J g^{\gamma\beta}\underline g_{SL} \\
& + \frac{1}{2}\bar R_{\alpha\beta}^S g^{\alpha\delta}\widehat C_{IK}^L\bar\mu_\delta^{\bar I} g^{A\beta}\underline g_{SL} \\
=& \frac{1}{2}\widehat C_{KJ}^S\bar\mu_\beta^J\bar R_S^{A\beta} + \frac{1}{2}\widehat C_{IK}^S\bar\mu_\alpha^I\bar R_S^{\alpha A} \\
& + \frac{1}{2}\bar R_L^{A\gamma}\widehat C_{KJ}^L\bar\mu_\gamma^J + \frac{1}{2}R_L^{\delta A}\widehat C_{IK}^L\bar\mu_\delta^{\bar I} \\
=& \frac{1}{2}\widehat C_{KJ}^S\bar\mu_\beta^J\bar R_S^{A\beta} + \frac{1}{2}\widehat C_{JK}^S\bar\mu_\beta^J R_S^{\beta A} \\
& + \frac{1}{2}\bar R_S^{A\beta}\widehat C_{KJ}^S\bar\mu_\beta^J + \frac{1}{2}\bar R_S^{\beta A}\widehat C_{JK}^S\bar\mu_\beta^J \\
=& [\widehat C_{KJ}^S\bar\mu_\beta^J, \bar R_S^{A\beta}]_+ + [\widehat C_{JK}^S\bar\mu_\beta^J, \bar R_S^{\beta A}]_+ \\
=& [\widehat C_{KJ}^S\bar\mu_\beta^J, \bar R_S^{A\beta}]_+ - (-1)^{|K||J|}[\widehat C_{KJ}^S\bar\mu_\beta^J, \bar R_S^{\beta A}]_+ .
\end{aligned}
$$

Moreover, one has:

$$
\begin{aligned}
(\partial\bar\mu_K^{AB}.L) =& (\partial\bar\mu_K^{AB}.\frac{1}{2}(\bar R_{\alpha\beta}^S\bar R_S^{\alpha\beta})) \\
=& \frac{1}{2}\{\delta_\beta^A\delta_\alpha^B\delta_K^S\bar R_S^{\alpha\beta} + \bar R_{\alpha\beta}^S(\partial\bar\mu_K^{AB}.\bar R_S^{\alpha\beta})\} \\
=& \frac{1}{2}\{\bar R_K^{\beta A} + \bar R_{\alpha\beta}^S g^{\alpha\delta}(\partial\bar\mu_K^{AB}.\bar R_{\delta\gamma}^L)g^{\gamma\beta}\underline g_{SL}\} \\
=& \frac{1}{2}\{R_K^{\beta A} + \bar R_{\alpha\beta}^S g^{\alpha\delta}\delta_\delta^\beta\delta_\gamma^A\delta_K^L g^{\gamma\beta}\underline g_{SL}\} \\
=& \frac{1}{2}\{\bar R_K^{\beta A} + \bar R_{\alpha\beta}^S g^{\alpha\beta}g^{A\beta}\underline g_{SK}\} \\
=& \frac{1}{2}\{\bar R_K^{\beta A} + R_K^{\beta A}\} = \bar R_K^{\beta A} .
\end{aligned}
$$

Therefore, it results: $(\partial\bar\mu_K^A.L) = [\widehat C_{KR}^H\bar\mu_C^R, \bar R_H^{[AC]}]_+$ and $(\partial\bar\mu_K^{AB}.L) = \bar R_K^{BA}$.

The situation is resumed in Tab. 2.26.

TAB.2.26 - Local expression of $(\widehat{YM}) \subset J\hat{D}^2(C)$ and Bianchi identity $(B) \subset J\hat{D}^2(\bar{C})$.

(Field equations) $E_K^A \equiv -(\partial_B.\bar{R}_K^{BA}) + [\widehat{C}_{KR}^H \bar{\mu}_C^R, \bar{R}_H^{[AC]}]_+ = 0 \ (\dagger)$	$(\widehat{YM})$
(Fields) $\bar{F}_{A_1 A_2}^K \equiv \bar{R}_{A_1 A_2}^K - \left[(\partial X_{A_1}.\bar{\mu}_{A_2}^K) + \frac{1}{2}\widehat{C}_{IJ}^K [\bar{\mu}_{A_1}^I, \bar{\mu}_{A_2}^J]_+ \right] = 0$	
(Bianchi identities) $B_{HA_1 A_2}^K \equiv (\partial X_H.\bar{R}_{A_1 A_2}^K) + \frac{1}{2}\widehat{C}_{IJ}^K [\bar{\mu}_H^I, \bar{R}_{A_1 A_2}^J]_+ = 0$	(B)

$(\dagger) \ [\widehat{C}_{KR}^H \bar{\mu}_C^R, \bar{R}_H^{[AC]}]_+ \equiv [\widehat{C}_{KR}^H \bar{\mu}_C^R, \bar{R}_H^{AC}]_+ - (-1)^{|K||R|} [\widehat{C}_{KR}^H \bar{\mu}_C^R, \bar{R}_H^{CA}]_+ .$

$$F_{A_1 A_2}^K : \Omega_1 \subset J\hat{D}(C) \to \overset{2}{\widehat{A}}; \quad B_{HA_1 A_2}^K : \Omega_2 \subset J\hat{D}^2(C) \to \overset{3}{\widehat{A}}; \quad E_K^A : \Omega_2 \subset J\hat{D}^2(C) \to \overset{3}{\widehat{A}}.$$

One can see that above equation $(\widehat{YM}) \subset J\hat{D}^2(\bar{C})$ is formally quantum integrable and also completely quantum superintegrable. (See also ref.[71].) In order to prove this, let us rewrite the quantum super Yang-Mills equation in a more explicit form. For this let us calculate the following:

$$(\partial_B.(\partial\bar{\mu}_K^{AB}.L))$$
$$= (\partial_B.\bar{R}_K^{BA})$$
$$= (\partial X_B.\bar{R}_K^{BA}) + (\partial\bar{\mu}_R^S.\bar{R}_K^{BA})\bar{\mu}_{SB}^R + (\partial\bar{\mu}_R^{ST}.\bar{R}_K^{BA})\bar{\mu}_{SBT}^R$$
$$= (\partial X_B.g^{B\delta})\bar{R}_{\delta\gamma}^L g^{\gamma A}\underline{g}_{SL} + g^{B\delta}\bar{R}_{\delta\gamma}^L(\partial X_B.g^{\gamma A})\underline{g}_{SL}$$
$$+ (\partial\bar{\mu}_R^S.\bar{R}_K^{BA})\bar{\mu}_{SB}^R + g^{B\delta}(\partial\bar{\mu}_R^{ST}.(\bar{\mu}_{\gamma\delta}^L + \widehat{C}_{\bar{I}\bar{J}}^L \bar{\mu}_\delta^{\bar{I}}\bar{\mu}_\gamma^{\bar{J}}))g^{\gamma A}\underline{g}_{KL}\bar{\mu}_{SBT}^R$$
$$= (\partial X_B.g^{B\delta})\bar{R}_{\delta\gamma}^L g^{\gamma A}\underline{g}_{SL} + g^{B\delta}\bar{R}_{\delta\gamma}^L(\partial X_B.g^{\gamma A})\underline{g}_{SL} + (\partial\bar{\mu}_R^S.\bar{R}_K^{BA})\bar{\mu}_{SB}^R$$
$$+ g^{B\delta}\delta_\gamma^S \delta_\delta^T \delta_R^L g^{\gamma A}\underline{g}_{KL}\bar{\mu}_{SBT}^R .$$

So we can write

$$(\partial\bar{\mu}_K^A.L) - (\partial_B.(\partial\bar{\mu}_K^{AB}.L))$$
$$= [\widehat{C}_{KJ}^S \bar{\mu}_\beta^J, \bar{R}_S^{A\beta}]_+ - (-1)^{|K||J|}[\widehat{C}_{KJ}^S \bar{\mu}_\beta^J, \bar{R}_S^{\beta A}]_+$$
$$- (\partial X_B.g^{BM})\bar{R}_{MN}^L g^{NA}\underline{g}_{KL} - g^{BM}\bar{R}_{MN}^L(\partial X_B.g^{NA})\underline{g}_{KL}$$
$$- (\partial\bar{\mu}_R^S.\bar{R}_K^{BA})\bar{\mu}_{SB}^R - g^{BT}g^{SA}\underline{g}_{KR}\bar{\mu}_{SBT}^R = 0.$$

Therefore we can write the equation $(\widehat{YM}) \subset J\hat{D}^2(\bar{C})$ in the following form:

$$(2.12) \quad g^{BT}g^{SA}\underline{g}_{KR}\bar{\mu}_{SBT}^R = [\widehat{C}_{KJ}^S \bar{\mu}_B^J, \bar{R}_S^{AB}]_+ - (-1)^{|K||J|}[\widehat{C}_{KJ}^S \bar{\mu}_B^J, \bar{R}_S^{BA}]_+$$
$$- (\partial X_B.\bar{R}_K^{BA}) - (\partial\bar{\mu}_R^S.\bar{R}_K^{BA})\bar{\mu}_{SB}^R,$$

where we have written on the left the term with the highest derivative, (i.e., second derivative).

TAB.2.27 - Local isomorphisms on $\widehat{(YM)}\subset J\hat{D}^2(\bar{C})$ and related fiber bundles.

$$M\cong A^m$$

$$\mathfrak{g}\cong A^{d|N_2}$$

$$\bar{C}\cong A^m\times(\overset{1}{\widehat{A}}_0)^{md|mN_2}\times(\overset{1}{\widehat{A}}_1)^{mN_2|md}$$

$$J\hat{D}(\bar{C})\cong\bar{C}\times(\overset{2}{\widehat{A}}_0)^{m^2d|m^2N_2}\times(\overset{2}{\widehat{A}}_1)^{m^2N_2|m^2d}$$

$$J\hat{D}^2(\bar{C})\cong J\hat{D}(\bar{C})\times(\overset{3}{\widehat{A}}_0)^{m^3d|m^3N_2}\times(\overset{3}{\widehat{A}}_1)^{m^3N_2|m^3d}$$

$$J\hat{D}^3(\bar{C})\cong J\hat{D}^2(\bar{C})\times(\overset{4}{\widehat{A}}_0)^{m^4d|m^4N_2}\times(\overset{4}{\widehat{A}}_1)^{m^4N_2|m^4d}$$

$$\widehat{(YM)}\cong J\hat{D}(\bar{C})\times(\overset{3}{\widehat{A}}_0)^{m(m^2-1)d|m(m^2-1)N_2}\times(\overset{3}{\widehat{A}}_1)^{m(m^2-1)N_2|m(m^2-1)d}$$

$$\widehat{(YM)}_{+1}\cong\widehat{(YM)}\times(\overset{4}{\widehat{A}}_0)^{m^2(m^2-1)d|m^2(m^2-1)N_2}\times(\overset{4}{\widehat{A}}_1)^{m^2(m^2-1)N_2|m^2(m^2-1)d}$$

$$(\dot{g}_2)_q\cong(\overset{3}{\widehat{A}}_0)^{m(m^2-1)d|m(m^2-1)N_2}\times(\overset{3}{\widehat{A}}_1)^{m(m^2-1)N_2|m(m^2-1)d}$$

$$((\dot{g}_2)_{+1})_q\cong(\overset{4}{\widehat{A}}_0)^{m^2(m^2-1)d|m^2(m^2-1)N_2}\times(\overset{4}{\widehat{A}}_1)^{m^2(m^2-1)N_2|m^2(m^2-1)d}$$

TAB.2.28 - Local isomorphisms on $\widehat{(YM)}_{+r}\subset J\hat{D}^{2+r}(\bar{C})$ and related fiber bundles.

$$J\hat{D}^k(\bar{C})\cong A^m\times\prod_{1\le s\le k}((\overset{s}{\widehat{A}}_0)^{m^sd|m^sN_2}\times(\overset{s}{\widehat{A}}_1)^{m^sN_2|m^sd})$$

$$\widehat{(YM)}_{+r}\cong A^m\times H_1\times H_2$$

$$H_1\equiv\prod_{1\le s\le 2}((\overset{s}{\widehat{A}}_0)^{m^sd|m^sN_2}\times(\overset{s}{\widehat{A}}_1)^{m^sN_2|m^sd})$$

$$H_2\equiv\prod_{3\le s\le 3+r}((\overset{s}{\widehat{A}}_0)^{m^{s-2}(m^2-1)d|m^{s-2}(m^2-1)N_2}\times(\overset{s}{\widehat{A}}_1)^{m^{s-2}(m^2-1)N_2|m^{s-2}(m^2-1)d})$$

$$Hom_Z(\dot{T}_0^{2+r}(T_pM);vT_q\bar{C})\cong(\overset{3+r}{\widehat{A}}_0)^{m^{3+r}d|m^{3+r}N_2}\times(\overset{3+r}{\widehat{A}}_1)^{m^{3+r}N_2|m^{3+r}d}$$

$$((\dot{g}_2)_{+r})_q\cong(\overset{3+r}{\widehat{A}}_0)^{m^{r+1}(m^2-1)d|m^{r+1}(m^2-1)N_2}\times(\overset{3+r}{\widehat{A}}_1)^{m^{r+1}(m^2-1)N_2|m^{r+1}(m^2-1)d}$$

$r\ge 1.$ $m=4,$ $(d|N_2)\equiv(8|4).$ $\overset{s}{\widehat{A}}_0\equiv\prod_{i_1+\cdots+i_s=0,\,i_r\in\mathbf{Z}_2}\overset{s}{\widehat{A}}_{i_1\cdots i_s}(A),\ \overset{s}{\widehat{A}}_1\equiv\prod_{i_1+\cdots+i_s=1,\,i_r\in\mathbf{Z}_2}\overset{s}{\widehat{A}}_{i_1\cdots i_s}(A).$

Taking into account the local isomorphisms reported in Tab.2.27 and Tab.2.28 we see that

$$\dim(\widehat{YM})_{+r} = \dim(\widehat{YM})_{+(r+1)} + \dim(\dot{g}_2)_{r+1})_q, \ \forall r \geq 1.$$

Therefore one has the following short exact sequences

$$(\widehat{YM})_{+(r+1)} \to (\widehat{YM})_{+r} \to 0.$$

So, taking into account that $(\widehat{YM})$ is quantum regular, that Z is a Noetherian **K**-algebra, that $(\widehat{YM})$ is δ-regular, and that $(\dot{g}_2)_{+r}$ is a bundle of Z-modules over $(\widehat{YM})$, we get, by using Theorem 2.22 and Theorem 2.23, that $(\widehat{YM})$ is formally quantumintegrable. Furthermore, by using Theorem 2.27, we are able to calculate the $(3|3)$-dimensional integral bordism group of $(\widehat{YM})$. One has: $\Omega_{3|3}^{(\widehat{YM})} = 0$. Hence we get existence of global Q_w^∞ solutions for any boundary condition of class Q_w^∞. Finally, the evaluation of $(\widehat{YM})$ on a macroscopic shell $i(M_C) \subset M$ is given by the equations reported in Tab.2.29.

TAB.2.29 - Local expression of $(\widehat{YM})[i] \subset J\hat{D}^2(i^*\bar{C})$ and Bianchi idenity $(B)[i] \subset J\hat{D}^2(i^*\bar{C})$.

(*Field equations*) $(\partial_\alpha.\tilde{R}^{K\alpha\beta}) + [\widehat{C}^K_{IJ}\tilde{\mu}^I_\alpha, \tilde{R}^{J\alpha\beta}]_+ = 0$	$((\widehat{YM})[i])$
(*Fields*) $\bar{R}^K_{\alpha_1\alpha_2} = (\partial\xi_{[\alpha_1}.\tilde{\mu}^K_{\alpha_2]}) + \frac{1}{2}\widehat{C}^K_{IJ}\tilde{\mu}^I_{[\alpha_2}\tilde{\mu}^J_{\alpha_1]}$ (*Bianchi identities*) $(\partial\xi_{[\gamma}.\tilde{R}^K_{\alpha_1\alpha_2]}) + \frac{1}{2}\widehat{C}^K_{IJ}\tilde{\mu}^I_{[\gamma}\tilde{R}^J_{\alpha_1\alpha_2]} = 0$	$((B)[i])$

This equation is also formally quantum integrable and completely quantum integrable. Furthermore, the 3-dimensional integral bordism group of $(\widehat{YM})[i]$ and its infinity prolongation $(\widehat{YM})[i]_{+\infty}$ are trivial: $\Omega_3^{(\widehat{YM})[i]} \cong \Omega_3^{(\widehat{YM})[i]_{+\infty}} \cong 0$. So equation $(\widehat{YM})[i] \subset \hat{J}^2_4(i^*\bar{C})$ admits global (smooth) solutions for any fixed time-like 3-dimensional (smooth) boundary conditions. Furthermore, with respect to coordinates $\{\xi^\alpha\}_{0 \leq \alpha \leq 3}$ on $N \equiv i(M_C)$, adapted to the frame, we can write the Hamiltonian, of such solutions, in the following form: $H = \frac{1}{2}(\tilde{R}^K_{\beta\alpha}\tilde{R}^{\beta\alpha}_K - \tilde{\mu}^K_{\alpha\beta}\tilde{R}^{\beta\alpha}_K) : N \to \widehat{A}$. Iff $H(p) \in G(\widehat{A}) \subset \widehat{A}$, $\forall p \in N$, the spectrum $Sp(H(p)) \subset \mathbf{K}$ does not contain the zero of $\mathbf{K}$. Here $G(\widehat{A})$ is the group of invertible elements of the **K**-algebra $\widehat{A}$, where the product is given by composition. Note that for a quantum algebra B, does not necessitate $G(B) = B$. (See Proposition 1.35.) Recall, also, that for any element $b \in B$, its spectrum is defined by $Sp(b) \equiv \mathbf{K} \setminus \rho(b)$, where $\rho(b)$ is the resolvent of b, i.e., $\rho(b) = -\{\lambda \in \mathbf{K} | (b - \lambda 1) \in G(B)\}$. In such a case we say that the solution has a *mass-gap*. Of course, if $G(\widehat{A}) = \widehat{A} \setminus \{0\}$, a solution with mass-gap is one such that $H \neq 0$. However, if $G(\widehat{A}) \subset \widehat{A}$, the condition $H \neq 0$ is necessary but not sufficient. Therefore, in general, the condition $H(p) \in G(\widehat{A})$, $\forall p \in N$, is necessary and sufficient in order to global solutions of $(\widehat{YM})[i]$ should be with mass-gap. This is equivalent to state that are satisfied the following conditions on matrices:

$$\mathrm{tr}\,(\tilde{R}^K_{\beta\alpha}\tilde{R}^{\beta\alpha}_H - \tilde{\mu}^K_{\alpha\beta}\tilde{R}^{\beta\alpha}_H) \neq 0;\ \exists\ \frac{1}{\mathrm{tr}\,(\tilde{R}^K_{\beta\alpha}\tilde{R}^{\beta\alpha}_H - \tilde{\mu}^K_{\alpha\beta}\tilde{R}^{\beta\alpha}_H)} \in \widehat{A}. \qquad \square$$

Corollary 2.5. *Trivial solutions, i.e. solutions with $\tilde{R}^K_{\alpha\beta} = 0$, of $(\widehat{YM})[i]$, cannot have mass-gap.*

Remark 2.29. (*Hamiltonian as distribution*). The Hamiltonian H of a solution of the Yang-Mills equation $(\widehat{YM})[i]$ can be identified with a section of the vector bundle $i^*E \to N$, where E is the trivial vector bundle $E \equiv M \times \widehat{A} \to M$. Therefore, by considering the *s-formal adjoint* $E'_s \equiv E^* \otimes_{\mathbf{K}} (\Lambda^s_0 M)^*$ of E, $s = 3, 4$, and the natural homomorphisms $C^\infty_w(E) \to (C^\infty_0(E'_s))'$, for any embedding $i_s : X \to M$, where X is a commutative s-dimensional manifold, we can represent the Hamiltonian as a distribution on the vector bundle $E \to M$. (See also ref.[63].) More precisely we have the following distribution: $< H[i_s], \alpha > = \int_X i^*_s < H, \alpha >\in \mathbf{K}$, $\forall \alpha \in C^\infty_0(E'_s)$, where $\alpha = \beta \otimes \eta$, $\beta \in C^\infty_0(E^*)$, with $\eta \in C^\infty_w((\Lambda^s_0 M)^*) \equiv \Omega^s(M)$, such that $i^*_s\eta$ is a volume form on X. We denote by $\Omega^s \subset \Omega^s(M)$ the corresponding subspace. Let us define *character-s-test functions*, $s = 3, 4$, test functions like before and with β identified by means of characters of the $\mathbf{K}$-algebra $\widehat{A}$, i.e., $\beta(p) = (p, \chi(p))$, where $\chi(p)$ is a character of $\widehat{A}$. Let us denote such subspace by $\mathcal{C}^s_{ar}(E) \subset C^\infty_0(E^*) \otimes \Omega^s$. Then we have the following theorem that characterizes in distributional sense, the requirement that a solution of the Yang-Mills equation is with mass-gap. $\qquad \blacksquare$

Theorem 2.35. *If a solution of the Yang-Mills equation $(\widehat{YM})[i]$ has mass-gap, the distributional value of H on the character-s-test functions, $s = 3, 4$, is different from zero for the frame $i : N \to M$, $(s = 4)$, or its space-like sections $i_t : N_t \subset N \to M$ with respect to the time-like flow defined on N, $(s = 3)$.*

Proof. We shall prove that the condition $E_s \equiv< H[i_s], \alpha > = \int_X i^*_s < H, \alpha > \neq 0$, $\forall \alpha \in \mathcal{C}^s_{ar}(E)$, $s = 3, 4$, with $i_4 = i : N \equiv X \to M$ and

$i_3 = i_t : N_t \equiv X \subset N \to M$, is a direct consequence of the condition $H(p) \in G(\widehat{A})$, $\forall p \in N$. In fact, we have the following.

Lemma 2.19. *Let B be a $\mathbf{K}$-algebra. Then $b \in G(B)$ iff $\chi(b) \neq 0$, $\forall \chi \in Ch(B)$, where $Ch(B)$ is the set of characters of B, i.e., the set of unitary, multiplicative linear functions $\chi : B \to \mathbf{K}$. (One has also $\chi(b) \in Sp(b)$, $\forall b \in B$.)*

Proof of Lemma 2.19. It is standard. (See e.g. ref.[5].) $\qquad\square$

As a by-product of above lemma we get that $i_s^* < H, \alpha > \neq 0$, $\forall \alpha \in \mathcal{C}_{ar}^s(E)$, whether $H(p) \in G(\widehat{A})$, $\forall p \in N$, or $\forall p \in N_t$. Then, for the continuity of the function $i_s^* < H, \alpha >$ on N or N_t, we get also that $E_s \neq 0$. So the proof is completed. (Of course, for the continuity of E_t, with respect to $t \in T$, we get also that $E_t \neq 0$, $\forall t \in T$. $\qquad\square$

Remark 2.30. (*Hamiltonian as quantum distribution*). Similarly to what made in Remark 2.29 we can represent also the Hamiltonian H, of a solution of the Yang-Mills equation, as a quantum distribution on the vector bundle $E \equiv M \times \widehat{A} \to M$. In fact, by considering the *s-formal quantum adjoint* $E_s^{\blacksquare} \equiv E^+ \otimes_A (\Lambda_0^s M)^\star$ of E, $s = 3, 4$, and the natural homomorphisms $C_w^\infty(E) \to (C_0^\infty(E_s^{\blacksquare}))^+$, for any embedding $i_s : X \to M$, like in Remark 2.29, we can represent the Hamiltonian as a A-valued distribution on the vector bundle $E \to M$. (See also ref.[65].) More precisely we have the following distribution: $< \hat{H}[i_s], \alpha >= \int_X i_s^* < H, \alpha >\in A$, $\forall \alpha \in C_0^\infty(E_s^{\blacksquare})$, where $\alpha = \beta \otimes \eta$, $\beta \in C_0^\infty(E^+)$, with $\eta \in C_w^\infty((\Lambda_0^s M)^\star) \equiv \hat{\Omega}^s(M)$, such that $i_s^* \eta$ is a $G(A)$-valued volume form on X. We denote by $\hat{\Omega}^s \subset \hat{\Omega}^s(M)$ the corresponding subspace. Let us define *quantum character-s-test functions*, $s = 3, 4$, test functions like before and with β identified by means of homomorphisms of the $\mathbf{K}$-algebra $\widehat{A} \to A$. Let us denote such subspace by $\hat{\mathcal{C}}_{ar}^s(E) \subset C_0^\infty(E^+) \otimes \hat{\Omega}^s$. Then we have the following theorem that characterizes in quantum distributional sense, the requirement that a solution of the Yang-Mills equation is with mass-gap. $\qquad\blacksquare$

Theorem 2.36. *If a solution of the Yang-Mills equation $(\widehat{YM})[i]$ has mass-gap, the quantum distributional value of H on the quantum character-s-test functions, $s = 3, 4$, belongs to $G(A)$, for the frame $i : N \to M$, $(s = 4)$, or its space-like sections $i_t : N_t \subset N \to M$ with respect to the time-like flow defined on N, $(s = 3)$.*

Proof. We shall prove that the condition $E_s \equiv < H[i_s], \alpha >= \int_X i_s^* < H, \alpha >\in G(A)$, $\forall \alpha \in \hat{\mathcal{C}}_{ar}^s(E)$, $s = 3, 4$, with $i_4 = i : N \equiv X \to M$ and

$i_3 = i_t : N_t \equiv X \subset N \to M$, is a direct consequence of the condition $H(p) \in G(\widehat{A})$, $\forall p \in N$. In fact, we have the following.

Lemma 2.20. *Let B and C be $\mathbf{K}$-algebras. Let $h \in Hom_{\mathbf{K}-algebra}(B;C)$. Then $h(G(B)) \subset G(C)$.*

Proof of Lemma 2.20. It is standard. (See e.g. ref.[4].) $\qquad\square$

As a by-product of above lemma we get that $i_s^* < H, \alpha > \in G(A)$, $\forall \alpha \in \hat{\mathcal{C}}_{ar}^s(E)$, whether $H(p) \in G(\widehat{A})$, $\forall p \in N$, or $\forall p \in N_t$. Then, we get also that $E_s \in G(A)$. So the proof is completed. $\qquad\square$

Note that for the proof of Theorem 2.34 we have assumed that the Yang-Mills equation is given by means of a Lagrangian density in reduced form. However, this results does not depend on the reduced form. In fact, we can consider also a Lagrangian density with Lagrangian defined by Lemma 2.17. Then the corresponding Euler-Lagrange equation it results a QSPDE yet, that is also formally quantum integrable. In order to understand this it is useful the following lemma.

Lemma 2.21. *The Bianchi identity is a formally quantum integrable QPDE.*

Proof of Lemma 2.21. For semplicity we shall consider only the case of quantum manifolds. (However the proof for quantum supermanifolds can be similarly done.) With reference to Theorem 2.7 one has the local isomorphisms reported in Tab.2.30. Then we get $\dim(B)_{+1} = \dim(B) + \dim((g_2)_{+1})$. Therefore, the projection $(B)_{+1} \to (B)$ is surjective. This property can be extended to any prolongation of (B), i.e., the canonical maps $(B)_{+r} \to (B)_{+(r-1)}$ are surjective, $\forall r \geq 1$, as $\dim(B)_{+r} = \dim(B)_{+(r-1)} + \dim((g_2)_{(r+1=)})$. In fact, for any r-prolongation, $r \geq 1$, one has the local isomorphisms reported in Tab.2.30. Therefore in quantum fibered coordinates $(X^A, Y^I, \dot{Y}_A^I, \dot{Y}_J^I) \equiv (X^A, Z^B)$, on $C \equiv Hom_Z(TP; \mathfrak{g}) \cong \widehat{\mathfrak{g}} \otimes_{\widehat{A}}(TP)^+$, fibered on M, we get that the vectors $\zeta \in ((\dot{g}_2)_{+r})_q$ iff

$$\zeta = \partial Z_B \otimes \zeta^{BA_1 \cdots A_{(2+r)}} d\dot{X}_{A_1 \cdots A_{(2+r)}}$$
$$= [\partial Y_I \otimes \zeta^{IA_1 \cdots A_{(2+r)}} + \partial \dot{Y}_I^A \otimes \zeta_A^{IA_1 \cdots A_{(2+r)}}$$
$$+ \partial \dot{Y}_I^J \otimes \zeta_J^{IA_1 \cdots A_{(2+r)}}] d\dot{X}_{A_1 \cdots A_{(2+r)}},$$

with $\zeta_A^{IA_1 \cdots A_{(2+r)}} = 0$. (These are $m^{(3+r)}n$ conditions.) Therefore, taking into account that (B) is quantum regular, that Z is a Noetherian $\mathbf{K}$-algebra, that (B) is δ-regular, and that $(\dot{g}_2)_{+r}$ is a bundle of Z-modules over (B), we can conclude, by using Theorem 2.6 and Theorem 2.7, that (B) is formally

TAB.2.30 - Local isomorphisms on $(B) \subset J\hat{D}^2(C)$ and related fiber bundles.

$M \cong A^m$
$P \cong A^{m+n}$
$C \cong A^{m+n} \times (\widehat{A})^{(m+n)n}$
$J\hat{D}(C) \cong A^{m+n} \times (\widehat{A})^{(2mn+n^2)} \times (\overset{2}{\widehat{A}})^{(m^2n+mn^2)}$
$J\hat{D}^2(C) \cong A^{m+n} \times (\widehat{A})^{(2mn+n^2)} \times (\overset{2}{\widehat{A}})^{(2m^2n+mn^2)} \times (\overset{3}{\widehat{A}})^{(m^3n+m^2n^2)}$
$J\hat{D}^3(C) \cong A^{m+n} \times (\widehat{A})^{(2mn+n^2)} \times (\overset{2}{\widehat{A}})^{(2m^2n+mn^2)} \times (\overset{3}{\widehat{A}})^{(2m^3n+m^2n^2)} \times (\overset{4}{\widehat{A}})^{(m^4n+m^3n^2)}$
$(B) \cong A^{m+n} \times (\widehat{A})^{(2mn+n^2)} \times (\overset{2}{\widehat{A}})^{(2m^2n+mn^2)} \times (\overset{3}{\widehat{A}})^{(m^2n^2)}$
$(B)_{+1} \cong A^{m+n} \times (\widehat{A})^{(2mn+n^2)} \times (\overset{2}{\widehat{A}})^{(2m^2n+mn^2)} \times (\overset{3}{\widehat{A}})^{(m^3n+m^2n^2)} \times (\overset{4}{\widehat{A}})^{(m^3n^2)}$
$(\dot{g}_2)_q \cong (\overset{2}{\widehat{A}})^{(m^2n)} \times (\overset{3}{\widehat{A}})^{(m^2n^2)}$
$((\dot{g}_2)_{+1})_q \cong (\overset{3}{\widehat{A}})^{(m^3n)} \times (\overset{4}{\widehat{A}})^{(m^3n^2)}$

TAB.2.31 - Local isomorphisms on $(B)_{+r} \subset J\hat{D}^{2+r}(C)$ and related fiber bundles.

$J\hat{D}^k(C) \cong A^{m+n} \times \prod_{1 \leq s \leq k} ((\overset{s}{\widehat{A}})^{(2m^sn+m^{(s-1)}n^2)} \times (\overset{k+1}{\widehat{A}})^{(m^{(k+1)}n+m^kn^2)})$
$(B)_{+r} \cong A^{m+n} \times (\prod_{1 \leq s \leq 2} (\overset{s}{\widehat{A}})^{(2m^sn+m^{(s-1)}n^2)}) \times (\prod_{3 \leq \bar{s} \leq 2+r} (\overset{\bar{s}}{\widehat{A}})^{(m^{\bar{s}}n+m^{\bar{s}-1}n^2)}) \times (\overset{3+r}{\widehat{A}})^{m^{(2+r)}n^2}$
$Hom_Z(\dot{T}_0^{2+r}(T_pM); vT_qC) \cong (\overset{(2+r)}{\widehat{A}})^{m^{(2+r)}n} \times (\overset{3+r}{\widehat{A}})^{(m^{(3+r)}n+m^{(2+r)}n^2)}$
$((\dot{g}_2)_{+r})_q \cong (\overset{(2+r)}{\widehat{A}})^{m^{(2+r)}n} \times (\overset{3+r}{\widehat{A}})^{m^{(2+r)}n^2}$

$r \geq 1$

quantum integrable. Note that we can also consider the Bianchi identities as a QPDE on the fiber bundle $\bar{\pi} : \bar{C} \equiv Hom_Z(TM; \mathfrak{g}) \to M$. In fact the pull-back of pseudoconnections $_1\mu : P \to C$ by means of sections $s : M \to P$ are sections $s^*{}_1\mu : M \to \bar{C}$. So we can consider quantum Yang-Mills equations $(\widehat{YM}) \subset J\hat{D}^\infty(\bar{C})$. In such case, the Bianchi identities are subbundles $(\bar{B}) \subset J\hat{D}^2(\bar{C})$. Then, we can similarly prove that $(\bar{B})$ is a formally quantum integrable QPDE, with $(\bar{B}) \cong J\hat{D}(\bar{C})$. (Its quantum symbol $\dot{g}_2 = 0$.)

Above lemma proves that Bianchi identity does not reduce the property of integrability of the variational field equation. $\qquad\qquad\square$

QUANTIZED PDE's III:
QUANTIZATIONS OF PDE's

Abstract - Here we consider a process that allows us to associate to a (super) PDE, defined in the category of (super)commutative manifolds, a quantum (super) PDE. This process is the *covariant quantization*. We described it in some steps. In fact, we first define quantizations of PDE's in the framework of the mathematical logic, by means of evaluations of the *logic of a PDE E_k*, that is the Boolean algebra of subsets of the classic limit $\Omega(E_k)_c$ of the *quantum situs* $\Omega(E_k)$ of E_k, into *quantum logics* $A \subset \mathcal{L}(\mathcal{H})$, that are algebras of (self-adjoint) operators on a locally convex topological vector (Hilbert) space $\mathcal{H}$, in such a way to define (pre-)spectral measures on $\Omega(E_k)_c$: $\Omega(E_k)_c \circ\!\!\!\to \mathcal{L}(\mathcal{H})$. We shaw that these quantizations can be obtained by means of a geometric process called *covariant quantization* of PDE's, (or *canonical quantization* of PDE's, that is, roughly speaking, the covariant quantization observed by a physical frame). In fact, in a purely geometric context, we prove that any physical observable deforms the classical PDE, $E_k \subset J\mathcal{D}^k(W)$, around its solutions. In this way we can associate to the Lie filtered (super)algebra of the (super)classical observables, $\mathcal{B}$, of E_k, a filtered quantum (super)algebra, $\hat{\mathcal{B}}$, defined by means of distributive kernels, $\widetilde{G}_q$, *propagators*, canonically associated to E_k. We characterize also the propagators of PDE's by means of their integral bordism groups. The final step is the relation between the formal properties $E_\infty \to \cdots \to E_{k+i} \to \cdots E_k \to \cdots$, of the classical equations E_k, with quantum ones $\hat{E}_\infty \to \cdots \to \hat{E}_{k+i} \to \cdots \hat{E}_k \to \cdots$. These are obtained in the category of QPDE's, where the quantum (super)algebra $\hat{\mathcal{B}}$, obtained as covariant quantization of E_k, identifies a quantum (super) PDE.

3.1 - INTEGRAL (CO)BORDISM GROUPS IN PDE's

In this section we resume some fundamental definitions and results on the geometry of PDE's in the category of commutative finite dimensional manifolds. (See also refs.[8,9,18–21,28–37,39,42,53,54,56–66,73,79,80,95].) In particular, we shall emphasize some our recents results on the algebraic topology of PDE's, that allowed us to characterize global solutions of PDE's. (See also refs.[62–66].)

Let W be a smooth manifold of dimension $m + n$. For any n-dimensional submanifold $N \subset W$ we denote by $[N]_a^k$ the *k-jet of N at the point $a \in N$*, i.e., the set of n-dimensional submanifolds of W that have in a a contact of order k. Set $J_n^k(W) \equiv \bigcup_{a \in W} J_n^k(W)_a$, $J_n^k(W)_a \equiv \{[N]_a^k | a \in W\}$. We call $J_n^k(W)$ the *space of all k-jets of submanifolds of dimension n* of W. $J_n^k(W)$ has the following natural structures of differential fiber bundles: $\pi_{k,s} : J_n^k(W) \to J_n^s(W)$, $s \leq k$, with affine fibers $J_n^k(W)_{\bar{q}}$, where $\bar{q} \equiv [N]_a^{k-1} \in J_n^{k-1}(W)$, $a \equiv \pi_{k,0}(\bar{q})$, with associated vector space $S^k(T_a^*N) \otimes \nu_a$, $\nu_a \equiv T_a W / T_a N$. For any n-dimensional submanifold $N \subset W$ one has the canonical embedding $j^k : N \to J_n^k(W)$, given by $j^k : a \mapsto j^k(a) \equiv [N]_a^k$. We call $j^k(N) \equiv N^{(k)}$ the *k-prolongation* of N. In the following we shall also assume that there is a fiber bundle structure on W, $\pi : W \to M$, where $\dim M = n$. Then there exists a canonical submanifold $J^k(W)$ of $J_n^k(W)$ that is called the *k-jet space for sections* of π. $J^k(W)$ is diffeomorphic to the k-jet-derivative space of sections of π, $JD^k(W)$ and has the same dimension of $J_n^k(W)$ [56]. Then, for any section $s : M \to W$ one has the following commutative diagram:

$$
\begin{array}{ccccccc}
JD^k(W) & \cong & J^k(W) & & \subset & J_n^k(W) & \\
D^k s \uparrow & & \uparrow & j^k(s) & & \downarrow & \pi_k \\
M & \cong & M & & \cong & M &
\end{array}
$$

where $D^k s$ is the k-derivative of s and $j^k(s)$ is the k-jet-derivative of s. If $s(M)^{(k)} \subset J_n^k(W)$ is the k-prolongation of $s(M) \subset W$, then one has $j^k(s)(M) \cong s(M)^{(k)} \cong s(M) \cong M$. Of course there are also n-dimensional submanifolds $N \subset W$ that are not representable as image of sections of π. As a consequence, in these cases, $N^{(k)} \cong N$ is not representable in the form $j^k(s)(M)$ for some section s of π. The condition that N is image of some (local) section s of π is equivalent to the following local condition: $s^*\eta \equiv s^* dx^1 \wedge \ldots \wedge dx^n \neq 0$, where $(x^\alpha, y^j)_{1 \leq \alpha \leq n, 1 \leq j \leq m}$, are fibered coordinates on W, with y^j vertical coordinates. In other words $N \subset W$ is locally representable by equations $y^j = y^j(x^1, \ldots, x^n)$. This is equivalent to saying that N is transversal to the fibers of π or that the tangent space TN identifies an horizontal distribution with respect to the vertical one $vTW|_N$ of the fiber bundle structure $\pi : W \to M$. Conversely, a completely integrable n-dimensional horizontal distribution on W determines a foliation of W by means of n-dimensional submanifolds that can be represented by images of sections of π.

Let $\Pi^0(W)$ denote the pseudogroup of local diffeomorphisms of the manifold W. Then, we have the following natural action $\Pi^0(W) \times J_n^k(W) \to J_n^k(W)$, $(\phi, [N]_n^k) \mapsto [\phi(N)]_{\phi(p)}^k$. We denote by $\phi^{(k)} : J_n^k(W) \to J_n^k(W)$ the action on $J_n^k(W)$ induced by $\phi \in \Pi^0(W)$, and we call $\phi^{(k)}$ the *k-th prolongation* of ϕ. One has $\pi_{k,s} \circ \phi^{(k)} = \phi^{(s)} \circ \pi_{k,s}$, $k > s$ and $(\phi \circ \psi)^{(k)} = \phi^{(k)} \circ \psi^{(k)}$, $(id)^{(k)} = 1$, $\forall \phi, \psi \in \Pi^0(W)$. Let $\Pi^k(W) \subset J_n^k(W,W)$ be the subbundle of $J_n^k(W,W)$ of k-jet derivative of local diffeomorphisms $W \to W$. We put $\Pi^k(W)_{(a,b)} \equiv \{[\phi]_a^k \in \Pi^k(W) | \phi(a) = b\}$. One has a natural pairing $\Pi^k(W)_{(b,c)} \times \Pi^k(W)_{(a,b)} \to \Pi^k(W)_{(a,c)}$, $([\psi]_b^k, [\phi]_a^k) \mapsto [\psi \circ \phi]_a^k$ that defines a structure of Lie groups on $\Pi^k(W)_a \equiv \Pi^k(W)_{(a,a)}$. One has the natural epimorphism of Lie groups

$$1 \to \Pi^k(W)_a \to \Pi^{k-1}(W)_a \to$$
$$\cdots \to \Pi^1(W)_a \equiv GL(T_a W) \to \Pi^0(W)_a \equiv 1, \ k \geq 1.$$

If $k \geq 2$ one has the following exact sequence

$$1 \to \Pi^{k+q,k}(W)_a \to \Pi^{k+q}(W)_a \to \Pi^k(W)_a \to 1.$$

This defines $\Pi^{k+q,k}(W)_a$ as the kernel of the bundle $\Pi^{k+q}(W)_a \to \Pi^k(W)_a$. One has the following diffeomorphism:

$$\Pi^{k+q,k}(W)_a \tilde{=} \bigoplus_{1 \leq i \leq q} S^{k+i}(T_a^* W) \bigotimes T_a W.$$

The group $\Pi^k(W)_a$ acts on the space $J_n^k(W)_a$ in a natural way : $[\phi]_a^k([N]_a^k) = [\phi(N)]_a^k$. Under this action the kernel $\Pi^{k,k-1}(W)_a$ acts transitively on the fibers: $\pi_{k,k-1}^{-1}(q) \subset J_n^k(W)_a$, $\forall q \equiv [N]_a^{k-1} \in J_n^k(W)_a$, $\pi_{k-1,0}(q) = a$. A stationary subgroup in $\Pi^{k,k-1}(W)_a$ of the element $u \in \pi_{k,k-1}^{-1}(q)$ under such action, $k \geq 2$, is

$$(Ann(T_a N)) \odot S^{k-1}(T_a^* W) \otimes T_a W \bigoplus S^k(T_a^* W) \otimes T_a N$$

where $(Ann(T_a N)) \subset T_a^* W$ is the annihilator of $T_a N$. Therefore, the fibers $\pi_{k,k-1}^{-1}(q)$, $k \geq 2$ are homogeneous spaces with respect to the action of the connected abelian group $\Pi^{k,k-1}(W)_a$ and thus carry an affine structure with associated vector sapces

$$S^k(T_a^* N) \bigotimes \nu_a, \ \nu_a \equiv T_a W / T_a N,$$

with $a = \pi_{k-1,0}(q)$. If $W \equiv \mathbf{R}^{n+m} \equiv \mathbf{R}^s$, we use the following notations [58,60]: $L_s^k \equiv \Pi^k(\mathbf{R}^s)_{(0,0)}$, $L_s^{k+q,k} \equiv \Pi^{k+q,k}(\mathbf{R}^s)_0$. The prolongations $\phi^{(k)}$ of any local diffeomorphism $\phi : W \to W$, for $k \geq 2$, preserves the affine structure described above. Jet bundles $\pi_{k,k-1} : J_n^k(W) \to J_n^{k-1}(W)$ are affine for $k \geq 2$. The prolongations $\phi^{(k)} : J_n^k(W) \to J_n^k(W)$ of the local diffeomorphism $\phi \in \Pi^0(W)$ are affine authomorphisms for $k \geq 2$, and are linear collineations of Grassmannians $G_{n,m+n}(T_aW)$, for $k = 1$.

In order to describe the contact structure of $J_n^k(W)$, let us, now, sketch some definitions, fundamental results and notations on ideals of differential forms and their associated distributions. Set $\Omega^k(M) \equiv$ module of exterior smooth differential forms of degree k on M; $\Omega^\bullet(M) \equiv \oplus_{k \geq 0}\Omega^k(M)$ is the graded algebra of exterior smooth differentiable forms over M; $\mathcal{I} \equiv <\alpha, \beta, \dots> \equiv$ ideal of $\Omega^\bullet(M)$ algebraically generated by the exterior differential forms $\alpha, \beta, \dots$. An ideal $\mathcal{I} \subset \Omega^\bullet(M)$ is said to be *closed* iff $d\mathcal{I} \subset \mathcal{I}$. If $\mathcal{I} \equiv <\alpha, \beta, \dots>$ we call *closure* $\overline{\mathcal{I}}$ of $\mathcal{I}$ the closed ideal $\overline{\mathcal{I}} \equiv <\alpha, d\alpha, \beta, d\beta, \dots>$. An ideal $\mathcal{I} \subset \Omega^\bullet(M)$ is said to be *stable* under transport by a vector field $\zeta : M \to TM$ iff $\mathcal{L}_\zeta\mathcal{I} \subset \mathcal{I}$, then ζ is called a (*infinitesimal*) *symmetry* of $\mathcal{I}$. We denote such a set by $\mathfrak{s}(\mathcal{I})$. ($\mathcal{L}_\zeta$ denotes Lie derivative with respect to ζ.) A vector field $\zeta : M \to TM$ is said to be a *Cauchy characteristic* of an ideal $\mathcal{I}$ iff $\zeta\rfloor\mathcal{I} \subset \mathcal{I}$. The set of Cauchy characteristics of $\mathcal{I}$ generates at each point $p \in M$ a vector space $\mathbb{D}_{\mathcal{I},p}$ that, if $\dim \mathbb{D}_{\mathcal{I},p}$ is locally constant, generates a distribution $\mathbb{D}_\mathcal{I}$ (*Cauchy characteristic distribution* of $\mathcal{I}$). $\mathbb{D}_\mathcal{I}$ is involutive iff $\mathcal{I}$ is a closed ideal. If $\mathcal{I} = <\omega^j>$, $1 \leq j \leq p$, where ω^j are differential 1-forms, (in this case $\mathcal{I}$ is called a *Pfaffian ideal*), then the Cauchy characteristic distribution $\mathbb{D}_\mathcal{I}$ is characterized by means of vector fields $\zeta : M \to TM$ such that $\zeta\rfloor\omega^j = 0$, $1 \leq j \leq p$. For any distribution $\mathbb{D} \subset TM$ on M we denote by $\mathfrak{d}(\mathbb{D})$ the $\Omega^0(M)$-module of all vector fields belonging to this distribution, and by $\mathcal{I}(\mathbb{D})$ the ideal generated by the following $\Omega^0(M)$-module

$$\Omega^1(\mathbb{D}) \equiv \{\omega \in \Omega^1(M) | \zeta\rfloor\omega = 0, \forall \zeta \in \mathfrak{d}(\mathbb{D})\}.$$

Then, one has: (a) $\mathcal{I}(\mathbb{D}_\mathcal{I}) \supset \mathcal{I}$, i.e., the ideal associated to the Cauchy characteristic distribution of $\mathcal{I}$ is larger than $\mathcal{I}$ itself; (b) $\mathbb{D}_{\mathcal{I}(\mathbb{D})} = \mathbb{D}$, i.e., the Cauchy characteristic distribution of the ideal associated to a distribution $\mathbb{D}$ coincides with $\mathbb{D}$. We define *characteristics* of $\mathcal{I}$ the vector fields $\zeta : M \to$

TM belonging to the following $\Omega^0(M)$-module:

$$\mathfrak{char}(\mathcal{I}) \equiv \{\zeta \in \mathfrak{d}(TM) | \zeta \in \mathfrak{d}(\mathbb{D}_{\mathcal{I}}), \zeta \rfloor d\mathcal{I} \subset \mathcal{I}\}.$$

If $\mathcal{I}$ is a Pfaffian ideal, then

$$\mathfrak{char}(\mathcal{I}) = \mathfrak{s}(\mathcal{I}) \cap \mathfrak{d}(\mathbb{D}_{\mathcal{I}}) \cong \mathfrak{d}(\mathbb{D}_{\overline{\mathcal{I}}}).$$

If $\dim \mathfrak{char}(\mathcal{I})_p = \text{constant}$, $\forall p \in M$, then $\mathfrak{char}(\mathcal{I})$ defines a distribution $\mathbb{C}_{har}(\mathcal{I})$ on M called *characteristic distribution* of $\mathcal{I}$. $\mathbb{C}_{har}(\mathcal{I})$ is a completely integrable distribution. If $\mathcal{I}$ is closed $\mathbb{C}_{har}(\mathcal{I}) = \mathbb{D}_{\mathcal{I}}$. If $\Phi : N \to M$ is a differentiable mapping one has the induced mapping $\Phi^* : \Omega^\bullet(M) \to \Omega^\bullet(N)$, such that: (a) $(\Phi \circ \Psi)^* \alpha = \Psi^*(\Phi^* \alpha)$, where $\Psi : N \to M$ is another differentiable mapping; (b) $\Phi^*(d\alpha) = d(\Phi^* \alpha)$; (c) $\Phi^*((\Phi_* \zeta) \rfloor \alpha) = \zeta \rfloor \Phi^* \alpha$ for any vector field $\zeta : N \to TN$. We say that Φ *solves* an ideal $\mathcal{I}$ iff $\Phi^* \mathcal{I} = 0$. We say also that $\Phi : N \to M$ is an *integral manifold* of $\mathcal{I}$. If Φ is a solving map of an ideal $\mathcal{I}$ and $\zeta \in \mathfrak{s}(\mathcal{I})$, then $\Phi_s \equiv \phi_s \circ \Phi$, $\partial\phi = \zeta$, is a solving map of $\mathcal{I}$ for all s in a neighbourhood of $s = 0$.

Theorem 3.1. *Let* $\mathbb{D} \subset TM$ *be a distribution on a finite dimensional manifold* M. *Set* $\Omega^1(\mathbb{D}) \equiv C^\infty(\mathbb{D}^*)$, $\Omega^1(M) \equiv C^\infty(T^*M)$, $C_\mathbb{D}\Omega^1(M) \equiv \{\omega \in \Omega^1(M) | \zeta \rfloor \omega = 0, \forall \zeta \in C^\infty(\mathbb{D})\}$. *Then, one has a canonical section* $\Omega_\mathbb{D} : M \to (TM/\mathbb{D} \bigotimes \Lambda^2(\mathbb{D}^*))\}$ *called the metasymplectic structure of* $\mathbb{D}$. *The characteristic distribution* $\mathbb{C}_{har}(\mathcal{I})$ *is a distribution of degeneration subspaces of the metasymplectic structure, i.e.,* $\mathbb{C}_{har}(\mathcal{I})_a = \ker(\Omega_\mathbb{D})_a$, $\forall a \in M$, $\mathcal{I} = \mathcal{I}_\mathbb{D}$. *Here, we denote by* $\ker(\Omega_\mathbb{D})_a$ *the kernel of the vector-valued 2-form:* $\ker(\Omega_\mathbb{D})_a = \bigcap_{\omega \in \Omega^1(\mathbb{D})} \ker(\Omega_a(\omega))$. *Note that* $\ker(\Omega_a(\omega))$ *is the kernel of the following homomorphism* $\mathbb{D}_a \to \mathbb{D}_a^*$ *canonically associated to* $\Omega_a(\omega)$, *i.e., the following short sequence*

$$0 \to \ker(\Omega_a(\omega)) \to \mathbb{D}_a \to \mathbb{D}_a^*$$

is exact.

Proof. See ref.[65]. □

The *k-order contact ideal* of W is the following Pfaffian ideal of differential forms $C_k(W) \equiv < C\Omega^1(J_n^k(W)) >$, where

$$C\Omega^1(J_n^k(W)) \equiv \left\{ \omega \in \Omega^1(J_n^k(W)) \left| \begin{array}{l} \omega|_{N^{(k)}} = 0, \\ \text{for any } n\text{-dimensional submanifold } N \subset W \end{array} \right. \right\}.$$

One has the following local representation: $C_k(W) = < \omega^j, \ldots, \omega^j_{\alpha_1,\ldots,\alpha_{k-1}} >$, where $\omega^j, \ldots, \omega^j_{\alpha_1\ldots\alpha_{k-1}}$ are differential 1-forms (*contact 1-forms*) given by the following expressions:

$$\left\{ \begin{array}{l} \omega^j \equiv dy^j - y^j_\alpha dx^\alpha \\ \omega^j_\alpha \equiv dy^j_\alpha - y^j_{\alpha\beta} dx^\beta \\ \ldots \\ \omega^j_{\alpha_1\ldots\alpha_{k-1}} \equiv dy^j_{\alpha_1\ldots\alpha_{k-1}} - y^j_{\alpha_1\ldots\alpha_{k-1}\beta} dx^\beta \end{array} \right\}.$$

Note that here we have used coordinates $(x^\alpha, y^j, y^j_\alpha, \ldots, y^j_{\alpha_1\ldots\alpha_k})$ in a neighborhood of a point $q \equiv [N]^k_a \in J^k_n(W)$, where $N \subset W$ is locally representable by equations $y^j = y^j(x^1, \ldots, x^n)$. $((x^\alpha, y^j)$ is not, in general, a fiber coordinate system of $\pi : W \to M)$. The *Cartan distribution* $\mathbb{E}^k_n(W)$ of $J^k_n(W)$ is defined by the following $\Omega^0(M)$-module

$$\mathfrak{d}(\mathbb{E}^k_n(W)) \equiv \{\zeta \in \mathfrak{d}(TJ^k_n(W))|\zeta\rfloor\omega^j = 0\}.$$

So

$$\mathbb{E}^k_n(W) = \mathbb{D}_{C_k(W)} \equiv \{\zeta \in \mathfrak{d}(TJ^k_n(W))|\zeta\rfloor C_k(W) \subset C_k(W)\}.$$

Therefore, the Cartan vector fields are the Cauchy characteristics of the k-order contact ideal of W. Of course $\mathcal{I}(\mathbb{E}^k_n(W)) \supset C_k(W)$. A Cartan vector field $\zeta \in \mathfrak{d}(\mathbb{E}^k_n(W))$ admits the following local representation

$$\zeta = X^\alpha(\partial x_\alpha + y^j_\alpha \partial y_j + \ldots + y^j_{\alpha_1\ldots\alpha_{k-1}\alpha}\partial y^{\alpha_1\ldots\alpha_{k-1}}_j) + Z^j_{\alpha_1\ldots\alpha_k}\partial y^{\alpha_1\ldots\alpha_k}_j,$$

where X^α, $Z^j_{\alpha_1\ldots\alpha_k} \in \Omega^0(J^k_n(W))$, $1 \le \alpha_1, \ldots, \alpha_k, \alpha \le n$, $1 \le j \le m$. The *closed k-order contact ideal* $\overline{C}_k(W)$ of W is the following closed ideal of differential forms on $J^k_n(W)$:

$$\overline{C}_k(W) \equiv < \omega^j, d\omega^j, \ldots, \omega^j_{\alpha_1\ldots\alpha_{k-1}}, d\omega^j_{\alpha_1\ldots\alpha_{k-1}} > .$$

It admits the reduced form

$$\overline{C}_k(W) \equiv < \omega^j, \ldots, \omega^j_{\alpha_1\ldots\alpha_{k-1}}, d\omega^j_{\alpha_1\ldots\alpha_{k-1}} > .$$

Of course $\overline{C}_k(W) \supset C_k(W)$ as a subideal. One can prove that the ideal generated by such q-forms, $q \ge 0$, coincides with the closure $\overline{C}_k(W)$ of

$\mathcal{C}_k(W)$. For $k = \infty$ one has $\overline{\mathcal{C}}_\infty(W) = \mathcal{C}_\infty(W)$. The characteristic distribution $\mathbb{C}_{har}(\mathcal{C}_k(W))$ of $\mathcal{C}_k(W)$, being $\mathcal{C}_k(W)$ a Pfaffian ideal, is related to the symmetry of $\mathfrak{s}(\mathcal{C}_k(W))$ by the following relation:

$$\mathfrak{char}(\mathcal{C}_k(W)) \cong \mathfrak{s}(\mathcal{C}_k(W)) \cap \mathfrak{d}(\mathbb{E}_n^k(W)) \cong \mathfrak{d}(\mathbb{D}_{\overline{\mathcal{C}}_k(W)}).$$

A vector field $\zeta \in \mathfrak{d}(TJ_n^k(W))$ has the following local representation

$$(3.1) \qquad \zeta = X^\alpha \partial x_\alpha + Y^j \partial y_j + Z_\alpha^j \partial y_j^\alpha + \ldots + Z_{\alpha_1 \ldots \alpha_k}^j \partial y_j^{\alpha_1 \ldots \alpha_k}.$$

If $\zeta \in \mathfrak{s}(\overline{\mathcal{C}}_k(W))$ it has the following local representation

$$\left\{ \begin{array}{l} Z_\alpha^j = (\zeta_\alpha^{(0)}.Y^j) - y_\beta^j(\zeta_\alpha^{(0)}.X^\beta) \\ Z_{\alpha_1 \ldots \alpha_k \beta}^j = (\zeta_\beta^{(k)}.Z_{\alpha_1 \ldots \alpha_k}^j) - y_{\alpha_1 \ldots \alpha_k \gamma}^j(\zeta_\beta^{(k)}.X^\gamma) \end{array} \right\},$$

where

$$\left\{ \zeta_\alpha^{(0)} \equiv \partial x_\alpha + y_\alpha^j \partial y_j, \ \zeta_\alpha^{(k)} \equiv \zeta_\alpha^{(k-1)} + y_{\alpha_1 \ldots \alpha_{k-1} \alpha}^j \partial y_j^{\alpha_1 \ldots \alpha_{k-1}} \right\}$$

and for $m > 1$: $X^\alpha = X^\alpha(x^\alpha, y^j)$, $Y^j = Y^j(x^\alpha, y^j)$ for any smooth choice of the $n + m$ functions $\{X^\alpha, Y^j\}$ of their indicated arguments; while for $m = 1$: $X^\alpha \equiv (\partial y^\alpha.\xi)$, $Y \equiv y_\alpha(\partial y^\alpha.\xi) - \xi$ for any smooth function $\xi = \xi(x^\alpha, y, y_\alpha)$. Therefore, for $m > 1$ any symmetry field of $\overline{\mathcal{C}}_k(W)$ is the kth order prolongation of a vector field on the manifold W, that is $\mathfrak{s}(\overline{\mathcal{C}}_k(W)) = \mathfrak{d}^{(k)}(TW)$; for $m = 1$ instead one has $\mathfrak{s}(\overline{\mathcal{C}}_k(W)) = \mathfrak{d}^{(k-1)}(TJ_n^1(W))$.

An *integral manifold* of $J_n^k(W)$ is an integral manifold $\Phi : P \to J_n^k(W)$ of the ideal $\mathcal{C}_k(W)$. Set $\Phi(P) \equiv Z \subset J_n^k(W)$. Then we can write $\mathcal{C}_k(W)|_Z = 0$. A *maximal integral manifold* of $J_n^k(W)$ is an integral manifold that is not properly contained into an integral manifold of higher dimension. Maximal integral manifolds are divided into types so that manifolds of type $n - p$ are of dimemsion $m\binom{p+k-1}{k} + n - p$. We shall consider the following cases. (A)(*Non-exceptional cases*). A maximal integral manifold $Z \subset J_n^k(W)$ of type $n-p$ can be represented as follows: $Z \setminus \Sigma(Z) = \bigcup_i V_i$, where $\Sigma(Z) \subset Z$ is a closed submanifold (with singularities) with $\dim \Sigma(Z) < \dim Z$, and V_i is a domain in

$$W(L_i) \equiv \{q \in J_n^k(W) | \pi_{k,k-1}(q) \in L_i; T_{\pi_{k,k-1}(q)} L_i \subset L_q\},$$

where L_i is an integal manifold of the Cartan distribution on $J_n^{k-1}(W)$, of dimension $n - p$, L_q is a n-dimensional subspace of $T_{\bar{q}} J_n^{k-1}(W)$ that is canonically identified by q, i.e., $L_q \equiv T_{\bar{q}} N^{(k-1)}$, where $[N]_a^k = q$ and $\bar{q} \equiv \pi_{k,k-1}(q)$. We call V_i the *regular components* of Z and $\Sigma(Z)$ the *set of singular points* of Z. In particular: (i) Maximal integral manifolds of $J_n^\infty(W)$ are of dimension n and are regular ones (without singular points). (ii) If $\dim V > \dim V' \Rightarrow \mathrm{type}V < \mathrm{type}V'$. (iii) The zero type integral manifolds coincide with the fibers of projections $\pi_{k,k-1}$. (iv) Maximal integral manifolds of type n are of dimension n. They have the representation $Z \setminus \Sigma(Z) = \bigcup_i V_i$, as said before, where the regular components V_i are the kth prolongations of n-dimensional submanifolds of W. We call such integral manifolds *solutions* of $J_n^k(W)$ and the corresponding set is denoted by $\mathcal{S}_{ol}(J_n^k(W))$. If the set of singular points is empty, we call such solutions *regular solutions* and the corresponding set is denoted by $\underline{\mathcal{S}}_{ol}(J_n^k(W))$. A regular solution can be characterized by means of an embedding $\Phi : P \to J_n^k(W)$, where P is a n-dimensional manifold, such that $\Phi^* \mathcal{C}_k(W) = 0$, $\Phi^* \eta|_U \neq 0$, where $\eta|_U \equiv dx^1 \wedge \ldots \wedge dx^n$ is the differential n-form on $J_n^k(W)$ defined on an open coordinate set $U \subset J_n^k(W)$, where the coordinates are $(x^\alpha, y^j, \ldots, y^j_{\alpha_1 \ldots \alpha_k})$ such that $Y \equiv \pi_{k,0}(\Phi(P)) \subset W$ is locally represented by the equations $y^j = y^j(x^\alpha)$. Then, a regular solution of $J_n^k(W)$ is locally represented by the equations:

$$(3.2) \qquad \left\{ \begin{array}{l} x^\alpha \circ \Phi = u^\alpha \\[2mm] y^j \circ \Phi = \psi^j(u^\alpha) \\[2mm] \ldots \\[2mm] y^j_{\alpha_1 \ldots \alpha_k} \circ \Phi = (\partial u_{\alpha_1} \ldots \partial u_{\alpha_k} . \psi^j) \end{array} \right\}.$$

Of course the same representation (3.2) holds for the regular components of any solution. Furthermore, forgetting condition $\Phi^* \eta|_U \neq 0$ (and also the one for Y) one has the following local representation of any solution:

$$(3.3) \qquad \left\{ \begin{array}{l} x^\alpha \circ \Phi = \phi^\alpha(u^\alpha) \\[2mm] y^j \circ \Phi = \psi^j(u^\alpha) \\[2mm] \ldots \\[2mm] y^j_{\alpha_1 \ldots \alpha_k} \circ \Phi = (\partial u_{\alpha_1} \ldots \partial u_{\alpha_k} . \psi^j) \end{array} \right\}.$$

Finally, if the coordinates (x^α) in the local differential n-form $\eta|_U \equiv dx^1 \wedge \ldots \wedge dx^n$ are that relative to the fibration $\pi : W \to M(x^\alpha)$, then the requirement $\Phi^*\eta|_U \neq 0$ almost every where, (resp. every where) implies that the n-dimensional integral manifold $Z \equiv \Phi(P) \subset J_n^k(W)$ is a solution (resp. regular solution) of $J^k(W) \subset J_n^k(W)$.

(B)(*Exceptional cases*). $(\alpha)(n = m = 1)$. Integral manifolds are one-dimensional and glued from pieces of type zero or type 1. A piece of type 0 is an open subset of the fibre $\pi_{k,k-1}^{-1}(\bar{q})$, $\bar{q} \in J_n^{k-1}(W)$; A piece of type 1 is an open subset of regular integral manifold (of dimension 1). $(\beta)(k = m = 1)$. In this case $J_n^1(W)$ becomes a classic contact manifold [63].

Theorem 3.2. *If* $\Phi \in \underline{S}_{ol}(J_n^k(W))$ *and* $\zeta \in \mathfrak{s}(\overline{C}_k(W))$ *then* $\phi_s \circ \Phi \in \underline{S}_{ol}(J_n^k(W))$ *for all s in some neighborhood* $]-\epsilon,\epsilon[$ *of $s = 0$, with $\partial\phi = \zeta$. Then, one has the following commutative diagram:*

$$
\begin{array}{ccc}
0 & \to & \underline{S}_{ol}(J_n^k(W)) & \to & S_{ol}(J_n^k(W)) \\
& & (\phi_s)_* \uparrow & & \uparrow (\phi_s)_* \\
0 & \to & \underline{S}_{ol}(J_n^k(W)) & \to & S_{ol}(J_n^k(W))
\end{array}
$$

for $s \in]-\epsilon,\epsilon[\subset \mathbb{R}$, induced by any infinitesimal symmetry $\zeta \in \mathfrak{s}(\overline{C}_k(W))$.

Proof. See ref.[65]. $\qquad\qquad\qquad\qquad\qquad\qquad\qquad\qquad\qquad\qquad\square$

The concept of solution of PDE's can be related to that of higher order connection.

Definition 3.1. 1) *A $(k+1)$-connection on W is a section* $] : J_n^k(W) \to J_n^{k+1}(W)$ *of* $\pi_{k+1,k}$. *(The restriction of* $]$ *to* $J^k(W) \subset J_n^k(W)$ *is also called $(k+1)$-connection).*

2) *Given a $(k+1)$-connection* $]$ *on W we define horizontal k-order contact ideal* $\mathcal{H}_k(])$ *of* $\Omega^\bullet(J_n^k(W))$ *the following one:* $\mathcal{H}_k(]) \equiv]^*\mathcal{C}_{k+1}(W)$.

Locally one can write $\mathcal{H}_k(]) = < \omega^j,\dots,\omega_{\alpha_1\dots\alpha_{k-1}}^j, H_{\alpha_1\dots\alpha_k}^j >$, where

$$
\begin{aligned}
H_{\alpha_1\dots\alpha_k}^j &\equiv]^*\omega_{\alpha_1\dots\alpha_k}^j =]^*(dy_{\alpha_1\dots\alpha_k}^j - y_{\alpha_1\dots\alpha_k\beta}^j dx^\beta) \\
&= dy_{\alpha_1\dots\alpha_k}^j -]_{\alpha_1\dots\alpha_k\beta}^j dx^\beta \in \Omega^1(J_n^k(W)),
\end{aligned}
$$

with $]_{\alpha_1\dots\alpha_k\beta}^j \equiv y_{\alpha_1\dots\alpha_k\beta}^j \circ] \in \Omega^0(J_n^k(W))$. Of course $\mathcal{C}_k(W)$ is a subideal of $\mathcal{H}_k(])$.

Definition 3.2. We call *horizontal k-order Cartan distribution* $\mathbb{H}_k(]) \subset TJ_n^k(W)$ (identified by a $(k+1)$-connection $]$) the Cauchy characteristic distribution associated to $\mathcal{H}_k(])$: $\mathbb{H}_k(]) = \mathbb{D}_{\mathcal{H}_k(])}$.

Remark 3.1. $\mathfrak{d}(\mathbb{H}_k(]))$ admits the following local basis (*canonical basis*)

$$
\begin{aligned}
\zeta_\alpha =&\, \partial x_\alpha + y_\alpha^j \partial y_j \\
&+ \cdots + y_{\alpha_1\dots\alpha_{k-1}\alpha}^j \partial y_j^{\alpha_1\dots\alpha_{k-1}} +]_{\alpha_1\dots\alpha_k\alpha}^j \partial y_j^{\alpha_1\dots\alpha_k}, \quad 1 \leq \alpha \leq n.
\end{aligned}
$$

Thus $\dim \mathbb{H}_k(\daleth) = n$. For any $(k+1)$-connection $\daleth$ on W, one has the following direct sum decompositions: $\mathbb{E}_n^k(W)_q \cong \mathbb{H}_k(\daleth)_q \bigoplus S^k(T_a^*N) \otimes \nu_a$, $\Omega^1(J_n^k(W)) \cong \Omega^1(J_n^k(W))_v \bigoplus \mathcal{H}_k(\daleth)^1$, with $a \equiv \pi_{k,0(q)} \in W$, $\daleth(q) = [N]_a^{k+1}$, and $\mathbb{H}_k(\daleth)_q \equiv T_q N^{(k)}$, $\mathcal{H}_k(\daleth)^1 \equiv \mathcal{H}_k(\daleth) \cap \Omega^1(J_n^k(W))$. So if $f : J_n^k(W) \to \mathbf{R}$ is any smooth function, one has the following local representation of its differential:

$$df = (\zeta_\alpha.f)dx^\alpha + (\partial y_j.f)\omega^j + (\partial y_j^\alpha.f)\omega_\alpha^j$$
$$+ \cdots + (\partial y_j^{\alpha_1\ldots\alpha_{k-1}}.f)\omega_{\alpha_1\ldots\alpha_{k-1}}^j + (\partial y_j^{\alpha_1\ldots\alpha_k}.f)H_{\alpha_1\ldots\alpha_k}^j$$
$$\equiv (\zeta_\alpha.f)dx^\alpha \quad \mathrm{mod}\,\mathcal{H}_k(\daleth).$$

Furthermore, for any horizontal ideal $\mathcal{H}_k(\daleth)$ we have:

$$d\omega^j \equiv 0 \, \mathrm{mod}\, \mathcal{H}_k(\daleth)$$

$$\cdots$$

$$d\omega_{\alpha_1\ldots\alpha_{k-1}}^j \equiv 0 \, \mathrm{mod}\, \mathcal{H}_k(\daleth)$$
$$\Omega_{\alpha_1\ldots\alpha_k}^j \equiv dH_{\alpha_1\ldots\alpha_k}^j$$
$$= \frac{1}{2}[(\zeta_\alpha.\daleth_{\alpha_1\ldots\alpha_k\beta}^j) - (\zeta_\beta.\daleth_{\alpha_1\ldots\alpha_k\alpha}^j)]dx^\alpha \wedge dx^\beta \, \mathrm{mod}\, \mathcal{H}_k(\daleth),$$

where $\{\zeta_\alpha\}_{1\leq\alpha\leq n}$ is the canonical basis of $\mathbb{H}_k(\daleth)$. $\blacksquare$

Definition 3.3. We call $\Omega_{\alpha_1\ldots\alpha_k\beta}^j$ the *curvature* of the $(k+1)$-connection $\daleth$.

Remark 3.2. In the next section we shall relate the curvature of a $(k+1)$-connection to the concept of complete integrability of PDEs. Note that all these relations are soldered and generalized by the concept of metasymplectic structure. (See below.)

Theorem 3.3. *The connection $\daleth$ is flat, i.e., with zero curvature, iff the differential ideal $\mathcal{H}_k(\daleth)$ is closed, or equivalently, iff $\mathbb{H}_k(\daleth)$ is involutive. Therefore flat $(k+1)$-connections on W induce involutive horizontal subdistributions of the Cartan distributions $\mathbb{E}_n^k(W)$ of $J_n^k(W)$.*

Proof. In fact $\mathcal{H}_k(\daleth)$ is closed iff

$$(\zeta_\alpha.\daleth_{\alpha_1\ldots\alpha_k\beta}^j) - (\zeta_\beta.\daleth_{\alpha_1\ldots\alpha_k\alpha}^j) = 0 \Leftrightarrow [\zeta_\alpha,\zeta_\beta] = 0.$$

This is equivalent to say that $\mathbb{H}_k(\daleth)$ is involutive and that $\daleth$ has zero curvature. $\square$

Corollary 3.1. *If $\daleth$ is a flat $(k+1)$-connection on W, then one has the following:*

1) $\overline{C}_k(W) \subset \mathcal{H}_k(\rceil)$ *as a closed subideal.*

2) $\mathbb{H}_k(\rceil) \cong \mathbb{C}_{har}(\mathcal{H}_k(\rceil))$; $\mathfrak{char}(\mathcal{H}_k(\rceil)) \subset \mathfrak{s}(\mathcal{H}_k(\rceil))$.

3) $J_n^k(W)$ *is foliated by regular solutions* Z *such that* $\mathcal{H}_k(\rceil)|_Z = 0$. *The leaves of the foliation are given in implicit form by the following equations:*

$$f^I(x^\alpha, y^j, \ldots, y^j_{\alpha_1 \ldots \alpha_k}) = \kappa^I, \quad 1 \le I \le r, \quad \dim J_n^k(W) - r = n,$$

where f^I *represent a complete independent system of primitive integrals of the linear system of PDEs* $(\zeta_\alpha . f) = 0$, $1 \le \alpha \le n$, *where* ζ_α *is a basis (e.g., the canonical basis) of the horizontal distribution* $\mathbb{H}_k(\rceil)$. *Furthermore, we can interpret each leaf that passes for a point* $q \in J_n^k(W)$ *as the orbit of* q *under* n-*parameter abelian Lie group* $\mathcal{G}_n$ *of point transformations of* $J_n^k(W)$ *that is generated by the abelian Lie algebra* $\mathfrak{d}(\mathbb{H}_k(\rceil)) = < \{\zeta_\alpha\}_{1 \le \alpha \le n} >$.

Example 3.1. Let us denote by $h(J_n^k(W))$ the set of all completely integrable horizontal k-order contact ideals of $\Omega^\bullet(J_n^k(W))$. This set is never locally empty. In fact, the local $(k+1)$-connection

$$\rceil^j_{\alpha_1 \ldots \alpha_{k+1}} \equiv (\partial x_{\alpha_1} \ldots \partial x_{\alpha_{k+1}} . s^j)$$

represents a local flat $(k+1)$-connection for every smooth choice of the functions $\{s^j(x^\alpha)\}_{1 \le \alpha \le n}$, (e.g., s^j could be a local representation of a section s of $\pi : W \to M$). ∎

Theorem 3.4. (Symmetries of horizontal k-order contact ideals). 1) *Any* $\zeta \in \mathfrak{s}(\mathcal{H}_k(\rceil))$ *has the following local representation:*

$$\zeta = X^\alpha \zeta_\alpha + Y^j \partial y_j + (\zeta_\alpha . Y^j) \partial y_j^\alpha + (\zeta_\alpha \zeta_\beta . Y^j) \partial y_j^{\alpha\beta}$$
$$+ \cdots + (\zeta_{\alpha_1} \ldots \zeta_{\alpha_k} . Y^j) \partial y_j^{\alpha_1 \ldots \alpha_k},$$

for any choice of $Y^j \in \Omega^0(J_n^k(W))$, $1 \le j \le m$, *such that*

$$(3.4) \qquad \begin{aligned} (\zeta_{\alpha_1} \ldots \zeta_{\alpha_k} . Y^j) &= (\partial y_i . \rceil^j_{\alpha_1 \ldots \alpha_k \beta}) Y^i + (\partial y_i^\gamma . \rceil^j_{\alpha_1 \ldots \alpha_k \beta})(\zeta_\gamma . Y^i) \\ &\quad + \cdots + (\partial y_i^{\gamma_1 \ldots \gamma_k} . \rceil^j_{\alpha_1 \ldots \alpha_k \beta})(\zeta_{\gamma_k} \ldots \zeta_{\gamma_1} . Y^i). \end{aligned}$$

2) *The space* $\mathfrak{s}(\mathcal{H}_k(\rceil))$ *admits the following direct sum decomposition:*

$$\mathfrak{s}(\mathcal{H}_k(\rceil)) \cong \mathfrak{d}(\mathbb{H}_k(\rceil)) \bigoplus \mathfrak{v}_k(\rceil),$$

where $\mathfrak{v}_k(\rceil)$ *is the collection of all vectors of the form*

$$\xi = \zeta - X^\alpha \zeta_\alpha = Y^j \partial y_j + (\zeta_\alpha . Y^j) \partial y_j^\alpha + (\zeta_\alpha \zeta_\beta . Y^j) \partial y_j^{\alpha\beta}$$
$$+ \cdots + (\zeta_{\alpha_1} \ldots \zeta_{\alpha_k} . Y^j) \partial y_j^{\alpha_1 \ldots \alpha_k},$$

for any choice of $Y^j \in \Omega^0(J_n^k(W))$, $1 \le j \le m$, such that conditions (3.4) are satisfied.

3) $\mathfrak{s}(\mathcal{H}_k(\daleth))$ *is a Lie algebra that admits the subalgebra $\mathfrak{o}(\mathbb{H}_k(\daleth))$ as an ideal; that is*

$$[\mathfrak{o}(\mathbb{H}_k(\daleth)), \mathfrak{o}(\mathbb{H}_k(\daleth))] \subset \mathfrak{o}(\mathbb{H}_k(\daleth)),$$
$$[\mathfrak{o}(\mathbb{H}_k(\daleth)), \mathfrak{v}_k(\daleth)] \subset \mathfrak{o}(\mathbb{H}_k(\daleth)),$$
$$[\mathfrak{v}_k(\daleth), \mathfrak{v}_k(\daleth)] \subset \mathfrak{v}_k(\daleth).$$

The subalgebra $\mathfrak{o}(\mathbb{H}_k(\daleth))$ is a module over $\Omega^0(J_n^k(W))$, and $\mathfrak{v}_k(\daleth)$ is a module over the associative algebra $\mathcal{P}_k(\daleth) \equiv \{f \in \Omega^0(J_n^k(W)) | (\zeta_\alpha . f) = 0, 1 \le \alpha \le n\}$. On $(\mathcal{P}_k(\daleth))^n$ we can recognize an algebraic structure ($\daleth$-Poisson algebra) with respect to the following bracket: $\{f_1, f_2\}^\alpha = f_1^\beta(\zeta_\beta . f_2^\alpha) - f_2^\beta(\zeta_\beta . f_1^\alpha)$. If $\mathcal{H}_k(\daleth) \in h(J_n^k(W))$ and $f \in \mathcal{P}_k(\daleth)$, then

$$\phi_s^* f = f, \quad \forall \zeta \in \mathfrak{o}(\mathbb{H}_k(\daleth)), \quad \zeta = \partial\phi,$$
$$\phi_s^* f = g \in \mathcal{P}_k(\daleth), \quad \forall \zeta \in \mathfrak{v}_k(\daleth), \quad \zeta = \partial\phi.$$

Thus

$$(\phi_s)_* : \mathcal{P}_k(\daleth) \to \mathcal{P}_k(\daleth), \quad \forall \zeta \in \mathfrak{s}(\mathcal{H}_k(\daleth)), \quad \partial\phi = \zeta.$$

4) *Let Z be any leaf of the distribution generated by $\mathbb{H}_k(\daleth)$. Then $\phi_s(Z) = Z$ for any $\partial\phi = \zeta \in \mathfrak{o}(\mathbb{H}_k(\daleth))$.*

5) *Let Z be any leaf of the distribution generated by $\mathbb{H}_k(\daleth)$. Then $\phi_s(Z) = Z'$, where Z' is another leaf of the same foliation to which belongs Z, for any $\partial\phi = \zeta \in \mathfrak{v}_k(\daleth)$.*

Theorem 3.5. (Metasymplectic structure of the Cartan distribution). The metasymplectic structure of the Cartan distribution $\mathbb{E}_n^k(W) \subset TJ_n^k(W)$ is a section $\Omega_k : J_n^k(W) \to [S^{k-1}(\tau^*) \otimes \nu] \bigotimes \Lambda^2(\mathbb{E}_n^k(W)^*)$, where $S^{k-1}(\tau^*) \equiv \bigcup_{q \in J_n^k(W)} S^{k-1}(\tau^*)_q$, with $S^{k-1}(\tau^*)_q \equiv S^{k-1}(T_a^* N)$, $\nu \equiv \bigcup_{q \in J_n^k(W)} \nu_q$, with $\nu_q \equiv (T_a W / T_a N)$, $[N]_a^k = q$.

Proof. See ref.[65]. $\square$

Definition 3.4. We say that vectors $X, Y \in \mathbb{E}_n^k(W)_q$ are in *involution* if

$$\Omega_k(q)(\lambda)(X, Y) = 0, \ \forall \lambda \in S^{k-1}(T_a N) \bigotimes \nu_a^*.$$

A subspace $P \subset \mathbb{E}_n^k(W)_q$ is called *isotropic* if any two vectors $X, Y \in P$ are in involution. We say that a subspace $P \subset \mathbb{E}_n^k(W)_q$ is a *maximal isotropic subspace* if P is not a proper subspace of any other isotropic subspace.

Theorem 3.6. 1) *Any maximal isotropic subspace* $P \subset \mathbb{E}_n^k(W)_q$ *has the form*

$$P \equiv \left\{ <x \otimes \theta> \;\middle|\; \begin{array}{l} x \in Ann(U) \subset T_a N, \\ \theta \in S^k U \bigotimes \nu_a \subset S^k(T_a^* N) \bigotimes \nu_a \end{array} \right\},$$

where $U \equiv Ann(T(\pi_{k,0})(P)) \subset T_a^* N$. *Now, let* $P \subset \mathbb{E}_n^k(W)_q$ *be an isotropic subspace for which the mapping* $T(\pi_{k,0}) : P \to T_a N$ *is an isomorphism. Then* P *is a maximal isotropic subspace and* $P = L_u$ *for some* $u \in \pi_{k+1,k}^{-1}(q)$.
2) *Let* $N \subset W$ *be a* n-*dimensional submanifold of* W *and let* $N_0 \subset N$ *be a submanifold in* N. *Set*

$$N_0^{(k)}(N) \equiv \left\{ q \in J_n^k(W) | \pi_{k,k-1}(q) \in N_0^{(k-1)}, L_q \supset T_{\pi_{k,k-1}(q)} N_0^{(k-1)} \right\},$$

where $N_0^{(k-1)} \equiv \left\{ [N]_a^{k-1} | a \in N_0 \right\} \subset J_n^{k-1}(W)$. *Then the tangent planes to* $N_0^{(k)}(N)$ *coincide with the maximal involutive subspaces described in the previous theorem. Therefore,* $N_0^{(k)}(N)$ *is a maximal integral manifold of the Cartan distribution.*
3) *If there is a fiber bundle structure* $\alpha : W \to M$, $\dim M = n$, *for the metasymplectic structure of* $J\mathcal{D}^k(W)$ *one has*

$$\Omega_k(q) \in \Lambda^2(\mathbb{E}_n^k(W)_q^*) \bigotimes S^{k-1}(T_b^* M) \bigotimes vT_a W,$$

with $a \equiv \pi_{k,0}(q) \in W, b \equiv \pi_k(q) \in M$. *If* α *is a trivial bundle* $\alpha : W \equiv M \times F \to M$, *then one has*

$$\Omega_k(q) \in \Lambda^2(\mathbb{E}_n^k(W)_q^*) \bigotimes S^{k-1}(T_b^* M) \bigotimes T_f F, \; \forall a \equiv (b, f).$$

Definition 3.5. A *partial differential equation* (PDE) for n-dimensional submanifolds of W is a submanifold $E_k \subset J_n^k(W)$. A *(regular) solution* of E_k is a (regular) solution of $J_n^k(W)$ that is contained into E_k. In particular, if $E_k \subset J^k(W) \subset J_n^k(W)$ we can talk about PDE for sections of $\pi : W \to M$. The *prolongation of order* l of $E_k \subset J_n^k(W)$ is the subset $(E_k)_{+l} \subset J_n^{k+l}(W)$ defined by $(E_k)_{+l} \equiv J_n^l(E_k) \cap J_n^{k+l}(W)$. A PDE $E_k \subset J_n^k(W)$ is called *formally integrable* if for all $l \geq 0$ the prolongations $(E_k)_{+l}$ are smooth submanifolds and the projections $\pi_{k+l+1,k+l} : (E_k)_{+(l+1)} \to (E_k)_{+l}, \pi_{k,0} : E_k \to W$ are smooth bundles. The *symbol* of the PDE $E_k \subset J_n^k(W)$ at the point $q \equiv [N]_a^k \in E_k$ is defined to be the following subspace: $g_k(q) \equiv T_q(E_k) \cap T_q(F_{\bar{q}})$, where $\bar{q} \equiv \pi_{k,k-1}(q)$ and $F_{bar q}$ is the fiber of $J_n^k(W) \to J_n^{k-1}(W)$ over

$\bar{q}$. Using the affine structure on the fibre $F_{\bar{q}}$, we can identify the symbol $g_k(q)$ with a subspace in $S^k(T_a^*N) \otimes \nu_a$: $g_k(q) \subset S^k(T_a^*N) \otimes \nu_a$. Suppose that all prolongations $(E_k)_{+l}$ are smooth manifolds, then their symbols at points $\check{q} \equiv [N]_a^{k+l}$ are lth prolongations of the symbol $g_k(q)$, hence $g_{k+l}(\check{q}) = g_{k+l}(q) \subset S^{k+l}(T_a^*N) \otimes \nu_a$ and $\delta(g_{k+l}(\check{q})) \subset g_{k+(l-1)}(q) \otimes T_a^*N$, $l = 1, 2, \ldots$, where by $\delta : S^{k+l}(T_a^*N) \otimes \nu_a \to T_a^*N \otimes S^{k+l-1}(T_a^*N) \otimes \nu_a$ we denote δ-*Spencer operator*. Therefore, at each point $q \in E_k$ the δ-Spencer complex

$$(3.5) \qquad 0 \to g_m(q) \xrightarrow{\delta} T_a^*N \otimes g_{m-1}(q) \xrightarrow{\delta} \Lambda^2(T_a^*N) \otimes g_{m-2}(q) \xrightarrow{\delta}$$
$$\ldots \xrightarrow{\delta} \Lambda^n(T_a^*N) \otimes g_{m-n}(q) \to 0$$

is defined, where $m \geq k$. We denote by $H^{m-j,j}(E_k, q)$ the cohomologies of this complex at the term $\Lambda^j(T_a^*N) \otimes g_{m-j}(q)$. They are called δ-*Spencer cohomologies* of PDE at the point $q \in E_k$. We say that g_k is *involutive* if the sequences (3.5) are exact and that g_k is *r-acyclic* if $H^{m-j,j}(E_k, q) = 0$ for $m - j \geq k$, $0 \leq j \leq r$. If $E_k \subset J_n^k(W)$ is a *2-acyclic* PDE , i.e., $H^{j,i}(E_k, q) = 0$, $\forall q \in E_k$, $0 \leq j \leq 2$, $m - j \geq k$, and $\pi_{k+1,k} : (E_k)_{+1} \to E_k$, $\pi_{k,0} : E_k \to W$ are smooth bundles, then E_k is formally integrable.

Example 3.2. Let $\mathbb{E} \subset TM$ be a r-distribution on M. Let $\{\zeta_\alpha\}_{1 \leq \alpha \leq r}$ be r-unconnected vector fields defined on $U \subset M$ which span the distribution at each point of $U \subset M$. Then, $\mathbb{E}$ is completely integrable, i.e., involutive, iff the following first order linear PDE $\{(\zeta_\alpha.\phi) \equiv \zeta_\alpha^\beta(\partial x_\beta.\phi) = 0, 1 \leq \alpha \leq r\}$, is involutive formally integrable, $(\phi \in C^\infty(U, \mathbf{R}))$. $\qquad\blacksquare$

A differential equation $E_k \subset J_n^k(W)$ is *completely integrable* if for any $q \in E_k$ there exists a n-dimensional submanifold $N \subset W$ in a neighborhood of $\pi_{k,0}(q) \equiv a \in W$ such that $[N]_a^k = q$, and $N^{(k)} \subset E_k$. This implies that the map $E_{k+l} \to E_k$, $\forall l \geq 0$, is surjective. The formal integrability assures the complete integrability in the analytic case.

Definition 3.6. A $(k + 1)$-connection on W can be seen as a $(k + 1)$-order PDE $C_{k+1} \equiv \rceil(J_n^k(W)) \subset J_n^{k+1}(W)$ diffeomorphic to $J_n^k(W)$ by means of the projection $\pi_{k+1,k}$. C_{k+1} is called *flat* if it is completely integrable. Then, to C_{k+1} we can associate the *curvature map* $R_\rceil : C_{k+1} \to \check{J}_n^{k+2}(W)/J_n^{k+2}(W)$ given by composition

$$C_{k+1} \subset J_n^{k+1}(W) \xrightarrow{\rceil^{(1)}} \check{J}_n^{k+2}(W) \xrightarrow{pr} \check{J}_n^{k+2}(W)/J_n^{k+2}(W),$$

where $\breve{J}_n^{k+2}(W)$ is the $(k+2)$-*order sesquiholonomic prolongation* of W defined by the kernel of the following double arrow:

$$
\begin{array}{ccccc}
0 & \to & \breve{J}_n^{k+2}(W) & \to & J_n^1(J_n^{k+1}(W)) & \to & J_n^1(J_n^k(W)) \\
& & & & \downarrow & & \cup \\
& & & & J_n^{k+1}(W) & = & J_n^{k+1}(W)
\end{array}
$$

A coordinate system on $\breve{J}_n^{k+2}(W)$ is the following

$$
\{x^\alpha, y^j, y^j_\alpha, \ldots, y^j_{\overline{\alpha_1\ldots\alpha_{k+1}}}, y^j_{\overline{\alpha_1\ldots\alpha_{k+1}}\beta}\},
$$

where the overline on the indexes denotes symmetry between the underlying indexes. The canonical projection $\breve{J}_n^{k+2}(W) \to \breve{J}_n^{k+1}(W)$ is an affine bundle with associated vector space $T_a^*N \otimes (S^{k+1}(T_a^*N) \otimes \nu_a)$, at each point $q \in \breve{J}_n^{k+2}(W)$, $\pi_{k+2,0}(q) = a$. Furthermore, $]^{(1)}$ is the first prolongation of the $(k+1)$-connection of $]$ and pr is the canonical projection on the quotient manifold.

Theorem 3.7. 1) *The definition of curvature of a $(k+1)$-connection given in above definition coincides with one given in Definition 3.3.*

2) *C_{k+1} is flat iff $C_{k+1} \subset \ker(R_])$, or equivalently, iff the corresponding horizontal distribution $\mathbb{H}_k(])$ is involutive.*

Proof. See last paper quoted in refs.[65]. □

Remark 3.3. Given a k-order PDE $E_k \subset J_n^k(W)$ we define the following subset $E_{k+r}^{(s)} \equiv \pi_{k+r+s,k+r}((E_k)_{+(r+s)}) \subset (E_k)_{+r}$. E_k is said to be *regular* if $E_{k+r}^{(s)}$ is a fibered manifold, for all $r, s \geq 0$. Furthermore, $E_k \subset J^k(W) \subset J_n^k(W)$ is said to be *strongly regular* if it is regular and the symbol $g_{k+r}^{(s)}$ of $E_{k+r}^{(s)}$ is induced from a vector bundle over M. Then, if E_k is a strongly regular PDE we can find two integers $\bar{r}, \bar{s} \geq 0$ such that the system $E_{k+\bar{r}}^{(\bar{s})}$ is involutive formally integrable with the same solutions as E_k. In particular, if E_k is strongly regular with the above integers $\bar{r} = \bar{s} = 0$, i.e., if there exists a sub-equation $E_k^{(0)} \subset E_k$ that is (involutive) formally integrable with the same solutions of E_k, we say that E_k is *universally regular*, or *universally formally integrable*. (We shall use use this important property in order to discuss the integrability of the Navier-Stokes equation.) ∎

Definition 3.7. The *contact ideal* of $E_k \subset J_n^k(W)$ is the ideal $\mathcal{C}(E_k)$ (or simply $\mathcal{C}_k$) of differential forms on E_k, defined by $\mathcal{C}_k \equiv \mathcal{C}_k(W)|_{E_k} \equiv i^*\mathcal{C}_k(W)$, where $i : E_k \to J_n^k(W)$ is the canonical inclusion.

Theorem 3.8. *Let W be a manifold (resp. $\pi : W \to M$ a fiber bundle) like above specified and let $E_k \subset J_n^k(W)$ (resp. $E_k \subset J\mathcal{D}^k(W)$) be a PDE. One has the following.*

1) Solutions V of $\hat{E}_k$ are branched coverings of some n-dimensional submanifold $N \subset W$, (resp. coverings of M) with branche points $p \subset \pi_k(\Sigma(V))$. Set $V' \equiv V \setminus \Sigma(V)$ and $N' \equiv N \setminus \pi_k(\Sigma(V))$, (resp. $M' \equiv M \setminus \pi_k(\Sigma(V))$). Then $\pi_k : V' \to N'$ (resp. $\pi_k : V' \to M'$) is a covering map.

2) In particular, if $\Sigma(V) = \emptyset$, then $\pi_k : V \to N$ (resp. $\pi_k : V \to M$) is a covering map. Furthermore, if $\pi_1(N) = 0$, (resp. $\pi_1(M) = 0$), i.e., N (resp. M) is simply connected, then V is globally diffeomorphic to N, (resp. M).

3) If the base space N (resp. M) is connected, then the number of sheets of the covering $V' \to N'$ (resp $V' \to M'$) is independent of the choice of the point $a \in N$ (resp. $p \in M$).

4) In the case of PDE's $E_k \subset J_n^k(W)$, if the manifolds V and N are closed, then the branched coverings (resp. coverings) are with a finite number of sheets. A similar proposition hold for the manifolds V and M in the case of solutions of PDE's $E_k \subset J\mathcal{D}^k(W)$.

Proof. The proof is similar to one for Theorem 2.8. $\square$

Theorem 3.9. *1) If E_k is locally characterized by the zero forms $h_a \in \Omega^0(J_n^k(W))$, then we can locally characterize C_k by means of the following Pfaffian ideal on $J_n^k(W)$: $C_k = < \omega^j, \ldots, \omega^j_{\alpha_1 \ldots \alpha_{k-1}}, dh_a >$. Then the closure $\overline{C}_k$ of C_k is the closed ideal of $\Omega^\bullet(E_k)$ that contains C_k as a subideal that can be locally characterized in the following reduced form:*
$$\overline{C}_k = < \omega^j, \ldots, \omega^j_{\alpha_1 \ldots \alpha_{k-1}}, d\omega^j_{\alpha_1 \ldots \alpha_{k-1}}, dh_a > .$$

2) The Cartan distribution $\mathbb{E}_k(E_k)$, or simply $\mathbb{E}_k$, of a PDE $E_k \subset J_n^k(W)$ is the Cauchy characteristic distribution of the contact ideal C_k of E_k. So we have:

$$\mathfrak{d}(\mathbb{E}_k) = \{\zeta \in \mathfrak{d}(TJ_n^k(W)) | \zeta \rfloor \omega^j = \ldots = \zeta \rfloor \omega^j_{\alpha_1 \ldots \alpha_{k-1}} = \zeta . h_a = 0\}.$$

Therefore, $\mathbb{E}_k \subset TE_k$. (We say that $q \in E_k$ is a regular point if in some neighborhood $U \subset E_k$ of this point the function $q' \mapsto \dim(\mathbb{E}_k)_{q'}$ is constant, $\forall q' \in U$. Otherwise, the point $q \in E_k$ is called singular. If all points $q \in E_k$ are regular then $\mathbb{E}_k \equiv \bigcup_{q \in E_k}(\mathbb{E}_k)_q$ is a distribution on E_k called Cartan distribution of E_k.

3) A solution (resp. regular solution) Z of E_k is a solution (resp. regular solution) of $J_n^k(W)$ such that $C_k|_Z = 0 \Leftrightarrow dh_a|_Z = 0$. Therefore, a (regular) solution of E_k is a (regular) solution of $J_n^k(W)$ that is contained into E_k. We denote by $\mathcal{S}_{ol}(E_k)$ (resp. $\underline{\mathcal{S}}_{ol}(E_k)$) the set of solutions (resp. regular solutions) of E_k. A solution (resp. regular solution) is representable by an embedding $\Phi : P \to J_n^k(W)$, with P a $n-$dimensional manifold, such that

$$\left\{ \begin{array}{l} \Phi^*\mathcal{C}_k = 0 \\ \Phi \in \mathcal{S}_{ol}(J_n^k(W)) \end{array} \right\}, \quad \left(resp. \left\{ \begin{array}{l} \Phi^*\mathcal{C}_k = 0 \\ \Phi \in \underline{\mathcal{S}}_{ol}(J_n^k(W)) \end{array} \right\} \right).$$

An alternative way to locally characterize a regular solution of E_k is by considering the following ideal of differential forms on E_k: $\mathcal{I}(E_k) \equiv <\omega^j, \ldots, \omega_{\alpha_1 \ldots \alpha_{k-1}}^j, B_a >$ where $B_a \equiv h_a \eta = h_a dx^1 \wedge \ldots \wedge dx^n$. Then a mapping $\Phi : P \to J_n^k(W)$ represents a regular solution of E_k iff the following conditions are verified:

$$\left\{ \begin{array}{l} \Phi \in \underline{\mathcal{S}}_{ol}(J_n^k(W)) \\ \Phi^*\mathcal{I}(E_k) = 0 \end{array} \right\} \Leftrightarrow \left\{ \begin{array}{l} \Phi^*\eta \neq 0 \\ \Phi^*\mathcal{I}(E_k) = 0 \end{array} \right\}.$$

Theorem 3.10. *One has the following propositions on the symmetries properties of PDEs:* (i) $\mathfrak{s}(\overline{\mathcal{C}}_k)$ *is a Lie subalgebra of* $\mathfrak{s}(\overline{\mathcal{C}}_k(W))$. (ii) $\mathfrak{s}(\overline{\mathcal{I}}(E_k))$ *is a Lie subalgebra of* $\mathfrak{s}(\overline{\mathcal{C}}_k(W))$. $\zeta \in \mathfrak{s}(\overline{\mathcal{I}}(E_k))$ *iff* $\zeta \in \mathfrak{s}(\overline{\mathcal{C}}_k(W)$ *and* $\mathcal{L}_\zeta B_a = N_a^b B_b \bmod \overline{\mathcal{C}}_k$, *where* N_a^b *are any* s^2 *elements of* $\Omega^0(J_n^k(W))$, $1 \leq a \leq s$. (iii) $\zeta \in \mathfrak{s}(\overline{\mathcal{C}}_k)$, $\Phi \in \mathcal{S}_{ol}(E_k) \Rightarrow \phi_s \circ \Phi \in \mathcal{S}_{ol}(E_k), \forall s \in]-\epsilon, \epsilon[\subset \mathbf{R}$. (iv) $\zeta \in \mathfrak{s}(\overline{\mathcal{C}}_k)$, $\Phi \in \underline{\mathcal{S}}_{ol}(E_k) \Rightarrow \phi_s \circ \Phi \in \underline{\mathcal{S}}_{ol}(E_k), \forall s \in]-\epsilon, \epsilon[\subset \mathbf{R}$. (v) $\zeta \in \mathfrak{s}(\overline{\mathcal{I}}(E_k))$, $\Phi \in \underline{\mathcal{S}}_{ol}(E_k) \Rightarrow \phi_s \circ \Phi \in \underline{\mathcal{S}}_{ol}(E_k), \forall s \in]-\epsilon, \epsilon[\subset \mathbf{R}$. (vi) *Let* $\mathcal{B}(E_k) \equiv \omega^i, \cdots, \omega_{\alpha_1 \ldots \alpha_{k-1}}^i, B_a, H_{\alpha_1 \ldots \alpha_k}^i >$ *be the ideal encoding* E_k. *Then, the relations between* $\mathfrak{s}(\mathcal{C}_k(W))$, $\mathfrak{s}(\mathcal{C}_k)$, $\mathfrak{s}(\mathcal{H}_k(\uparrow))$, $\mathfrak{s}(\mathcal{I}(E_k))$, *and* $\mathfrak{s}(\mathcal{B}(E_k))$ *are given by the following canonical isomorphisms* $\mathfrak{s}(\overline{\mathcal{C}}_k) \cong \mathfrak{s}(\mathcal{C}_k) \cong \mathfrak{s}(\mathcal{I}(E_k))$, $\mathfrak{s}(\mathcal{B}(E_k)) \cong \mathfrak{s}(\mathcal{H}_k(\uparrow))$ *and by the following commutative diagram:*

$$\begin{array}{ccccc} 0 & \to & \mathfrak{s}(\mathcal{C}_k(W)) & \to & \mathfrak{s}(\mathcal{H}_k(\uparrow)) \\ & & \uparrow & & \uparrow \\ & & \mathfrak{s}(\mathcal{C}_k(W))_\uparrow & = & \mathfrak{s}(\mathcal{C}_k(W))_\uparrow \\ & & \uparrow & & \uparrow \\ & & 0 & & 0 \end{array}$$

where $\mathfrak{s}(\mathcal{C}_k(W))_\uparrow \equiv \{\zeta \in \mathfrak{s}(\mathcal{C}_k(W)) | d(\zeta \rfloor H_\alpha^J) = 0 = \zeta \rfloor dH_\alpha^J\} \neq 0$. *So, in general, existence of relations between* $\mathfrak{s}(\mathcal{C}_k)$ *(or* $\mathfrak{s}(\mathcal{I}(E_k))$*) and* $\mathfrak{s}(\mathcal{B}(E_k))$ *depends on the structure of* $\mathfrak{s}(\mathcal{C}_k(W))_\uparrow$.

Proof. See ref.[65]. $\qquad\qquad\qquad\qquad\qquad\qquad\qquad\qquad\qquad\qquad$ $\square$

Corollary 3.2. *If $E_k \subset J_n^k(W)$ is a PDE such that hold the following propositions:*

(i) $(E_k)_{+1} \to E_k$ is a fiber bundle.

(ii) There exists a flat $(k+1)$-connection $\rceil$ on W such that $\rceil|_{E_k}$ is a section of above fiber bundle.

Then the characteristic distribution $\mathbb{C}_{har}(E_k)$ of E_k, i.e., the characteristic distribution $\mathbb{C}_{har}(C_k)$ of the ideal C_k, is non-zero.

Proof. In fact under our assumptions one has:

$$\left.\begin{cases} \partial(\mathbb{H}_k(\rceil)) \subset \mathfrak{s}(C_k) \cap \mathfrak{s}(C_k(W))_{\rceil} \neq 0 \\ \mathbb{H}_k(\rceil)|_{E_k} \subset \mathbb{E}_k \end{cases}\right\} \Rightarrow \mathbb{H}_k(\rceil)|_{E_k} \subset \mathbb{C}_{har}(C_k).$$

$\square$

Theorem 3.11. *Let $E_k \subset J_n^k(W)$ be a PDE such that $\pi_{k+1,k} : (E_k)_{+1} \to E_k$ is a smooth bundle. Then $\mathcal{C}har(E_k)_q$ is the set of such vectors $v = \zeta - \theta \in L_{\hat{q}}$, $\zeta \in (\mathbb{E}_n^k)_q \cong L_{\hat{q}} \bigoplus [S^k(T_a^*N) \otimes \nu_a]$ such that $v \rfloor \delta(\theta) = 0$ for any $\theta \in (g_k)_q \subset S^k(T_a^*N) \otimes \nu_a$, where $L_{\hat{q}}$ is the subspace determined for any point $\hat{q} \in (E_k)_{+1}$ such that $\pi_{k+1,k}(\hat{q}) = q$. The local expression of v is the following:*

$$\begin{aligned} v = &X^\alpha(\partial x_\alpha + y_\alpha^j \partial y_j + y_{\alpha\beta}^j \partial y_j^\beta \\ &+ \ldots + y_{\alpha\beta_1\ldots\beta_{k-1}}^j \partial y_j^{\beta_1\ldots\beta_{k-1}} + y_{\alpha\beta_1\ldots\beta_k}^j \partial y^{\beta_1\ldots\beta_k}) \end{aligned}$$

where $y_{\alpha\beta_1\ldots\beta_k}^j$ are functions on $J_n^{k+1}(W)$ that satisfy the following equations:

$$(E_k)_{+1} : \left.\begin{cases} F_\alpha^I(x^\alpha, y^j, \ldots, y_{\alpha\beta_1\ldots\beta_{k-1}}^j, y_{\alpha\beta_1\ldots\beta_k}^j) = 0 \\ F^I(x^\alpha, y^j, \ldots, y_{\beta_1\ldots\beta_k}^j) = 0 \end{cases}\right\}$$

and the functions X^α on $J_n^k(W)$ are determined by the equations:

$$\left\{ (\partial y_j^{\alpha_1\ldots\alpha_k}.F^I)Y_{\alpha_1\ldots\alpha_k}^j = 0; \; v\rfloor\delta(\theta) = 0, \; \theta = Y_{\alpha_1\ldots\alpha_k}^j \, dx^{\alpha_1} \odot \ldots \odot dx^{\alpha_k} \right\}.$$

Proof. The proof follows directly from Theorem 3.1 and above considerations on PDEs. (See also refs.[42,65].) $\square$

Definition 3.8. A *Cartan connection* on E_k is a n-dimensional subdistribution $\mathbb{H} \subset \mathbb{E}_k$ such that $T(\pi_{k,k-1})(\mathbb{H}_q) = L_q \equiv T_{\pi_{k,k-1}(q)}N^{(k-1)}$, $[N]_q^k \equiv q$, $\forall q \in E_k$. Let us assume that $(E_k)_{+1} \to E_k$ is a smooth subbundle of $J_n^{k+1}(W) \to J_n^k(W)$. Then any section $H : E_k \to (E_k)_{+1}$ is called a *Bott connection*. We call *curvature* of the Cartan connection $\mathbb{H}$ on $E_k \subset J_n^k(W)$ the field of geometric objects on E_k:

$$\Omega_{\mathbb{H}} : q \mapsto \Lambda^2(\mathbb{H}_q^*) \bigotimes [S^{k-1}(T_a N) \otimes \nu_a^* / Ann(g_{k-1})]^*$$

$$\cong \Lambda^2(T_a^* N) \bigotimes [S^{k-1}(T_a N) \otimes \nu_a^* / Ann(g_{k-1})]^*$$

obtained by restriction on $\mathbb{H}$ of the metasymplectic structure on the distribution $\mathbb{E}_n^k$.

Theorem 3.12. 1) *A Cartan connection $\mathbb{H}$ is a Bott connection iff $\Omega_{\mathbb{H}} = 0$.*
2) *A Cartan connection $\mathbb{H}$ gives a split of the Cartan distribution*

$$\mathbb{E}_n^k \cong g_k \bigoplus \mathbb{H}.$$

Two Cartan connections $\mathbb{H}$, $\mathbb{H}'$ on E_k identify a field of geometric objects λ on E_k called soldering form: $\lambda \equiv \lambda_{\mathbb{H},\mathbb{H}'} : E_k \to \mathbb{H}^* \bigotimes g_k$, $\lambda(q) \in T_a^* N \otimes g_k(q)$. One has: $\Omega_{\mathbb{H}'} = \Omega_{\mathbb{H}} + \delta\lambda$, (Bianchi identity) $\delta\Omega_{\mathbb{H}} = 0$, $\Omega_{\mathbb{H}}(q) \bmod \delta(T_a^* N \otimes g_k(q))) \in H^{k-1,2}(E_k)_q$. We call such δ-cohomology class of $\Omega_{\mathbb{H}}$ the Weyl tensor of E_k at $q \in E_k$: $W_k(q) \equiv [\Omega_{\mathbb{H}}(q)]$. Then, there exists a point $u \in (E_k)_{+1}$ over $q \in E_k$ iff $W_k(q) = 0$.
3) *Suppose that g_{k+1} is a vector bundle over $E_k \subset J_n^k(W)$. Then if the Weyl tensor W_k vanishes the projection $\pi_{k+1,k} : (E_k)_{+1} \to E_k$ is a smooth affine bundle.*
4) *If g_{k+l} are vector bundles over E_k and $W_{k+l} = 0$, $l \geq 0$, then E_k is formally integrable.*
5) *If the system E_k is of finite type, i.e., $g_{k+l}(q) = 0$, $\forall q \in E_k$, $l \geq l_0$, then $W_{k+l} = 0$, $0 \leq l \leq l_0$, is a sufficient condition for integrability.*
Theorem 3.13. *Given a Cartan connection $\mathbb{H}$ on E_k, for any regular solution $N^{(k)} \subset E_k$ we identify a section $_{\mathbb{H}}\nabla \in C^\infty(T^* N \bigotimes g_k)$ called covariant differential of $\mathbb{H}$ of the solution N. Furthermore, for any vector field $\zeta : N \to TN$ we get a section $_{\mathbb{H}}\nabla\zeta \in C^\infty(g_k|_{N^{(k)}})$.*
Remark 3.4. Let $E_k \subset J_n^k(W)$ be a PDE. A *symmetry* of E_k is a local diffeomorphism $\phi : W \to W$ such that $\phi^{(k)}(E_k) \subset E_k$. An *infinitesimal symmetry* of E_k is a vector field $X \in C^\infty(TW)$ such that $X^{(k)} = \partial\phi^{(k)}$, where $\phi_t^{(k)} : J_n^k(W) \to J_n^k(W)$, $t \in \mathbf{R}$, is the 1-paramenter group of transformations on $J_n^k(W)$ corresponding to the flow ϕ_t associated to X on W. In particular, if $E_k \subset J^k(W)$, where $\alpha : W \to M$ is a fiber bundle, a local

authomorphism of α, defined by the following commutative digram

$$
\begin{array}{ccc}
W & \xrightarrow{\phi} & W \\
\alpha \downarrow & & \downarrow \alpha \\
M & \xrightarrow[\phi]{} & M
\end{array}
$$

is a symmetry of E_k if $\phi^{(k)}(E_k) \subset E_k$. Furthermore, a projectable vector field on W, $X \in \widehat{C}^\infty(TW)$, defined by the following commutative diagram

$$
\begin{array}{ccc}
W & \xrightarrow{X} & TW \\
\alpha \downarrow & & \downarrow T\alpha \\
M & \xrightarrow[\overline{X}]{} & TM
\end{array}
$$

where $\overline{X} \in C^\infty(TM)$, is an infinitesiml symmetry of E_k if $X^{(k)}$ is tangent to E_k. Let $E_k \subset J_n^k(W)$ be a PDE. Then the linear PDE $R_k \subset J^k(TW)$ defined by

$$
R_k \equiv \left\{ [X]_a^k, X \in C^\infty(TW) | X^{(k)}(q) \in T_q E_k, \pi_k(q) = a \right\}
$$

is called the *infinitesimal Lie equation* of E_k. The set of regular solutions of R_k is the set of infinitesimal symmetries of E_k. If there exists a fiber bundle structure $\alpha : W \to M$ and $E_k \subset J^k(W)$ then one has

$$
R_k \equiv \left\{ [X]_a^k, X \in \widehat{C}^\infty(TW) | X^{(k)}(q) \in T_q E_k, \pi_k(q) = a \right\}.
$$

The set of regular solutions of R_k is the set of infinitesiml symmetries of E_k. The pseudogroup G of symmetries of E_k is the set of solutions of a Lie equation $\mathcal{R}_k \subset \Pi^k(W)$ such that $J^k(G) \subseteq \mathcal{R}_k$, where $J^k(G)$ is the k-prolongation of the groupoid associated to G. The equality holds if $\mathcal{R}_k$ is completely integrable. The linearized equation R_k of $\mathcal{R}_k$ is the corresponding infinitesimal Lie equation of E_k. $\blacksquare$

Definition 3.9. We call a PDE $E_k \subset J_n^k(W)$ *degenerate* at the point $q \in E_k$ if there is a subspace $\Xi(q) \subset T_a^* N$ such that $g_k(q) \subset S^k(\Xi(q)) \otimes \nu_a$. The subspace $\Xi(q)$ is called *subspace of degeneration*. Let $P \subset E_k$ be a $(n-1)$-dimensional integral manifold (*Cauchy data*). Let $\Sigma_l(P)$ denote the set of Thom-Bordman singular points of P of type Σ_l, i.e., $\ker(\pi_{k,0}|_P)_{*,q} = l$. We say that $q \in \Sigma_l(P)$ is a *frozen singularity* if $L_0 \equiv \ker(\pi_{k,0}|_P)_{*,q} \subset S^k(\Xi) \otimes \nu_a$ is nondegenerate subspace in $S^k(\Xi) \otimes \nu_a$.

Theorem 3.14. 1) *There are no smooth n-dimensional integral manifolds containing Cauchy data passing through frozen singularities.*

2)(*Cauchy problem and characteristic regular solutions of PDE's*) *Let $E_k \subset J_n^k(W)$ be a PDE encoded by an ideal $\mathcal{I} \equiv\, <\Omega_I>$, $1 \le I \le r$. Let $\zeta \in \mathfrak{s}(\overline{\mathcal{I}})$ and let $N \subset E_k$ be a regular integral submanifold of E_k of dimension $n-1$ characterized by a mapping $\psi : P \to E_k$, $(P \equiv (n-1)-$dimensional manifold), such that the following conditions are satisfied:*

$$(3.6) \qquad \left\{ \begin{array}{l} \text{(a)}(\textit{transversality conditions}) \quad \psi^*(\zeta\rfloor\eta) \neq 0 \\[2ex] \text{(b)}(\textit{initial conditions}) \quad \left\{ \begin{array}{l} \psi^*\Omega_I = 0 \\[1ex] \psi^*(\zeta\rfloor\Omega_I) = 0 \end{array} \right\}, 1 \le I \le r \end{array} \right\}.$$

Furthermore, let us assume that $X \equiv \pi_{k,0}(N) \subset W$ admits a cylindrical neighbourhood $X \times I$, $I \subset \mathbf{R}$, or collar $X \times [0,\infty)$, in W. Then, a regular solution Z of E_k exists that passes through the Cauchy data N. More precisely $Z \subset E_k$ is characterized by means of a mapping $\Psi : P \times J \to E_k$, where J is a neighborhood of $0 \in \mathbf{R}$ such that $\Psi^*(\zeta\rfloor\Omega_I) = 0$, $1 \le I \le r$. We call such a solution a characteristic solution *of E_k.*

Proof. 1) See ref.[42]. 2) See ref.[65]. $\qquad\qquad\qquad\qquad\qquad\qquad$ $\square$

Theorem 3.15. (Horizontal contact ideals and regular solutions of PDE's).
1) *Let $E_k \subset J_n^k(W)$ be a k-order PDE encoded by the differential ideal*

$$\mathcal{C}_k \equiv\, <\omega^j, \ldots, \omega^j_{\alpha_1\ldots\alpha_{k-1}}, dF_a>,$$

where $\omega^j, \ldots, \omega^j_{\alpha_1\ldots\alpha_{k-1}}$ are contact 1-forms and $F_a \in \Omega^0(J_n^k(W))$ such that $F_a = 0$ are the local representation of E_k. Then any regular solution of E_k can be locally represented in the form $\Psi : \mathcal{G}_n \to J_n^k(W)$ of a leaf mapping of a closed horizontal k-order contact ideal $\mathcal{H}_k(\rceil) \subset \Omega^\bullet(J_n^k(W))$, ($\mathcal{G}_n$ is the abelian Lie group that has $\mathfrak{d}(\mathbb{H}_k(\rceil))$ as Lie algebra), such that:

$$\{\Psi^* B_a = 0 \Leftrightarrow F_a \circ \Psi = 0\}, \quad B_a \equiv F_a\eta.$$

Resuming:

$$\left\{ \begin{array}{l} \Psi^*\eta \neq 0 \\[1ex] \Psi^*\omega^j = \ldots = \Psi^*\omega^j_{\alpha_1\ldots\alpha_{k-1}} = 0 \\[1ex] \Psi^* B_a = 0 \\[1ex] \Psi^* H^j_{\alpha_1\ldots\alpha_k} = 0 \end{array} \right\} \Leftrightarrow \left\{ \begin{array}{l} \Psi^*\eta \neq 0 \\[1ex] \Psi^*\mathcal{B}(E_k) = 0 \\[1ex] \mathcal{B}(E_k) \equiv\, <\omega^j, \ldots, \omega^j_{\alpha_1\ldots\alpha_{k-1}}, H^j_{\alpha_1\ldots\alpha_k}, B_a> \end{array} \right\},$$

where $H^j_{\alpha_1\ldots\alpha_k} = dy^j_{\alpha_1\ldots\alpha_k} - \rceil^j_{\alpha_1\ldots\alpha_{k-1}\alpha} dx^\alpha \in \Omega^1(J^k_n(W))$ is the connection 1-form.

2) If $\zeta \in \mathfrak{s}(\overline{C}_k) \Rightarrow \phi_s \circ \Psi$ is a regular solution of E_k if Ψ is so, $\partial\phi = \zeta$.

3) (Theorem of existence of regular solutions). Let $E_k \subset J^k_n(W)$ be a PDE locally characterized by the equations $F_a = 0$, $F_a \in \Omega^0(J^k_n(W))$. Let $\mathcal{H}_k(\rceil)$ be an element of $h(J^k_n(W))$ such that the following conditions are verified: $(\zeta_\alpha.F_a) = 0$, where $(\zeta_\alpha)_{1\le\alpha\le n}$ is a canonical basis of $\mathfrak{d}(\mathbb{H}_k(\rceil))$. Let $\mathcal{L}(q)$ be the leaf of the foliation of $J^k_n(W)$ that contains the point $q \in E_k$, i.e., $F_a(q) = 0$. If Ψ is the map $\mathcal{G}_n \to J^k_n(W)$ that is constructed by sequential integration of the orbital equation of (ζ_α) starting from q then $\Psi \in \underline{S}_{ol}(E_k)$.

Proof. See ref. [65]. $\qquad\qquad\qquad\qquad\qquad\qquad\qquad\qquad\qquad\qquad\quad$ $\square$

Theorem 3.16. *Let be given a PDE $E_k \subset J^k_n(W)$ locally given by the equations $F^J = 0$. Let $(E_k)_{+1} \to E_k$ be a subbundle of $J^{k+1}_n(W) \to J^k_n(W)$. Then the horizontal distribution $\mathbb{H}_k(\rceil)$ induced by a flat $(k+1)$-connection $\rceil : J^k_n(W) \to J^{k+1}_n(W)$ satisfies the condition*

$$(3.7) \qquad\qquad\qquad \mathbb{H}_k(\rceil)|_{E_k} \subset TE_k$$

iff $\rceil|_{E_k}$ is a section of the fiber bundle $(E_k)_{+1} \to E_k$.

Proof. In fact we can write condition (3.7) in the following way:

$$(3.8) \qquad\qquad\qquad (\zeta_\alpha.F^J) = 0, \quad 1 \le J \le r,$$

for any vector field ζ_α of the canonical basis $\{\zeta_\alpha\}_{1\le\alpha\le n}$ of $\mathfrak{d}(\mathbb{H}_k(\rceil))$. Let us explicitly write equations (3.8):

$$((\partial x_\alpha + y^j_\alpha \partial y_j + \ldots + y^j_{\alpha_1\ldots\alpha_{k-1}\alpha}\partial y_j^{\alpha_1\ldots\alpha_{k-1}}).F^J) = -\rceil^j_{\alpha\alpha_1\ldots\alpha_k}(\partial y_j^{\alpha_1\ldots\alpha_k}.F^J).$$

These can be also rewritten in the following form:

$$(3.9) \qquad F^J_\alpha + (\partial y_j^{\alpha_1\ldots\alpha_k}.F^J)[-y^j_{\alpha\alpha_1\ldots\alpha_k} + \rceil^j_{\alpha\alpha_1\ldots\alpha_k}] = 0 \quad 1 \le J \le r,$$

where F^J_α is the first prolongation of F^J. (3.9) are conditions on the points of $J^{k+1}_n(W)$. So, as we are interested that conditions (3.8) should be verified on the points of E_k we require that conditions (3.9) should be verified on the points of $(E_k)_{+1}$. These are characterized by the equations $\{F^J_\alpha = 0, \quad F^J = 0\}$. Then (3.9) on $(E_k)_{+1}$ gives $\rceil^j_{\alpha\alpha_1\ldots\alpha_k} = y^j_{\alpha\alpha_1\ldots\alpha_k}$. In other words $\rceil$ is a section of $(E_k)_{+1} \to E_k$. Vice versa if $\rceil|_{E_k}$ is a section of $(E_k)_{+1} \to E_k$ we can made the converse road and to prove that the condition (3.7) is verified. Note also that the flatness of $\rceil$ is a consequence of Theorem 3.12(1). In fact, if $\mathbb{H}_k(\rceil)|_{E_k} \subset TE_k$, then it is a Cartan connection on E_k. $\qquad\square$

Theorem 3.17. (Cauchy problem with singular Cauchy data and characteristic solutions). *Let $E_k \subset J_n^k(W)$ be a PDE determined by an ideal $\mathcal{I} \equiv\; <\Omega_I>,\ 1 \le I \le r$. Let $\zeta \in \mathfrak{s}(\overline{\mathcal{I}})$ and let $N \subset E_k$ be an integrable submanifold of dimension $n-1$ without frozen singularities, represented by a mapping $\psi : P \to E_k \subset J_n^k(W)$, such that the following conditions are satisfied:*

$$\left\{ \begin{array}{l} \zeta \notin \mathfrak{d}(TN), \quad \psi^*(\zeta\rfloor\eta) \ne 0 \text{ on } N \setminus \Sigma(N) \\[2mm] \psi^*\Omega_I = 0 \\[2mm] \psi^*(\zeta\rfloor\Omega_I) = 0, \quad 1 \le I \le r \end{array} \right\}.$$

Furthermore, let us assume that $X \equiv \pi_{k,0}(N \setminus \Sigma(N)) \subset W$ admits a cylindrical neighborhood $X \times I,\ I \subset \mathbf{R}$, (or collar $X \times [0,\infty)$) in W. Then, a solution Z of E_k exists that passes through the Cauchy data N. More precisely $Z \subset E_k$ is expressed by means of a mapping $\Psi : P \times J \to E_k$, where J is a neighborhood of $0 \in \mathbf{R}$, such that $\Psi^*(\zeta\rfloor\Omega_I) = 0,\ 1 \le I < r$. We call such a solution a characteristic solution of E_k.

Proof. It is enough to forget the condition $\psi^*(\zeta\rfloor\eta) \ne 0$ on $\Sigma(N)$ in the proof of Theorem 3.14(2), taking into account however that the resulting integral manifold has dimension n and type n. In fact, the set $\Sigma(N)$ of singular points of N is of dimension $\dim\Sigma(N) < \dim N = n - 1$, and as a consequence $\dim\Sigma(Z) < \dim Z = n$. $\qquad\square$

Theorem 3.18. (Characteristic strips and solutions). *Let $E_k \subset J_n^k(W)$ be a PDE such that for any $q \in E_k$ passes a i-dimensional characteristic strip, $\mathbb{S}_q \subset (\mathbb{E}_n^k)_q$, i.e., there exists some $\theta \in (g_k)_q$ such that $v\rfloor\delta(\theta) = 0$, where $v \in L_{\hat{q}} \subset (\mathbb{E}_n^k)_q \cong (g_k)_q \oplus L_{\hat{q}}$, where $\hat{q} \in (E_k)_{+1}$, and $L_{\hat{q}}$ is the n-dimensional subspace of $(\mathbb{E}_n^k)_q$ uniquely identified by $\hat{q}$. Let us assume that the set of such $\theta \in (g_k)_q$ is a s-dimensional sub-symbol $\bar{g}_k$ of g_k. Then, we can identify a subequation $\bar{E}_k \subset E_k$ such that $(\bar{g}_k)_q$ is its symbol at q. Furthermore, for such an equation $\bar{E}_k$ the characteristic sub-distribution $\mathbb{C}\mathrm{har}(\bar{E}_k)$ is non empty, and one can build solutions of $\bar{E}_k$, hence solutions of E_k, by means of characteristics of $\bar{E}_k$, (even if $\mathbb{C}\mathrm{har}(E_k) = 0$).*

Proof. In fact, let $\Phi^p(Y_{\alpha_1\dots\alpha_k}^j) = 0$ be the local equations that define the subspaces $(\bar{g}_k)_q \subset (g_k)_q,\ \forall q \in E_k$, if $\theta = Y_{\alpha_1\dots\alpha_k}^j \partial y_j^{\alpha_1\dots\alpha_k} \in (g_k)_q$. Then, if

$F^I(x^\alpha, \ldots, y^j_{\alpha_1 \ldots \alpha_k}) = 0$ are the local expressions for E_k, it results that

$$\left\{ F^I(x^\alpha, \ldots, y^j_{\alpha_1 \ldots \alpha_k}) = 0,\ \Phi^p(y^j_{\alpha_1 \ldots \alpha_k}) = 0 \right\}$$

represents a sub-equation $\bar{E}_k \subset E_k$ with symbol $\bar{g}_k$. For such an equation we get the condition $v \rfloor \delta(\theta) = 0$ is satisfied for any $\theta \in \bar{g}_k$ and $v \in \mathbb{S}$. Therefore, $\mathbb{C}har(\bar{E}_k) = \mathbb{S} \subset \mathbb{E}_n^k$. Then, we can use the method of characteristics to obtain solutions of $\bar{E}_k$. On the other hand as $\bar{E}_k \subset E_k \subset J_n^k(W)$, and $\mathbb{S} \subset \mathbb{E}_n^k$, we get that such solutions are also solutions of E_k. $\square$

Theorem 3.19. (Horizontal contact ideals and regular solutions of PDE's) [65]. 1) *Any regular solution of $E_2 \subset J_n^2(W)$ where $\mathcal{I}(E_2) \equiv < \omega^j, B_j >$, with*

$$\omega^j \equiv dy^j - y^j_\alpha dx^\alpha,\ B_j \equiv h_j \eta - dw^\alpha_j \wedge \eta_\alpha,\ 1 \leq j \leq m,\ n_j, \omega^\alpha_j \in \Omega^0(J_n^1(W)),$$

can be realized as a leaf of a foliation of a closed horizontal 1-contact ideal $\mathcal{H}_1(\rceil)$, $\Psi : \mathcal{G}_n \to J_n^1(W)$ such that

$$\left\{ \Psi^* B_j = \Psi^* F_j du^1 \wedge \ldots \wedge du^n = 0 \Leftrightarrow F_j \circ \Psi = 0 \right\}$$

with $F_j \equiv h_j - (\zeta_\alpha . w^\alpha_j)$, $1 \leq j \leq m$, or equivalently

$$\left\{ \begin{array}{l} \Psi^* \eta \neq 0 \\ \Psi^* \omega^j = 0 \\ \Psi^* H^j_\alpha = 0 \\ \Psi^* B_j = 0 \end{array} \right\}.$$

Definition 3.10. It is useful to define some subsets of $h(J_n^1(W))$:

(a) $h_s(J_n^1(W)) \equiv \{ \mathcal{H}_1(\rceil) \in h(J_n^1(W)) | F_a \equiv h_a - (\zeta_\alpha . w^\alpha_a) = 0,\quad 1 \leq a \leq r \}$.

(b) $h_r(J_n^1(W)) \equiv \left\{ \begin{array}{l} \mathcal{H}_1(\rceil) \in h(J_n^1(W)) - h_s(J_n^1(W)) : \\ (\zeta_\alpha . F_a) \equiv (\zeta_\alpha . (h_a - (\zeta_\beta . w^\beta_a))) = 0,\quad 1 \leq a \leq r \end{array} \right\}$.

(c) $h_g(J_n^1(W)) \equiv \left\{ \begin{array}{l} \mathcal{H}_1(\rceil) \in h(J_n^1(W)) - h_s(J_n^1(W)) - h_r(J_n^1(W)) : \\ (\zeta_\alpha . F_a) = L^b_{a\alpha} F_b,\quad L^b_{a\alpha} \in \Omega^0(J_n^1(W)) \end{array} \right\}$.

Theorem 3.20. (Existence theorems of regular solutions). 1) *Let us consider quasi-linear PDEs of second order $E_2 \subset J_n^2(W)$. Let $\mathcal{I}(E_2) \equiv < \omega^j, B_a >$ be the fundamental ideal where $B_a \equiv h_a \eta - dw^\alpha_a \wedge \eta_\alpha \equiv F_a \eta \bmod \mathcal{H}_1(\rceil)$*

with $F_a \equiv h_a - (\zeta_\alpha.w_a^\alpha)$. Then, a leaf map Ψ of $\mathbb{H}_1(\daleth)$ is a regular solution of E_2 if

$$\Psi^* F_a = \Psi^*(h_a - (\zeta_\alpha.w_a^\alpha)) = 0.$$

2) Let $\mathcal{H}_1(\daleth) \in h_s(J_n^1(W)) \neq \emptyset$, then every leaf of the foliation generated by $\mathbb{H}_1(\daleth)$ is a regular solution of E_2, i.e., $J_n^1(W)$ is foliated by regular solutions of E_2. The conditions that $\daleth$ must satisfy in these circumstances are

$$\left\{ \begin{array}{l} \daleth_{\alpha\beta}^j = \daleth_{\beta\alpha}^j \\ (\zeta_\alpha.\daleth_{\beta\gamma}^j) - (\zeta_\beta.\daleth_{\alpha\gamma}^j) = 0 \\ (\zeta_\alpha.w_a^\alpha) = h_a \end{array} \right\},$$

with at least one of the $w's$ not constant on $J_n^1(W)$ and $\zeta_\alpha = \partial x_\alpha + y_\alpha^j \partial y_j + \daleth_{\alpha\beta}^j \partial y_j^\beta$.

3) Let $\mathcal{H}_1(\daleth) \in h_r(J_n^1(W)) \neq \emptyset$, then on any leaf of the folitaion on $J_n^1(W)$ generated by $\mathbb{H}_1(\daleth)$ the F_a are constant. Thus, any leaf of this foliation that touches the points of $\mathcal{F}(\daleth) \equiv \{q \in J_n^1(W)|F_a(q) = 0\}$ is a regular solution of E_2. The conditions that $\daleth$ must satisfy in these circumstances are

$$\left\{ \begin{array}{l} \daleth_{\alpha\beta}^j = \daleth_{\beta\alpha}^j \\ (\zeta_\alpha.\daleth_{\beta\gamma}^j) - (\zeta_\beta.\daleth_{\alpha\gamma}^j) = 0 \\ 0 = (\zeta_\alpha.F_a) = \zeta_\alpha.(h_a - (\zeta_\beta.w_a^\beta)) \end{array} \right\}, \quad \zeta_\alpha = \partial x_\alpha + y_\alpha^j \partial y_j + \daleth_{\alpha\beta}^j \partial y_j^\beta.$$

4) (Cartan-Kähler Theorem). The Cartan-Kähler theorem for first order PDEs is based on the closed fundamental ideal $\overline{\mathcal{I}}(E_1) \equiv < \omega^j, d\omega^j, h_a, dh_a >$. In order to describe regular solutions we can use the following closed ideal $\mathcal{B}(E_1) \equiv < \omega^j, d\omega^j, h_a\eta >$ that with respect to the situation of previous theorem has $w_a^\alpha = 0$. Then, we can use the existence theorem for the case $\mathcal{H}_1(\daleth) \in h_r(J_n^1(W))$.

5) Let $\mathcal{H}_1(\daleth) \in h_g(J_n^1(W)) \neq \emptyset$. Then the F_a are constants in value on the leaves of the foliation generated by $\mathbb{H}_1(\daleth)$ that intersect $\mathcal{F}(\daleth) \equiv \{q \in J_n^1(W)|F_a(q) = 0\}$. So any of such a leaf is a regular solution of E_2. The conditions that $\daleth$ must satisfy under these circumstances are

$$\left\{ \begin{array}{l} \daleth_{\alpha\beta}^j = \daleth_{\beta\alpha}^j \\ (\zeta_\alpha.\daleth_{\beta\gamma}^j) - (\zeta_\beta.\daleth_{\alpha\gamma}^j) = 0 \\ (\zeta_\alpha.F_a) = \zeta_\alpha.(h_a - (\zeta_\beta.w_a^\beta)) = L_{a\alpha}^b(h_b - (\zeta_\beta.w_b^\beta)) \end{array} \right\},$$

for some $L_{a\alpha}^b$ not all zero and $\zeta_\alpha = \partial x_\alpha + y_\alpha^j \partial y_j +]_{\alpha\beta}^j \partial y_j^\beta$.

Example 3.3. The Euler-Lagrange equation of the variational calculus, for Lagrangians of first order, is a PDE of the type given in Theorem 3.20. ∎

Theorem 3.21. A system of balance equations with generating 1-form $w \equiv w_\alpha dx^\alpha + h_j dy^j + w_j^\alpha dy_\alpha^j$ is the system of Euler-Lagrange equations of variational calculus, for Lagrangians of first order, iff the fundamental $(n+1)$-form $\Omega(w) \equiv w \wedge \eta + dJ(w)$, with $J(w) \equiv \omega^j \wedge w_j^\beta \eta_\beta = \omega^j \wedge (\partial y_j^\beta \rfloor w)\eta_\beta$, $\eta_\beta \equiv \partial x_\beta \rfloor \eta$, belongs to the zero class of $H^{n+1}(J_n^1(W))$.

Proof. We get

$$d\Omega(w) = dw \wedge \eta = (dh_j \wedge dy^j + dw_j^\alpha \wedge dy_\alpha^j)\eta$$
$$= d(dL \wedge \eta) = 0$$

if $h_j = (\partial y_j . L)$, $w_j^\alpha = (\partial y_j^\alpha . L)$ for some $L \in \Omega^0(J_n^1(W))$. Therefore, $[\Omega(\omega)] \in 0 \in H^{n+1}(J_n^1(W))$. On the other hand if $[\Omega(w)] \in 0 \in H^{n+1}(J_n^1(W))$, we have $\Omega(w) = d\rho$ for some $\rho \in \Omega^n(J_n^1(W))$, and hence we have $w \wedge \eta = d(\rho - J(w))$. Since $w \wedge \eta$ is a simple $(n+1)$-form, $d(\rho - J(w))$ must be a simple $(n+1)$-form that is divisible by the closed, simple n-form η. This can be the case iff there exists an element $L \in \Omega^0(J_n^1(W))$ such that $d(\rho - J(w)) = d(L\eta)$. We thus have

$$w \wedge \eta = \{h_j dy^j + w_j^\alpha dy_\alpha^j\} \wedge \eta = d(L\eta) = [(\partial y_j . L)dy^j + (\partial y_j^\alpha . L)dy_\alpha^j] \wedge \eta$$

and the result is established. □

Remark 3.5. In order to characterize the global properties of solutions of PDE's we shall characterize the integral bordism groups of PDE's. Here we shall report on our some results given in refs.[62–65]. ∎

Let $V \subset J_n^k(W)$ be an integral manifold, $\dim V = p$, $p \leq n - 1$. We denote by $\Sigma(V) \subset V$ the set of all singular points of the mapping $\pi_{k,0} : V \to W$, and $\Sigma_l(V) \equiv \{q \in V \mid \dim(\ker(\pi_{k,0}|_V)_{*,q}) = l\} \subset \Sigma(V)$. Set $L_{0,q} \equiv \ker(\pi_{k,0}|_V)_{*,q} \subset T_qV$. Then, one has $L_{0,q} \subset S^k(\Xi_q) \otimes \nu_a$ for some subspace $\Xi_q \subset L_q^*$. We will consider only the minimal subspace Ξ_q with the above property.

Definition 3.11. An integral manifold $V \subset J_n^k(W)$ will be called *admissible* if the following properties hold:

(i) $\Sigma(V) \subset V$ has no open subsets and has no *frozen singularities*, i.e., for any $q \in \Sigma(V)$, $L_{0,q}$ is a degenerate subspace in $S^k(\Xi_q) \otimes \nu_a$, with respect to the exterior 2-form $\Omega(q)(\lambda) \in C^\infty(\Lambda^2(\mathbb{E}_n^k(W)_q^*))$. Recall that

$$ker\,\Omega(q)(\lambda) = \left\{\theta \in S^k(T_a^*N) \otimes \nu_a \,|\, \delta\theta \in T_a^*N \otimes g_\lambda\right\} \tilde{=} g_\lambda^{(1)}$$

where g_λ is a symbol space of tensor $\lambda \in S^{k-1}(T_aN) \otimes \nu_a^*$: $g_\lambda \equiv \{\gamma \in S^{k-1}(T_a^*N) \otimes \nu_a \,|\, <\lambda, \gamma> = 0\}$.

(ii) There is a vector bundle $e : H \to V$ such that the fibers $H_q \supset \Xi_q$ for any point $q \in V$, satisfy the following conditions:

$$\Xi_q \subset H_q \subset L_q^*, \quad \dim H_q \leq l.$$

(iii) The family $L_0 : q \mapsto L_{0,q}$ over $\Sigma(V)$ can be prolonged to some subbundle $L_0 \subset H \otimes \nu$.

(iv) If V is a compact closed integral manifold of dimension p, $0 \leq p \leq n-1$, for V passes at least one integral manifold of dimension $(p+1)$ that satisfies above conditions (i), (ii), (iii). More precisely, we shall assume that an integral manifold V of dimension $p \leq n - 1$, is *admissible* if its set of singular points $\Sigma(V)$ can be solved by means of integral deformations. Furthermore, we say that V is *integral admissible* with respect to a PDE E_k, if V is admissible in the above sense, is contained into E_k, and the $(p+1)$−dimensional integral manifold mentioned in the above point (iv) is also contained into E_k.

Definition 3.12. 1) Let $E_k \subset J_n^k(W), k \geq 0$, be a k-order PDE, $\dim W = n + m$. If $f_i : P_i \to E_k$, $i = 1, 2$, are C^∞ mappings that represent p-dimensional integral admissible manifolds $N_i \subset E_k$ respectively, then $N_1 \sim_{E_k} N_2 \Leftrightarrow \exists$ a C^∞ mapping $f : R \to E_k$ such that R is a $(p + 1)$-dimensional smooth manifold with $\partial R = P_1 \bigcup P_2$, $f|_{P_i} = f_i$, $i = 1, 2$, and the following condition is verified: (i) $f(R) \equiv V \subset E_k$ is a $(p + 1)$-dimensional piecewise admissible integral manifold of E_k. Then we say that N_1 and N_2 are E_k-*bordant*. We call $\Omega_p^{E_k}$, the *integral p-bordism group* of E_k. Note that if $N = \partial V$ is nonorientable, V cannot be simply connected. In fact, a compact nonorientable p-manifold without boundary cannot be embedded in a simply connected $(p + 1)$-manifold [26]. The empy set $\emptyset$ will be regarded as a p-dimensional compact closed admissible integral manifold for all $p \geq 0$. $\sim_{E_k}$ is an equivalence relation. We write $\Omega_p^{E_k}$ for the set of all E_k-bordism classes $[N]_{E_k}$ of compact closed p-dimensional admissible integral manifolds of $E_k, 0 \leq p \leq n - 1$. The operation of taking disjoint union $\bigcup$ defines a sum $+$ on $\Omega_p^{E_k}$ such that $\Omega_p^{E_k}$ becomes an abelian group. The class $[\emptyset]_{E_k}$ defines the zero element. For $k = 0$ we set $\Omega_p^{E_0} = \underline{\Omega}_p(E_0)$, where $E_0 \subset J_n^o(W) \equiv W$, and $\underline{\Omega}_p(E_0)$ is the p-bordism group of E_0 [17, 65, 66]. We shall denote by Ω_p, $p \in \mathbf{N}$, the p-bordism groups of unoriented smooth compact p-dimensional manifolds.

2) For any couple $(J_n^k(W), E_k)$, where $E_k \subset J_n^k(W)$ is a PDE, we will denote by $\Omega_p(E_k)$, $p \in \{0, \ldots, n - 1\}$, the corresponding relative integral p-bordism groups, and we call them the *quantum p-bordism groups* of E_k.

Remark 3.6. In the following we shall consider integral and quantum bordism groups of PDEs E_k only for integral admissible submanifolds of E_k, in the sense of Definition 3.11. Then, standing the definition of integral admissibility, we shall simply assume that such manifolds are regular ones, hence diffeomorphic to C^∞ submanifolds of W. Emphasize also that if a k-order PDE $E_k \subset J_n^k(W)$ is such that $E_k \subset JD^k(W) \subset J_n^k(W)$ for some fiber bundle structure $\pi : W \to M$, where M is a n-dimensional manifold, then regular integral manifolds V of $J_n^k(W)$ are not necessarily regular integral manifolds of $JD^k(W)$. In other words the fact that $\pi_{k,0}(V)$ is diffeomrphic to some submanifold of W does not necessarily imply that $\pi_{k,0}(V)$ is also diffeomorphic to a submanifold of M. Furthermore, theorems just proved in refs.[64,65] will be stated without proofs. Let us note that if $N' \in [N]_{E_k}$ then there exists a piecewise smooth admissible integral manifold $V \subset E_k$ of dimension $p + 1$ such that $\partial V = N \bigcup N'$. However, for any $N' \in [N]_{E_k}$ there exists also at least another $N'' \in [N]_{E_k}$ such that $N' \bigcup N''$ is the boundary of a smooth admissible integral manifold $X \subset E_k$ of dimension $p + 1$: $\partial X = N' \bigcup N''$. In general, for two integral admissible manifolds $N', N'' \in [N]_{E_k}$ there exists a smooth admissible integral manifold $V \subset E_k$ of dimension $p + 1$ such that $\partial V = N' \bigcup N''$, iff there exists a nonsingular vector field ζ on V, transversal to N' and N'' (for example interior normal on N' and exterior normal on N''). Of course, there are also cases, (PDEs), where the integral admissible manifolds, bording admissible integral manifolds, are globally smooth. (See, e.g., the Hamilton-Jacobi equation [65].)∎

Theorem 3.22. (Extension theorem for integral bordism groups)[63–65].
1) *The integral bordism groups $\Omega_p^{E_k}, 0 \leq p < n$, are extensions of the corresponding quantum bordism groups $\Omega(E_k)_p, 0 \leq p < n$. One has the*

following exact commutative diagrams

$$
\begin{array}{ccccccccc}
 & & & & & & 0 & & \\
 & & & & & & \downarrow & & \\
 & & & & & & K_p(E_k) & & \\
 & & & & & & \downarrow & & \\
0 & \to & K_p^{E_k} & \to & \Omega_p^{E_k} & \xrightarrow{\widehat{i_p}} & \Omega_p(E_k) & \to & 0 \\
 & & & & & & \downarrow & & \\
 & & & 0 & \to & \widehat{\Omega}_p(E_k) & \to & \Omega_p \cong \mathbf{Z}_2 \oplus \cdots_q \cdots \oplus \mathbf{Z}_2 \\
 & & & & & & \downarrow & & \\
 & & & & & & 0 & &
\end{array}
$$

for $0 \le p \le n-1$. So we can write

$$\Omega_p^{E_k} \in H^2(\Omega_p(E_k), K_p^{E_k}) \cong Hom_{\mathbf{Z}}(H_2(\Omega_p(E_k), \mathbf{Z}), K_p^{E_k})$$
$$\Omega_p(E_k) \in H^2(\widehat{\Omega}_p(E_k), K_p(E_k)) \cong Hom_{\mathbf{Z}}(H_2(\widehat{\Omega}_p(E_k), \mathbf{Z}), K_p(E_k)).$$

One has:

$$
K_p^{E_k} = \left\{ [N]_{E_k} \in \Omega_p^{E_k} \;\middle|\; \begin{array}{l} N - \partial V, \\[4pt] \text{for some } (p+1) - \text{dimensional admissible} \\[4pt] \text{integral manifold } V \subset J_n^k(W) \end{array} \right\},
$$

$$
K_p(E_k) = \left\{ [N]_{\overline{E_k}} \in \Omega_p(E_k) \;\middle|\; \begin{array}{l} N = \partial V, \text{ for some} \\[4pt] (p+1) - \text{dimensional manifold } V \end{array} \right\},
$$

$$\widehat{\Omega}_p(E_k) \equiv \Omega_p(E_k)/K_p(E_k), \quad \widehat{\Omega}_p^{E_k} \equiv \Omega_p^{E_k}/K_p^{E_k}.$$

We call $\widehat{\Omega}_p^{E_k}$ *reduced p-integral bordism group* of E_k, and $\widehat{\Omega}_p(E_k)$ *reduced p-quantum bordism group* of E_k. Furthermore, if $H^2(\Omega_p(E_k), K_p^{E_k}) = 0$ we say that E_k is *p-integral bording reducible*. If $H^2(\widehat{\Omega}_p(E_k), K_p(E_k)) = 0$ we say that E_k is *p-quantum bording reducible*.

2) The reduced p-integral bordism group of E_k is isomorphic to the corresponding p-quantum bordism group: $\widehat{\Omega}_p^{E_k} \cong \Omega_p(E_k), \quad 0 \le s \le n-1$.

3) Stiefel-Whitney numbers completely characterize $\widehat{\Omega}_p(E_k)$. More precisely, we have:

$$
\widehat{\Omega}_p(E_k) = \left\{ [N] \in \Omega_p \;\middle|\; \begin{array}{l} \exists X \equiv \text{compact closed } p\text{-dimensional admissible} \\[4pt] \text{integral manifold of } E_k : w[X] = w[N] \end{array} \right\}
$$

4) (Structure of the quantum bordism groups). $\Omega_p(E_k)$ is an extension of a finite abelian torsion group of the form $\mathbf{Z}_2 \oplus \cdots_q \cdots \oplus \mathbf{Z}_2$. Recall that $\mathbf{Z}_2 \cong \{0,1\}$ is an additive abelian group with "addition" given by: $0+0 = 0$, $0+1 = 1$, $1+1 = 0$, $-1 = 1$. One has the following exact commutative diagram

$$
\begin{array}{ccccccccc}
 & & & & & & 0 & & \\
 & & & & & & \downarrow & & \\
 & & & & & & K_p(E_k) & & \\
 & & & & & & \downarrow & & \\
0 & \to & K_p^{E_k} & \to & \Omega_p^{E_k} & \stackrel{\widehat{i_p}}{\to} & \Omega_p(E_k) & \to & 0 \\
 & & & & & & \downarrow & & \\
 & & & & & & \mathbf{Z}_2 \oplus \cdots_p \cdots \oplus \mathbf{Z}_2 & & \\
 & & & & & & \downarrow & & \\
 & & & & & & 0 & &
\end{array}
$$

So we can write

$$
\begin{aligned}
\Omega_p(E_k) &\in H^2(\mathbf{Z}_2 \oplus \cdots_p \cdots \oplus \mathbf{Z}_2, K_p(E_k)) \\
&\cong Hom_{\mathbf{Z}}(H_2(\mathbf{Z}_2 \oplus \cdots_q \cdots \oplus \mathbf{Z}_2, \mathbf{Z}); K_p(E_k)) \\
&\cong \bigoplus_{s \in \{1,\dots,r\}} Hom_{\mathbf{Z}}(\mathbf{Z}_2 \otimes \mathbf{Z}_2; K_p(E_k)),
\end{aligned}
$$

where $r = r(q)$. One has $rank(\Omega_p(E_k)) = rank(K_p(E_k))$. We can write $K_p(E_k) = \{[N]_{\overline{E_k}} | w[N] = 0\}$.

5) (Cohomology structure of $\Omega_p^{E_k}$ and $\Omega_p(E_k)$). One has the following isomorphisms:

$$
\begin{aligned}
\Omega_p^{E_k} &\cong H^2(\Omega_p(E_k), f_p)(u_p), \\
\Omega_p(E_k) &\cong H^2(\mathbf{Z}_2 \oplus \cdots \oplus \mathbf{Z}_2, h_p)(v_p),
\end{aligned}
$$

where

$$
\begin{aligned}
u_p &\in H^2(\Omega_p(E_k), H_2(\Omega_p(E_k), \mathbf{Z})), \\
v_p &\in H^2(\mathbf{Z}_2 \oplus \cdots \oplus \mathbf{Z}_2, H_2(\mathbf{Z}_2 \oplus \cdots \oplus \mathbf{Z}_2, \mathbf{Z})),
\end{aligned}
$$

and f_p, and h_p are uniquely determined mappings $f_p : H_2(\Omega_p(E_k), \mathbf{Z}) \to K_p^{E_k}$, $h_p : H_2(\mathbf{Z}_2 \oplus \cdots \oplus \mathbf{Z}_2, \mathbf{Z}) \to K_p(E_k)$.

6) If E_k is p-integrable reducible, (resp. p-quantum reducible), one has the following isomorphism:

$$
\Omega_p^{E_k} \cong K_p^{E_k} \bigoplus \Omega_p(E_k), \quad (\text{resp. } \Omega_p(E_k) \cong K_p(E_k) \bigoplus \widehat{\Omega}_p(E_k))).
$$

7) (Characteristic numbers for classes in integral and quantum bordism groups).[34] *If N_0 and N_1 are two p-dimensional compact closed admissible integral submanifolds of E_k and $N_0 \bigcup N_1 = \partial V$, for some integral submanifold V of E_k, (or $J_n^k(W)$), then one has: $w[N_0] = w[N_1]$. Furthermore, if on V there exists a nonsingular vector field ζ interior normal on N_0 and exterior normal on N_1, (hence V is smooth), then one has also: $\chi(N_0) = \chi(N_1)$, where $\chi(-)$ denotes Euler number. In particular, $\chi(\partial V) = 0$.*

Remark 3.7. Taking into account that $\widehat{\Omega}_p(E_k)$ are subgroups of Ω_p, and the classic results on the bordism groups [10,77,81,83] we can directly calculate the groups $\widehat{\Omega}_p(E_k)$ in some cases, and in other ones we have some important informations on their structure. See Tab.3.1. Of course, when there are more that one possibilities it is necessary to look more to the structure of the particular PDE in order to decide about. ∎

TAB.3.1 - Examples of calculated groups $\widehat{\Omega}_p(E_k)$.

$\widehat{\Omega}_p$	$\widehat{\Omega}_p(E_k)$	p
$\mathbf{Z}_2$	$0, \quad \mathbf{Z}_2$	$0, \quad 2$
0	0	$1, \quad 3$
$\mathbf{Z}_2 \oplus \mathbf{Z}_2$	$0, \quad \mathbf{Z}_2, \quad \mathbf{Z}_2 \oplus \mathbf{Z}_2$	4

There are important classes of differential relations where it is possible to give some further general informations of the structure of integral and quantum bordism groups. For this, let us give the following definition that generalizes previous one given by Gromov [21] for holonomic situations.

Definition 3.13. A PDE $E_k \subset J_n^k(W)$ satisfies the *(p)-homotopy principle*, (or E_k is of *(p)-h-type*), if for any compact smooth $(p+1)$-dimensional manifold Y, $0 < p+1 \leq n$, and continuous mapping $f : Y \to E_k$, such that $\partial V = f(\partial Y) = N \subset E_k$, $V \equiv f(Y) \subset E_k$, such that N is a compact closed admissible integral manifold of dimension p, there exists a $(p+1)$-dimensional admissible integral manifold X of E_k with $\partial X = N$, $X \simeq_N V$ (X homotopic to V with relation to N in the class of $(p+1)$-dimensional singular submanifolds of E_k). Here $(p+1)$-dimensional singular submanifold of

---

[34] In order to characterize more integral and quantum bordism groups, we shall emphasize the fact that on such bordism manifolds there are characteristic vector fields. This allows us to relate the bordism in PDEs to the bordism with vectors as introduced by Reinhart.[65].

E_k means, of course, a pair (Z, h) consisting of a smooth $(p+1)$-dimensional manifold Z and continuous mapping $h : Z \to E_k$.

Theorem 3.23. [63,65] 1) *If E_k is a PDE that satisfies the (p)-h-principle then*

$$\Omega_p^{E_k} \lhd \underline{\Omega}_p(E_k) \cong \bigoplus_{r+s=p} H_r(E_k; \mathbf{Z}_2) \bigotimes_{\mathbf{Z}_2} \Omega_s,$$

where we denote by $\underline{\Omega}_p(X)$ (resp. $\underline{\Omega}_p(X,Y)$) the p-bordism group (resp. relative p-bordism group) of the space X (resp. couple (X,Y), $Y \subset X$). In particular, for $p = 1$ one has: $\Omega_1^{E_k} \lhd H_1(E_k; \mathbf{Z}_2)$.

2) If, in addition, E_k is contractible to a point one has: $\Omega_p^{E_k} \lhd \Omega_p$, $p \geq 0$.[35]

Example 3.4. If $E_k \subset J_n^k(W)$ satisfies the $(n-1)$-h-principle then it satisfies also the e-principle [18]. However, if we consider admissible only regular manifolds, the two above principles coincide. So the examples given in ref.[60] about differential relations of immersions, Lagrangian immersions, and Legendrian immersions fit above theorem, as there we have considered regular manifolds only. ∎

Similar considerations can be made for the quantum bordism. More precisely, we can give the following.

Definition 3.14. A PDE $E_k \subset J_n^k(W)$ satisfies the *quantum (p)-homotopy principle*, (or E_k is of $q(p)$-h-type), if for any compact smooth $(p + 1)$-dimensional manifold Y, $0 < p + 1 \leq n$, and continuous mapping $f : Y \to J_n^k(W)$, such that $\partial V = f(\partial Y) = N \subset E_k$, $V \equiv f(Y) \subset J_n^k(W)$, such that N is a p-dimensional compact closed admissible integral manifold of E_k, there exists a $(p + 1)$-dimensional admissible integral manifold $X \subset J_n^k(W)$, with $\partial X = N, X \simeq_N V$, ($X$ homotopic to V with relation to N in the class of $(p + 1)$-dimensional singular submanifolds of $J_n^k(W)$). Here $(p+1)$-dimensional singular submanifold of $J_n^k(W)$ means, of course, a pair (Z, h) consisting of a smooth $(p+1)$-dimensional manifold Z and continuous mapping $h : Z \to J_n^k(W)$.

Theorem 3.24. [63,65] 1) *If E_k is a PDE that satisfies the $q(p)$-h-principle then one has*

$$\Omega_p(E_k) \cong \mathbf{Z}_2 \bigoplus \cdots_q \cdots \bigoplus \mathbf{Z}_2 \lhd \Omega_p.$$

---

[35] Here and in the following we shall write $A \lhd B$, to denote that the group A is a subgroup of the group B.

2) If E_k satisfies the q(p)-h-principle $\Omega_p^{E_k}$ is an extension of an abelian finite torsion group of the form $\mathbf{Z}_2 \oplus \cdots_q \cdots \oplus \mathbf{Z}_2$, i.e., one has the following short exact sequence:

$$0 \to K_p^{E_k} \to \Omega_p^{E_k} \to \mathbf{Z}_2 \oplus \cdots_q \cdots \oplus \mathbf{Z}_2 \to 0.$$

So we can write

$$\Omega_p^{E_k} \in H^2(\mathbf{Z}_2 \oplus \cdots_q \cdots \oplus \mathbf{Z}_2; K_p^{E_k})$$

$$\cong Hom_{\mathbf{Z}}(H_2(\mathbf{Z}_2 \oplus \cdots_q \cdots \oplus \mathbf{Z}_2; \mathbf{Z}); K_p^{E_k})$$

$$\cong \bigoplus_{s \in \{1,\ldots,r\}} Hom_{\mathbf{Z}}(\mathbf{Z}_2 \otimes \mathbf{Z}_2; K_p^{E_k}).$$

3) If $E_k \subset J_n^k(W)$ is a PDE of q(p)-h-type, then it is also of q(r)-h-type, for $0 \le r \le p$, if any compact closed admissible integral manifold of dimension s of E_k is contained into a compact closed admissible integral manifold of dimension $s + 1$, $0 \le s \le p - 1$.

4) If W is p-connected, $p \in \{0, 1, \ldots, n-1\}$, $\dim W = n + m$, one has the following short exact sequence: $0 \to \Omega_p^{J_n^k(W)} \xrightarrow{j} \Omega_p$, $k \ge 1$, $0 \le p < n$.

5) If W is p-connected, $p \in \{0, 1, \ldots, n-1\}$, $\dim W = n + m$, and any compact closed smooth p-dimensional manifold can be embedded into W up to bordisms, (for example if $\dim W \ge 2p+1$), then one has the following short exact sequence

$$0 \to \Omega_p^{J_n^k(W)} \xrightarrow{j} \Omega_p \to 0, \quad k \ge 1.$$

6) Let W be a manifold of dimension $n + m$ such that the following conditions are satisfied:

(i) $\dim W \ge 2p + 1$.

(ii) The q-Hurewicz homotopy groups $\pi_q(W)$ are trivial for $0 \le q \le p - 1$, i.e., W is $(p-1)$-connected.

Then one has the following exact commutative diagram:

$$
\begin{array}{ccccccccc}
 & & & & 0 & & & & \\
 & & & & \downarrow & & & & \\
0 & \to & K_p^{J_n^k(W)} & \to & \Omega_p^{J_n^k(W)} & \to & \Omega_p & \to & 0 \\
 & & & & \downarrow & & & & \\
 & & & & \pi_p(W) & & & & \\
 & & & & \downarrow & & & & \\
 & & & & 0 & & & &
\end{array}
$$

Furthermore, if W is p-connected one has the following isomorphism:

$$\Omega_p^{J_n^k(W)} \cong \Omega_p.$$

7) *Let $E_k \subset J_n^k(W)$, $k \geq 1$, be a PDE. Let us assume that W is p-connected, $0 \leq p \leq n-1$. Then $\Omega_p(E_k)$ is a finite group of the form*

$$\Omega_p(E_k) \cong \mathbf{Z}_2 \bigoplus \cdots_q \cdots \bigoplus \mathbf{Z}_2 \triangleleft \Omega_p, \quad 0 \leq p < n, \quad (\text{i.e., } K_p(E_k) = 0).$$

8) *Let $E_k \subset J_n^k(W)$, $k \geq 1$, be a PDE. Let us assume that W is p-connected, $0 \leq p \leq n-1$. Then $\Omega_p^{E_k}$ is an abelian group such that one has a short exact sequence*

$$0 \to K_p^{E_k} \to \Omega_p^{E_k} \to \mathbf{Z}_2 \bigoplus \cdots_q \cdots \bigoplus \mathbf{Z}_2 \to 0,$$

with $K_p^{E_k} \equiv \{ [N]_{E_k} , N = \partial V : V = (p+1)-\text{dimensional admissible integral manifold of } J_n^k(W) \}$. *So we can write*

$$\Omega_p^{E_k} \in H^2(\mathbf{Z}_2 \bigoplus \cdots_q \cdots \bigoplus \mathbf{Z}_2, K_p^{E_k})$$
$$\cong Hom_{\mathbf{Z}}(H^2(\mathbf{Z}_2 \bigoplus \cdots_q \cdots \bigoplus \mathbf{Z}_2, \mathbf{Z}), K_p^{E_k}).$$

Definition 3.15. If $\widehat{\Omega}_p(E_k) = 0$ we say that E_k is *reduced-q(p)-cobording free*. Furthermore, if $\Omega_p(E_k) = 0$ we say that E_k is *q(p)-cobording free*.

Proposition 3.1. 1) *If E_k is reduced-q(p)-cobording free we have: $\Omega_p(E_k) \cong K_p(E_k)$. This is equivalent to say that all the classes $[N]_{\overline{E_k}} \in \Omega_p(E_k)$, have zero Stiefel-Whitney and Euler numbers. If E_k is also of q(p)-h-type, we have $\Omega_p(E_k) = 0$. So in this case E_k results of q(p)-cobording free also. Hence, $\Omega_p^{E_k} = K_p^{E_k}$. So we have the following implication: reduced q(p)-cobording free+q(p)-h-type$\Rightarrow$ q(p)-cobording free.*

2) If E_k is q(p)-cobording free we have: $\Omega_p^{E_k} \cong K_p^{E_k}$. This is equivalent to say that all the classes $[N]_{E_k} \in \Omega_p^{E_k}$, have zero Stiefel-Whitney and Euler numbers.

We shall relate, now, the formal integrability properties of PDEs to their integral bordism groups. More precisely our main theorem is the following.

Theorem 3.25. *Let us assume that the piecewise admissible integral manifolds, bording integral admissible manifolds, are smooth regular integral manifolds for a k-order PDE $E_k \subset J_n^k(M)$. (This condition is surely verified if we consider the infinity prolongation E_∞ of E_k.) Then we have the following.*

1) *If $E_k \subset J_n^k(W)$ is a formally integrable PDE, then one has the following isomorphisms: $\Omega_p^{E_k} \cong \Omega_p(E_k)$, $\Omega_p^{E_\infty} \cong \Omega_p^{E_{k+s}} \cong \Omega_p^{E_k}$, $\Omega_p(E_\infty) \cong \Omega_p(E_{k+s}) \cong \Omega_p(E_k)$, for $p \in \{0, \dots, n-1\}$ and $s \geq 0$.*

2) *If W is p-connected, $p \in \{0, \dots, n-1\}$, and the PDE $E_k \subset J_n^k(W)$ is formally integrable, then E_k satisfies the quantum p-homotopy principle, i.e., one has the following monomorphism: $\Omega_p^{E_k} \triangleleft \Omega_p$, $p \in \{0, \dots, n-1\}$. Furthermore, if any compact closed smooth p-dimensional manifold can be embedded into W up to bordisms, then E_k satisfies the p-homotopy principle and one has the following isomorphism: $\Omega_p^{E_k} \cong \Omega_p$, $p \in \{0, \dots, n-1\}$.*

Proof. See ref.[65]. $\qquad\qquad\qquad\qquad\qquad\qquad\qquad\qquad\qquad\qquad\quad$ $\square$

In the following, when we will refer to Theorem 3.25, we will understand that the "admissible integral manifolds" considered are the regular smooth integral manifolds of $J_n^k(W)$ or $E_k \subset J_n^k(W)$.

Example 3.5. Here we will consider some examples of PDE's that are formally integrable, as well as other that are not so, but for all we will apply our previous methods to calculate their integral bordism groups. The results are resumed in Tab.3.2 and Tab.3.3. More precisely we will consider the following equations:

(1)(*D'Alembert equation*).

$$(d'A) \subset J\mathcal{D}^2(\mathbf{R}^2, \mathbf{R}) : \{F \equiv uu_{xy} - u_x u_y = 0\}.$$

It is a 2-acyclic formally integrable equation [65] for which we can apply Theorem 3.25.

TAB.3.2 - Quantum and integral bordism groups of $(d'A)$ [†].

p	$\Omega_p(d'A)$	$\Omega_p^{(d'A)}$	Ω_p
0	$\mathbf{Z}_2$	$\mathbf{Z}_2$	$\mathbf{Z}_2$
1	0	0	0

(†) $(d'A) \subset J\mathcal{D}^2(\mathbf{R}^2, \mathbf{R}) \subset J_2^2(W)$.

2)(*Euler equation*). The Euler equation for isothermic incompressible fluids can be described as a subbundle $(E) \subset J\mathcal{D}(W)$, where $\pi : W \to M$ is the trivial affine fiber bundle $W \equiv M \times \mathbf{I} \times \mathbf{R} \to M$, over the affine 4-dimensional Galilean space-time M, with $\mathbf{I} \equiv \{v \in \mathbf{M}| <\underline{\tau}, v> = 1\}$, where

M is the space of free vectors of M and $\underline{\tau} \in \mathbf{M}^*$ is the 1-form induced by differentiation of the affine time-function $\tau : M \to \mathbf{R}$. A section $s \equiv (v, p)$ of π represents the vector field velocity v of the flow and the hydrostatic pressure p. We can write (E) in the following form

$$(E) \subset J\mathcal{D}(W) : \left\{ \begin{array}{l} G^j_{jk}\dot{x}^k + \dot{x}^i_i = 0 \\ \rho(G^j_{ik}\dot{x}^i\dot{x}^k + \dot{x}^\alpha\dot{x}^j_\alpha) + y_i g^{ij} - \rho B^j = 0 \end{array} \right\},$$

where ρ is a constant (mass density), $g : M \to vS^0_2 M$ is the vertical (with respect to the projection $\tau : M \to \mathbf{R}$) metric field, and B is the density of the body force. Furthermore, $\{x^\alpha, \dot{x}^i, y, \dot{x}^i_\alpha, y_\alpha\}$, $0 \le \alpha \le 3$, $1 \le i \le 3$, are frame-adapted coordinates on $J\mathcal{D}(W)$, and G^i_{jk} are functions of x^p, $1 \le p \le 3$, that represent the spatial components of the connection coefficients of the canonical connection on the Galilean space-time M. For conservative forces, we can see that (E) is an involutive formally integrable non-linear PDE of first order on the affine bundle $\pi : W \to M$. Also for this equation we can apply Theorem 3.25.

3) Similar results can be obtained for Klein Gordon, Maxwell, Dirac, Einstein, Yang-Mills equations over a p-connected, $0 \le p \le 3$, 4-dimensional space-time. In fact these are examples of formally integrable PDEs where the fiber bundle W is also p-connected. Furthermore, we can embedd into W all the possible compact closed smooth manifolds of dimension $p \in \{0, \ldots, 3\}$ up to bordisms. Therefore, by means of Theorem 3.25 we can determine the groups $\Omega^{E_k}_p$ and $\Omega^{E_k}_p$.

4)(*Navier-Stokes equation*). We can describe the non-isothermal Navier-Stokes equation (NS) on the Galilean space-time M too. Then, (NS) results a 74-dimensional submanifold of the second jet-derivative space $J\mathcal{D}^2(W)$, where $W \equiv J\mathcal{D}(M) \times_M T^0_0 M \times_M T^0_0 M$, with $J\mathcal{D}(M) \equiv$ first order jet-derivative space for motions, i.e., first derivative of sections $m : T \to M$ of the Galilean affine fiber bundle $\tau : M \to T$, where T represents the time axis. In the following we shall identify T with $\mathbf{R}$, i.e., we shall assume that an origin and a unity for the times are chosen. (For informations on the geometric structure of M see refs.[56,63].) $T^0_0 M \equiv M \times \mathbf{R}$. A section $s : M \to W$, of $\pi : W \to M$ is a triplet $s = (v, p, \theta)$, where $v \equiv$ velocity vector field, $p \equiv$ isotropic pressure field, $\theta \equiv$ temperature field. More precisely, with respect to an inertial frame ψ, assuming the body force conservative, with potential f, the thermal conductivity ν, the viscosity χ and the specific heat constants, we can write (NS) in the following form:

$$(NS)\begin{cases} \text{(a)} \left\{ F^0 \equiv \dot{x}^k G^j_{jk} + \dot{x}^i_s \delta^s_i = 0 \right\} \\[2ex] \text{(b)} \begin{cases} F^j \equiv \dot{x}^s R^j_s + \dot{x}^s \dot{x}^i \rho G^j_{is} + \dot{x}^s \dot{x}^j_s \rho + \rho \dot{x}^j_0 \\[1ex] \qquad + \dot{x}^k_s S^{js}_k + \dot{x}^k_{is} T^{jis}_k + p_i g^{ij} + \rho (\partial x_i . f) g^{ij} = 0 \end{cases}_{1 \leq j \leq 3} \\[2ex] \text{(c)} \begin{cases} F^4 \equiv \theta_0 \rho C_p + \dot{x}^k \theta_k \theta_{is} \bar{E}^{is} \rho C_p + \dot{x}^k \dot{x}^p W_{kp} \\[1ex] \qquad + \dot{x}^k \dot{x}^s_p \overline{W}^p_{ks} + \dot{x}^k_i \dot{x}^s_p Y^{ip}_{ks} = 0 \end{cases} \end{cases},$$

where we have the following coordinates:

$$M : (x^\alpha), \quad 0 \leq \alpha \leq 3$$
$$W : (x^\alpha, \dot{x}^i, p, \theta), \quad 1 \leq i \leq 3$$
$$J\mathcal{D}^2(W) : (x^\alpha, \dot{x}^i, p, \theta, \dot{x}^i_\beta, p_\beta, \theta_\beta, \dot{x}^i_{\beta\alpha}, p_{\beta\alpha}, \theta_{\beta\alpha}), \quad 0 \leq \beta \leq 3$$

and

$$\begin{cases} R^i_s \equiv -2\chi[G^i_{ik}(\partial x_s . g^{kj}) + G^j_{ki}(\partial x_s . g^{ki}) + (\partial x_i . \partial x_s . g^{ij})] \\[1ex] S^{js}_k \equiv -2\chi[-G^i_{ik} g^{js} - G^i_{ip} g^{ps} \delta^j_k - 2G^j_{ik} g^{is} + \delta^j_k (\partial x_p . g^{ps})] \\[1ex] T^{jis}_k \equiv 2\chi(g^{sj} \delta^i_k + g^{is} \delta^j_k) \\[1ex] \bar{E}^{is} \equiv -\nu g^{is} \\[1ex] W_{kp} \equiv -2\chi G^j_{ik} g_{js} (\partial x_p . g^{si}) \\[1ex] \overline{W}^p_{ks} \equiv 2\chi[G^j_{ik} g_{js} g^{ip} + G^p_{sk} + (\partial x_k . g^{pe}) g_{se}] \\[1ex] Y^{ip}_{ks} \equiv 2\chi[g_{ks} g^{ip} + \delta^p_k \delta^i_s]. \end{cases}_{1 \leq i,j,k,p,s \leq 3}$$

Here G^i_{ip} are the spatial components of the connection coefficients of the canonical connection of the Galilean spce-time. One can prove that (NS) is an involutive PDE, but not formally integrable [56,65].) Moreover, we can associate to the Navier-Stokes equation (NS) an involutive formally integrable second order PDE $(\widehat{NS}) \subset (NS)$. The local expression of $(\widehat{NS})$

is the following:

(3.10)

$$(\widehat{NS}) \left\{ \begin{array}{l} \text{(a)} \left\{ F^0 \equiv \dot{x}^k G^j_{jk} + \dot{x}^i_s \delta^s_i = 0 \right\} \\[2mm] \text{(b)} \left\{ F^0_\alpha \equiv \dot{x}^k (\partial x_\alpha . G^j_{jk}) + \dot{x}^k_\alpha G^j_{jk} + \dot{x}^i_{s\alpha} \delta^s_i = 0 \right\}_{0 \le \alpha \le 3} \\[2mm] \text{(c)} \left\{ \begin{array}{l} F^j \equiv \dot{x}^s R^j_s + \dot{x}^s \dot{x}^i \rho G^j_{is} + \dot{x}^s \dot{x}^j_s \rho + \rho \dot{x}^j_0 \\[2mm] \quad + \dot{x}^k_s S^{js}_k + \dot{x}^k_{is} T^{jis}_k + p_i g^{ij} + \rho(\partial x_i . f) g^{ij} = 0 \end{array} \right\}_{1 \le j \le 3} \\[2mm] \text{(d)} \left\{ \begin{array}{l} F^4 \equiv \theta_0 \rho C_p + \dot{x}^k \theta_k \theta_{is} \bar{E}^{is} \rho C_p + \dot{x}^k \dot{x}^p W_{kp} \\[2mm] \quad + \dot{x}^k \dot{x}^s_p \overline{W}^p_{ks} + \dot{x}^k_i \dot{x}^s_p Y^{ip}_{ks} = 0 \end{array} \right\} \end{array} \right\}.$$

In fact, (NS) is an universally formally integrable PDE, with $(NS)_{+\infty} = (\widehat{NS})_{+\infty}$. Therefore, we can apply Theorem 3.25 to obtain the integral and quantum bordism groups of (NS). However, in ref.[65] we apply our general methods directly to (NS), in order to calculate its integral and quantum bordism groups, without using Theorem 3.25.

TAB.3.3 - Quantum and integral bordism groups of PDE's E_k [†] of Examples 3.5(2,3,4).

p	$\Omega_p(E_k)$	$\Omega_p^{E_k}$	Ω_p
0	$\mathbf{Z}_2$	$\mathbf{Z}_2$	$\mathbf{Z}_2$
1	0	0	0
2	$\mathbf{Z}_2$	$\mathbf{Z}_2$	$\mathbf{Z}_2$
3	0	0	0

(†) $E_k \subset JD^k(W) \subset J^k_4(W)$. $k=1$ for Euler and Dirac eqs. and $k=2$ for the other eqs.

For all these equations we see that the integral and quantum bordism groups coincide with the corresponding usual bordism groups for smooth manifolds. Furthermore, from such table of integral bordism groups we can state that in set of p-dimensional, $2 \le p \le 4$, integral manifolds of E_k one has ones that change their sectional topology. More precisely one has the following. Let V be a p-dimensional compact integral manifold of E_k such that $\partial V = N_0 \bigcup N_1$, where N_0 and N_1 are admissible closed compact smooth integral manifolds of E_k. Then, there exists an admissible Morse function [26] $f : V \to [a,b]$ such that $f^{-1}(a) = N_0$, $f^{-1}(b) = N_1$, and such that: (A) (*Simple tunnel effect*). If f has a critical point q of index k, then there exists a k-cell $e^k \subset V - N_1$ and a $(p-k)$–cell $e_*^{p-k} \subset V - N_0$ such that: (i) $e^k \cap N_0 = \partial e^k$; (ii) $e_*^{p-k} \cap N_1 = \partial e_*^{p-k}$; (iii) there is a deformation retraction of V onto $N_0 \cup e^k$; (iv) there is a deformation retraction of V onto $N_1 \cup e_*^{p-k}$; (v) $e_*^{p-k} \cap e^k = q$; $e_*^{p-k} \pitchfork e^k$. (Here $\pitchfork$ denotes transverse). (B) (*Multi tunnel effect*). If f is of type $(\nu_0, \ldots, \nu_n)$, where ν_k denotes the number of critical points with index k, such that f has just one critical value c, $a < c < b$, then there are disjoint k-cells $e_i^k \subset V - N_1$, and disjoint $(p-k)$-cells $(e_*)_i^{p-k} \subset V - N_0$, $1 \le i \le \nu_k$, $k = 0, \ldots, n$, such that: (i) $e_i^k \cap N_0 = \partial e_i^k$; (ii) $(e_*)_i^{p-k} \cap N_1 = \partial (e_*)_i^{p-k}$; (iii) there is a deformation retraction of V onto $N_0 \cup \{ \underset{i,k}{\cup} e_i^k \}$; (iv) there is a deformation retraction of V onto $N_1 \cup \{ \underset{i,k}{\cup} (e_*)_i^{p-k} \}$; (v) $(e_*)_i^{p-k} \cap e_i^k = q_i \in V$; $(e_*)_i^{p-k} \pitchfork e_i^k$. (C) (*No topology transition*). If f has no critical point then $V \cong N_0 \times I$, where $I \equiv [0,1]$. ∎

In the following we will characterize integral and quantum bordism groups by means of differential forms.

Definition 3.16. In Tab.3.4 we define some important spaces associated to a PDE $E_k \subset J_n^k(W)$.

TAB.3.4 - Important spaces associated to PDE E_k.

$\left(Space\ of\ characteristic\ q\text{-}forms,\ q{=}1,2,\cdots\right)$ $Ch\Omega^q(E_k) \equiv \{\beta \in \Omega^q(E_k) \mid \beta(\zeta_1,\cdots,\zeta_q)(p){=}0, \zeta_i(p) \in \mathbb{C}har(E_k)_p,\ \forall p \in E_k\};$ $Ch\Omega^0(E_k){=}0$
$\left(Space\ of\ Cartan\ q\text{-}forms,\ q{=}1,2,\cdots\right)$ $C\Omega^q(E_k) \equiv \{\beta \in \Omega^q(E_k) \mid \beta(\zeta_1,\cdots,\zeta_q)(p){=}0, \zeta_i(p) \in (\mathbb{E}_k),\ \forall p \in E_k\};$ $C\Omega^0(E_k){=}0$
$\left(Space\ of\ p\text{-}characteristic\ q\text{-}forms,\ q{=}1,2,\cdots\right)$ $Ch^p\Omega^q(E_k) \equiv \{\beta \in \Omega^q(E_k) \mid \beta(\zeta_1,\cdots,\zeta_q){=}0$ if at least $q{-}p{+}1$ of the fields $\zeta_1,\cdots,\zeta_q$ are characteristic$\}$
$\left(Space\ of\ p\text{-}Cartan\ q\text{-}forms,\ q{=}1,2,\cdots\right)$ $C^p\Omega^q(E_k) \equiv \{\beta \in \Omega^q(E_k) \mid \beta(\zeta_1,\cdots,\zeta_q){=}0$ if at least $q{-}p{+}1$ of the fields $\zeta_1,\cdots,\zeta_q$ are Cartan$\}$

Remark 3.8. If the fiber dimension of $\mathbb{C}har(E_k)_p$ is s one has: $Ch\Omega^q(E_k) =$

$\Omega^q(E_k)$, $q > s$. If the fiber dimension of $\mathbb{E}_k$ is r one has: $C\Omega^q(E_k) = \Omega^q(E_k)$, $q > r$. If $k = \infty$ one has: $Ch\Omega^q(E_k) = C\Omega^q(E_\infty)$, $Ch^p\Omega^q(E_k) = C^p\Omega^q(E_\infty)$, $C\Omega^q(E_\infty) = \Omega^q(E_\infty) = Ch\Omega^q(E_k)$, $q > n$. $\alpha \in Ch\Omega^q(E_k)$, iff $\alpha|_V = 0$, for all the characteristic integral manifolds of E_k. If $E_\infty \subset J_n^k(W)$, then $\mathbb{C}har(E_\infty) = \mathbb{E}_\infty$, and $Ch\Omega^q(E_\infty) = C\Omega^q(E_\infty)$. Furthermore, even if for any $p \in E_\infty$, one has an infinity number of maximal integral manifolds (of dimension n) passing for p, one has that all these integral manifolds have at p the same tangent space $(\mathbb{E}_\infty)_p$. Hence, a differential q-form on E_∞ is Cartan iff it is zero on all the integral manifolds of E_∞. One has the following natural differential complex:

(3.11)
$$0 \to Ch\Omega^1(E_k) \xrightarrow{d} Ch\Omega^2(E_k) \xrightarrow{d} \cdots \to Ch\Omega^s(E_k) \xrightarrow{d} Ch\Omega^{s+1}(E_k) \xrightarrow{d} \cdots \to Ch\Omega^r(E_k) \xrightarrow{d} 0,$$

where s = fiber dimension of $\mathbb{C}har(E_k)$, and $r = \dim E_k$, with $k \le \infty$. In particular, if $k = \infty$ we can write above complex by fixing $Ch\Omega^q(E_\infty) = C\Omega^q(E_\infty)$. One has: $d : Ch^p\Omega^q(E_k) \to Ch^p\Omega^{q+1}(E_k)$, $k \le \infty$. In particular, for $k = \infty$ we can write $d : C^p\Omega^q(E_\infty) \to C^p\Omega^{q+1}(E_\infty)$. One has the following filtration compatible with the exterior differential:

$$Ch^0\Omega^q(E_k) \equiv \Omega^q(E_k) \supset Ch^1\Omega^q(E_k) \equiv Ch\Omega^q(E_k) \supset Ch^2\Omega^q(E_k) \supset \cdots \supset Ch^q\Omega^q(E_k) \supset 0,$$

for $k \le \infty$. As a consequence we have associated a spectral sequence (*characteristic spectral sequence of E_k*): $\{E_r^{p,q}(E_k), d_r^{p,q}\}$. In particular, if E_∞ is the infinity prolongation of a PDE $E_k \subset J_n^k(W)$ above spectral sequence applied to E_∞ coincides with the $\mathcal{C}$-spectral sequence of E_k [39, 61]. Of particular importance is the following spectral term:

$$E_1^{0,q}(E_k) = \frac{\Omega^q(E_k) \cap d^{-1} Ch\Omega^{q+1}(E_k)}{d\Omega^{q-1}(E_k) \oplus Ch\Omega^q(E_k)}$$

that for $k = \infty$ can be also written

$$E_1^{0,q}(E_k) = \frac{\Omega^q(E_k) \cap d^{-1} C\Omega^{q+1}(E_k)}{d\Omega^{q-1}(E_k) \oplus C\Omega^q(E_k)}.$$

Set: $\bar{\Omega}^q(E_k) = \Omega^q(E_k)/Ch\Omega^q(E_k)$, $\le \infty$. Then, one has the following differential complex associated to E_k (*bar de Rham complex of E_k*):

$$0 \to \bar{\Omega}^0(E_k) \xrightarrow{\bar{d}} \bar{\Omega}^1(E_k) \xrightarrow{\bar{d}} \bar{\Omega}^2(E_k) \xrightarrow{\bar{d}} \cdots \to \bar{\Omega}^q(E_k) \xrightarrow{\bar{d}} \cdots \to \bar{\Omega}^s(E_k) \xrightarrow{\bar{d}} 0.$$

We call *bar de Rham cohomology* of E_k the corresponding homology $\bar{H}^q(E_k)$. One has the following canonical isomorphism: $E_1^{0,q}(E_k) \cong \bar{H}^q(E_k)$, $\quad k \leq \infty$.

Definition 3.17. Set

$$\mathcal{I}(E_k)^p \equiv \frac{\Omega^p(E_k) \cap d^{-1}(C\Omega^{p+1}(E_k))}{d\Omega^{p-1}(E_k) \oplus \{C\Omega^p(E_k) \cap d^{-1}(C\Omega^{p+1}(E_k))\}},$$
$$\mathcal{Q}_n^k(W)^p \equiv \mathcal{I}(J_n^k(W))^p.$$

Remark 3.9. For $k = \infty$ one has: $\mathcal{I}(E_\infty)^p \cong E_1^{0,p}(E_\infty) \cong \bar{H}^p(E_\infty)$.   ∎

Theorem 3.26. [63,65] 1) *Let us assume that $\mathcal{I}(E_k)^p \neq 0$. One has a natural group homomorphism:*

$$j_p : \Omega_p^{E_k} \to (\mathcal{I}(E_k)^p)^*$$

$$[N]_{E_k} \mapsto j_p([N]_{E_k}), \quad j_p([N]_{E_k})([\alpha]) = \int_N \alpha \equiv <[N]_{E_k}, [\alpha]> .$$

We call $i[N] \equiv <[N]_{E_k}, [\alpha]>$ integral characteristic numbers of N for all $[\alpha] \in \mathcal{I}(E_k)^p$. Then a necessary condition that $N' \in [N]_{E_k}$ is the following

$$(3.12) \qquad\qquad i[N'] = i[N], \ \forall [\alpha] \in \mathcal{I}(E_k)^p.$$

Above condition is also sufficient for $k = \infty$ in order to identify elements belonging to the same singular integral bordism classes of $\Omega_{p,s}^{E_\infty}$. In fact, one has the following exact commutative diagram:

$$
\begin{array}{ccc}
& & 0 \\
& & \downarrow \\
\Omega_p^{E_\infty} & \xrightarrow{i_p} & \Omega_{p,s}^{E_\infty} \equiv \bar{H}_p(E_\infty;\mathbf{R}) \\
\downarrow{\scriptstyle j_p} & & \downarrow \\
0 \to \ (\mathcal{I}(E_\infty)^p)^* & \to & \bar{H}^p(E_\infty;\mathbf{R})^* \qquad \to 0
\end{array}
$$

2) For any $k \leq \infty$ one has the following exact commutative diagram:

$$
\begin{array}{ccccc}
& & & 0 & \\
& & & \downarrow & \\
0 \to \overline{K_p^{E_k}} & \to & \Omega_p^{E_k} & \to & \overline{\Omega_p^{E_k}} & \to 0 \\
& & & {\scriptstyle j_p}\searrow & \downarrow & \\
& & & & (\mathcal{I}(E_k)^p)^* &
\end{array}
$$

Therefore, we can write

$$\overline{K_p^{E_k}} \equiv \{[N]_{E_k}| < [\alpha], [N]_{E_k} >= 0, \forall [\alpha] \in \mathcal{I}(E_k)^p\}$$

$$N' \in \overline{[N]_{E_k}} \in \overline{\Omega_p^{E_k}} \Leftrightarrow \int_{N'} \alpha = \int_N \alpha \,, \forall [\alpha] \in \mathcal{I}(E_k)^p.$$

3) *Let us assume that* $\mathcal{Q}_n^k(W)^p \neq 0$. *One has a natural group homomorphism*

$$\bar{j}_p : \Omega_p(E_k) \to (\mathcal{Q}_n^k(W)^p)^*$$

$$[N]_{\overline{E_k}} \mapsto \bar{j}_p([N]_{\overline{E_k}})$$

$$\bar{j}_p([N]_{\overline{E_k}})([\alpha]) = \int_N \alpha \equiv < [N]_{\overline{E_k}}, [\alpha] > .$$

We call $q[N] \equiv < [N]_{\overline{E_k}}, [\alpha] >$ *quantum characteristic numbers of* N, *for all* $[\alpha] \in \mathcal{Q}_n^k(W)^p$. *Then, a necessary condition that* $N' \in [N]_{\overline{E_k}}$ *is that*

$$(3.13) \qquad\qquad q[N'] = q[N] \,, \forall [\alpha] \in \mathcal{Q}_n^k(W)^p.$$

4) *(Criterion in order condition (3.12) should be sufficient). Let us assume that* $E_k \subset J_n^k(W)$ *is such that all its p-dimensional compact closed admissible integral submanifolds are orientable and* $\mathcal{I}(E_k)^p \neq 0.$[36] *Then,* $\ker(j_p) = 0$, *i.e.,*

$$N' \in [N]_{E_k} \Leftrightarrow \int_{N'} \alpha = \int_N \alpha \,, \forall [\alpha] \in \mathcal{I}(E_k)^p.$$

In particular, for $k = \infty$, *one has* $\Omega_p^{E_\infty} \cong \Omega_{p,s}^{E_\infty}$ *as* $\mathcal{I}(E_\infty)^p \cong \bar{H}^p(E_\infty)$.
5) *Under the same hypotheses of above theorem one has*

$$N' \in [N]_{\overline{E_k}} \Leftrightarrow \int_{N'} \alpha = \int_N \alpha, \quad \forall [\alpha] \in \mathcal{Q}_n^k(W)^p.$$

Proof. See refs.[63,65].   □

Remark 3.10. In above criterion $\Omega_p^{E_k}$ (resp. $\Omega_p(E_k)$) does not necessarily coincides with the oriented version of the integral (resp. quantum) bordism groups. In fact, the Möbius band is an example of non orientable manifold B with $\partial B \cong S^1$, that, instead, is an orientable manifold.   ■

---

[36] It is important to note that can be $\mathcal{I}(E_k)^p \neq 0$ even if E_k is p-cohomologically trivial, i.e., $H^p(E_k; \mathbf{R}) = 0$. This, for example, can happen if E_k is contractible to a point.

Remark 3.11. The oriented version of integral and quantum bordism can be similarly obtained by substituting the groups Ω_p with the corresponding groups $^+\Omega_p$ for oriented manifolds. We will not go in to details. ∎

Let us give, now, a full characterization of singular integral and quantum (co)bordism groups by means of suitable characteristic numbers.

Definition 3.18. 1) Let $E_k \subset J_n^k(W)$ be a PDE. We call *bar singular chain complex, with coefficients into an abelian group G,* of E_k the chain complex $\{\bar{C}_p(E_k;G),\bar{\partial}\}$, where $\bar{C}_p(E_k;G)$ is the G-module of formal linear combinations, with coefficients in G, $\sum \lambda_i c_i$, where c_i is a singular p-chain $f : \triangle^p \to E_k$ that extends on a neighborhood $U \subset \mathbf{R}^{p+1}$, such that f on U is differentiable and $Tf(\triangle^p) \subset \mathbb{E}_k$. Denote by $\bar{H}_p(E_k;G)$ the corresponding homology (*bar singular homology with coefficients in G*) of E_k. Let $\{\bar{C}^p(E_k;G) \equiv Hom_{\mathbf{Z}}(\bar{C}_p(E_k;\mathbf{Z});G),\bar{\delta}\}$ be the corresponding dual complex and $\bar{H}^p(E_k;G)$ the associated homology spaces (*bar singular cohomology, with coefficients into G of E_k*).

2) A *G-singular p-dimensional integral manifold* of $E_k \subset J_n^k(W)$, is a bar singular p-chain V with $p \le n$, and coefficients into an abelian group G, such that $V \subset E_k$.

3) Set $\bar{B}_\bullet(E_k;G) \equiv \mathrm{im}\,(\bar{\partial})$, $\bar{Z}_\bullet(E_k;G) \equiv \ker(\bar{\partial})$. Therefore, one has the following exact commutative diagram:

$$
\begin{array}{ccccccccc}
& & 0 & & 0 & & & & \\
& & \downarrow & & \downarrow & & & & \\
0 & \to & \bar{B}_\bullet(E_k;G) & \to & \bar{Z}_\bullet(E_k;G) & \to & \bar{H}_\bullet(E_k;G) & \to & 0 \\
& & \downarrow & & \downarrow & & & & \\
& & \bar{C}_\bullet(E_k;G) & = & \bar{C}_\bullet(E_k;G) & & & & \\
& & \downarrow & & \downarrow & & & & \\
0 & \to & {}^G\Omega_{\bullet,s}^{E_k} & \to & \bar{Bor}_\bullet(E_k;G) & \to & \bar{Cyc}_\bullet(E_k;G) & \to & 0 \\
& & & & \downarrow & & \downarrow & & \\
& & & & 0 & & 0 & &
\end{array}
$$

where $\bar{Bor}_\bullet(E_k;G) \equiv$ *bordism group*; $b \in {}^G[a]_{E_k} \in \bar{Bor}_\bullet(E_k;G) \Rightarrow \exists c \in \bar{C}_\bullet(E_k;G) : \bar{\partial}c = a - b$; $\bar{Cyc}_\bullet(E_k;G) \equiv$ *cyclism group*; $b \in {}^G[a]_{E_k} \in \bar{Cyc}_\bullet(E_k;G) \Rightarrow \bar{\partial}(a - b) = 0$; ${}^G\Omega_{\bullet,s}^{E_k} \equiv$ *closed bordism group*; $b \in {}^G[a]_{E_k} \in {}^G\Omega_{\bullet,s}^{E_k} \Rightarrow \left\{ \begin{array}{l} \bar{\partial}a = \bar{\partial}b = 0 \\ a - b = \bar{\partial}c \end{array} \right\}.$

Theorem 3.27. 1) *One has the following canonical isomorphism:* ${}^G\Omega_{\bullet,s}^{E_k} \cong \bar{H}_\bullet(E_k;G).$

2) If $^G\Omega^{E_k}_{\bullet,s} = 0$ one has: $\bar{Bor}_\bullet(E_k;G) \cong \bar{Cyc}_\bullet(E_k;G)$.

3) If $\bar{Cyc}_\bullet(E_k;G)$ is a free G-module, then the bottom horizontal exact sequence, in above diagram, splits and one has the isomorphism:

$$\bar{Bor}_\bullet(E_k;G) \cong {}^G\Omega(E_k)_{\bullet,s} \bigoplus \bar{Cyc}_\bullet(E_k;G).$$

Remark 3.12. In the following we shall consider only closed bordism groups ${}^G\Omega(E_k)_{s,\bullet}$. So, we will omit the term "closed".

Similar definitions and results can be obtained in dual form by using the cochain complex $\{\bar{C}^\bullet(E_k;G);\bar{\delta}\}$.   ■

Definition 3.19. A *G-singular p-dimensional quantum manifold* of E_k is a bar singular p-chain $V \subset J^k_n(W)$, with $p \le n$, and coefficients into an abelian group G, such that $\partial V \subset E_k$. Let us denote by ${}^G\Omega_{p,s}(E_k)$ the corresponding (closed) bordism groups in the singular case. Let us denote also by ${}^G[N]_{\overline{E_k}}$ the equivalence classes of quantum singular bordisms respectively.

NOTE. In the following, for $G = \mathbf{R}$ we will omit the apex G in the symbols ${}^G\Omega^{E_k}_{p,s}$, ${}^G\Omega^{p,s}_{E_k}$ and ${}^G\Omega_{p,s}(E_k)$.

Theorem 3.28. (Bar de Rham theorem for PDEs). *One has a natural bilinear mapping: $<,>: \bar{C}_p(E_k;\mathbf{R}) \times \bar{C}^p(E_k;\mathbf{R}) \to \mathbf{R}$ such that: (bar Stokes formula) $< \bar{\delta}\alpha, c > +(-1)^p < \alpha, \bar{\partial}c >= 0$. One has the canonical isomorphism: $\bar{H}^p(E_k;\mathbf{R}) \cong Hom_\mathbf{R}(\bar{H}_p(E_k;\mathbf{R});\mathbf{R}) \equiv \bar{H}_p(E_k;\mathbf{R})^*$, and a nondegenerate mapping:*

$$<,>: \bar{H}_p(E_k;\mathbf{R}) \times \bar{H}^p(E_k;\mathbf{R}) \to \mathbf{R}.$$

Hence one has the following short exact sequence $0 \to \bar{H}_p(E_k;\mathbf{R}) \to \bar{H}^p(E_k;\mathbf{R})^*$. This means that if c is a $\bar{\partial}$-closed bar singular p-chain ($\bar{\partial}c = 0$) of E_k, c is the boundary of a bar-singular $(p+1)$-chain c' of E_k ($\bar{\partial}c' = c$), iff $< c, \alpha >= 0$, for all the $\bar{\delta}$-closed bar singular p-cochains α of E_k. Furthermore, if α is a $\bar{\delta}$-closed bar singular p-cochain of E_k, α is $\bar{\delta}$-exact, ($\alpha = \bar{\delta}\beta$) iff $< c, \alpha >= 0$, for all the $\bar{\partial}$-closed bar singular p-chains c of E_k.

Proof. The full proof has been given in refs.[63,65].   □

Remark 3.13. 1) Similarly to the classical case, we can also define the *relative (co)homology spaces* $\bar{H}^p(E_k, X;\mathbf{R})$ and $\bar{H}_p(E_k, X;\mathbf{R})$, where $X \subset E_k$ is a bar singular chain.

2) One has the following exact sequence:

$$\cdots \to \bar{H}_p(X;\mathbf{R}) \to \bar{H}_p(E_k;\mathbf{R}) \to \bar{H}_p(E_k,X;\mathbf{R}) \to \bar{H}_{p-1}(X;\mathbf{R}) \to$$
$$\cdots \bar{H}_0(X;\mathbf{R}) \to \bar{H}_0(E_k;\mathbf{R}) \to \bar{H}_0(E_k,X;\mathbf{R}) \to 0.$$

3) One has the following isomorphisms: $\bar{H}_p(E_k,*;\mathbf{R}) \cong \bar{H}_p(E_k;\mathbf{R})$, with $p > 0$; $\bar{H}_0(E_k,*;\mathbf{R}) = 0$ if E_k is arcwise connected. ∎

Theorem 3.29. *Let us assume that $E_k \subset J_n^k(W)$ is a formally integrable PDE.*

1) As $\pi_\infty : E_\infty \to E_k$ is surjective, one has the following short exact sequence of chain complexes: $\bar{C}_\bullet(E_\infty;\mathbf{R}) \to \bar{C}_\bullet(E_k;\mathbf{R}) \to 0$, $\bar{C}^\bullet(E_\infty;\mathbf{R}) \leftarrow \bar{C}^\bullet(E_k;\mathbf{R}) \leftarrow 0$. These induce the following homomorphisms of vector spaces: $\bar{H}_p(E_\infty;\mathbf{R}) \to \bar{H}_p(E_k;\mathbf{R})$, $\bar{H}^p(E_\infty;\mathbf{R}) \leftarrow \bar{H}^p(E_k;\mathbf{R})$.

2) One has the following isomorphisms: $\Omega_{p,s}^{E_k} \cong \bar{H}_p(E_k;\mathbf{R})$, $\quad k \le \infty$, $\Omega_{p,s}(E_k) \cong \bar{H}_p(J_n^k(W),E_k;\mathbf{R})$.

3) One has the following exact sequences of vector spaces:

$$\Omega_{n-1,s}^{E_k} \xrightarrow{a_{n-1}} \Omega_{n-1,s}^{J_n^k(W)} \xrightarrow{b_{n-1}} \Omega_{n-1,s}(E_k) \xrightarrow{c_{n-1}} \Omega_{n-2,s}^{E_k} \xrightarrow{a_{n-2}}$$
$$\cdots \xrightarrow{a_0} \Omega_{0,s}^{E_k} \xrightarrow{b_0} \Omega_{0,s}^{J_n^k(W)} \xrightarrow{c_0} \Omega_{0,s}(E_k) \to 0.$$

Therefore, one has unnatural splits:

$$\Omega_{p,s}(E_k) \cong \underline{\Omega_{p,s}^{J_n^k(W)}} \times \overline{\Omega_{p-1,s}^{E_k}} ; \quad \Omega_{p,s}^{J_n^k(W)} \cong \underline{\Omega_{p,s}^{E_k}} \times \overline{\Omega_{p,s}(E_k)},$$

where

$$\underline{\Omega_{p,s}^{J_n^k(W)}} \equiv \mathrm{im}\,(b_p) \cong \ker(c_p),$$

$$\overline{\Omega_{p-1,s}^{E_k}} \equiv \mathrm{im}\,(c_p) \cong \mathrm{coim}\,(c_p) \equiv \Omega_{p,s}(E_k)/\ker(c_p) \cong \mathrm{coker}\,(b_p) \equiv \Omega_{p,s}(E_k)/\mathrm{im}\,(b_p),$$

$$\underline{\Omega_{p,s}^{E_k}} \equiv \mathrm{im}\,(a_p) \cong \ker(b_p),$$

$$\overline{\Omega_{p,s}(E_k)} \equiv \mathrm{im}\,(b_p) \cong \mathrm{coim}\,(b_p) \equiv \Omega_{p,s}^{J_n^k(W)}/\ker(b_p) \cong \mathrm{coker}\,(a_p) \equiv \Omega_{p,s}^{J_n^k(W)}/\mathrm{im}\,(a_p).$$

4) One has a natural homomorphism: $\pi_{\infty,k*} : \Omega_{p,s}^{E_\infty} \to \Omega_{p,s}^{E_k}$.

Example 3.6. After the result by Thom [84] it easy to prove the following isomorphism: ${}^G\Omega_{p,s}^{J_n^k(W)} \cong H_p(W;G)$. ∎

Example 3.7. Let $E_k \subset J_1^k(W)$ be a ODE. Then, one has the following isomorphism: $\Omega_{0,s}^{E_k} \cong \mathbf{Z}_2$. ∎

Definition 3.20. 1) We call *singular integral characteristic numbers* of a p-dimensional $\bar{\partial}$-closed singular integral manifold $N \subset E_k \subset J_n^k(W)$ the numbers $i[N] \equiv\, <N, \alpha >\,\in \mathbf{R}$, where α is a $\bar{\delta}$-closed bar singular p-cochain of E_k.

2) We call *singular quantum characteristic numbers of a p-dimensional $\bar{\partial}$-closed singular integral manifold* $N \subset E_k \subset J_n^k(W)$, the numbers $q[N] \equiv\, <N, \alpha >\,\in \mathbf{R}$, where α is a $\bar{\delta}$-closed bar singular p-cochain of $J_n^k(W)$.

Theorem 3.30. 1) $N' \in [N]_{E_k}^s \Leftrightarrow N'$ and N have equal all the singular integral characteristic numbers: $i[N'] = i[N]$.

2) $N' \in [N]_{\overline{E_k}}^s \Leftrightarrow N'$ and N have equal all the singular quantum characteristic numbers: $q[N'] = q[N]$.

Proof. It follows from the bar de Rham theorem that one has the following short exact sequences: $0 \to \Omega_{p,s}^{E_k} \to \bar{H}^p(E_k; \mathbf{R})^*$, $0 \to \Omega_{p,s}(E_k) \to \bar{H}^p(J_n^k(W), E_k; \mathbf{R})^*$. $\qquad\square$

Theorem 3.31. *The relation between singular integral (quantum) bordism groups and homolohy is given by the following exact commutative diagrams:*

$$
\begin{array}{ccccccc}
& & & & 0 & & \\
& & & & \downarrow & & \\
0 \to & K\bar{H}_p(E_k;\mathbf{R}) & \to & \Omega_{p,s}^{E_k} & \to & \widehat{\bar{H}}_p(E_k;\mathbf{R}) & \to 0 \\
& & & & \downarrow & & \\
& & & & H_p(E_k;\mathbf{R}) & &
\end{array}
$$

where

$$
\begin{aligned}
K\bar{H}_p(E_k;\mathbf{R}) &\equiv \{[N]_{E_k}^s \,|\, N = \partial V, V = \text{singular } p\text{-chain in } E_k\} \\
&\equiv \{[N]_{E_k}^s \,|\, <[\alpha]|[N]_{E_k}^s> = 0, \forall [\alpha] \in H^p(E_k;\mathbf{R})\}.
\end{aligned}
$$

We call $s[N] \equiv\, <[\alpha]|[N]_{E_k}^s> \equiv$ *singular characteristic numbers of* $[N]_{E_k}^s$.

$$
\begin{array}{ccccccc}
& & & & 0 & & \\
& & & & \downarrow & & \\
0 \to & K\bar{H}_p(J_n^k(W),E_k;\mathbf{R}) & \to & \Omega_{p,s}(E_k) & \to & \widehat{\bar{H}}_p(J_n^k(W),E_k;\mathbf{R}) & \to 0 \\
& & & & \downarrow & & \\
& & & & H_p(J_n^k(W),E_k;\mathbf{R}) & &
\end{array}
$$

where

$$
\begin{aligned}
& K\bar{H}_p(J_n^k(W), E_k; \mathbf{R}) \\
&\equiv \{[N]_{\overline{E_k}}^s \,|\, N = \partial V, V = \text{ singular } p\text{-chain in } J_n^k(W)\} \\
&\equiv \{[N]_{\overline{E_k}}^s \,|\, <[\alpha]|[N]_{\overline{E_k}}^s> = 0, \forall [\alpha] \in H^p(J_n^k(W), E_k; \mathbf{R})\}.
\end{aligned}
$$

We call *singular characteristic numbers* of $[N]^s_{\overline{E_k}}$ the numbers $s_q[N] \equiv\ <[\alpha]|[N]^s_{\overline{E_k}}>$.

Theorem 3.32. 1) *The integral bordism group* $\Omega_p^{E_k}, 0 \leq p \leq n-1$, *is an extension of a subgroup* $\hat{\Omega}_{p,s}^{E_k}$ *of the singular integral bordism group* $\Omega_{p,s}^{E_k}$.

2) *The quantum bordism group* $\Omega_p(E_k), 0 \leq p \leq n-1$, *is an extension of a subgroup* $\hat{\Omega}_{p,s}(E_k)$ *of the singular quantum bordism group* $\Omega_{p,s}(E_k)$.

Proof. 1) In fact, one has a canonical group-homomorphism $j_p : \Omega_p^{E_k} \to \Omega_{p,s}^{E_k}$, that generates the following exact commutative diagram:

$$
\begin{array}{ccccccc}
 & & & & 0 & & \\
 & & & & \downarrow & & \\
0 \to K_{p,s}^{E_k} \to & \Omega_p^{E_k} & \xrightarrow{\ i_p\ } & \hat{\Omega}_{p,s}^{E_k} & \to 0 & & \\
 & & {}_{j_p}\searrow & \downarrow & & & \\
 & 0 & \to & \Omega_{p,s}^{E_k} & \to & \bar{H}_p(E_k;\mathbf{R}) & \to 0
\end{array}
$$

where $K_{p,s}^{E_k} \equiv \ker(j_p)$ and $\hat{\Omega}_{p,s}^{E_k} \equiv \Omega_p^{E_k}/K_{p,s}^{E_k}$. Furthermore, $K_{p,s}^{E_k}$ can be characterized by means of characteristic numbers. In fact we get

$$K_{p,s}^{E_k}$$
$$= \left\{ [N]^s_{E_k} \,|\, \exists (p+1) - \text{dimensional singular integral submanifold } V \subset E_k, \text{ with } \partial V = N \right\}$$
$$= \left\{ [N]^s_{E_k} \,|\, i[N] = 0 \text{ for all singular integral characteristic numbers} \right\}.$$

2) In fact, one has a canonical group homomorphism $\bar{j}_p : \Omega_p(E_k) \to \Omega_{p,s}(E_k)$, hence one has the following exact commutative diagram:

$$
\begin{array}{ccccccc}
 & & & & 0 & & \\
 & & & & \downarrow & & \\
0 \to K_{p,s}(E_k) \to & \Omega_p(E_k) & \to & \hat{\Omega}_{p,s}(E_k) & \to 0 & & \\
 & & {}_{j_p}\searrow & \downarrow & & & \\
 & 0 & \to & \Omega_{p,s}(E_k) & \to & \bar{H}_p(J_n^k(W),E_k;\mathbf{R}) & \to 0
\end{array}
$$

where

$$K_{p,s}(E_k)$$
$$\equiv \{ [N]^s_{\overline{E_k}} \,|\, q[N] = 0 \quad \text{for all singular quantum characteristic numbers}\}.$$

$\square$

In ref.[65] we have also related integral (co)bordism groups of PDEs to some spectrum in such a way to generalize also to PDEs the Thom-Pontrjagin construction usually adopted for bordism theories. In fact we have the following theorem.

Theorem 3.33. (Integral spectrum of PDEs). *1) Let $E_k \subset J_n^k(W)$ be a PDE. Then there is a spectrum $\{\Xi_s\}$ (singular integral spectrum of PDEs), such that $\Omega_{p,s}^{E_k} = \lim_{r \to \infty} \pi_{p+r}(E_k^+ \wedge \Xi_r)$, $\Omega_{E_k}^{p,s} = \lim_{r \to \infty} [S^r E_k^+, \Xi_{p+r}]$, $p \in \{0, 1, \ldots, n-1\}$.*

2) There exists a spectral sequence $\{E_{p,q}^r\}$, (resp. $\{E_r^{p,q}\}$), with $E_{p,q}^2 = H_p(E_k, E_q(*))$, (resp. $E_2^{p,q} = H^p(E_k, E^q(*)))$, converging to $\Omega_{\bullet,s}^{E_k}$, (resp. $\Omega_{E_k}^{\bullet,s}$). We call the spectral sequences $\{E_{p,q}^r\}$ and $\{E_r^{p,q}\}$ the integral singular spectral sequences of E_k.

Proof. See refs.[65]. $\qquad\qquad\square$

Remark 3.14. Note that the integral bordism groups do not define a homology theory on the category of differential equations. In fact the quantum bordism groups can be different from zero even if the inclusion $E_k \to J_n^k(W)$ is a homotopy equivalence. (See e.g., the case of the Navier-Stokes equation.) $\qquad\blacksquare$

Let us, now, relate integral bordism to the spectral term $E_1^{0,n-1}$ of the $\mathcal{C}$-spectral sequence, that represents the space of conservation laws of PDEs. In fact we represent $E_1^{0,n-1}$ into Hopf algebras that give the true full meaning of conservation laws of PDEs.

Definition 3.21. We define *conservation law* of a PDE $E_k \subset J_n^k(W)$, any differential $(n-1)$-form β belonging to the following quotient space:

$$Cons(E_k) \equiv \frac{\Omega^{n-1}(E_\infty) \cap d^{-1}C\Omega^n(E_\infty)}{C\Omega^{n-1}(E_\infty) \oplus d\Omega^{n-2}(E_\infty)},$$

where $\Omega^q(E_\infty), q = 0, 1, 2, \ldots$, is the space of differential q-forms on E_∞, $C\Omega^q(E_\infty)$ is the space of all Cartan $q$forms on E_∞, $q = 1, 2, \ldots$, (see Tab.3.4), and $C\Omega^o(E_\infty) \equiv 0$, $C\Omega^q(E_\infty) \equiv \Omega^q(E_\infty)$, for $q > n$, $\Omega^{-1}(E_\infty) = 0$. Thus a conservation law is a $(n-1)$-form on E_∞ non trivially closed on the (singular) solutions of E_k.

The space of conservation laws of E_k can be identified with the spectral term $E_1^{0,n-1}$ of the $\mathcal{C}$-spectral sequence associated to E_k. One can see that locally we can write

$$Cons(E_k) = \frac{\{\omega \in \Omega^{n-1}(E_\infty) | \partial\omega = 0\}}{\{\omega = \partial\theta | \theta \in \Omega^{n-2}(E_\infty)\}},$$

where

$$\partial\omega = \sum_{\mu_0,\dots,\mu_{n-1}} (\partial_{[\mu_0}\omega_{\mu_1\dots\mu_{n-1}]})dx^{\mu_0}\wedge\cdots\wedge dx^{\mu_{n-1}},$$

with

$$\omega = \sum_{\mu_1,\dots,\mu_{n-1}} \omega_{\mu_1\dots\mu_{n-1}}(x^\mu,y^j)dx^{\mu_1}\wedge\cdots\wedge dx^{\mu_{n-1}}\,mod\,C\Omega^{n-1}(E_\infty)$$

and

$$\partial_\mu \equiv \partial x_\mu + \sum_{i\in I} A^i_\mu(x,y)\partial y_i\,,\ \mu = 1,\dots,n,$$

basis Cartan fields of $\mathbb{E}_\infty$, where $\{x^\mu,y^j\}_{1\le\mu\le k,j\in I}$ are adapted coordinates.

Theorem 3.34. 1) *One has the canonical isomorphism:* $\mathcal{I}(E_\infty)^{n-1} \cong Cons(E_\infty)$. *So that integral numbers of E_∞ can be considered as conserved charges of E_k.*

2) *One has the following homomorphism of vector spaces*

$$(3.14).\qquad\qquad j : E_1^{0,n-1} \to \mathbf{R}^{\Omega^{E_\infty}_{n-1}}.$$

Then $E_1^{0,n-1}$ identifies a subspace $E^{0,n-1}$ of $\mathbf{R}^{\Omega^{E_\infty}_{n-1}}$, where

$$E^{0,n-1} \equiv im(j) = \Big\{\phi\in\mathbf{R}^{\Omega^{E_\infty}_{n-1}}\,|\,\exists\beta\in E_1^{0,n-1},\ \phi([N]_{E_\infty}) = \int_N \beta|_N\Big\}.$$

Proof. 1) It is a direct consequence of previous definitions and results.

2) In fact, to any conservation law $\beta : E_\infty \to \Lambda^o_{n-1}(E_\infty)$ we can associate a function $j(\beta) \equiv \phi : \Omega^{E_\infty}_{n-1} \to \mathbf{R}$, $\phi([N]) = \int_N \beta|_N$. This definition has sense as it does not depend on the representative used for $[N]_{E_\infty}$. In fact, if β is a conservation law, then $\forall V\in\Omega(E_\infty)_c$, with $\partial V = N_0\bigcup N_1$, we have

$$\int_{\partial V}\beta|_{\partial V} = \int_V d\beta|_V = 0 \Rightarrow \int_{N_0}\beta|_{N_0} = \int_{N_1}\beta|_{N_1}.$$

Furthermore, the mapping j is not necessarily injective. Indeed one has

$$(3.15)\quad \ker(j) = \left\{\beta\in E_1^{0,n-1}\,\left|\,\begin{array}{l}\displaystyle\int_N\beta|_N = 0\\[4pt]\text{for all }(n-1)\text{-dimensional admissible}\\[4pt]\text{integral manifolds of }E_\infty\end{array}\right.\right\}.$$

So $\ker(j)$ can be larger than the zero-class $[0] \in Cons(E_k)$.[37] $\square$

Remark 3.15. Note that one has the following short exact sequence:

$$0 \to \mathbf{R}^{\widehat{\Omega}^{E\infty}_{n-1,s}} \xrightarrow{i_*} \mathbf{R}^{\Omega^{E\infty}_{n-1}}$$

where i_* is the mapping $i_* : \phi \mapsto \phi \circ i$, $\forall \phi \in \mathbf{R}^{\widehat{\Omega}^{E\infty}_{n-1,s}}$, and $i \equiv i_{n-1}$ is the canonical mapping defined in the following commutative diagram:

$$
\begin{array}{ccccc}
 & \mathbf{R} & = & \mathbf{R} & \\
i_*(\phi) & \uparrow & & \uparrow & \phi \\
 & \Omega^{E\infty}_{n-1} & \xrightarrow[i_{n-1}]{} & \hat{\Omega}^{E\infty}_{n-1,s} & \to \quad 0
\end{array}
$$

As i_{n-1} is surjective it follows that i_* is injective. So any function on $\Omega^{E\infty}_{n-1,s}$ can be identified with a function on $\Omega^{E\infty}_{n-1}$. In particular, if $\hat{\Omega}^{E\infty}_{n-1,s} \cong \Omega^{E\infty}_{n-1,s}$ then any function on $\Omega^{E\infty}_{n-1,s}$ can be identified with a function on $\Omega^{E\infty}_{n-1}$.   ∎

By means of Theorem 3.34 we are able to represent $E_1^{0,n-1}$ by means of a Hopf algebra. (For informations on Hopf algebras see section 1.2 and, e.g. ref.[1].) In the following Ω can be considered indifferently one of previously considered "bordism groups".

Lemma 3.1. *Denote by* $\mathbf{K}\Omega$ *the free* $\mathbf{K}$*-module generated by* Ω. *Then,* $\mathbf{K}\Omega$ *has a natural structure of* $\mathbf{K}$*-bialgebra (group* $\mathbf{K}$*-bialgebra). (Here* $\mathbf{K} = \mathbf{R}$).

Proof of Lemma 3.1. In fact define on the free $\mathbf{K}$-module $\mathbf{K}\Omega$ the moltiplication

$$\left(\sum_{x\in\Omega} a_x x\right)\left(\sum_{y\in\Omega} b_y y\right) = \sum_{z\in\Omega}\left(\sum_{xy=z} a_x b_y\right)z.$$

Then, $\mathbf{K}\Omega$ becomes a ring. The map $\eta_{\mathbf{K}\Omega} : \mathbf{K} \to \mathbf{K}\Omega$, $\eta_{\mathbf{K}\Omega}(\lambda) = a1$, where 1 is the unit in Ω, makes $\mathbf{K}\Omega$ an $\mathbf{K}$-algebra. Furthermore, if we define $\mathbf{K}$-linear maps $\triangle : \mathbf{K}\Omega \to \mathbf{K}\Omega \bigotimes_{\mathbf{K}} \mathbf{K}\Omega$, $\triangle(s) = s \otimes s$ and $\epsilon : \mathbf{K}\Omega \to \mathbf{K}$, $\epsilon(s) = 1$, then $(\mathbf{K}\Omega, \triangle, \epsilon)$ becomes a $\mathbf{K}$-coalgebra. $\square$

Lemma 3.2. *The dual linear space* $(\mathbf{K}\Omega)^*$ *of* $\mathbf{K}\Omega$ *can be identified with the set:* $\mathbf{R}^{\Omega} \equiv Map(\Omega, \mathbf{K})$, *where the dual* $\mathbf{K}$*-algebra structure of* $\mathbf{K}\Omega$ *is given by*

$$
\left\{
\begin{array}{l}
(f+g)(s) = f(s) + g(s) \\
(fg)(s) = f(s)g(s) \\
(af)(s) = af(s), \; \forall f, g \in Map(\Omega, \mathbf{K}), s \in \Omega, a \in \mathbf{K}
\end{array}
\right\}.
$$

---

[37] For example for the d'Alembert equation one can see that for any conservation law ω one has $<\omega,N>=0$, where N is any admissible 1-dimensional compact integral manifold of $(d'A)$, but $\omega \notin [0] \in E_1^{0,n-1}$.

Lemma 3.3. *If Ω is a finite group $A \equiv Map(\Omega, \mathbf{K})$ has a natural structure of $\mathbf{K}$-bialgebra $(\mu, \eta, \triangle, \epsilon)$, with*

$$(a) \qquad \mu : A \bigotimes_{\mathbf{K}} A \to A , \quad \mu(f \otimes g) = f.g;$$

$$(b) \qquad \eta : \mathbf{K} \to A , \quad \eta(\lambda)(s) = \lambda , \ \forall s \in \Omega;$$

$$(c) \qquad \triangle : A \to A \bigotimes_{\mathbf{K}} A , \quad \triangle(f)(x, y) = f(xy);$$

$$(d) \qquad \epsilon : A \to \mathbf{K} , \quad \epsilon(f) = f(1).$$

Lemma 3.4. $\mathbf{K}\Omega$ *has a natural structure of $\mathbf{K}$-Hopf algebra.*

Proof of Lemma 3.4. Define the $\mathbf{K}$-linear map $S : \mathbf{K}\Omega \to \mathbf{K}\Omega$, $S(x) = x^{-1}$, $\forall x \in \Omega$. Then, $(1 * S)(x) = xS(x) = xx^{-1} = 1 = \epsilon(x)1 = \eta \circ \epsilon(x), x \in \Omega$. Then, S is the antipode of $\mathbf{K}\Omega$ so that $\mathbf{K}\Omega$ becomes a $\mathbf{K}$-Hopf algebra. $\square$

Lemma 3.5. *If Ω is a finite group $A \equiv Map(\Omega, \mathbf{K})$ has a natural structure of $\mathbf{K}$-Hopf algebra. If Ω is not a finite group $Map(\Omega; \mathbf{K})$ has a structure of Hopf algebra in extended sense, i.e., an extension of an Hopf $\mathbf{K}$-algebra K contained into $Map(\Omega; \mathbf{K})$. More precisely, $K = R_{\mathbf{K}}(\Omega)$ is the Hopf $\mathbf{K}$-algebra of all the representative functions on Ω. In fact, one has the following short exact sequence: $0 \to R_{\mathbf{K}}(\Omega) \to Map(\Omega; \mathbf{K}) \to H \to 0$, where H is the quotient algebra. (If Ω is a finite group then $H = 0$.) Therefore, $< E^{1,0} >$ is, in general, an Hopf algebra in this extended sense.*

Proof of Lemma 3.5. In fact one can define the antipode $S(f)(x) = f(x^{-1})$, $\forall f \in A, x \in \Omega$. It satisfies the equalities: $\mu(1 \otimes S)\triangle = \mu(S \otimes 1)\triangle = \eta \circ \epsilon.\square$

Theorem 3.35. *The space of conservation laws $E_1^{0,n-1}$ of a PDE identifies in a natural way a $\mathbf{K}$-Hopf algebra: $< E^{0,n-1} > \subset \mathbf{H}(E_\infty) \equiv Map(\Omega_{n-1}^{E_\infty}, \mathbf{R})$. If $E^{0,n-1} = 0 \in Map(\Omega_{n-1}^{E_\infty}, \mathbf{R})$, we put for definition $< E^{0,n-1} > \equiv \mathbf{H}(E_\infty)$. We call $< E^{0,n-1} >$ the Hopf algebra of E_k.*

Proof. It is an immediate consequence of Theorem 3.34 and above lemmas, and taking into account the following commutative diagram

$$
\begin{array}{ccc}
\mathbf{R}^{\Omega_{n-1}^{E_\infty}} \times \mathbf{R}^{\Omega_{n-1}^{E_\infty}} & \longrightarrow & \mathbf{R}^{\Omega_{n-1}^{E_\infty}} \\
\uparrow & & \uparrow \\
E^{0,n-1} \times E^{0,n-1} & \longrightarrow & <E^{0,n-1}>
\end{array}
$$

where $< E^{0,n-1} >$ is the Hopf subalgebra of $\mathbf{R}^{\Omega_{n-1}^{E_\infty}}$ generated by $E^{0,n-1}$. We denote by f_α the image of the conservation law $\alpha \in E_1^{0,n-1}$ into $\mathbf{R}^{\Omega_{n-1}^{E_\infty}}$. So in $< E^{0,n-1} >$ we have the following product: $< E^{0,n-1} > \times <$

$E^{0,n-1} > \to < E^{0,n-1} >$, $(f_\alpha, f_\beta) \mapsto f_\alpha \cdot f_\beta$. Furthermore, we can explicitly write

$$\overline{\eta} : \mathbf{K} \to < E^{0,n-1} >, \quad \overline{\eta}(\lambda)(s) = \lambda,$$

$$\triangle : < E^{0,n-1} > \to < E^{0,n-1} > \bigotimes_{\mathbf{K}} < E^{0,n-1} >, \quad \triangle(f)(x,y) = f(xy),$$

$$\epsilon : < E^{0,n-1} > \to \mathbf{K}, \quad \epsilon(f) = f(1),$$

$$S : < E^{0,n-1} > \to < E^{0,n-1} >, \quad S(f)(x) = f(x^{-1}).$$

So the proof is complete. $\qquad\qquad\qquad\qquad\qquad\qquad\qquad\qquad\qquad\square$

Definition 3.22. 1) We call *full p-Hopf algebra* of $E_k \subset J_n^k(W)$ the following Hopf algebra: $\mathbf{H}_p(E_k) \equiv \mathbf{R}^{\Omega_p^{E_k}}$. In particular for $p = n - 1$ we write $\mathbf{H}(E_k) \equiv \mathbf{H}_{n-1}(E_k)$ and we call it *full Hopf algebra* of E_k.

2) If $< E^{0,n-1} > \cong \mathbf{H}(E_\infty) \equiv \mathbf{R}^{\Omega_{n-1}^{E_\infty}}$, we say that E_k is *wholly Hopf-bording*. Furthermore, we say also that $\mathbf{R}^{\Omega_p^{E_k}} \equiv \mathbf{H}_p(E_k)$ is the *space of the full p-conservation laws* of E_k.

Theorem 3.36. 3 If $\Omega_{n-1}^{E_\infty}$ is trivial then E_k is wholly Hopf-bording. Furthermore, in such a case $E^{0,n-1} = 0$.

Proof. In fact, in such a case one has $\int_N \omega = \int_{\partial V} = \int_V d\omega = 0$, $\forall [\omega] \in E_1^{0,n-1}$, $[N] \in \Omega_{n-1}^{E_\infty}$, and $V = n$-dimensional admissible integral manifold contained into E_∞. Hence, for definition one has $< E^{0,n-1} > \cong \mathbf{H}(E_\infty)$.$\square$

Example 3.8. If $E_\infty = J_n^\infty(W)$, with $n = 2, 4$, then E_k is wholly Hopf-bording. In fact, in such cases, we have $\Omega_{n-1}^{E_\infty} \cong \Omega_{n-1} = 0$. $\qquad\qquad\blacksquare$

Example 3.9. Let us consider $E_k \equiv JD^k(\mathbf{R}, \mathbf{R})$. Therefore the infinity prolongation of E_k is just $E_\infty \equiv JD^\infty(\mathbf{R}, \mathbf{R})$. Then, from Theorem 3.25 we get $\Omega_0^{E_\infty} = \Omega_0^{E_k} \cong \Omega_0 \cong \mathbf{Z}_2 \Rightarrow \mathbf{R}^{\Omega_0^{E_\infty}} \cong (\mathbf{R})^{\mathbf{Z}_2} \cong \mathbf{R}^2$. On the other hand the Cartan distribution $\mathbb{E}_\infty$ corresponding to the infinity prolongation E_∞ of E_k is a 1-dimensional distribution on $JD^\infty(W)$, endowed with coordinates $(x, y, y^{(i)})$, $i \in \mathbf{N}$, generated by the following vector field $\partial_x = \partial x + \sum_{i \in \mathbf{N}} y^{(i)} \partial y_{(i)}$. Then, $Cons(E_k) = E_1^{0,0} \cong \{f(x, y^{(i)}) | (\partial_x \cdot f) = 0\} \cong \mathbf{R}$. Therefore, j is the canonical monomorphism $\mathbf{R} \to \mathbf{R}^2$. One has $< E^{0,0} > = \mathbf{R} \subset \mathbf{H}(E_\infty) \cong \mathbf{R}^2$. It follows that E_k is not wholly Hopf-bording. $\qquad\qquad\blacksquare$

Example 3.10. If $E_k \subset J_n^k(W)$ is a q(n-1)-cobording free PDE, one has that $E^{0,0}$ is a subspace of $\mathbf{R}^{K_{n-1}^{E_k}}$. $\qquad\qquad\blacksquare$

Example 3.11. Let us consider the following equation:

$$(E_1) \subset JD(\mathbf{R}, \mathbf{R}) : \{\dot{u} = f(t, u)\},$$

with $f \in C^\infty(\mathbf{R}^2, \mathbf{R})$. This PDE is diffeomorphic to $\mathbf{R}^2$ by means of the canonical projection $\pi_{1,0} : J\mathcal{D}(\mathbf{R}, \mathbf{R}) \to \mathbf{R}^2$, $(t, u, \dot{u}) \mapsto (t, u)$. Then, $H_q(E; G) = 0$ for $q > 0$, $q \neq 2$. The Cartan distribution $\mathbb{E}_1$ of E_1 is 1-dimensional and generated by the following vector field $\partial_t \equiv \partial t + f(t, u)\partial u$. Then, one has $\mathcal{I}(E_1)^0 = \{F(t, u) \in C^\infty(\mathbf{R}^2, \mathbf{R}) | \partial_t.F = 0\}$. As one has $\partial_t.F = (\partial t.F) + f(t, u)(\partial u.F) = (\partial t.F) + \dot{u}(\partial u.F)$, we see that on any integral manifold V, that locally is the image of a first derivative Ds of a C^∞ function $s : \mathbf{R} \to \mathbf{R}$, $s(t)$, one has $0 = \partial_t F \Leftrightarrow dF|_V = 0 \Rightarrow F|_V = const$. In other words $\mathcal{I}(E_1)^0$ coincides with the space of first integrals of E_1. Note that the number of functionally independent first integrals of E_1 is 1. So taking into account that $\dim(E_1) = 2$ and that the manifold $N_c \equiv \{F = c\}$ is 2-dimensional, we get that $\dim(E_1 \cap N_c) = 1$, i.e., is an integral curve of the Cartan distribution $\mathbb{E}_1$. So we conclude that two points $a, b \in E_1$ belong to the same integral curve of E_1 iff $F(a) = F(b)$. This means that we can identify each integral curve by means of a real number. Finally note that $\Omega_0 \cong \mathbf{Z}_2 \cong \Omega_0(E_1)$. So we have the following short exact sequence: $0 \to K_0^{E_1} \to \Omega_0^{E_1} \to \mathbf{Z}_2 \to 0$. Of course as E_1 is an ODE one has $\Omega_{0,s}^{E_1} = \mathbf{Z}_2$. Then, $E_1^{0,0} \cong \mathbf{R}$ and $\mathbf{H}(E_\infty) \cong \mathbf{R}^2$. So, also this equation is not wholly Hopf-bording.   ∎

Example 3.12. The *Korteweg de Vries equation*: $(KdV) \subset J\mathcal{D}^3(W)$: $\{u_{xxx} + 6uu_x - u_t = 0\}$, $\pi : W \equiv \mathbf{R}^2 \times \mathbf{R} \to \mathbf{R}^2$, and the *wave equation*: $(Wa) \subset J\mathcal{D}^2(W)$: $\{u_{tt} - u_{xx} = 0\}$, $\pi : W \equiv \mathbf{R}^2 \times \mathbf{R} \to \mathbf{R}^2$, are two examples where $H^1(E_k) = 0$, and $\mathcal{I}(E_k)^1 \neq 0$. Furthermore, all the compact closed admissible integral 1-dimensional manifolds are orientable and diffeomorphic to S^1. So we can apply Theorem 3.26. (In ref.[58] we have explicitly calculated differential forms (conservation laws) for such two equations that identify non zero vectors of $\mathcal{I}(E_k)^1$.)   ∎

Example 3.13. The *Euler equation* for incompressible fluids has been considered in refs.[56,65]. All the compact closed admissible integral manifolds of dimension p are orientable for $p = 1, 3$. Instead, if $p = 2$ are orientable only that of the type S^2 with hands. So, for $p = 2$, in order to apply Theorem 3.26 we shall restrict the class of admissible integral manifolds to that of type S^2 with hands. Note that $\mathcal{I}(E)^1 \neq 0$ for $p = 1, 2, 3$. In fact, in ref.[58] it is proved that for $p = 3$ there are non trivial conservation laws. These induce, by restriction, also conservation laws for $p = 1, 2$. Finally note that as for the physical point of view are interesting only admissible integral manifolds that are defined as submanifolds of 3-dimensional affine spaces $M_t \subset M$, $t \in T \equiv$ time-axis, representing equal time events, then only oriented admissible manifolds can be considered. In fact, non orientable compact p-dimensional manifolds cannot be embedded into $(p+1)$-dimensional ones. So in these cases we can apply Theorem 3.26. ∎

Example 3.14. *Field equations of mathematical physics.* Similar considerations can be made for the standard equations of the field theory (e.g., Klein-Gordon, Maxwell, Dirac, Einstein, Yang-Mills) over p-connected, $0 \le p \le 3$, 4-dimensional space-time. ∎

Example 3.15. *D'Alembert equation.* Let us return on the d'Alembert equation. The conservation laws for this equation are the following differential 1-forms: $\omega = f(x, I, D_x I, \ldots)dx + g(y, J, D_y J, \ldots)dy$ where $\equiv u_x/u$ and $J \equiv u_y/u$. Let us show how these conservation laws are related to the bordism group $\Omega_1^{(d'A)}$. A curve $\gamma : A \subset \mathbf{R} \to J\mathcal{D}^2(\mathbf{R}^3)$ is an integral curve of $(d'A)$ iff $\gamma^*\omega_0 = \gamma^*\omega_1 = \gamma^*\omega_2 = \gamma^*\omega_3 = 0$, where ω_0, ω_1, ω_2, ω_3, are the differential 1-forms encoding the equation $(d'A)$. Then we can see that if γ is an integral closed curve, i.e., $\gamma : S^1 \to (d'A) \subset J\mathcal{D}^2(\mathbf{R}^3)$, then $\oint \gamma^*\omega = 0$. This agrees with the fact that $\Omega_1^{(d'A)} = 0$ and with Theorem 3.26. Let us explicitly prove that $\omega = Idx + Jdy$ is a conservation law for $(d'A)$, In fact, $d\omega = dI \wedge dx + dJ \wedge dy$. On the other hand $dI = d(\frac{u_x}{u}) = \frac{du_x}{u} - I\frac{du}{u}$. On any 2-dimensional integral manifold $V \subset (d'A)$ one has $\frac{du}{u} = Idx + Jdy$, $\frac{du_x}{u} = \frac{u_{xx}}{u}dx + IJdy$, $\frac{du_y}{u} = IJdx + \frac{u_{yy}}{u}dy$. Hence, $dI = (\frac{u_{xx}}{u} - I^2)dx$, and $dJ = (\frac{u_{yy}}{u} - J^2)dy$. As a consequence on V one has $d\omega|_V = 0$. This proves that ω is a conservation law. In the general case, i.e., for conservation laws $\omega = fdx + gdy$, as above specified, the proof can be similarly obtained observing that $df = \kappa dx$ and $dg = \bar\kappa dy$ respectively. Furthermore, if $\omega = Idx + Jdy$, on an integral curve $\gamma \subset (d'A)$ one has $\gamma^*\omega = (I\dot{x} + J\dot{y})dt = \frac{\dot{u}}{u}dt$. Therefore, on a closed curve we get $\oint \gamma^*\omega = \oint \frac{du}{u} = [\log u]_{u_0}^{u_0} = \log 1 = 0$. Same conclusions hold if $\omega = fdx + gdy$. In fact, let $N = \gamma(S^1) \subset V \subset (d'A)$, where V is an admissible 2-dimensional integral manifold. We can choice V such that $H^1(V) = 0$. Since $d\omega|_V = 0$ it follows that $\omega|_V$ is exact. Therefore, $\oint \gamma^*\omega = 0$. In this case $\Omega_1^{(d'A)\infty} \cong \Omega_1^{(d'A)} \cong \Omega_1 = 0$. Instead $E_1^{0,1}$ is different from zero, but its image into the Hopf algebra $\mathbf{H}((d'A)_\infty) \cong \mathbf{R}$ is zero. Therefore, $< E^{0,1} >= \mathbf{H}((d'A)_\infty)$. Hence $(d'A)$ is wholly Hopf-bording. ∎

Example 3.16. *Burger's equation:* $(BE) \subset J\mathcal{D}^2(\mathbf{R}^3) :$ $\quad u_{xx} + uu_x - u_t =$

0. This is a completely integrable PDE. In fact one has the following short exact sequences: $(BE)_{+m} \to (BE)_{+(m-1)} \to 0$, $\dim(BE)_{+m} = \dim(BE)_{+(m-1)} + \dim(g_2)_{+m} = 7 + 2(m-1) + 2 = 7 + 2m$. Furthermore as (BE) is analytic, it follows that it is also formally integrable. Therefore, from Theorem 3.25 we get: $\Omega_1^{(BE)} \cong \Omega_1^{(BE)\infty} \cong 0$. Moreover, $E_1^{0,1} \cong \mathbf{R}$ [72], hence the homomorphism $E_1^{0,1} \to \mathbf{H}((HE)_\infty) \cong \mathbf{R}$ is zero. So the Burger's equation is wholly Hopf-bording. Note that $E_1^{0,1}$ is generated by $\omega = c[udx + (u_x + \frac{u^2}{2})dt]$, $c \in \mathbf{R}$. In fact, on any 2-dimensional admissible integral manifold $V \subset (BE)$, one has $d\omega|_V = c[du \wedge dx + du_x \wedge dt + udu \wedge dt] = c[(u_x dx + u_t dt) \wedge dx + (u_{xx} dx + u_{xt} dt) \wedge dt + u(u_x dx + u_t dt) \wedge dt] = c[-u_t + u_{xx} + uu_x]dx \wedge dt = 0$. The proof that $\oint \gamma^*\omega = 0$ for any $[\omega] \in E_1^{0,1}$, where $\gamma : S^1 \to (BE)$ is an admissible integral manifold, is similar to one given in Example 3.15 for the d'Alembert equation. ∎

Example 3.17. *Heat equation:* $(HE) \subset J\mathcal{D}^2(\mathbf{R}^3)$: $\quad u_{xx} - u_t = 0$. Also this is a completely integrable PDE. In fact one has the following short exact sequences: $(HE)_{+m} \to (HE)_{+(m-1)} \to 0$, $\dim(HE)_{+m} = \dim(HE)_{+(m-1)} + \dim(g_2)_{+m} = 7 + 2(m-1) + 2 = 7 + 2m$. As the heat equation is analytic, it follows that it is also formally integrable. Therefore, from Theorem 3.25 we get: $\Omega_1^{(HE)} \cong \Omega_1^{(HE)\infty} \cong \Omega_1 \cong 0$. Hence $\mathbf{H}((HE)_\infty) \cong \mathbf{R}$. On the other hand one has $E_1^{0,1} \cong \mathbf{R}^3$. In fact one can see [95] that $E_1^{0,1} \cong \{\varphi \in C^\infty(\mathbf{R}^2, \mathbf{R}) | \varphi_{xx} + \varphi_t = 0\}$. On the other hand the general solution of $\varphi_{xx} + \varphi_t = 0$ is $\varphi = C_1 e^{\sqrt{c}x - ct} + C_2 e^{-(\sqrt{c}x + ct)}$, C_1, C_2, $c \in \mathbf{R}$. So the manifold of solutions of $\varphi_{xx} + \varphi_t = 0$ is a 3-dimensional manifold X fibered on $\mathbf{R}$, with fibers $X_c \cong \mathbf{R}^2$. Furthermore, X has also a global structure of vector space (i.e., $X \cong \mathbf{R}^3$), endowed with the following addition: $(C_1, C_2, c) + (\bar{C}_1, \bar{C}_2, \bar{c}) = (C_1 + \bar{C}_1, C_2 + \bar{C}_2, \frac{c+\bar{c}}{2})$. In this case the homomorphism $E_1^{0,1} \to \mathbf{H}((HE)_\infty) \cong \mathbf{R}$ is zero. So the heat equation is wholly Hopf-bording. ∎

Example 3.18. *Modified heat equation:* $(MHE) \subset J\mathcal{D}^2(\mathbf{R}^3)$: $\quad u_{xx}^2 - u_t = 0$. Also this is a completely and formally integrable PDE. In fact, one has the following short exact sequences: $(MHE)_{+m} \to (MHE)_{+(m-1)} \to 0$, $\dim(MHE)_{+m} = \dim(MHE)_{+(m-1)} + \dim(g_2)_{+m} = 7 + 2(m-1) + 2 = 7 + 2m$. Therefore, from Theorem 3.25 we get: $\Omega_1^{(MHE)} \cong \Omega_1^{(MHE)\infty} \cong \Omega_1 \cong 0$. Hence the Hopf algebra $\mathbf{H}((MHE)_\infty)$ of (MHE) is $\mathbf{R}$. Instead for the space of conservation laws one has $E_1^{0,1} \cong 0$ [95]. Hence this a good example where the full conservation laws exist, but the conservation laws are zero.

Of course (MHE) is wholly Hopf-bording: $< E^{0,1} >= \mathbf{H}((NHE)_\infty) \cong \mathbf{R}.$ ∎

Example 3.19. *Navier-Stokes equation.* An orientable closed compact admissible p-dimensional, $0 \le p \le 3$, integral manifold $N = N_0 \bigcup N_1$, $N_0, N_1 \subset (NS)$, is the boundary of a $(p + 1)$-dimensional integral manifold V iff $\int_{N_0} \alpha = \int_{N_1} \alpha$, $\forall [\alpha] \in \mathcal{I}(NS)^p \ne 0$. So it remains to prove that for the Navier-Stokes equation one has $\mathcal{I}(NS)^p \ne 0$. For $p = 3$ it is surely $\mathcal{I}(NS)^3 \ne 0$, as one can see that the symmetry vector field ∂x_0 generates a conservation law. The proof is similar to one given in ref.[58] for the Euler equation. For example, a conservation law induced by the symmetry vector field ∂x_0 is the following

$$\omega = \sqrt{\det(g_{ij})}[(G^j_{jk}\dot{x}^k + \dot{x}^s_s)dx^1 \wedge dx^2 \wedge dx^3 + \dot{x}^1_0 dx^0 \wedge dx^2 \wedge dx^3$$
$$- \dot{x}^2_0 dx^0 \wedge dx^1 \wedge dx^3 + \dot{x}^3_0 dx^0 \wedge dx^1 \wedge dx^3].$$

In fact, on any 4-dimensional admissible integral manifold $V \subset (NS)$ one has

$$\omega|_V = \sqrt{\det(g_{ij})}[(\partial x_0.v^1)dx^0 \wedge dx^2 \wedge dx^3$$
$$- (\partial x_0.v^2)dx^0 \wedge dx^1 \wedge dx^3 + (\partial x_0.v^3)dx^0 \wedge dx^1 \wedge dx^3]$$

and

$$d(\omega|_V) = - (\partial x_k.\sqrt{\det(g_{ij})})[(\partial x_0.v^k)dx^0 \wedge dx^1 \wedge dx^2 \wedge dx^3$$
$$- \sqrt{\det(g_{ij})})(\partial x_0.\mathrm{div}\,(v))dx^0 \wedge dx^1 \wedge dx^2 \wedge dx^3 = 0.$$

Of course ω is trivial on 3-dimensional integral manifolds N contained into the hypersurfaces $x^0 =$cost. But as N are not, in general, such manifolds, the equivalence class $[\omega]$, corresponding to the above differential 3-form, are contained into $\mathcal{I}(NS)^3$ and does not coincide with its zero element. On the other hand, from Tab.3.3 it is clear that such a test is not necessary for the Navier-Stokes equation. Let us recall that the *full p-conservation laws*, $0 \le p \le 3$, of (NS) are functions $f : \Omega_p^{(NS)} \to \mathbf{R}$. For $p = 3$ we simply call *full conservation law* of (NS) a function $f : \Omega_3^{(NS)} \to \mathbf{R}$. Then, the full p-Hopf algebras, $\mathbf{H}_p(NS) \equiv \mathbf{R}^{\Omega_p^{(NS)}}$, $0 \le p \le 3$, of (NS) are given by the following table:

TAB.3.5 - Full p-Hopf algebras of (NS).

p	$\mathbf{H}_p(N)$
0	$\mathbf{R}^2$
1	$\mathbf{R}$
2	$\mathbf{R}^2$
3	$\mathbf{R}$

If, in particular, the bordism class $[N]_{(NS)} \in \Omega_3^{(NS)}$ is represented by an orientable closed 3-dimensional manifold N, and a full conservation law is represented by a differential 3-form $\omega \in \mathcal{I}(NS)^3$, then we can write $f : \Omega_3^{(NS)} \to \mathbf{R}$, $f([N]_{(NS)}) = \int_N \omega = 0$. The proof is obtained directly from the definition of full p-Hopf algebra and from Tab.3.3 and taking into account observations similar to ones given for the d'Alembert equatiom. ∎

3.2 - ALGEBRAIC GEOMETRY OF PDE's

In this section we shall consider a relation between the algebraic localization of modules and the geometry of PDE's. This algebraic characterization of PDE's is principally useful to describe singularities in PDE's given by means of polynomial functions.

In this section we will start by considering commutative rings. (See also, e.g., refs.[20,22,23,28–37,48].)

Definition 3.23. (*Distinguished elements in a ring*). A *zero-divisor* in a ring A is an element $a \in A$ such that there exists $b \in A$, $b \neq 0$, such that $ab = 0$. An element $a \in A$ is *nilpotent* if $a^n = 0$, for some $n > 0$. (A nilpotent element is a zero-divisor (unless $A = 0$), but not conversely.)[38] The *identity element* $1 \in A$ is defined by $a1 = 1a = a$, for all $a \in A$. A *unit* in A is an element $a \in A$ that divides 1, i.e., $\exists b \in A$ such that $ab = 1$. Then b is called the *inverse* of a and denoted by a^{-1}. (a^{-1} is uniquely determined by a.)

Definition 3.24. An *integral domain* is a ring with non-zero-divisors $\neq 0$.

Proposition 3.2. (Distinguished subsets of a ring). 1) *The set of units forms a multiplicative, abelian group in A, that we denote by $G(A)$.*

---

[38] A ring A, such that the set of its nilpotent elements is reduced to $\{0\}$, is called a *reduced ring.*

2) *For any subset $S \subset A$, the smallest ideal $\mathfrak{a} \subset A$ containing S is called the ideal generated by S and denoted by $< S >$. Any ideal $\mathfrak{p} =< a >$, $a \in A$, is called a principal ideal. A ring is called principal ideal ring if every ideal is principal. In particular we denote $\mathfrak{a} + \mathfrak{b} =< \{a + b\}_{a \in \mathfrak{a}, b \in \mathfrak{b}} >$, and $\mathfrak{ab} =< \{ab\}_{a \in \mathfrak{a}, b \in \mathfrak{b}} >$.*

3) *If $S \subset A$ is a subset and $\mathfrak{a} \subset A$ is an ideal, then the quotient $\frac{\mathfrak{a}}{S} \equiv \{a \in A | aS \subset \mathfrak{a}\}$ is an ideal of A.*

4) *If $S \subset A$ is a multiplicatively closed subset of A, we call a maximal ideal with respect to S, a maximal member $\mathfrak{m} \subset A$ among the set of ideals do not meet S. In particular, if $S = \{1\}$, then $\mathfrak{m}$ is called a maximal ideal of A. An ideal $\mathfrak{m} \subset A$ is maximal iff $A/\mathfrak{m}$ is a field.*

5) *An ideal $\mathfrak{p} \subset A$ is prime if $\mathfrak{p} \neq < 1 >$ and $xy \in \mathfrak{p}$ implies $x \in \mathfrak{p}$ or $y \in \mathfrak{p}$. $\mathfrak{p}$ is prime iff $A/\mathfrak{p}$ is any integral domain. A maximal ideal $\mathfrak{m}$ of A is prime.*

6) *If $\mathfrak{a}$ is any ideal of A, the radical of $\mathfrak{a}$ is the following ideal*

$$\sqrt{\mathfrak{a}} \equiv r(\mathfrak{a}) \equiv \{x \in A | x^n \in \mathfrak{a} \text{ for some } n > 0\}.$$

An ideal $\mathfrak{a}$ such that $\mathfrak{a} = \sqrt{\mathfrak{a}}$ is called radical-ideal (or perfect). One has the following properties:

(i) *$\sqrt{\{0\}}$ is the ideal consiting of all nilpotent elements of A and is denoted also by $nil(A)$: $\sqrt{\{0\}} = nil(A)$. If $\{0\}$ is a radical ideal, or equivalently, when $nil(A) = \{0\}$, A is said to be reduced.*

(ii) *$r(\mathfrak{a}) \supseteq \mathfrak{a}$;*

(iii) *$r(r(\mathfrak{a})) = r(\mathfrak{a})$;*

(iv) *$r(\mathfrak{ab}) = r(\mathfrak{a} \cap \mathfrak{b}) = r(\mathfrak{a}) \cap r(\mathfrak{b})$;*

(v) *$r(\mathfrak{a}) =< 1 > \Leftrightarrow \mathfrak{a} =< 1 >$;*

(vi) *$r(\mathfrak{a} + \mathfrak{b}) = r(r(\mathfrak{a}) + r(\mathfrak{b}))$;*

(vii) *If $\mathfrak{p}$ is prime, $r(\mathfrak{p}^n) = \mathfrak{p}$ for all $n > 0$;*

(viii) *One has*

$$r(\mathfrak{a}) = \bigcap_{\mathfrak{p} = \text{ prime ideals containing } \mathfrak{a}} \mathfrak{p};$$

(ix) *If $\mathfrak{a}$ is a radical-ideal, $A/\mathfrak{a}$ is reduced, and in particular $A/nil(A)$ is reduced.*

(x) *The radical $nil(A)$ of a commutative ring A is the intersection of all prime ideals of A:*

$$nil(A) = \bigcap_{\mathfrak{p} = \text{ prime ideals of } A} \mathfrak{p}.$$

In particular $nil(A) \subset rad(A)$, where $rad(A)$ is the *radical* of A defined by:

$$rad(A) = \bigcap_{\mathfrak{m} \,=\, maximal\ ideals\ of\ A} \mathfrak{m}.$$

$rad(A)$ is an ideal of A. One has: $a \in rad(A)$ iff for all $x \in A$, one has that $1 - xa$ has a left inverse.

6) An ideal $\mathfrak{q} \subset A$ is *primary* if $\mathfrak{q} \neq A$ and if $xy \in \mathfrak{q}$ implies $x \in \mathfrak{q}$ or $y^n \in \mathfrak{q}$ for some $n > 0$. One has the following properties:

(i) $\mathfrak{q}$ is primary iff $A/\mathfrak{q} \neq 0$ and every zero-divisor in $A/\mathfrak{q}$ is nilpotent;

(ii) Every prime ideal is primary;

(iii) Let $\mathfrak{q}$ be a primary ideal in a ring A. Then $r(\mathfrak{q})$ is the smallest prime ideal containing $\mathfrak{q}$. If $\mathfrak{p} = r(\mathfrak{q})$ then $\mathfrak{q}$ is said to be $\mathfrak{p}-primary$.

Definition 3.25. The *prime spectrum* of a ring A is the set $S_{pec}(A)$ of all prime ideals of A.

Proposition 3.3. Let A be a ring and let $\mathfrak{p} \in S_{pec}(A)$. Then one has the following properties:

(i) If $\mathfrak{a}_1, \cdots, \mathfrak{a}_r$ are ideals of A and $\mathfrak{a}_1 \cdots \mathfrak{a}_r \subset \mathfrak{p}$, then $\mathfrak{a}_i \subset \mathfrak{p}$ for one $i \in \{1, \cdots, r\}$.

(ii) If $\mathfrak{a}_1, \cdots, \mathfrak{a}_r$ are ideals of A and $\mathfrak{a}_1 \cap \cdots \cap \mathfrak{a}_r \subseteq \mathfrak{p}$, then $\mathfrak{a}_i \subset \mathfrak{p}$ for one $i \in \{1, \cdots, r\}$. In particular if the equality holds, then $\mathfrak{a}_i = \mathfrak{p}$.

Proof. (i) Let us assume that $\mathfrak{a}_i \not\subset \mathfrak{p}$ for all $i = 1, \cdots, r$. Then, we can find elements $a_i \in \mathfrak{a}_i$ such that $a_i \notin \mathfrak{p}$, $i = 1, \cdots, r$. Taking into account that $\mathfrak{p}$ is prime, it should be $a_1 \cdots a_r \notin \mathfrak{p}$. But this does not agree with the fact that must be $a_1 \cdots a_r \in \mathfrak{a}_1 \cdots \mathfrak{a}_r$.

(ii) Let us first take $\mathfrak{a}_1 \cap \cdots \cap \mathfrak{a}_r \subset \mathfrak{p}$. Let us assume that there exist $\mathfrak{a}_i \not\subset \mathfrak{p}$ for all $i = 1, \cdots, r$. Then, we can find $a_i \in \mathfrak{a}_i$, $i = 1, \cdots, r$, such that $a_1 \cdots a_r \in \mathfrak{a}_1 \cap \cdots \cap \mathfrak{a}_r \subset \mathfrak{p}$. But this does not agree with the fact that $\mathfrak{p}$ is prime. Let us, now, assume that $\mathfrak{a}_1 \cap \cdots \cap \mathfrak{a}_r = \mathfrak{p}$. Then, as it is $\mathfrak{a}_1 \cap \cdots \cap \mathfrak{a}_r \subset \mathfrak{p}$ it follows that there exist some $\mathfrak{a}_i \subset \mathfrak{p}$. On the other hand, for assumption, must be $\mathfrak{p} \subseteq \mathfrak{a}_j$, for all $j = 1, \cdots, r$. Then must exists one $\mathfrak{a}_i$ such that $\mathfrak{a}_i = \mathfrak{p}$. $\square$

Definition 3.26. (*Localization*). 1) Let $S \subset A$ be a multiplicatively closed subset of A with $1 \in S$. Let $S^{-1}A \equiv A \times S/ \sim$, where $\sim$ is the following equivalence relation: $(a, s) \sim (a', s') \Leftrightarrow as' - a's = 0$. We denote by $\frac{a}{s}$ the equivalence class of (a, s) and call $S^{-1}A$ the set of *fractions* of A by S.

2) In particular, if $\mathfrak{p} \subset A$ is a prime ideal we call $A_{\mathfrak{p}} \equiv S^{-1}A$, $S \equiv A - \mathfrak{p}$, the *localization* of A at $\mathfrak{p}$.[39]

3) Let M be a A-module. Set $S^{-1}M \equiv M \times S/\sim$, where $\sim$ is the following equivalence relation: $(p, s) \sim (p', s') \Leftrightarrow \exists t \in S : t(sp' - s'p) = 0$. We denote by $\frac{p}{s}$ the equivalence class of (p, s) and call $S^{-1}M$ the set of *fractions* of M by S. This is a $S^{-1}A$-module.

4) In particular, if $\mathfrak{p} \subset A$ is a prime ideal and $S \equiv A - \mathfrak{p}$, then $M_{\mathfrak{p}} \equiv S^{-1}M$ is called the *localization* of M at $\mathfrak{p}$.

Proposition 3.4. (Localization). 1) *One has the following properties:*

(i) S^{-1} *is an exact functor;*

(ii) $S^{-1}(M \oplus P) = (S^{-1}M) \oplus (S^{-1}P)$; *In particular* $(M \oplus P)_{\mathfrak{p}} = M_{\mathfrak{p}} \oplus P_{\mathfrak{p}}$, *for any* $\mathfrak{p} \in S_{pec}(A)$;

(iii) $S^{-1}(M + P) = (S^{-1}M) + (S^{-1}P)$; *In particular* $(M + P)_{\mathfrak{p}} = M_{\mathfrak{p}} + P_{\mathfrak{p}}$, *for any* $\mathfrak{p} \in S_{pec}(A)$;

(iv) $S^{-1}(M \cap P) = (S^{-1}M) \cap (S^{-1}P)$; *In particular* $(M \cap P)_{\mathfrak{p}} = M_{\mathfrak{p}} \cap P_{\mathfrak{p}}$, *for any* $\mathfrak{p} \in S_{pec}(A)$;

(v) $S^{-1}(M/N) \cong (S^{-1}M)/(S^{-1}N)$, *(as* $(S^{-1}A)$*-module); In particular* $(M/N)_{\mathfrak{p}} \cong M_{\mathfrak{p}}/N_{\mathfrak{p}}$, *for any* $\mathfrak{p} \in S_{pec}(A)$;

(vi) $S^{-1}A \otimes_A M \cong S^{-1}M$, *for any* $\mathfrak{p} \in S_{pec}(A)$; *In particular* $A_{\mathfrak{p}} \otimes_A M \cong M_{\mathfrak{p}}$, *for any* $\mathfrak{p} \in S_{pec}(A)$;

(vii) $S^{-1}A$ *is a flat* A*-module; In particular* $A_{\mathfrak{p}}$ *is a flat* A*-module, for any* $\mathfrak{p} \in S_{pec}(A)$;

(viii) $S^{-1}M \otimes_{S^{-1}A} S^{-1}N \cong S^{-1}(M \otimes_A N)$; *In particular* $M_{\mathfrak{p}} \otimes_{S^{-1}A} N_{\mathfrak{p}} \cong (M \otimes_A N)_{\mathfrak{p}}$, *with* $S \equiv A \setminus \mathfrak{p}$ *and , for any* $\mathfrak{p} \in S_{pec}(A)$.

2) *If S is the set of non-zero-divisors of A, one denotes $S^{-1}A \equiv Q(A)$ and calls it the full ring of fractions of A. In that case, A can be considered as a subring of $Q(A)$. One has the following exact sequence:* $0 \to t_S(M) \to M \to S^{-1}M$, *where the map $M \to S^{-1}M$ is $x \mapsto \frac{x}{1}$ and $t_S(M) \equiv \{x \in M \mid sz = 0,$ for some $s \in S\}$ is called the S-torsion submodule of M.*[40]

3) *When A is an integral domain and $S = A \setminus \{0\}$ then $K \equiv Q(A)$ is a field called the field of fractions of A. If $S = S_a$, then $S^{-1}A = A[\frac{1}{a}]$.*

4) *One has the following local properties:*

(i) $M = 0$ *iff $M_{\mathfrak{p}} = 0$ for all prime ideals $\mathfrak{p} \subset A$;*

---

[39] If $a \in A$ and $S = \{a^n\}_{n \geq 0}$, then one writes $A_a \equiv S^{-1}A$.

[40] A module is called *torsion-free* if it has no *torsion element*, i.e., $t_S(M) = 0$. For example the **Z**-module **Q/Z** is torsion free.

(ii) $M = 0$ iff $M_{\mathfrak{m}} = 0$ for all maximal ideals $\mathfrak{m} \subset A$:

(iii) $\phi : M \to N$ injective iff $\pi_{\mathfrak{p}} : M_{\mathfrak{p}} \to N_{\mathfrak{p}}$ is so, for each prime ideal $\mathfrak{p} \subset A$;

(iv) $\phi : M \to N$ injective iff $\pi_{\mathfrak{m}} : M_{\mathfrak{m}} \to N_{\mathfrak{m}}$ is so, for each maximal ideal $\mathfrak{m} \subset A$;

(v) M flat iff $M_{\mathfrak{p}}$ is so, for each prime ideal $\mathfrak{p} \subset A$;

(vi) M flat iff $M_{\mathfrak{m}}$ is so, for each maximal ideal $\mathfrak{m} \subset A$;

(vii) A sequence $M \to N \to F$ of A-modules and homomorphisms is exact iff the corresponding *localized sequences*: $M_{\mathfrak{m}} \to N_{\mathfrak{m}} \to F_{\mathfrak{m}}$ is exact, for all maximal ideals $\mathfrak{m} \subset A$.

(viii) A short exact sequence $0 \to M \to N \to F \to 0$ of A-modules and homomorphisms, where M is finitely presentable *splits* iff the the corresponding localized exact sequences: $0 \to M_{\mathfrak{m}} \to N_{\mathfrak{m}} \to F_{\mathfrak{m}} \to 0$ splits for all maximal ideals $\mathfrak{m} \subset A$.

(ix) If M is a finitely presentable A-module and $N \subset M$ is a finitely generable submodule, then N is a direct summand of M iff $N_{\mathfrak{m}}$ is a direct summand of $M_{\mathfrak{m}}$, for any maximal ideals $\mathfrak{m} \subset A$.

(x) If A is an integral domain, and M is an A-module, then M is torsion-free iff $M_{\mathfrak{m}}$ is torsion-free for any maximal ideal $\mathfrak{m}$ of A.

(xi) If M is a finitely generated module over a Noetherian integral domain A, then M is reflexive, i.e., $M \cong Hom_A(Hom_A(M; A); A)$, iff $M_{\mathfrak{m}}$ is reflexive for any maximal ideal $\mathfrak{m}$ of A.

(xii) If M is a torsion-free module over an integral domain A and $S = A \setminus \{0\}$, then we have an exact sequence $0 \to M_{\mathfrak{m}} \to S^{-1}N$, for any maximal ideal $\mathfrak{m}$ of A and $N = \bigcap_{\mathfrak{m}} M_{\mathfrak{m}}$.

5) If $\phi : A \to A/\mathfrak{a}$ is the canonical homomorphism, it follows that

$$r(\mathfrak{a}) = \phi^{-1}(nil(A)_{A/\mathfrak{a}}).$$

6) A ring with a unique maximal ideal $\mathfrak{m}$ is called a local ring. $A_{\mathfrak{p}}$ is a local ring, where $\mathfrak{p}$ is a prime ideal of A.

Theorem 3.37. 1) One has a natural structure of topological space (Zariski topology) on $S_{pec}(A)$. The open sets in this topology are the sets $V(E) \equiv \{\mathfrak{p} \supset E | E \subset A, \mathfrak{p}$ prime ideals$\}$. With respect to such a topology, $S_{pec}(A)$ is a quasi-compact space, that is every open covering of $S_{pec}(A)$ has a finite subcovering.

2) If the ring A is Noetherian, $S_{pec}(A)$ is a Noetherian space. (The converse of this is false.)

3) $S_{pec}\left(\prod_{1\leq i\leq n} A_i\right) \cong \bigcup_{1\leq i\leq n} S_{pec}(A_i)$.

4) Let $\mathfrak{a}$ be an ideal of A. The ideals of $A/\mathfrak{a}$ are in correspondence one-to-one with the ideals of A that contain $\mathfrak{a}$, hence $S_{pec}(A/\mathfrak{a}) \cong V(\mathfrak{a})$, where $V(\mathfrak{a})$ is the set of prime ideals of A that contain $\mathfrak{a}$. This is just a closed subspace of $S_{pec}(A)$.

5) S_{pec} is a contravariant functor from the category of rings and ring-homomorphisms to the category of topological spaces and continuous maps. In particular, if $\phi : A' \to A$ is surjective, then $S_{pec}(\phi) : S_{pec}(A) \to S_{pec}(A')$ is a closed embedding, i.e., a homeomorophism of $S_{pec}(A)$ onto a closed subset of $S_{pec}(A')$. Furthermore, if ϕ is injective, then $S_{pec}(\phi)$ is dominant, i.e., $S_{pec}(\phi)(S_{pec}(A))$ is dense in $S_{pec}(A')$.

6) Let $a \in A$. Then $S_{pec}(A_a) = \{x \in S_{pec}(A)|a \notin x\}$.

7) Let $\phi : A \to A_a$ be the canonical homomorphism associated to an element $a \in A$. Then the corresponding mapping $S_{pec}(\phi)$ is a homeomorphism of $S_{pec}(A_a)$ onto the open set: $D(a) \equiv S_{pec}(A) - V(a) \equiv \{x \in S_{pec}(A)|a(x) \neq 0\} \equiv$ support of a. Here $a(x)$ denotes the class of $a \bmod x$ in A/x. Thus $a(x) = 0$ iff $a \in x$. ($D(a)$ is an open set.)

In the following table we report some distinguished objects associated to the points of a prime spectrum of a ring. (For the definition of affine variety see Definition 3.34.)

TAB.3.6 - Objects associated to $x \in S_{pec}(A)$.

Name	Definition
local ring at x	$A_x \equiv S^{-1}A,\ S \equiv A\backslash x$
residue field of A_x	$\kappa(x) \equiv A_x/\mathfrak{m}_x,\ \mathfrak{m}_x \subset A_x$ maximal ideal. $(\star)$
value of $a \in A$ at x	$a(x) \in A/x \subseteq \kappa(x),\ A \to A/x,\ (a(x)=0 \Leftrightarrow a \in x).$ $(\star\star)$
tangent space at x	$T_x S_{pec}(A) \equiv Hom_{\kappa(x)}(\mathfrak{m}_x/\mathfrak{m}_x^2;\kappa(x)).$ $(\star\star\star)$

$(\star)$ One has the exact sequence: $0 \to x \to A \to \kappa(x)$

$(\star\star)$ If $A \subset \kappa[X]$, $X \equiv$ affine variety, $x \in X$ determines a maximal ideal of A

$\kappa(x) = \kappa$, $a(x) \equiv$ value of the function a at x.

$(\star\star\star)$ $\mathfrak{m}_x/\mathfrak{m}_x^2 \equiv \kappa(x)$-vector space.

If A_x is a Noetherian ring (e.g. if A is Noetherian) then $\mathfrak{m}_x/\mathfrak{m}_x^2$ is finite dimensional.

Definition 3.27. The *support* of a A-module M is defined by the following:

$$supp_A(M) \equiv \{\mathfrak{p} \in S_{pec}(A) : M_\mathfrak{p} \neq 0\} \subseteq S_{pec}(A).$$

Proposition 3.5. (Properties of module support). 1) $M \neq 0 \Leftrightarrow supp_A(M) \neq \emptyset$.

2) $V(\mathfrak{a}) = supp_A(A/\mathfrak{a})$, where $V(\mathfrak{a})$ is the set of prime ideals containing $\mathfrak{a}$.

3) If $0 \rightarrow M' \rightarrow M \rightarrow M'' \rightarrow 0$ is an exact sequence, then $supp_A(M) = supp_A(M') \cup supp_A(M'')$.

4) If $M = \sum_i M_i \Rightarrow supp_A(M) = \bigcup_i supp_A(M_i)$.

5) If M is finitely generated A-module, it follows that

$$supp_A(M) = V(ann_A(M))$$

and therefore is a closed subset of $S_{pec}(A)$.

6) If M and N are finitely generated, then $supp_A(M \otimes_A N) = supp_A(M) \cap supp_A(N)$.

7) If M is finitely generated and $\mathfrak{a}$ is an ideal of A, then $supp_A(M/\mathfrak{a}M) = V(\mathfrak{a} + ann_A(M))$.

8) If $f : A \rightarrow B$ is a ring homomorphism and M is a finitely generated A-module, then $supp_A(B \otimes_A M) = f^{*-1}(supp_A(M))$, where $f^* : S_{pec}(B) \rightarrow S_{pec}(A)$ is the map induced by f.

Proposition 3.6. Let $A \equiv \bigoplus_{0 \leq n \leq \infty} A_n$ be a Noetherian graded ring. Then A_0 is a Noetherian ring and $A = A_0[x_1, \cdots, x_s]$, with $|x_j| > 0,$[41] i.e., A is a finitely generated A_0-algebra. Let $\mathcal{C}$ be a class of A-modules and $\lambda : \mathcal{C} \rightarrow \mathbf{Z}$. The function λ is additive if, for each short exact sequence $0 \rightarrow M' \xrightarrow{f} M \xrightarrow{g} M'' \rightarrow 0$ in $\mathcal{C}$, we have $\lambda(M) = \lambda(M') + \lambda(M'')$. Let $0 \rightarrow M_0 \rightarrow M_1 \rightarrow \cdots \rightarrow M_n \rightarrow 0$ be an exact sequence of A-modules in which all the modules M_i and the kernels of all the homomorphisms belong to $\mathcal{C}$. Then for any additive function λ on $\mathcal{C}$ we have $\sum_{0 \leq i \leq n} \lambda(M_i) = 0$.

Definition 3.28. We call *Hilbert polynomial* $P(M,t) = \sum_{0 \leq n \leq \infty} \lambda(M_n) t^n \in \mathbf{Z}[[t]]$, where λ is an additive $\mathbf{Z}$-valued function.

Theorem 3.38. (Hilbert-Serre). $P(M,t)$ is a rational function in t of the form $P(M,t) = \dfrac{f(t)}{\prod_{1 \leq i \leq s}(1 - t^{|x_i|})}$, where $f(t) \in \mathbf{Z}[t]$, and s is the number of generators of A over A_0.

---

[41] Here $|x_j| = n$ iff $x_j \in A_n$.

Proof. By induction on s. Start with $s = 0$; this means that $A_n = 0$ for all $n > 0$ and that M is a finitely generated A_0-module, hence $M_n =$ for some $n \le n_0$. In such a case $P(M,t)$ is a polynomial. Let us assume, now, $s > 0$, and the theorem true for $s - 1$. Multiplication by x_s is an A-module homomorphism of M_n into $M_{n+|x_s|}$ hence it gives the following exact sequence:

$$(3.16) \qquad 0 \to K_n \to M_n \xrightarrow{x_s} M_{n+|x_s|} \to L_{n+|x_s|} \to 0.$$

Let $K \equiv \oplus_n K_n$, $L \equiv \oplus_n L_n$; these are both finitely-generated A-modules (because K is a submodule and L a quotient module of M, that is a Noetherian A-module), and both are annihilated by x_s, hence are $A_0[x_1, \cdots, x_{s-1}]$-modules. Applying λ to (3.16) we have

$$\lambda(K_n) - \lambda(M_n) + \lambda(M_{n+|x_s|}) - \lambda(L_{n+|x_s|}) = 0.$$

Multiplying by $t^{n+|x_s|}$ and summing with respect to n we get

$$(1 - t^{|x_s|})P(M,t) = P(L,t) - t^{|x_s|}P(K,t) + g(t),$$

where $g(t)$ is a polynomial. Applying the inductive hypotheses then we can conclude the proof. $\qquad\qquad\square$

Definition 3.29. We denote by $d(M)$ the order of the pole of $P(M,t)$ at $t = 1$. It provides a measure of the "size" of M (relatively to λ). In particular $d(A)$ is defined.

Corollary 3.3. *If each $|x_i| = 1$, then for all sufficiently large n, $\lambda(M_n)$ is a polynomial in n (with rational coefficients) of degree $d - 1$. (We adopt the convention that the degree of the zero polyniomial is -1; also the binomial coefficient $\binom{n}{-1} = 0$ for $n \ge 0$, and $n = 1$ for $n = -1$.) This polynomial is called* Hilbert polynomial *of M (with respect to λ).*

Example 3.20. If $x \in A_\kappa$ is not a zero-divisor of M, then $d(M/xM) = d(M) - 1$. $\qquad\qquad\blacksquare$

Example 3.21. If A_0 is an Artinian ring (in particular, a field),

$$A = A_0[x_1, \cdots, x_s],$$

then A_n is a free A_0-module generated by the monomials $x_1^{m_1} \cdots x_s^{m_s}$, with $\sum_i m_i = n$; there are $\binom{s-n-1}{s-1}$ of these, hence $P(A,t) = (1 - t)^{-s}$. $\qquad\blacksquare$

Definition 3.30. We define *dimension* of the $Supp(M)$ the degree of the Hilbert polynomial.

Definition 3.31. We define *(Krull) dimension of a ring A* the sup of lengths of $(n+1)$-chains of prime ideals $\mathfrak{p}_0 \subset \mathfrak{p}_1 \subset \cdots \subset \mathfrak{p}_n$ of length n:

$$\dim A = sup_n\{\mathfrak{p}_0 \subset \mathfrak{p}_1 \subset \cdots \subset \mathfrak{p}_n\}.$$

Theorem 3.39. (Properties of dimension of a ring). *One has the following properties.*

1) $\dim A = sup_{\mathfrak{p} \in S_{pec}(A)} \dim(A_\mathfrak{p}).$[42]

2) $\dim A \geq 0, +\infty.$

3) *A field has dimension 0; the ring $\mathbf{Z}$ has dimension 1.*

4) *A ring A is Artin iff A is Noetherian and $\dim A = 0$.*

5) *Let A be a Noetherian local ring, $\mathfrak{m}$ its maximal ideal. Then, exactly one of the following two statements is true: i) $\mathfrak{m}^n \neq \mathfrak{m}^{n+1}$ for all n; ii) $\mathfrak{m}^n = 0$ for some n, in which case A is an Artinian local ring.*

6) $\dim A = d(A) = \delta(A)$, *where $\delta(A)$ is the least number of generators of an $\mathfrak{m}$-primary ideal of A, Noetherian local ring.*

7) *Let M be an A-module. Then*

$$\dim_A(M) = \dim(A/ann_A(M)) = sup_{\mathfrak{p} \in supp_A(M)}\{\dim(A/\mathfrak{p})\}.$$

Definition 3.32. Let κ be a field of *characteristic zero*, that is a field containing the field $\mathbf{Q}$ of rational numbers as s subfield, and set $A = \kappa[\chi] \equiv \kappa[\chi_1, \cdots, \chi_n]$ the ring of polynomials in the indeterminates χ_i with coefficients in κ. For r given polynomials $P_1, \cdots, P_r \in \kappa[\chi]$, we define *algebraic set X* determined by the ideal $\mathfrak{a} \equiv < P_1, \cdots, P_r >$ of $\kappa[\chi]$ the following one:

$$X \equiv \left\{ x = (x_1, \cdots, x_n) \,\middle|\, \begin{array}{l} x_i, 1 \leq i \leq n, \text{ belong to an extension of } \kappa \\[4pt] \text{(for example its algebraic closure)} \\[4pt] \text{such that } P_j(x) = 0, 1 \leq j \leq r. \end{array} \right\}.$$

We define *irreducible algebraic set* (or *variety*) the set X of points in $\bar{\kappa}^n$, where $\bar{\kappa}$ is an algebraic closure of κ, which are zeros of solutions of a prime ideal $\mathfrak{p} \in S_{pec}(\kappa[\chi])$. We shall write $X = Z(\mathfrak{p})$ and we shall introduce the field $K \equiv \kappa(X) \equiv Q(\kappa[\chi]/\mathfrak{p})$ as the quotient field of the integral domain

---

[42] For a Noetherian ring A one has $\dim(A_\mathfrak{p}) < +\infty$, but can be also $\dim(A) = +\infty$.

$A \equiv \kappa[X] \equiv \kappa[\chi]/\mathfrak{p}$ of polynomial functions on X. (We have the inclusions $\kappa \subseteq A \subseteq K$.)[43] We call $A \equiv \kappa[X]$ the *coordinates ring* of X. $K \equiv \kappa(X)$ is called the *field of rational functions* on X.

Definition 3.33. A ring B is *integral* over a ring A if each element of B is *integral* over A, that is to say if each element of B is a root of a *unitary polynomial* with coefficients in A.[44] If $A \subset B$ are two rings, the *integral closure* of A in B is the subring of B consisting of all elements of B that are integral over A. Let us consider a field extension K/κ.[45] Elements $a_1, \cdots, a_n \in K$ are said to be *algebraically independent* over κ in K if one cannot find a polynomial $P \in \kappa[\chi_1, \cdots, \chi_n]$ such that $P(a_1, \cdots, a_n) = 0$.[46] A maximal subset of K which is algebraically independent over κ is called a *transcendence basis* of K/κ. The number of elements of a transcendence basis is uniquely defined and only depends on the extension K/κ. This numebr is called the *transcendence degree* of K/κ and denoted by $trd(K/\kappa)$. If A is an integral domain containing a field κ, we shall define $trd(A(\kappa)) = trd(K/\kappa)$ with $K = Q(A)$. Since $K = \kappa(A)$, transcendence basis exist that are subsets of A and any such transcendence basis will be called a transcendence basis of A/κ. If K/κ is a finitely generated extension, say $K = \kappa(a_1, \cdots, a_n)$ we may find a transcendence basis among the a_i, say $a_1, \cdots, a_r$ and $\kappa(a_1, \cdots, a_r)$, called a *purely* transcendental extension of κ-module K, is a finite algebraic extension of $\kappa(a_1, \cdots, a_r)$.

Proposition 3.7. (Noether normalization lemma). *If $A = \kappa[a_1, \cdots, a_n]$ is a finitely integral domain with $trd(A/\kappa) = r$, then there exist r algebraically*

---

[43]   We may consider A as a vector space over κ and K as a vector space over A or over κ. If we put $|A/\kappa| = \dim_\kappa A$, $|K/A| = \dim_A K$, $|K/\kappa| = \dim_\kappa K$, then we get $|K/\kappa| = |K/A| \cdot |A/\kappa|$.

[44]   A unitary polynomial is a polynomial where the coefficient of the term of highest degree is equal to 1.

[45]   For simplicity we shall denote by K/κ an extension of fields $0 \to \kappa \to K \to K/\kappa \to 0$.

[46]   Otherwise we say that $a_1, \cdots, a_n \in K$ are *algebraic* over κ. An extension K/κ is called an *algebraic (field) extension* if every element $a \in K$ is *algebraic* over κ, i.e., we can find a polynomial $P \in \kappa[\chi]$, such that $P(a) = 0$. One has the following. (*Existence of primitive element theorem*): *Every finitely generated algebraic extension is generated by a single element. More precisely, if $L = K(\eta^1, \cdots, \eta^m) \equiv Q(K[\eta^1, \cdots, \eta^m])$, then we can find $c_1, \cdots, c_m \in K$ such that $L = K(\zeta) \equiv Q(K[\zeta])$, with $\zeta = c_1\eta^1 + \cdots + c_m\eta^m$.* An extension L/K is called *regular* if K is *algebraically closed* in L, that is no element of L is algebraic over K. Let us consider the following extensions of fields: $\kappa \subset K \subset L$. If L/κ is regular, then $K \otimes_\kappa L$ is an integral domain.

independent linear combinations $b_1, \cdots, b_r$ of the a_i such that A is integral over $B = \kappa[b_1, \cdots, b_r]$ and $K = Q(A)$ is algebraic over $\kappa(b_1, \cdots, b_r)$.

Lemma 3.6. *If κ is an infinite field and $P \in \kappa[\chi_1, \cdots, \chi_n]$ is a non-zero polynomial, then we can find elements $\alpha_1, \cdots, \alpha_n \in \kappa$ such that $P(\alpha_1, \cdots, \alpha_n) \neq 0$ in an infinite number of fashions.*

Definition 3.34. If X is a variety defined over the ground field κ we define the *dimension* of X to be the integer $\dim_\kappa(X) = trd(\kappa(X)/\kappa)$. If X is an *affine variety*, that is to say $X = Z(\mathfrak{p})$ with $\mathfrak{p} \in S_{pec}(\kappa[\chi_1, \cdots, \chi_n])$, then we define the *codimension* of X by the integer $\text{codim}\,(X) = n - \dim(X) \equiv n - d(X)$.

Definition 3.35. Let X be an algebraic set defined as the zero of an ideal $\mathfrak{a}$, that is $X = Z(\mathfrak{a})$. Let us consider the decomposition of the ideal $I(X) = rad(\mathfrak{a})$ into prime ideals. Then, we define *dimension of algebraic set* (or of the ideal) to be the maximum of the dimensions of its irreducible components.

Example 3.22. An *irreducible plane curve* is a variety of dimension 1.   ∎

Example 3.23. A *hypersurface* is an algebraic set which is defined by principal ideal in $\kappa[\chi_1, \cdots, \chi_n]$ and all its irreducible components are of codimension 1. Conversely, any algebraic set whose components all have codimension 1 is a hypersurface and its defining ideal is principal.   ∎

Example 3.24. If a polynomial P does not vanish identically on a variety X, then the irreducible components of the intersection of X with the hypersurface defined by P all have dimension equal to $\dim(X) - 1$.   ∎

Theorem 3.40. (Hilbert theorem of zeros.) *A polynomial $P \in \kappa[\chi]$ vanishes on all the zeros of $P_1, \cdots, P_r \in \kappa[\chi]$ iff $P \in rad < P_1, \cdots, P_r >$.*

Corollary 3.4. *The polynomials $P_1, \cdots, P_r \in \kappa[\chi]$ have no-common zero iff*

$$< P_1, \cdots, P_r >= A \equiv \kappa[\chi_1, \cdots, \chi_n].$$

Theorem 3.41. *1) Let us denote by $X = Z(\mathfrak{a})$ the algebraic set dteremined by $\mathfrak{a}$ and by $I(X) \subset A$ the ideal of all polynomials vanishing on X. One has: $I(Z(\mathfrak{a})) = rad(\mathfrak{a})$. One has the following properties:*

(i) $X \subset Y \Rightarrow I(X) \supset I(Y)$.

(ii) $\mathfrak{a} \subset \mathfrak{b} \Rightarrow Z(\mathfrak{a}) \supset Z(\mathfrak{b})$.

(iii) $I(X \cup Y) = I(X) \cap I(Y)$.

(iv) $Z(\mathfrak{a} + \mathfrak{b}) = Z(\mathfrak{a}) \cap Z(\mathfrak{b})$.

(v) $Z(\mathfrak{a}) = \emptyset \Rightarrow \mathfrak{a} = A$.

(vi) Let X and Y be algebraic sets defined over κ with $I(X) = \mathfrak{a} \subset \kappa[u_1, \cdots, u_r]$ and $I(Y) = \mathfrak{b} \subset \kappa[v_1, \cdots, v_s]$. Then one has $\kappa[X \times Y] \cong \kappa[X] \otimes_\kappa \kappa[Y]$, where $\kappa[X] \equiv \kappa[u]/\mathfrak{a}$, $\kappa[Y] \equiv \kappa[v]/\mathfrak{b}$ and $\kappa[X \times Y] = \kappa[u,v]/(\mathfrak{a}, \mathfrak{b})$.

(vii) The ring $\kappa[X] = \kappa[\chi]/\mathfrak{a}$ is reduced, that is does not contain any nilpotent element. Such a ring is called the ring of rational fractions on X. $\mathfrak{a}$ is an intersection of prime ideals according to the decomposition theorem, say $\mathfrak{a} = \bigcap_i \mathfrak{p}_i$, then $X = \bigcup_i X_i$, with each $X_i = Z(\mathfrak{p}_i)$ a variety and $\kappa[X]$ is contained in a direct sum of integral domains $A_i \equiv \kappa[X]/\mathfrak{p}_i$: $\kappa[X] \subset \bigoplus_i A_i$. $\kappa(X)$ is contained in a direct sum of fields $K_i \equiv Q(\kappa[\chi]/\mathfrak{p}_i)$: $\kappa(X) \subset \bigoplus_i K_i$. Each field K_i is a finitely-generated extension of κ and so has a finite transcendence degree over κ, i.e., the maximum number of algebraically independent elements. Each variety X_i coincides with the set of maximal ideals of A_i. Then, the local dimension of X_i at $p \in X_i$ is $\dim(A_i)_{\mathfrak{m}_i}$, where $\mathfrak{m}_i$ is the maximal ideal corresponding to p. The local dimension of X at $p \in \bigcap_j X_j$ is $\dim_p X = \sup_j \{\dim(A_j)_{\mathfrak{m}_j}\}$.[47]

(viii) Let X and Y be two varieties defined over the field κ. One has: $\mathrm{codim}_\kappa(X \cap T) \leq \mathrm{codim}_\kappa(X) + \mathrm{codim}_\kappa(Y)$.

Theorem 3.42. *Let A be a Noetherian ring then one has: $\dim A[x_1, \cdots, x_n] = n + \dim A$.*

Definition 3.36. *Let Y be any variety. Y is nonsingular at a point $p \in Y$ if the local ring $A_p \equiv \kappa[Y]_p$ is a regular local ring. Y is nonsingular it it is nonsingular at every point. Y is regular if it is not nonsingular.*

Proposition 3.8. *If A is a Noetherian local ring with maximal ideal $\mathfrak{m}$ and residue field κ, then $\dim_\kappa \mathfrak{m}/\mathfrak{m}^2 \geq \dim A$. If A is regular one has $\dim_\kappa \mathfrak{m}/\mathfrak{m}^2 = \dim A$.*

Theorem 3.43. *Let Y be a variety. Then the set, $Sing(Y)$, of singular*

---

[47] If A is an integral algebra of finite type on a field κ, then for any maximal ideal $\mathfrak{m} \subset A$, one has $\dim A_\mathfrak{m} = trd(Q(A)/\kappa)$. If A is a Noetherian local ring with $\dim A = d$, the minimal number of generators of the maximal ideal $\mathfrak{m}$ is always $\geq d$. We say that A is *regular* if the equality holds. A Noetherian ring A is called *regular* if $A_\mathfrak{m}$ is regular for any maximal ideal $\mathfrak{m}$ of A. In particular, the ring of polynomials $A[\chi]$ and formal series $A[[\chi]]$ are regular. One has the following property for A-modules (*syzygies theorem of Hilbert-Serre*): The A-modules M of finite type, with A regular, are characterized by the fact to admit a projective resolution: $0 \to L_d \to L_{d-1} \to \cdots \to L_1 \to M \to 0$. A regular local ring is integral and integrally closed.

points of Y, is a proper closed subset of Y.

Theorem 3.44. (Topologies on a ring). *Let (A_n) be a filtration of a commutative ring A. Then (A_n) defines a topology on A compatible with the structure of ring, where (A_n) is a fundamental system of neighbourhoods of $0 \in A$. Set $A_\infty = \bigcap_n A_n$. Then A/A_∞ is a metrizable topological ring and its completion $\hat{A}$ is called the (separated) completion of A. The adherence in $\hat{A}$ of A_n/A_∞ forms a fundamental system of neighborhoods of 0. Then $\hat{A} = \varprojlim A/A_n$.*

Example 3.25. ($\mathfrak{a}$-*adic topology*). 1) Let $\mathfrak{a} \subset A$ be an ideal of A and let us consider the filtration $(A_n \equiv \mathfrak{a}^n)$ of A. Then the corresponding topology is called $\mathfrak{a}$-*adic*.

2) Furthermore, let us assume that $\bigcap_n \mathfrak{a}^n = \{0\}$ and that $\mathfrak{a}$ is of finite type. Then, one has $\overline{\mathfrak{a}^n} = (\overline{\mathfrak{a}})^n = \mathfrak{a}^n \hat{A}$, where $\overline{\mathfrak{a}}$ is the adherence of $\mathfrak{a}$ with respect to the $\mathfrak{a}$-adic topology.

3) In particular, if $A = R[\chi_1, \cdots, \chi_n]$ and $\mathfrak{a} \equiv < \chi_1, \cdots \chi_n >$, then one has $\hat{A} = R[[\chi_1, \cdots, \chi_n]]$, the ring of formal poweseries in n indeterminates on R, that is $\sum_\alpha c_\alpha \chi^\alpha$, $\chi^\alpha \equiv \chi_1^{\alpha_1} \cdots \chi_n^{\alpha_n}$, $c_\alpha \in R$, $\alpha \equiv (\alpha_1, \cdots, \alpha_n) \in \mathbf{N}^n$, converging in the topology of $\hat{A}$. If R is integer, then so is $\hat{A} = R[[\chi_1, \cdots, \chi_n]]$. ∎

Theorem 3.45. *Let A be a Noetherian local ring with maximal ideal $\mathfrak{m}$, and let $\hat{A}$ be its completion with respect to the $\mathfrak{m}$-adic topology. Then one has the following properties.*

(i) *$\hat{A}$ is a local Noetherian ring with maximal ideal $\hat{\mathfrak{m}} = \mathfrak{m}\hat{A}$, and there is a natural injective homomorphism $A \to \hat{A}$. The residue field of $\hat{A}$ is canonically isomorphic to that of A. Furthermore, $\hat{A}$ is a flat A-module such that for any $\mathfrak{p} \in S_{pec}(A)$, one has $\mathfrak{p} = A \cap \mathfrak{p}'$, for some $\mathfrak{p}' \in S_{pec}(\hat{A})$, (that is the A-module $\hat{A}$ is fidelly flat).*

(ii) *If M is a finitely generated A-module, its completion $\hat{M}$ with respect to its $\mathfrak{m}$-adic topology is isomorphic to $M \otimes_A \hat{A}$.*

(iii) *$\dim A = \dim \hat{A}$.*

(iv) *A is regular iff $\hat{A}$ is regular. (Furthermore, if A is regular any localized ring $A_\mathfrak{p}$ is also regular.)*

Theorem 3.46. (Cohen structure theorem). *If A is a complete regular local ring of dimension n containing some field, then $A \cong \kappa[[x_1, \cdots, x_n]]$, the ring of formal power series over the residue field κ of A.*

Definition 3.37. *We say two points $p \in X$ and $q \in Y$ are analytically* *isomorphic* if there is an isomorphism $\hat{A}(X)_p \cong \hat{A}(Y)_q$ as κ-algebras. Here $A(X) \equiv \kappa[X]$ and $A(Y) \equiv \kappa[Y]$.

Theorem 3.47. (Elimination theorem). *Let $f_1, \cdots, f_r$ be homogeneous polynomials in $x_0, \cdots, x_n$ having indeterminates coefficients a_{ij}. Then there is a set $g_1, \cdots, g_t$ of polynomials in the a_{ij}, with integer coefficients, which are homogeneous in the coefficients of each f_i separately, with the following property: for any field κ, and for any set of spacial values of the $a_{ij} \in \kappa$, a necessary any sufficient condition for the f_i to have a common zero different from $(0, 0, \cdots, 0)$ is that the a_{ij} are a common zero of the polynomials g_j.*

Example 3.26. (*Examples of singular points*). In the following table are reported some examples of singular points of curves and surfaces.

TAB.3.7 - Examples of singular points

Name	Equation
Singular points of curves in κ^2 ($char\kappa \neq 2$).	
node	$x^4 + y^4 - x^2 = 0$
triple point	$x^6 + y^6 - xy = 0$
cusp	$x^4 + y^4 - x^3 + y^2 = 0$
tacnode	$x^4 + y^4 - x^2 y - xy^2 = 0$
Singular points of surfaces in κ^3 ($char\kappa \neq 2$).	
conical double point	$z^2 - xy^2 = 0$
double line	$z^2 - x^2 - y^2 = 0$
pinch point	$z^3 + y^3 + xy = 0$

Let $Y \subseteq \kappa^2$ be a curve defined by the equation $f(x, y) = 0$. Let $p = (a, b)$ be a point of κ^2. Make a linear change of coordinates so that p becomes the point $(0, 0)$. Then, write f as a sum $f = f_0 + \cdots + f_d$, where f_i is a homogeneous polynomial of degree i in x and y. Then, we define *multiplicity* of p on Y, denoted $\mu_p(Y)$, the least r such that $f_r \neq 0$. (Note that $p \in Y$ iff $\mu_p(Y) > 0$.) The linear factors of f_r are called the *tangent directions* at p. Then one has the following: $\mu_p(Y) = 1$ iff p is a nonsingular point of Y. If $Y, Z \subset \kappa^2$ are two distinct curves, given by equations $f = 0$ and $g = 0$, respectively, and if $p \in Y \cap Z$, we define the *intersection multiplicity*, $(Y.Z)_p$, of Y and Z at p to be the length of A_p-module $A_p/ < f, g >$.

Then one has the following: (a) $(Y.Z)_p$ is finite and $(Y.Z)_p \geq \mu_p(Y).\mu_p(Z)$; (b) If $p \in Y$ for almost all lines L trough p (i.e., all but a finite number) $(L.Y)_p = \mu_p(Y)$. ∎

Definition 3.38. The fields, the ring $\mathbf{Z}$, the complete Noetherian local rings, are called *excellent rings*.

Theorem 3.48. (Properties of excellent rings). 1) *Let A be an excellent ring. Then the permanence properties hold:*

(i) *Any localized ring $S^{-1}A$ is an excellent ring.*

(ii) *Any A-algebre of finite type is an excellent ring.*

2) *Let A be an excellent local ring. The following properties hold:*

(i) *A is a reduced ring iff $\hat{A}$ is so.*

(ii) *A is integral and integrally closed iff $\hat{A}$ is so.*

(iii) *If A is integral, its integral closure A' is a finite A-algebra.*

Proposition 3.9. (Derivations and localization). *Let A be a ring containing a field κ.*

1) *Let $S \subset A$ be a subset of a ring A. Then $\delta \in Der_\kappa(A, M)$ induces a unique derivation $\delta \in Der_\kappa(S^{-1}A; S^{-1}M)$. More precisely $\delta(\frac{a}{s}) = \frac{s\delta a - a\delta s}{s^2}$. In particular, δ induces a unique derivation $\delta \in Der_\kappa(A_\mathfrak{p}; M_\mathfrak{p})$, $\forall \mathfrak{p} \in S_{pec}(A)$.*

2) *There are the following isomorphisms:*

$$S^{-1}A \bigotimes_A \Omega^1(A) \cong \Omega^1(S^{-1}A),$$

$$S^{-1}A \bigotimes_A Der_K(A, M) \cong Der_K(S^{-1}A; S^{-1}M).$$

In particular, one has the following isomorphisms:

$$A_\mathfrak{p} \bigotimes_A \Omega^1(A) \cong \Omega^1(A_\mathfrak{p}),$$

$$A_\mathfrak{p} \bigotimes_A Der_K(A, M) \cong Der_K(A_\mathfrak{p}; M_\mathfrak{p}),$$

for any $\mathfrak{p} \in S_{pec}(A)$.

Example 3.27. Let us consider a finitely generated integral domain $A = \kappa[\chi_1, \cdots, \chi_n]$ over a field κ, with quotient field $K \equiv Q(A)$ and extension of finitely elements $y_1, \cdots, y_r \in A$ such that A is *integral* over $\kappa[y] =$

$\kappa[y_1, \cdots, y_r]$, that is each element $a \in A$ satisfies a polynomial equation of the form

$$P(a) \equiv a^d + c_1 a^{d-1} + \cdots + c_d = 0,$$

where $c_i \in \kappa[y]$. In that case K is an *algebraic extension* of $\kappa(y)$, that is each element of K satisfies a polynomial equation with coefficients in $\kappa(y) = Q(\kappa[y])$ and $(y_1, \cdots, y_r)$ is called a *trascendence basis* of the extension K/κ of degree r. Then we have that there exists $0 \neq c \in \kappa[y] \subseteq A$ such that $\Omega(A)$ is a free A_c-module with basis $\{dy_i\}$ and $Der_\kappa(A_c)$ is a free A_c-module with basis $\{\partial y_i\}$.   ∎

Proposition 3.10. *Let $\kappa \subset K \subset L$ be a chain of extensions of fields. We have the following short exact sequence of vector spaces over L:*

$$0 \to L \bigotimes_K \Omega^1(K)_\kappa \to \Omega^1(L)_\kappa \to \Omega^1(L)_K \to 0.$$

Here we denote $\Omega^1(A)_\kappa \equiv \Omega^1(A)$, (resp. $\Omega^1(A)_K \equiv \Omega^1(A)$) to emphasize that we are talking about κ-derivations, (resp. K-derivations). If L/K is any algebraic extension, we have of course $\Omega(L)_K = 0$ because any derivation of K can be extended uniquely to L and we get the following isomorphism: $L \bigotimes_K \Omega(K)_\kappa \cong \Omega(L)_\kappa$.

Definition 3.39. A *differential ring* is a ring A with a finite number n of commutating derivations $d_1, \cdots, d_n$, $d_i d_j - d_j d_i = 0$, $\forall i, j = 1, \cdots, n$. A *differential ideal* is an ideal $\mathfrak{a} \subset A$ which is stable by each d_i, $i = 1, \cdots, n$.

Proposition 3.11. 1) *A differential ring $(A, \{d_j\}_{1 \leq j \leq n})$ identifies a subring (subring of constants): $C \equiv cst(A) \equiv \{a \in A | d_j a = 0, \forall j = 1, \cdots, n\} \subset A$.*
2) *We may extend each d_i to a derivation of $Q(A)$, still denoted by d_i and such that $d_i(a/r) = (r d_i a - a d_i r)/r^2$, for any $0 \neq r$, $a \in A$.*

Example 3.28. If K is a differential field with derivations $\partial_1, \cdots, \partial_\mu$ and y^k, $k = 1, \cdots, m$, are indeterminates over K, we set $y_0^k = y^k$. Then the polynomial ring $K[y]_d = K[y_\mu^k, k = 1, \cdots, m, \mu = \mu_1 \cdots \mu_s, |\mu| \geq 0]$, can be endowed with a structure of differential ring by defining the *formal derivations* $d_i \equiv \partial_i + y_{\mu+1i}^k \partial y_k^\mu$. Of course $K[y]_d$ is not a Noetherian ring. We write $K[y_q]_d = K[y_\mu^k | k = 1, \cdots, m; 0 \leq |\mu| \leq q]$ and one has $K(y_q)_d = Q(K[y_q]_d)$. We set also $K(y)_d = Q(K[y]_d)$.   ∎

Definition 3.40. A *differential subring* A of a differential ring B is a subring which is stable under the derivations of B. Similarly we can define a *differential extension* L/K of differential fields, and such an extension is said to be *finitely generated* if one can find elements $\eta^1, \cdots, \eta^m \in L$ such that $L = K(\eta^1, \cdots, \eta^m)$. Then the *evaluation epimorphism* is defined by $K[y]_d \to K[\eta]_d \subset L$, $y^k \mapsto \eta^k$. Its kernel is a prime differential ideal.

Definition 3.41. Let $< S >_d$ denote the differential ideal generated by the subset $S \subset A$, where A is a differential ring.

Lemma 3.7. *If A is a differential ring and $a, b \in A$, then one has the following:*

(i) $a^{|\mu|+1} d_\mu b \in < d_\nu(ab) || \nu | \leq |\mu| >$.

(ii) $(d_i a)^{2r-1} \in < a^r >_d$.

(iii) *If $\mathfrak{a}$ is a radical differential ideal of the differential ring A and S is any subset of A, then $\mathfrak{a} : S \equiv \{a \in A | aS \subset \mathfrak{a}\}$ is again a radical differential ideal of A.*

(iv) *If $\mathfrak{a}$ is a differential ideal of a differential ring A, then $rad(\mathfrak{a})$ is a differential ideal too.*

(v) *One has the following inclusion: $a\, rad < S >_d \subset rad < aS >_d$, $\forall a \in A$, and for all subset $S \subset A$.*

(vi) *If S and T are two subsets of a differential ring A, then*

$$rad < S >_d . rad < T >_d \subset rad < ST >_d = rad < S >_d \cap rad < T >_d .$$

(vii) *If S is any subset of a differential ring A, then we have:*

$$rad < S, a_1, \cdots, a_r >_d = rad < S, a_1 >_d \cap \cdots \cap rad < S, a_r >_d .$$

Definition 3.42. A *differential vector space* is a vector space V over a differential field $(K, \partial_i)_{1 \leq i \leq n}$ such that are defined n homomorphisms d_i, $i = 1, \cdots, n$, of the additive group V such that: $d_i(av) = (\partial_i a)v + a(d_i v)$, $\forall a \in K$, $\forall v \in V$. Then we say that K is a *differential field of definition*.

Proposition 3.12. *Let V be a differential vector space over a differential field K, with derivations d_i, $i = 1, \cdots, n$, and let $\{e_j\}_{j \in I}$ be a basis of V. Then the field of definition κ of a differential subspace $W \subset V$ is a differential subfield of K if it contains the field of definition of each $d_1 e_i, \cdots, d_n e_i$ with respect to $\{e_i\}$.*

Definition 3.43. A family $\eta = (\eta^1, \cdots, \eta^m)$ of elements in a differential extension of the differential field K is said to be *differentially algebraically independent* (or a *family of differential indeterminates*) over K, if the kernel of the evaluation epimorphism $K[y]_d \to K[\eta]_d$ is zero. Otherwise the family is said to be *differentially algebraically dependent* (or *differentially algebraic*) over K.

Proposition 3.13. *If K/κ and L/κ are two given differential extensions with respective derivations d_K and d_L, there always exists a differential free composite field of K and L over κ.*

Proof. The ring $K \otimes_\kappa L$ has a natural differential structure given by $d(a \otimes b) = (d_K a) \otimes b + a \otimes (d_L b)$, as $d_K|_\kappa = d_L|_\kappa = \partial$. On the other hand there is a finite number of prime ideals $\mathfrak{p}_i \subset K \otimes_\kappa L$ such that $\bigcap_i \mathfrak{p}_i = 0$ and $\mathfrak{p}_i + \mathfrak{p}_j =< 1 >$, $\forall i \neq j$. Now we have the following lemma.

Lemma 3.8. *If $\mathfrak{a}_1, \cdots, \mathfrak{p}_r$ are ideals of a differential ring A such that $\mathfrak{a}_i + \mathfrak{a}_j = A$, $\forall i \neq j$, and $\mathfrak{a}_1 \cap \cdots \cap \mathfrak{a}_r$ is a differential ideal of A, then each $\mathfrak{a}_i$ is a differential ideal too.*

Therefore we can conclude that each $\mathfrak{p}_i$ is a differential ideal, hence the proposition is proved. $\qquad\square$

Lemma 3.9. *A family η is differentially algebraic over K iff a differential polynomial $P \in \mathfrak{p}$ exists such that $(\partial y_P.P) \notin \mathfrak{p}$, where y_P is the highest power of y_p appearing in P. $S_P \equiv (\partial y_P.P)$ is called the separout of P. (The initial of P is the coefficient of the highest power of y_P appearing in P and it is denoted by I_P. More precisely one has $P = I_P(y_P)^r +$ terms of lower degree.)*

Definition 3.44. If S is any subset of a differential ring A and $r \geq 0$ is any integer, we call r-prolongation of S, the ideal

$$(S)_{+r} =< d_\nu a | a \in S,\ 0 \leq |\nu| \leq r >\subset A.$$

Proposition 3.14. *One has the following properties:*

(i) $(S)_{+(r+s)} = ((S)_{+s})_{+r}.$

(ii) $(S)_{+\infty} =< S >_d.$

Proposition 3.15. *Let $\mathfrak{a}$ be a differential ideal of the differential ring $K[y]_d$. We set $\mathfrak{a}_q = \mathfrak{a} \cap K[y_q]_d$, $\mathfrak{a}_0 = \mathfrak{a} \cap K[y]_d$, $\mathfrak{a}_\infty = \mathfrak{a}$. We call the r-prolongation of $\mathfrak{a}_q$, the following ideal:*

$$(\mathfrak{a}_q)_{+r} =< d_\nu P | P \in \mathfrak{a}_q,\ 0 \leq |\nu| \leq r >\subset K[y_{q+r}]_d.$$

One has:

$$\left\{ \begin{array}{l} (\mathfrak{a}_q)_{+r} \subseteq \mathfrak{a}_{q+r}, \\[2mm] (\mathfrak{a}_q)_{+\infty} \subseteq \mathfrak{a}, \\[2mm] (\mathfrak{a}_q)_{+r} \cap K[y_q]_d = \mathfrak{a}_q, \end{array} \right\} \quad \forall q, r \geq 0.$$

Remark 3.16. With algebraic sets it is better to consider radical ideals. Hence if $\mathfrak{r} \subset K[y]_d$ is a radical differential ideal, then $\mathfrak{r}_q$ is a radical ideal of $K[y_q]_d$, for all $q \geq 0$. Then if $E_q = Z(\mathfrak{r}_q)$ is the algebraic set defined over K by $\mathfrak{r}_q = I(E_q)$, we call *r-prolongation* of E_q the following algebraic set: $(E_q)_{+r} = Z((\mathfrak{r}_q)_{+r})$. In general one has $(\mathfrak{r}_q)_{+r} \subseteq \mathfrak{r}_{q+r}$, hence $rad((\mathfrak{r}_q)_{+r}) \subseteq \mathfrak{r}_{q+r}$. Therefore, in general one has: $E_{q+r} \subseteq (E_q)_{+r}$. ∎

Proposition 3.16. *Let $\mathfrak{p} \subset K[y]_d$ be a prime differential ideal. Then we can identify each field $L_q = Q(K[y_q]_d/\mathfrak{p}_q)$ with a non-differential subfield of $L = Q(K[y]_d/\mathfrak{p})$ and we have: $K \subseteq L_0 \subseteq \cdots \subseteq L_\infty = L$. Then there are vector spaces R_q over L_q or L defined by the following linear system:*

$$(\partial y_k^\mu . P_\tau)(\eta)v_\mu^k = 0, \quad \left\{\begin{array}{l} 1 \leq \tau \leq t, \\ 1 \leq k \leq m, \\ |\mu| = q, \end{array}\right\},$$

where η is a generic solution of $\mathfrak{p}$ and $P_1, \cdots, P_t$ are generating $\mathfrak{p}_q$. Such result does not depend on the generating polynomials. We can also define the vector space g_q (symbol) over L_q or L, by means of the linear system:

$$(\partial y_k^\mu . P_\tau)(\eta)v_\mu^k = 0, \quad \left\{\begin{array}{l} 1 \leq \tau \leq t, \\ 1 \leq k \leq m, \\ 0 \leq |\mu| \leq q, \end{array}\right\}.$$

For the prolongations $(g_q)_{+r}$ one has, in general, $g_{q+r} \subseteq (g_q)_{+r}$, $\forall q, r \geq 0$.

Definition 3.45. We say that R_q or g_q is *generic* over E_q, if one can find a certain number of maximum rank determinants D_α that cannot be all zero at a generic solution of $\mathfrak{p}$.

Proposition 3.17. *R_q or g_q is generic if we may find polynomials $A_\alpha, B_\tau \in K[y_q]_d$ such that:*

$$\sum_\alpha A_\alpha D_\alpha + \sum_\tau B_\tau P_\tau = 1.$$

Furthermore, R_q or g_q are projective modules over the ring $K[y_q]_d/\mathfrak{p}_q \subset K[y]_d/\mathfrak{p}$.

Proof. It follows directly from the Hilbert theorem of zeros. □

Theorem 3.49. (Primality criterion). *Let $\mathfrak{p}_q \subset K[y_q]_d$ and $\mathfrak{p}_{q+1} \subset K[y_{q+1}]_d$ be prime ideals such that $\mathfrak{p}_{q+1} = (\mathfrak{p}_q)_{+1}$ and $\mathfrak{p}_{q+1} \cap K[y_q]_d = \mathfrak{p}_q$. If the* symbol g_q of the variety R_q defined by $\mathfrak{p}_q$ is 2-acyclic and its first prolongation g_{q+1} is generic over E_q, then $\mathfrak{p} = (\mathfrak{p}_q)_{+\infty}$ is a prime differential ideal with $\mathfrak{p} \cap K[y_{q+r}]_d = (\mathfrak{p}_q)_{+r}$, for all $r \geq 0$.

Corollary 3.5. *Let $\mathfrak{r}_q \subset K[y_q]_d$ and $\mathfrak{r}_{q+1} \subset K[y_{q+1}]_d$ be radical ideals such that $\mathfrak{r}_{q+1} = (\mathfrak{r})_{+1}$ and $\mathfrak{r}_{q+1} \cap K[y_q]_d = \mathfrak{r}_q$. If the symbol g_q of the algebraic set E_q defined by $\mathfrak{r}_q$ is 2-acyclic and its first prolongation g_{q+1} is generic over E_q, then $\mathfrak{r} = (\mathfrak{r}_q)_{+\infty}$ is a radical differential ideal with $\mathfrak{r} \cap K[y_{q+r}]_d = (\mathfrak{r}_q)_{+r}$, for all $r \geq 0$.*

Theorem 3.50. *(Differential basis). If $\mathfrak{r}$ is a differential ideal of $K[y]_d$, then $\mathfrak{r} = rad((\mathfrak{r}_q)_{+\infty})$ for q sufficiently large.*

Proof. In fact one has the following lemma.

Lemma 3.10. *If $\mathfrak{p}$ is a prime ideal of $K[y]_d$, then for q sufficiently large, there is a polynomial $P \in K[y_q]_d$ such that $P \notin \mathfrak{p}_q$ and $P\mathfrak{p}_{q+r} \subset rad((\mathfrak{p}_q)_{+r}) \subset \mathfrak{p}_{q+r}$, for all $r \geq 0$.*

After above lemma the proof follows directly. $\qquad\square$

Proposition 3.18. *1) Every radical differential ideal of $K[y]$ can be expressed in a unique way as the non-redundant intersection of a finite number of prime differential ideals.*

2) The smallest field of definition κ of a prime differential ideal $\mathfrak{p} \subset K[y]$ is a finitely generated differential extension of $\mathbf{Q}$.

Example 3.29. With $n = 2$, $m = 2$, $q = 1$. Let us consider the differential polynomial $P = y_1^1 y_2^2 - y_1^2 y_2^1 - 1$. We obtain for the symbol g_1: $y_1^1 v_2^2 + y_2^2 v_1^1 - y_2^1 v_1^2 - y_1^2 v_2^1 = 0$. Setting $v_i^k = y_l^k w_i^l$ we obtain $(y_1^1 y_2^2 - y_1^2 y_2^1)(w_2^2 + w_1^1) = 0$ and thus $w_2^2 + w_1^1 = 0$ on E_1. Hence g_1 is generic. One can also set $P_1 = y_2^1$, $P_2 = y_1^2$ and we get the relation: $y_2^2 P_1 - y_2^1 P_2 - P \equiv 1$. A similar result should hold for E_1. g_1 is involutive and the differential ideal generted by P in $\mathbf{Q} < y^1, y^2 >$ is therefore a prime ideal. $\qquad\blacksquare$

Definition 3.46. A *differentially algebraic extension L over* of a differential field K is a differential extension over K where every element of L is differentially algebraic over K.

Definition 3.47. The *differential transcendence degree* of a differential extension L/K is the number of elements of a maximal subset S of elements of L that are differentially transcendental over K and such that L becomes differentially algebraic over $K(S)$. We shall denote such number by $trd_d(L/K)$.

Theorem 3.51. *One has the following formula:*

$$\dim(\mathfrak{p}_{q+r}) = \dim(\mathfrak{p}_{q-1}) + \sum_{1 \leq i \leq n} \frac{(r+i)!}{r!\,i!}\alpha_q^i, \quad \forall r \geq 0,$$

where α_q^i is the character of the corresponding system of PDE's.[48] The character α_q^n and the smallest non-zero character only depend on the differential extension L/K and not on the generators. In particular, one has: $trd_d(L/K) = \alpha_q^n$.

Proposition 3.19. *If ζ is differentially algebraic over $K(\eta)_d$ and η is differentially algebraic over K, then ζ is differentially algebraic over K.*

Corollary 3.6. *If L/K is a differential extension and $\xi, \eta \in L$ are both differentially algebraic over K, then $\xi + \eta$, $\xi\eta$, ξ/η, $(\eta \neq 0)$, and $d_i\xi$ are differentially algebraic over K.*

Definition 3.48. *If L/K is a differential extension, the set of elements of L that are algebraic over K is an intermediate field K_0, called the* algebraic closure *of K in L. Furthermore, the set of elements of L that are differentially algebraic over K is an intermediate differential field K', called the* differential algebraic closure *of K in L. (K' is differentially algebraic closed in L.)*

Definition 3.49. *If K is a differential field with derivations $\partial_1, \cdots, \partial_n$, we say that the derivative operators $\{\partial_\mu\}_{|\mu| \geq 0}$ are* algebraically independent *over K if there does not exist a differential indeterminate z over K and a nontrivial differential polynomial in $K[z]_d$ vanishing on any element of K.*

Proposition 3.20. *The derivative operators $\{\partial_\mu\}_{|\mu| \geq 0}$ are algebraically independent over K iff one of the following equivalent propositions are verified:*

(i) The derivative operators $\{\partial_\mu\}_{|\mu| \geq 0}$ are linearly independent over K.

(ii) The derivatives $\partial_1, \cdots, \partial_n$ are linearly independent over K. (In such a case we simply say that $\partial_1, \cdots, \partial_n$ are independent over K.)

Theorem 3.52. (Differentially primitive element). *If K is a differential field with independent derivations, then every finitely generated differentially algebraic extension of K can be generated by a simple element.*

Proposition 3.21. *1) If L/K is a finitely generated differentially algebraic extension with derivations $d_1, \cdots, d_n$, then L/K can be considered as a*

---

[48] The character α_q^i of a q-order PDE $E_q \subset JD^q(W)$, $\pi{:}W \to M$, $\dim M = n$, with symbol g_q, is the integer $\alpha_q^i \equiv \dim(g_q^{(i-1)})_p - \dim(g_q^{(i)})_p$, $p \in E_q$, where $(g_q^{(i)})_p \equiv \{\zeta \in (g_q)_p | \zeta(v_1) = \cdots = \zeta(v_i) = 0\}$, where $(v_1, \cdots, v_n)$ is the natural basis in $T_{\pi_k(p)}M$.

finitely generated extension with derivations $d'_1, \cdots, d'_{n-1}$ such that $d'_j = c^i_j d_i$ for certain $c^i_j \in C = cst(K)$.

2) If L/K is a finitely generated differential extension, then one has the following propositions:

(i) Any intermediate differential field between K and L is also finitely generated over K.

(ii) If $C = cst(K)$, $D = cst(L)$, then D/C is a finitely generated extension.

(iii) If K_0 is the algebraic closure of K in L, then K_0 is a finitely generated extension of K and $|K_0/K| < \infty$.

(iv) If K' is the differential algebraic closure of K in L, then K'/K is a finitely generated differential extension.

Proposition 3.22. If ζ is differentially algebraic over $K(\eta)_d$ but η is not differentially algebraic over $K(\zeta)_d$, then ζ is differentially algebraic over K.

Theorem 3.53. If $K \subset L \subset M$ are differential fields and S is a differential transcendence basis of L/K while T is a differential transcendence basis of $M(L)$, then $S \cap T = \emptyset$ and $S \cup T$ is a differential transcendence basis of M/K Furthermore

$$trd_d(M/K) = trd_d(M/L) + trd_d(L/K).$$

Proposition 3.23. Let $(A, \{d_j\}_{1 \leq j \leq n}$ be a differential ring such that A is an integral domain with field quotients $K = Q(A)$, then $cst(A)$ is integrally closed in A while $cst(K)$ is algebraically closed in K. In general one has $cst(A) \subseteq cst(K)$.

Proposition 3.24. Let M and N be filtered modules over the filtered ring A. This means that there are increasing filtrations, $\cdots \subset M_{q-1} \subset M_q \subset \cdots \subset M$ and $\cdots \subset N_{q-1} \subset N_q \subset \cdots \subset M$, such that $A_r M_q \subseteq M_{q+r}$ and $A_r N_q \subseteq N_{q+r}$ respectively, for all $q, r \geq 0$. To such filtered modules we can associate graded modules $gr(M) \equiv G = \bigoplus_{0 \leq q \leq \infty} G_q$ and $gr(N) \equiv H = \bigoplus_{0 \leq q \leq \infty} H_q$ respectively, such that $G_q \equiv M_q/M_{q-1}$ and $H_q \equiv N_q/N_{q-1}$.[49] Then any morphism $f : M \to N$, compatible with above filtrations, that is $f(M_q) \subset N_q$, induces by restriction morphisms $f_q : M_q \to N_q$ that pass to quotient inducing morphisms $gr_q(f) : G_q \to H_q$, hence a morphism

---

[49] Note that G and H are modules over the graded ring $gr(A)$.

$gr(f) : G \to H$, such that the following diagram

$$
\begin{array}{ccccccc}
0 & \to & M_{q-1} & \to & M_q & \to & G_q & \to & 0 \\
& & f_{q-1} \downarrow & & f_q \downarrow & & \downarrow gr_q(f) & & \\
0 & \to & N_{q-1} & \to & N_q & \to & H_q & \to & 0
\end{array}
$$

is commutative with exact horizontal lines. In general $f(M) \cap N_q \neq f_q(M_q)$, for all $q \geq 0$.

Definition 3.50. A *strict* morphism $f : M \to N$ between filtered modules is a compatible morphism such that the following equivalent propositions are verified:

(i) $f(M) \cap N_q = f_q(M_q)$, for all $q \geq 0$.

(ii) *The following short exact sequences are exact* $0 \to \mathrm{coker}\,(f_q) \to \mathrm{coker}\,(f)$, *for all* $q \geq 0$.

Proposition 3.25. *For filtered modules and strict compatible morphisms hold the following propositions:*

(i) *A short sequence*

$$
0 \to M' \xrightarrow{\,f\,} M \xrightarrow{\,g\,} M'' \to 0
$$

is exact iff the associated sequence

$$
0 \to gr(M') \xrightarrow{\,gr(f)\,} gr(M) \xrightarrow{\,gr(g)\,} gr(M'') \to 0
$$

is exact.

(ii) *Let* $M' \xrightarrow{\,f\,} M \xrightarrow{\,g\,} M''$ *be an exact sequence then the associated sequence* $gr(M') \xrightarrow{\,gr(f)\,} gr(M) \xrightarrow{\,gr(g)\,} gr(M'')$ *is exact too.*

(iii) *If the sequence* $gr(M') \xrightarrow{\,gr(f)\,} gr(M) \xrightarrow{\,gr(g)\,} gr(M'')$ *is exact (without assuming that f and g are strict), and $M = \bigcup_{q \in \mathbf{Z}} M_q$ with $M_q = 0$ for $q \ll 0$, then f and g are strict and the filtered sequence* $M' \xrightarrow{\,f\,} M \xrightarrow{\,g\,} M''$ *is exact.*

Proposition 3.26. 1) *If the sequence of filtered modules* $M' \xrightarrow{\,f\,} M \xrightarrow{\,g\,} M''$ *is exact does not imply that sequences* $M'_q \xrightarrow{\,f_q\,} M_q \xrightarrow{\,g_q\,} M''_q$ *are exact for all $q \geq 0$.*

2) *Let*

$$
M' \xrightarrow{\,f\,} M \xrightarrow{\,g\,} M''
$$

be an exact sequence of Noetherian filtered modules and compatible morphisms. If the associated sequence

$$G'_q \xrightarrow{gr_q(f)} G_q \xrightarrow{gr_q(g)} M''_q$$

are exact for sufficiently large q, then the sequence

$$M'_q \xrightarrow{f_q} M_q \xrightarrow{g_q} M''_q$$

are exact for sufficiently large q.

Proposition 3.27. *Let M be a filtered module over a filtered ring A. One has the following propositions:*

1) M *is finitely generated iff $gr(M)$ is finitely generated over $gr(A)$. This is equivalent to say that there exist homogeneous elements $\bar{x}_1, \cdots, \bar{x}_r \in G$, with $\bar{x}_i \in G_{q_i}$ being the canonical projection of $x_i \in M_{q_i}$ such that any element of G_q may be written as finite sum $\sum_{1 \leq i \leq r} \bar{a}_i \bar{x}_i$ where $\bar{a}_i \in gr_{q-q_i}(A)$.*

2) M *is Noetherian iff $gr(M)$ is Noetherian.*

3) *If A is a Noetherian filtered ring, the following three conditions are equivalent:*

(i) M *is a finitely generated A-module.*

(ii) G *is a finitely generated $gr(A)$-module.*

(iii) *There exist q_0 such that $A_1 M_q = M_{q+1}$ for any $q \geq q_0$.*

Theorem 3.54. *Let $(A, \{\partial_j\}_{1 \leq j \leq n})$ be a differential ring. The set $D(A)$ of differential operators over $(A, \{\partial_j\}_{1 \leq j \leq n})$ is a non-commutative filtered ring and a filtered bimodule over A.*

Proof. If y is a differential indeterminate over A, we may introduce the *formal derivatives* $d_1, \cdots, d_n$ which are such that $d_i d_j - d_j d_i = 0$, $\forall i, j = 1, \cdots, n$, and are defined by: $d_i(ay) = (\partial_i a)y + a(d_i y)$. We shall write $d_i y = y_i$, $d_i y_\mu = y_{\mu+1i}$, where μ is the multi-index $\mu = (\mu_1, \cdots, \mu_n)$ with length $|\mu| = \mu_1 + \cdots + \mu_n$. If $y = (y^1, \cdots, y^m)$, we set $d_\mu = (d_1)^{\mu_1} \cdots (d_n)^{\mu_n}$ and $d_\mu y^k = y^k_\mu$. Any differential operator of order q over A can be written in the form $P = \sum_{0 \leq \mu \leq q} a^\mu d_\mu$, $a^\mu \in A$. Set $ord(P) = q$. Then, we can write $D(A) \cong A[d_1, \cdots, d_n] \equiv A[d]$ the *ring of partial differential operators* over A with derivatives $d_1, \cdots, d_n$. The addition rule is clear. The multiplication rule comes from the Leibniz formula:

$$\left\{\begin{array}{l} \partial_\nu(ab) = \displaystyle\sum_{\lambda+\mu=\nu} \frac{\nu!}{\lambda!\mu!}(\partial_\lambda a)(\partial_\mu b) \\[3mm] d_\nu(ay) = \displaystyle\sum_{\lambda+\mu=\nu} \frac{\nu!}{\lambda!\mu!}(\partial_\lambda a)d_\mu y \end{array}\right\} \Rightarrow d_\nu a = \sum_{\lambda+\mu=\nu} \frac{\nu!}{\lambda!\mu!}(\partial_\lambda a)d_\mu.$$

Here we have put $\mu! = \mu_1!\cdots\mu_n!$. With these rules $D(A)$ becomes a non-commutative ring and a bimodule over A. In fact, the previous formula defines the right action of A on $D(A)$. The left action of A on $D(A)$ is simply the multiplication on the left by A, that is $aP = a(\sum_{0\le\mu\le q} a^\mu d_\mu) = \sum_{0\le\mu\le q} aa^\mu d_\mu$. Now, the filtration of $D(A)$ is naturally induced by filtration of spaces of differential operators. More precisely $D_q(A) = \{P \in D(A)|ord(P) \le q\}$, where $ord(P) = sup\{|\mu||a^\mu \ne 0\}$. We set $D_{-1}(A) = 0$ and $D_0(A) = A$. Then, $D_q(A) \subset D_{q+1}(A)$, $D(A) = \bigcup_{q\ge 0} D_q(A)$ and $D_q(A)D_p(A) \subseteq D_{p+q}(A)$. $\qquad\qquad\square$

Definition 3.51. If $\chi_1,\cdots,\chi_n$, are indeterminates over A, we define the *(prinipal) symbol* of $P \in D(A)$, with respect to χ, by setting: $\sigma_\chi(P) = \sum_{|\mu|=ord(P)} a^\mu \chi_\mu$.

Definition 3.52. If $f,g \in A[\chi]$, we define the *Poisson bracket*:

$$\{f,g\} = \sum_{1\le i\le n} (\partial_{\chi_i}.f)(\partial_i.g) - (\partial_{\chi_i}.g)(\partial_i.f) \in A[\chi],$$

where $\partial_i.f$ is the polynomial obtained by applying ∂_i to the coefficients of the polynomial f.

Proposition 3.28. *For any $P,Q \in D(A)$ with $ord(P) = p$, $ord(Q) = q$, we have:*

1) $\sigma_\chi(PQ) = \sigma_\chi(P)\sigma_\chi(Q)$;

2) $ord([P,Q]) \le p + q - 1$, *with* $[P,Q] = PQ - QP$;

3) $\{\sigma_\chi(P),\sigma_\chi(Q)\} \ne 0 \Rightarrow ord([P,Q]) = p + q - 1$ *and* $\sigma_\chi([P,Q]) = \{\sigma_\chi(P),\sigma_\chi(Q)\}$.

Theorem 3.55. *1) $D_q(A)$ identifies a representable functor D_q on the category of modules. More precisely, for any A-module E one has the following A-module*

$$D_q(E) = Hom_A(\mathcal{I}^q(E); A),$$

where $\mathcal{I}^q(E) \equiv \mathcal{I}^q(A) \bigotimes_A E$ and $\mathcal{I}^q(A) \equiv Hom_A(D_q(A); A)$. One has the isomorphism: $D_q(E) \cong Hom_A(\mathcal{I}^q(A); E^*)$, where $E^* \equiv Hom_A(E; A)$.

2) If E is finitely generated and projective or free one has the isomorphism $D_q(E) \cong D_q(A) \bigotimes_A E^*$.

3) $D(A)$ identifies a functor from the category of modules to the category of filtered modules. More precisely, for any A-module E one has: $D(E) = \bigcup_{q \geq 0} D_q(E)$.

4) If E is finitely generated projective module over A, one has the isomorphism: $D(E) \cong D(A) \bigotimes_A E^*$.

5) $D(A)$ identifies a representable functor of two variables, controvariant in the first and covariant in the second. More precisely one has; $D_q(E; F) = Hom_A(\mathcal{I}^q(E); F)$.

Proof. 1) Let $\mathcal{I}^q(A) \equiv Hom_A(D_q(A); A) \equiv D_q(A)^*$. One has the isomorphism $D_q(A) \cong Hom_A(\mathcal{I}^q(A); A) \equiv \mathcal{I}^q(A)^*$. Then, if E is a module over A, setting $\mathcal{I}^q(E) = \mathcal{I}^q(A) \bigotimes_A E$, we get on $\mathcal{I}^q(E)$ a left-module structure over A. If $\{e_k\}$ is a set of generators for E, typical terms of $\mathcal{I}^q(E)$ are $\xi_\mu^k \otimes e_k$, $0 \leq |\mu| \leq q$, $\xi_\mu^k \in A$. One has also the q-jet operator: $j_q : E \to \mathcal{I}^q(E)$, $j_q(\xi^k e_k) = (\partial_\mu \xi^k) \otimes e_k$, $0 \leq |\mu| \leq q$. One has the isomorphism:

$$D_q(E) \equiv Hom_A(\mathcal{I}^q(E); A) \cong Hom_A(\mathcal{I}^q(A); E^*).$$

2) If E is finitely genertaed and projective or free one has the following isomorphisms:

$$Hom_A(\mathcal{I}^q(A); E^*) \cong Hom_A(\mathcal{I}^q(A); A) \bigotimes_A E^*$$
$$\cong \mathcal{I}^q(A)^* \bigotimes_A E^*$$
$$\cong D_q(A) \bigotimes E^*.$$

3) It is enough to consider that to the inclusions $D_q(A) \subset D_{q+1}(A)$ there correspond the projection $\mathcal{I}^{q+1}(A) \to \mathcal{I}^q(A)$. Then by taking the inductive limit on the filtration of $D(A)$ we get $D(A) = \lim_{\leftarrow} \mathcal{I}^q(A)^* \equiv \mathcal{I}^\infty(A)^*$ and the corresponding projective limit gives $\mathcal{I}^\infty(A) = \lim_{\rightarrow} \mathcal{I}^q(A)$. So $D(E) = \bigcup_{q \geq 0} D_q(E)$.

4) This follows directly from the points 2) and 3).

5) In fact,

$$D(E,F) = F \bigotimes_A D(A) \bigotimes_A E^* \cong F \bigotimes_A D(E)$$

$$\cong F \bigotimes_A Hom_A(\mathcal{I}^\infty(E); A)$$

$$\cong Hom_A(\mathcal{I}^\infty(E); F).$$

$\square$

Proposition 3.29. 1) *The ring A can be considered a left-module over $D(A)$: $D(A) \times A \to A$, $(P, a) \mapsto P(a)$. Then one has the isomorphisms:*
(i) *$Hom_{D(A)}(D(A); A) \cong A$;*
(ii) *$A \cong D(A)/I(A)$, $D(A) \cong A \bigoplus I(A)$, where $I(A) \equiv \{P \in D(A)|P(1) = 0\}$.*
(iii) *$Hom_{D(A)}(A; A) = C = cst(A)$.*
(iv) *$T(A) \equiv D_1(A)/A \cong \{P \in D_1(A)|P(1) = 0\} \subset Der_C(A)$; $D_1(A) \cong A \bigoplus T(A)$.*
2) *If A is an integral domain then $D(A)$ is also an integral domain.*
3) *Identifying $T(A) \subset D(A)$ as a submodule, the commutator in $D(A)$ restricts to $T(A)$ to produce the standard bracket for vector fields.*
4) *We have the following short exact sequences:*

$$0 \to S_q T^*(A) \to \mathcal{I}^q(A) \to \mathcal{I}^{q-1}(A) \to 0$$

$$0 \to D_{q-1}(A) \to D_q(A) \to S_q T(A) \to 0$$

where $T^*(A) \equiv Hom_A(T(A); A)$.
5) *For the graded module $gr(D(A))$ associated with the filtered module $D(A)$, i.e., $gr_q(D(A)) = D_q(A)/D_{q-1}(A)$, we have: $gr_q(D(A)) \cong S_q T(A)$. Furthermore, introducing indeterminates $\chi_1, \cdots, \chi_n$ over A, we get: $gr(D(A)) \cong A[\chi_1, \cdots, \chi_n]$.*
6) *If A is a Noetherian ring, then $D(A)$ is also a Noetherian ring.*

Proof. 1) The left action of $D(A)$ on A coincides with the evaluation of the differential operators, i.e., $(P, a) \mapsto P(a)$. This induces the isomorphism (i). $I(A)$ is a left ideal of $D(A)$ generated by $d_1, \cdots, d_n$. This induces the isomorphism (ii). By means of the inclusions $A = D_0(A) \subset D(A)$, we can identify A with a subring of $D(A)$, hence any $D(A)$-module can be considered also as a A-module, just by forgetting about the differential structure. Then, one has $Hom_{D(A)}(A; A) = cst(A) \equiv C$. Furthermore, for the A-module $D_1(A)/D_0(A) = D_1(A)/A$ one has the natural short exact sequence: $0 \to D_1(A)/A \to Der_C(A)$, as any element $\delta \in D_1(A)/A$ is of the form $\delta = a^1 d_1 + \cdots + a^n d_n$, hence can be identified with a vector $\delta = a^1 \partial_1 + \cdots + a^n \partial_n \in Der_C(A)$. Therefore, $T(A) \equiv D_1(A)/A$ is a free module over A. Furthermore, as $T(A)$ can be considered also an element of $D_1(A)$, we get isomorphism (iv).

2), 3), 4), 5), 6) are natural consequences of previous results. $\qquad\square$

Proposition 3.30. 1) *The module $D(E)$ can be considered also a left-module over $D(A)$: $D(A) \times D(E) \to D(E)$, $(P,Q)(e) = P(Q(e))$.*

2) *In particular, if E is a finitely generated projective module over A, one has the isomorphism $Hom_{D(A)}(D(E); A) \cong E$.*

3) *One has a canonical C-linear map (adjoint involution):*

$$ad \in Hom_C(D(A); D(A)),$$

$$ad(P) = ad(a^\mu d_m) = (-1)^{|\mu|} d_\mu a^\mu.$$

One has the following properties:

(i) $ad^2 = id_{D(A)}$;

(ii) $ad|_{D_0 = A} = id_A$;

(iii) $ad(d_i) = -d_i$;

(iv) $ad(PQ) = ad(Q).ad(P)$;

Proof. 1) If $P \in D(A)$ is a differential operator of order q and $Q \in D(E)$ is a differential operator of order r, $(PQ) \in D(E)$ is a differential operator of order $q + r$, such that for any $e \in E$ one has $P(Q(e))$.

2) In fact, under our hypotheses one has:

$$Hom_{D(A)}(D(E); A) \cong Hom_{D(A)}(D(A) \bigotimes_A E^*; A)$$

$$\cong Hom_A(E^*; A) \cong E^{**} \cong E.$$

3) It follows directly from above considerations. $\qquad\square$

Theorem 3.56. (Localization in non-commutative ring). 1) *Let A be a non-commutative ring and let S be a multiplicative set in A, that is $1 \in S$ and S is closed under multiplication. Let us assume that the following two conditions are satisfied:*

(i) $Sa \cap As \neq 0$, $\forall a \in A$, $s \in S$. *(Left Ore set condition).*

(ii) *If $s \in S$ and $a \in A$ are such that $as = 0$, then there is $t \in S$ such that $ta = 0$.*

Then there exists a ring $S^{-1}A$ (left ring fractions or left localization of A with respect to S) and a homomorphism $\theta = \theta_S : A \to S^{-1}A$, with the following properties:

(iii) $\theta(s)$ is invertible in $S^{-1}A$;

(iv) Each element of $S^{-1}A$ or fraction has the form $\theta(s)\theta(a)$ for some $s \in S$ and $a \in A$;

(v) $\ker(\theta) = \{a \in A | \exists s \in S,\ sa = 0\}$.

More precisely $S^{-1}A = S \times A/ \sim$, where $\sim$ is the following equivalence relation: $(s, a) \sim (t, b) \Leftrightarrow \exists u, v \in A$ such that $us = vt \in S$, $ua = vb$. We will denote the equivalence class of (s, a) by $s^{-1}a \in S^{-1}A$.[50]

2) For symmetry we have a similar theorem for right ring fractions or right localization of A with respect to S.

3) If A is Noetherian and $S \subset A$ is simply a left Ore set in A, then the existence of $S^{-1}A$ is assured.

Proof. We shall use the following lemmas.

Lemma 3.11. *If there exists a left localization of A with respect to S, that is a homomorphism $\theta_S = \theta : A \to S^{-1}A$ such that the above conditions* (iii), (iv), (v) *are satisfied, then also conditions* (i) *and* (ii) *must be satisfied.*

Lemma 3.12. *If S is a left Ore set in a ring A, then one has the following properties:*

(a) *If $s, t \in S$ then $A_s \cap A_t \cap S \neq 0$;*

(b) *Two fractions $\theta(s)^{-1}\theta(a)$ and $\theta(t)^{-1}\theta(b)$, $a, b \in A$, can be reduced to the same common denominator $\theta(us) = \theta(vt)$, where $u, v \in A$ such that $us = vt \in S$.*

Now, in order to conclude the proof it is enough to prove that the above equivalence relation is well defined in $S \times A$. This can be down just by using above lemmas. $\qquad\square$

2) The proof can be conduced for similarity.

3) It is enough to prove that from the fact that A is Noetherian, the assumption that S satisfies the left Ore condition (i) implies also condition (ii). Now let $s \in$ and $a \in A$ such that $as = 0$. Let us denote yet by ann the left annihilator in A. Then one has $ann(s^n) \subseteq ann(s^{n+1})$, $\forall n \in \mathbf{N}$. As A is Noetherian, we have $ann(s^n) = ann(s^{n+1})$ for $n \gg 0$. From the left

---

[50] Let us emphasize that for a non-commutative ring A we cannot write $\frac{a}{s}$ as it does not distinguish between $s^{-1}a$ and as^{-1}.

Ore condition we may find $t \in S$, $b \in A$ such that $ta = bs^n$ and we get $bs^{n+1} = tas = 0$. Hence, $b \in ann(s^{n+1}) = ann(s^n)$. Therefore $ta = 0$. $\square$

Theorem 3.57. (Localization of module over non-commutative ring). 1) *Let M be a left module over the non-commutative ring A and let $S \subset A$ be a set in A that satisfies conditions (i) and (ii) in Theorem 3.124(1). Then there exists a left module $S^{-1}M$ over $S^{-1}A$ (left module of fractions or left localization of M with respect to S) and a homomorphism $\theta_S = \theta : M \to S^{-1}M$ with the following properties:*

(iii) Each element of $S^{-1}M$ has the form $s^{-1}\theta(x)$ for $s \in S$, $x \in M$;

(iv) $\ker(\theta_S) = t_S(M) \equiv \{x \in M | \exists s, \, sx = 0\} \equiv S$-torsion submodule of M. More precisely $S^{-1}M = S \times M/ \sim$, where $\sim$ is the following equivalence relation: $(s,x) \sim (t,y) \Leftrightarrow \exists u, v \in A$ such that $us = vt \in S$, $ux = vy$.

2) For symmetry we have a similar theorem for right module fractions or right localization of M with respect to S.

3) One has the following isomorphism of modules over $S^{-1}A$:

$$(S^{-1}A) \bigotimes_A M \cong S^{-1}M.$$

Proof. 1) The proof can be conduced for similarity with one of the previous Theorem 3.124(1). Let us simply emphasize, here, that $t_S(M)$ is just a sub-module of M.

2) For symmetry with respect to the point 1).

3) This isomorphism is induced by the *multiplication map $S^{-1}A \times M \to S^{-1}M$*. $\square$

Proposition 3.31. (Internal operations). *Let A be a differential ring with n derivations $\{\partial_i\}_{1 \leq i \leq n}$. A vector $\zeta \in T(A)$ can be written $\zeta = \zeta^i \partial_i$, $\zeta^i \in A$, and a r-form $\alpha \in \Lambda^r T^*(A)$, $T^*(A) \equiv Hom_A(T(A); A)$ can be written in the form $\alpha = \sum_{1 \leq i_1 < \cdots < i_r \leq n} \alpha_{i_1,\dots,i_r} \delta^{i_1} \wedge \cdots \wedge \delta^{i_r} \equiv \alpha_I \delta^I$, where $\alpha_{i_1,\dots,i_r}, \alpha_I \in A$ and $I \equiv (i_1 < \cdots < i_r)$ is a multi-index and $\delta^I \equiv \delta^{i_1} \wedge \cdots \wedge \delta^{i_r}$.[51] One has the following distinguished properties:*

(1) (Exterior differential). $d : \Lambda^r T^*(A) \to \Lambda^{r+1} T^*(A)$, $d\alpha = (\partial_i a_I)\delta^i \wedge \delta^I$, $a_I.(\partial_i a_I) \in A$. One has the following property:

$$d(\alpha \wedge \beta) = (d\alpha) \wedge \beta + (-1)^r \alpha \wedge d\beta, \quad \alpha \in \Lambda^r T^*(A), \; \beta \in \Lambda^s T^*(A).$$

---

[51] For example if $A = \mathbf{Q}[\chi^1, \cdots, \chi^n]$, one has $\partial_i = \partial x_i$, $\delta^i = dx^i$, $i = 1, \cdots, n$.

(2) (Interior multiplication by a vector $\xi \in T(A)$). $\xi\rfloor : \Lambda^{r+1}T^*(A) \to \Lambda^r T^*(A)$, $\xi\rfloor\alpha = (\xi^i \alpha_{ii_1 \cdots i_r})\delta^{i_1} \wedge \cdots \wedge \delta^{i_r}$. One has the following properties:

(i) $(\xi\rfloor d + d\xi\rfloor)a = \xi\rfloor da = \xi(a) = \xi^i(\partial_i a) \in A$, $a \in A$.

(ii) $\xi\rfloor d + d\xi\rfloor d = d(\xi\rfloor)d + d\xi\rfloor$, (as $d^2 = 0$).

(iii) $\xi\rfloor(\alpha \wedge \beta) = (\xi\rfloor\alpha) \wedge \beta + (-1)^r \alpha \wedge (\xi\rfloor\beta)$, $\alpha \in \Lambda^r T^*(A)$, $\beta \in \Lambda^s T^*(A)$.

(3) (Lie derivative with respect to a vector $\xi \in T(A)$). $\mathcal{L}_\xi : \Lambda^r T^*(A) \to \Lambda^r T^*(A)$, $\mathcal{L}_\xi \alpha = \xi^i\rfloor(\alpha) + d(\xi\rfloor\alpha)$, $\alpha \in \Lambda^r T^*(A)$. One has the following property: $[\mathcal{L}_\xi, \mathcal{L}_\zeta] = \mathcal{L}_{[\xi,\zeta]}$, $\forall \xi, \zeta \in T(A)$.

Theorem 3.58. *Let A be Noetherian integral domain. Then $D(A)$ is also a Noetherian integral domain and if $S \subset D(A)$ is a left Ore domain in $D(A)$, then one has the following isomorphisms: $S^{-1}D(A) \cong D(A)S^{-1}$.*

Proposition 3.32. (Adjoint operation). 1) *The adjoint operation identifyies a functor from the category of left $D(A)$-modules to the category of right $D(A)$-modules. More precisely $m.P = ad(P)m$, for any $m \in M$, where M is a A-module, and $P \in D(A)$.*

2) *One has $m.PQ = ad(PQ)m = ad(Q)ad(P)m = (m.P).Q$.*

3) *$ad(ad(P)) = P$.*

Proposition 3.33. *$\Lambda^n T^*(A)$ has a right module structure over $D(A)$, if A is a differential ring with derivatives $\{\partial_i\}_{1 \le i \le n}$.*

Proof. If $\alpha = a\delta^1 \wedge \cdots \wedge \delta^n$, $a \in A$, one has $\alpha.P = ad(P)(a)\delta^1 \wedge \cdots \wedge \delta^n$. An alternative proof can be given also by taking into account that $D(A)$ is generated by $D_1(A) = A \oplus T(A)$. Then, any P can be considered as linear combinations of iterated operators of first order. Such operators can be written as linear combinations of term of the form $a + \xi$, with $a \in A$, $\xi \in T(A)$. Then, it is enough define the right action $\alpha.(a + \xi) = \alpha.a + \alpha.\xi$. The action $\alpha.a$ is defined by $\alpha(\zeta)a$, $\forall \zeta \in T(A)$. Furthermore, one has: $\alpha.\xi = -\mathcal{L}_\xi \alpha$. $\qquad\square$

Proposition 3.34. 1) *If M and N are left $D(A)$-modules, then $M \otimes_A N$ and $Hom_A(M; N)$ become left $D(A)$-modules.*

2) *If M is a left $D(A)$-module and N is a right $D(A)$-module, then $M \otimes_A N$ and $Hom_A(M; N)$ become right $D(A)$-modules.*

3) *If M and N are right $D(A)$-modules, $Hom_A(M; N)$ becomes a left $D(A)$-modules.*

4) *If M is a left $D(A)$-module and N is a right $D(A)$-module, then one has the following properties:*

(i) $M_r \equiv \Lambda^n T^*(A) \bigotimes_A M$ is a right $D(A)$-module, *called the converted right* $D(A)$-*module of* M.

(ii) $N_l \equiv Hom_A(\Lambda^n T^*(A); N) \cong \Lambda^n T(A) \bigotimes_A N$ is a left $D(A)$-*module, called the converted left* $D(A)$-*module of* N.

Above conversions identify corresponding functors that are inverse to each other.

Definition 3.53. A *linear differential system of order* q, on the A-module E, is a submodule over A, $E_q \subset \mathcal{I}^q(E)$ of the A-module $\mathcal{I}^q(E)$. We say that $E_q \subset \mathcal{I}^q(E)$ is a *normal differential system* if $F \equiv \mathcal{I}^q(E)/E_q$ and E_q are free or projective modules over A. We call r-*prolongation of* E_q, $E_{q+r} \equiv \mathcal{I}^r(E_q) \cap \mathcal{I}^{q+r}(E)$. We say that E_q is *regular* if E_{q+r}, $r \geq 0$, is a normal differential system of order $q + r$. We say that E_q is *formally integrable* if it is regular and $E_{q+r+1} \to E_{q+r} \to 0$ are short exact sequences for all $r \geq 0$.

Theorem 3.59. *Let* $E_q \subset \mathcal{I}^q(E)$ *be a differential system of order* q. *The following properties hold:*

(i) *The following sequences are exact:*

$$(a)\ 0 \to E_q \to \mathcal{I}^q(E) \to F \to 0;$$

$$(b)\ 0 \to F^* \to D_q(A) \bigotimes_A E^* \to E_q^*;$$

$$(c)\ 0 \to E_{q+r} \to \mathcal{I}^{q+r}(E) \to \mathcal{I}^r(F) \to Q_r \to 0;$$

$$(d)\ 0 \to Q_r^* \to D_r(F) \to D_{q+r}(E).$$

Furthermore, the exact sequence $D(F) \to D(E) \to M \to 0$, *defines* M *equipped with the quotient filtration* $M_q \subset \cdots \subset M_{q+r} \subset \cdots \subset M$. *In general the induced sequences* $D_r(F) \to D_{q+r}(E) \to M_{q+r} \to 0$ *are not necessarily exact.*

(ii) *If* E_q *is normal then the exact sequence* (a) *splits and the exact sequence* (b) *can be completed obtaining the following one:*

$$(bb)\ 0 \to F^* \to D_q(A) \bigotimes_A E^* \to E_q^* \to 0.$$

(iii) *If* E_{q+r} *is a regular system, then the sequence* (c) *splits and* (d) *can be completed giving the following exact sequence:*

$$(dd)\ 0 \to Q_r^* \to D_r(F) \to D_{q+r}(E) \to E_{q+r}^* \to 0.$$

(iv) If the morphism $D(F) \to D(E)$ is strict the sequences

$$D_r(F) \to D_{q+r}(E) \to M_{q+r} \to 0$$

are exact. In this case we have the exact sequences: $0 \to E^*_{q+r} \to M$, $0 \to E^*_{q+r} \to E^*_{q+r+1}$, $E_{q+r+1} \to E_{q+r} \to 0$, for all $r \geq 0$, where E_q is regular. Therefore, in such a case E_q is formally integrable.

(v) If $E_q \subset \mathcal{I}^q(E)$ is a formally integrable linear differential system, one has

$$E_{q+r} \subseteq (E_q)_{+r} \Leftrightarrow D_r M_q \subseteq M_{q+r}, \ \forall q, r \geq 0.$$

More precisely one has: $E_{q+r} = (E_q)_{+r}$ for $r \geq 0$ and $E_q \subseteq (E_p)_{+(q-p)}$, for all $p \leq q$. One has the filtration of $M = E^*_\infty \Leftrightarrow E_\infty = M^*$ by means of $M_q = E^*_q \Leftrightarrow E_q = M^*_q$. The corresponding gradiation is $gr(M) \equiv G$, with $g^*_q = G_q \Leftrightarrow g_q = G^*_q$. g_q is the symbol of E_q. One has $g_q \subseteq (g_p)_{+(q-p)}$, for all $p \leq q$, and $g_{q+r} = (g_q)_{+r}$, for all $r \geq 0$.

Definition 3.54. A *solution* of a q-order *linear differential system* $E_q \subset \mathcal{I}^q(E)$, is an element $e \in E$ such that $j^q(e) \in E_q$. The set of solutions is a submodule $\underline{Sol}(E_q) \equiv \Theta \subset E$, over A.

Theorem 3.60. Given a formally integrable q-order linear differential system $E_q \subset \mathcal{I}^q(E)$, one has the following isomorphism:

$$\Theta \cong Ext^0_{D(A)}(M; A) \cong Hom_{D(A)}(M; A).$$

Proof. As E and F are projective modules of finite rank over A, we obtain the exact sequence of filtered left $D(A)$-modules:

$$(3.17) \qquad\qquad D(F) \to D(E) \to 0.$$

If N is another left $D(A)$-module, any sequence $D(F) \to D(E) \to N$ induces a morphism $M \to N$ of left $D(A)$-modules. We can define a *solution* of M in N such a morphism $M \to N$. We denote the set of such solutions by $\underline{Sol}_N(M)$. Let us apply the controvariant functor $Hom_{D(A)}(-; N)$ to the sequence (3.17), and taking into account the following isomorphisms:

$$Hom_{D(A)}(D(E); N) \cong Hom_{D(A)}\left(D(A) \bigotimes_A E^*; N\right)$$

$$\cong Hom_A(E^*; N) \cong E^{**} \bigotimes_A N \cong E \bigotimes_A N$$

we get the following exact sequence:

$$0 \to Hom_{D(A)}(M;N) \to E \bigotimes_A N \to F \bigotimes_A N.$$

So we get thet $\underline{Sol}_N(M) \cong Hom_{D(A)}(M;N) \cong Ext^0_{D(A)}(M;N)$. In the case that $N = A$ we get that $\Theta \cong Ext^0_{D(A)}(M;A)$. $\qquad\qquad\square$

Theorem 3.61. (Algebraic criterion for formal integrability). *Let $Z_q = Z(\mathfrak{p}_q)$ be the variety defined by means of ideal $\mathfrak{p}_q \subset K[y_q]_d$ such that the following conditions are verified:*

(i) $(\mathfrak{p}_q)_{+1} = \mathfrak{p}_{q+1} \subset K[y_{q+1}]_d$ is also a prime ideal.

(ii) $\mathfrak{p}_{q+1} \cap K[y_q]_d = \mathfrak{p}_q$.

(iii) g_{q+1} is generic over E_q.

(iv) g_q is 2-acyclic.

Then $(\mathfrak{p}_q)_{+\infty} = \mathfrak{p} \subset K[y]_d$ is a prime differential ideal, where $\mathfrak{p}$ is the differential ideal generated by a finite number of differential polynomials $P_1, \cdots, P_t$, defining E_q, and E_q is formally integrable. If one of these conditions is not satisfied we get that $\mathfrak{p}$ is not a prime ideal, hence we have a factorization of $\mathfrak{p}$. In other words the PDE is not formally integrable.

Proof. It follows from above propositions. $\qquad\qquad\square$

Example 3.30. (*D'Alembert equation*). Let us consider the d'Alembert equation $(d'A) \subset J\mathcal{D}^2(W)$, defined by means the following differential polynomial $F \equiv u_{xy}u - u_x u_y$, over $J\mathcal{D}^2(W)$, where $\pi : W \equiv \mathbf{R}^3 \to M \equiv \mathbf{R}^2$, $(x, y, u) \mapsto (x, y)$. The ideal $< F >$ is not prime in $\mathbf{R}[u, u_x, u_y, u_{xy}]$. One irreducible component $\mathfrak{p}_1$ is generated by (u_x, u_y), while the other is described by the system in solved form: $(d'A)'\, u_{xy} = \frac{u_x u_y}{u}$, with the localization $u \neq 0$, that is formally integrable. The intersection of these two components is described by the system $\{u_x = 0, u_y = 0, u_{xy} = 0\}$. $\qquad\blacksquare$

3.3 - SPECTRAL MEASURES OF PDE's

The set of quantum fluctuations around classical configurations is represented by the *quantum situs* [58,59,61,63] of the PDE $E_k \subset J\mathcal{D}^k(W)$ where the classical configurations are represented by hypersurfaces Cauchy data. The quantum situs $\Omega(E_k)$ of E_k is the set of admissible integral manifolds, contained into $J\mathcal{D}^k(W)$, bording hypersurfaces Cauchy data contained into E_k. In particular, the quantum situs contains the set $\Omega(E_k)_c$ of solutions of E_k, i.e., the *classic limit* of $\Omega(E_k)$. $\Omega(E_k)_c$ and $\Omega(E_k)$ are related to the integral $(n-1)$-bordism groups $\Omega_{n-1}^{E_k}$ and quantum $(n-1)$-bordism groups $\Omega(E_k)_{n-1}$ respectively of E_k, where n is the dimension of the integral manifolds representing the solutions of E_k. (See above chapters and refs.[63–65] for such bordism groups.) Observables can be represented by means of numerical functions on $\Omega(E_k)$, or $\Omega(E_k)_c$, and conservation laws are numerical functions on $\Omega_{n-1}^{E_\infty}$. Spectral measures on $\Omega(E_k)_c$, or $\Omega(E_k)$, allow us to pass from the logic of a PDE to a quantum logics that are represented by means of noncommutative algebras of operators on locally convex topological vector spaces. So in this section we shall consider first in some detail the geometrical theory of measures and spectral sequences on topological spaces and them we will specialize on PDE's.

Definition 3.55. Let $\mathcal{M}$ be the *category of measurable spaces*, that is the set of objects, $Ob(\mathcal{M})$, of $\mathcal{M}$ is given by all abstract measurable spaces $(X, \mathcal{T})$, where X is a set and $\mathcal{T}$ is a σ-algebra, that is a non-void collection of subsets of X, closed for complementation and countable union (and intersection) and containing the empty set. (A σ-algebra on X results in a *monotone class* of sets of X, such that for every monotone sequence $\{A_k\}$ of sets in X, we have $\lim_{k} A_k \in X$.) The set of morphisms, $Hom(\mathcal{M})$, of $\mathcal{M}$ is given by all measurable transformations of $(X, \mathcal{T})$ into $(Y, \mathcal{R})$, $f : X \to Y$, that is $f^{-1}(B) \in \mathcal{T}, \forall B \in \mathcal{R}$. A subset $A \subset X$ of a measurable space $(X, \mathcal{T})$ is called *measurable* if A belongs to $\mathcal{T}$. If X is a topological space there is a canonical σ-algebra on X : that generated by the open (or closed) subsets of X. This is called *Borel σ-algebra* on X and denoted by $\mathcal{B}(X) \equiv \mathcal{B}$. Let $\mathcal{P}(X)$ denote the collection of all subsets of X. Then, for any subset $L \subset \mathcal{P}(X)$ the σ-algebra *generated* by L is the smallest σ-algebra, $\sigma(L)$, containing L. Of course $\mathcal{P}(X)$ is also the Borel σ-algebra with respect to the finest topology on X, i.e., the *discrete topology*, namely the topology where the open sets are all subsets of X, $\emptyset$ and X itself. If $f : X \to Y$ is a mapping between two sets X and Y, and $\mathcal{T}$ is a σ-algebra on Y, then $f^{-1}(\mathcal{T}) \equiv \{A \subset X | A = f^{-1}(B), B \in \mathcal{T}\}$ is a σ-algebra on X. Of course $f : (X, f^{-1}(\mathcal{T})) \to (Y, \mathcal{T})$ is a measurable mapping. If $\mathcal{T} = \sigma(L)$, then $f^{-1}(\mathcal{T})$ is the σ-algebra generated by $f^{-1}(L)$. A *Borel mapping* $f : X \to Y$ between two topological spaces X and Y is a measurable map $(X, \mathcal{B}(X)) \to (Y, \mathcal{B}(Y))$. Any continuous map is Borel one.

Definition 3.56. A *measure* is a covariant functor $\mathbf{M}^+ : \mathcal{M} \to \mathcal{I}(\mathbf{R}) \equiv$ category of $\mathbf{R}$-valued set functions given by: (a) $\mathbf{M}^+(X, \mathcal{T}) \equiv \mu \equiv$ *measure* on $\mathcal{T}$, that is a non-negative valued set function $\mu : \mathcal{T} \to \overline{\mathbf{R}}$ (possibly taking the value $+\infty$), equal to 0 on the empty set (monotone increasing), and countably additive (or σ-additive); (b) if $f \in Hom((X,\mathcal{T}),(Y,\mathcal{R}))$, $\mathbf{M}^+(f) : \mathbf{M}^+(X,\mathcal{T}) \equiv \mu \mapsto \mathbf{M}^+(f)(\mu) \equiv \nu = \mu \circ f^{-1}$. A *finite measure* μ on $(X,\mathcal{T})$ is a measure such that $\mu(X) < +\infty$. A *complete measure* μ on $(X,\mathcal{T})$ is a mesare such if $E \in \mathcal{T}$, $F \subset E$, $\mu(E) = 0$, then $F \in \mathcal{T}$. A σ-*finite measure* μ on (X,T) is a measure such that if for all $E \in \mathcal{T}$ there is a uniquely sequence $C_i (i = 1, 2, \cdots) \in \mathcal{T}$, such that $E = \bigcup_{i=1}^{\infty} C_i$ and $\mu(E_i) < +\infty$. Given a measure μ on $(X,\mathcal{T})$ there is a uniquely complete measure $\bar{\mu}$ on $(X,\overline{\mathcal{T}})$, given by: (a) $\overline{\mathcal{T}}$ is the σ-algebra of all sets of the form $E \triangle N$, with $E \in \mathcal{T}$ and $N \subset F \in \mathcal{T}$ with $\mu(F) = 0$; (b) $\bar{\mu}(E \triangle N) = \mu(E)$. Here $E \triangle N \equiv (E - N) \cup (N - E)$ is the *symmetric difference* of two sets E and N. (One has $E \triangle N = N \triangle E$.) $\bar{\mu}$ is called the *completion* of μ. $\overline{\mathcal{T}}$ is called the *completion of $\mathcal{T}$ with respect to μ*. A measure on a set X can also be defined with respect to a ring: $\mu : R \to \mathbf{R}^+$ such that $X = \bigcup_{n=1}^{\infty} E_n$ for some sequence of sets $\{E_n\}$ in R. Then, there is an extension of μ for a measure ν defined on $\mathcal{S}(R)$, the σ-ring generated by R. If μ is σ-finite on R, then the extension is unique, and is σ-finite on $\mathcal{S}(R)$.

Example 3.31. The *Lebesgue measure* on $\mathbf{R}$ is obtained by the measure μ on the ring E of elementary figures: $E \equiv \{E = \bigcup_{i=1}^{n}]a_i, b_i]\}$, with $b_i < a_{i+1}$, putting $\mu(]a, b]) = b - a$. The corresponding σ-ring $\mathcal{S}(E)$ is the class of Borel sets $\mathcal{B}(\mathbf{R})$ in $\mathbf{R}$. The unique extension of μ to $\mathcal{B}(\mathbf{R})$ is called the *Lebesgue measure* on $\mathbf{R}$.   ∎

Example 3.32. A similar example can be obtained considering the *Stieltjes measures* on $\mathbf{R}$ associated to a monotone increasing function $F : \mathbf{R} \to \mathbf{R}$, that is everywhere continuous on the right, putting $\mu(]a, b]) = F(b) - F(a)$. F is called *Stieltjes measure function*. This can be extended to $\mathbf{R}^n$ by considering an n-dimensional Stieltjes measure function $F : \mathbf{R}^n \to \mathbf{R}$ satisfying the following properties: (i) F is continuous on the right in each variable; (ii) $\mu_F(I) = \sum_{i=1}^{2^n} y_i F(v_i) \geq 0$, where v_i are the 2^n vertices of the set: $I \in P^n \equiv \{\prod_{i=1}^{n}]a_i, b_i]\}$ and $y_i = +1$ for the vertex in which each coordinate is the largest and $y_i = (-1)^r$ if the vertex v_i is such that r of its coordinates are at the lower bound; (iii) $F(x_1, \cdots, x_n) \to 0$ as any one of $x_1, \cdots, x_n \to -\infty$; $F(x_1, \cdots, x_n) \to 1$ as all of $x_1, \cdots, x_n \to +\infty$.   ∎

Definition 3.57. A triplet $(X, \mathcal{T}, \mu)$ is called a *measure space*. The set of measures on $(X, \mathcal{T})$ is denoted by $M^+(X, \mathcal{T})$. The *support of the measure* $\mu = M^+(X, \mathcal{T})$ is the complement of the largest (open) set $V \subset X$ such that $\mu(V) = 0$. Let $\mathcal{M}_\bullet$ be the canonical sub-category of $\mathcal{M}$ of measurable spaces like $(X, \mathcal{P}(X))$, where $\mathcal{P}(X)$ is the collection of all subsets of X, (i.e., $\mathcal{M}$ is the category of discrete topological spaces). Any measure μ on $(X, \mathcal{T}) \in Ob(\mathcal{M})$ identifies an *outer measure* μ^* (resp. *inner measure* μ_*) on $(X, \mathcal{P}(X)) \in Ob(\mathcal{M}_\bullet)$ given by $\mu^*(A) = \inf \mu(B)$ for all $B \supset A$ and $B \in \mathcal{T}$, (resp. $\mu_*(A) = \sup \mu(B)$ for all $B \subset A$ and $B \in \mathcal{T}$). Then, μ^* satisfies the *Fatou property*: if $A_n \uparrow A \Rightarrow \mu^*(A_n) \uparrow \mu^*(A)$.

Definition 3.58. If X is a topological space a *Borel measure* on X is a measure on $(X, \mathcal{B}(X))$. The set of all Borel measures on X is denoted by $M_{bor}^+(X)$. Let $\overline{\mathcal{M}}$ be the *category of measure spaces*:

$$f \in Hom_{\overline{\mathcal{M}}}((X, \mathcal{T}, \mu), (Y, \mathcal{R}, \nu))$$

iff $f \in Hom_{\mathcal{M}}((X, \mathcal{T}), (Y, \mathcal{R}))$ and $\nu = \mu \circ f^{-1}$.

Remark 3.17. On $\overline{\mathbf{R}} \equiv \mathbf{R} \sqcup \{-\infty, +\infty\}$ we say that a set $B \subset \overline{\mathbf{R}}$ is a Borel set in $\overline{\mathbf{R}}$ if it is the union of a set $B' \in B(\mathbf{R})$ with any subset of $\overline{\mathbf{R}} - \mathbf{R} = \{-\infty, +\infty\}$. In practice in order to know if a function $f : (X, \mathcal{T}) \to \mathbf{R}$ is measurable, it is sufficient to check that $f^{-1}(E) \in \mathcal{T}$ for a suitable class of subsets which generates the σ-field $\mathcal{B}(\mathbf{R})$. More precisely, the most important class is given by the following: (a) $\{x : f(x) \leq c\}, \forall c \in \mathbf{R}$; (b) $\{x : f(x) > c\}, \forall c \in \mathbf{R}$; (c) $\{x : f(x) \geq c\}, \forall c \in \mathbf{R}$; (d) $\{x : f(x) < c\}, \forall c \in \mathbf{R}$. Note that any continuous function $\Omega \to \mathbf{R}$ on a topological space Ω is Borel measurable.   ∎

Definition 3.59. We have the following controvariant functor: $F : \overline{\mathcal{M}} \to \mathcal{B}_{an}$, where $\mathcal{B}_{an}$ is the category of Banach spaces such that $\forall (X, \mathcal{T}, \mu) \in Ob(\overline{\mathcal{M}})$, one has $F(X, \mathcal{T}, \mu) \equiv$ Banach space of $\overline{\mathbf{R}}$-valued measurable functions on $(X, \mathcal{T}, \mu)$; $\forall f \in Hom_{\overline{\mathcal{M}}}((X, \mathcal{T}, \mu), (Y, \mathcal{R}, \nu))$, one has $F(f) \in Hom_{\mathcal{B}an}(F(Y, \mathcal{R}, \nu), (X, \mathcal{T}, \mu))$, $F(f) = h \circ f$, $\forall h \in F(Y, \mathcal{R}, \nu)$. Then, $F(\overline{\mathcal{M}})$ is a sub-category of $\mathcal{B}_{an}$: $F(\overline{\mathcal{M}}) \subset \mathcal{B}_{an}$.

Definition 3.60. The *integral* is a $\mathbf{R}$-linear function: $\int : F(X, \mathcal{T}, \mu) \to \mathbf{R}$,
$$\int : f \mapsto \mu(f) \equiv \int f d\mu = \int f_+ d\mu - \int f_- d\mu = \lim_{n \to \infty} \left(\int f_{+,n} d\mu - \int f_{-,n} d\mu \right),$$
where $f = f_+ - f_-$, and $f_{+,n}$, $f_{-,n}$ are monotone increasing sequences of simple functions such that $f_{+,n} \to f_+$ and $f_{-,n} \to f_-$ (that surely

exist). Furthermore, if $f_{\pm,n}(x) = \sum_{i=1}^{s_n} {}^{\pm}c_i^{(n)}\chi_{E_i}(x)$, ${}^{\pm}c_i^{(n)} \geq 0$, where χ_{E_i} is the characteristic functions of $\{E_i\}$, that form a T-dissection of X; we put: $\int f_{\pm,n}d\mu = \sum_{i=1}^{s_n} {}^{\pm}c_i^{(n)}\mu(E_i)$. (This definition does not depend on the particular sequence $f_{+,n}$ or dissection $\{E_i\}$.) A function $f \in F(X,T,\mu)$ is called μ-*integrable* if $|\int f d\mu| < +\infty$.

Definition 3.61. The *indefinite integral* on $F(X,T,\mu)$ is a map:

$$N : F(X,T,\mu) \to M(X,T), \; N(f) \equiv \nu : A \to \nu(A) = \int_A f d\mu = \int \chi_A f d\mu.$$

One has that if $f \in Hom_{F(\overline{M})}(F(X,T,\mu), F(Y,R,\nu))$ then $\int_X (g \circ f)d\mu = \int_Y g d\nu$, $\forall g \in F(Y,R,\nu)$.

Definition 3.62. A set $A \subset X$ is said to be μ-*integrable* if its characteristic function χ_A is μ-integrable.

Definition 3.63. A set $A \subset X$ is said to be μ-*measurable* if its intersection with every μ-integrable set C is μ-integrable.

Definition 3.64. The μ-measurable sets form a σ-algebra T_μ containing T and μ is a measure on T_μ *extending* μ, (*extension of a measure*).

Example 3.33. The set of *Lebesgue measurable sets* of $\mathbf{R}$ forms a class $\mathcal{L}$ that is larger than $\mathcal{B}(\mathbf{R}^n) \equiv \mathcal{B}$. If μ is the Lebesgue measure on $\mathbf{R}$ then $\mu^* : \mathcal{L} \to \mathbf{R}^+$ is also called Lebesgue measure on $\mathbf{R}$. Similarly, the class $\mathcal{L}_F^n$ of sets in $\mathbf{R}^n$ which are Lebesgue-Stieltjes measurable, for a function F, is larger than $\mathcal{B}(\mathbf{R}^n)$. Then $\mu_F^* : \mathcal{L}_F^n \to \mathbf{R}$ is also called *Lebesgue-Stieltjes measure* on $\mathbf{R}^n$. This measure on the σ-algebra $\mathcal{L}_F^n$ satisfies the property $\mu_F(\mathbf{R}^n) = 1$ and it is complete. ∎

Proposition 3.35. 1) *If* $f \in Hom_{\overline{M}}((X,T,\mu),(Y,R,\nu))$, *then* f *is said to be* (μ,R)-*measurable if* $f^{-1}(B)$ *belongs to* T_μ *for every* $B \in R$. *So, in the category* $\overline{M}$ *we can distinguish a sub-category* $\overline{\overline{M}} \subset \overline{M}$, *where* $Ob(\overline{\overline{M}}) = Ob(\overline{M})$ *but the set of morphisms* $Hom_{\overline{\overline{M}}}((X,T,\mu),(Y,R,\nu))$ *is the subset of all* $f \in Hom_{\overline{M}}((X,T,\mu),(Y,R,\nu))$ *such that they are* (μ,R)-*measurable.*

2) *Let* $Y \equiv \prod_{i\in I} Y_i$, *let* T_i *be the* σ-*algebra on* $Y_i, i \in I$, *and let* $f \equiv (f_i)_{i\in I} : X \to Y$ *be a map. Then the* σ-*algebra* $\sigma(f_i)_{i\in I}$ *on* X *generated by* $\bigcup_{i\in I} f_i^{-1}(T_i)$ *is called the* σ-*algebra generated by* $(f_i)_{i\in I}$. *In particular, if* $X = Y$ *and* $f_i \equiv pr_i \equiv$ *projection on* i-*th factor, we call product* σ-*algebra the* σ-*algebra on* Y *generated by* $(pr_i)_{i\in I}$ *and we write* $\otimes_{i\in I} T_i \equiv \sigma(pr_i)_{i\in I}$.

A map $f : (X, \mathcal{T}) \to (\prod_{i \in I} Y_i, \otimes_{i \in I} \mathcal{T}_i)$ between measurable spaces X and $Y \equiv \prod_{i \in I} Y_i$ is measurable iff the mappings $pr_i \circ f : (X, \mathcal{T}) \to (Y_i, \mathcal{T}_i)$ are measurable $\forall i \in I$. If $(X_n)_{n \in \mathbf{N}}$ is a countable family of topological spaces with countable basis then $\mathcal{B}(X) = \otimes_{n \in \mathbf{N}} \mathcal{B}(X_n)$, where $X = \prod_{n \in \mathbf{N}} X_n$.

3) Let $\overline{\dot{\mathcal{M}}} \subset \overline{\mathcal{M}}$ be the *sub-category of σ-finite measure spaces*. Then, there is a canonical functor: $\otimes : \overline{\dot{\mathcal{M}}} \times \overline{\dot{\mathcal{M}}} \to \overline{\dot{\mathcal{M}}}$, $\otimes((X, \mathcal{T}, \mu), (Y, \mathcal{R}, \nu)) = (X \times Y, \mathcal{T} \otimes \mathcal{R}, \mu \otimes \nu)$, where $\mathcal{T} \otimes \mathcal{R}$ is the canonical σ-algebra in $X \times Y$ associated to $\mathcal{T}$ and $\mathcal{R}$, and $\mu \otimes \nu(A \times B) = \mu(A).\nu(B), A \in \mathcal{T}, B \in \mathcal{R}$, with the usual convention: $0\infty = \infty 0 = 0$. For each $E \in \mathcal{T} \otimes \mathcal{R}$ set $E_x = pr_1^{-1}(x) \cap E$, $E_y = pr_2^{-1}(y) \cap E$, where pr_i, $i = 1, 2$, are the canonical projections $X \times Y \to X, Y$. Then, one also has $\mu \otimes \nu(E) = \int \nu(E_x) d\mu(x) = \int \nu(E_y) d\nu(y)$. (*For examples the Lebesgue measure on $\mathbf{R}^n$ is obtained by tensor product of the Lebesgue measure on $\mathbf{R}$.*)

4) Let $(X, \mathcal{T}, \mu) \in Ob(\overline{\mathcal{M}})$ and let $Y \subset X$ be a μ-measurable set. We have: $B(Y) = \{A \cap Y; A \in B(X)\} \subset B(X)$ and $\mu|B(Y)$ is the *induced measure* on Y by μ.

Definition 3.65. A *signed measure* is a covariant functor $\mathbf{M} : \mathcal{M} \to \mathcal{I}(\mathbf{R})$ such that $\mu \equiv \mathbf{M}(X, \mathcal{T})$ is a real valued, countably additive set function such that $\mu(\emptyset) = 0$ and assuming at most one of the values $+\infty$ or $-\infty$. Let us denote by $\mathbf{M}(X, \mathcal{T})$ the set of signed measures on $(X, \mathcal{T})$. If $\nu \in \mathbf{M}(X, \mathcal{T})$ there exist two disjoint sets A and B where $A \cup B = X$ such that $\mu(A) > 0$ and $\mu(B) < 0$, (*Hann decomposition*). Set $\mu_+(E) \equiv \mu(E \cap A)$ and $\mu_-(E) = -\mu(E \cap B)$, for every measurable set E of X. Then, we have $\mu(E) = \mu_+(E) - \mu_-(E)$ for $\mu \in M(X, \mathcal{T})$, (*Jordan decomposition*). If μ is finite then so are μ_+ and μ_-, at least one of the measures μ_+ and μ_- is always finite. We call μ_+ (resp. μ_-) the *upper variation* (resp. *lower variation*) of μ. So, we have the canonical maps: $p_\pm : M(X, \mathcal{T}) \to M^+(X, \mathcal{T}), p_\pm(\mu) = \mu_\pm$.

Definition 3.66. Let $\mathcal{M}_H$ be the sub-category of $\mathcal{M}$ of measurable spaces $(X, \mathcal{B})$ where X is a Hausdorff space and $\mathcal{B}(X) \equiv B$ is the σ-algebra of all Borel sets of X. A *Radon measure* is a Borel measure $\mathbf{M}_{Ra} : \mathcal{M}_H \to \mathcal{I}(\mathbf{K})$ such that:[52] (a) $\mu(C) < \infty$ for each compact subset $C \subseteq X$; (b)

$\mu(B) = \sup\{\mu(C)|C \subseteq B, compact\}$, for each $B \in \mathcal{B}(X)$. The set of all-Radon measures on X is denoted by $M_{Ra}^+(X)$ and the set of signed Radon measures is denoted $M_{Ra}(X)$. A *finite Radon measure* μ (i.e., $\mu(X) < \infty$) satisfies: $\mu(B) = \inf\{\mu(G)|B \subseteq G, G \text{ open}\}$, $B \in \mathcal{B}(X)$. A *molecular Radon measure* μ has $\text{supp}(\mu)$ finite.

Definition 3.67. Let X be any infinite space and let $(X, \mathcal{F})$ be a measurable space where $\mathcal{F}$ contains all single points sets. Let $x_1, x_2, \cdots$, be an enumerable sequence of distinct points of X and assume that $p_1, p_2, \cdots$, is a sequence of real numbers such that $\sum_{i=1}^{\infty} p_i$ either converges absolutely or is properly divergent to $+\infty$ or $-\infty$. Put: $\mu(E) = \sum_{x_i \in E} p_i$, $\quad \forall E \in \mathcal{F}$. Then μ is called a *discrete measure*.

Definition 3.68. 1) Let $\mathcal{C}$ be a set of subsets of X. The set $E_0 \in \mathcal{C}$ is said to be an *atom* of a set function $\phi : \mathcal{C} \to \overline{\mathbf{R}}$ if: (a) $\phi(E_0) \neq 0$ and; (b) for every $E \subset E_0$, $E \in \mathcal{C}$, $\phi(E) = 0$ or $\phi(E) = \phi(E_0)$.

2) A set function $\phi : \mathcal{C} \to \overline{\mathbf{R}}$ is said to be *non-atomic* if it has no atom.

Proposition 3.36. *Assume that $\phi : \mathcal{F} \to \overline{\mathbf{R}}$ is σ-additive on the σ-field $\mathcal{F}$, non-atomic, and finite valued. Then, for any $A \in \mathcal{F}$, ϕ takes every real value between $-\phi_-(A)$ and $\phi_+(A)$ for some subset $E \subset A$.*

Definition 3.69. Given any measure space $(X, \mathcal{F}, \mu)$ in which $\mathcal{F}$ contains all single points sets, the point $p \in X$ is said to be an *atom* for the measure μ if $\mu(p) > 0$. A measure μ with no atoms is said to be *non-atomic*.

Theorem 3.62. *1) If μ is σ-finite, the set of atoms of μ is countable.*

2) In this case if we put: $\nu(E) = \sum_{p \in E, \mu(p) \neq 0} \mu(p)$, we obtain a new measure ν defined on all subsets of X, and ν is a discrete measure.

3) Given a σ-finite measure space $(X, \mathcal{F}, \mu)$ in which $\mathcal{F}$ contains all single point sets there is a unique decomposition of μ: $\mu = \nu + \tau$ for which ν is a discrete measure on X and τ is a non-atomic measure on $\mathcal{F}$.

Note that the theory of Radon measures can be generalized to spaces which are not necessarily Hausdorff if we replace the word "compact" by "closed, quasi-compact". Recall that a subset E of a topological space X is said to be *quasi-compact* if every open covering of E has a finite subcovering. (A quasi-compact set in a non-Hausdorff space is not in general closed!)

---

space or a separable space (i.e. it has a countable base), then so is its one-point compactification. Recall also that a metric space with the usual topology is Hausdorff, and a topological space is metrizable iff it is Hausdorff.

Definition 3.70. Let $\mathcal{K} \equiv \mathcal{K}(X)$ denote the family of all compact subsets of X. A *Radon content* is a set function: $\lambda : \mathcal{K} \to [0, \infty]$, $\lambda(C_2) - \lambda(C_1) = \sup\{\lambda(C)|C \subseteq C_2 - C_1, C \in \mathcal{K} \,|\, \forall\, C_1, C_2 \in \mathcal{K}, C_1 \subseteq C_2\}$.

Proposition 3.37. *1) If μ is a Radon measure, its restriction to $\mathcal{K}$ is a Radon content. Vice versa, any Radon content on a Hausdorff space has a unique extension to a Radon measure.*

2) If X is a Hausdorff space and $B \in \mathcal{B}(X)$, then B is again a Hausdorff space and any Radon measure $\mu \in M_{Ra}^+(X)$ restricted to $\mathcal{B}(B)$ gives a Radon measure $\mu|_B$ on B. ($\mu|_B$ is called the restriction Radon measure of μ to B.)

3) If $\mu \in M_{Ra}^+(X)$ and $f : X \to [0, \infty]$ is Borel measurable, then $\int f d\mu = \sup_{K \in \mathcal{K}} \int_K f d\mu$.

4) If $f : X \to [0, \infty[$ is continuous, then $\mathcal{B}(X) \to [0, \infty]$ defined by $\nu(B) = \int_B f d\mu$ is again a Radon measure.

5) If $\mu \in M_{Ra}^+(X), \nu \in M_{Ra}^+(Y)$, with $X, Y =$ Hausdorff spaces, then there is a uniquely determined Radon measure on $X \times Y$ called product measure of μ and ν denoted by $\mu \otimes \nu$, with the property $\mu \otimes \nu(K \times L) = \mu(K).\nu(L), \forall\, K \in \mathcal{K}(X), L \in \mathcal{K}(Y)$. For all Borel sets $A \subseteq X, B \subseteq Y$ we have $\mu \otimes \nu(A \times B) = \mu(A).\nu(B)$. In particular the restriction of $\mu \otimes \nu$ to the product σ-algebra $\mathcal{B}(X) \otimes \mathcal{B}(Y)$ is a product measure of μ and ν in the usual sense. Furthermore, if $f : X \times Y \to [0, \infty]$ is lower semicontinuous, (resp. Borel measurable), one has (Fubini's theorem): $\int_{X \times Y} f d(\mu \otimes \nu) = \int_X \int_Y f(x, y) d\nu(y) d\mu(x) = \int_Y \int_X f(x, y) d\mu(x) d\nu(y)$.

6) Let $(G_\alpha)_{\alpha \in D}$ be an open covering of the Hausdorff space of X and on each G_α, let a Radon measure μ_α be given such that $\mu_\alpha(B) = \mu_\beta(B)$ for each pair of indices $\alpha, \beta \in D$ and for each Borel set $B \subseteq G_\alpha \cap G_\beta$. Then there is a uniquely determined Radon measure μ on X such that $\mu(B) = \mu_\alpha(B)$ if B is a Borel set in G_α.

7) If X is strongly Lindelöf, every Radon measure μ on X is σ-finite.

Theorem 3.63. (Riesz representation theorem). *1) Let X be a Hausdorff space, let C denote a point separating convex cone of continuous functions $f : X \to [0, \infty[$ and let $T : C \to [0, \infty]$ together with C fulfil the following conditions: (a) if $f, g \in C \Rightarrow f \wedge g \in C, f - g^+ \in C$ and $f \wedge 1 \in C$; (b) $T(f + g) = T(f) + T(g), \forall f, g \in C$; (c) For each $f \in C \Rightarrow T(f) = \sup\{T(g)|f \geq g \in C, g$ bounded$, T(g) < \infty\}$; (d) For each compact set $K \subseteq X$ there is some $f \in C$ with $1_K \leq f$ and $T(f) < \infty$. (e) Given*

$f \in C$ such that $T(f) < \infty$ and given $\epsilon > 0$ there is a compact set $K \subseteq X$ such that $g \in C$, $g \leq f$ and $g|_K = 0$ implies $T(g) < \epsilon$. (As a consequence we have: monotonicity of T: $T(f) \leq T(g)$ if $f \leq g$ and $T(\alpha f) = \alpha T(f)$ for all $f \in C$ and $\alpha \in \mathbf{R}^+$.) Then, there is a uniquely determined Radon measure μ on X such that $T(f) = \int f d\mu$ for all $f \in C$. The measure μ is furthermore locally finite.

2) If X is locally compact there is a bijection: $C_0^0(X)' \to M_{Ra}(X)$, $T \mapsto \mu : T(f) = \int f d\mu = \int f d\mu_1 - \int f d\mu_2$, $\mu = \mu_1 - \mu_2$, $\mu_i \in M_{Ra}^+(X)$.

3) If X is compact, $C^0(X)$ with the sup-norm is a Banach space and $C(X)'$ is also a Banach space with the norm $\|T\| = \sup_{\|f\| \leq 1} |T(f)|$. The set $M_{Ra}(X)$ of all signed Radon measure on X is also a Banach space with norm $\|\mu\| = \sup\{|\mu(A)| + |\mu(CA)| \| A \in B(X)\}$. Then, one has the identification $M_{Ra}(X) \equiv C^0(X)'$ by means of the isometry: $C^0(X)' \to M_{Ra}(X)$, $T \mapsto \mu = \mu_1 - \mu_2$, $T(f) = \int f d\mu = \int f d\mu_1 - \int f d\mu_2, \mu_i \in M_{Ra}^+(X)$.

Definition 3.71. Let $\mathcal{CM}$ be the category of cylindrical measurable locally convex vector Hausdorff spaces over $\mathbf{R}$. The set $Ob(\mathcal{CM})$ is given by all *cylindrical measurable spaces* that are couples $(E, \mathcal{T})$ where E is a convex Hausdorff vector space over the field $\mathbf{R}$ and $\mathcal{T}$ is the set of *cylindrical subsets A of E*, that is there exists a closed subspace of finite codimension $F \subset E$ and a subset $B \subset E/F$ such that $A = \pi^{-1}_{E/F}(B)$, where $\pi_{E/F}$ is the canonical projection $E \to E/F$. B is called the *basecylindrical set* and F a *generator subspace of the cylindrical set A*. If $G \subset F$ has a finite codimension, here $B_0 \equiv \pi^{-1}_{E/F, E/G}(B)$ is also a base and G a generator subspace of A finer $A = \pi^{-1}_{E/F}(B) = \pi^{-1}_{E/G}(\pi^{-1}_{E/F, E/G}(B)) = \pi^{-1}_{E/G}(B)$, where $\pi_{E/F, E/G}$ is the canonical map $E/G \to E/F$, such that $\pi_{E/F, E/G} \circ \pi_{E/G} = \pi_{E/F}$. Then, $f \in Hom_{\mathcal{CM}}((E, \mathcal{T}), (H, \mathcal{R}))$ iff $f : E \to H$ is continuous and linear and $f^{-1}(B) \in \mathcal{T}, \forall B \in \mathcal{R}$. A *cylindrical measure* is a covariant functor $\mathbf{CM} : \mathcal{CM} \to \mathcal{I}(\mathbf{R})$, such that: (a) $\mathbf{CM}(E, \mathcal{T}) \equiv \mu \equiv$ *cylindrical measure on E* defined by $\mu(A) = \mu_{E/F}(B)$, $\forall A \in \mathcal{T}$ with base B and generator subspace F, where $\mu_{E/F}$ is a probability on the finite dimensional space E/F. (This definition is independent of the particular choice of F and B.) (b) If $f \in Hom_{\mathcal{CM}}((E, \mathcal{T}), (H, \mathcal{R}))$, $\mathbf{CM}(f) : \mu \mapsto \mathbf{CM}(f)(\mu) = \nu \equiv \mu \circ f^{-1}$. Set $\overline{\mathcal{CM}} =$ *category of cylindrical measure spaces*; $\overline{\mathcal{M}}(X) =$ subcategory of $\overline{\mathcal{M}}$ restricted to a set X, fixed. $\overline{\mathcal{CM}}(E) =$ subcategory of $\overline{\mathcal{CM}}$ restricted to a space E.

Proposition 3.38. 1) *We have a canonical injective functor*

$$J : \overline{\mathcal{CM}}(F) \to \overline{\mathcal{M}}(\prod_F (E/F)), \ \mu \mapsto \check{\mu},$$

where $(E/F)^\vee$ is the Cech-compactification of E/F. (Note that we have a canonical injection $j : E \to \prod_F (E/F)^\vee$ but it can happen that E has measure zero in the measure $\check{\mu}$.)

2) *Let $\overline{\mathcal{M}}_{Ra}$ be the subcategory of $\overline{\mathcal{M}}$ of Radon measure spaces and $\mathcal{M}_{Ra}(X) = \overline{\mathcal{M}}_{Ra} \cap \overline{\mathcal{M}}(X)$. We have a canonical functor: $j : \overline{\mathcal{M}}_{Ra}(E) \to \overline{\mathcal{M}}(\prod_F (E/F)^\vee)$. Then, we say that $j(\mu)$ comes from the Radon measure μ. So, we can identify $\overline{\mathcal{M}}_{Ra}(E)$ with a subset of $\overline{\mathcal{CM}}(E)$.*

Remark 3.18. If E is a reflexive Banach space and μ is a cylindrical measure on E, concentrated cylindrically on the balls of E, then μ is a Radon measure on E concentrated on the strongly compact subsets of E. (Recall that a measure μ is said to be concentrated on a subset $A \subset E$ if $\mu(A) = 1$, i.e., if $\mu(CA) = 0$, where CA is the complementary of A. Furthermore, a cylindrical measure μ is cylindrically concentrated on a subset $A \subset E$ if $\mu^*(CA) = 0$.

Theorem 3.64. 1) (Sazonov theorem). *Let E and F be two Hilbert spaces and $u : E \to F$ a continuous linear map. The image $\mu \circ u^{-1}$ of every cylindrical measure μ scalarly concentrated on the balls of E is a Radon measure on F (i.e., u is radonifying) iff u is a Hilbert-Schmidt operator (that is belongs to the canonical image of $\overline{E} \widehat{\otimes} F$ into $\mathcal{L}(E, F)$, where $\overline{E}$ is the anti-space of E and $\overline{E} \widehat{\otimes} F$ is the Hilbert-Schmidt tensor product of $\overline{E}$ and F).[53]*

2) (Minlos theorem). *Let E be a locally convex Hausdorff vector space. Let $\mathcal{G}$ be a collection of convex, completing subsets of E, such that if A is a balanced, convex completing subsets of E contained in some $B \in \mathcal{G}$, then $A \in \mathcal{G}$. Assume that E is $\mathcal{G}$-conuclear. [A $\mathcal{G}$-conuclear space is a locally convex Hausdorff space E such that there exists a subset collection $\mathcal{G}$ of balanced closed subsets of E, such that for every $A \in \mathcal{G}$ there exists $B \in \mathcal{G}$, $B \supset A$, such that the canonical map $\widehat{E}_A \to \widehat{E}_B$ is nuclear.] If μ a cylindrical measure on E concentrated scalarly on the sets of $\mathcal{G}$, then μ is a Radon measure on E concentrated on the sets of $\mathcal{G}$.*

---

[53] Recall that a cylindrical measure μ is scalarly concentrated on $A \subset E$ if for every closed hyperplane $H \subset E$ we have $\mu_{E/H}(\pi_{E/H}(A)) \geq 1$.

Definition 3.72. The *Fourier transform of a Radon measure* on a locally convex Hausdorff vector space E endowed with a finite Radon measure μ is a function $\phi \equiv F(\mu) : E' \to \mathbf{K}$ given by: $\phi(xi) = \int_E e^{-2i<x,\xi>} d\mu(x)$.

Theorem 3.65. (Sazonov-Badrikian theorem). *Let E be a quasi-complete locally convex Hausdorff vector space and let E'_c be its dual equipped with the topology of uniform convergence on compact subsets of E. Assume that every compact subset of E is contained in a balanced, convex, compact Hilbert set. A function of positive type ϕ on E'_c is the Fourier transform of a Radon measure on E iff it is continuous for Hilbert-Schmidt topology associated with the topology of E'_c.*

Definition 3.73. A *probability space* (or *sample space*) is a triplet $(\Omega, \mathcal{T}, \mu)$ consisting of a measurable space $(\Omega, \mathcal{T})$ and probability μ, i.e., a measure $\mu : \mathcal{T} \to [0,1]$, such that $\mu(\Omega) = 1$. An *event* is a μ-measurable subset $E \subset \Omega$; an *elementary event* is a point $\omega \in \Omega$. The *probability $P(E)$* of an event E, is its measure: $P(E) = \mu(E)$. Two events E and F are said to be *independent events* if $P(E \cap F) = P(E).P(F)$. The *conditional probability* of an event $A \in \Omega$ given the event B (with $\mu(B) > 0$), is the probability measure μ^B on Ω given by $\mu^B(A) = P(A|B) = \mu(A \cap B)/\mu(B)$.

Example 3.34. We recognize on $\mathbf{R}^n$ a structure of probability space

$$(\mathbf{R}^n, \mathcal{L}_F^n, \mu_F),$$

where F is a Stieltjes function and μ_F is the covariant measure on the σ-algebra $\mathcal{L}_F^n$. ∎

Example 3.35. A particular important example of probability measure on Hilbert spaces is the Gauss measure. Let $\mathcal{E}^\bullet(\mathbf{R})$ be the category of isometric finite dimensional real Euclidean vector spaces; we put

$$Hom_{\mathcal{E}^\bullet(\mathbf{R})}((E,g),(F,f)) = \emptyset$$

if (E,g) and (F,f) are not isometric spaces. Then, there exists a canonical measure functor (*Gauss measure functor*) $\mathbf{M}_G^\bullet : \mathcal{E}^\bullet(\mathbf{R}) \to \mathcal{I}(\mathbf{R})$, such that $\gamma \equiv \mathbf{M}_G^\bullet(E,g)$ is the measure (*Gauss measure*) given by $\gamma(A) = \int_A e^{-\pi|x|^2} d\mu(x)$, $A \subseteq E$, where $d\mu(x) = *1$ is the canonical volume form associated to g. We simply write $\gamma = e^{-\pi|x|^2} d\mu(x)$. More generally a Gauss measure on (E,g) can be defined with respect to a vector $x_0 \in E$ and a positive definite isomorphism $B \in GL(E)$ as

$$\gamma_{x_0'B} = \frac{1}{\sqrt{2}} \frac{1}{n\sqrt{|B|}} e^{-\frac{1}{2}<B^{-1}(x-x_0),(x-x_0)>} d\mu(X),$$

where $n = \dim E$, $|B| = \det(B)$. For example the Gauss measure on the real line $\mathbf{R}$ is $\gamma = e^{-\pi|x|^2}dx$ and one has $\gamma(\mathbf{R}) = \int_{\mathbf{R}} e^{-\pi|x|^2}dx = 1$. Let $\mathcal{E}(\mathbf{R})$ be the category of isometric Hilbert spaces. Then, there exists a canonical measure-functor (*Gauss measure functor*) $\mathbf{M}_G : \mathcal{E}(\mathbf{R}) \to \mathcal{I}(\mathbf{R})$ such that $\gamma \equiv \mathbf{M}_G(E)$ is the measure on E (*Gauss measure*) that is the cylindrical measure on E such that for any finite-dimensional subspace F of E one has $\gamma_F = \gamma \circ p^{-1}$, where $p : E \to F$ is the orthogonal projection and γ_F is the Gauss measure on F for the Euclidean structure induced on F by E. (The Gauss measure γ on E is scalarly concentrated on the balls of E since for any $\epsilon > 0$ there exists $R > 0$ such that $\int_{|x| \geq R} e^{-\pi|x|^2}d\mu(x) \leq \epsilon$.) Note also that the measure induced in the finite-dimensional Hilbert subspace $F \subset E$ by the Gauss measure γ on E coincides with the Gauss measure γ_F on F. In fact one has: (a) $\gamma \circ p^{-1} = \gamma_F$; (b) $p \circ j = id_F \Rightarrow \gamma \circ p^{-1} \circ p \circ j = \gamma_F \Rightarrow \gamma \circ j = \gamma_F$ i.e., $\gamma|F = \gamma_{F'}$ where $j : F \to E$ is the canonical injection. Furthermore, the Fourier transform $\phi = \mathcal{F}(\gamma)$ of the Gauss measure γ on E is given by $\phi(\xi) = e^{-\pi|\xi|^2}$, where $|\xi|$ is the norm corresponding to the canonical Euclidean structure of E'.   ∎

Example 3.36. Let $\mathcal{H}_{l.c.}(\mathbf{R})$ be the category of the couples (E, H) where E is a locally convex Hausdorff $\mathbf{R}$-vector space and H is a Hilbert subspace of E. The set of morphisms $Hom_{\mathcal{H}_{l.c.}(\mathbf{R})}((E, H), (\mathcal{F}, K))$ is the set of continuous maps $f : E \to F$ such that $f(H) = K$ and $f|H$ is an isometry between H and K. This set can be eventually empty. Then, there exists a canonical measure-functor (*relative Gauss measure-functor*) $\mathbf{M}_G^r : \mathcal{H}_{l.c.}(\mathbf{R}) \to \mathcal{I}(\mathbf{R})$, such that $\mu \equiv \mathbf{M}_G^r(E, H)$ is the image of the Gauss measure γ_H on the Hilbert subspace with respect to the canonical injection $j : H \to E, \mu = \gamma_H \circ j^{-1}$. For example, the *Dirac measure* δ on any locally convex Hausdorff vector space E is a Gauss measure relative to the Hilbert subspace $H \equiv \{0\} \subset E$. One has $\delta(E) = 1$. The simplest type of probability on a measurable space $(\Omega, \mathcal{A})$ is the σ-measure μ_{δ_x} at $x \in \Omega$, defined as follows: for all $\triangledown \in \mathcal{A}$: $\mu_{\delta_x}(\triangledown) = \{0 \text{if } x \neq \triangledown, 1 \text{if } x \in \triangledown\}$, where $\mathcal{A} = $ Boolean σ-algebra of subsets of Ω. Another example of relative Gauss measure is the *Wiener-Levy measure* of Brownian motion on $C^0([0, 1])$ with respect to the Hilbert subspace $H \equiv H^1([0, 1])$ of the real-valued continuous functions on $[0, 1]$ which vanish at the origin and whose derivative in the sense of distributions in the open interval $[0, 1]$ belongs to $L^2[0, 1]$.

The inner product in H is $(f,g) = \int_0^1 f'(t)g'(t)dt$. In fact, H is the image of the canonical continuous linear map $\mathcal{I} : L^2[0,1] \to C^0([0,1])$ given by $\phi \mapsto \mathcal{I}(\phi) \equiv \Phi = $ primitive of ϕ, i.e., $\Phi(x) = \int_0^x \phi(t)dt$. *Wiener's probability space* is the couple $([0,1],\mu)$ where $[0,1] \subset \mathbf{R}$ is endowed with the class of all Lebesgue-measurable subsets of $[0,1]$ and μ is just the Lebesgue measure. This probability space describes the random choice of a point from the interval $[0,1]$. ∎

Let us quote the following theorems that relate Gauss measure and Radon measure.

Theorem 3.66. 1) *Let E be a separated locally convex vector space. Let H be a Hilbert space and let $u : H \to E$ be a linear continuous map. Let $\mu = \gamma_H \circ u^{-1}$ be the cylindrical probability image of γ_H by u. If μ is a Radon measure on E one has $\mathrm{supp}(\mu) = u(H)$.*

2) *Let E be a Fréchet vector space and let μ be a Radon measure on E. There exists a Hilbert space H, a cylindrical measure ν on H scalar concentrated on the balls of H and a linear continuous map $u : H \to E$ such that $\nu \circ u^{-1} = \mu$.*

Definition 3.74. Let $(X,\mathcal{R})$ be a measurable space. We say that two μ-measurable maps $\Omega \to X$ are *equivalent measurable maps* if they coincide almost everywhere with respect to μ.

1) A *random variable* with value in X, on the probability space $(\Omega,\mathcal{T},\mu)$, is an equivalent class of μ-measurable maps $\Omega \to X$. We shall denote by $Mes(\Omega,\mu,X)$ the set of all random variables $\Omega \to X$. If $(X \equiv \mathbf{R}, \mathcal{R} \equiv \mathcal{B}(\mathbf{R}))$ we shall speak about *real random variables*.

2) Let $x : \Omega \to X$ be a random variable. The *probability distribution* of x is the probability $\mu \circ x^{-1} \equiv \nu$ on X extended for completion. (In fact, it should not be, in general, complete!) If $A \subset X$ is a $\nu \equiv \mu \circ x^{-1}$-measurable subset of X, then its measure $\nu(A)$ will be denoted by $P\{x \in A\}$ and considered to be the *probability* that $x(\omega)$ should lie in A.

Definition 3.75. Two random variables $x \in Mes(\Omega,\mu,X), y \in Mes(\Omega,\mu,Y)$ define a random variable $(x,y) \in Mes(\Omega,\nu;X \times Y)$, where $\nu \equiv \mu \circ (x \times y)^{-1}$. Then, x and y are said to be *independent random variables* if $\nu \equiv (\mu \circ x^{-1}) \otimes (\mu \circ y^{-1})$, or equivalently if $P\{(x \times y) \in A \times B\} = P\{x \in A\}P\{y \in B\}$ for arbitrary measurable $A \subset X$ and $B \subset Y$.

Remark 3.19. This definition can be extended to any countable finite or infinite family $\{x_i\}_{i \in I}$ of random variables $x_i : \Omega \to X_i$. If the family is not countable, the definition is meaningless since the map $(x_i) : \omega \mapsto (x_i(\omega))$ from ω into $\prod_{i \in I} x_i$ is not necessarily μ-measurable. However, in this case we say that $(x_i)_{i \in I}$ is *independent random variables* if every finite or countable subfamily is independent. $\blacksquare$

Definition 3.76. 1) Let $x \in Mes(\Omega, \mu, X), x' \in Mes(\Omega', \mu', X)$. We say that x and x' are *isonomous random variables* if they have the same probability distribution in $X : \mu \circ x^{-1} = \mu' \circ x'^{-1}$.

2) Let $(X_i)_{i \in I}$ be a family of completely regular spaces and let $(x_i)_{i \in I}$ be a family of random variables $x_i : \Omega \to X$. The family (x_i) is *canonical random variable* (or *regular random variable*) if there exists a random variable

$$x : \Omega \to X \equiv \prod_{i \in I} \overset{\vee}{X}_i,$$

$(\overset{\vee}{X}_i \equiv$ Cech-compactification of X_i, such that $x_i = \pi_i \circ x$, for each $i \in I$, where π_i denotes the canonical projection $X \to \overset{\vee}{X}_i$.

3) Let $x \in Mes(\Omega, \mu, E)$, where E is a separable Banach space. The *mean value* of x is the vector of E (if it exists) given by

$$E(x) \equiv < x > = (B) \int_{\Omega} x d\mu,$$

where $(B) \int$ denotes the Bochner integral.

4) Let E be a topological vector space, over $\mathbf{K}(\equiv \mathbf{C}, \mathbf{R})$, let $x \in Mes(\Omega, \mu, E)$, we define *characteristic function* of x the map $\phi : E' \to \mathbf{C}$ given by

$$\phi(\theta) = E e^{i<\theta,x>} = \int_{E} e^{i<\theta,y>} d\nu, \ \nu \equiv \mu \circ x^{-1}, \ y \equiv id_E.$$

Here ϕ is also called the *characteristic functional* of x, or characteristic functional of the measure ν.

TAB.3.8 - Useful definitions and properties for real random variables

Name	Definitions and Properties
Expectation ($\dagger+$)	$E(x)\equiv\int_\Omega x\,d\mu$ $E(x+y)=E(x)+E(y)$ $E(cx)=cE(x),\ c\in\mathbf{R}$
Generating function	$E(e^{tx}),\ t\in\mathbf{R}$ $E(xy)=E(x)E(y)$ if x and y are indep.
Moments	$E(x^n),(n=1,2,\cdots)$
Variance	$\sigma^2\equiv var(x)=E[(x-E(x))^2]$ (Thebichef inequality)$\left(^{+*}\right)$: $P\{|x-E(x)|\geq\lambda\sigma\}\leq\lambda^{-2},\ \forall\lambda\geq1$
Covariance of two rand. var	$cov(x,y)=E\{(x-E(x))(y-E(y))\}$ $cov(x,x)=var(x)$ $var(x+y)=var(x)+var(y)+2cov(x,y)$ If x,y are indep. $cov(x,y)=0$ (Schwartz inequality): $(cov(x,y))^2\leq var(x)var(y)$
Corrl.coeff.two rand.var.	$\rho(x,y)\equiv\dfrac{cov(x,y)}{(var(x)var(y))^{1/2}}$ $\rho(x,y)\in[-1,1];\ \rho(x,x)=1;\ \rho(x,-x)=-1$

$(\dagger+)$($E(x)$ is also called *mean value* of x.)

$((^{+*}))$(If $E(x)$ and $\sigma>0$ are finite.)

Remark 3.20. Let E be a separable Banach space and let $L : E \to E$ be a linear continuous operator. If $x \in Mes(\Omega,\mu,E)$, then $E(L\circ x) = L(E(x))$. Let $(H,<,>)$ be a separable Hilbert space. Let $x \in Mes(\Omega,\mu,H)$, then

$$< E(x),h' >= \int_H < h,h' > d\nu,$$

where $\nu \equiv \mu \circ x^{-1}$. ∎

The following proposition gives a relation between random variables and random measures on a topological space X.

Proposition 3.39. *Let X be a uniform Hausdorff space. The map*

$$Mes(\Omega, \mu, X) \to M_{Ra}^{+1}(X),$$

where $M_{Ra}^{+1}(X) \equiv$ set of finite positive Radon measures on X, $x \mapsto \mu \circ x^{-1}$, is continuous if $Mes(\Omega, \mu, X)$ is equipped with the topology of convergence in probability and $M_{Ra}^{+1}(X)$ with the narrow topology.

Remark 3.21. *Let $\xi \in Mes(\Omega, \mu, X)$ and $f : X \to \mathbf{R}^I$. Then $f \circ \xi \equiv f(\xi) : \Omega \to \mathbf{R}^I$ is a random variable and the expectation value $E(f(\xi))$ of $f(\xi)$ is defined by $E(f(\xi)) = \int_\Omega f(\xi)d\mu = \int_X f d\nu$, where $\nu \equiv \mu \circ f^{-1}$.* ■

In Tab.3.8 we summarize some useful definitions and properties for real random variables.

Proposition 3.40. (Distribution function of real random variable). *1) Assume that $x \in Mes(\Omega, \mu, \mathbf{R})$ is almost certainly finite, the probability distribution $\nu \equiv \mu \circ x^{-1}$ is a probability on $\mathbf{R}$, i.e., $\nu(\mathbf{R}) = 1$. (In this case we say that the random variable x has probability 1.) Then, there exist: (a) monotone non-decreasing, (b) right-continuous function $F : \mathbf{R} \to [0, 1]$ with (c) $\lim\limits_{x \to -\infty} F(x) = 0$, $\lim\limits_{x \to \infty} F(x) = 1$, such that $L_F \subset \mathcal{P}(\mathbf{R}) \equiv \sigma$-field of subsets on $\mathbf{R}$; (L_F contains all Borel subsets of $\mathbf{R}$, and such that ν coincides on L_F with the Lebesgue-Stieltjes measure μ_F determined by F). More precisely, one has: $F(t) = P\{\omega; x(\omega) \le t\} = \mu_F(-\infty, x] = \nu(-\infty, x]$. Furthermore, for any Borel set $B \subseteq \mathbf{R}$, $P\{\omega; x(\omega \in B)\} = \int_B dF(t)$. Then, F is called distribution function of $x \in Mes(\Omega, \mu, \mathbf{R})$. Furthermore, F characterizes ν in the sense that any function $F : \mathbf{R} \to [0, 1]$ satisfying the above three properties defines a probability measure on $\mathbf{R}$. (In fact we can put $\nu((a, b]) = F(b) - F(a)$, $a < b$, this can be extended uniquely to a probability measure ν on $(\mathbf{R}, \mathcal{B}(\mathbf{R}))$.*

2) If the random variable $x \in Mes(\Omega, \mu, \mathbf{R})$ is finite with probability one, and if $f : \mathbf{R} \to \mathbf{R}$ is Borel measurable, then $E\{f \circ x\} = \int_{\mathbf{R}} f(t)dF(t)$. Furthermore, the distribution function $f \circ x \in Mes(\Omega, \mu, \overline{\mathbf{R}})$ is $G(t) = F(\gamma(t))$, where $\gamma(t) = \sup\{t'; f(t') \le t\}$.

3) Let x be any (finite, real) random variable $x \in Mes(\Omega, \mu, \mathbf{R})$. Then, there exists a function $\phi : \mathbf{R} \to \mathbf{R}$, such that: (a) it is positive definite (for any finite sets $\{t_1, \cdots, t_n\} \in \mathbf{R}$ and $\{\alpha_1, \cdots, \alpha_n\} \in \mathbf{C}$, $\sum_{j,k} \alpha_j \bar{\alpha}_k (z_j - z_k) \ge 0$); (b) uniformly continuous; (c) $\phi(0) = 1$. ϕ is uniquely determined by the distribution F. More precisely, $\phi(t) = E(e^{itx}) = \int_{\mathbf{R}} e^{it\lambda}dF(\lambda) = \int_{\mathbf{R}} e^{it\lambda}d\nu$, so ϕ is the Fourier-Stieltjes transform of F. ϕ is called the characteristic *function of x. If $x, y \in Mes(\Omega, \mu, \mathbf{R})$ are independent random variables, then the characteristic function ϕ_{x+y} of $x + y$ is $\phi_{x+y} = \phi_x \phi_y$.*

4) (P.Levy). Let ϕ_n be characteristic functions of distributions F_n. Let $\phi_n \to \phi$ uniformly on some neigbourhood of $z = 0$. Then, ϕ is also a characteristic function, and one has $F_n \to F$, where F is the distribution associated with ϕ. It is assumed that $F_n \to F$ means that for all $f \in \widetilde{C}^0(\mathbf{R}) =$ Banach space of all continuous functions on $\mathbf{R}$ vanishing at $+\infty$ and $-\infty$ equipped with the topology of uniform convergence,

$$\lim_{n \to \infty} \left(\int_{-\infty}^{+\infty} f dF_n \right) = \int_{-\infty}^{+\infty} f dF.$$

5) The space R of distribution functions, $R \equiv \{F : \mathbf{R} \to [0,1]\}$, F monotone, non-decreasing, right-continuous function, such that $\lim_{n \to -\infty} F(x) = 0$, $\lim_{n \to \infty} F(x) = 1$, is a complete metric space with respect to the Levy's metric:

$$d(F, G) \equiv \inf\{\epsilon > 0 | F(x - \epsilon) - \epsilon \leq G(x) \leq F(x + \epsilon) + \epsilon\}, \ \forall x \in \mathbf{R}, \ F, G \in R.$$

Example 3.37. A *discrete random variable* is one that takes on at most countably many values. A such random variable is characterized by a jump function as distribution. Assume that x is a real random variable which takes on only (positive and negative) integer values and write $p_j = \mu(x = j)$, $j = 0, \pm 1, \pm 2, \cdots$. Then x has the following characteristic function $\phi(t) = P(e^{it})$, where $P(z) = \sum_{j=-\infty}^{+\infty} p_j z^j$, $z \in \mathbf{C}$. The series for $P(z)$ converges at least when $|z| = 1$, the resulting function is called the *probability generating function*. One has the following properties: (i) if $E(x) < \infty \Rightarrow E(x) = P'(1)$; (ii) if $E(x^2) < \infty \Rightarrow var(x) = P''(1) + P'(1) - \{P'(1)\}^2$. ∎

Example 3.38. A *continuous random variable* is one such that the distribution function F can be expressed as follows: $F(t) = \int_{-\infty}^{t} f(\lambda) d\lambda$, where $f : \mathbf{R} \to \mathbf{R}$ is a measurable function such that $f(\lambda) \geq 0$, $\int_{-\infty}^{+\infty} f(\lambda) d\lambda = 1$. Then, f is called *a probability density*. This is equivalent to saying that F is *absolutely continuous*, (which in turn occurs iff its probability distribution is absolutely continuous with respect to Lebesgue measure). Notice that if f_1 and f_2 are probability densities of the same random variable, then $f_1 = f_2$ at most everywhere. Furthermore $F'(t) = f(t)$ at most everywhere. The relation between probability density and characteristic function is the following: $\phi(\lambda) = \frac{\lambda}{2\pi} \int\limits_{-\infty}^{+\infty} \phi(t)e^{-it\lambda}dt.$ ∎

Example 3.39. (*Examples of real random variables:* $x \in Mes(\Omega, \mu, \mathbf{R})$).
1) *Binomial:* $B(n,p)$: (Discrete). Distribution function: $\Sigma_{r \leq t}\binom{n}{r}p^r(1 - p)^{n-r}$. Probability generating function: $(1-p+pz)^n$. Mean: np. Variance: $np(1-p)$. In the binomial case $B(n,p)$, the probability that x takes value r is $P(x = r) = \binom{n}{r}p^r(1-p)^{n-r} \equiv b(n,p;r)$.
2) *Poisson:* $P(\mu)$: (Discrete). Distribution function: $\Sigma_{r \leq t}\mu^r e^{-\mu}/r!$. Characteristic function: $e^{\mu(e^{it}-1)}$. Probability generating function: $e^{\mu(z-1)}$. Mean $= \mu =$ variance.
3) *Geometric:* $G(q), q \in [0,1]$: (Discrete). Probability generating function: $\frac{1-q}{1-qz}$. Mean: $\frac{q}{1-q}$. Variance: $\frac{q}{(1-q)^2}$. In the geometric case $G(q)$, the probability that x takes value $r = 0,1,2,\cdots$ is $P(x = r) = (1-q)q^r$.
4) *Gaussian:* $N(0,1)$: (Continuous). Characteristic function: $e^{-\frac{1}{2}t^2}$. Probability density: $(2\pi)^{-\frac{1}{2}}e^{-\frac{1}{2}\lambda^2}$. Mean: 0. Variance: 1. In Gaussian case $N(0,1)$, the moments $E(x^k)$ of a Gaussian random variable are 0 for k odd, and related to integral gamma function for k even.
5) *Random variable with normal distribution random variable:* $N(\mu, \sigma^2)$: (Continuous). Charactersitic function: $e^{i\mu t - \frac{1}{2}\sigma^2 t^2}$. Probability density: $\sigma^{-1}(2\pi)^{-\frac{1}{2}}e^{-\frac{(\lambda-\mu)^2}{2\sigma^2}}$. Mean: μ. Variance: σ^2.
6) *Random variable with uniform distributionrandom variable:* $U(a,b)$: (Continuous). Characteristic function: $\frac{e^{ibt}-e^{iat}}{i(b-a)t}$. Probability density: $f(\lambda) = (b-a)^{-1}$ if $a \leq \lambda \leq b$, 0 otherwise.
7) *Random variable with exponential distributionrandom variable:* $E(\xi)$: (Continuous) Characteristic function: $\xi/(\xi-it)$. Probability density: $f(\lambda) = \xi e^{-\xi\lambda}$ if $\lambda \geq 0$, 0 if $\lambda < 0$.
8) *Random variable with Cauchy distributionrandom variable:* (Continuous). Characteristic function: $e^{-|t|}$. Probability density: $[\pi(1 + \lambda^2)]^{-1}$.∎

Proposition 3.41. (Conditional probability). *Let* $(\Omega, \mathcal{T}, \mu)$ *be a probability space. Let A be a fixed element of $\mathcal{T}$. Let $\mathcal{G}$ be a sub-σ-field of $\mathcal{T}$, not necessarily containing A. Then, there exists a unique function $\mu|_\mathcal{G}$-measurable $\phi \equiv \mu(A|\mathcal{G}) : \Omega \to \mathbf{R}$ such that $\mu(A \cap E) \equiv \mu_A(E) = \int_E \phi d\mu|E = \int_E \phi d\mu$, $\forall E \in \mathcal{G}$. Then, $\mu(A|\mathcal{G})$ is called the probability of A conditioned on the σ-field $\mathcal{G}$. (Note that, in general, $\mu(\bullet|\mathcal{G})$ is not a measure on $(\Omega, \mathcal{T})$.) One has: (a) If $A \in \mathcal{G} \Rightarrow \mu(A|\mathcal{G}) = \chi_A =$ indicator function of A; (b) If A and $\mathcal{G}$ are independent, then $\mu(A|\mathcal{G}) = \mu(A)$; (c) If $\mathcal{S}$ is the σ-field containing* *only* $\emptyset$ and $\Omega \Rightarrow \mu(A|\mathcal{S}) = \mu(A)$.

Definition 3.77. If $x \in Mes(\Omega, \mu, X)$ is a random variable on the probability space $(\Omega, \mathcal{T}, \mu)$, we call *conditional probability* of $A \in \mathcal{T}$ with respect to x the following: $\mu(A|\mathcal{G} \equiv x^{-1}(R)) \equiv \mu(A|x)$.

Definition 3.78. 1) A *stochastic process* (*random function* or *random process*) is a map $f : T \to Mes(\Omega, \mu, X)$, where T is an arbitrary set. The map $f_\omega : T \to X$, for $\omega \in \Omega$ fixed, is called a *sample function* (or *sample path* if $T \equiv \mathbf{R}$). The map $\tilde{f} : T \times \Omega \to X, \tilde{f}(t, \omega) = f(t)(\omega)$ is called a *random map*.

2) Two stochastic processes $f : T \to Mes(\Omega, \mu, X), g : S \to Mes(\Omega, \mu, Y)$ corresponding to the same sample space (Ω, μ), are said to be *independent stochastic processes* if for arbitrary finite sequence, $t_1, \cdots, t_p \in T$ and $s_1, \cdots, s_p \in S$ the random variables $(f(t_1), \cdots, f(t_p)) : \Omega \to X \times \cdots_p \cdots \times X \equiv X^p$ and $(g(s_1), \cdots, g(s_p)) : \Omega \to Y \times \cdots_p \cdots \times Y \equiv Y^p$ are independent.

3) Let X be a completely regular space. We say that the stochastic process $f : T \to Mes(\Omega, \mu, X)$ is *canonical stochastic process* (or *regular stochastic process*) if there exists a random variable $\check{f} : \Omega \to \check{X}^T \equiv F(T, \check{X})$, such that for every $t \in T, f(t)(\omega) = \check{f}(\omega)(t)$ for almost every $\omega \in \Omega$. (Every stochastic process is isonomous to a canonical stochastic process, so we shall often assume that a given random function is canonical.)

4) Let T be a topological space. A canonical stochastic process $f : T \to Mes(\Omega, \mu, X)$ is *almost surely continuous* if $\check{f}(\omega) : T \to X$ is an injective continuous map for almost every $\omega \in \Omega$.

5) If T and X are vector spaces, a canonical stochastic process $f : T \to Mes(\Omega, \mu, X)$ is *linear stochastic process* if $\check{f}(\omega) : T \to X$ is a linear map for almost every $\omega \in \Omega$. f is one-to-one iff $\tilde{f}(t, \omega) = 0$ a.e. implies $t = 0$.

6) If $T \equiv \mathbf{Z}$ or $T \equiv \mathbf{N}$ the stochastic process is called a *discrete stochastic process*.

Remark 3.22. If Y is a topological space and $\mathcal{B}$ is its Borellian σ-algebra, for any $n \in \mathbf{N}, B^n \equiv \overset{n}{\underset{i=1}{\otimes}} B_i, B_i \equiv \mathcal{B}$, is the product σ-algebra of Y^n. Furthermore, a *cylindrical set* in Y^X with *basis cylindrical set* $B_1 \times \cdots \times B_n \in B^n$, (over coordinates $x_1, \cdots, x_n \in X$), is the set $C_{x_1 \cdots x_n}(B_1 \times \cdots \times B_n) \equiv \{f \in Y^X | f(x_i) \in B_i, i = \overline{1, n}\}$. The class $\mathcal{C}$ of all cylindrical sets over a finite number of coordinates is an algebra of sets in Y^X. Denote by $\mathcal{A}$ the σ-algebra of sets in Y^X generated by $\mathcal{C}$. $\tilde{f} : X \times \Omega \to Y$ is a random map from X to Y iff the map $\check{f} : (\Omega, \mathcal{T}) \to (Y^X, \mathcal{A})$ defined by $f(\omega) = \widetilde{f}_\omega$ is a measurable map. Then, $\nu \equiv \mu \circ \check{f}^{-1}$ is a probability measure on $(Y^X, \mathcal{A})$ called *distribution of the random process $\widetilde{f}$*. For a random map $\widetilde{f} : X \times \Omega \to (X, \mathcal{B})$, $n \in \mathbf{N}$, $x_1, \cdots, x_n \in T$, $B_1, \cdots, B_n \in \mathcal{B}$ the probabilities $P\{\omega | \widetilde{f}_{x_1}(\omega) \in B_1, \cdots, \widetilde{f}_{x_n}(\omega) \in B_n\}$ are called the *finite dimensional distributions* of $\widetilde{f}$. One has: $P\{\omega | \widetilde{f}_{x_1}(\omega) \in B_1, \cdots, \widetilde{f}_{x_n}(\omega) \in B_n\} = \nu(C_{x_1 \cdots x_n}(B_1 \times \cdots \times B_n))$, where $\nu \equiv \mu \circ \check{f}^{-1}$. The corresponding measures $\nu_{x_1 \cdots x_n}$ on (X^n, B^n) are probability measures, also called finite dimensional distributions of the stochastic process $\widetilde{f}$. In particular if $X \equiv E$ is a separable Banach space, for any measurable map $h : (E^n, B^n) \to E$ one has $E(h(\widetilde{f}_{x_1}, \cdots, \widetilde{f}_{x_n})) = (B) \int_{E^n} h \, d\nu_{x_1 \cdots x_n}$. The *characteristic functional* of $f : T \to Mes(\Omega, \mu, E)$ is the map $\phi : T \times E' \to \mathbf{C}$, $\phi(t, \theta) = \phi_t(\theta) = Ee^{i<\theta, f_t>} = \int_E e^{i<\theta, y>} d\nu_t$, where $\nu_t \equiv \mu \circ \widetilde{f}_t^{-1}$, $y \equiv id_E$. $\blacksquare$

Theorem 3.67. *Let (E, F) be a duality couple of vector spaces on $\mathbf{R}$. Then, there exists a bijection between cylindrical measures on E and the set of stochastic processes on F.*

Example 3.40. Set $T \equiv [0, 1] \subset \mathbf{R}$. As $(C^0(T), M_{Ra}(T))$ is a duality couple of vector spaces on $\mathbf{R}$, we have that for a cylindrical measure w on $C^0(T)$ there corresponds a linear stochastic process on $M_{Ra}(T)$. For example, the stochastic process $f : M_{Ra}(T) \to Mes(C^0(T), w; \mathbf{R})$ to $\mu \in M_{Ra}(T)$ associates the w-class of functions $f(\mu) : C^0(T) \to \mathbf{R}$, $x \mapsto \mu(x) \equiv \int_T x \, d\mu$ is a canonical linear stochastic process associated to the Wiener (cylindrical) measure w. $\blacksquare$

Definition 3.79. Let $X \equiv E$ be a topological vector space and $\{T, d\}$ a compact metric space. A map $f : T \to Mes(\Omega, \mu; E)$ is *μ-Riemannian integrable* if the base of the filter $\{B_\epsilon(t, \mu)\}_{\epsilon > 0}$, where $B_\epsilon(t, \mu)$ is the set of all elements x of $Mes(\Omega, \mu; E)$ like $x = \sum_{i \in I} \mu(A_i) x_i$, where I is finite, $\{A_i\}_{i \in I}$ is a partition of T in borelians with diameter ϵ and $x_i \in f(A_i)$, converges in $Mes(\Omega, \mu; E)$. This limit is called the *μ-Riemannian integral* of f. (If f is continuous, f is μ-Riemannian integrable for any $\nu \in Ob(\overline{\mathcal{M}}(T))$.)

Theorem 3.68. *Let (T, d) be a compact metric space and let*

$$L : Ob(\overline{\mathcal{M}}_c(T)) \to Mes(\Omega, \mu; E)$$

be a linear random process, where $Ob(\overline{\mathcal{M}}_c(T))$ is the vector space $Ob(\overline{\mathcal{M}}(T))$ endowed with the topology of uniform convergence on the compact parts of $C^0(T)$. Then, $L(\mu)$ is the μ-Riemannian integral of the function $f : T \to Mes(\Omega, \mu; E)$, $f(t) = L(\delta_t)$. Furthermore, the stochastic process f admits a version g with continuous trajectories $\omega \mapsto \int_T \tilde{f}(t,\omega)\mu(dt)$, is an element of the class $L(\mu)$.

Of particular importance are the real random processes.

Definition 3.80. 1) A *real stochastic process* is a map $f : T \to Mes(\Omega, \mu; \mathbf{R})$. The *covariance of a stochastic process* $f : T \to Mes(\Omega, \mu; \mathbf{R})$ is the function $\Gamma : T \times T \to \mathbf{R}$, $\Gamma(s,t) = E(f_s f_t) - E(f_s)E(f_t)$.

2) Let $\tilde{f} : T \times \Omega \to \mathbf{R}$ be a random function. The *mean value* of $\tilde{f}$ is the map $< \tilde{f} > : T \to \mathbf{R}$, $< \tilde{f} > (t) = E(\tilde{f}_t)$. The *correlation function* of $\tilde{f}$ is the map $\rho(\tilde{f}) : T \times T \to \mathbf{R}$, $\rho(\tilde{f})(t_1, t_2) = \rho(\tilde{f}_{t_1}, \tilde{f}_{t_2})$. The *variance* of $\tilde{f}$ is the map $var(\tilde{f}) : T \to \mathbf{R}$, $var(\tilde{f})(t) = var(\tilde{f})$.

Proposition 3.42. (Properties of real stochastic processes). 1) *The covariance Γ of a real stochastic process $f : T \to Mes(\Omega, \mu; \mathbf{R})$ identifies a metric $d : T \times T \to \mathbf{R}^+$, $d(s,t) = \|f_s - f_t\| = (\Gamma(t,t) + \Gamma(s,s) - 2\Gamma(s,t))^{1/2}$.*

2) *The correlation function has the following properties:* (a) $\rho(\tilde{f})(t_1, t_2) = \rho(\tilde{f})(t_2, t_1)$, $\forall t_1, t_2 \in T$. (b) $\sum_{i,j=1}^{n} \rho(\tilde{f})(t_i, t_j)\alpha^i \alpha^j \geq 0$, *for any* $n \in \mathbf{N}$, $t_i \in T$, $\alpha^i \in \mathbf{R}$.

Theorem 3.69. (Finite dimensional distributions of real stochastic processes). 1) *Let $T \equiv E$ be a vector space and let $f : E \to Mes(\Omega, \mu, \mathbf{R})$ be a linear real stochastic process. Then f is characterized by a measurable map $\tilde{f} : (\Omega, T) \to (E^*, A)$, where A is the σ-algebra generated by cylindrical sets in E^*. More precisely, a cylindrical set in E^* is a subset A of E^* such that there exists a family of n-vectors of E, $(x_1, \cdots, x_n), n \in \mathbf{N}$, such that $A = \{\alpha \in E^* | (< \alpha, x_1 >, \cdots, < \alpha, x_n >) \in B(\mathbf{R}^n)\}$. Furthermore, the finite dimensional distributions of $\tilde{f}$ are probabilities measures $\nu_{x_1 \cdots x_n}$ on $(\mathbf{R}^n, \mathcal{B}(\mathbf{R}^n))$, $n \in \mathbf{N}$.*

2) *If E is a separable Banach space, for any measurable map $h : (E^n, B^n) \to (E, B)$ one has*

$$E(h(\tilde{f}_{x_1}, \cdots, \tilde{f}_{x_n})) = (B) \int_{E^n} h \, d\nu_{x_1 \cdots x_n}.$$

The mean value of $\tilde{f}$ is an element of E^* :$< \tilde{f} > \in E^*$. The characteristic functional of $\tilde{f}$ is: $\phi : E \times \mathbf{R} \to \mathbf{C}$, $\phi(x,t) = Ee^{it\tilde{f}_x} = \int_{\mathbf{R}} e^{it} d\nu_x$, $\nu_x \equiv \mu \circ \tilde{f}_x^{-1}$.

3) (Inverse theorem). *A system $(\nu_{x_1 \cdots x_n})$, indexed by the set of finite families $(x_1, \cdots, x_n)$ of vectors belonging to a vector space E, is the finite dimensional distributions of a linear real stochastic process $f : E \to Mes(\Omega, \mu, \mathbf{R})$*

iff the following conditions are satisfied: (*Kolmogorov consistency conditions*): (a) For any $(x_1, \cdots, x_n), \nu_{x_1 \cdots x_n}$ is a probability measure on $\mathbf{R}^n$; (b) For any $(x_1, \cdots, x_n)$, any linear map $u \in L(\mathbf{R}^n, \mathbf{R}^m)$, and any $A \in \mathcal{B}(\mathbf{R}^m)$, one has $\nu_{u(x_1) \cdots u(x_n)}(A) = \nu_{x_1 \cdots x_n}(u^{-1}(A))$.

4) Let E be a space, inductive limit of a sequence $(E_n)_{n \in \mathbf{N}}$ of Fréchet subspaces of countable type. A system $(\nu_{x_1 \cdots x_n})$ like before, is the system of finite dimensional distributions of probability measure on $(E', \mathcal{B}(E'))$ iff the consistency conditions of Kolmogorov and the following are satisfied: (C) For any $\epsilon > 0$, there exists a neighbourhood U of 0 in E such that, for any $p \in \mathbf{N}$ and any element $(x_1, \cdots, x_p) \in U^p$, one has $\nu_{x_1 \cdots x_p}(U^p) \geq 1 - \epsilon$. Furthermore, if E_n are nuclear spaces, then it is sufficient that the following are fulfilled: the Kolmogorov consistency conditions and that the map $\phi_t : E \to \mathbf{C}, \phi_t(x) = \int_{\mathbf{R}} e^{it} d\nu_x$, should be continuous in E or E_n.

Example 3.41. (*Real stochastic processes*). 1) Let $T \equiv B([0, 1])$. Then, there exists a real stochastic process $f : T \to Mes(\Omega, \mu, \mathbf{R})$ such that: (a) $f(t)$ has a Poisson distribution whose mean is the Lebesgue measure of t; (b) if $t_1, \cdots, t_n$ are disjoint, $f(t_1), \cdots, f(t_n)$ are independent, and

$$f(t_1, \cdots, t_n) = f(t_1) + \cdots + f(t_n).$$

Let (Ω, T, μ) be a probability space. A vector subspace H of $Mes(\Omega, \mu, \mathbf{R})$ is *Gaussian* if each element of H is a Gaussian random variable: (We can see that the closure $\overline{H}$ of H in $Mes(\Omega, \mu, \mathbf{R})$ is also Gaussian and that $\overline{H}$ is a closed subspace of $Mes(\Omega, \mu, \mathbf{R})$. A *Gaussian process* is a real stochastic process $f : T \to Mes(\Omega, \mu, \mathbf{R})$ such that the closed vector subspace generated by $\{f_t\}_{t \in T}$ is Gaussian. For any Gaussian process $f : T \to Mes(\Omega, \mu, \mathbf{R})$ one has $E(f_t) = 0$ and the covariance Γ is a symmetric function $\Gamma : T \times T \to \mathbf{R}$ such that for any finite subset $\{t_1, \cdots, t_n\} \in T$ and any real numbers $z_1, \cdots, z_n$ not all zero $\sum_{1 \leq i,j \leq n} \Gamma(t_i, t_j) z_i z_j > 0$ (i.e., Γ is *positive definite*). Furthermore, a Gaussian random function $\tilde{f} : T \times \Omega \to \mathbf{R}$ is characterized by definite dimensional distributions given by $F\{\omega | (\tilde{f}_{t_1}(\omega), \cdots, \tilde{f}_{t_k}(\omega)) \in A\} = \int_A \gamma_{x_0, B}$, where $k \in \mathbf{N}, t_i \in T, i = \overline{1, k}$, A is Borel set in $\mathbf{R}^k, x_0 \equiv (<\tilde{f}>(t_1), \cdots, <\tilde{f}>(t_k))$. $B \equiv (a_{ij}), i, j = \overline{1, k}, a_{ij} \equiv \rho(\tilde{f})(t_i, t_j)$, and $\gamma_{x_0, B}$ is the Gaussian measure on $\mathbf{R}^k$ relative to the vector x_0 and the isomorphism B.

2) A *Wiener process* (or *Brownian motion*) is a real stochastic Gaussian process $f : T \to Mes(\Omega, \mu, \mathbf{R})$ such that: (a) $T \equiv [0, 1]$; (b) The increments

$f(s,t) = f(s) - f(t)$ are independent; (c) $f(0) = 0$; (d) the process $g :$ $T \times T \to Mes(\Omega, \mu, \mathbf{R})$ given by $g(s,t) = f(s,t)$ is a Gaussian process with variance $var(g)(s,t) = |t - s|$. The distribution of Wiener process is a measure μ on $(\mathbf{R}^T, \mathcal{B}^T)$ supported by $C^0(T)$, (i.e., $\mu(C^0(T)) = 1$), so μ identifies a measure w on $C^0(T)$ endowed with the collection $\mathcal{B}'(C^0(T))$, trace of $\mathcal{B}^T$ on $C^0(T)$. w is just the Wiener measure. Further, one has: (a) $C^0(T) \neq \mathcal{B}^T$; (b) $\mathcal{B}(C^0(T)) = \mathcal{B}'(C^0(T))$. Furthermore, one can prove that almost all sample paths of a Brownian motion are nowhere differentiable.

3) A *Markov process* is a real stochastic process $f : T \to Mes(\Omega, \mu, \mathbf{R})$ where: (a) $T \subset \mathbf{R}$; (b) if for any subset $S \subset T, F_S$ denotes the smallest σ-field of event with respect to which the function $f(s) : \Omega \to \mathbf{R}$ is measurable for any $s \in S$, then $((\mu(A \cap B)|f(t)) = \mu(A|f(t))\mu(B|f(t))$ with probability one with $A \in F_{\{s \in T; s < t\}}, B \in F_{\{s \in T; s > t\}}$. In the following we shall consider some fundamental properties of Markov process and a relation with Brownian motion. Let $f : T \to Mes(\Omega, \mu, \mathbf{R})$ be a Markov process. Then, there exists a function $P : T\widehat{\times}T \times \Omega \times \mathcal{T} \to \mathbf{R}$, where $T\widehat{\times}T \equiv \{(s,t) \in T \times T | s < t\}$, such that: (i) $P(s,t,\bullet,A)$ is a Borel-measurable function on Ω and $P(s,t,\lambda,\bullet)$ is a probability measure on $\mathcal{T} \equiv \sigma$-algebra of the probability space; (ii) $P(s,s,\lambda,A) = \chi_A(\lambda)$; (iii) the following *Chapman-Kolmogorov equation* holds for almost all $\lambda \in \Omega$: $P(s,t,\lambda,A) = \int P(s,t,\lambda,dy)P(t,u,y,A), s < t < u$. Any such function $P(s,t,\lambda,A)$ is called a *transition probability* of the *Markov process* f. The distribution of a Markov process f is uniquely determined by the initial distribution μ_0 of f_0 and the system $\{P(s,t,\lambda,A)\}$ of transition probabilities. Then finite dimensional distributions of a Markov process $f : T \to Mes(\Omega, \mu, \mathbf{R})$ may be defined so that to satisfy the relation

$$E(\psi(f(t_0), f(t_1), \cdots, f(t_n)))$$

$$= \int_\Omega \cdots \int_\Omega \psi(\lambda_0, \lambda_1, \cdots \lambda_n) d\mu_0 \prod_{j-1}^n dP(t_{j-1}, t_j, \lambda_{j-1}, \bullet),$$

for any measurable function $\psi : \Omega \times \cdots_{n+1} \cdots \times \Omega \to \mathbf{R}$. Any Brownian motion $f : T \to Mes(\Omega, \mu, \mathbf{R})$ is a Markov process.   ∎

Definition 3.81. 1) A *partially ordered system* $(E, \leq)$ is a non-empty set E, together with a relation $\leq$ on E, such that: (a) if $a \leq b$ and $b \leq c \Rightarrow a \leq c$; (b) $a \leq a$. The relation $\leq$ is called an *order relation* in E. The notation $y \geq x$ is sometimes used in place of $\leq$.

2) A *totally ordered subset F* of partially ordered system $(E, \leq)$ is a subset of E such that for every pair $x, y \in F$ either $x \leq y$ or $y \leq x$.

3) If F is a subset of a partially ordered system $(E, \leq)$ then an element x in E is said to be an *upper bound* for F if every f in F has the property $f \leq x$. An upper bound for F is said to be a *least upper bound* of F ($\sup F$) if every upper bound y of F has the property $x \leq y$.

4) In a similar fashion, the terms *lower bound* and *greatest lower bound* (in F) may be defined.

5) An element $x \in E$ is said to be *maximal* if $x \leq y$ implies $y \leq x$.

Example 3.42. Let X be a set and $\mathcal{P}(X)$ be the set of all subsets of X. The couple $(\mathcal{P}(X), \subseteq)$ is a partially ordered system where the order relation is the inclusion relation $\subseteq$ between the sets contained in X. An upper bound for a subfamily $B \subseteq \mathcal{P}(X)$ is any set containing $\cup B$, and $\cup B$ is the only least upper bound of B. Similarly $\cap B$ is the only greatest lower bound of B. The only maximal element of $\mathcal{P}(X)$ is X. ∎

Definition 3.82. Let $(E, \leq)$ be a non-void partially ordered system. A subset $B \subset E$ with the following three properties will be called *admissible* with respect to a function $f : E \to E$ such that $f(x) \geq x$; and with respect to an element $a \in E$: (a) $a \in B$; (b) $f(B) \subseteq B$; (c) *Every least upper bound of a totally ordered subset of B is in B*.

Theorem 3.70. 1) (Hausdorff maximality theorem). *Every partially ordered system contains a maximal totally ordered subsystem.*

2) (Zorn's lemma). *A partially ordered system has a maximal element if every totally ordered subsystem has an upper bound.*

3) (Zermelo well-ordering theorem). *Every set E may be well-ordered that is a partially ordered system $(E, \leq)$ such that: (i) $a \leq b$, $b \leq a \Rightarrow a = b$; (ii) any non-void subset of E contains a lower bound for itself.*

Example 3.43. $\mathbf{N} \equiv$ the set of positive integers in the usual order is a familiar example of well-ordered system. ∎

Definition 3.83. A partially ordered system $(E, \leq)$ is said to be *complete* if: (i) $a \leq b$ and $b \leq a \Rightarrow a = b$; (ii) Every non-void subset has a least upper bound and a greatest lower bound.

Proposition 3.43. (Tarski). *If $(E, \leq)$ is a complete partially ordered system , $f : E \to E$, and $x \leq y$ implies $f(x) \leq f(y)$, then f has a fixed element $x_{o,}$, i.e., $f(x_0) = x_0$, and the set of all fixed elements contains its least upper bound and its greatest lower bound.*

Definition 3.84. An element x in a ring R is said to be *idempotent* if $x^2 = x$, and to be *nilpotent* if $x^n = 0$ for some positive integer n. A *Boolean ring* is one in which every element is idempotent.

Proposition 3.44. 1) *Every Boolean ring is a commutative ring where the identity $x + x = 0$ (or equivalently $x = -x$) holds.*

2) *In a Boolean ring, any prime ideal is maximal.*

Example 3.44. The smallest Boolean ring with unit is $\mathbf{Z}_2 \equiv \{0,1\}$, i.e., the set of integers modulo 2. $\mathbf{Z}_2$ is actually a field. Conversely every Boolean ring with unit which is also a field is necessarily isomorphic to the field $\mathbf{Z}_2$. If a Boolean ring A does not contain zero divisor it is reduced to 0 or it is isomorphic to $\mathbf{Z}_2$. The ring $\mathcal{P}(X)$ of subsets of a fixed set X is a Boolean ring with respect to the following definitions of multiplication and addition of arbitrary subsets of X: (a) *multiplication*: $EF = E \cap F, E, F \in \mathcal{P}(X))$; (b) *addition*: $E + F = (E \cap CF) \cup (CE \cap F) = (E - F) \cup (F - E) \equiv E \triangle F$ ($\equiv$ *symmetric difference*). X is the unity element and $\emptyset$ is the zero element. The Boolean ring $\mathcal{P}(X)$ is isomorphic to the ring $\mathbf{Z}_2{}^X$ of the applications X into the ring $\mathbf{Z}_2$: $h : \mathcal{P}(X) \to \mathbf{Z}_2{}^X$, $h : A \mapsto h(A) \equiv \chi_A$ where χ_A is the characteristic function of A. Every Boolean ring is isomorphic to a sub-ring of the ring $\mathbf{Z}_2{}^X$. Every finite Boolean ring is of the form $\mathbf{Z}_2{}^n$.   ■

Definition 3.85. A topological space is said to be *totally disconnected* if its topology has a base consisting of sets which are simultaneously open and closed.

Example 3.45. A set X with the discrete topology is a totally disconnected topological space.   ■

Theorem 3.71. (Stone representation theorem). *Every Boolean ring with unit is isomorphic to the Boolean ring of all open and closed subsets of a totally disconnected compact Hausdorff space.*

Definition 3.86. A *lattice* is a partially ordered set L such that every pair x, y of elements in L has a least upper bound and a greatest lower bound, denoted by $x \vee y$ and $x \wedge y$ respectively. The lattice L has a *unit* if there exists an element 1 such that $x \leq 1$, $x \in L$. The lattice L has a *zero* if there exists an element 0 such that $0 \leq x$, $x \in L$. The lattice is called *distributive lattice* if $x \wedge (y \vee z) = (x \wedge y) \vee (x \wedge z)$, $x, y, z \in L$. The lattice is called *complemented lattice* if for every $x \in L$, there exists an $x' \in L$ such that $x \vee x' = 1$, $x \wedge x' = 0$. A lattice L is said to be *complete lattice* if every subset bounded above has a least upper bound, or equivalently, every subset bounded below has a greatest lower bound. A lattice L is said to be σ-*complete lattice* if this is true for denumerable subsets of L.

Definition 3.87. A *Boolean algebra* is a lattice with unit and zero which is distributive and complemented.

Proposition 3.45. *The concepts of Boolean algebra and Boolean ring with unit are equivalent.*

Proof. In fact let B be a Boolean algebra and define multiplication and addition as: $xy = x \wedge y$, $x + y = (x \wedge y') \vee (x' \wedge y)$. Then, it may be verified that B is a Boolean ring with 1 as unit. On the other hand, if B is a Boolean ring with unit denoted by 1, then if $x \leq y$ is defined to mean $x = \lambda y$ and $x' = 1 + x$, then B is a Boolean algebra and $x \vee y = x + y$, $x \wedge y = xy$. $\square$

Definition 3.88. If B and C are Boolean algebras and $h : B \to C$, then h is said to be a *Boolean algebra homomorphism*, if $h(x \wedge y) = h(x) \wedge h(y)$, $h(x \vee y) = h(x) \vee h(y)$, $h(x') = h(x)'$. If h is one-to-one, it is called an *isomorphism Boolean algebra*. If h is an isomorphism and $h(B) = C$, then we say that B is *isomorphic* to C.

Example 3.46. The following are examples of Boolean algebras: 1) $\mathbf{Z}_2 \equiv \{0,1\}$; 2) $\mathcal{P}(X)$, where $\leq$ is taken to be the inclusion, and $\wedge$ and $\vee$ are taken as intersection and union respectively. $\blacksquare$

Theorem 3.72. *Every Boolean algebra is isomorphic with the Boolean algebra of all open and closed subsets of a totally disconnected compact Hausdorff space.*

Definition 3.89. 1) A *projection* in V means a linear operator $E \in L(V)$ with $E^2 = E$.

2) The *intersection* $A \wedge B$ and the *union* $A \vee B$ of two commuting projections A and B in V are projections: (a) *intersection*: $A \wedge B = A \circ B$; (b) *union*: $A \vee B = A + B - (A \circ B)$. The ranges of the intersection and union of two commuting projections are given by: $(A \wedge B)(V) = A(V) \cap B(V)$; $(A \vee B) = A(V) \oplus B(V) = S\bar{p}(A(V), B(V))$, where $S\bar{p}(A(V), B(V))$ means the closed linear manifold spanned by the sets $A(V)$ and $B(V)$.

3) The *natural ordering* $A \leq B$ between two commuting projections A and B has the geometrical significance that $A \leq B$ is equivalent to $A(V) \subseteq B(V)$.

4) A *Boolean algebra of projections* in V is a set of projections in V which is a Boolean algebra under the operations $A \vee B$ and $A \wedge B$ which has for

its zero and unit elements the operators 0 and $id_V \equiv I \in V$.

5) The notions of abstract σ-completeness, and completeness can be extended.

Definition 3.90. Let V be a locally convex topological vector space, X a set and $\mathcal{R}$ a ring of subsets of X. A *pre-spectral measure* on V is a map $E : \mathcal{R} \to \mathcal{L}(V)$ with the following properties: (a) E is countable additive in the weak operator topology, that is an $E(\bigcup_n A_n) = \sum_n E(A_n)$, whenever A_n are disjoint sets in $\mathcal{R}$. (b) E is *continuous spectral measure* in the sense that $E(M_n) \to E(M)$ weakly, whenever M_n is an increasing sequence of sets in $\mathcal{R}$ whose union M is also in $\mathcal{R}$.[54] A pre-spectral measure $E : \mathcal{R} \to \mathcal{L}(V)$ is *normalized spectral measure* if $E(X) = id_V$, (if $X \in \mathcal{R}$). $E : \mathcal{R} \to \mathcal{L}(V)$ is said to be *equicontinuous spectral measure* if the values of E on $\mathcal{R}$ are. $E : \mathcal{R} \to \mathcal{L}(V)$ is said to be *Baire* or *Borel* if X is a (compact) Hausdorff space and $\mathcal{R} \equiv \mathcal{B}_a(X)$ or $\mathcal{R} \equiv \mathcal{B}(X)$, its σ-algebra of Baire or Borel sets respectively. We call *sub-spectral* (or *quasi-spectral*) a normalized pre-spectral measure. A *spectral measure* on V is a pre-spectral measure $E : \mathcal{R} \to \mathcal{L}(V)$ such that $E(A_1 \cap A_2) = E(A_1) \circ E(A_2)$ for any $A_1, A_2 \in \mathcal{R}$.

Remark 3.23. For a spectral measure, normalization can always be achieved by throwing away an irrelevant part of the underlying locally convex space V. Furthermore, a spectral measure in V can be regarded as a homomorphic map of Boolean algebra of sets into a σ-complete or complete Boolean algebra of (equicontinuous) projections in V, which has the additional property that it maps the unit into the identity operator $I \equiv id_V$.   ■

Definition 3.91. Let H be a nonzero complex Hilbert space. A *pre-spectral* (resp. *sub-spectral, spectral*) *measure spectral* on H is a pre-spectral (resp. sub-spectral, spectral) measure $E : \mathcal{R} \to \mathcal{L}(H)$, where H is considered as a locally convex space such that E is *positive spectral measure*, that is $E(M) \geq 0$ for each $M \in \mathcal{R}$; (so $E(M)$ is also hermitian).

Theorem 3.73. 1) *Let* $(X, \mathcal{R}, E)$ *be a triplet where* $\mathcal{R}$ *is a ring of subsets of* X *and* $E : \mathcal{R} \to \mathcal{L}(V)$ *is a set function with* V *a locally convex space. Then,* E *is a pre-spectral measure on* V *iff* $E_{x,\alpha} : \mathcal{R} \to \mathbf{R}$, $E_{x,\alpha}(A) = <E(A)(x), \alpha >$, *is a signed measure on* $(X, \mathcal{R})$ *for any* $x \in V, \alpha \in V'$.

2) *If* $V \equiv H$ *is a hermitian vector space, then* E *is a pre-spectral measure*

---

[54] If $A, A_j \in \mathcal{L}(V)$, where V is a locally convex space, writing "$A_j \to A$ weakly", we mean that is: $<A_j(x), \alpha> = <A(x), \alpha>$, for each $\alpha \in V'$ ($\equiv$ topological dual of V), $x \in V$.

on H iff $E_{x,x} : \mathcal{R} \to \mathbf{R}$, $E_{x,x}(A) = < E(A)(x), x >$, is a measure on $(X, \mathcal{R})$.

Proof. 1) Note that a pre-spectral measure must take its values in a weakly normal cone in $\mathcal{L}(V)$, namely the convex cone generated by its values: if $x \in V$ and $\alpha \in V'$, then the set function $A \mapsto < E(A)(x), \alpha >$ is a finite-valued signed measure on $\mathcal{R}$ and hence there is a set $B \in \mathcal{R}$ for which $A \mapsto < E(A \cap B)(x), \alpha >$ is a non negative finite valued measure and $A \mapsto < E(A \sim B)(x), \alpha >$ is a non-positive finite valued measure (*Hahn-decomposition*). Thus, for every $A \in \mathcal{R}$,

$$< E(A)(x), \alpha > = < E(A)(x), E(B)'(\alpha) > + < E(A)(x), E(A \sim B)'(\alpha) >,$$

where $'$ denotes the transposition, and $A \sim B \equiv C_A(A \cap B) = A \setminus (A \cap B)$.
2) The proof is similar. $\qquad\qquad\qquad\qquad\qquad\qquad\qquad\qquad\qquad\qquad\quad\square$

Definition 3.92. A Borel spectral measure $E : \mathcal{B}(X) \to \mathcal{L}(V)$ is said to be *regularBorel spectral measure* if $< E(\bullet)(x), \alpha >$ is a regular Borel measure for each $x \in V$ and $\alpha \in V'$.

Theorem 3.74. (Naimark-Nagy). *Let $(X, \mathcal{R}, E)$ be a quasi-spectral measure, $E : \mathcal{R} \to \mathcal{L}(H)$, where H is a Hermitian space and $\mathcal{R}$ is a σ-ring. Then, there exists an extended hermitian space $\widetilde{H}$ such that E can be extended to a spectral-measure $\widetilde{E} : \mathcal{R} \to \mathcal{L}(\widetilde{H})$. More precisely, the projections $\widetilde{E}(\omega)$, $\omega \in \mathcal{R}$, are such that $\widetilde{E}(\omega)(f)$, $f \in H$, $\omega \in \mathcal{R}$, identify the space $\widetilde{H}$.*

Proof. Let $E(\mathcal{R})$ be the space of complex numerical valued $\mathcal{R}$-sets functions and let $E(\mathcal{R}, H) \equiv E(\mathcal{R}) \otimes H$ be the space of H-valued, complex numerical $\mathcal{R}$-set functions. We can define a scalar semiproduct in $E(\mathcal{R}, H)$ by putting: $(\phi|\psi) = \sum_{k,e}(E(A_k \cap B_e)(x_k)|y_e)$; if $\phi = \sum_k 1_{A_k} \otimes x_k$, $\psi = \sum_e 1_{B_e} \otimes y_e$. We can prove that, thank to the additivity of E, $(\phi|\psi)$ is independent of the representation chosen for ϕ and ψ. Let $\widetilde{H} \equiv E(\mathcal{R})\widehat{\otimes}H$ be the completion of $E(\mathcal{R}) \otimes H$ with respect to the above scalar semi-product. One has the isometric monomorphism $j : H \to \widetilde{H}$ given by $j : x \mapsto 1 \otimes x$. By means of j we can identify H with a closed subspace of $\widetilde{H}$. Furthermore, we have the orthogonal projection on $H : \pi_H : \widetilde{H} \to H$ given by $\pi_H(1_A \otimes x) = E(A)(x)$, $\forall A \in \mathcal{R}$, $x \in H$. We define $\widetilde{E} : \mathcal{R} \to \mathcal{L}(E(\mathcal{R}, H))$ as $\widetilde{E}(A)(\phi) = \widetilde{E}(A)(1_A \otimes x) = 1_A \otimes E(A)(x)$. One verifies that $\|\widetilde{E}(A)(\phi)\| \le \|\phi\|$ and $\widetilde{E}(A)$ canonically extends to an operator $\widetilde{E}(A) : \widetilde{H} \to \widetilde{H}$. Then, we can see that $\widetilde{E}$ is a spectral measure on H such that $E(\bullet) = \pi_H \circ \widetilde{E}(\bullet) \circ \pi_H$. Finally, one can prove that $\widetilde{H}$ is a *minimal extension* of H, that is $\widetilde{H}$ is generated by the vectors $\widetilde{E}(A)(x)$, $A \in \mathcal{R}$, $x \in H$. $\qquad\qquad\square$

Theorem 3.75. *Any (pre-)spectral measure $E : (X, \mathcal{R}) \circ\!\!\to \mathcal{L}(V)$ identifies a map $\phi : M(X, \mathcal{R}) \to \mathcal{L}(V)$, where $M(X, \mathcal{R})$ is the algebra of measurable functions $X \to \mathbf{K}$ on $(X, \mathcal{R})$, such that $< \phi(f)(x), \alpha > = \int_X f dE_{x,\alpha}$, $\forall x \in V$, $\alpha \in V'$. If $V \equiv H$ is a hermitian space then one has $< \phi(f)(x), y > = \int_X f dE_x$, $\forall x, y \in H$. If E is a spectral measure the map ϕ is a morphism of algebras.*

Definition 3.93. 1) A *(pre-)spectral measure* on a PDE $E_k \subset J\mathcal{D}^k(W)$ is any (pre-)spectral measure $E : (\Omega(E_k), \Sigma) \circ\!\!\to \mathcal{L}(\mathcal{H})$, where $\Omega(E_k)$ is the *quantum situs* of E_k, i.e. the set of admissible integral manifolds bording hypersurfaces Cauchy data of E_k. Furthermore, $\mathcal{H}$ is a locally convex topological vector space, obtained by means of the space $C_0^\infty(H)$ of the sections with support compact of a hermitian bundle $\pi_H : H \to J\mathcal{D}^p(W)$, $-1 \le p \le \infty$, and such that there is a map $\eta : C^\infty(W) \to C^\infty(\Lambda_n^0 M)$, $\quad (n = \dim M)$, such that $\eta(s) : M \to \Lambda_n^0 M$ is a volume form on the base manifold M of $\pi : W \to M$. So $\mathcal{H}(s) \subset \mathcal{H}$ is the completion of $C_0^\infty(D^p s^* H)$ with respect to the inner product $(\psi | \psi')_s = \int_M h(s)(\psi | \psi')\eta(s)$ where $h(s)$ is the scalar product induced by the hermitian structure of $D^p s^* H$. Furthermore, $(H, W, \mathbb{B})$ is a super-bundle of geometric objects over M, with respect to a covariant functor $\mathbb{B} : \mathcal{C}(W) \to \mathcal{C}(H)$.

2) A *weak spectral measure (pre-)spectral measure* on E_k is a (pre-)spectral measure where H is not a hermitian bundle, but simply a vector bundle over $J\mathcal{D}^p(W)$, $\quad -1 \le p \le \infty$. So $\mathcal{H}(s)$ are not necessarily Hilbert spaces.[55]

Theorem 3.76. *Any spectral measure $E : (\Omega(E_k), \Sigma) \circ\!\!\to \mathcal{L}(\mathcal{H})$ of a differential equation $E_k \subset J\mathcal{D}^k(W)$ identifies a representation of random variables of E_k, i.e., functions $f : \Omega(E_k) \to \mathbf{R}$.*

Proof. In fact we have the following morphism: $E : M(\Omega(E_k), \Sigma) \to \mathcal{L}(\mathcal{H})$, such that $E(f)$ is the linear operator on $\mathcal{H}$ defined by the following: $< E(f)(\psi), \psi' > = \int_{\Omega(E_k)} f dE_{\psi,\psi'}$, $\forall f \in M(\Omega(E_k), \Sigma)$, and $\forall \psi \in \mathcal{H}$, $\psi' \in \mathcal{H}'$. This morphism becomes an homomorphism of algebras if $E : (\Omega(E_k), \Sigma) \circ\!\!\to \mathcal{L}(\mathcal{H})$ is just a spectral measure and not only a prespectral measure. $\qquad\square$

Theorem 3.77. *A spectral measure on $E_k \subset J\mathcal{D}^k(W)$ identifies a spectral measure on any differential equation $\overline{E}_k \subset J\mathcal{D}^k(\overline{W})$ equivalent to E_k by*

---

[55] In the following, whether no confusion can arise, for sake of simplicity, we shall say spectral measures to denote also prespectral measures and weak (pre)spectral measures.

means of a fiber bundle diffeomorphism $f \equiv (f_W, f_M) : W \to \overline{W}$ over a diffeomorphism $f_M : M \to \overline{M}$.

Proof. In fact, for functorial property, we have the following commutative diagram:

$$
\begin{array}{ccc}
(\Omega(E_k), \Sigma) & \overset{E}{\multimap} & \mathcal{L}(\mathcal{H}) \\
f_* \quad \uparrow & & \downarrow \quad Q(f) \\
(\Omega(\overline{E}_k), \overline{\Sigma}) & \underset{\overline{E}}{\multimap} & \mathcal{L}(\overline{\mathcal{H}})
\end{array}
$$

where $\overline{\mathcal{H}} \equiv C_0^\infty((f^{-1})^* H)$ and $Q(f)$ is the natural transformation $\mathcal{L}(\mathcal{H}) \to \mathcal{L}(\overline{\mathcal{H}})$ induced for functoriality. $\qquad\square$

Corollary 3.7. (Quantum covariance). *Under the hypotheses of the above theorem we have that for any function* $h : \Omega(\overline{E}_k) \to \mathbf{K}$ *one has:*

$$
< \overline{E}(h)(\bar{\alpha}), \bar{y} > = \int_{\Omega(\overline{E}_k)} h \, d\overline{E}_{\bar{x}, \bar{\alpha}}
$$

$$
= \int_{\Omega(E_k)} h \circ (f^{-1})_* dE_{f_\bullet(\bar{x}), f'_\bullet(\bar{\alpha})}, \ \forall \bar{x} \in \overline{\mathcal{H}}, \forall \bar{\alpha} \in \overline{\mathcal{H}}',
$$

and with $f_\bullet : \overline{\mathcal{H}} \to \mathcal{H}$, $f'_\bullet : \mathcal{H}' \to \overline{\mathcal{H}}'$ *the canonical homomorphisms induced for functoriality by* f.

3.4 - QUANTIZATIONS OF PDE's

In this section we shall justify on the ground of the mathematical logic what we mean by the proposition: *"quantization of PDE "*. Let us start with some definitions and fundamental results of mathematical logic. (See also e.g., ref.[3].)

Definition 3.94. An *n-ary operation* on the set A is a function $t : A^n \to A$, $n \in \mathbf{N}$; the number n is called *arity* of t. A *0-ary* operation on a set A is a function $t : A^0 \equiv \emptyset \to A$. Hence, it can be regarded as a distinguished element of A. (For example, if A is a group, the 0-ary operation e gives the identity element of the group.) A *type* $\mathcal{T}$ is a couple $\mathcal{T} \equiv (T, ar)$, where T is a set and ar is a map $ar : T \to \mathbf{N}$. We denote $T_n \equiv \{t \in T \mid ar(t) = n\}$. An *algebra A of type $\mathcal{T}$*, or *$\mathcal{T}$-algebra*, is a triplet $(A, \mathcal{T}, \sigma)$, where A is a set, $\mathcal{T}$ is a type and σ is a section of the bundle $\bigcup_{t \in T} Map(A^{ar(t)}; A) \to T$, i.e., $\sigma(t) \equiv t_A : A^{ar(t)} \to A$. The elements $t \in T$ are called *n-ary $\mathcal{T}$-algebra operations*. $\mathcal{T}$-algebras A and B are *equal* iff $A = B$ and $t_A = t_B, \forall t \in T$. A *$\mathcal{T}$-subalgebra* of A is a subset $B \subset A$ such that it is a $\mathcal{T}$-algebra with respect to the operation restrictions to B of those in A.

Remark 3.24. 1) Any intersection of subalgebras is a subalgebra.

2) $\emptyset$ is a subalgebra iff $T_0 = \emptyset$.

3) Given any subset $X \subset A$, there is a unique smallest subalgebra containing X-namely, the subalgebra $\cap\{U \mid U = \text{subalgebra of } A, U \supseteq X\}$. We call this the *subalgebra generated by* X and denote it by $< X >_T$ or simply $< X >$.

4) *For all T, every T-algebra has a unique smallest subalgebra.*   ∎

Example 3.47. (*Rings*). A ring may be considered as an algebra of type $T \equiv (\{0, -, +, \cdot\}, ar)$ with $ar(0) = 0$, $ar(-) = 1$, $ar(+) = 2$, $ar(\cdot) = 2$.   ∎

Example 3.48. (*R-Modules*). An R-module ($R \equiv$ ring) may be regarded as algebra of type $T \equiv (\{0, -, +, \cdot\} \cup R, ar)$ with $ar(0) = 0$, $ar(-) = 1$, $ar(+) = 2$, $ar(\lambda) = 1$, $\forall \lambda \in R$.   ∎

Example 3.49. (*Groups*). Any group G can be considered as an algebra of type $T \equiv (\{*\}, ar)$, with $ar(*) = 2$, or an algebra of type $T' \equiv (\{e, i, *\}, ar)$ with $ar(e) = 0$, $ar(i) = 1$, $ar(*) = 2$. Then any T'-subalgebra of a finite group is a group, but this is not true for any non-empty T-subalgebra, if G is not a finite group.   ∎

Definition 3.95. A *homomorphism* between T-algebras A and B is a function $\phi : A \to B$ such that the following diagram is commutative:

$$
\begin{array}{ccc}
A^n & \xrightarrow{\ t_A\ } & A \\
\phi^n \downarrow & & \downarrow \phi, \\
B^n & \xrightarrow[\ t_B\ ]{} & B
\end{array}
$$

$$\forall t \in T, \text{ i.e., } \phi(t_A(a_1, \cdots, a_n)) = t_B(\phi(a_1), \cdots, \phi(a_n)).$$

An *isomorphism* of T-algebras is an invertible homomorphism.

Remark 3.25. The set of T-algebras and homomorphisms of T-algebras is a category.

Definition 3.96. A *free T-algebra* is a couple (F, σ), where F is a T-algebra, and σ is a mapping $\sigma : X \to F$, such that for every T-algebra A and function $\tau : X \to A$, there exists a unique homomorphism $\phi : F \to A$ such that $\tau = \phi \circ \sigma$, i.e., the following diagram is commutative:

$$
\begin{array}{ccc}
X & \xrightarrow{\ \sigma\ } & F \\
\tau \downarrow & \swarrow \phi & \\
A & &
\end{array}
$$

Then, we say that X is a set of *free generators* of the T-algebra F (σ is necessarily injective).

Theorem 3.78. 1) *For any set X and any type $\mathcal{T}$, there exists a free $\mathcal{T}$-algebra on X. This free $\mathcal{T}$-algebra on X is unique up to isomorphisms.*

2) *Let F be the free $\mathcal{T}$-algebra on the set $X_n \equiv \{x_1, \cdots, x_n\}$. Let A be any $\mathcal{T}$-algebra. Then, we have a canonical mapping (associated to $w \in F$) $w_A : A^n \to A$, $w_A(a_1, \cdots, a_n) = w_A(a_1, \cdots, a_n) = \phi(w)$, where ϕ is the unique homomorphism $\phi : F \to A$, $\phi(x_i) = a_i$, $i = 1, \cdots, n$. In particular, if $A = F$, $a_i = x_i$, $i = 1, \cdots, n$, then ϕ is the identity and $w_A(x_1, \cdots, x_n) = w$.*

Definition 3.97. A $\mathcal{T}$-*algebra variable* is an element of the free generating set of a free $\mathcal{T}$-algebra. A $\mathcal{T}$-*word* in the variables $x_1, \cdots, x_n$, is an element of the free $\mathcal{T}$-algebra on the set $X_n \equiv \{x_1, \cdots, x_n\}$ of free generators. A *word* in the elements $a_1, \cdots, a_n$ of a $\mathcal{T}$-algebra A, is an element $w_A(a_1, \cdots, a_n) \in A$, where w is a $\mathcal{T}$-word in the variables $x_1, \cdots, x_n$. Among the words in the variables $x_1, \cdots, x_n$ there are the words $x_i, (i = 1, \cdots, n)$, having the property that $x_i(a_1, \cdots, a_n) = a_i$. These variables may also be regarded as coordinate functions.

Definition 3.98. 1) An *identical relation* on $\mathcal{T}$-algebra A is a pair $(u, v) \in F \times F$. There is an n for which u, v are in the free algebra on $X_n = \{x_1, \cdots, x_n\}$, and we say that (u, v) is an n-*variable identical relation* (u, v) for any such n.

2) We say that $(u, v) \in F \times F$ is a *law* of the $\mathcal{T}$-algebra A, (or that A *satisfies the n-variable identical relation* (u, v), if $\phi(u) = \phi(v)$ for every homomorphism $\phi : F \to A$, (or equivalently $u(a_1, \cdots, a_n) = v(a_1, \cdots, a_n)$ for all $a_1, \cdots, a_n$).

3) Let L be a set of identical relations on $\mathcal{T}$-algebra. We call *variety of $\mathcal{T}$-algebra defined by L* the class V of all $\mathcal{T}$-algebras which satisfies all the identical relations satisfied by every algebra in V. (This set includes L, but may be larger.)

Example 3.50. (*Semigroups*). Let us consider the type $\mathcal{T} \equiv (\{*\}, ar)$ with $ar(*) = 2$. Let L be the set of identical relations on $\mathcal{T}-$algebra given by $L \equiv \{x_1 * (x_2 * x_2), (x_1 * x_2) * x_3\}$. If A satisfies this identical relation, then $a * (b * c) = (a * b) * c$, for all $a, b, c \in A$. Thus the operation on A is associative and A is a semigroup. Hence, the variety defined by L in this case is the class of all semigroups. ∎

Example 3.51. (*Groups*). Let us consider the type $\mathcal{T} \equiv (\{e, i, *\}, ar)$ with $ar(e) = 0$, $ar(i) = 1$, $ar(*) = 2$. Let L be the set of identical relations on $\mathcal{T}-$algebra given by $L \equiv \{x_1 * (x_2 * x_2), (x_1 * x_2) * x_3, (e * x_1, x_1), (i(x_1) *$

$x_1, e)\}$. Thus, the first law ensures that $*$ is an associative operation in every algebra of the variety defined by L. The second law shows that the distinguished element e is always a left identity. The third law guarantees that $i(a)$ is a left inverse of the element e. Hence, the algebras of the variety are groups. $\blacksquare$

Definition 3.99. Let V be the variety of $\mathcal{T}$-algebras defined by the set L of laws. A $\mathcal{T}$-algebra R in V is *(relatively) free algebra of V* on the set X of *(relatively) free generators*, (where a function $\sigma : X \to R$ is given, usually an inclusion), if, for any algebra A in V and every function $\tau : X \to A$, there exists a unique homomorphism $\phi : R \to A$ such that $\phi \circ \tau$, i.e., one has the following commutative diagram:

$$
\begin{array}{ccc}
X & \xrightarrow{\ \sigma\ } & R \\
& \phi \searrow \quad & \downarrow \phi \\
& & A
\end{array}
$$

An algebra is *relatively free* if it is a free algebra of some variety.

Theorem 3.79. *For any type $\mathcal{T}$, and any set L of laws, let V be the variety of $\mathcal{T}$-algebras defined by L. For any set X, there exists a free $\mathcal{T}$-algebra of V on X.*

Definition 3.100. Let V be a variety of $\mathcal{T}$-algebras. Then, we call V-*variables* an element of the free generating set of a free algebra of V, and V-*word* in the V-variables $x_1, \cdots, x_n$, an element of the free algebra of V on the free generators $\{x_1, \cdots, x_n\}$.

Example 3.52. (*Variety of abelian groups*). Let us consider the type $\mathcal{T} \equiv (\{e, i, *\}, ar)$ with $ar(e) = 0$, $ar(i) = 1$, $ar(*) = 2$. Let L be the set of identical relations on $\mathcal{T}-$algebra given by $L \equiv \{x_1 * (x_2 * x_2), (x_1 * x_2) * x_3, (e*x_1, x_1), (i(x_1)*x_1, e)\}$. Then, the veriety V defined by L is the variety of abelian groups. In this case, the relatively free algebra on $\{x_1, \cdots, x_n\}$ is the set of all $x_1^{r_1} x_2^{r_2} \cdots, x_n^{r_n}$, with $r_i \in \mathbf{Z}$. $\blacksquare$

Example 3.53. (*Variety of algebras*). Let κ be a field. Then, the vector spaces over κ form a variety V of algebras, and every vector space over κ is a free algebra of L. $\blacksquare$

Example 3.54. (*Variety of commutative rings*). Let R be a commutative ring with unity 1, and let V be the free variety of commutative rings S which contain R as a subring and in which 1_R is a multiplicative identity in S. Then, the free algebra of V on the set X of variables is the polynomial ring over R in the elements of X. $\blacksquare$

Definition 3.101. Let $\mathcal{T} \equiv (\{F, \Rightarrow\}, ar)$ be a type with $ar(F) = 0$, $ar(\Rightarrow) = 0$. Then, any $\mathcal{T}$-algebra is called a *proposition algebra*. The *proposition algebra* $\mathcal{P}(X)$ *of the proposition calculus on the set* X (*set of proposition variables*) is the free $\mathcal{T}$-algebra on X. In any proposition algebra, we introduce the following further operations:

TAB.3.9 - Operations defined in proposition algebra.

$$
\begin{array}{ll}
\sim: & \sim p = p \Rightarrow F \; (ar(\sim)=1) \\[2mm]
\vee: & p \vee q = (\sim) \Rightarrow q \quad (ar(\vee)=2) \\[2mm]
\wedge: & p \wedge q = \sim(\sim p \vee \sim q) \quad (ar(\wedge)=2) \\[2mm]
\Leftrightarrow: & p \Leftrightarrow q = (p=q) \wedge (q \Rightarrow p) \quad (ar(\Leftrightarrow)=2)
\end{array}
$$

(We shall often omit brackets in writing $\sim p \vee \sim q$ for $(\sim p) \vee (\sim q)$.) A *valuation of* $\mathcal{P}(X)$ *is a proposition algebra homomorphism* $v : \mathcal{P}(X) \to \mathbf{Z}_2$. We say that $p \in \mathcal{P}(X)$ *is true with respect to* v *if* $v(p) = 1$, *and that* p *is false with respect to* v *if* $v(p) = 0$. Since X is a set of free generators of $\mathcal{P}(X)$ the values $v(x)$ for $x \in X$ may be assigned arbitrary. These values, once assigned, determine the homomorphism v uniquely and so determine $v(p)$ for all $p \in \mathcal{P}(X)$. Let $A \subseteq \mathcal{P}(X)$ and $q \in \mathcal{P}(X)$. *We say that* q *is a consequence of the set* A *of assumptions, or that* A *semantically implies* q, *if* $v(q) = 1$ *for every valuation* v *such that* $v(p) = 1$ *for all* $p \in A$. *We write:* $A \models q$. Set: $Con(A) \equiv \{p \in \mathcal{P}(X) \,|\, A \models p\} \equiv$ *set of all consequences of* A. Let $p \in \mathcal{P}(X)$. *We say that* p *is valid, or is a tautology, if* $v(p) = 1$, *for every valuation* v *of* $\mathcal{P}(X)$.

Proposition 3.46. 1) *The proposition* p *is a tautology if* $\varnothing \models p$. (*We shall write this simply as* $\models p$.)

2) *Con is a closure operation on* $\mathcal{P}(X)$, *i.e., has the following properties:* (i) $A \subseteq Con(A)$; (ii) *If* $A_1 \subseteq A_2$, *then* $Con(A_1) \subseteq Con(A_2)$; (iii) $Con(Con(A)) = Con(A)$.

Definition 3.102. (Proof in propositional calculus). For the propositional calculus on the set X, we take as axioms all elements of the subset: $\mathcal{A} \equiv \mathcal{A}_1 \cup \mathcal{A}_2 \cup \mathcal{A}_3 \subset \mathcal{P}(X)$, where: $\mathcal{A}_1 \equiv \{p \Rightarrow (q \Rightarrow p) \,|\, p, q \in \mathcal{P}(X)\}$, $\mathcal{A}_2 \equiv \{(p \Rightarrow (q \Rightarrow r)) \Rightarrow ((p \Rightarrow q) \Rightarrow (p \Rightarrow r)) \,|\, p, q, r \in \mathcal{P}(X)\}$, $\mathcal{A}_3 \equiv \{\sim\sim p \Rightarrow \,|\, p \in \mathcal{P}(X)\}$.

1) Let $q \in \mathcal{P}(X)$ and let $A \subseteq \mathcal{P}(X)$. In the propositional calculus on the set X, a *proof of* q *from the assumption* A is a finite sequence $p_1, p_2, \cdots, p_n$

of elements $p_i \in \mathcal{P}(X)$ such that $p_n = q$ and for each i, either $p_i \in \mathcal{A} \cup A$ or for some $j, k < i$, we have: $p_k = (p_j \Rightarrow p_i)$.

2) Let $q \in \mathcal{P}(X)$ and let $A \subseteq \mathcal{P}(X)$. We say that q is a *deduction* from A, or q is *prevable* from A, or that A *syntactically implies* q, if there exists a proof of q from A. We shall write this by: $A \vdash q$. Set: $Ded(A) \equiv$ set of all deductions from A.

3) Let $p \in \mathcal{P}(X)$. We say that p is a *theorem* of the propositional calculus on X if there exists a proof of p from $\emptyset$ and we write simply: p is a theorem if $\vdash p$.

Theorem 3.80. 1) *If* $q \in Ded(A)$, *then* $q \in Ded(A')$ *for some finite subset* A' *of* A.

2) *Ded is a closure operation on* $\mathcal{P}(X)$.

3) *(The substitution theorem). Let* X *and* Y *be any two sets, and let* $\phi : \mathcal{P}(X) \to \mathcal{P}(Y)$ *be a homomorphism of the (free) proposition algebra on* X *into the (free) proposition algebra on* Y. *Let* $w = w(x_1, \cdots, x_n)$ *be any element of* $\mathcal{P}(X)$ *and let* A *be any subset of* $\mathcal{P}(X)$. *Put* $a_i = \phi(x_i)$. *Then, we have the following: (a)* $A \vdash w$, *then* $\phi(A) \vdash w(a_1, \cdots, a_n)$, *(b)* $A \models w$, *then* $\phi(A) \models w(a_1, \cdots, a_n)$.

Example 3.55. The algebra $\mathbf{Z}_2$ can be made into a proposition algebra by defining $F_{\mathbf{Z}_2} = 0$ and $w \Rightarrow u = 1 + m(1+n)$. Furthermore, one can easily see the expressions $\sim, \vee, wedge$ in $\mathbf{Z}_2$ in terms of multiplications and addition. ∎

Example 3.56. $Ded(A)$ is the smallest subset D of $\mathcal{P}(X)$ such that $D \supseteq \mathcal{A}(X) \cup A$ and such that if $p, p \Rightarrow \in D$, then $q \in D$ also. ∎

Definition 3.103. A *logic* $\mathcal{L}$ is a triplet $\mathcal{L} \equiv (P, \mathcal{V}, \mathcal{P})$, where: (i) P is a set of elements (called *propositions*); (ii) $\mathcal{V}$ is a set of functions (called *valuations*) $P \to W$ ($\equiv$ *value set*); (iii) $\mathcal{P} \equiv (\{p_1, \cdots, p_{nA}\})$. For each subset $A \subset P$, $\{p_1, \cdots, p_{nA}\}$ is a set of finite sequences of elements of P (called *proofs from the assumptions*).

Theorem 3.81. *The propositional calculus on the set* X *is a logic:* $\mathcal{L} = Prop(X)$ *such that: (i)* $P = \mathcal{P}(X)$; *(ii)* $\mathcal{V} \equiv$ *set of all homomorphisms* $\mathcal{P}(X) \to \mathbf{Z}_2$; *(iii) For each subset* $A \subseteq \mathcal{P}(X)$, *the set of proofs are defined above.*

Now, we are ready to give the following important definitions that clarify the meaning of quantization of a physical theory.

Definition 3.104. 1) The *logic of a PDE* $E_k \subset J\mathcal{D}^k(W)$, $\mathcal{L}(E_k)$, is the

Boolean algebra of subsets of $\Omega(E_k)_c \equiv$ set of solutions of E_k: $\mathcal{L}(E_k) \equiv \mathcal{P}(\Omega(E_k)_c)$.

2) A *quantum logic*, $\mathcal{L}_q$, is an algebra, A, of (self-adjoint) operators on a locally convex, (or Hilbert), space $\mathcal{H}$: $\mathcal{L}_q \equiv A \subset \mathcal{L}(\mathcal{H})$.

3) *Quantize a PDE* $E_k \subset JD^k(W)$, means to define a map $\mathcal{L}(E_k) \to \mathcal{L}_q$, that identifies a spectral measure on $\Omega(E_k)_c$: $\Omega(E_k)_c \circ\!\!\to L(\mathcal{H})$.

Theorem 3.82. *A quantization $\mathcal{L}(E_k) \to \mathcal{L}_q$ of a PDE $E_k \subset JD^k(W)$, identified by means of a spectral measure $\Omega(E_k)_c \circ\!\!\to \mathcal{L}(\mathcal{H})$, identifies an homomorphism of Boolean algebras $q : \mathcal{P}(\Omega(E_k)_c) \to \mathcal{P}_r(\mathcal{H})$, where $\mathcal{P}_r(\mathcal{H})$ is a Boolean algebra of projections on $\mathcal{H}$.*

Proof. The proof is a direct consequence of Definition 3.45, Definition 3.27 and Remark 3.7.   □

Theorem 3.83. *A quantization of $\mathcal{L}(E_k) \to \mathcal{L}_q$, identified by means of a spectral measure $E : \Omega(E_k)_c \circ\!\!\to \mathcal{L}(\mathcal{H})$, identifies an homomorphism of algebras $E : M(\Omega(E_k)_c, \Sigma) \to \mathcal{L}(\mathcal{H})$. We say that $\hat{f} \equiv E(f)$ is the quantized function corresponding to $f \in M(\Omega(E_k)_c, \Sigma)$.*

Proof. It is a direct consequence of Theorem 3.15.   □

Theorem 3.84. (Heisenberg uncertainty relation). *Let $f, f' : \Omega(E_k) \to \mathbf{R}$ be two random variables associated to a PDE $E_k \subset JD^k(W)$, quantized by means of two spectral measures $E, F : \Omega(E_k) \circ\!\!\to \mathcal{L}(\mathcal{H})$. Then one has the following relation between the corresponding variances:*

(3.18)(*Heisenberg uncertainty relation*)

$$\sigma^2(f)_\psi . \sigma^2(f')_\psi \geq \frac{1}{4} | < [\hat{f}, \hat{f}'](\psi)|\psi > |^2.$$

Proof. Let us consider the following definitions and results. A *quantum state* of a PDE $E_k \subset JD^k(W)$, quantized by means of a quantization $\mathcal{L}(E_k) \to \mathcal{L}_q \equiv A \subset \mathcal{L}(\mathcal{H})$ is an element $T \in \mathcal{L}(\mathcal{H})$ such that $\mathrm{tr}(T) = 1$. If $T = \psi' \otimes \psi$, with $\psi' \in \mathcal{H}'$, $\psi \in \mathcal{H}$, T is called a *pure quantum state*. If the Dirac quantization is proper, we call *proper quantum state* a positive Hermitian nuclear operator $T \in \mathcal{L}_1(\mathcal{H})$ such that $\mathrm{tr}(T) = 1$, where $\mathcal{L}_1(\mathcal{H})$ is the space of nuclear operators on $\mathcal{H}$. If $T = \psi \otimes \psi, \psi \in \mathcal{H}$, and $\|\psi\| = 1$, T is called a *proper pure quantum state*. The space E of proper quantum states can be identified with a convex part of the unit bubble in $K(\mathcal{H})'$ where $K(\mathcal{H})$ is the space of compact operators on $\mathcal{H}$. In fact, from the canonical isomorphism $\mathcal{L}_1(\mathcal{H}) \cong K(\mathcal{H})'$ we can define a proper quantum state by using the following. A *state* of a C^* algebra $\mathcal{L}(\mathcal{H})$ is a self-adjoint

complex-valued linear functional E on $\mathcal{L}(\mathcal{H})$ which is non-negative or non-negative self-adjoint, and normalized by the constraint: $\sup_{\|A\|\leq 1} E(A^*A) = 1$.
A *pure state* is one which is not a linear combination with positive coefficients of two other states, otherwise it is called *mixed state*. A set of states is called *complete state* if the only element of $\mathcal{L}(\mathcal{H})$ which vanishes in every state of the set is zero. Any vector $\psi \in \mathcal{H}$ identifies a pure state: $E(A) = (A(\psi), \psi)$ (called the *vertex state*). When $\mathcal{H}$ is finite-dimensional, all pure states have this form; furthermore in this case the general state has the form $E(A) = \text{tr}(AD)$, where D is a positive semidefinite operator of unit trace (known as the "density operator" in statistical mechanics). Note that the normalized condition is equivalent to the condition $E(e) = 1$. The set of all pure states is always a Borel subset of state space of $\mathcal{L}(\mathcal{H})$. The *variance* of a self-adjoint element A of $\mathcal{L}(\mathcal{H})$ in a state E is defined as $E(A^2) - E(A)^2$. A is said to *have the exact value operator $E(A)$* in the state E in case its variance vanishes in this state. A state of $\mathcal{L}(\mathcal{H})$ in which every self-adjoint element has an exact value is called an *observation*.

Lemma 3.13. *If F is a pure state of a C^*-subalgebra B, containing the identity of $\mathcal{L}(\mathcal{H})$, then there exists a pure state E of $\mathcal{L}(\mathcal{H})$, such that $E|_B = F$. Any observation on $\mathcal{L}(\mathcal{H})$ is pure. The variance of a self-adjoint operator of $A \in \mathcal{L}(\mathcal{H})$ in a state E of $\mathcal{L}(\mathcal{H})$ is identical to the variance in the usual probabilistic sense of the random variable which represents A on the regular compact measure space obtained by imposing on the spectrum of C^*-algebra $B(A) \subset \mathcal{L}(\mathcal{H})$ generated by A the measure representing $E|_{B(A)}$.*

Proof of Lemma 3.13. See e.g., ref.[94(a)]. $\qquad\qquad\square$

If G is a group of automorphisms of the C^*-algebra $\mathcal{L}(\mathcal{H})$, a *stationary state* (with respect to G) is defined as one with the property that: $E(A) = E(g(A))$, $\forall A \in \mathcal{L}(\mathcal{H})$, $g \in G$. The state E is called *ergodic state* in case it is an extreme point of the set of all stationary states.

Lemma 3.14. *Every stationary state is w^*-limit of convex linear combinations of ergodic states.*

Proof of Lemma 3.14. See e.g., ref.[94(a)]. $\qquad\qquad\square$

A *quantum value random variable* of a random variable $f : \Omega(E_k) \to \mathbf{R}$ is a point λ of the spectrum of the quantized function $\hat{f}$, with respect to a spectral measure $E : \Omega(E_k) \multimap \mathcal{L}(\mathcal{H})$: $\lambda \in Sp(\hat{f})$. The probability to find a quantum value of f in the Borel set $\omega \subset \mathbf{R}$, if the system is in the quantum state T, is given by the following formula $p(f; T; \omega)\text{tr}(E(\omega)T) =$

$\mathrm{tr}\,(TE(\omega))$, where we have written $E \equiv E \circ f^{-1}$. If $T \equiv \psi \otimes \psi$ we have $p(f;T;\omega) = \int_\omega dE_\psi(\lambda)$. The corresponding *mean value* is:

$$<< f >>_T \equiv \int_{S_p} (\hat{f})\lambda dp(f;T;\lambda) = \mathrm{tr}\,\int \lambda dE(\lambda)T$$

$$= \mathrm{tr}\,(\hat{f}T) = \mathrm{tr}\,(T\hat{f}) = \sum \alpha_n(\hat{f}(e_\alpha), e_\alpha).$$

If $T \equiv \psi \otimes \psi$ we have: $<< f >>_T = \int \lambda dE_\psi(\lambda)$. The *variance* is:

$$\sigma^2(f)_T = \int_{S_p} (\hat{f})(\lambda - << f >>_T)^2 dp(f;T;\lambda)$$

$$= \mathrm{tr}\,(\int \lambda^2 dE(\lambda)T - << f >>_T^2$$

$$= \mathrm{tr}\,(\hat{f}^2 T)[\mathrm{tr}\,(\hat{f})].$$

If $T \equiv \psi \otimes \psi$ one has:

$$\sigma^2(f)_T = \int (\lambda - << f >>_T)^2 dE_\psi(\lambda) = \|\hat{f}(\psi)\|^2 - (\hat{f}(\psi)|\psi)^2.$$

Now, let $\hat{f}$ and $\hat{f}'$ be two quantized random variables of $E_k \subset J\mathcal{D}^k(W)$. If $[\hat{f}, \hat{f}'] = 0$ we say that f and f' are *simultaneously observable*. Then the probability to find the quantum value λ (resp. λ') of f (resp. f') in the Borel set $\omega \subset \mathbf{R}$ (resp. $\omega' \subset \mathbf{R}$) if the system is in the pure quantum state is given by $p(f;f';\psi;\omega,\omega') = \int_\omega \int_{\omega'} dG_\psi(\lambda,\lambda')$, where $G_\psi(\bullet,\bullet) \equiv E_\psi(\bullet) \otimes F_\psi(\bullet)$ is the spectral measure on $\mathbf{R}^2$ obtained by product of the spectral measures E and F respectively. From above considerations it follows directly the Heisenberg uncertainty relation. $\qquad\square$

Theorem 3.85. (Bohr Sommerfield quantization conditions). *Let the sub-manifold $L \subset J\mathcal{D}^k(W)$ be an n-dimensional integral manifold solution of a PDE $E_k \subset J\mathcal{D}^k(W)$. Let L lead the charge $Q = \int_a i_L{}^*\beta$, where β is an $(n-1)$-differential form on $J\mathcal{D}^k(W)$; $i_L : L \hookrightarrow J\mathcal{D}^k(W)$ is the canonical inclusion and a is an $(n-1)$-chain on L. If $i_L{}^*\beta = k\theta$, where $k \in \mathbf{R}$ and θ belongs to the image of the canonical map*

$$\left\{ \begin{array}{ccc} \pi^{n-1}(L) & \xrightarrow{\;j_{n-1}\;} & H^{n-1}(L) \\[4pt] \| & & \|\wr \\[4pt] [L; S^{n-1}] & \to Hom(H^{n-1}(S^{n-1}), H^{n-1}(L)) \cong & Hom(\mathbf{R}, H^{n-1}(L)) \\[4pt] [f] & \mapsto [f^*\mu] & \end{array} \right\} ,$$

where μ is the canonical volume form on S^{n-1}, then Q is quantized, that is $Q = k.n$, $k \in \mathbf{R}$, $n \in \mathbf{Z}$.

Proof. Recall that any $(n-1)$-differential form α on S^{n-1} can be written as $\alpha = b\mu + d\gamma$, where $b \in \mathbf{R}$ and γ is an $(n-2)$-differential form on S^{n-1}. In fact, $H^{n-1}(S^{n-1}) \cong \mathbf{R}$. Then, we get $\int_a i_L{}^*\beta = \int_a \theta = k \int_a f^*\mu = k\deg(\bar{f}) \int_{S^{n-1}} \mu$, where $f : L \to S^{n-1}$ is the representative of the cohomology class $[f] \in \pi^{n-1}(L)$, $\bar{f} \equiv f|_\triangle$, where $\triangle$ is the support of a. So we get, $\int_a i_L{}^*\beta = \bar{k}.n$, where $\bar{k} \equiv k.vol(S^{n-1})$ and $n \equiv \deg(\bar{f}) \in \mathbf{Z}$ is the Brouwer degree of $\bar{f}$. $\qquad\qquad\square$

3.5 - COVARIANT AND CANONICAL QUANTIZATIONS OF PDE's

The "canonical quantization" of a classical mechanical system (system of particles or field theory), consists into reformulate the classical dynamical PDE (or ODE) as a differential equation of operators on a suitable vector space, with an additional constraint on the initial data, given by "canonical" commutation relations. Even if in the linear cases this approach is satisfactory enough, for non-linear equations a direct interpretation of the classical equations as operatorial ones is, in general, very problematic. In the context of the quantization of PDEs, in the sense of Definition 3.184(3), we have, instead, a natural way to consider classical entities as operatorial ones, whether in linear or non linear cases. Scope of this section is, just, to show that our quantization of PDEs is able to interpret the concept of canonical quantization also for nonlinear PDE's. Furthermore, we characterize integral bordisms of (nonlinear) PDEs by means of geometric Green kernels and prove that these are invariant for the classic limit of statistical sets of formally integrable PDEs.[56] Such geometric characterization of Green kernels is related to the geometric approach of covariant and canonical quantizations of (nonlinear) PDEs, previously introduced by us [59,61,63,70]. In particular we will prove that to any classic-limit-statistical set of PDEs we can associate a geometric Green kernel. By means of such objects we get the covariant and canonical quantizations of PDE's, i.e., representations of their observable by means of operator algebras. Further-

---

[56] Recall that for any two admissible $(n-1)$-dimensional closed compact integral manifolds N_1, N_2 contained into a k-order PDE $E_k \subset J_n^k(W)$ of n-dimensional submanifolds of the $(n+m)$-dimensional manifold W, the classic-limit statistical set $\Omega_c(N_1, N_2)$ is the set of all solutions V of E_k such that $\partial V = N_1 \bigcup N_2$.

more, we prove that covariant quantizations of PDE's identify quantum PDE's in algebraic sense. Some applications are given, also, where particle fields on curved space-times having physical or unphysical masses, (i.e., bradions, luxons and massive neutrinos) are covariantly quantized respecting microscopic causality.

Definition 3.105. Set, for any vector fiber bundle $\pi : F \to M$, over a n-dimensional manifold M, $F' \equiv F^* \otimes \Lambda_n^0 M$, where F^* is the dual of the fiber bundle F and $\Lambda_n^0 M$ is the fiber bundle of n-forms on M. We call F' the *formal adjoint* of F. If $\bar{\pi} : E \to N$ is another vector fiber bundle over N we set $E \boxed{\times} F \equiv \bigcup_{(p,q)\in N\times M} E_p \bigotimes F_q$. This is a vector fiber bundle over $N \times M$ of dimension $n + m + r.s$, if $\dim E_p = r$, $\forall p \in N$, and $\dim F_q = s$, $\forall q \in M$. For any vector fiber bundle $E \to N$ we denote by $C_0^\infty(E)$ the space of smooth sections with compact support over N.

Theorem 3.86. (Green kernels and sistribution solutions of affine PDE's). *Let $\kappa : J\mathcal{D}^r(E) \to F$ be a morphism of vector fiber bundles on a manifold M of dimension n such that it defines a linear differential operator of order r, $\kappa. : C^\infty(E) \to C^\infty(F)$. Let $E_r \equiv \ker_f \kappa \subset J\mathcal{D}^r(E)$ be an affine equation, where $f \in C^\infty(F)$. Assume that $R \subset J\mathcal{D}^r(E) \subset J_n^r(E)$ is an integral manifold of dimension n, such that in some neighborhood of all its points $q \in R$, (except for a nowhere dense subset $\Sigma(R) \subset R$), it can be represented as the r-holonomic prolongation $N^{(r)}$ of some n-dimensional submanifold $N \subset E$. Furthermore, we assume that $\pi_{r,0}|_R : R \to E$ is a proper application, and $\omega_1{}^{(r)}(R) = 0$, where $\omega_1{}^{(r)}(R) = i_R^* \omega_1^{(r)}$, with $i_R : R \to J_n^r(W)$ the natural inclusion mapping and $\omega_1^{(r)}$ is the first generator of the cohomology algebra $H^\bullet(I_r(W); \mathbf{Z}_2) \cong \mathbf{Z}_2[\omega_1^{(r)}, \ldots, \omega_m^{(r)}]$, where $I_r(W)$ is the Grassmannian bundle over $J_n^r(W)$ of oriented integral planes of $J_n^r(W)$. If r is orientable then $\omega_1^{(r)}(R) = 0$, as it coincides with the first Stiefel-Whitney characteristic class of R. This is a direct consequence of the following short exact sequence:*

$$0 \to\; <w_1> \;\to H^\bullet(BO(n); \mathbf{Z}_2) \overset{j^\bullet}{\to} H^\bullet(BSO(n); \mathbf{Z}_2) \to 0,$$

where $w_1 \in H^1(BSO(n); \mathbf{Z}_2)$ is the first Stiefel-Whitney class, $<w_1>$ the corresponding ideal and $j^\bullet$ is induced by the natural mapping $j : BSO(n) \to BO(n)$ which forgets the orientation on the oriented planes of dimension n representing the points $\hat{G}_{\infty,n} = BSO(n)$. Then, we can associate to R a distribution $F[R] \in (C_0^\infty(E'))'$. Therefore, $F[R]$ is a distributive solution of E_r, i.e., a solution of the distributional extension $\widetilde{E}_r$ of E_r, iff R is a multivalued solution of $E_r \subset J\mathcal{D}^r(E)$. Furthermore, if R is such that $\partial R = R_1 \bigcup R_2$ with $R_1 \in [R_2]_{E_r} \in \Omega_{n-1}^{E_r}$, [57] then we say that $F[R]$ satisfies the boundary condition $\partial R = R_1 \bigcup R_2$ and it can be represented in the following form: $F[R] = {}_f\mathbb{G} + \omega_0$, with $\omega_0 \in Sol(\widetilde{\overline{E}}_r)$, where $Sol(\widetilde{\overline{E}}_r)$ is the space of solutions of the linear equation $\widetilde{\overline{E}}_r$ associated to $\widetilde{E}_r$, and $\mathbb{G}$ is the Green kernel of κ. Then, ${}_f\mathbb{G}$ can be considered an invariant of the classic-limit-statistical set $\Omega_c(R_1, R_2)$.

Proof. We shall introduce some definitions and lemmas. Let $E \xrightarrow{\pi} M \xleftarrow{\bar{\pi}} F$ be two vector fiber bundles over a n-dimensional manifold M. We call *formal adjoint* of a r-order linear differential operator $\kappa : C^\infty(E) \to C^\infty(F)$ the linear operator $\kappa^* : C^\infty(F') \to C^\infty(E')$ defined by $< \kappa^*(\alpha), e >=< \alpha, \kappa(e) >$, $\forall e \in C^\infty(E)$, $\alpha \in C^\infty(F')$. The restriction $\kappa^* : C_0^\infty(F') \to C_0^\infty(E')$ satisfies the following formula: $\int_M < e, \kappa^*(\alpha) >= \int_M < \kappa(e), \alpha >$, $\forall e \in C^\infty(E)$, $\alpha \in C_0^\infty(F')$. The *Dirac kernel* $\mathbb{D}$ of the vector fiber bundle F on a n-dimensional manifold M, $\dim F = n + m$, is the distributive kernel of local character $\mathbb{D} \in (C_0^\infty(F \boxtimes F'))' \cong (C_0^\infty(F' \boxtimes F))'$ defined by $\mathbb{D}(f \otimes \alpha) = \int_M < f, \alpha >$, $\forall f \in C_0^\infty(F)$, $\alpha \in C_0^\infty(F')$. The Dirac kernel admits the following local representation: $\mathbb{D}_{U,i}^j(x, x') = \delta_i^j \delta(x, x')$, where $\delta(x, x')$ is the Dirac function. In fact we can write:

$$\mathbb{D}(f \otimes \alpha) = \sum_U \int_U g_U < f_U, \alpha_U >$$

$$= \sum_U \int_{U \times U} g_{U \times U} f_U^i(x) \alpha_{U,j}(x') \delta_i^j \delta(x, x') d\mu(x'),$$

where g_U is a partition of unity on M, subordined to the open covering $\{U\}$ of M. Furthermore, the *Green kernel* of κ is a kernel $\mathbb{G} \in (C_0^\infty(E' \boxtimes F))' \cong (C_0^\infty(F \boxtimes E'))'$, such that it satisfies the following equation: $(\widetilde{\kappa} \otimes \mathbf{1})(\mathbb{G}) = (\mathbf{1} \otimes \widetilde{\kappa})(\mathbb{G}) = \mathbb{D}$, where $\widetilde{\kappa} : (C_0^\infty(E'))' \to (C_0^\infty(F'))'$ is the linear mapping such that the following diagram is commutative:

$$
\begin{array}{ccccc}
0 & \to & C^\infty(F) & \xrightarrow{j} & (C_0^\infty(F'))' \\
 & & \kappa \uparrow & & \uparrow \widetilde{\kappa} \\
0 & \to & C^\infty(E) & \xrightarrow[j]{} & (C_0^\infty(E'))'
\end{array}
$$

---

[57] $\Omega_{n-1}^{E_r}$ denotes the integral bordism group for $(n-1)$-dimensional closed admissible manifolds of $E_r \subset J_n^r(E)$. (See also refs.[61–66] for more informations about.)

where with the same symbol j we denote the canonical inclusions. We call $\widetilde{\kappa}$ the *distributional extension* of κ. More precisely $\widetilde{\kappa}$ is given by $\widetilde{\kappa}(\omega)(\alpha) = <\omega, \kappa^*(\alpha)>$, $\forall \omega \in (C_0^\infty(E'))'$, $\alpha \in C_0^\infty(F')$ where κ^* is the formal adjoint of κ. Furthermore, one has:

$$\mathbf{1} \otimes \widetilde{\kappa} : (C_0^\infty(F \boxtimes E'))' \to (C_0^\infty(F \boxtimes F'))', \quad (\mathbf{1} \otimes \widetilde{\kappa})(\Xi)(f \otimes \alpha) =_f \Xi(\kappa^*(\alpha)),$$

$$\widetilde{\kappa} \otimes \mathbf{1} : (C_0^\infty(E' \boxtimes F))' \to (C_0^\infty(F' \boxtimes F))', \quad (\widetilde{\kappa} \otimes \mathbf{1})(\Xi)(\alpha \otimes f) = \Xi_f(\kappa^*(\alpha)).$$

Lemma 3.15. (Solutions of affine PDE's and Green kernels). 1) *Let* $E_r \equiv \ker_f \kappa \subset J\mathcal{D}^r(E)$, *be an affine PDE identified by a linear differential operator* $\kappa : C^\infty(E) \to C^\infty(F)$, *of order* r, *and a section* $f \in C^\infty(F)$. *Let us denote by* $Sol(E_r)$ *the set of solutions of* E_r. *Let* $(\widetilde{E}_r)$: $\widetilde{\kappa}(\omega) = \widetilde{f}$ *be the distributional equation corresponding to* E_r. *Then,* $Sol(\widetilde{E}_r)$ *is related to* $Sol(E_r)$ *by means of the following exact sequence:* $0 \to Sol(E_r) \overset{j}{\to} Sol(\widetilde{E}_r)$, *where* $j : C^\infty(E) \to (C_0^\infty(E'))'$ *is the canonical inclusion.*

2) *Let us denote by* $Sol(\overline{E}_r)$ *and* $Sol(\overline{\widetilde{E}}_r)$ *respectively the set of solutions of the following linear equations:* $\kappa(e) = 0$, $e \in C^\infty(E)$, *and* $\widetilde{\kappa}(\omega) = 0$, $\omega \in (C_0^\infty(E'))'$. *Then one has:* $j(Sol(\overline{E}_r)) \subset Sol(\overline{\widetilde{E}}_r)$.[58]

3) *Any solution* $\omega \in Sol(\widetilde{E}_r)$ *can be represented by a distribution* $\omega \in (C_0^\infty(E'))'$ *written in the following form:* $(\heartsuit)$: $\omega = {}_f\mathbb{G} + \omega_0$, *where* $\omega_0 \in \ker(\widetilde{\kappa})$, *and* $\mathbb{G}$ *is the Green kernel of* κ.

4) *Let us assume that* ${}_f\mathbb{G}$ *can be identified with a section* ${}_fG \in C^\infty(E)$ *by means of the canonical embedding* $j : C^\infty(E) \to (C_0^\infty(E'))'$. *Then, any solution* $e \in Sol(E_r)$ *can be written in the form:* $e = {}_fG \bmod Sol(\overline{E}_r)$.

Proof of Lemma 3.15. 1) If $\kappa(e) = f$, one has also

$$\widetilde{\kappa}(j(e))(\alpha) = <j(e), \kappa^*(\alpha)> = \int_M <\kappa^*(\alpha), e> = \int_M <\kappa(e), \alpha> = \int_M <f, \alpha> = \widetilde{j}(\alpha),$$

for all $\alpha \in C_0^\infty(F')$. Therefore, we conclude that $\widetilde{\kappa}(j(e)) = \widetilde{f}$.

2) One has: $\omega \in Sol(\overline{\widetilde{E}}_r) \Leftrightarrow \widetilde{\kappa}(\omega)(\alpha) = 0$, $\forall \alpha \in C_0^\infty(F')$ iff $\omega(\kappa^*(\alpha)) = 0$, $\forall \alpha \in C_0^\infty(F')$. Then, if $e \in Sol(\overline{E}_r)$, and $\omega = j(e)$, one has also, $\forall \alpha \in C_0^\infty(F')$:

$$\omega(\kappa^*(\alpha)) = \int_M <e, \kappa^*(\alpha)> = \int_M <\kappa(e), \alpha> = 0 \Rightarrow j(Sol(\overline{E}_r) \subset Sol(\overline{\widetilde{E}}_r).$$

---

[58] In general $j(Sol(\overline{E}_r))$ is properly contained into $Sol(\overline{\widetilde{E}}_r)$. However, there are some equations, (e.g., elliptic equations), for which $j(Sol(\overline{E}_r))=Sol(\overline{\widetilde{E}}_r)$.

Vice versa, if $\omega = j(e) \in j(C^\infty(E))$ and $\omega(\kappa^*(\alpha)) = 0$, $\forall \alpha \in C_0^\infty(F')$, one has also: $0 = \int_M < \kappa(e), \alpha > \Rightarrow \kappa(e) = 0 \Rightarrow e \in Sol(\overline{E}_r)$. Therefore, $Sol(\overline{\overline{E}}_r) \bigcap j(C^\infty(E)) \cong Sol(\overline{E}_r)$.

3) We must prove that $\widetilde{\kappa}(_fG) = \widetilde{f}$, where $\widetilde{f} = j(f) \in (C_0^\infty(F'))'$. In fact, for any $\alpha \in C_0^\infty(F')$ we have: $\widetilde{\kappa}(_fG)(\alpha) = {}_fG(\kappa^*(\alpha)) = (\kappa \otimes \mathbf{1})(G)(f \otimes \alpha) = \mathbb{D}(f \otimes \alpha) = \int_M < f, \alpha >= \widetilde{f}(\alpha)$. Vice versa, let $\omega \in Sol(\widetilde{E}_r)$. Then, one has: $\widetilde{\kappa}(\omega)(\alpha) = \widetilde{f}(\alpha) = \int_M < f, \alpha >= \mathbb{D}(f \otimes \alpha) = \widetilde{\kappa}(_fG)(\alpha) = {}_fG(\kappa^*(\alpha))$, $\forall \alpha \in C_0^\infty(F')$. On the other hand $\widetilde{\kappa}(\omega)(\alpha) = \omega(\kappa^*(\alpha))$, $\forall \alpha \in C_0^\infty(F')$. Then, as $\widetilde{E}_r$ is an affine equation, we have $\omega = {}_fG + \omega_0$, where $\omega_0 \in \ker(\widetilde{\kappa})$, i.e., any distribution $\omega_0 \in (C_0^\infty(E'))'$ such that $\omega_0|_{\mathrm{im}\,(\kappa^*)} = 0$. In other words $\omega_0 \in Sol(\overline{\overline{E}}_r)$.

4) In fact as $j(e) \in Sol(\widetilde{E}_r)$, we can write $j(e) = {}_fG + \omega_0$, $\forall \omega_0 \in Sol(\overline{\overline{E}}_r)$. On the other hand, if $_fG = j(_fG)$ and $\omega_0 = j(e_0)$, where $e_0 \in Sol(\overline{E}_r)$, (from point (2)), it follows that $e = {}_fG + e_0$ is a solution of E_r. In fact, one has $\kappa(e) = \kappa(_fG) + \kappa(e_0) = \kappa(_fG) = f$. $\qquad\square$

Lemma 3.16. (Determination of Green kernels of linear differential operators). 1) *Let* $G \in (C_0^\infty(E' \boxtimes F))'$ *be a Green kernel of a linear differential operator* $\kappa : C^\infty(E) \to C^\infty(F)$ *of order* r. *Then also* $\widetilde{G} \equiv G + \omega_0 \otimes \beta$, $\forall \beta \in (C_0^\infty(F))'$, $\forall \omega_0 \in Sol(\overline{\overline{E}}_r)$ *is a Green kernel of* κ.

2) *A kernel* $G \in (C_0^\infty(E' \boxtimes F))' \cong (C_0^\infty(E'))' \bigotimes (C_0^\infty(F))'$ *can be identified with a section* $G : M \times M \to J\mathcal{D}^\infty(E')^* \otimes \Lambda_n^0(M) \boxtimes J\mathcal{D}^\infty(F)^* \otimes \Lambda_n^0(M)$ *such that* $G(\alpha \otimes f) = \int_{M \times M} < G, D^\infty \alpha \otimes D^\infty f >$. *One has the following local representation:*

$$G(\alpha \otimes f)$$
$$= \sum_U \sum_{|\gamma|,|\beta| \leq \infty} \int_{U \times U} g_{U \times U} G_i^{\beta\gamma j}(x, x')(\partial_\gamma.\alpha_j)(x)(\partial_\beta.f^j)(x')d\mu(x)d\mu(x'),$$

where $G_i^{\beta\gamma j}$ *are the local components of* G, *and* $g_{U \times U}$ *is a partition of unity subordined to the covering* $\{U \times U\}$ *of* $M \times M$. *Furthermore, if* G *is identified with a section of* $(E' \boxtimes F)' \to M \times M$ *then* G *is identified with local functions* G_i^j *on* $M \times M$ *that are called Green functions. Furthermore, if* G_i^j *are Green functions of the linear differential operator of order* r, $\kappa. : C^\infty(E) \to C^\infty(F)$, *they must satisfy the following local equations:*

$$(3.19) \qquad \sum_{|\gamma| \leq r} \kappa_i^{\gamma j}(x)(\partial_{x,\gamma}.G_u^i)(x, x') = \delta_u^j \delta(x, x'),$$

where $\sum_{|\gamma|\le r} \kappa_i^{\gamma j}(x)\partial_{x,\gamma}$ is the local representation of κ. Taking into account point (1) we get also that the general solution of (3.19) can be written

$$(3.20) \qquad G_i^j(x,x') \equiv \underline{G}_i^j(x,x') + u^j(x)g_i(x'),$$

where $\underline{G}_i^j(x,x')$ is any particular solution of (3.19), (u^j) are the local components of any solution of the linear equation $\sum_{|\gamma|\le r} \kappa_i^{*\gamma j}(x)(\partial_{x,\gamma}.u^j)(x) = 0$ and g_i are local arbitrary functions on M. In particular, if κ is an hyperbolic differential operator over a global hyperbolic space-time, there exists a unique Green function $G^+(x,x')$, (resp. $G^-(x,x')$), supported in $\mathcal{E}^+(x')$, i.e. the future of x', (resp. $\mathcal{E}^-(x')$, i.e. the past of x'), $\forall x' \in M$. Then, $\widetilde{G}(x,x') \equiv G^+(x,x') - G^-(x,x')$ is called the *propagator* of κ and satisfies the following Cauchy problem:

$$\left\{\begin{array}{l} \sum_{|\gamma|} \kappa_i^{\gamma j}(x)(\partial x_\gamma.\widetilde{G}_u^i(x,x'))=0 \\[2mm] \widetilde{G}_u^i(x,x')|_{x'^0=x^0}=0 \\[2mm] (\partial x_0'.\widetilde{G}_u^i(x,x'))|_{x'^0=x^0}=-\delta(x,x')|_{x'^0=x^0} \end{array}\right\},$$

where $(x^\alpha) = (x^0,x^1,x^2,x^3)$ is any adapted coordinate system on M, i.e., ∂x_0 is a time-like vector field.

Proof of Lemma 3.16. 1) First let us note that $\omega_0 \otimes \beta$ belongs to the same space to which belongs $\mathbb{G}$. In fact, one has the following commutative diagram:

$$\begin{array}{ccc} (C_0^\infty(E'\boxed{\times}F))' & \cong & (C_0^\infty(E'))'\bigotimes(C_0^\infty(F))' \\ \uparrow & & \uparrow\, j\otimes 1 \\ C^\infty(E)\bigotimes(C_0^\infty(F))' & = & C^\infty(E)\bigotimes(C_0^\infty(F))' \\ \uparrow & & \uparrow \\ 0 & & 0 \end{array}$$

Furthermore, $_f(\omega_0 \otimes \beta)(\kappa^*(\alpha)) = 0$, $\omega_0 \in Sol(\overline{\overline{E}}_r)$, $\forall \alpha \in C_0^\infty(F')$, $\forall f \in C_0^\infty(F)$. In fact, one has: $_f(\omega_0 \otimes \beta)(\kappa^*(\alpha)) = <\omega_0 \otimes \beta, f \otimes \kappa^*(\alpha) >= \beta(f).\omega_0(\kappa^*(\alpha)) = \beta(f).\widetilde{\kappa}(\omega_0)(\alpha) = 0$. Then, we also have: $(1\otimes\widetilde{\kappa})(\mathbb{G}+\omega_0\otimes\beta) = (1\otimes\widetilde{\kappa})(\mathbb{G}) + (1\otimes\widetilde{\kappa})(\omega_0\otimes\beta) = \mathbb{D}$. Thus we have proved the theorem.

2) In local coordinates we must have:

$$0=\sum_U \int_{U\times U} g_{U\times U} \sum_{|\omega|,|\epsilon|\le r} [G_u^{\omega\epsilon i}(x,x')(\partial_\epsilon.\kappa^*(\alpha)_i)(x)(\partial_\omega.f^u)(x')-\alpha_i(x)f^i(x')]d\mu(x)d\mu(x').$$

Now, if G_i^j are Green functions, the above expression can be written:

$$0 = \sum_U \int_{U\times U} g_{U\times U}[G_u^i(x,x')\kappa^*(\alpha)_i(x)f^u(x')$$
$$- \alpha_j(x)f^u(x')\delta_u^j\delta(x,x')]d\mu(x)d\mu(x').$$

As this expression must hold for any f we can write: $G_u^i(x,x')\kappa^*(\alpha)_i(x) = \alpha_j(x)\delta_u^j\delta(x,x')$. Taking into account that $\kappa^*(\alpha)_i = \sum_{|\gamma|\le r}\kappa_i^{*\gamma j}(\partial_{x,\gamma}.\alpha_j)$, we can also write

$$\sum_{|\gamma|\le r} G_u^j(x,x')\kappa_j^{*\gamma i}(x)(\partial_{x,\gamma}.\alpha_i)(x) = \alpha_j(x)\delta_u^j\delta(x,x')\,.$$

From the definition of adjoint of κ and taking into account that the above equation holds for any α we get equations (3.19). $\qquad\square$

Definition 3.106. Let κ_1 be a differential operator on M between the vector fiber bundles $E\to M\leftarrow F$, of order $\le r$ and class C^h, $r\le h\le s-r$. Let κ_2 be another differential operator between F' and E'. Define *Green operator* of (κ_1,κ_2) any differential operator G of order $\le r-1$ (and class C^h, $h\ge 1$) between $E\bigotimes F'$ and $\Lambda_{n-1}^0(M)$, such that $<\kappa_1(u),v> - <u,\kappa_2(v)> = dG(u\otimes v)$. If $\kappa_2 = \kappa_1{}^*$, we say that G is a *Green operator* for κ_1.

Example 3.57. The Green operator of a second order differential operator κ, locally written as follows

$$\kappa(s) = [a_j^{i\alpha\beta}(\partial x_\alpha\partial x_\beta.s^j) + b_j^{i\alpha}(\partial x_\alpha.s^j) + c_j^i s^j]e_i,\ \forall s\in C^\infty(E),$$

where $\{e_i\}$ is a local basis for $C^\infty(E)$, can be locally written in the following way:

$$G(s\otimes\gamma) = \sum_\alpha(-1)^{\alpha-1}\left\{a_i^{j\alpha\beta}[\gamma_j(\partial x_\beta.s^i) - s^i(\partial x_\beta.\gamma_j)] + \bar{b}_i^{j\alpha}s^i\gamma_j\right\}$$
$$dx^1\wedge\ldots\wedge\widehat{dx^\alpha}\wedge\ldots\wedge dx^n,$$

with $\bar{b}_i^{j\alpha}\equiv b_i^{j\alpha} - (\partial x_\beta.a_i^{j\alpha\beta})$, $\forall\gamma\in C^\infty(F')$, $\gamma = \gamma_j\epsilon^j\otimes dx^1\wedge\ldots\wedge dx^n$, where $\{\epsilon^j\}$ is a local basis for $C^\infty(F^*)$. In fact,

$$\kappa^*(\gamma) = \left[(\partial x_\alpha\partial x_\beta.(a_i^{j\alpha\beta}\gamma_j) - (\partial x_\alpha.(b_i^{j\alpha}\gamma_j)) + c_i^j\gamma_j\right]\epsilon^i\otimes dx^1\wedge\ldots\wedge dx^n.$$

Furthermore:

$$<\kappa(s),\gamma> - <s,\kappa^*(\gamma)> = [a_i^{j\alpha\beta}(\partial x_\alpha\partial x_\beta.s^i)\gamma_j - (\partial x_\alpha\partial x_\beta(a_i^{j\alpha\beta}\gamma_j))s^i$$
$$+ b_i^{j\alpha}(\partial x_\alpha.s^i)\gamma_j - (\partial x_\alpha(b_i^{j\alpha}\gamma_j))s^i]dx^1\wedge\ldots\wedge dx^n$$
$$= (\partial x_\alpha.\sum_\alpha\{a_i^{j\alpha\beta}[\gamma_j(\partial x_\beta.s^i) - s^i(\partial x_\beta.\gamma_j)] + \bar{b}_i^{j\alpha}s^i\gamma_j\})dx^1\wedge\ldots\wedge dx^n.$$

Lemma 3.17. *Let κ be a linear differential operator as given in above lemma. Let $A \subset M$ be a compact oriented domain in M. Then, one has*

(3.21)

$$\int_A < \kappa(e), v > - \int_A < e, \kappa^*(v) >= \int_{\partial A} G(e{\otimes}v), \ \forall e \in C^\infty(E), v \in C^\infty(F').$$

In particular , if $\partial A = \emptyset$, one has: $\int_A < \kappa(e), v >= \int_A < e, \kappa^*(v) >$. *Furthermore, if M is a compact manifold formula (3.21) can be written in the following way:*

$$(3.22) \qquad \int_M < \kappa(e), v > - \int_M < e, \kappa^*(v) >= \int_{\partial M} G(e \otimes v).$$

The Green operator of a linear differential operator of order r identifies a section

$$G \in C^\infty \left(J\mathcal{D}^{r-1}(E \otimes F')^* \bigotimes \Lambda^0_{n-1}M \right).$$

In particular, if $r = 1$, one has $G \in C^\infty \left(E^* \otimes \Lambda^0_{n-1}M \bigotimes(F')^*) \right)$.
Proof of Lemma 3.17. In fact, one has the canonical isomorphism:

$$Hom \left(J\mathcal{D}^{r-1}(E \bigotimes F'); \Lambda^0_{n-1}M \right) \cong J\mathcal{D}^{r-1}(E \bigotimes F')^* \bigotimes \Lambda^0_{n-1}M.$$

So, for $r = 1$ we get

$$Hom \left(E \bigotimes F'; \Lambda^0_{n-1}M \right) \cong E^* \bigotimes(F')^* \bigotimes \Lambda^0_{n-1}M$$

$$\cong E^* \bigotimes \Lambda^0_{n-1}M \bigotimes(F')^*.$$

$\square$

Lemma 3.18. *Let $\kappa. : C^\infty(E) \to C^\infty(F)$ be a linear differential operator of first order between vector fiber bundles $\pi : E \to M$, and $\pi' : F \to M$ on a differential manifold M of dimension n. Then, one has the following relation between the symbol $\sigma(\kappa)$ of κ and Green operator G:* $< \sigma(\kappa)(\lambda \otimes e), v >= \lambda \wedge G(e \otimes v), \ \forall \lambda \in \Omega^1(M) \equiv C^\infty(\Lambda^0_1 M)$.
Proof of Lemma 3.18. If $\kappa_1. : C^\infty(E) \to C^\infty(F), \kappa_2. : C^\infty(F') \to C^\infty(E')$ are two differential operators of the first order, such that the symbols $\sigma_\lambda(\kappa_1)$ and $\sigma_\lambda(\kappa_2)$, for any differential 1-form $\lambda \in \Omega^1(M)$ are skew-symmetric: $\sigma_\lambda^*(\kappa_1)+\sigma_\lambda(\kappa_2) = 0$, then the operator $\gamma(\kappa_1, \kappa_2) : C^\infty(E \bigotimes F') \to \Omega^n(M)$,

given by $\gamma(\kappa_1, \kappa_2)(u \otimes v) = <\kappa_1(u), v> - <u, \kappa_2(v)>$ is a differential operator of the first order. The symbol of this operator determines the morphism $\omega : C^\infty(E \otimes F') \to \Omega^{n-1}(M)$ such that $\sigma_\lambda(\gamma(\kappa_1, \kappa_2))(u \otimes v) = \lambda \wedge \omega(u \otimes v)$. Thus, the operators $\gamma(\kappa_1, \kappa_2)$ and ω have the same symbol, and, hence, differ by an operator of zero order. By substituting, if necessary, the operator κ_2 with $\kappa_2 + \kappa_2'$, where $\kappa_2' \in Hom(C^\infty(F'), C^\infty(E'))$, we get for each operator $\kappa_1 \in Diff_1(C^\infty(E), C^\infty(F))$, the operator $\kappa_1^* \in Diff_1(C^\infty(F'), C^\infty(E'))$, that is the adjoint of the operator κ_1 for which $\gamma(\kappa_1, \kappa_1^*) = d\omega$. Thus $\omega = G$. $\qquad\square$

Lemma 3.19. *Let* $\kappa. : C^\infty(E) \to C^\infty(F)$ *be a linear differential operator of order* r *over a compact manifold* M *of dimension* n. *Then the distributional extension* $\widetilde{\kappa} : (C_0^\infty(F'))' \to (C_0^\infty(E'))'$ *of* κ *is related to its formal adjoint* κ^* *by the following formula:*

$$(3.23) \qquad \widetilde{\kappa}(\omega)(\alpha) = \omega(\kappa^*(\alpha)) + \widetilde{G}|_{\partial M}(\omega)(\alpha),$$

where $\widetilde{G}|_{\partial M}(\omega) \in (C_0^\infty(F'))'$ *is the distribution supported on* ∂M *identified by the* ω *and the Green operator of* κ. *Furthermore, if* $r = 1$ *above equation can be written as follows:* $\widetilde{\kappa}(\omega)(\alpha) = \omega(\kappa^*(\alpha)) + \widetilde{\sigma}_1(\kappa)|_{\partial M}(\omega)(\alpha)$, *where* $\widetilde{\sigma}_1(\kappa)|_{\partial M}(\omega)$ *is the distribution supported on* ∂M *identified by the symbol* $\sigma_1(\kappa)$ *of* κ.

Proof of Lemma 3.19. One has the following Green formula: $<\kappa(e), \alpha> - <e, \kappa^*(\alpha)> = dG(e \otimes \alpha)$, $\forall e \in C^\infty(E)$, $\alpha \in C_0^\infty(F')$. Then, taking into account the canonical immersion $j : C^\infty(E) \to (C_0^\infty(E'))'$, we get:

$$\widetilde{\kappa}(j(e), \alpha) = \int_M <\kappa(e), \alpha> = \int_M <e, \kappa^*(\alpha)> + \int_M dG(e \otimes \alpha)$$

$$= j(e)(\kappa^*(\alpha)) + \int_{\partial M} G(e \otimes \alpha) = j(e)(\kappa^*(\alpha)) + \widetilde{G}|_{\partial M}(\omega)(\alpha),$$

where $\widetilde{G}|_{\partial M}(e) \in (C_0^\infty(F'))'$ is the distribution with support on ∂M, identified by G, i.e., $\widetilde{G}|_{\partial M}(e)(\alpha) = \int_{\partial M} G(e \otimes \alpha)$. Then, we get: $\widetilde{\kappa}(j(e))(\alpha) = j(e)(\kappa^*(\alpha)) + \widetilde{G}|_{\partial M}(e)(\alpha)$. On the other hand, as $C^\infty(E)$ is dense in

$(C_0^\infty(E'))'$, for any $\omega \in (C_0^\infty(E'))'$ we can write

$$\widetilde{\kappa}(\omega)(\alpha) = \widetilde{\kappa}(\sum_s j(e_s))(\alpha) = \sum_s \int_M < \kappa(e_s), \alpha >$$

$$= \sum_s \int_M < e_s, \kappa^*(\alpha) > + \sum_s \int_M dG(e_s \otimes \alpha)$$

$$= \sum_s j(e_s)(\kappa^*(\alpha)) + \sum_s \int_{\partial M} G(e_s \otimes \alpha).$$

Thus, we have $\omega(\kappa^*(\alpha)) + \widetilde{G}|_{\partial M}(\omega)(\alpha) = j(\sum_s e_s)(\kappa^*(\alpha)) + \sum_s \widetilde{G}|_{\partial M}(e_s)(\alpha). \square$

Lemma 3.20. *As particular cases we have:*

(a) *For the Green kernel $\mathbb{G}$ of κ we have, $_f\mathbb{G} \in (C_0^\infty(E'))'$, $\forall f \in C^\infty(F)$.*
Thus we get

$$(3.24) \qquad \widetilde{\kappa}(_f\mathbb{G})(\alpha) = {}_f\mathbb{G}(\kappa^*(\alpha)) + \widetilde{G}|_{\partial M}(_f\mathbb{G})(\alpha), \quad \alpha \in C_0^\infty(F').$$

If $r = 1$ one can also write:

$$(3.25) \qquad \widetilde{\kappa}(_f\mathbb{G})(\alpha) = {}_f\mathbb{G}(\kappa^*(\alpha)) + \widetilde{\sigma}_1|_{\partial M}(_f\mathbb{G})(\alpha), \quad \alpha \in C_0^\infty(F').$$

(b) *If the Green kernel $\mathbb{G}$ is such that it can be identified with a section $G \in C^\infty(E\boxtimes F)$, then we can write:*

$$(3.26) \qquad \int_M < \kappa(_fG), \alpha >= \int_M < {}_fG, \kappa^*(\alpha) > + \int_{\partial M} G(_fG \otimes \alpha).$$

If $r = 1$ we can also write:

$$(3.27) \qquad \int_M < \kappa(_fG), \alpha >= \int_M < {}_fG, \kappa^*(\alpha) > + \int_{\partial M} < \sigma_1(\kappa)(_fG), \alpha > .$$

Furthermore the Green kernel satisfies the following equivalent equations:
(a) $\widetilde{\kappa}(_f\mathbb{G}) = {}_f\mathbb{D}$, $\forall f \in C^\infty(F)$; (b) $_f\mathbb{G}(\kappa^*(\alpha)) + \widetilde{G}|_{\partial M}(_f\mathbb{G})(\alpha) = \int_M < f, \alpha >$, $\forall f \in C^\infty(F)$, $\alpha \in C_0^\infty(F')$; (c) *If $\mathbb{G}$ is identified by a section $G \in C^\infty(E\boxtimes F)$ the above equations become:* $\kappa(_fG) = {}_f\mathbb{D}$, $\forall f \in C^\infty(F)$, $\int_M < {}_fG, \kappa^*(\alpha) > + \int_{\partial M} < \widetilde{G}|_{\partial M}(_fG), \alpha >= \int_M < f, \alpha >$. *If $r = 1$ we can also write:* (a) $\widetilde{\kappa}(_f\mathbb{G}) = {}_f\mathbb{D}$, $\forall f \in C^\infty(F)$; (b) $_f\mathbb{G}(\kappa^*(\alpha)) + \widetilde{\sigma}_1(\kappa)|_{\partial M}(_f\mathbb{G})(\alpha) = \int_M < f, \alpha >$, $\forall f \in C^\infty(F)$, $\alpha \in C_0^\infty(F')$; (c) *If $\mathbb{G}$ is identified by a section $G \in C^\infty(E\boxtimes F)$ the above equations become:*

$\kappa(_f G) = {_f}\mathbb{D}, \forall f \in C^\infty(F); \int_M < {_f}G, \kappa^*(\alpha) > + \int_{\partial M} < \sigma_1(\kappa)(_f G), \alpha >= \int_M < f, \alpha >.$

Proof of Lemma 3.20. The proof follows from above results.   □

Now, from our assumptions on R and using results given in refs.[60], we get that we can associate to R a distribution $F[R] \in (C_0^\infty(E'))'$, that is a distributive solution of E_r iff R is a multivalued solution of $E_r \subset JD^r(E) \subset J_n^r(E)$. Moreover, from above lemmas, we get that the relation between the Green kernel of κ and $F[R]$ is the following: (♣): $F[R] = {_f}G + \omega_0$, where $\omega_0 \in \mathcal{S}ol(\widetilde{\widetilde{E}}_r)$, with $\widetilde{\widetilde{E}}_r$ the linear equation associated to $\widetilde{E}_r$. ($\widetilde{E}_r$ is the distributional equation associated to E_r.) Furthermore, if $\partial R = R_1 \bigcup R_2$, with $R_1 \in [R_2]_{E_r} \in \Omega_{n-1}^{E_r}$, then we say that $F[R]$ satisfies the boundary conditions $R_1 \bigcup R_2 \subset E_r$. If $R' \in \Omega_c(R_1, R_2)$ is another solution of E_r that satisfies the same boundary conditions than R, then $F[R'] = {_f}G + \omega_0'$, where $\omega_0' \in \mathcal{S}ol(\widetilde{\widetilde{E}}_r)$. Let us denote by $[F[R]]$ the element belonging to the space $[\mathcal{S}ol(\widetilde{E}_r)]$ corresponding to $F[R]$ by means of the following commutative diagram:

$$
\begin{array}{ccc}
0 \;\;\rightarrow & \mathcal{S}ol(\widetilde{E}_r) & \rightarrow & (C_0^\infty(E'))' \\
& \downarrow & & \downarrow \tilde{\pi} \\
0 \;\;\rightarrow & [\mathcal{S}ol(\widetilde{E}_r)] & \rightarrow & (C_0^\infty(E'))'/\mathcal{S}ol(\widetilde{\widetilde{E}}_r) \\
& \downarrow & & \downarrow \\
& 0 & & 0
\end{array}
$$

Then, the equivalence class $[F[R]]$ of $F[R]$ in the space $[\mathcal{S}ol(\widetilde{E}_r)]$ is identified by a unique distribution: $_f G$. This proves that $_f G$ is an invariant of the classic-limit statistical set $\Omega_c(R_1, R_2)$.   □

Corollary 3.8. (Generalized Green kernels and solutions of affine PDE's). *Any solution R of an affine PDE $E_r \equiv \ker_f \kappa \subset JD^r(E) \subset J_n^r(E)$ that satisfies the boundary conditions $\partial R = R_1 \bigcup R_2$, where R_1 and R_2 are such that: (i) $\pi_{r,0}|_R : R \to E$ is a proper application; (ii) $\omega_1^{(r)}(R) = 0$; identifies a distributive kernel: (♠): $\mathbb{F}[R] = F[R] \otimes \tilde{f} \in (C_0^\infty(E' \boxtimes F'))'$, given by $\mathbb{F}[R](\alpha \otimes \phi) = F[R](\alpha) \int_M < \phi, f >$. Define $\mathbb{F}[R]$ the generalized Green kernel of the singular solution $R \subset E_r$ that satisfies the boundary conditions $\partial R = R_1 \bigcup R_2$. The relation between $\mathbb{F}[R]$ and the Green kernel $\mathbb{G}$ of κ is given by the following formula: $\mathbb{F}[R] = {_f}\mathbb{G}[R] \otimes \tilde{f} \mod \mathcal{S}ol(\widetilde{\widetilde{E}}_r) \otimes \tilde{f}$.*
Proof. It is a direct consequence of Theorem 3.154.   □

Theorem 3.87. (Green kernels and propagators for nonlinear PDE's). *1) Let $E_r = \ker_\chi \kappa \subset JD^r(W)$ be a PDE given as a kernel of a differential* operator of order r: $\kappa : J\mathcal{D}^r(W) \to K$, with respect to a section C^∞: $\chi : M \to K$ of the fiber bundle $\pi : K \to M$. Then, for any section $s \in C^\infty(W)$, solution of E_r, and $j[\chi] = \partial\widetilde{\chi}$, where $\widetilde{\chi}$ is a deformation of χ, we can associate to E_r an affine equation $E_r[s] = \ker_{j[\chi]} J[s] \subset J\mathcal{D}^r(s^*vTW)$, where $J[s] : C^\infty(s^*vTW) \to C^\infty(\chi^*vTK)$ is the linearized of κ at the section s. Define $E_r[s]$ Jacobi equation of E_r at the solution s. Furthermore, define Green kernels, (resp.propagator), $G[s]$ of E_r, at the section s, the Green kernels, (resp. propagator) of $E_r[s]$.

2) Assume that $E_r = \ker_f \kappa \subset J\mathcal{D}^r(W)$ is a PDE as given in the above point (1). Then, the Green kernels (resp. propagator), $G[s]$ identifies an integral manifold (integral bordism) $R \in \Omega_c(R_1, R_2)[s]$, i.e., belonging to the classic limit of a statistical set of $E_r[s]$, if $G[s]$ satisfies to the boundary condition $\partial R = R_1 \bigcup R_2 \subset E_r[s]$.

3) Let R_1 and R_2 be two admissible integral compact closed manifolds of dimension $(n-1)$ contained into E_r such that the following conditions are satisfied: (i) $R_1 \in [R_2]_{E_r} \in \Omega^{E_r}_{n-1}$; (ii) There exists a vector fiber bundle neighborhood $\overline{E}_r[s] \subset E_r$ such that $R_1, R_2 \subset E_r[s]$. Then the equivalence class $[G[s]]$, identified by the Green kernels, (resp. propagator), $G[s]$, is invariant for $\Omega_c(R_1, R_2)[s]$, i.e., the set of solutions V of E_r with $\partial V = R_1 \bigcup R_2$ and such that $V \subset E_r[s]$.

4) Furthermore, if E_r is a formally integrable PDE and $\Omega_c(R_1, R_2)$ is restricted to the regular solutions of $E_r \subset J^r_n(W)$, then $G[s]$ is an invariant of $\Omega_c(R_1, R_2)$.

Proof. For the full proof of 1) and 2) see refs.[42,61]. Here, let us emphasize only that $E_r[s]$ is an affine equation, that is an affine bundle over M, with associated vector bundle the linearized equation of E_r at s: $(D^r s)^*vTE_r \subset (D^r s)^*vTJ\mathcal{D}^r(W) \cong J\mathcal{D}^r(s^*vTW)$. Therefore, with respect to a solution of $E_r[s]$, one has the identification of $E_r[s]$ with $(D^r s)^*vTE_r \equiv \overline{E}_r[s]$, hence one has the identification of $E_r[s]$ with a submanifold of E_r.

3) It is a direct consequence of above two points and taking into account above lemmas.

4) In fact, one has the following short exact sequence:

$$0 \to \Omega_c(R_1, R_2)[s] \to \Omega_c(R_1, R_2).$$

Moreover, if E_r is formally integrable, taking into account of Theorem 4.5 in the last paper quoted in ref.[65], that relates bordism to formal integrability, we get also the following short exact sequence: $\Omega_c(R_1, R_2) \rightarrow \Omega_c(R_1, R_2)[s] \rightarrow 0$. Hence, from the above point, it follows that we can characterize $\Omega_c(R_1, R_2)$ also by means of the Green kernel, (resp. propagator), $G[s]$. $\qquad\square$

Definition 3.107. A *physical observables*, associated to a PDE $E_k \subset J\mathcal{D}^k(W)$, is any random variable: $A : \Omega(E_k)_c \rightarrow \mathbf{R}$ given as $A(s) = \int_M f \circ D^q s d\mu$, where μ is a suitable measure on M, $f : J\mathcal{D}^q(W) \rightarrow \mathbf{R}$ is a numerical function on the q-jet derivative space of $\pi : W \rightarrow M$, $q \geq 0$ and s is a solution of E_k. Local coordinates on $J\mathcal{D}^q(W)$ identify random variables of this type. We denote by $\mathcal{P}_{hys}(E_k)_q$ the set of physical observables coming from functions on $J\mathcal{D}^q(W)$, $q \geq 0$, and we call $\mathcal{P}_{hys}(E_k)_q$ the *space of q-order physical observables*.

We have the following important theorem.

Theorem 3.88. (Covariant quantization). *Let $E_k = \ker_\chi \kappa \subset J\mathcal{D}^k(W)$ be a formally integrable PDE such that $K = vT^*W$. Let κ be an hyperbolic differential operator on the fiber bundle $\pi : W \rightarrow M$, and let M be a global hyperbolic space-time. Then the space of physical observables $\mathcal{P}_{hys}(E_k)$ of E_k has a natural structure of filtered Lie algebra, denoted $\mathcal{B}_0 \subset \cdots \subset \mathcal{B}_q \subset \cdots \subset \mathcal{B}_\infty \equiv \mathcal{B}$. Furthermore, to such Lie algebras $\mathcal{B}$ we canonically associate a filtered quantum algebra $\hat{\mathcal{B}}_0 \subset \cdots \subset \hat{\mathcal{B}}_q \subset \cdots \subset \hat{\mathcal{B}}_\infty \equiv \hat{\mathcal{B}}$ that we call the covariant quantization of E_k.*[59]

Proof. Let us be given a function $f_1 : W \rightarrow \mathbf{R}$ on the fiber bundle W. This identifies a physical observable characterized by a function, yet denoted by f_1, $f_1 : C^\infty(W) \rightarrow C^\infty(M, \mathbf{R}) \rightarrow \mathbf{R}$ (0-order physical observable of E_k). To f_1 we can associate a *current* $jf_1 : C^\infty(W) \rightarrow \bigcup_{s \in C^\infty(W)} C^\infty(s^* vT^*W)$, $jf_1(s) \equiv (vdf) \circ s \in C^\infty(s^* vT^*W)$. If $f_2 : C^\infty(W) \rightarrow \mathbf{R}$ is another physical observable, we can consider the corresponding current of f_2 calculated in correspondence of a solution of E_k for any vector field $\nu \in T_s C^\infty(W)$ belonging to the set of solutions of the Jacobi equation of E_k corresponding to the current $j[\chi.s] = -jf_1(s) : J[s].\nu = -jf_1(s)$. Then, we get:
$$< \nu^\pm, jf_2(s) \otimes \eta > = -_{jf_1(s)} G^\pm(jf_2(s) \otimes \eta) = G^\pm(jf_1(s) \otimes jf_2(s) \otimes \eta).$$
Set: $(D_{f_1} f_2)(s) \equiv < \nu^-, jf_2(s) \otimes \eta > = G^-(jf_1(s) \otimes jf_2(s) \otimes \eta)$. Thus, in $\mathcal{P}_{hys}(E_k)_0$, the set of 0-order physical observables on E_k, we can define the

---

[59] For abuse of language we will refer to covariant quantization of 0-order physical observables of E_k as to covariant quantization too.

following bracket: $(f_1, f_2)(s) \equiv (D_{f_1} f_2 - D_{f_2} f_1)(s) = G^-(jf_2(s) \otimes jf_1(s) \otimes \eta) - G^-(jf_1(s) \otimes jf_2(s) \otimes \eta) = G^+(jf_1(s) \otimes jf_2(s) \otimes \eta) - G^-(jf_1(s) \otimes jf_2(s) \otimes \eta) \equiv \tilde{G}(jf_1(s) \otimes jf_2(s) \otimes \eta)$. We can easily see that this bracket satisfies the following relations: $(f_1, f_2) = -(f_2, f_1)$, $((f_1, f_2), f_3) + ((f_2, f_3), f_1) + ((f_3, f_1), f_2) = 0$. So $\mathcal{P}_{hys}(E_k)_0$ becomes a Lie algebra, denoted by $\mathcal{B}_0$. This process can be generalized to any order $q > 0$. In fact, if the physical observable f is of order $q \geq k$ we can consider the $s \equiv (q + 1 - k)$-prolongation $(E_k)_{+s}$ of E_k, considered as a first order PDE on the fiber bundle $JD^q(W) \to M$, via the canonical embeddings $(E_k)_{+s} \subset JD^{q+1}(W) \subset JD(JD^q(W))$. So we can recognize on $\mathcal{P}_{hys}(E_k)_q$, $q \geq k$, a structure of Lie algebra, denoted by $\mathcal{B}_q$. Furthermore, if f is a physical observable of order $0 < q < k$, we can consider E_k, as a $(k - q)$-order PDE on $JD^q(W) \to M$, via the embeddings $E_k \subset JD^{k-q}(JD^q(W))$. So we can recognize on $\mathcal{P}_{hys}(E_k)_q$, $0 < q < k$, a structure of Lie algebra, denoted by $\mathcal{B}_q$, with $0 < q < k$. As $E_k \subset J_n^k(W)$ is formally integrable it follows that $(E_k)_{+s} \subset J_n^{k+s}(W)$ is equivalent to E_k, from the point of view of regular solutions, hence we get $\Omega(E_k)_c = \Omega((E_k)_{+(s)})_c$. Moreover, taking into account the short exact sequence: $0 \to C^\infty(E_{k+s}; \mathbf{R}) \to C^\infty(E_{k+s+1}; \mathbf{R})$, we get also the natural inclusions. $\mathcal{P}_{hys}(E_k)_q \subset \mathcal{P}_{hys}(E_k)_{q+1}$, $q \geq 0$. Therefore, we recognize a filtered Lie algebra structure on the set of physical observables of E_k: $\mathcal{B}_0 \subset \cdots \subset \mathcal{B}_q \subset \cdots \subset \mathcal{B}_\infty \equiv \mathcal{B}$. Then, we call *covariant quantization* of E_k the quantum logics $\mathcal{L}_q \equiv \hat{\mathcal{B}}_q$, $q \geq 0$, associated to E_k defining on $\mathcal{P}_{hys}(E_k)_q$ the following Lie algebra structure: $[\hat{f_1}, \hat{f_2}](s) = (\hat{f_1} \circ \hat{f_2} - \hat{f_2} \circ \hat{f_1})(s) = i\hbar \tilde{G}_q(jf_1(s) \otimes jf_2(s) \otimes \eta) \hat{\otimes} id_{\mathcal{H}(s)}$ $(\triangle)$, where $\mathcal{H}(s)$ is a suitable locally convex topological vector space, and $\tilde{G}_q$ are the propagators of order $q \geq 0$ associated to $E_k \subset JD^{k-q}(JD^q(W))$, for $0 \geq q \leq k$, and to $(E_k)_{+(s+1)} \subset JD^{k+s+1}(W) \subset JD(JD^q(W))$, for $q = k + s > k$. The corresponding expectation value for any $\psi \in \mathcal{H}(s)$, $\psi' \in \mathcal{H}(s)'$, is as follows:

$$< \psi' |[\hat{f_1}, \hat{f_2}](s)| \psi > = i\hbar \tilde{G}_q(jf_1(s) \otimes jf_2(s) \otimes \eta) < \psi', \psi > .$$

Note also that any p-order physical observable f can be identified with a q-order physical observable $f^{(q)}$, $p < q$, via the filtration $\mathcal{B}_p \subset \cdots \subset \mathcal{B}_q$. On the other hand, any p-order propagator $\tilde{G}_p$ identifies a q-order propagator $\tilde{G}_{q(p)}$, via the canonical inclusion $C_0^\infty(JD^p(E) \boxed{\times} JD^p(E)')' \to C_0^\infty(JD^q(E) \boxed{\times} JD^q(E)')'$ for $E \equiv s^* vTW$. This means that quantum bracket between two p-order physical observables f_i, $i = 1, 2$, can be also written as follows: $[\widehat{f}_1, \widehat{f}_2](s) = i\hbar\tilde{G}_{q(p)}(jf_1^{(q)}(s) \otimes jf_2^{(q)}(s) \otimes \eta)\widehat{\otimes}id_{\mathcal{H}(s)} = i\hbar\tilde{G}_p(jf_1(s) \otimes jf_2(s) \otimes \eta)\widehat{\otimes}id_{\mathcal{H}(s)}$. Furthermore, if f_1 is a p-order physical observable and f_2 is a q-order physical observable, $q > p$, we get $[\widehat{f}_1, \widehat{f}_2](s) = i\hbar\tilde{G}_q(jf_1^{(q)}(s) \otimes jf_2(s) \otimes \eta)\widehat{\otimes}id_{\mathcal{H}(s)}$. Let us emphasize, also, that the structure of $\tilde{G}_q$ assures the local causality. In other words, for any two observable fields f_1 and f_2, if the support of $jf_1(s)$ is separated from every point of the support of $jf_2(s)$ by spacelike interval, we have $[\widehat{f}_1, \widehat{f}_2](s) = 0$. The choice of the quantum commutator allows us to also recognize quantum spectral measures $E : (\Omega(E_k)_c, \Sigma)\circ\!\!\!\to\mathcal{L}(\mathcal{H})$ associated to random functions $\Omega(E_k)_c \to \mathbf{R}$. In fact if a scalar measure μ is recognized on $\Omega(E_k)_c$ one has $< \phi'|E(A)|\phi > = \int_A \hat{f}_{\phi,\phi'}d\mu$ for any $S \in \Sigma \equiv$ Borel σ-algebra of $\Omega(E_k)_c$, where $\hat{f} : \Omega(E_k)_c \to \mathcal{L}(\mathcal{H})$ is the map associated to the canonical quantization of f. $\hat{f}(s)$ is determined by means of its spectral measure on $\mathbf{R}$. In conclusion we get a filtered quantum algebra $\hat{\mathcal{B}}_0 \subset \cdots \subset \hat{\mathcal{B}}_q \subset \cdots \subset \hat{\mathcal{B}}_\infty \equiv \hat{\mathcal{B}}$ that just represents the covariant quantization of E_k. This completes the proof. $\qquad\square$

Definition 3.108. We call *Dirac covariant quantization*, (or simply *Dirac quantization*), a covariant quantization of a PDE $E_k \subset J\mathcal{D}^k(W)$, made with respect to physical observables performed with respect to Dirac measures on the base manifold M of the fiber bundle $\pi : W \to M$.

Theorem 3.89. (Canonical quantization). *Let $\pi : W \to M$ be a bundle of geometric objects, and let $(\tau : M \to \mathbf{R}, \phi : \mathbf{R} \times M \to M) \equiv \psi$ be a frame on M, i.e., τ is a function of constant rank 1 (time function) and ψ is a 1-parameter group of diffeomorphisms with time-like velocity, i.e., $< \dot{\phi} \equiv \partial\phi, d\tau > = 1$. Let $E_k \subset J\mathcal{D}^k(W)$ be a PDE like in above Theorem 3.88. Let us denote by $\mathcal{P}_{hys}^{can}(E_k)$ the physical observables of E_k obtained by the physical observables f, associated with Dirac measure on M, and $\dot{f} \equiv \frac{df}{dt}$, where the time derivative is performed with respect to the frame ψ. Then the covariant quantization of E_k identifies a filtered quantum algebra $\hat{\mathcal{B}}_t^{can}$, at any istant t, called canonical quantization of E_k, at the time t, with respect to the frame ψ.*

Proof. Let us start with 0-order physical observables. Then by the covariant quantization we get the following commutator: $[\widehat{f}_1(x), \widehat{f}_2(x')] = i\hbar\tilde{G}_0(x, x')$. By considering composition with the flow ϕ and deriving with respect to the first variable, and by restriction on the space-like submanifold $M_t \equiv \tau^{-1}(t) \subset M$, we get: $[\widehat{f}_1(t, \mathbf{x}), \widehat{f}_2(t, \mathbf{x}')] = 0$, $[\widehat{f}_1(t, \mathbf{x}), \widehat{f}_2(t, \mathbf{x}')] = i\hbar\delta(x, x')|_{t=t'}$. Here we have used the properties of the propagator. (See Lemma 3.16.) (Note also that the frame determines the following split $x = (t, \mathbf{x})$ of any point $x \in M$.) This is just the quantum algebra $(\hat{\mathcal{B}}_0)_t^{can}$ called *canonical quantization* of E_k at the 0-order and at the instant t, with respect the frame ψ. By extending this process to all the orders, we get a filtered quantum algebra $\hat{\mathcal{B}}_t^{can}$, called the *canonical quantization* of E_k at the instant t, with respect to the frame ψ. $\qquad\square$

The relation between quantized PDEs and quantum PDEs can be obtained by considering an algebraic characterization of the category of (quantum) PDE's. The following theorem just gives a relation between quantized PDEs and quantum PDEs.

Theorem 3.90. (Dirac quantized PDE's vs. quantum PDE's). *The Dirac covariant quantization of a formally integrable PDE $E_k \subset J\mathcal{D}^k(W)$, identifies a quantum PDE.*

Proof. Let $\mathcal{A}$ be a $\mathbf{K}$−algebra filtered by the $\mathbf{K}$−algebras $\mathcal{A}_i$, $\mathcal{A}_i \subset \mathcal{A}_{i+1}$, $\bigcup_i \mathcal{A}_i = \mathcal{A}$. A *filtered $\mathcal{A}$−module* structure in a $\mathcal{A}$−module Q is given by fixing $\mathcal{A}_i$−modules $Q_i \subset Q$, where $Q_i \subset Q_{i+1}$, $\bigcup_i Q_i = Q$. A *filtered differential operator* $\triangle \in Diff_\bullet(P, Q)$, $P \equiv \{P_i\}$, $Q \equiv \{Q_i\}$, is a differential operator such that $\triangle(P_i) \subset Q_{i+s}$, for all i. A filtered $\mathcal{A}$−module $Q \equiv \{Q_i\}$ is called *geometric* if, for any i, the $\mathcal{A}_i$−module Q_i is geometric, i.e., $\bigcap_m mQ_i = 0$, where $m \in Spec(\mathcal{A}_i)$. For example, if $\mathcal{A} \equiv C^\infty(M; \mathbf{R})$, with M a smooth differentiable manifold, geometric modules are the ones whose elements are defined by their values at the points of the manifold M. The functor $Diff_s(P; -)$ of differentiable operators of order s on the module P, is representable in the category of geometric modules by means of the objects $\mathcal{I}^s(P)$. More precisely, $Diff_s(P; Q) \cong Hom_\mathcal{A}(\mathcal{I}^s(P); Q)$. The *FG-category* over $\mathcal{A}$ is the category whose objects are filtered geometric $\mathcal{A}$−modules and whose morphisms are filtered homomorphisms. The *general category of differential equations*, $\mathcal{DE}_g$, is defined by the following: (i) $\mathcal{A} \in Ob(\mathcal{DE}_g)$ iff $\mathcal{A}$ is a filtered algebra $\mathcal{A} \equiv \{\mathcal{A}_i\}$, $\mathcal{A}_i \subset \mathcal{A}_{i+1}$, such that in the differential calculus in the FG−category over $\mathcal{A}$ it is defined a natural operation $\mathcal{C}$ that satisfies $\mathcal{C}\Omega^1 \wedge \Omega^\bullet = \mathcal{C}\Omega^\bullet$, where $\Omega^i \equiv \mathcal{A} \wedge \cdots_i \cdots \wedge \mathcal{A}$ are the representative objects of the functor D_i in the FG−category over $\mathcal{A}$, where $D_i \equiv D \cdots_i \cdots D$, being $D(P)$ the *$\mathcal{A}$-module of all differentiations of algebra $\mathcal{A}$ with values in module P*; (ii) $f \in Hom(\mathcal{DE}_g)$ iff f is a homomorphism of filtered algebras preserving operation $\mathcal{C}$. If $\mathcal{A}_i$ are commutative algebras, we take for example $\mathcal{A}_i \equiv C^\infty(E_{k+i}; \mathbf{K})$, where E_{k+i} is the i-prolongation of a PDE E_k. Then we have the canonical inclusion $j_i : E_{k+i} \to Spec(\mathcal{A}_i) = \{F \in C^\infty(\mathcal{A}_i; \mathbf{K}) | F(f_1 f_2) = F(f_1)F(f_2), \forall f_i \in \mathcal{A}_i\}$, $x \mapsto j_i(x) \equiv e_x \equiv$ evaluation map at $x \in E_{k+i}$. Then, to the inclusion $\mathcal{A}_i \subset \mathcal{A}_{i+1}$ corresponds the projection $E_{k+i+1} \to E_{k+i}$. So we set $E_\infty = \varprojlim E_{k+i}$. We can see that $\overline{E_\infty} = Spec(\mathcal{A}_\infty)$. Similar considerations can be made by starting with quantum algebras. Therefore, in the *general category of quantum differential equations* $\hat{\mathcal{D}\mathcal{E}}_g$ a quantum PDE is represented by a filtered quantum algebra $\hat{\mathcal{A}} \equiv \{\hat{\mathcal{A}}_i\}$. Now, by considering the Dirac covariant quantization of PDE's, with respect to physical observables identified by means of Dirac measures on the base manifolds M, we get that the corresponding filtered quantum algebras $\hat{\mathcal{B}}$, identify filtered quantum modules. Therefore, the corresponding FG-category is well identified and, as a by-product, it follows that the quantum algebras $\hat{\mathcal{B}}$ identify quantum PDE's in the category $\hat{\mathcal{D}\mathcal{E}}_g$. Therefore, we get also a functor $Q : \underline{\mathcal{D}\mathcal{E}} \to \hat{\mathcal{D}\mathcal{E}}_g$, that we call *Dirac covariant quantization* too, (or simply *Dirac quantization*), with $\underline{\mathcal{D}\mathcal{E}}$ the category where the objects are infinity prolongations of formally integrable PDE's. We say that two QPDE's $\hat{E}_k$, $\hat{E}'_{k'}$ are *algebraic equivalent* if their filtered quantum algebras $\hat{\mathcal{B}}$, $\hat{\mathcal{B}}'$, are isomorphic. Let us denote by $[\hat{E}_k]_g$ the corresponding equivalence class in the category of formally integrable quantum differential equations $\underline{\hat{\mathcal{D}\mathcal{E}}}$. We call $[\hat{E}_k]_g$ the *general equivalence class* of $\hat{E}_k$. Then the Dirac covariant quantization of a PDE E_k identifies an equivalence class $[\hat{E}_k]_g$ in $\underline{\hat{\mathcal{D}\mathcal{E}}}$. $\qquad\qquad\square$

Remark 3.26. Note, that the method of covariant quantization given in this section, by means of Jacobi equation clarifies the linking between quantification and deformation of original structure (i.e., PDE). Furthermore, taking into account the relation between generalized Green kernels and singular solutions, we see that multivalued solutions of Jacobi equation govern the covariant and canonical quantizations. Finally, note that the knowledge of Bäcklunds of PDEs can aid the procedure of quantization. (For details see ref.[61].) $\qquad\qquad\blacksquare$

Example 3.58. In ref.[61] the canonical quantization of unharmonic oscillator is shown. More generally it also produces the canonical quantization of Newton-equation. Note that the condition in order to apply the covariant and canonical quantizations of PDE's $E_k \subset J\mathcal{D}^k(W)$ is that the corresponding 0-order propagators $G_0[s]$, belong to $C_0^\infty(E\boxtimes E')'$, with $E \equiv s^*vTW$, for any solution s. These conditions are surely verified for the PDE's of the field theory, where W is an Euclidean vector bundle or an Euclidean affine bundle, and PDE's are obtained by means of differential operators $\kappa : J\mathcal{D}^k(W) \to W$. (See, e.g., the next examples.) In particular, our geometric method allows us to reproduce the canonical quantizations of the usual linear PDEs of field theories, like Klein-Gordon, Dirac, Maxwell equations. ∎

Example 3.59. (*Quantum fluids: Canonical quantization of Euler equaton and Navier-Stokes equation*). In this example we shall simply state that also for the classic nonlinear equations (E) and (NS) we can recognize a procedure of covariant, or canonical, quantization and to obtain a description of quantum fluids. Let us recall that (E) is formally integrable, but (NS) is not so. (See Example 3.5(4) and refs.[56,65].) So we can apply Theorem 3.89 directly to (E). Furthermore, equation (NS) is universally formally integrable, i.e., there exists a subequation $\widehat{(NS)} \subset (NS)$, that is formally integrable and that has the same regular solutions of (NS), i.e., $\widehat{(NS)}_{+\infty} \cong (NS)_{+\infty}$. Therefore, we can apply above theorems on the covariant and canonical quantizations of PDE's just to $\widehat{(NS)}$ in order to quantize (NS). The Jacobi equations $(E)[s]$ and $\widehat{(NS)}[s]$ that govern the covariant and canonical quantizations of the Euler equation and Navier-Stokes equation respectively, are reported in Tab.3.10 and Tab.3.11. Note that for the (E) the section $s = (v,p)$ and the corresponding perturbed term is (ν^j,μ), $1 \le j \le 3$. Furthermore for (NS) the section $s = (v,p,\theta)$ and the corresponding perturbed term is (ν^j,μ,ϵ), $1 \le j \le 3$. Then, the covariant quantization of physical observables representing the components of a section s is obtained by means of the following quantum bracket: $[\widehat{s}^i(x,t),\widehat{s}^j(x',t')] = -i\hbar\widetilde{G}^{ij}(x,t;x',t';s)$, where $\widetilde{G}^{ij}(x,t;x',t';s)$ is the propagator of $(E)[s]$ or $\widehat{(NS)}[s]$. The derivation with respect to t' and taking $t' = t$ gives: $[\widehat{s}^i(x,t),\dot{\widehat{s}}^j(x',t)] = -i\hbar(\partial t'.\widetilde{G}^{ij}(x,t;x',t';s))|_{t'=t}$. Furthermore, taking into account that the propagators satisfy the boundary conditions:

$$(\partial t'.\widetilde{G}^{ij}(x,t;x',t';s))|_{t'=t} = \delta(x,x')\delta^{ij},$$

we get: $[\widehat{s}^i(x,t),\dot{\widehat{s}}^j(x',t)] = -i\hbar\delta(x,x')\delta^{ij}$ that is the desired canonical quantization of (E) and (NS) respectively.   ∎

TAB.3.10 - Jacobi equation of (E)

$$
\text{(a) } G^{j}_{jk}\nu^{k}+(\partial x_{i}.\nu^{i})=J[s]
$$

$$
\begin{aligned}
\text{(b) } &\rho[2G^{j}_{ik}v^{k}+(\partial x_{i}.v^{j})-(\partial v_{i}.(\rho B^{j}))]\nu^{i} \\
&+[\rho\delta^{j}_{k}-(\partial v^{0}_{k}.(\rho B^{j}))](\partial x_{0}.\nu^{k}) \\
&+[\rho v^{i}\delta^{j}_{k}-(\partial v^{i}_{k}.(\rho B^{j}))](\partial x_{i}.\nu^{k})+(\partial x_{i}.\mu)g^{ij}=J^{j}[s]
\end{aligned}
$$

TAB.3.11 - Jacobi equation of (NS)

$$
\text{(a) } G^{j}_{jk}\nu^{k}+(\partial x_{i}.\nu^{i})=J[s]
$$

$$
\text{(b) } (\partial x_{\alpha}.G^{j}_{jk})\nu^{k}+G^{j}_{jk}(\partial x_{\alpha}.\nu^{k})+(\partial x_{\alpha}(\partial x_{i}.\nu^{i}))=J_{\alpha}[s]
$$

$$
\begin{aligned}
\text{(c) } &\nu^{s}\{2\rho G^{i}_{sk}v^{k}+(\partial x_{s}.v^{j})-2\chi[G^{i}_{ik}(\partial x_{s}.g^{kj}) \\
&+G^{j}_{ik}(\partial x_{s}.g^{ki})(\partial v_{s}.\rho B^{j})+(\partial x_{i}.\partial x_{s}.g^{ij})]\} \\
&+(\partial x_{0}.\nu^{i})\{\rho\delta^{j}_{i}-(\partial v_{0}.\rho B^{j})\}+(\partial x_{s}..\nu^{k}) \\
&\{v^{s}\delta^{j}_{k}-2\chi[-G^{i}_{ik}g^{js}-G^{i}_{ip}g^{ps}\delta^{j}_{k}-2G^{i}_{ik}g^{is}-(\partial x_{i}.g^{is})\delta^{j}_{k}] \\
&-(\partial v^{s}_{k}.\rho B^{j})\}+(\partial x_{i}.\mu)g^{ij}+(\partial x_{i}.\partial x_{s}.v^{k})\{2\chi[g^{js}\delta^{i}_{k}-g^{is}\delta^{j}_{k}]=J[s]^{j}
\end{aligned}
$$

$$
\begin{aligned}
\text{(d) } &\nu^{s}\{\rho C_{p}(\partial x_{s}.\theta)-2\chi[G^{j}_{ik}(\partial x_{s}.g^{pi})v^{k}g_{jp}-G^{j}_{is}(\partial x_{p}.v^{k})g_{kq}g^{qp} \\
&-G^{j}_{is}(\partial x_{j}.v^{i})+(\partial x_{i}.v^{k})(\partial x_{s}.g^{pi})g_{kp}]\} \\
&+(\partial x_{s}.\nu^{k})\{-2\chi[-G^{j}_{iq}v^{q}g_{kj}g^{is}-2(\partial x_{p}.v^{i})g_{ki}g^{sp} \\
&-2(\partial x_{k}.v^{s})-G^{s}_{ik}v^{i}+v^{p}(\partial x_{p}.g^{is})g_{ki}]\} \\
&+(\partial x_{0}.\nu^{i})\rho C_{p}+(\partial x_{k}.\epsilon)\{2\rho C_{p}v^{k}-\phi[G^{j}_{ji}g^{ki}+(\partial x_{i}.g^{ik})]\} \\
&-(\partial x_{i}.\partial x_{s}.\epsilon)\phi g^{is}=\overline{J}[s]
\end{aligned}
$$

Example 3.60. (*Generalized Klein-Gordon equation*). In this example we shall consider scalar fields. In order to include in our formulation also scalar massive neutrinos we will introduce a generalized Klein-Gordon equation on an hyperbolic space-time and, after studied its geometrical structure, we will obtain the covariant and canonical quantizations. We shall emphasize that our geometric approach preserves the microscopic causality even if particles are massive neutrinos. Let M be a globally hyperbolic space-time. The *generalized Klein-Gordon* equation is the submanifold $(GKG)_{\chi} \subset J\mathcal{D}^{2}(E^{c})$, $\pi : E^{c} \equiv M \times \mathbf{C} \to M$, obtained as the kernel of the following fiber bundle morphism: $\mathcal{K}_\chi \equiv (\Box + \chi) : J\mathcal{D}^2(E^c) \to E^c$, where $\Box$ is the d'Alembertian for scalar fields with respect to the metric g on M, $\chi \equiv \xi R + \bar\chi$, with $\xi \in \mathbf{R}$, R is the Ricci scalar curvature and $\bar\chi \in \mathbf{R}$, (square of *mass*). The numerical factor ξ has two values of particular interest: the so-called *minimallly complet case*, $\xi = 0$, and the *conformally complete case*, $\xi = \frac{1}{6}$. In this latter case, if $m = 0$ the field equation $\left[\Box + \frac{R}{6}\right]\phi_{\pi_0} = 0$ for free pion π_0^+ is conformal invariant. Note that the parameter χ can be, in general, considered a costant-coupling of self-interaction. In the particular case of bradions and luxons, is $\bar\chi \geq 0$. In the case of massive neutrinos, instead, one has $\bar\chi < 0$. We call the numerical factor χ the *geometric mass* of the field. We can have $\chi \gtrless 0$. In adapted coordinates $(GKG)_\chi$ can be locally written as follows:

$$(3.28) \qquad \left\{ F \equiv g^{\alpha\beta} z_{\alpha\beta} - \lceil^{\alpha}_{\beta\gamma} g^{\gamma\beta} z_\alpha + \chi z = 0 \right\},$$

where $(x^\alpha, z, z_\alpha, z_{\alpha\beta})$ are coordinates on $J\mathcal{D}^2(E^c)$ induced by fibered coordinates (x^α, z) on $\pi : E^c \to M$. Furthermore, $\lceil^{\alpha}_{\beta\gamma} : U \subset M \to \mathbf{R}$ denote the connection coefficients induced by the metric g. Note that equation (3.28) is equivalent to the product two times of a same equation: $(GKG)_\chi \cong (GKG)^{\mathbf{R}}_\chi \times (GKG)^{\mathbf{R}}_\chi \subset J\mathcal{D}^2(E) \times J\mathcal{D}^2(E) \cong J\mathcal{D}^2(E^c)$, where $E = M \times \mathbf{R} \to M$, ($E^c$ is just the complexification of E: $E^c = \mathbf{C} \otimes_{\mathbf{R}} E$), and $(GKG)^{\mathbf{R}}_\chi = \ker(\mathcal{K}^{\mathbf{R}}_\chi) \subset J\mathcal{D}^2(E)$, with $\mathcal{K}^{\mathbf{R}}_\chi \equiv (\Box + \chi) : J\mathcal{D}^2(E) \to E$. Thus in order to discuss the equation $(GKG)_\chi$ it is enough to consider $(GKG)^{\mathbf{R}}_\chi \subset J\mathcal{D}^2(E)$, that, in real coordinates $(x^\alpha, y, y_\alpha, y_{\alpha\beta})$ on $J\mathcal{D}^2(E)$, looks like $F^{\mathbf{R}} \equiv g^{\alpha\beta} y_{\alpha\beta} - \lceil^{\alpha}_{\beta\gamma} g^{\gamma\beta} y_\alpha + \chi y = 0$. As $\mathcal{K}^{\mathbf{R}}_\chi$ is an epimorhism of constant rank 5, it follows that $(GKG)^{\mathbf{R}}_\chi$ is a vector sub-bundle of $J\mathcal{D}^2(E)$ of dimension $19 - 1 = 18$. The Cartan distribution $\mathbb{E}_2$ of $J\mathcal{D}^2(E) \cong M \times \mathbf{R} \times \mathbf{R}^4 \times \mathbf{R}^{10} \cong M \times \mathbf{R}^{15}$, is a distribution of dimension 14: $\mathbb{E}_2 \subset TJ\mathcal{D}^2(E)$. The Cartan distribution of $(GKG)^{\mathbf{R}}_\chi$ is a sub-distribution $\mathbb{E}_2(GKG)^{\mathbf{R}}_\chi$ of $\mathbb{E}_2$ of dimension 13 generated by the vector fields $\zeta = X^\alpha \left(\partial x_\alpha + y_\alpha \partial y + y_{\alpha\beta} \partial y^\beta \right) + Y_{\alpha\beta} \partial y^{\alpha\beta}$ on $J\mathcal{D}^2(E)$ such that: $X^\omega \left[g^{\alpha\beta}_{,\omega} y_{\alpha\beta} - (\lceil^{\alpha}_{\beta\gamma,\omega} g^{\gamma\beta} + \lceil^{\alpha}_{\beta\gamma} g^{\gamma\beta}_{,\omega}) y_\alpha \right] + \chi(y_\alpha X^\alpha) - \lceil^{\alpha}_{\beta\gamma} g^{\gamma\beta}(y_{\alpha\delta} X^\delta) + g^{\alpha\beta} Y_{\alpha\beta} = 0$. Then the solutions of $(GKG)^{\mathbf{R}}_\chi$ are 4-dimensional integral manifolds of $\mathbb{E}_2(GKG)^{\mathbf{R}}_\chi$ that, except for a subset of dimension lower than 4, are diffeomorphic projected on 4-dimensional submanifolds of E. As each $(GKG)^{\mathbf{R}}_\chi$ is an involutive formally integrable, completely integrable PDE, it follows that in the neighbourhood of each point of $(GKG)^{\mathbf{R}}_\chi$ we can build a regular solution, i.e., a local 4-dimensional integral submanifold, diffeomorphic to an open set of M, by means of the projection $\pi_2 : J\mathcal{D}^2(E) \to M$. For the equation $(GKG)_\chi^{\mathbf{R}}$ we can solve the Cauchy problem by means of characteristics. In fact, let us rewrite the equation $(GKG)_\chi^{\mathbf{R}}$ in orthogonal coordinates: $F \equiv u_{00} - u_{xx} - u_{yy} - u_{zz}\chi u = 0$. Therefore, $\dim(GKG)_\chi^{\mathbf{R}} = \dim J\mathcal{D}^2(E) - 1 = 19 - 1 = 18$. A characteristic vector field of $(GKG)_\chi^{\mathbf{R}}$ is a vector field $\zeta = X^\alpha \left(\partial x_\alpha + u_\alpha \partial u + u_{\alpha\beta}\partial u^\beta + u_{\alpha\beta\gamma}\partial u^{\beta\gamma} \right)$, where X^α are functions defined by the following equations:

$$\left\{ \begin{array}{l} Y_{00} - Y_{11} - Y_{22} - Y_{33} = 0 \\[2mm] 0 = \zeta \rfloor \delta(\theta) \equiv \zeta \rfloor \frac{1}{2} Y_{\alpha\beta} \left(dx^\alpha \otimes dx^\beta + dx^\beta \otimes dx^\alpha \right) \end{array} \right\}$$

and the other coordinates functions in the expression of ζ are constrained to satisfy the first prolongation of $(GKG)_\chi^{\mathbf{R}}$:

$$((GKG)_\chi^{\mathbf{R}})_{+1} : \left\{ \begin{array}{l} u_{00\alpha} - u_{11\alpha} - u_{22\alpha} - u_{33\alpha} + \chi u_\alpha = 0 \\[2mm] u_{00} - u_{11} - u_{22} - u_{33} + \chi u = 0 \end{array} \right\}.$$

Then we can see that a characteristic strip can be the following: $X^1 = X^2 = X^3 = 0$, $X^0 \neq 0$. In other words:

(3.29)

$$\zeta = X^0 \left[\partial x_0 + u_0 \partial u + (u_{11} + u_{22} + u_{33} - \chi u)\partial u^0 + u_{0k}\partial u^k + (u_{11\gamma} + u_{22\gamma} + u_{33\gamma} - \chi u_\gamma)\partial u^{0\gamma} + u_{0\chi\gamma}\partial u^{\chi\gamma} \right].$$

One can also directly see that ζ is tangent to $(GKG)_\chi^{\mathbf{R}}$ and it belongs to the Cartan distribution!! Furthermore, we can see also that ζ can be considered the characteristic distribution for the following subequation $(E_2) \subset (GKG)_\chi^{\mathbf{R}}$:

$$(E_2) \subset (GKG)_\chi^{\mathbf{R}} : \left\{ \begin{array}{l} u_{00} - (u_{11} + u_{22} + u_{33}) + \chi u = 0 \\[2mm] u_{11} + u_{22} + u_{33} = 0 \\[2mm] u_{\alpha\beta} = 0, \quad \alpha \neq \beta. \end{array} \right\}.$$

For any Cauchy data N of (E_2), transversal to ζ, given by (3.29) we can generate a 4-dimensional integral manifold of (E_2) that is contained into $(GKG)_\chi^{\mathbf{R}}$, hence it is a solution of $(GKG)_\chi^{\mathbf{R}}$. In particular if $N = D^2 s(N_0)$, where $N_0 \subset M$ is a space-like submanifold of M, and s is a solution of $(GKG)_\chi^{\mathbf{R}}$, then $Y \equiv \bigcup_{t \in \mathbf{R}} \phi_t(N)$ is a regular solution of $(GKG)_\chi^{\mathbf{R}} \subset$

$J\mathcal{D}^2(E)$, where ϕ is the flow generated by ζ on $(GKG)^{\mathbf{R}}_\chi$. Furthermore, let ζ be a vector field on $(GKG)^{\mathbf{R}}_\chi$ that represents an infinitesimal symmetry of this equation. If $N \subset (GKG)^{\mathbf{R}}_\chi$ is a Cauchy hypersurface trasnversal to ζ, then $Y \equiv \bigcup_{t\in J} \phi_t(N)$, $\partial\phi = \zeta$, is a solution of $(GKG)^{\mathbf{R}}_\chi$, for a suitable neighborhood J of $0 \in \mathbf{R}$. (This is a direct application of a general property of PDEs.) Furthermore, by using the same calculations given in ref.[56] to obtain the infinitesimal symmetry algebra for the Klein-Gordon equation, we can see that also for the generalized Klein-Gordon equation the infinitesimal symmetry algebra $\mathfrak{s}((GKG)^{\mathbf{R}}_\chi)$ of $(GKG)^{\mathbf{R}}_\chi$ is generated by the second order holonomic prolongation of the following vector fields $\check\zeta : X^\alpha\partial x_\alpha + fy\partial y : E|_U \to T(E|_U)$, where X^α and f are local numerical functions on M solutions of the linear PDE reported in Tab.3.5.

TAB.3.12 - Equations for (X^α)

$$\Box f - \frac{\chi}{\operatorname{tr}(g)}\operatorname{tr}(\mathcal{L}_{\bar\zeta}g)=0$$

$$-(\partial x_\alpha\partial x_\beta.X^\omega)g^{\alpha\beta}-(\partial x_\alpha.X^\omega)h^\alpha-X^\alpha(\partial x_\alpha.h^\omega)+2fh^\omega+2(\partial x_\sigma.f)g^{\omega\sigma}+\frac{h^\omega\chi}{\operatorname{tr}(g)}\operatorname{tr}(\mathcal{L}_{\bar\zeta}g)=0$$

$$\bar\zeta\equiv X^\alpha\partial x_\alpha:U\subset M\to TM,\quad h^\omega\equiv\Gamma^\omega_{\beta\gamma}g^{\gamma\beta}:U\subset M\to\mathbf{R}$$

Then $(GKG)^{\mathbf{R}}_\chi$ admits the following covariant quantization for real scalar fields: ($\spadesuit\spadesuit$): $[\hat s(x), \hat s(x')] = i\hbar\widetilde{G}(x;x'|\chi)\,\mathbf{1}$, where $\mathbf{1} = id_\mathcal{H}$, for some suitable locally convex vector space $\mathcal{H}$. The propagator $\widetilde{G}$ is the Green kernel solution of the following Cauchy problem:

$$(3.30)\qquad\left\{\begin{array}{l}(\Box_{x'}+\chi)\widetilde{G}(x;x'|\chi)=0\\[4pt]\widetilde{G}(x;x'|\chi)_{x^0=x'^0}=0\\[4pt]\left(\partial x'_0.\widetilde{G}\right)(x;x'|\chi)|_{x^0=x'^0}=-\delta(x;x')|_{x^0=x'^0}\end{array}\right\}.$$

Note that the second condition in (3.30) implies that we exlcude instantaneous propagations, so to conserve the usual microscopic causality. One has the following equal time commutator:

$$(3.31)\qquad\left\{\begin{array}{l}[\hat s(x),\hat s(x')]|_{x^0=x'^0}=0\\[4pt][\hat s(x),\dot{\hat s}(x')]|_{x^0=x'^0}=-i\hbar\delta(x;x')|_{x^0=x'^0}\,\mathbf{1}\end{array}\right\}.$$

Moreover, the PDE $(GKG)_\chi$ admits the following covariant quantization of the complex scalar fields $s = s_1 + is_2$: $[\hat s_i(x), \hat s_j(x')] = i\hbar\delta_{ij}\widetilde{G}(x;x'|\chi)\,\mathbf{1}$,

$i, j = 1, 2$; hence we get the following canonical quantization: $[\hat{s}_i(x), \dot{\hat{s}}_j(x')] = -i\hbar\delta_{ij}\delta(x; x')|_{s^0 = x'^0}\,\mathbf{1}$, $i, j = 1, 2$. Furthermore, we get also for the full complex field $s = s_1 + is_2$ and its c.c. $\bar{s} = s_1 - is_2$ the following commutation relations:

$$\left\{ \begin{array}{l} [\hat{s}(x), \hat{s}(x')] = [\widehat{\bar{s}}(x), \widehat{\bar{s}}(x')] = 0 \\[2mm] [\hat{s}(x), \widehat{\bar{s}}(x')] = 2i\hbar\widetilde{G}(x; x'|\chi)\,\mathbf{1} \end{array} \right\} \;\Rightarrow\; \left\{ \begin{array}{l} [\hat{s}(x), \dot{\hat{s}}(x')] = [\widehat{\bar{s}}(x), \dot{\widehat{\bar{s}}}(x')] = 0 \\[2mm] [\hat{s}(x), \dot{\widehat{\bar{s}}}(x')]|_{x^0 = x'^0} = -2i\hbar\delta(x; x')|_{x^0 = x'^0}\,\mathbf{1} \end{array} \right\}.$$

The microscopic causality is conserved also for scalar massive neutrinos (takions) having a geometric mass $\chi \equiv \xi R + \bar{\chi} < 0$. Note that the Green kernel $G(x; x'|\chi)$ of the generalized Klein-Gordon equation is a solution of the following equation $(\Box_{x'} + \chi)\,G(x; x'|\chi) = \delta(x, x')$. In a geodesically convex domain, there exist two fundamental solutions G^+ and G^- which vanish outside the future cone $\mathcal{E}_x^+$ and outside the past cone $\mathcal{E}_x^-$ respectively. Then, the propagator is given by $\widetilde{G}(x; x'|\chi) = G^+(x'; x|\chi) - G^-(x'; x|\chi)$. $\widetilde{G}(x; x'|\chi)$, $G^+(x; x'|\chi)$, $G^-(x; x'|\chi)$ are real and $G^+(x; x'|\chi) = G^-(x'; x|\chi)$. It follows that $\widetilde{G}(x; x'|\chi) = -\widetilde{G}(x'; x|\chi)$. The solution of (3.29), in the Minkowsky space-time, is the following for $\chi \neq 0$,

$$\widetilde{G}(x; x'|\chi) = \tfrac{1}{4\pi r}(\partial r. J_0(\sigma(\chi)\sqrt{|\chi|(t^2 - r^2)}))H(t+r) - \tfrac{1}{4\pi r}(\partial r. J_0(\sigma(\chi)\sqrt{\chi(t^2 - r^2)}))H(r-t)$$

and for $\chi = 0$,

$$\widetilde{G}(x; x'|\chi) = -\frac{1}{4\pi r}[\delta(r - t) - \delta(r + t)],$$

with $r \equiv |\mathbf{x}' - \mathbf{x}|$, $t \equiv x'^0 - x^0$, J_0 the regular Bessel function and $H(\xi)$ is the Heaviside function. (Here we use the notation $(x^\alpha) = (x^0, \mathbf{x}) = (x^0, x^1, x^2, x^3)$.) Furthermore, $\sigma(\chi) = +1$ if $\chi > 0$ and $\sigma(\chi) = -1$ if $\chi < 0$. Let us emphasize, also, that, even if the geometric mass χ is negative, the generalized Klein-Gordon operator $\Box + \chi$ remains of hyperbolic type, so it admits a propagator $\widetilde{G}(x, x'|\chi)$ with support in $\mathcal{E}^-(x') \cup \mathcal{E}^+(x')$, $\forall x' \in M$. Therefore, the quantum commutator $[\hat{s}(x), \hat{s}(x')] = i\hbar\widetilde{G}(x, x'|\chi)$ respects the microscopic causality. Finally note that the integral bordism group $\Omega_3^{(GKG)_\chi^{\mathbf{R}}}$ of the generalized Klein-Gordon equation is trivial: $\Omega_3^{(GKG)_\chi^{\mathbf{R}}} = 0$. Therefore are admmissible tunnel effects, i.e., solutions with change of sectional topology. The proof can be directly obtained appling some general theorems given by us in order to calculate integral bordism groups in PDEs, and taking into account that $_f(GKG)_\chi^{\mathbf{R}}$ is a conic equation with trivial cohomoly: $H^s((GKG)_\chi^{\mathbf{R}}) = 0$, $\forall s \neq 0$, $H^0((GKG^{\mathbf{R}})_\chi) = \mathbf{R}$. If the scalar field is in interaction with some other field such that it produces a current $f \in C^\infty(E)$, i.e., the PDE considered is the following affine PDE:

$$_f(GKG)_\chi^{\mathbf{R}} \subset J\mathcal{D}^2(E) : \quad \left\{ g^{\alpha\beta} y_{\alpha\beta} - \lceil_{\beta\gamma}^\alpha g^{\gamma\beta} y_\alpha + \chi y = f \right\}.$$

We can associate a generalized propagator to any singular solution V of $_f(GKG)_\chi^{\mathbf{R}}$, (that realizes a tunnel effect, if $\partial V = N_1 \bigcup N_2$, with $\pi_n(N_1) \neq \pi_n(N_2)$, for some $n \geq 0$, $n \in \mathbf{N}$, where $\pi_n(-)$ are the Hurewitz homotopy group-functors). More precisely, if V is the integral manifold (quantum cobord) cobording two 3-dimensional admissible integral manifolds N_0 and N_1 contained into $_f(GKG)_\chi^{\mathbf{R}}$, such that the mapping $\pi_2|_V : V \to M$ is a proper application and $\omega_1^{(2)}(V) = 0$, where $\omega_1^{(2)}(V)$ is the characteristic of Stiefel-Whitney of V, then the propagator (generalized Green kernel) $G[V]$ between N_0 and N_1 is identified with the following kernel

$$G[V] \in (C_0^\infty(E' \boxtimes E'))',$$

given by

$$G[V](\alpha \otimes \phi) = F[V](\alpha) \int_M < \phi, f >,$$

where

$$E' \equiv E^* \bigotimes \Lambda_4^0 M,$$

and

$$F[V] \in C_0^\infty(E')$$

is the distribution associated to V. The physical intepretation for the generalized propagator $G[V]$, $\partial V = N_0 \bigcup N_1$, is that it represents the amplitude probability $< N_0|N_1 >$ for the transition from an extendon N_0 to another N_1. (Compare, e.g., with the exposition of extendons given in ref.[96(i)].) More precisely $< N_0|N_1 >= G[V](\alpha \otimes \phi)$, where α and ϕ represent the physical states of N_0 and N_1 respectively. ∎

Example 3.61. (*Generalized Dirac equation*). Here we want to formulate a generalized Dirac equation in order to include also fermionic massive neutrinos-takions. Let (M, g) be as before a 4-dimensional space-time and let $\pi : E \to M$ be a vector bundle over M identified with the complexificated Clifford bundle over M: $E \equiv \bigcup_{p \in M} E_p$,

$$E_p \equiv \bigoplus_{s \geq 0} T_0^s(\mathbf{C} \bigotimes_{\mathbf{R}} T_p M)/\mathbb{I}_p,$$

where $\mathbb{I}_p$ is the subspace of $\bigoplus_{s \geq 0} T_0^s(\mathbf{C} \otimes_{\mathbf{R}} T_p M)$ generated by elements of the type $v \otimes v - g^c(v, v)\mathbf{1}$, where $\mathbf{1}$ is the unity of $\mathbf{R}$ and g^c the scalar product on $\mathbf{C} \otimes_{\mathbf{R}} T_p M$ defined by $g^c(z_1 \otimes v_1, z_2 \otimes v_2) = z_1 z_2 \otimes g(v_1, v_2)$. Then, each fiber E_p becomes a complexificated Clifford algebra, hence it is possible to define a product (Clifford product) on E_p. The Levi-Civita connection on M lifts on E. The Clifford connection is flat iff the space-time manifold M is flat. This is equivalent to say that C_1 is completely integrable iff (M, g) is flat. In fact, let us recall that the Clifford connection is a first order PDE on the fiber bundle $\pi : E \to M$, $C_1 \subset J\mathcal{D}(E)$, identified by a section $\rceil_C : E \to J\mathcal{D}(E)$ of $\pi_{1,0} : J\mathcal{D}(E) \to E$, such that the following diagram is commutative:

$$
\begin{array}{ccccccccc}
& & C_1 & & \cong & & E & & \\
& & \| & & & & \| & & \\
& & & & & & \rceil_C & & \\
JD(\gamma) & & C_1 & \subset & J\mathcal{D}(E) & \overset{\longleftarrow}{} & E & \to & M \\
& & \uparrow & & \uparrow J\mathcal{D}(\gamma) & & \uparrow \gamma & & \| \\
& & (LC)_1 & \subset & J\mathcal{D}(TM) & \underset{\rceil_{LC}}{\longleftarrow} & TM & \to & M \\
& & \| & & & & \| & & \\
& & (LC)_1 & & \cong & & TM & &
\end{array}
$$

where $(LC)_1$ is the canonical connection on TM identified by means of the metric g. Furthermore, γ is the canonical monomorphism of vector bundles over M given by composition: $TM \to TM \otimes \mathbf{C} \to E$. If $\{x^\alpha\}$ is a coordinate system on M and ∂x_α is the natural basis induced on the tangent bundle, the monomorphism γ induces a set of sections of E denoted by $\gamma_\alpha \equiv \gamma(\partial x_\alpha)$. Set $\gamma_{\alpha_1 \dots \alpha_p} \equiv \gamma_{\alpha_1} \dots \gamma_{\alpha_p}$. We get the following relation

$$\gamma_{\alpha_1 \dots \alpha_p} = (-1)^q \gamma_{\alpha_1} \dots \gamma_{\alpha_{p-q-1}} \gamma_{\alpha_p} \gamma_{\alpha_{p-q}} \dots \gamma_{\alpha_{p-1}}$$
$$+ \sum_{1 \leq r \leq q} (-1)^{r+1} 2 g_{\alpha_p \alpha_{p-r}} \gamma_{\alpha_1} \dots \widehat{\gamma_{\alpha_{p-r}}} \dots \gamma_{\alpha_{p-1}}.$$

The set $\{1, \gamma_{\alpha_1 \dots \alpha_p}\}_{0 \leq \alpha_1 < \dots < \alpha_p \leq 3, 1 \leq p \leq 4}$ is a local basis for sections of $\pi : E \to M$. Hence, if $\psi : M \to E$ is such a (local) section, we get the following local representation in the natural basis induced by the coordinate system $\{x^\alpha\}$ on M:

$$\psi = \phi 1 + \psi^\alpha \gamma_\alpha$$

$$+ \cdots + \sum_{0 \le \alpha_1 < \ldots < \alpha_p \le 3} \psi^{\alpha_1 \ldots \alpha_p} \gamma_{\alpha_1 \ldots \alpha_p} + \cdots + \psi^{0123} \gamma_{0123}$$

$$\equiv \psi^B \gamma_B,$$

where $(\gamma_B) \equiv (1, \gamma_\alpha, \ldots, \gamma_{\alpha_1 \ldots \alpha_p}, \ldots, \gamma_{0123})$, $0 \le \alpha_1 < \alpha_2 < \ldots < \alpha_p \le 3$. $(\psi^B) \equiv (\phi, \psi^\alpha, \ldots, \psi^{0123})$ are $\mathbf{C}$-valued local functions on M. Therefore $\dim_{\mathbf{C}} E_p = 16$, where E_p is the fiber of E over $p \in M$. As a consequence we get the following induced fibered system of coordinates on E: $\{x^\alpha, z, z^B\} \equiv \{x^\alpha, z, z^\alpha, \ldots, z^{\alpha_1 \ldots \alpha_p}, \ldots, z^{0123}\}$, where x^α are $\mathbf{R}$-valued and the other ones are $\mathbf{C}$-valued. If we denote by $\{x^\alpha, \dot{x}^\alpha\}$ the induced fibered coordinates system on TM, the monomorphism γ has the local expression given in Tab.3.13.

TAB.3.13 - Local expression of γ

$$\boxed{x^\alpha \circ \gamma = x^\alpha; \ z \circ \gamma = 0; \ z^\alpha \circ \gamma = \dot{x}^\alpha; \ \cdots; \ z^{\alpha_1 \cdots \alpha_p} \circ \gamma = 0; \ \cdots; \ z^{0123} \circ \gamma = 0}$$

We will denote respectively by $\{x^\alpha, \dot{x}^\alpha, \dot{x}^\alpha_\beta\}$ and $\{x^\alpha, z, z^B, z_\beta, z^B_\beta\}$ the induced coordinates on $J\mathcal{D}(TM)$ and $J\mathcal{D}(E)$. Then, the local expression of $J\mathcal{D}(\gamma)$ can be written as given in Tab.3.14.

TAB.3.14 - Local expression of $J\mathcal{D}(\gamma)$

$$\boxed{x^\alpha \circ J\mathcal{D}(\gamma) = x^\alpha; \ z \circ J\mathcal{D}(\gamma) = 0; \ z^\alpha \circ J\mathcal{D}(\gamma) = \dot{x}^\alpha; \ \cdots; \ ; z^{\alpha_1 \cdots \alpha_p} \circ J\mathcal{D}(\gamma) = 0; \ \cdots; \ z^{0123} \circ J\mathcal{D}(\gamma) = 0}$$

Furthermore, the Levi-Civita connection can be written as follows:

$$(LC)_1 \subset J\mathcal{D}(TM): \quad \{\dot{x}^\alpha_\beta + \dot{x}^\gamma \lceil^\alpha_{\beta\gamma} = 0\},$$

where $\lceil^\alpha_{\beta\gamma} = g^{\alpha\delta}[\beta\gamma, \delta] = \frac{1}{2} g^{\alpha\delta}[(\partial x_\gamma . g_{\beta\delta}) + (\partial x_\beta . g_{\gamma\delta}) - (\partial x_\delta . g_{\gamma\beta})]$ are local numerical functions on M representing the usual Levi-Civita connection symbols. Then, the induced Clifford connection $C_1 \subset J\mathcal{D}(E)$, is locally written as follows:

$$C_1 \subset J\mathcal{D}(E): \quad \left\{ \begin{array}{l} z^B_\lambda + \underset{C}{\lceil}^B_{\lambda A} z^A = 0 \\[2ex] \underset{C}{\lceil}^B_{\lambda A} \equiv \underset{C}{\lceil}^{\beta_1 \ldots \beta_p}_{\lambda \alpha_1 \ldots \alpha_p} \equiv \lceil^{\beta_1}_{\lambda \alpha_1} \delta^{\beta_2 \ldots \beta_p}_{\alpha_2 \ldots \alpha_p} + \cdots + \lceil^{\beta_p}_{\lambda \alpha_p} \delta^{\beta_1 \ldots \beta_{p-1}}_{\alpha_1 \ldots \alpha_{p-1}} \end{array} \right\},$$

where $\Gamma^{B}_{C\lambda A} \equiv -(z^{B}_{\lambda}\circ]) \circ \gamma_{A} \equiv -]^{B}_{C\lambda}\circ\gamma_{A} \equiv -]^{B}_{C\lambda A}$ are the connection coefficients given as **R**-valued local functions on M. Then, the absolute differential $\nabla_{C}\psi \equiv \Gamma_{C}\circ D\psi : M \to J\mathcal{D}(E) \to T^{*}M\otimes E$ of a section $\psi : M \to E$, locally written $\psi = \phi 1 + \psi^{\alpha}\gamma_{\alpha} + \ldots + \psi^{\alpha_1\ldots\alpha_p}\gamma_{\alpha_1\ldots\alpha_p} + \ldots + \psi^{0123}\gamma_{0123} \equiv \psi^{B}\gamma_{B}$, has the following local expression:

$$(3.32) \qquad \nabla_{C}\psi = (\nabla_{C\lambda}\psi)^{B}dx^{\lambda}\otimes\gamma_{B} = [(\partial x_{\lambda}.\psi^{B}) + \Gamma^{B}_{C\lambda A}\psi^{A}]dx^{\lambda}\otimes\gamma_{B}.$$

By using results on the curvature of PDEs, we can state that the curvature R_{C} of the Clifford connection can be identified with a section $R_{C} : M \to \Lambda^{0}_{2}M\otimes E$ locally written as follows:

$$\left\{\begin{array}{l} R_{C} = \displaystyle\sum_{0\leq\alpha,\beta\leq 3, A, B} R_{C\alpha\beta}{}^{B}{}_{.A}dx^{\alpha}\wedge dx^{\beta}\otimes\gamma_{B}\otimes\gamma^{A} \\[2mm] R_{C\alpha\beta}{}^{B}{}_{.A} = (\partial x_{\alpha}.\Gamma^{B}_{C\beta A}) - (\partial x_{\beta}.\Gamma^{B}_{C\alpha A}) + \Gamma^{B}_{C\alpha D}\Gamma^{D}_{C\beta A} - \Gamma^{B}_{C\beta D}\Gamma^{D}_{C\alpha A} \end{array}\right\}.$$

Here $R_{C\alpha\beta}{}^{B}{}_{.A}$ are **R**-valued local functions on M. It is easy to prove the following formula for any section $\psi = \psi^{B}\gamma_{B}$: $<R,\psi> = R_{C\alpha\beta}{}^{B}{}_{.D}\psi^{D}dx^{\alpha}\wedge dx^{\beta}\otimes\gamma_{B}$, with $R_{C\alpha\beta}{}^{B}{}_{.D}\psi^{D}\gamma_{B} = (\nabla_{C\alpha}\nabla_{C\beta} - \nabla_{C\beta}\nabla_{C\alpha})(\psi^{B}\gamma_{B})$ that relates the curvature of the Clifford connection to the Clifford covariant derivative. Now the curvature of a connection is zero iff the equation representing the connection is completely integrable. On the other hand we can easily compute $R_{C\alpha\beta}{}^{B}{}_{.A}$ and we get the following formula:

$$R_{C\alpha\beta}{}^{B}{}_{.C} \equiv R_{C\alpha\beta}{}^{\delta_1\ldots\delta_p}{}_{.\gamma_1\ldots\gamma_p} = R^{\delta_1}_{\alpha\beta.\gamma_1}\delta^{\delta_2\ldots\delta_p}_{\gamma_2\ldots\gamma_p} + \ldots + R_{\alpha\beta}{}^{\delta_p}{}_{.\gamma_p}\delta^{\delta_1\ldots\delta_{p-1}}_{\gamma_1\ldots\gamma_{p-1}}.$$

Therefore, $R_{C\alpha\beta}{}^{B}{}_{.C} = 0$ iff the Levi-Civita curvature $R_{\alpha\beta}{}^{\delta}{}_{.\gamma} = 0$. Thus the theorem is proved.

Let us, now, M be a space-time such that one has the following isomorphism: $E \cong S^{*}\otimes_{C} S$, for a suitable vector bundle $S \to M$. It is well known [57,63] that if M is endowed with an almost complex structure J such that g is Hermitian, (this appens e.g. if M is an almost Hermitian manifold or an almost Kählerian manifold), and such that it admits a $Spin^{c}$-structure (i.e., the second Stiefel-Whitney class $w_2 \in H^{2}(M;\mathbf{Z}_2)$ is zero), then the bundle

E is canonically isomorphic to $Hom_{\mathbf{C}}(S; S) \cong S^* \otimes_{\mathbf{C}} S$, where $S \equiv \Lambda E_1$, with $E_1 = \ker(\omega - id_E)$, where ω is the involution canonically induced by J. One has $\dim_{\mathbf{C}} S = 2^2 = 4$. $S \to M$ is called the *spinors bundle*. The sections of S over M represent spinor fields with spin $s = \frac{1}{2}$. Then the Clifford connection induces the following linear connection on S:

$$(3.33) \qquad C_1^S \subset J\mathcal{D}(S) : \quad \left\{ \begin{array}{l} y_\lambda^r + \underset{S_\lambda}{\lceil} {}^r_{s} y^s = 0 \\[2mm] \underset{S}{\lceil} {}_\lambda {}^r_s = C^{-1}{}^{rb}_{sa} D_{\lambda}{}^a_b \end{array} \right\}.$$

Here $\underset{S}{\lceil} {}_\lambda {}^r_s$ are $\mathbf{C}$-valued local functions on M and

$$(3.34)$$
$$\left\{ \begin{array}{l} C^{sa}_{rb} \equiv 4\delta^{as}_{rb} - \gamma_{\alpha r}{}^a \gamma^{\alpha}{}^s_b \in L(M(4; \mathbf{C})) \\[2mm] D_{\lambda}{}^a_b \equiv \underset{\lambda}{\lceil}{}^{\beta\delta} \gamma_{\beta s}{}^a \gamma_{\delta b}{}^s - (\partial x_\lambda . \gamma_{\alpha r}{}^a)\gamma^{\alpha}{}^r_b \in M(4; \mathbf{C}), \quad \forall \lambda = 0, 1, 2, 3 \end{array} \right\}.$$

One has the following consistency condition between Levi-Civita connection and γ_α-matrices:

$$(3.35) \qquad 2\lceil^{\beta}_{\lambda\beta} \mathbf{1} = (\partial x_\lambda . \gamma_\alpha)\gamma^\alpha + \gamma^\alpha(\partial x_\lambda . \gamma_\alpha) \in M(4; \mathbf{C}).$$

Moreover by using equation (3.35) we can also write:

$$(3.36) \qquad \underset{S}{\lceil}{}_\lambda = \frac{1}{4}\lceil^{\beta\delta}_{\lambda} \gamma_\beta \gamma_\delta \in M(4; \mathbf{C}).$$

In fact, under our hypotheses there is a unique first order linear connection on S, $C_1^S \subset J\mathcal{D}(S)$, such that the following diagram is commutative:

$$
\begin{array}{ccccccc}
E & \overset{j}{\cong} & S^* \otimes S & \cong & C_1^{S^*} \otimes C_1^S & \cong & C_1 \\
& & \| & & \cap & & \cap \\
& & \overset{*}{\lceil} \otimes \lceil_{\;S\;\;S} & & & & \\
& & S^* \otimes S & \longrightarrow & J\mathcal{D}(S^* \otimes S) & \cong & J\mathcal{D}(E) \\
& & \uparrow \bar{\pi} & & \uparrow J\mathcal{D}(\bar{\pi}) & & \\
& & S^* \times S & \longrightarrow & J\mathcal{D}(S^*) \times J\mathcal{D}(S) & & \\
& & \lceil_{\;S} \times \lceil_{\;S} & & & & \\
& & \| & & \cup & & \\
& & S^* \times S & \cong & C_1^{S^*} \times C_1^S & &
\end{array}
$$

In order to give a local representation of the isomorphism $C_1 \cong C_1^{S^*} \otimes C_1^S$, let us first locally characterize the isomorphism $j : C \cong L(S) \cong S^* \otimes S$.

348   *Quantized Partial Differential Equations*

We get $j(\psi = \phi 1 + \phi^B \gamma_B) = \phi j(1) + \phi^B j(\gamma_B)$, where $j(1) = id_S \in L(S)$, $j(\gamma_B) = j(\gamma_{\alpha_1 \ldots \alpha_p}) = j(\gamma_{\alpha_1}) \ldots j(\gamma_{\alpha_p})$, with $j(\gamma_\alpha) \in L(S)$. Hence if $\{y_b \otimes y^a\}_{1 \le a,b \le 4}$ is a coordinate system on $S^* \otimes S$, we obtain a representation of $j(\gamma_\alpha)$, (and as a consequence of any operator $j(\gamma_B)$), as a 4×4 matrix-function $(y_b \otimes y^a \circ j(\gamma_\alpha)) \equiv (\gamma_{\alpha\,b}^{\ a})$ such that $j(\gamma_\alpha)(p) \in M(4;\mathbf{C})$ for any $p \in M$.[60] (For abuse of notation we shall simply write $\gamma_B = j(\gamma_B)$.) The matrices $(\gamma_{B\,b}^{\ a})$, $B = \alpha_1 \ldots \alpha_p$, $0 \le \alpha_i \le 3$, $1 \le a,b \le 4$, are called *Dirac matrices*. For example, if M is the Minkowsky space-time and x^α is a system of cartesian coordinates, we can identify an isomorphism $j : E \cong L(S)$ such that if $j(\gamma_\alpha) = \gamma_{\alpha\,b}^{\ a} \theta^b \otimes e_a$, where $\{\theta^b\}$ and $\{e_a\}$ are the corresponding local bases on S^* and S respectively, one has the following constant matrices for $(\gamma_{\alpha\,b}^{\ a})$:

$$
\gamma_0 = \begin{pmatrix} 1 & 0 & 0 & 0 \\ 0 & 1 & 0 & 0 \\ 0 & 0 & -1 & 0 \\ 0 & 0 & 0 & -1 \end{pmatrix}, \gamma_1 = \begin{pmatrix} 0 & 0 & 0 & 1 \\ 0 & 0 & 1 & 0 \\ 0 & -1 & 0 & 0 \\ -1 & 0 & 0 & 0 \end{pmatrix},
$$

$$
\gamma_2 = \begin{pmatrix} 0 & 0 & 0 & -i \\ 0 & 0 & i & 0 \\ 0 & i & 0 & 0 \\ -i & 0 & 0 & 0 \end{pmatrix}, \gamma_3 = \begin{pmatrix} 0 & 0 & 1 & 0 \\ 0 & 0 & 0 & -1 \\ -1 & 0 & 0 & 0 \\ 0 & 1 & 0 & 0 \end{pmatrix}.
$$

One usually puts $\gamma_5 \equiv \gamma_{0123}$. If one takes anoter coordinate system $\bar{x}^\alpha$ the corresponding matrices $(\bar\gamma_{\alpha\,b}^{\ a})$ are related to $\gamma_{\alpha\,b}^{\ a}$ by the following transformations:

$$
\bar\gamma_{\alpha l}^{\ \ m} = A_\alpha^\beta \Lambda_l^r \Lambda_s^m \gamma_{\beta r}^{\ \ s},
$$
$$
(A_\alpha^\beta) = (\partial \bar{x}_\alpha . x^\beta) \in SO(1,3),
$$
$$
(\Lambda_p^s) = (\partial \bar{y}_p . y^s) \in Spin(1,3).
$$

Of course $(\bar\gamma_{\alpha\,b}^{\ a})$, are not more, in general, constant matrices. Now, recall that a linear connection on $S \to M$, $C_1^S \subset J\mathcal{D}(S)$: $\{y_\alpha^a + \lceil_{\alpha b}^a y^b = 0\}$, identifies a unique linear connection on the dual bundle $S^* \to M$, such

---

[60] We use small greek indexes for space-time indexes, (they run from 0 to 3), and small italic indexes for spinor indexes, (they run from 1 to 4).

that the following diagram is commutative:

$$
\begin{array}{ccc}
C_1^S \times C_1^{S^*} & \subset & J\mathcal{D}(S \times S^*) \\
\|\wr & & \|\wr \\
& \underset{\underset{S\ \ S}{1 \times 1^*}}{\longrightarrow} & \\
S \times S^* & & J\mathcal{D}(S) \times J\mathcal{D}(S^*) \\
0 \downarrow & & \downarrow J\mathcal{D}(<,>) \\
J\mathcal{D}(M \times \mathbf{C}) & = & J\mathcal{D}(M \times \mathbf{C})
\end{array}
$$

More precisely: $C_1^{S^*} \subset J\mathcal{D}(S^*)$: $\{y_{k\alpha} - \underset{S}{\lceil}{}_{\alpha k}^j y_j = 0\}$. Then we have:

$$
\underset{C}{\nabla}{}_\lambda \gamma_\alpha = \lceil{}_{\lambda\alpha}^\beta \gamma_\beta = \lceil{}_{\lambda\alpha}^\beta \gamma_{\beta}{}_s^r \theta^s \otimes e_r
$$

$$
= \underset{S^* \otimes S}{\nabla}{}_\lambda (\gamma_\alpha{}_s^r \theta^s \otimes e_r)
$$

$$
= (\partial x_\lambda . \gamma_\alpha{}_s^r) \theta^s \otimes e_r + \gamma_\alpha{}_s^r (\underset{S^*}{\nabla}{}_\lambda \theta^s) \otimes e_r + \gamma_\alpha{}_s^r \theta^s \otimes (\underset{S}{\nabla}{}_\lambda e_r)
$$

$$
= (\partial x_\lambda . \gamma_\alpha{}_s^r) \theta^s \otimes e_r - \gamma_\alpha{}_s^r \underset{S}{\lceil}{}_{\lambda p}^s \theta^p \otimes e_r + \gamma_\alpha{}_s^r \theta^s \otimes \underset{S}{\lceil}{}_{\lambda r}^q e_q
$$

$$
= \left[(\partial x_\lambda . \gamma_\alpha{}_s^r) - \gamma_\alpha{}_p^r \underset{S}{\lceil}{}_{\lambda s}^p + \gamma_\alpha{}_s^q \underset{S}{\lceil}{}_{\lambda q}^r \right] \theta^s \otimes e_r .
$$

Therefore, we must have

$$
\lceil{}_{\lambda\alpha}^\beta \gamma_{\beta}{}_s^r = (\partial x_\lambda . \gamma_\alpha{}_s^r) - \gamma_\alpha{}_p^r \underset{S}{\lceil}{}_{\lambda s}^p + \gamma_\alpha{}_s^q \underset{S}{\lceil}{}_{\lambda q}^r .
$$

This relation can be also rewritten in matrix form in the following way:

$$
(3.37) \qquad \lceil{}_{\lambda\alpha}^\beta \gamma_\beta = (\partial x_\lambda . \gamma_\alpha) + \underset{S}{\lceil}{}_\lambda \gamma_\alpha - \gamma_\alpha \underset{S}{\lceil}{}_\lambda ,
$$

where $\gamma_\beta(p), (\partial x_\lambda . \gamma_\alpha)(p), \underset{S}{\lceil}{}_\lambda(p) \in M(4; \mathbf{C})$, $\forall p \in M$. Set $\gamma^\alpha = g^{\alpha\beta} \gamma_\beta$.
Then one has $\gamma^\alpha \gamma_\alpha = 4\,\mathbf{1}$. (This follows directly by contracting both side
of the relation $\gamma_\alpha \gamma_\beta + \gamma_\beta \gamma_\alpha = 2 g_{\alpha\beta}\,\mathbf{1}$, by $g^{\alpha\beta}$ and taking into account that
$g^{\alpha\beta} g_{\alpha\beta} = 4$.) From (3.37) we get also the following equations:

$$
\lceil{}_{\lambda\alpha}^\beta \gamma_\beta \gamma^\alpha = (\partial x_\lambda . \gamma_\alpha) \gamma^\alpha + \underset{S}{\lceil}{}_\lambda \gamma_\alpha \gamma^\alpha - \gamma_\alpha \underset{S}{\lceil}{}_\lambda \gamma^\alpha
$$

$$
\lceil{}_{\lambda\alpha}^\beta \gamma^\alpha \gamma_\beta = \gamma^\alpha (\partial x_\lambda . \gamma_\alpha) + \gamma^\alpha \underset{S}{\lceil}{}_\lambda \gamma_\alpha - \gamma^\alpha \gamma_\alpha \underset{S}{\lceil}{}_\lambda .
$$

Therefore, we get also:

$$
(3.38) \qquad \left\{
\begin{array}{l}
\lceil{}_\lambda^{\beta\delta} \gamma_\beta \gamma_\delta = (\partial x_\lambda . \gamma_\alpha) \gamma^\alpha + 4 \underset{S}{\lceil}{}_\lambda - \gamma_\alpha \underset{S}{\lceil}{}_\lambda \gamma^\alpha \\[2mm]
\lceil{}_\lambda^{\beta\delta} \gamma_\delta \gamma_\beta = \gamma^\alpha (\partial x_\lambda . \gamma_\alpha) + \gamma^\alpha \underset{S}{\lceil}{}_\lambda \gamma_\alpha - 4 \underset{S}{\lceil}{}_\lambda
\end{array}
\right\} .
$$

Of course the second of above equations is equivalent to the first one. Note also that the first term on the left in above equation can be written as follows: $\lceil_{\lambda}^{\beta\delta}\gamma_\beta\gamma_\delta = \lceil_{\lambda\beta}^{\beta}1 + \lceil_{\lambda}^{[\beta\delta]}\gamma_\beta\gamma_\delta$. In fact we have: $\lceil_{\lambda}^{\beta\delta}\gamma_\beta\gamma_\delta = \lceil_{\lambda}^{\beta\delta}\frac{1}{2}(\gamma_\beta\gamma_\delta + \gamma_\delta\gamma_\beta) + \lceil_{\lambda}^{\beta\delta}\frac{1}{2}(\gamma_\beta\gamma_\delta - \gamma_\delta\gamma_\beta) = g_{\beta\delta}\lceil_{\lambda}^{\beta\delta}1 + \lceil_{\lambda}^{[\beta\delta]}\gamma_\beta\gamma_\delta = \lceil_{\lambda\beta}^{\beta}1 + \lceil_{\lambda}^{[\beta\delta]}\gamma_\beta\gamma_\delta$. By addition of both equations (3.38) we get equation (3.35). Taking into account that $\gamma^\alpha\gamma_\alpha = \gamma_\alpha\gamma^\alpha = 4\,\mathbf{1}$, we get that the consistency condition (3.35) can be also written in the following way: $2\lceil_{\lambda\beta}^{\beta}\mathbf{1} = (\partial x_\lambda.\gamma_\alpha)\gamma^\alpha - (\partial x_\lambda.\gamma^\alpha)\gamma_\alpha = \gamma^\alpha(\partial x_\lambda.\gamma_\alpha) - \gamma_\alpha(\partial x_\lambda.\gamma^\alpha)$. In order to obtain the explicit expression for $\lceil_{\lambda}^{}$ in the general case, let us rewrite equation (3.37) as an

equation in $M(4;\mathbf{C})$ in the following way: $C_{rb}^{as}\lceil_{S}{}_{\lambda}^{r}{}_{s} = D_{\lambda}{}_{b}^{a}$, $\forall\lambda = 0, 1, 2, 3$,

with $D_{\lambda}{}_{b}^{a} \equiv \lceil_{\lambda}^{\beta\delta}\gamma_{\beta}{}_{s}^{a}\gamma_{\delta}{}_{b}^{s} - (\partial x_\lambda.\gamma_{\alpha}{}_{r}^{a})\gamma^{\alpha}{}_{b}^{r}$ and $C \equiv 4\mathbf{1} \otimes \mathbf{1} - \gamma_\alpha \otimes \gamma^\alpha$. So, assuming that the matrix $(C_{rb}^{sa}) \in L(M(4;\mathbf{C}))$, that represents a linear application of the 16-dimensional vector space $M(4;\mathbf{C})$, over $\mathbf{C}$, is invertible, we get for $\lceil_{S\lambda}^{}$ the following solution:

$$(3.39) \qquad \lceil_{S}{}_{\lambda}{}_{s}^{r} = C^{-1}{}_{sa}^{rb}D_{\lambda}{}_{b}^{a}.$$

Therefore, the general expression of the connection $C_1^S \subset J\mathcal{D}(S)$, induced from the Clifford connection on E, has the local expression given in (3.32). Now let us prove that we can write $\lceil_{S}{}_{\lambda}$ in the Lychnerowicz's form (3.36). In fact, let us substitute (3.36) in the first equation (3.38). We get:

$$(3.40) \qquad \lceil_{\lambda}^{\beta\delta}\gamma_\beta\gamma_\delta = (\partial x_\lambda.\gamma_\alpha)\gamma^\alpha + \lceil_{\lambda}^{\beta\delta}\gamma_\beta\gamma_\delta - \frac{1}{4}\gamma_\alpha\lceil_{\lambda}^{\beta\delta}\gamma_\beta\gamma_\delta\gamma^\alpha.$$

On the other hand we can see that

$$(3.41) \qquad \gamma_\alpha\gamma_\beta\gamma_\delta\gamma^\alpha = g_{\beta\delta}.$$

Therefore, from (3.40) we get

$$(3.42) \qquad \frac{1}{4}\lceil_{\lambda\beta}^{\beta}\mathbf{1} = (\partial x_\lambda.\gamma_\alpha)\gamma^\alpha.$$

Now, let us substitute (3.36) in the second equation of (3.38). We similarly get:

$$(3.43) \qquad \frac{7}{4}\lceil_{\lambda\beta}^{\beta}\mathbf{1} = \gamma^\alpha(\partial x_\lambda.\gamma_\alpha).$$

Therefore, we see that (3.36) is a solution of both (3.38) iff (3.42) and (3.43) are both respected. On the other hand by adding these equations we get (3.35). This proves that conditions (3.42) and (3.43) are not new requirements but are automatically satisfied thanks to the consitency condition (3.35).

The absolute differential of a controvariant spinor field $\psi^a e_a : M \to S$, is $\nabla_S \psi = \lceil \circ D\psi = (\nabla_S{}_\lambda \psi)^r dx^\lambda \otimes e_r$, where the spinor-covariant derivative $(\nabla_S{}_\lambda \psi)^r$ is given by

$$(3.44) \qquad (\nabla_S{}_\lambda \psi)^r = (\partial x_\lambda \psi^r) + \lceil{}_{\lambda s}^r \psi^s = (\partial x_\lambda \psi^r) + C^{-1rb}_{\ \ \ sa} D^a_{\lambda b} \psi^s.$$

The absolute differential for the covariant spinor field $\varphi = \varphi_a \theta^a : M \to S^*$ is given by: $\nabla_{S^*} \varphi = \lceil \circ D\varphi = (\nabla_{S^*}{}_\rho \varphi)_s dx^\rho \otimes \theta^s$, where the spinor-covariant derivative $(\nabla_{S^*}{}_\rho \varphi)_s$ is given by:

$$(3.45) \qquad (\nabla_{S^*}{}_\rho \varphi)_s = (\partial x_\rho . \varphi_s) - \lceil{}_{\rho s}^r \varphi_r = (\partial x_\rho . \varphi_s) - C^{-1rb}_{\ \ \ sa} D^a_{\rho b} \varphi_r.$$

The curvature of the spinor connection $C_1^S \subset J\mathcal{D}(S)$ on $\pi : S \to M$, is given by a morphism of vector fiber bundles over M, $R_S : C_1^S \to \Lambda_2^0 M \otimes S$ locally written, as a section over M, as follows:

$$(3.46) \qquad \left\{ \begin{aligned} &R_S = \sum_{0 \le \alpha, \beta \le 3, 1 \le a, b \le 4} R_S{}_{\alpha\beta}{}^a{}_{.b} dx^\alpha \wedge dx^\beta \otimes e_a \otimes \theta^b \\ &R_S{}_{\alpha\beta}{}^a{}_{.b} = (\partial x_\alpha . \lceil{}_{\beta b}^a) - (\partial x_\beta . \lceil{}_{\alpha b}^a) + \lceil{}_{\alpha c}^a \lceil{}_{\beta b}^c - \lceil{}_{\beta s}^a \lceil{}_{\alpha b}^c \end{aligned} \right\},$$

where $R_S{}_{\alpha\beta}{}^a{}_{.b}$ are **C**-valued local functions on M. If we use the Lichnerowicz's formula we get that (3.46) gives the following expression for the spinor curvature:

$$(3.47) \qquad R_S{}_{\alpha\beta}{}^a{}_{.b} = \frac{1}{4} R_{\alpha\beta}{}^\mu{}_{.\lambda} (\gamma_\mu \gamma^\lambda)^a_b,$$

where $R_{\alpha\beta}{}^\mu{}_{.\lambda}$ is the Levi-Civita curvature. Similarly to the case of the Clifford connection, we have the following formula:

$$(3.48) \qquad \left\{ \begin{aligned} &< R_S, \psi > = R_S{}_{\alpha\beta}{}^b{}_{.c} \psi^c dx^\alpha \wedge dx^\beta \otimes e_b \\ &R_S{}_{\alpha\beta}{}^b{}_{.c} \psi^c e_b = \left(\nabla_S{}_\alpha \nabla_S{}_\beta - \nabla_S{}_\beta \nabla_S{}_\alpha \right) (\psi^b e_b) \end{aligned} \right\}$$

that relates the spinor-curvature to the spinor-covariant derivative.

The *Dirac operator* on the Clifford bundle $\pi : E \to M$ is a first order linear differential operator $\underset{C}{P} : C^\infty(E) \to C^\infty(E)$, given by $\underset{C}{P} = i\underset{C}{P_0}$, where i is the imaginary unity and $\underset{C}{P_0}$ is defined by means of the following homomorphism of vector bundles over M:

$$
\begin{array}{ccccc}
J\mathcal{D}(E) & \xrightarrow{\ \underset{C}{\Gamma}\ } & T^*M\otimes E & \overset{g'\otimes 1}{\cong} & TM\otimes E \\[4pt]
\underset{C}{P_0}\downarrow & & & & \downarrow\,\gamma\otimes 1 \\[4pt]
E & & \underset{a}{\longleftarrow} & & E\otimes E
\end{array}
$$

where a is the morphism induced by the Clifford multiplication. Therefore, if $\psi = \psi^B\gamma_B : M \to E$ is a section of $\pi : E \to M$, we get the following local expression of $\underset{C}{P}.\psi$:

$$
\begin{aligned}
\underset{C}{P}.\psi &= ig^{\alpha\epsilon}\left[(\partial x_\alpha.\psi^B) + \underset{C}{\Gamma}{}^{B}_{\alpha D}\psi^D\right]\gamma_\epsilon\gamma_B = i\gamma^\alpha\left[(\partial x_\alpha.\psi^B) + \underset{C}{\Gamma}{}^{B}_{\alpha D}\psi^D\right]\gamma_B \\
&= i\gamma^\alpha(\underset{C}{\nabla}_\alpha\psi)^B\gamma_B.
\end{aligned}
$$

If there is an isomorphism $j : E \cong S^* \otimes S$, where $S \to M$ is a vector bundle over M, then we can define also a Dirac operator on S, (*spinor Dirac operator*), as the first order linear differential operator $\underset{S}{P} : C^\infty(S) \to C^\infty(S)$, given by $\underset{S}{P} = i\underset{S}{P_0}$, where i is the imaginary unity and $\underset{S}{P_0}$ is defined by means of the following homomorphism of vector bundles over M:

$$
\begin{array}{ccccc}
J\mathcal{D}(S) & \xrightarrow{\ \underset{S}{\Gamma}\ } & T^*M\otimes S & \overset{g'\otimes 1}{\cong} & TM\otimes S \\[4pt]
\underset{S}{P_0}\downarrow & & & & \downarrow\,\gamma\otimes 1 \\[4pt]
S & \longleftarrow & S^*\otimes S\otimes S & \cong & E\otimes S
\end{array}
$$

Of course there is also a Dirac operator for covariant spinors, i.e., defined on S^*, $\underset{S^*}{P} : C^\infty(S^*) \to C^\infty(S^*)$, with $\underset{S^*}{P} = i\underset{S^*}{P_0}$, where $\underset{S^*}{P_0}$ is defined by means of the following homomorphism of vector bundles over M:

$$
\begin{array}{ccccc}
J\mathcal{D}(S^*) & \xrightarrow{\ \underset{S^*}{\Gamma}\ } & T^*M\otimes S^* & \overset{g'\otimes 1}{\cong} & TM\otimes S^* \\[4pt]
\underset{S^*}{P_0}\downarrow & & & & \downarrow\,\gamma\otimes 1 \\[4pt]
S^* & \longleftarrow & S^*\otimes S\otimes S^* & \cong & E\otimes S^*
\end{array}
$$

The covariant local expression for controvariant spinor field $\psi^a e_a : M \to S$ and covariant spinor field $\varphi = \varphi_a \theta^a : M \to S^*$ are respectively:

$$\underset{S}{P} \cdot \psi = i g^{\lambda \epsilon} \left[(\partial x_\lambda \cdot \psi^r) + \underset{S}{\Gamma^r_{\lambda s}} \psi^s \right] \gamma_{\epsilon r}{}^a e_a,$$

$$\underset{S^*}{P} \cdot \varphi = i g^{\lambda \epsilon} \left[(\partial x_\lambda \cdot \varphi_s) - \underset{S}{\Gamma^r_{\lambda s}} \varphi_r \right] \gamma_{\epsilon b}{}^s \theta^b.$$

In matrix form we can write:

$$\underset{S}{P} \cdot \psi = i \gamma^\lambda (\underset{S}{\nabla}_\lambda \cdot \psi), \quad \underset{S^*}{P} \cdot \varphi = i \gamma^\lambda (\underset{S^*}{\nabla}_\lambda \cdot \varphi).$$

The *Clifford Laplace operator* is the second order linear differential operator on the Clifford bundle $\underset{C}{\Delta} : C^\infty(E) \to C^\infty(E)$ defined as the square of the Dirac operator: $\underset{C}{\Delta} = \underset{C}{P}^2$. So if $\psi = \psi^B \gamma_B$ is the local expression of a section of the fiber bundle $\pi : E \to M$, then the local expression of $\underset{C}{\Delta} \cdot \psi$ is as follows:

$$
\begin{aligned}
(3.49) \quad \underset{C}{\Delta} \cdot \psi &= \underset{C}{P} \, i ((\underset{C}{\nabla}_\alpha \cdot \psi)^B \gamma^\alpha \gamma_B) \\
&= -(\underset{C}{\nabla}_\beta (\underset{C}{\nabla}_\alpha \cdot \psi))^B \gamma^\beta \gamma^\alpha \gamma_B = -\gamma^\beta \gamma^\alpha (\underset{C}{\nabla}_\beta (\underset{C}{\nabla}_\alpha \cdot \psi))^B \gamma_B.
\end{aligned}
$$

The Clifford Laplace operator is related to the curvature $\underset{C}{R}$ of the Clifford connection, (hence also to the Levi-Civita curvature). In fact one has

$$
\begin{aligned}
\underset{C}{\Delta} \cdot \psi &= -\frac{1}{2} (\gamma^\beta \gamma^\alpha + \gamma^\alpha \gamma^\beta)(\underset{C}{\nabla}_\beta \underset{C}{\nabla}_\alpha \cdot \psi)^B \gamma_B \\
&\quad - \frac{1}{2} (\gamma^\beta \gamma^\alpha - \gamma^\alpha \gamma^\beta)(\underset{C}{\nabla}_\beta \underset{C}{\nabla}_\alpha \cdot \psi)^B \gamma_B \\
&= g^{\beta \alpha} (\underset{C}{\nabla}_\beta \underset{C}{\nabla}_\alpha \cdot \psi)^B \gamma_B - \frac{1}{2} \gamma^\beta \gamma^\alpha \left[(\underset{C}{\nabla}_\beta \underset{C}{\nabla}_\alpha - (\underset{C}{\nabla}_\alpha \underset{C}{\nabla}_\beta) \cdot \psi)^B \right] \gamma_B.
\end{aligned}
$$

On the other hand we have

$$\left[\left(\underset{C}{\Delta}_\beta \underset{C}{\Delta}_\alpha - \underset{C}{\Delta}_\alpha \underset{C}{\Delta}_\beta \right) \cdot \psi)^B \right] \gamma_B = \underset{C}{R}_{\beta \alpha}{}^B{}_{.D} \psi^D \gamma_B.$$

Therefore we can also write:

$$(3.50) \quad \underset{C}{\Delta} \cdot \psi = - \left[g^{\beta \alpha} (\underset{C}{\nabla}_\beta \underset{C}{\nabla}_\alpha \cdot \psi)^B + \underset{C}{R}_{\beta \alpha}{}^B{}_{.D} \psi^D \gamma^\beta \gamma^\alpha \right] \gamma_B.$$

Similarly we can define the *spinor Laplace operator* as follows: $\underset{S}{\Delta} \equiv \underset{S}{P}{}^2$: $C^\infty(S) \to C^\infty(S)$. Then for any spinor field ψ, locally written as $\psi = \psi^a e_a$, we get:

$$(3.51) \qquad \underset{S}{\Delta}\cdot\psi = -\gamma^\beta\gamma^\alpha(\underset{S}{\nabla}{}_\beta(\underset{S}{\nabla}{}_\alpha\cdot\psi))^a e_a.$$

Similarly we can find the expression of $\underset{S}{\Delta}\cdot\psi$ in terms of curvature of the spinor connection on $\pi : S \to M$. We have:

$$(3.52) \qquad \underset{S}{\Delta}\cdot\psi = -\left[g^{\beta\alpha}(\underset{S}{\nabla}{}_\beta\underset{S}{\nabla}{}_\alpha\cdot\psi)^a + \frac{1}{2}\underset{S}{R}{}_{\beta\alpha}{}^a{}_{.b}\gamma^\beta\gamma^\alpha\psi^b\right]e_a.$$

Moreover if we use for the spinor connection the Lichnerowicz's formula, we can write (3.52) in the following form:

$$\underset{S}{\Delta}\cdot\psi = -\left[g^{\beta\alpha}(\underset{S}{\nabla}{}_\beta\underset{S}{\nabla}{}_\alpha\cdot\psi)^a + \frac{1}{8}R_{\beta\alpha\mu\lambda}(\gamma^\beta\gamma^\alpha\gamma^\mu\gamma^\lambda)^a_b\psi^b\right]e_a.$$

Then by using the relation $R_{\beta\alpha\mu\lambda}(\gamma^\beta\gamma^\alpha\gamma^\mu\gamma^\lambda)^a_b = 2\delta^a_b R$, where R is the scalar curvature of the Levi-Civita connection, we get:[61]

$$(3.53) \qquad \underset{S}{\Delta}\cdot\psi = -\left[g^{\beta\alpha}(\underset{S}{\nabla}{}_\beta\underset{S}{\nabla}{}_\alpha\cdot\psi)^a + \frac{1}{4}R\psi^a\right]e_a.$$

Furthermore, for covariant spinor field $\varphi = \varphi_a\theta^a$ we similarly get:

$$(3.54) \qquad \left\{\begin{aligned} \underset{S^*}{\Delta}\cdot\varphi &= -\gamma^\beta\gamma^\alpha(\underset{S^*}{\nabla}{}_\beta(\underset{S^*}{\nabla}{}_\alpha\cdot\varphi))_a\theta^a \\ &= -\left[g^{\beta\alpha}(\underset{S^*}{\nabla}{}_\beta\underset{S^*}{\nabla}{}_\alpha\cdot\varphi)_a + \frac{1}{2}\underset{S}{R}{}_{\beta\alpha}{}^b{}_{.a}\gamma^\beta\gamma^\alpha\varphi_b\right]\theta^a \\ &= \left[g^{\beta\alpha}(\underset{S^*}{\Delta}{}_\beta\underset{S^*}{\Delta}{}_\alpha\cdot\varphi)_a + \frac{1}{4}R\varphi_a\right]\theta^a \end{aligned}\right\}.$$

One has two generalized Dirac-Clifford equations:

$$(3.55)$$

$$(DC)_\mp \equiv \ker(\underset{C}{P}\mp m) \subset J\mathcal{D}(E) : \quad \left\{\left[i\gamma^\alpha(\underset{C}{\nabla}\cdot\psi)^B \mp m\psi^B\right]\gamma_B = 0\right\},$$

---

[61] This expression coincides with the well known formula given by A. Lichnerowicz [96(g)].

where $m \in \mathbf{R} \bigcup i\mathbf{R}^+$. An equivalent expression of (3.55) is the following:

$$(3.56) \qquad i\gamma^\alpha \left[(\partial x_\alpha.\psi^B) + \lceil^{\;B}_{\alpha C}\psi^C \right] \mp m\psi^B = 0.$$

Similarly one has two generalized Dirac-spinor equations:

$$(3.57) \qquad \left\{ \begin{array}{l} (DS)_\mp \equiv \ker(\underset{S}{P}\mp m) \subset J\mathcal{D}(S) \\[2mm] (\underset{S}{P}\mp m)\psi = 0 \Leftrightarrow \left\{ i\gamma^{\lambda a}_{\;r} \left[(\partial x_\lambda.\psi^r) + \lceil^{\;r}_{\lambda s}\psi^s \right] \mp m\psi^a = 0 \right\} \end{array} \right\}.$$

We set $(GM)_m \equiv (DC)_-$ or $(GM)_m \equiv (DS)_-$. The corresponding dual Dirac adjoint equation results:

$$(3.58) \qquad \left\{ \begin{array}{l} (DS^*)_\mp \equiv \ker(\underset{S^*}{P}\mp m) \subset J\mathcal{D}(S^*) \\[2mm] (\underset{S^*}{P}\mp m)\varphi = 0 \Leftrightarrow \left\{ i\gamma^{\lambda s}_{\;b} \left[(\partial x_\lambda.\varphi_s) - \lceil^{\;r}_{\lambda s}\varphi_r \right] \mp m\varphi_b = 0 \right\} \end{array} \right\}.$$

The *generalized Klein-Gordon Clifford Equation)* is the following equation:

$$(3.59) \qquad \left\{ \begin{array}{l} (KGC) \subset J\mathcal{D}^2(E) \\[2mm] (\underset{C}{P^2} - m^2)\psi = 0 \Leftrightarrow (\underset{C}{P}-m)(\underset{C}{P}+m)\psi = 0 \Leftrightarrow (\underset{C}{\Delta}-m^2)\psi = 0 \end{array} \right\}.$$

Tle local expression of (KGC) is the following:

$$(3.60) \qquad g^{\beta\alpha}(\underset{C}{\nabla}_\beta \underset{C}{\nabla}_\alpha \psi)^B + \left(\underset{C}{R}_{\beta\alpha}{}^B{}_{.D}\gamma^\beta\gamma^\alpha + \delta^B_D m^2 \right) \psi^D = 0.$$

Similarly the *generalized Klein-Gordon Spinor Equation* is the following equation:

$$(3.61) \qquad \left\{ \begin{array}{l} (KGS) \subset J\mathcal{D}^2(S) \\[2mm] (\underset{S}{P^2} - m^2)\psi = 0 \Leftrightarrow (\underset{S}{P}-m)(\underset{S}{P}+m)\psi = 0 \Leftrightarrow (\underset{S}{\Delta}-m^2)\psi = 0 \end{array} \right\}$$

Tle local expression of (KGS) is the following:

$$(3.62) \qquad g^{\beta\alpha}(\underset{S}{\nabla}_\beta \underset{S}{\nabla}_\alpha \psi)^a + \left(\underset{S}{R}_{\beta\alpha}{}^a{}_{.b}\gamma^\beta\gamma^\alpha + \delta^a_b m^2 \right) \psi^b = 0.$$

In particular by using the Lichnerowicz's formula for the spinor connection we get the following local expression for (KGS):

$$(3.63) \qquad g^{\beta\alpha}(\underset{S}{\nabla}_\beta \underset{S}{\nabla}_\alpha \psi)^a + \left(\frac{1}{4}R + m^2\right)\psi^a = 0.$$

Note that

$$g^{\beta\alpha}(\underset{S}{\nabla}_\beta \underset{S}{\nabla}_\alpha \psi)^a = g^{\beta\alpha}\{(\partial x_\beta \partial x_\alpha . \psi^a) + (\partial x_\alpha . \psi^b)\,\underset{S}{\Gamma}{}^a_{\beta b} + (\partial x_\beta . \psi^b)\,\underset{S}{\Gamma}{}^a_{\alpha b} + \psi^c[(\partial x_\beta . \underset{S}{\Gamma}{}^a_{\alpha c}) + \underset{S}{\Gamma}{}^b_{\alpha c}\,\underset{S}{\Gamma}{}^a_{\beta b}]\}.$$

The corresponding expression for covariant spinor field is the following:

$$(3.64) \qquad g^{\beta\alpha}(\underset{S^*}{\nabla}_\beta \underset{S^*}{\nabla}_\alpha \varphi)_a + \left(\frac{1}{4}R + m^2\right)\varphi_a = 0.$$

The *Dirac conjugated*, (or h.c.), of a spinor field $\psi = \psi^a e_a : M \to S$ is the covariant field $\bar\psi = \varphi\gamma^0 : M \to S^*$, where $\varphi = \varphi_a \theta^a$ with $\varphi_a = (\psi^a)^*$, ($* = $ complex conjugated). Then the generalized Dirac eqution for $\bar\psi$, (*generalized Dirac h.c.*), is the following (written in matrix form): $i\underset{S^*}{\nabla}_\mu \bar\psi\gamma^\mu + m\bar\psi = 0$. It is important to note that the Dirac equation and its conjugated, must be considered together as they together are the Euler-Lagrange equation of a first order Lagrangian $\mathcal{L} : J\mathcal{D}(S^* \otimes S) \to \mathbf{C}$, defined on $S^* \otimes S$, i.e., defined on the Clifford bundle. More precisely the Lagrangian $\mathcal{L}$, in matrix form, can be written in the following way:

$$(3.65) \qquad \mathcal{L} = \frac{1}{2}i\left[\bar\psi\gamma^\mu \underset{S}{\nabla}\psi - (\underset{S^*}{\nabla}\bar\psi)\gamma^\mu\psi\right] - m\bar\psi\psi.$$

The variation of the action integral with respect to $\bar\psi$, (resp. ψ), yields the generalized Dirac equation for ψ, (resp. $\bar\psi$). So the Euler-Lagrange equation, $E[\mathcal{L}] \subset J\mathcal{D}^2(S^* \otimes S)$, of $\mathcal{L}$ is the following:

$$E[\mathcal{L}]\subset J\mathcal{D}^2(S^*\otimes S): \left\{ \begin{array}{ll} i\gamma^\mu \underset{S}{\nabla}_\mu\psi - m\psi = 0 & (\textit{Generalized Dirac-equation}) \\[2ex] i(\underset{S^*}{\nabla}_\mu\bar\psi)\gamma^\mu - m\bar\psi = 0 & (\textit{Generalized Dirac-h.c.- equation}) \end{array} \right\}.$$

Above remark shows that a theory for spinor fields could be written on the Clifford bundle instead then on spinor bundle only. Furthermore, if we consider also that the isomorphism $E \cong S^* \otimes S$ is conditioned to some very particular categories of space-times, (e.g., almost Hermitian or almost Kählerian space-times), it follows that it should be more convenient to implement any physical theory for "spinor fields" on Clifford bundles. Therefore the local expression of $(GD)_m$ with respect to local coordinates $\{x^\alpha, z^A, z^A_\alpha\}$ on $J\mathcal{D}(E)$ is the following:[62]

$$(3.66) \qquad (GM)_m \subset J\mathcal{D}(E): \left\{ F^A \equiv i\gamma^\alpha \left[z^B_\alpha + z^A \lceil^B_{\alpha A} \rceil_C \right] - m\delta^B_A z^A = 0 \right\}.$$

For the generalized Dirac equation $(GD)_m$ we have a structure similar to $(GKG)_\chi$, even if it is of the first order. In fact, $(GD)_m$ is an involutive formally integrable, as well completely integrable PDE, hence it is equivalent to its first prolongation $((GD)_m)_{+1} \subset J\mathcal{D}^2(E)$. This can be easily seen by rewriting equation (3.66) in the following split form:

$$(3.67) \qquad \left\{ \begin{array}{l} i\gamma^\alpha \left[\xi^B_\alpha + \lceil^B_{\alpha A} \xi^A \right]_C - m\delta^B_A \xi^A = 0 \\[2mm] i\gamma^\alpha \left[\eta^B_\alpha + \lceil^B_{\alpha A} \eta^A \right]_C - m\delta^B_A \eta^A = 0 \end{array} \right\},$$

where $z^A \equiv \xi^A + i\eta^A$ and $z^A_\alpha \equiv \xi^A_\alpha + i\eta^A_\alpha$.

If there is the isomorphism $E \cong S^* \otimes S$, then $(GD)_m$ is isomorphic to the tensor product of two equations: $(GD)_m \cong \underline{(GD)_m} \otimes \overline{(GD)_m}$, where $\overline{(GD)_m}$ is the Dirac hermitiam conjugated of $\underline{(GD)_m}$ such that the following diagram is commutative:

$$\begin{array}{ccc} ((GD)_m \times \overline{(GD)_m} & \subset & J\mathcal{D}(S^* \times S) \cong J\mathcal{D}(S^*) \times J\mathcal{D}(S) \\ \downarrow & & \downarrow \\ (GD)_m \quad \cong \quad \underline{(GD)_m} \otimes \overline{(GD)_m} & \subset & J\mathcal{D}(S^* \otimes S) \end{array}$$

Then similarly to what made for the equation $(GKG)_\chi$, we can prove that Cauchy problem can be solved for the equation $(GD)_m$ by the method of characteristics. Furthermore we can also prove that the integral bordism group $\Omega_3^{(GD)_m}$ of $(GD)_m$ is trivial: $\Omega_3^{(GD)_m} = 0$. Therefore, tunnel effects can be observed and propagators associated to these calculated. In fact, if there is an isomorphism $E \cong S^* \otimes S$ we can represent $(GM)_m$ in the form $(GM)_m \cong ((GM)_m \otimes \overline{(GD)_m}$.

---

[62] Note that in the particular case of fermionic bradions and luxons the parameter m is a real number, instead for massive neutrinos it is an imaginary number.

In the following we shall specialize on the Dirac equation for spin 1/2, on globally hyperbolic 4-dimensional space-time. Thus we shall consider the following equations:

$$(3.68) \qquad \left\{ \begin{array}{l} (i\gamma^\mu \nabla_\mu - m)\psi = 0 \\ i\nabla_\mu \bar\psi \gamma^\mu - m\bar\psi = 0 \end{array} \right\}.$$

The canonical quantization of the generalized Dirac equation $(GD)_m \subset JD^2(E)$ is given by the following anticommutation relations:

$$(3.69) \qquad \left\{ \begin{array}{l} [\widehat{s}_A(x), \widehat{s}_B(x')]_+ = [\widehat{\bar s}_A(x), \widehat{\bar s}_B(x')]_+ = 0 \\ [\widehat{s}_A(x), \widehat{\bar s}_B(x')]_+ = i\hbar \widetilde{G}_{AB}(x; x'|m) \end{array} \right\},$$

where

$$\widetilde{G}_{AB}(x; x'|m) = \left(\blacksquare_{x'} + m\right)_{AB} \widetilde{G}(x; x'|m), \quad (\blacksquare_{x'})_{AB} \equiv (i\gamma^\mu \nabla_\mu)_{AB}$$

where $\widetilde{G}(x; x'|m)$ is the propagator of the generalized Klein-Gordon spinor equation. Then, the microscopic causality is conserved also for fermionic massive neutrinos having a *geometric mass* $\chi \equiv \frac{1}{4}R + m^2 < 0$. In fact, the propagator $\widetilde{G}$ is the Green kernel solution of the following Cauchy problem:

$$(3.70) \qquad \left\{ \begin{array}{l} \left((\blacksquare_{x'})_A^C - m\delta_A^C\right)\widetilde{G}_B^A(x; x'|m) = 0 \\ \widetilde{G}_B^A(x; x'|m)\big|_{x^0 = x'^0} = 0 \\ \left(\partial x_0'.\widetilde{G}_B^A\right)(x; x')\big|_{x^0 = x'^0} = -\delta_B^A \delta(x; x')\big|_{x^0 = x'^0} \end{array} \right\} + \text{h.c.}.$$

Let us find solutions of the type

$$\widetilde{G}(x; x'|m) = \left(\blacksquare_{x'} + m\right)\widetilde{G}(x; x'|m).$$

Then we find that this is a solution iff $\widetilde{G}_B^A(x; x'|m)$ satisfies the equation

$$(\underset{S}{\Delta} - m^2)\widetilde{G}(x; x'|m) = 0.$$

This means that $\widetilde{G}(x; x'|m)$ is the propagator of the generalized Klein-Gordon spinor equation. Then we get the usual anticommutation relations:

$$[\widehat{\psi}_A(x), \widehat{\psi}_B(x')]_+ = [\widehat{\bar\psi}_A(x), \widehat{\bar\psi}_B(x')]_+ = 0,$$

$$[\widehat{\psi}_A(x), \widehat{\bar\psi}_B(x')]_+ = i\hbar \widetilde{G}_{AB}(x; x'|m).$$

On the other hand, as the operator $\underset{S}{\Delta} - m^2$ is hyperbolic its propagator $\widetilde{G}_{AB}(x; x'|m)$ exists and it is unique and with support in $\mathcal{E}^-(x') \cup \mathcal{E}^+(x')$, $\forall x' \in M$. So the quantum commutator $[\hat{\psi}^A(x), \hat{\psi}^B(x')]_+ = i\hbar\widetilde{G}_{AB}(x; x'|m)$ respects the microscopic causality even if the geometric mass of the field is negative. $\blacksquare$

Example 3.62. (*Fields in interaction*). The field equations, considered in the above two examples, are linear PDEs, and represent free fields on a general curved space-time background identified with a globally hyperbolic 4-dimensional manifold. For interacting fields the corresponding PDEs are not more linear. However, the general geometric method developed by us to quantize PDEs works well also for interacting fields. As an example let us consider the following.

$\square$ (*Decay of scalar pion*). We shall study the following decay: $\pi_0^+ \to \mu^+ + \nu_\mu$. Of course the neutrino path is not directly observed. We will assume that the neutrino is massive. In this case we have scalar particles (π_0^+) and Dirac particles (μ^+, ν_μ). So the process is described by a second order PDE $(E_2) \subset J\mathcal{D}^2(E)$, where the fiber bundle E is defined by $E \equiv E_{(\pi_0)} \times_M E_{(\mu)} \times_M E_{(\nu)}$, given by:

$$(3.71) \quad \left\{ \begin{array}{l} \left(\Box_{x'} + \chi_{(\pi_0)}\right)\varphi = j_{(\pi_0)} \equiv \lambda\bar{\psi}_{(\nu)}\psi_{(\mu)} \\[4pt] \left(i\gamma^\alpha\nabla_\alpha - m_{(\mu)}\right)\psi_{(\mu)} = j_{(\mu)} \equiv \lambda\psi_{(\nu)}\varphi \\[4pt] \left(i\gamma^\alpha\nabla_\alpha - m_{(\nu)}\right)\psi_{(\nu)} = j_{(\nu)} \equiv \lambda\bar{\varphi}\psi_{(\mu)} \end{array} \right\} \left\{ \begin{array}{l} i\nabla_\alpha\bar{\psi}_{(\mu)}\gamma^\alpha - m_{(\mu)}\bar{\psi}_{(\mu)} = \bar{j}_{(\mu)} \\[4pt] i\nabla_\alpha\bar{\psi}_{(\nu)}\gamma^\alpha - m_{(\nu)}\bar{\psi}_{(\nu)} = \bar{j}_{(\nu)} \end{array} \right\}.$$

There we have put: $\left\{ \chi_{(\pi_0)} \equiv \xi_{(\pi_0)}R + m^2_{(\pi_0)} \right\}$. The local expression of (3.71) is the following:

$$\left\{ \begin{array}{l} g^{\alpha\beta}\varphi_{\alpha\beta} - \Gamma^\alpha_{\beta\gamma}g^{\gamma\beta}\varphi_\alpha + \chi_{(\pi_0)}\varphi = \lambda\bar{\psi}_{(\nu)j}\psi^j_{(\mu)} \\[6pt] \left(i\gamma^{\alpha}{}^i_j\nabla_\alpha - m_{(\mu)}\delta^i_j\right)\psi^j_{(\mu)} \equiv i\gamma^{\alpha}{}^i_j\left[(\partial x_\alpha.\psi^j_{(\mu)}) + \underset{S}{\Gamma}{}^j_{\alpha s}\psi^s\right] - m_{(\mu)}\delta^i_j\psi^j_{(\mu)} = \lambda\varphi\psi^i_{(\nu)} \\[6pt] \left(i\gamma^{\alpha}{}^i_j\nabla_\alpha - m_{(\nu)}\delta^i_j\right)\psi^j_{(\nu)} \equiv i\gamma^{\alpha}{}^i_j\left[(\partial x_\alpha.\psi^j_{(\nu)}) + \underset{S}{\Gamma}{}^j_{\alpha s}\psi^s\right] - m_{(\nu)}\delta^i_j\psi^j_{(\nu)} = \lambda\bar{\varphi}\psi^i_{(\mu)} \end{array} \right\} + \text{h.c.}$$

Here we used the following notation: $\bar{\varphi}$ is the c.c. of φ and $\bar{\psi} = \psi^*\gamma^0$, with ψ^* the c.c. of ψ. Then, the corresponding Jacobi operator has the following components:

$$\left\{ \begin{array}{l} (J[s].\nu)_{(\pi_0)} \equiv [\chi_{(\pi_0)}]\nu_{(\pi_0)} - [\Gamma^\alpha_{\beta\gamma}g^{\gamma\beta}]\bullet\nu_{(\pi_0),\alpha} + [g^{\alpha\beta}]\bullet\nu_{(\pi_0),\alpha\beta} - \delta_{ij}[\lambda\bar{\psi}^i_{(\nu)}]\nu^j_{(\mu)} - \delta_{ij}[\lambda\psi^j_{(\mu)}]\bullet\bar{\nu}^i_{(\nu)} \\[6pt] (J[s].\nu)^i_{(\mu)} \equiv [i\gamma^i_{\beta l}g^{\alpha\beta}\Gamma^l_{\alpha j} - m_{(\mu)}\delta^i_j]\bullet\nu^j_{(\mu)} + i[\gamma^i_{\alpha j}g^{\alpha\beta}]\bullet\nu^j_{(\mu),\beta} - [\lambda\psi^i_{(\nu)}]\bullet\nu_{(\pi_0)} - [\lambda\varphi\delta^i_j]\bullet\nu^j_{(\nu)} \\[6pt] (J[s].\nu)^i_{(\nu)} \equiv [i\gamma^i_{\beta l}g^{\alpha\beta}\Gamma^l_{\alpha j} - m_{(\nu)}\delta^i_j]\bullet\nu^j_{(\nu)} + i[\gamma^i_{\alpha j}g^{\alpha\beta}]\bullet\nu^j_{(\nu),\beta} - [\lambda\bar{\varphi}]\bullet\nu^i_{(\mu)} - [\lambda\psi^i_{(\mu)}]\bullet\bar{\nu}_{(\pi_0)} \end{array} \right\} + \text{h.c.}$$

Here and in the following the bracket $[-]_\bullet$ denotes "evaluation at a solution of the equation (3.71)". The equations for the Green kernel are given by the following Cauchy problem:

(3.72)

$$
\left\{
\begin{aligned}
& \chi_{(\pi_0)} G_{(\pi_0)}(x;x') - [\lceil^\alpha_{\beta\gamma} g^{\gamma\beta}]_\bullet (\partial x_\alpha . G_{(\pi_0)})(x;x') \\
& \qquad\qquad + [g^{\alpha\beta}]_\bullet (\partial x_\alpha \partial x_\beta . G_{(\pi_0)})(x;x') \\
& \qquad\qquad - \delta_{ij}[\lambda \bar\psi^i_{(\nu)}]_\bullet G^j_{(\mu)}(x;x') \\
& \qquad\qquad - \delta_{ij}[\lambda \psi^i_{(\mu)}]_\bullet \bar G^j_{(\nu)}(x;x') = 0 \\[4pt]
& [i\gamma^i_{\beta l} g^{\alpha\beta} \lceil^l_{\alpha j} - m_{(\mu)}\delta^i_j]_\bullet G^j_{(\mu)}(x;x') + [i\gamma^i_{\alpha j} g^{\alpha\beta}]_\bullet (\partial x_\beta . G^j_{(\mu)})(x;x') \\
& \qquad\qquad - [\lambda \psi^i_{(\nu)}]_\bullet G_{(\pi_0)}(x;x') \\
& \qquad\qquad - [\lambda \varphi \delta^i_j]_\bullet G^j_{(\nu)}(x;x') = 0 \\[4pt]
& [i\gamma^i_{\beta l} g^{\alpha\beta} \lceil^l_{\alpha j} - m_{(\nu)}\delta^i_j]_\bullet G^j_{(\nu)}(x;x') + [i\gamma^i_{\alpha j} g^{\alpha\beta}]_\bullet (\partial x_\beta . G^j_{(\nu)})(x;x') \\
& \qquad\qquad - [\lambda \bar\varphi]_\bullet G^i_{(\mu)}(x;x') \\
& \qquad\qquad - [\lambda \psi_{(\mu)}]_\bullet \bar G_{(\pi_0)}(x;x') = 0 \\[4pt]
& \left\{
\begin{aligned}
& G(x^0,\mathbf{x};x^0,\mathbf{x}') = 0 \\
& (\partial x'_0 . G^{IJ})(x;x')|_{x^0=x'^0} = \delta^{IJ}\delta(x;x')|_{x^0=x'^0}
\end{aligned}
\right\}
\end{aligned}
\right\} + \text{h.c.}.
$$

Set

$$
G(x;x') = \begin{pmatrix} X \\ Y^j \\ Z^j \end{pmatrix}; \qquad \bar G(x;x') = \begin{pmatrix} \bar X \\ \bar Y^j \\ \bar Z^j \end{pmatrix}.
$$

Then, we get the following expression for the system (3.72):

(3.73)

$$
\left\{
\begin{aligned}
& [\Box_{x'} + \chi_{(\pi_0)}]_\bullet X - [\lambda \bar\psi_{(\nu)}]_\bullet Y - [\lambda \psi_{(\mu)}]_\bullet \bar Z = 0 \\
& [\blacksquare_{x'} - m_{(\mu)}]_\bullet Y - [\lambda \psi_{(\nu)}]_\bullet X - [\lambda \varphi]_\bullet Z = 0 \\
& [\blacksquare_{x'} - m_{(\nu)}]_\bullet Z - [\lambda \bar\varphi]_\bullet Y - [\lambda \psi_{(\mu)}]_\bullet \bar X = 0 \\
& +\text{Cauchy conditions:}
\left\{
\begin{aligned}
& X|_{x^0=x'^0} = Y|_{x^0=x'^0} = Z|_{x^0=x'^0} = 0 \\
& (\partial x'_0,X)|_{x^0=x'^0} = \delta(x;x')|_{x^0=x'^0} \\
& (\partial x'_0,Y)|_{x^0=x'^0} = \delta(x;x')|_{x^0=x'^0} \\
& (\partial x'_0,Z)|_{x^0=x'^0} = \delta(x;x')|_{x^0=x'^0}
\end{aligned}
\right\}
\end{aligned}
\right\} + \text{h.c.}
$$

Then the propagator $\widetilde{G} = G^+ - G^-$ is given by means of the retarded and advanced solutions

$$G^\pm(x;x') = \begin{pmatrix} X^\pm \\ Y^{\pm,j} \\ Z^{\pm,j} \end{pmatrix}, \quad \bar{G}^\pm(x;x') = \begin{pmatrix} \bar{X}^\pm \\ \bar{Y}^{\pm,j} \\ \bar{Z}^{\pm,j} \end{pmatrix}, \quad \widetilde{G}(x;x') = \begin{pmatrix} X^+(x;x') - X^-(x;x') \\ Y^+(x;x') - Y^-(x;x') \\ Z^+(x;x') - Z^-(x;x') \end{pmatrix}.$$

As a consequence the quantization of the dynamic equation (3.71) is obtained by considering the following quantum bracket:

$$\left[\widehat{\phi^I}(x), \widehat{\phi^J}(x') \right]_\pm = i\hbar \widetilde{G}^{IJ}(x;x')\mathbf{1},$$

where $(\phi^I) = \begin{pmatrix} \varphi \\ \psi^i_{(\mu)} \\ \psi^i_{(\nu)} \end{pmatrix}$. Note that $[A,B]_- \equiv AB - BA$ is the commutator or simply $[A,B]$, instead $[A,B]_+ \equiv AB + BA$ is the anticommutator. If both A and B are fermionic one has the anticommutator otherwise one uses the commutator. By derivation of above commutator with respect to x'^0, and taking $x^0 = x'^0$, and by considering also equation (3.72), we get:

$$\left[\widehat{\phi^I}(x), \dot{\widehat{\phi^J}}(x') \right]_\pm \Big|_{x^0=x'^0} = i\hbar\delta^{IJ}(x;x')\big|_{x^0=x'^0}\mathbf{1}.$$

This completes the procedure to obtain the canonical quantization of our dynamical equation for the decay $\pi_0^+ \to \mu^+ + \nu_\mu$. Note that equation (3.71) admits the zero section $(\varphi, \psi_{(\mu)}, \psi_{(\nu)}) = (0,0,0)$ as a solution. The canonical quantization of (3.71) at the zero section, reduces to the following Jacobi equation:

$$\left\{ \left\{ \begin{array}{l} \left(\Box_{x'} + \chi_{(\pi)} \right)X = 0 \\ \left(\blacksquare_{x'} - m_{(\mu)} \right)Y = 0 \\ \left(\blacksquare_{x'} - m_{(\nu)} \right)Z = 0 \end{array} \right\} + \text{Cauchy conditions:} \left\{ \begin{array}{l} X|_{x^0=x'^0} = Y|_{x^0=x'^0} = Z|_{x^0=x'^0} = 0 \\ (\partial x'_0, X)|_{x^0=x'^0} = \delta(x;x')|_{x^0=x'^0} \\ (\partial x'_0, Y)|_{x^0=x'^0} = \delta(x;x')|_{x^0=x'^0} \\ (\partial x'_0, Z)|_{x^0=x'^0} = \delta(x;x')|_{x^0=x'^0} \end{array} \right\} \right\} + \text{h.c.}$$

Then the propagator of (3.71) at the zero section is directly identified by the propagator of the independent fields. So we get the following (anti)commutation relations:

$$[\hat{\varphi}(x),\hat{\varphi}(x')]=[\widehat{\bar{\varphi}}(x),\widehat{\bar{\varphi}}(x')]=0$$

$$[\hat{\varphi}(x),\widehat{\bar{\varphi}}(x')]=i\hbar\widetilde{G}_{(\pi_0)}(x;x'|\chi_{(\pi_0)})\ \mathbf{1}$$

$$[\hat{\varphi}(x),\dot{\hat{\varphi}}(x')]=[\widehat{\bar{\varphi}}(x),\dot{\widehat{\bar{\varphi}}}(x')]=0$$

$$[\hat{\varphi}(x),\dot{\widehat{\bar{\varphi}}}(x')]|_{x^0=x'^0}=-i\hbar\delta(x;x')|_{x^0=x'^0}\ \mathbf{1}$$

$$[\widehat{\psi_{(\mu)A}}(x),\widehat{\psi_{(\mu)B}}(x')]_+=[\widehat{\bar{\psi}_{(\mu)A}}(x),\widehat{\bar{\psi}_{(\mu)B}}(x')]_+=0$$

$$[\widehat{\psi_{(\mu)A}}(x),\widehat{\bar{\psi}_{(\mu)B}}(x')]_+=i\hbar\widetilde{G}_{(\mu)AB}(x;x'|m_{(\mu)})$$

$$[\widehat{\psi_{(\mu)}}(x),\dot{\widehat{\psi_{(\mu)}}}(x')]_+=[\widehat{\bar{\psi}_{(\mu)}}(x),\dot{\widehat{\bar{\psi}_{(\mu)}}}(x')]_+=0$$

$$[\widehat{\psi_{(\mu)A}}(x),\dot{\widehat{\bar{\psi}_{(\mu)B}}}(x')]_+|_{x^0=x'^0}=i\hbar\delta_{AB}(x;x')|_{x^0=x'^0}$$

$$[\widehat{\psi_{(\nu)A}}(x),\widehat{\psi_{(\nu)B}}(x')]_+=[\widehat{\bar{\psi}_{(\nu)A}}(x),\widehat{\bar{\psi}_{(\nu)B}}(x')]_+=0$$

$$[\widehat{\psi_{(\nu)A}}(x),\widehat{\bar{\psi}_{(\nu)B}}(x')]_+=i\hbar\widetilde{G}_{(\nu)AB}(x;x'|m_{(\nu)})$$

$$[\widehat{\psi_{(\nu)}}(x),\dot{\widehat{\psi_{(\nu)}}}(x')]_+=[\widehat{\bar{\psi}_{(\nu)}}(x),\dot{\widehat{\bar{\psi}_{(\nu)}}}(x')]_+=0$$

$$[\widehat{\psi_{(\nu)A}}(x),\dot{\widehat{\bar{\psi}_{(\nu)B}}}(x')]_+|_{x^0=x'^0}=i\hbar\delta_{AB}(x;x')|_{x^0=x'^0}$$

where $\widetilde{G}_{(\pi_0)}$, (resp. $\widetilde{G}_{(\mu)}$, $\widetilde{G}_{(\nu)}$), is the propagator for the Klein-Gordon, (resp. Dirac μ-particle, resp. Dirac ν-particle), equation. As an application of Theorem 1.2 we get that $\widetilde{G}[s]$ is an invariant of $\Omega_c(R_1,R_2)$. Furthermore, the integral bordism groups of the dynamic equation (3.73) E_2 are given by $\Omega_p^{E_2}\cong \mathbf{Z}_2$, $p=0,2$, and $\Omega_p^{E_2}\cong 0$, $p=1,3$. In particular we get existence of tunnel effects for global solutions of (3.73). ∎

Example 3.63. (*Quantum particles in interaction with the gravitational field*). Emphasize that we can consider also the interaction of above particles with the gravitational field. In fact our general method to obtain the covariant and canonical quantizations of PDEs can be applied also to this case. In particular we shall reconsider the decay $\pi_0^+ \to \mu^+ + \nu_\mu$ by inserting the interaction with gravitons, i.e., the interaction with the gravitational field as a quantum field. So we will consider the following fiber bundle $\pi : E \equiv E_{(\pi_0)} \times_M E_{(\mu)} \times_M E_{(\nu)} \times_M S_2^0 M \to M$. A section $s = (\varphi, \psi_{(\mu)}, \psi_{(\nu)}, g)$ of π must be a solution of the following second order

PDE on E:

(3.74)
$$\left\{\begin{array}{l} (\Box+\chi_{(\pi_0)})\varphi=j_{(\pi_0)} \\ (\blacksquare-m_{(\mu)})\psi_{(\mu)}=j_{(\mu)} \\ (\blacksquare-m_{(\nu)})\psi_{(\nu)}=j_{(\nu)} \\ G=-kT,\quad \mathrm{div}\,(T)=0 \end{array}\right\}+\text{h.c.,} \quad \left\{\begin{array}{l} G_{\alpha\beta}=R_{\alpha\beta}-\frac{1}{2}g_{\alpha\beta}R \\ R_{\alpha\beta}=R^{\lambda}_{\beta\alpha\lambda}=(\partial x_{\alpha}\lceil^{\lambda}_{\beta\lambda})-(\partial x_{\lambda}\lceil^{\lambda}_{\alpha\beta})+\lceil^{\lambda}_{\alpha\sigma}\lceil^{\sigma}_{\beta\lambda}-\lceil^{\lambda}_{\sigma\lambda}\lceil^{\sigma}_{\alpha\beta} \\ R=R^{\alpha}_{\alpha} \end{array}\right\}.$$

where the currents are as before and the stress-tensor T is given by $T = T_{(\pi_0)} + T_{(\mu)} + T_{(\nu)}$ with

$$\left\{\begin{array}{l} T_{(\pi_0)\,\alpha\beta}=(1-2\xi_{(\pi_0)}\varphi_{/\alpha}\varphi_{/\beta}+(2\xi_{(\pi_0)}-\frac{1}{2})g_{\alpha\beta}g^{\rho\sigma}\varphi_{/\rho}\varphi_{/\sigma}-2\xi_{(\pi_0)}\varphi_{/\alpha\beta}\varphi \\[4pt] \qquad +\frac{1}{2}\xi_{(\pi_0)}g_{\alpha\beta}\varphi\Box\varphi-\xi_{(\pi_0)}[R_{\alpha\beta}-\frac{1}{2}Rg_{\alpha\beta}+\frac{3}{2}\xi_{(\pi_0)}Rg_{\alpha\beta}]\varphi^2 \\[4pt] \qquad +\frac{1}{2}[1-3\xi_{(\pi_0)}]m^2 g_{\alpha\beta}\varphi^2 \\[4pt] T_{(\mu)\,\alpha\beta}=\frac{i}{2}[\bar{\psi}_{(\mu)}\gamma_{(\alpha}\Delta_{\beta)}\psi_{(\mu)}-(\Delta_{(\alpha}\bar{\psi}_{(\mu)})\gamma_{\beta)}\psi_{(\mu)}] \\[4pt] T_{(\nu)\,\alpha\beta}=\frac{i}{2}[\bar{\psi}_{(\nu)}\gamma_{(\alpha}\Delta_{\beta)}\psi_{(\nu)}-(\Delta_{(\alpha}\bar{\psi}_{(\nu)})\gamma_{\beta)}\psi_{(\nu)}] \end{array}\right\}.$$

The canonical quantization of the dynamic equation (3.74) gives

$$(3.75) \qquad [\widehat{s^I}(x),\widehat{s^J}(x')]_{\pm} = i\hbar\widetilde{G}^{IJ}(x;x'|s)\,\mathbf{1},$$

where $\widetilde{G}^{IJ}$ is the propagator, solution of the following Cauchy problem:

$$(3.76) \qquad \left\{\begin{array}{l} (J[s].\widetilde{G})^{IJ} = 0 \\ \widetilde{G}^{IJ}(x^0,\mathbf{x};x'^0,\mathbf{x}'|s) = 0 \\ (\partial x'_0.\widetilde{G}^{IJ})(x;x'|s)|_{x^0=x'^0} = \delta^{IJ}\delta(x;x')|_{x^0=x'^0} \end{array}\right\},$$

where $J[s]$. is the Jacobi operator of equation (3.74) at the solution s. Then, by derivation of (3.75) with respect to x'^0, and taking $x^0 = x'^0$, and by considering also equation (3.76), we get:

$$(3.77) \qquad [\widehat{s^I}(x),\widehat{\dot{s}^J}(x')]_{\pm}|_{x^0=x'^0} = i\hbar\delta^{IJ}(x;x')|_{x^0=x'^0}\,\mathbf{1}.$$

Then, similarly to previous examples, as the Jacobi operator is hyperbolic we get that the microscopic causality is conserved even if the geometric mass of the neutrino is negative. Furthermore, as the integral bordism groups $\Omega_3^{E_2} = 0$, we get also the existence of global solutions of the dynamic equation with tunnel effects. $\blacksquare$

Example 3.64. (*Canonical quantization of super PDE's*). In this example we consider the covariant and canonical quantizations of super PDE's as a generalization in supersense of the corresponding quantizations of PDE's. Furthermore, we will show as such quantizations identify quantum super PDE's. Here we work in the category of super manifolds as defined in refs.[62]. To a super PDE $\underline{E}_k \subset J\underline{D}^k(W)$, we associate its *quantum supersitus* $\underline{\Omega}(\underline{E}_k)$, that is the set of admissible integral supermanifolds in $J\underline{D}^k(W)$, that cobord two Cauchy hypersurface data belonging to $\underline{E}_k$. The *classic limit* $\underline{\Omega}(\underline{E}_k)_c$ of $\underline{\Omega}(\underline{E}_k)$ is the set of solutions of $\underline{E}_k$, and the *covariant quantization* of $\underline{E}_k$ can be formulated by associating a quantum superalgebra to the set of *physical superobservables* of $\underline{E}_k$, that are functions on $\Omega(\underline{E}_k)_c$, by means of a method that is the direct extension to the category of super PDE's of the one just considered for PDE's. Then we will show that singular solutions of super PDE's describe whether quantum tunneling effects or covariant quantization. A costructive method to obtain quantum cobords, cobording fixed boundary $X \equiv X_1 \bigcup X_2$, is given.

Let A be a BG **K**-superalgebra (in the sense of ref.[62]) with $\mathbf{K} = \mathbf{R}, \mathbf{C}$. The *general category of superdifferential equations*, $\mathcal{SDE}_g$, is defined by the following: 1) $\mathcal{A} \in Ob(\mathcal{SDE}_g)$ iff $\mathcal{A}$ is a filtered A-algebra $\mathcal{A} \equiv \{\mathcal{A}_i\}, \mathcal{A}_i \subset \mathcal{A}_{i+1}$ such that in the differential calculus in the SFG-category over $\mathcal{A}$ is defined a natural operation $\mathcal{C}$ that satisfies $\mathcal{C}\widetilde{\Omega}^1 \wedge \widetilde{\Omega}^\bullet = \mathcal{C}\widetilde{\Omega}^\bullet$, where $\widetilde{\Omega}^i \equiv \mathcal{A} \wedge \cdots_i \cdots \wedge \mathcal{A}$ are the representative objects of the functor $\underline{D}_i$ in the SFG-category over $\mathcal{A}$, where $\underline{D}_i \equiv \underline{D} \cdots_i \cdots \underline{D}$, being $\underline{D}(P)$ the $\mathcal{A}$-module of all superdifferentiations of algebra $\mathcal{A}$ with values in module P. 2) $f \in Hom(\mathcal{SDE}_g)$ iff f is a homomorphism of filtered algebras preserving operation $\mathcal{C}$. In practice we shall take $\mathcal{A} \equiv \{\mathcal{A}_i \equiv G_w^\infty(M_i; A)\}$, where M_i is a supermanifold. Then, we have a canonical inclusion: $j_i : M_i \to Sp(\mathcal{A}_i), x \mapsto j_i(x) \equiv e_x \equiv$ evaluation map at $x \in M_i$. Then to the inclusion $\mathcal{A}_i \subset \mathcal{A}_{i+1}$ corresponds the smooth map $M_{i+1} \to M_i$. So we set $M_\infty = \underleftarrow{lim} M_i$. We can see that $\overline{M}_\infty = Sp(\mathcal{A}_\infty)$. However, as M_∞ contains all the "nice" points of $Sp(\mathcal{A}_\infty)$, we shall use the space M_∞ to denote an object of *the category of superdifferential equations*. The *category of superdifferential equations* $\underline{SDE}$ is defined by the following: 1) $X \in Ob(\underline{SDE})$ iff X is a supermanifold equipped with a finite dimensional Frobenius graded superdistribution $\widetilde{C}(X) \subset \widetilde{T}X \equiv Hom_{A_0}(A; TX)$, which is locally the same as $\underline{\widetilde{\mathbb{E}}}_\infty \equiv$ graded Cartan superdistribution of $\underline{E}_\infty$ for some SPDE $\underline{E}_k \subset J\underline{D}^k(W)$. We set:

$sdimX \equiv \dim \widetilde{C}(X) = m + n \equiv$ *Cartan superdimension of* $X \in Ob(\underline{SDE})$. $f \in Hom(\underline{SDE})$ iff it is a supersmooth map $f : X \to Y$, where $X, Y \in Ob(\underline{SDE})$, which conserves the corresponding Frobenius graded superdistributions: $\widetilde{T}(f) : \widetilde{C}(X) \to \widetilde{C}(Y)$, $f \in Hom_{\underline{SDE}}(X, Y)$, $sdimX = m + n$, $sdimY = m' + n'$, $srankf = r = \dim(\widetilde{T}(f)_x(\widetilde{C}(X)_x))$, $x \in X$. Then the fibers $f^{-1}(y)$, $y \in im(f) \subset Y$, are $(m + n - r)$-superdimensional objects of $\underline{SDE}$. *Isomorphisms of* $\underline{SDE}$: supermorphisms with fibres consisting of separate points. *Covering maps of* $\underline{SDE}$: supermorphims with zero-superdimensional fibres. Let $A(J\underline{D}^k(W))$ be the sheaf of germs of superdifferentiable functions $J\underline{D}^k(W) \to A$. It is a sheaf of rings, but also a sheaf of A-modules. A subsheaf of ideals $\mathcal{B}$ of $A(J\underline{D}^k(W))$ that is also a subsheaf of A-modules is a *super PDE of order k* on the fiber bundle $\pi : W \to M$. A *regular solution* of $\mathcal{B}$ is a section $s : M \to W$ such that $f \circ D^k s = 0, \forall f \in \mathcal{B}$. The set of *integral points* of $\mathcal{B}$ (i.e., the zeros of $\mathcal{B}$ on $J\underline{D}^k(W)$) is denoted by $J(\mathcal{B})$. The first *prolongation* $\mathcal{B}_{+1}$ of $\mathcal{B}$ is defined as the system of order $k + 1$ on $W \to M$, defined by the $f \circ \pi_{k,k-1}$ and $f^{(1)}$, where $f^{(1)}$ on $D^{k+1}s(p)$ is defined by $f^{(1)}(D^{k+1}s(p)) = (\partial x_\alpha.(f \circ D^k s(p)))$. In local coordinates $(x^\alpha, y^j, y^j_\alpha)$ the *formal derivative* $f^{(1)}$ is given by $f^{(1)}(x^\alpha, y^j, y^j_\alpha) = (\partial x_\alpha.f) + \sum_{[\beta] \le k} y^j_{\beta\alpha}(\partial y^\beta_j.f)$. The system $\mathcal{B}$ is said to be *superinvolutive* at an integral point $q \in J\underline{D}^k(W)$ if the following two conditions are satisfied: (i) $\mathcal{B}$ is a *regular local equation* for the zeros of $\mathcal{B}$ at q (i.e., there are local sections $F_1, ..., F_t \in \Gamma(U, \mathcal{B})$ of $\mathcal{B}$ on an open neighbourhood U of q such that the integral points of $\mathcal{B}$ in U are precisely the points q' for which $F_j(q') = 0$ and $dF_1 \wedge \cdots \wedge dF_t(q) \ne 0$, that is $F_1, \cdots, F_t$ are linearly independent at q; and (ii) there is a neighbourhood U of q such that $\pi^{-1}_{k+1,k}(U) \bigcap J(\mathcal{B}_{+1})$ is a fibered manifold over $U \bigcap J(\mathcal{B})$ (with projection $\pi_{k+1,k}$). For a system $\mathcal{B}$ generated by linearly independent Pfaffian forms $\theta^1, \cdots, \theta^k$ (i.e., a Pfaffian system) this is equivalent to the superinvolutiveness defined for superdistributions. Let $\mathcal{B}$ be a system defined on $J\underline{D}^k(W)$, and suppose that $\mathcal{B}$ is involutive at $q \in J(\mathcal{B})$. Then, there is a neighbourhood U of q satisfying the following. If $\widetilde{q} \in J(\mathcal{B}_{+s})$ and $\pi_{k+s,k}(\widetilde{q})$ is in U, then there is a regular solution s of $\mathcal{B}$ defined on a neighbourhood $p = \pi_{k+s,-1}(\widetilde{q})$ of such that $D^{k+s}s(p) = \widetilde{q}$. (*Cartan-Kuraniski superprolongation theorem*). Suppose that there exists a sequence of integral points $q^{(s)}$ of $\mathcal{B}_{+s}, s = 0, 1, \cdots$, projecting onto each other, $\pi_{k+s,k+s-1}(q^{(s)}) = q^{(s-1)}$, such that: (a) $\mathcal{B}_{+s}$ is a regular local equation for $J(\mathcal{B}_{+s})$ at $q^{(s)}$; and (b) there is a neighborhood $U^{(s)}$ of $q^{(s)}$ in $J(\mathcal{B}_{+s})$ such that its projection under $\pi_{k+s,k+s-1}$ contains a neighborhood of $q^{(s-1)}$ in $J(\mathcal{B}_{+(s-1)})$ and such that $\pi_{k+s,k+s-1} : U^{(s)} \to \pi_{k+s,k+s-1}(U^{(s)})$ is a fibered manifold. Then, $\mathcal{B}_{+s}$ is superinvolutive at $q^{(s)}$ for s large enough. For any $X \in Ob(\underline{SDE})$ set: $\widetilde{\Omega}^i \equiv G_w^\infty(\widetilde{\Lambda}_i^0 X)$; $C\widetilde{\Omega}^1 \subset \Omega^1$: annulator of the graded superdistribution $\widetilde{C}(X)$; $C\widetilde{\Omega}^i \equiv C\widetilde{\Omega}^1 \wedge \widetilde{\Omega}^{i-1}$; $C\widetilde{\Omega} \equiv \bigoplus_{i \geq 0} C\widetilde{\Omega}^i$; $\overline{\widetilde{\Omega}}^i \equiv \widetilde{\Omega}^i / C\widetilde{\Omega}^i$. $C\widetilde{\Omega}$ is a differential ideal of the Grassmann algebra $\widetilde{\Omega} \equiv \bigoplus_{i \geq 0} \widetilde{\Omega}^i$, i.e. $C\widetilde{\Omega}$ is an ideal of $\widetilde{\Omega}$ such that $dC\widetilde{\Omega} \subset C\widetilde{\Omega}$. The operator d induces the factor-operator $\overline{d} : \overline{d}_i : \overline{\widetilde{\Omega}}^i \to \overline{\widetilde{\Omega}}^{i+1}$. In the case $X = J\underline{D}^\infty(W)$, elements $\omega \in \overline{\widetilde{\Omega}}^{m+n}$ are horizontal superdifferential forms locally written as follows: $\omega = L dx^1 \wedge \cdots \wedge dx^{m+n}$, $L : J\underline{D}^\infty(W) \to A$. For any $X \in Ob(\underline{SDE})$ there exists a spectral sequence $\{\widetilde{E}_r^{p,q}, d_r\}$, associated to it such that: $\widetilde{E}_0^{0,i} \equiv \overline{\widetilde{\Omega}}^i$, $d_0^{0,i} = \overline{d}_i$; $\widetilde{E}_1^{0,i} = \overline{\widetilde{H}}^i \equiv \ker(\overline{d}_i)/im(\overline{d}_{i-1})$. In fact, $\{\widetilde{E}_r^{p,q}, d_r\}$ is induced by the natural filtration $\widetilde{\Omega} \equiv C^0\widetilde{\Omega} \supset C\widetilde{\Omega} \equiv C^1\widetilde{\Omega} \supset \cdots \supset C^k\widetilde{\Omega} \supset \cdots$, where $C^k\widetilde{\Omega}$ is the k-th degree of the superdifferential ideal $C\widetilde{\Omega} \subset \widetilde{\Omega}$, p is the filtration index and $p + q$ is the full degree. If $X = \underline{E}_\infty \subset J\underline{D}^\infty(W)$ one has the following: 1) $d_1^{0,m} : \overline{\widetilde{H}}^m \to \widetilde{E}_1^{1,m}$ is the body-Euler-Lagrange superoperator for variational problems constrained by the SPDE $\underline{E}_k$. 2) $\widetilde{E}_1^{0,m-1} = \overline{\widetilde{H}}^{m-1}$ may be interpreted as the group of all higher body-conservation superlaws of $\underline{E}_k$. (See e.g. refs.[62]). 3) Every finite dimensional super-manifold M, when equipped with the zero-superdimensional superdistribution $\widetilde{C}(M) \equiv 0 \subset \widetilde{T}M$, may be regarded as a zero-superdimensional object of $\underline{SDE}$. In this case one has the following: 1) $C\widetilde{\Omega} \equiv \bigoplus_{i \geq 1} \widetilde{\Omega}^i$, $\widetilde{\Omega}^i \equiv G_w^\infty(\widetilde{\Lambda}_i^0 M)$; 2) $C^k\widetilde{\Omega} = \bigoplus_{i \geq k} \widetilde{\Omega}^i$; 3) $\widetilde{E}_0^{p,0} \equiv \widetilde{\Omega}^p$; 4) $\widetilde{E}_0^{p,q} = 0$ if $q > 0$; 5) Since $d_0^{p,q} : \widetilde{E}_0^{p,q} \to \widetilde{E}_0^{p,q+1}$ it follows that $d_0^{p,q} = 0$ for every p, q; 6) $\widetilde{E}_0^{p,q} = \widetilde{E}_1^{p,q}$; 7) $d_1^{p,0} : \widetilde{E}_1^{p,0} = \widetilde{\Omega}^p \to \widetilde{E}_1^{p+1,0} = \widetilde{\Omega}^{p+1} \Rightarrow d_1^{p,0} = d$; 8) $d_2^{p,0} : \widetilde{E}_2^{p,0} = \widetilde{H}^p(M) \to \widetilde{E}_2^{p+1,0} = \widetilde{H}^{p+1}(M)$. Let $\pi : R \to X, \rho : R \to Y$ be epimorphisms in $\underline{SDE}$. Then to each integral sub-supermanifold of X corresponds a set of local integral sub-supermanifolds of Y parametrized by auxiliary coordinates in R, and conversely. [This is called *Bäcklund correspondence* in $\underline{SDE}$]. In fact, let $\pi : R \to X$ be an epimorphism in $\underline{SDE}$. Then for each point $b \in X$ the inverse image $\pi^{-1}(b)$ is an object of $\underline{SDE}$, having Cartan superdimension $l = (r + s) - (m + n)$, where $r + s = sdimR, m + n = sdimX$. Moreover, let $\pi : R \to X$ be an epimorphism in $\underline{\mathcal{SDE}}$. Then any integral sub-supermanifold P of X gives rise to a reduction S of R, where $S = \pi^{-1}(P)$ and $\widetilde{C}(S) = \widetilde{C}(R)|_S$. Finally, let $\rho : R \to Y$ be an epimorphism. Then, locally the image, by means of ρ, of every integral sub-supermanifold of R is an integral sub-supermanifold of Y. Taking into account above considerations we can easily conclude the proof of the theorem. Now, the problem to find a superconservation law for a super PDE can be reconduced to a *Bäcklund correspondence* in $\underline{\mathcal{SDE}}$. Let us, now, consider superdistributions for fiber bundles of A_0-modules on supermanifolds and as these are related to solutions of super PDE's. Furthermore, we will show as to associate to any singular solution L of an affine SPDE a superdistributive kernel (*generalized Green Kernel of L*). Let M be a (m, n)-dimensional supermanifold. Let $\pi : E \to M$ be a fiber bundle of (m', n')-dimensional vector superspaces. E can be considered whether a fiber bundle of A_0-modules or fiber bundle of $\mathbf{K}$-vector spaces over M. So we can define the following structures, reported in Tab.3.15, for any section $i : M_B \to M$, where M_B is the body of M.

TAB.3.15 - Definitions associated to $\pi : E \to M$

$C_0^\infty(i^*E) \equiv \mathbf{K}$-vector space of C^∞ sections of $\pi|i{:}i^*E \to M_B$ with support compact.(†)

$(i^*E)' \equiv (i^*E)^* \otimes_{\mathbf{K}} \Lambda_m^0 M_B \equiv$ *formal adjoint of i^*E.* $((i^*E)'' \cong i^*E)$

$(C_0^\infty(i^*E))' \equiv \mathcal{L}_{\mathbf{K}}(i^*E, \mathbf{K}) \equiv$ *space of distribution sections of i^*E* (†).

$(C_0^\infty((i^*E)'))' \equiv$ *space of distribution-sections of $(i^*E)'$*

$E' \equiv E^* \bigotimes_{\mathbf{K}} \Lambda_m^0 M \equiv$ *formal adjoint of E.*

$E^\phi \equiv E^+ \otimes_A \widetilde{\Lambda}_m^0 M \equiv$ *formal superadjoint of E.*

$C_0^\infty(E)^+ \equiv L_{A_0}(C_0^\infty(E); A) \equiv$ *space of superdistributions on E.*

(†) It is a nuclear LF-and Montel, Lindelof locally convex topological vector space

hence paracopmact and normal. It is also reflexive.

(†)It is nuclear.

One has the following exact sequences of $\mathbf{K}$-vector spaces:

$$0 \to C^\infty((i^*E)') \xrightarrow{\gamma} (C_0^\infty(i^*E))', \gamma : \sigma \mapsto \gamma(\sigma) : f \mapsto \int_{M_B} , <\sigma, f>;$$

$$0 \to C^\infty(i^*E) \xrightarrow{\omega} (C_0^\infty((i^*E)'))', \omega : f \mapsto \omega(f) : \sigma \mapsto \int_{M_B} <\sigma, f> .$$

One has the following exact sequence of vector bundles over M_B: $i^*E' \to (i^*E)' \to 0$. As a consequence we have also the following ones: $C_0^\infty(i^*E') \to C_0^\infty((i^*E)') \to 0$; $(C_0^\infty(i^*E'))' \leftarrow (C_0^\infty((i^*E)'))' \leftarrow 0$. For any section i of b we have the following mappings: (a) $l[i] : C^\infty(E) \times C_0^\infty(E') \to A$; (b) $l[i] : C^\infty(E) \times C_0^\infty(E^\phi) \to A$, given by $l[i](s,\alpha) = \int_{M_B} i^* <\alpha, s>$, $\forall \alpha \in C_0^\infty(E')$, $C_0^\infty(E^\phi)$, $s \in C^\infty(E)$, where the integral belongs to $\mathbf{K}$ in the case (a) and to A in the case (b). Recall that $i^*\eta : M_B \to A \bigotimes_{\mathbf{K}} \Lambda_m^0 M_B$, for any graded m-form η on M. So, for any section i of b we have the following natural homomorphisms of $\mathbf{K}$-vector spaces and A_0-modules respectively: $C^\infty(E) \overset{\underline{l}[i]}{\to} (C_0^\infty(E'))^+$; $C^\infty(E) \overset{\underline{l}[i]}{\to} (C_0^\infty(E^\phi))^+$, given by $\underline{l}[i](s)(\alpha) = \int_{M_B} i^* <\alpha, s>$. Let M and N be (m,n)-dimensional supermanifolds. Let $h \equiv (h_F, h_N)$ be a proper homomorphism of fiber bundles of A_0-modules, i.e. the following diagram:

$$
\begin{array}{ccc}
F & \overset{h_F}{\to} & E \\
\pi_F \downarrow & & \downarrow \pi_E \\
N & \overset{h_N}{\to} & M
\end{array}
$$

is commutative. Proper means that the inverse image of a compact in M is a compact also. Then h identifies the following homomorphism for any section i of $b : N \to N_B$, where N_B is the body of N: $\omega_h[i] : C^\infty(F) \to (C_0^\infty(E^\phi))^+$, $\omega_h[i](f) : \sigma \mapsto \int_{N_B} i^* <h^*\sigma, f>$. So, for any section $f \in C^\infty(F)$ we get a superdistribution of $E^\phi : \omega_h[i](f) \in (C_0^\infty(E^\phi))^+$. Really σ is a A-linear combination of objects of E^ϕ like this $\alpha \otimes \eta \in C_0^\infty(E^\phi)$, so it is well defined the pull-back $h^*(\alpha \otimes \eta)$ by means of the following commutative diagram

$$
\begin{array}{ccc}
F^\phi \equiv F^+ \otimes \widetilde{\Lambda}_m^0 N & \overset{h_F^+ \otimes \widetilde{\Lambda}_m^0(h_N)}{\longleftarrow} & E^+ \otimes \widetilde{\Lambda}_m^0 M \equiv E^\phi \\
\uparrow & & \uparrow \\
N & \underset{h}{\longrightarrow} & M
\end{array}
$$

with $h^*(\alpha \otimes \eta)$ on the left and $\alpha \otimes \eta$ on the right.

As h is proper it follows that $h^*(\alpha \otimes \eta)$ (as well $h^*\sigma$) belongs to $C_0^\infty(F^\phi)$. So has sense the integral $\int_{N_B} i^* <h^*\sigma, f>$. Let L be an *integral supermanifold* (maximal (m,n)-dimensional integral manifold of the Cartan superdistribution on $J\underline{\mathcal{D}}^k(E)$, such that: (i) $\pi_k|L : L \to M$ is a proper map; (ii) $\omega_1^{(k)}(L_B) = 0$, where L_B is the body of L, and $\omega_1^{(k)}(L_B) \equiv i^*_{L_B}\omega_1^{(k)} \in H^1(L_B; \mathbf{Z}_2)$, being $\omega_1^{(k)} \in H^1(I_k(E_B); \mathbf{Z}_2)$ the first generator of $H^1(I_k(E_B); \mathbf{Z}_2) \cong \mathbf{Z}_2[\omega_1^{(k)}, \cdots, \omega_m^{(k)}]$, where $I_k(E_B)$ is the bundle of

Grassmannian of oriented integral planes on $J\mathcal{D}^k(E_B)$, being E_B the body of E. Then L identifies an element of $C_0^\infty(E^\phi)^+$ for any section i of $L \to L_B$, where L_B is the body of L. (For a proof see ref.[57].) Let E and F be bundles of vector superspaces over a (m,n)-dimensional supermanifold M. Let $\kappa : C^\infty(E) \to C^\infty(F)$ be a superdifferential operator over M of order $\leq k$. Then to κ we can associate its *formal superadjoint* $\kappa^\phi : C^\infty(F^\phi) \to C^\infty(E^\phi)$ that is a superdifferential operator of order $\leq k$ such that

(3.78)
$$\int_{M_B} i^* <\kappa(e), \alpha> = \int_{M_B} i^* <e, \kappa^\phi(\alpha)>, \ \forall e \in C_0^\infty(E), \alpha \in C^\infty(F^\phi),$$

and for any section $i : M_B \to M$ of $b : M \to M_B$. More precisely, one has:

(3.79) $\quad <\kappa^\phi(\alpha), e> = <\alpha, \kappa(e)> - dG(e \otimes \alpha), \ \forall e \in C_0^\infty(E), \alpha \in C^\infty(F^\phi),$

where G is a $(k-1)$-order superdifferential operator
(3.80)(*Green superoperator*)
$$G : C^\infty(E \bigotimes F^\phi) \to \underline{\widetilde{\Omega}}^{m-1}(M), \ \underline{\widetilde{\Omega}}^{m-1}(M) \equiv C^\infty(\widetilde{\Lambda}^0_{m-1}M).$$

One has the following properties: (a) The mapping $\kappa \mapsto \kappa^\phi$ is A_0-linear. (b) Let κ_1, κ_2 be two superdifferential operators then $(\kappa_2 \circ \kappa_1)^\phi = \kappa_1^\phi \circ \kappa_2^\phi$. (c) In particular, for trivial bundles $E \cong F \cong M \times A_0$ one has $E^+ \cong M \times A$, $E^\phi \cong F^\phi \cong \widetilde{\Lambda}^0_m M$ and the operation $\kappa \mapsto \kappa^\phi$ transforms scalar differential operators into superdifferential operators between graded m-differential forms on M. Let N be a compact domain of M. Endowe N (and hence ∂N) with the induced orientation from M. Let κ be a superdifferential operator on M between E and F, of order $\leq k$. Let G be a Green superoperator for κ. Let $e \in C_0^k(E)$ and $\alpha \in C_0^k(F^\phi)$ such that $supp(e) \cap N \neq \emptyset \neq supp\alpha \cap N$. Then one has:
(3.81)(*Green's formula*)
$$\int_{N_B} i^* <\kappa(e), \alpha> - \int_{N_B} i^* <e, \kappa^\phi(\alpha)> = \int_{\partial N_B} i^* G(e \otimes \alpha),$$

where N_B is the body of N, and i is any section of $b : N \to N_B$. In particular, if $\partial N = \emptyset$, one has

(3.82)
$$\int_{N_B} i^* <\kappa(e), \alpha> = \int_{N_B} i^* <e, \kappa^\phi(\alpha)>.$$

(For a proof see ref.[60].) Let $\kappa|U$, where U is an open subset of M, defined by

$$\kappa(e)^B = \sum_{c \in \mathbf{I}^m \times \mathbf{I}^n, 1 \leq A \leq m'+n'} \kappa^B_{Ac}(D^c e^A), \ B = 1, ..., m'+n',$$

$$\mathbf{I} \equiv \{1, 2, ..k\} \subset \mathbf{N}$$

$$e \equiv (e^A) \equiv (e^1, \cdots, e^{m'+n'}) \in \mathcal{A}^{m'+n'}, \ \mathcal{A} \equiv C^\infty(U; A)$$

$$c \equiv (c_1, \cdots, c_m, c_{m+1}, \cdots, c_{m+n}), \ D^c \equiv \partial_1^{c_1} \cdots \partial_m^{c_m} \partial_{m+1}^{c_{m+1}} \cdots \partial_{m+n}^{c_{m+n}},$$

(where ∂_ν^S denotes the sth power of partial derivative ∂x_ν), $\kappa^B_{Ac} \in \mathcal{A}$, and only finitely many of these coefficients are nonzero, so that the sum over c is finite. Then, the formal superadjoint of $\kappa|_U$ is the local operator defined by

$$(\kappa^\phi(\alpha))_{AA_1 \cdots A_m} = \sum_{c \in \mathbf{I}^m \times \mathbf{I}^n, 1 \leq B \leq m''+n''} (-1)^{[c]} D^c(\kappa^B_{Ac} \alpha_{BA_1 \cdots A_m}),$$

$$1 \leq A \leq m'+n',$$

where $\alpha_{BA_1 \cdots A_m} \in \mathcal{A}$, $[c] = c_1 + \cdots + c_m + c_{m+1} + \cdots + c_{m+n}$. Now, if $\{U_j\}_{j \in J}$ is an open covering of M, the pull-backs of $\{\kappa|U_j\}_{j \in J}$ by means of i define local A-valued differential m-forms on M_B that can be soldered together into global one by using a partition of unity of M_B subordined to the open covering $\{b(U_j)\}_{j \in J}$. Then, one can easy see that such global differential form on M_B satisfies equation (3.82). Let $\kappa : C^\infty(E) \to C^\infty(F)$ be a k-order A_0-linear differential operator, where E and F are fiber bundles of A_o-modules over a (m, n)-dimensional supermanifold M. Then for any section i of $b : M \to M_B$ we have a natural extension of κ to superdistributional spaces:

$$
\begin{array}{ccc}
C_0^\infty(E^\phi)^+ & \xrightarrow{\ \widetilde{\kappa}[i]\ } & C_0^\infty(F^\phi)^+ \\
\cup & & \cup \\
C^\infty(E) & \xrightarrow[\kappa]{} & C^\infty(F)
\end{array}
$$

One has the following formula for linear differential operators of first order

$$(3.83) \qquad \widetilde{\kappa}[i](\omega)(\alpha) = \omega(\kappa[i]^\phi(\alpha)) + \widetilde{\sigma}_1(\kappa)[i]|_{\partial M}(\omega)(\alpha),$$

where $\omega \in (C_0^\infty(E^\phi))^+, \alpha \in C_0^\infty(F^\phi), \widetilde{\kappa}$ is the distributive extension of $\kappa. : C^\infty(E) \to C^\infty(F)$, $\widetilde{\sigma}_1(\kappa)[i]|_{\partial M}(\omega)$ is the distribution supported on ∂M

identified by the supersymbol $\dot{\sigma}_1(\kappa)$ of κ. 3) Let $\kappa : C^\infty(E) \to C^\infty(F)$ be a k-order A_0-linear differential operator. Let $\underline{E}_k \equiv ker_f \kappa \subset J\underline{D}^k(E)$ be an affine superequation, where $f \in C^\infty(F)$. Suppose that $L \subset J\underline{D}^{k+s}(E)$ is a (m,n)-dimensional integral supermanifold such that $\pi_{k+s}|_L : L \to M$ is a proper map, and $\omega_1{}^{(k+s)}(L_B) = 0$. Then, for any section i of $b : M \to M_B$, $F[L|i] \in (C_0^\infty(E^\phi)^+$ is a superdistribution-solution of $\underline{E}_k$ iff L is a multivalued supersolution. Furthermore, for any section i of b we have that the superdistribution solution of $\underline{E}_k$ can be written as follows: $\nu = G[i]_f$, where $G[i]$ is the *Green's kernel* of κ at the section i, that is $G[i]$ is the superdistributive kernel $G[i] \in C_0^\infty(E^\phi\boxtimes F)^+ \cong C_0^\infty(F\boxtimes E^\phi)^+$ such that $(\widetilde{\kappa}[i] \otimes 1)(G[i]) = (1 \otimes \widetilde{\kappa}[i])(G[i]) = \widetilde{\mathbb{D}}[i]$, being $\widetilde{\mathbb{D}}[i] \in C_0^\infty(F\boxtimes F^\phi)^+$ the *Dirac superkernel* of F, at the section i, i.e., $\widetilde{\mathbb{D}}[i](f\otimes\alpha) = \int_{M_B} i^* < f, \alpha >$, $\forall f \in C_0^\infty(F)$, $\alpha \in C_0^\infty(F^\phi)$. In the particular case that κ is a first order differential superoperator we get: (a) As for the Green's kernel $G[i]$ of the operator $\widetilde{\kappa}[i]$, $_fG[i] \in (C_0^\infty(E^\phi))^+$, for any $f \in C^\infty(F)$, we have:

$$\widetilde{\kappa}[i](_fG[i])(\alpha) =_f G[i](\kappa^\phi(\alpha)) + \widetilde{\sigma}_1(\kappa)[i]|_{\partial M}(_fG[i])(\alpha), \forall\alpha \in C_0^\infty(F^\phi).$$

b) In the case where the Green-kernel $G[i]$ is such tat it can be identifed with a section $G[i] \in C^\infty(E\boxtimes F)$, then we can write:

$$\int_{M_B} i^* < \widetilde{\kappa}[i](_fG[i]), \alpha > = \int_{M_B} i^* <_f G[i], \kappa^\phi(\alpha) >$$
$$+ \int_{\partial M_B} i^* < \dot{\sigma}_1(\kappa)(_fG[i]), \alpha > .$$

The Green kernel satisfies the following equivalent equations:

    (a) $\widetilde{\kappa}[i](_fG[i]) =_f \mathbb{D}[i]$, $\forall f \in C^\infty(F)$.

    (b) $_fG[i](\kappa^\phi(\alpha)) + \widetilde{\sigma}_1(\kappa)[i]|_{\partial M}(_fG)(\alpha) = \int_{M_B} i^* < f, \alpha >,$

    $\forall f \in C^\infty(F)$, $\alpha \in C_0^\infty(F^\phi)$.

If $G[i] \in C^\infty(E\boxtimes F)$ above equations become:

    $(\overline{a})$ $\widetilde{\kappa}[i](_fG[i]) =_f \mathbb{D}[i]$, $\forall f \in C^\infty(F)$.

    $(\overline{b})$ $\int_{M_B} i^* <_f G[i], \kappa^\phi(\alpha) >$
$$+ \int_{\partial M_B} i^* < \dot{\sigma}_1(\kappa)(_fG[i]), \alpha >= \int_{M_B} i^* < f, \alpha > .$$

(For a proof see ref.[48–49].) Finally, we are ready to give the following important theorem that relates singular solutions of affine SPDE's with superdistributive kernels. Any singular supersolution L of an affine SPDE $\underline{E}_k \equiv ker_f\kappa \subset J\underline{D}^k(E)$, that satisfies some boundary conditions, i.e., $\partial L = X_1 \bigcup X_2$, where X_1 and X_2 are some fixed hypersurfaces Cauchy data, and such that : (i) $\pi_{k+s}|_L : L \to M$ is a proper map;(ii) $\omega_1{}^{(k+s)}(L_B) = 0$; identifies, for any section $i : M_B \to M$, a superdistributive kernel $G[L|i] \in C_o^\infty(E^\phi\boxed{\times}F^\phi)^+$. In fact let $F[L|i]$ be the superdistribution associated to L (and i) by means of above theorem. Then, $G[L|i] \in C_0^\infty(E^\phi\boxed{\times}F^\phi)^+$ is defined by means of the following: $G[L|i](\alpha\otimes\phi) = F[L|i](\alpha) \int_{M_B} i^* < \phi, f >$. We call $G[L|i]$ the *generalized Green kernel* of the singular solution $L \subset \underline{E}_{k+s}$ satisfying the boundary condition $\partial L = X_1 \bigcup X_2$, where $\bigcup$ denotes disjoint union. In particular, if L is the singular solution corresponding to the Green kernel $G[i]$ of K, i.e., $F[L|i] = G[i]_f$, then the corresponding generalized Green kernel is defined by $G[L|i](\alpha\otimes\phi) = G[i]_f(\alpha) \int_{M_B} i^* < \phi, f >$. In this case, we call L *Green singular solution* of the boundary value problem $\partial L = X_1 \bigcup X_2$. Let $\mathcal{SDE}$ be the sub-category of $\underline{\mathcal{SDE}}$, where $X \in Ob(\mathcal{SDE})$ iff $X = \underline{E}_\infty = \lim_s \underline{E}_{k+s}$, where $\underline{E}_k \subset J\underline{D}^k(W)$ is a SPDE. There exists a canonical covariant functor, *formal superquantization*, $\underline{\Omega} : \mathcal{SDE} \to \mathcal{M}$, where $\mathcal{M}$ is the category of measurable spaces, (see Definition 3.55), such that $\forall \underline{E}_\infty \in Ob(\mathcal{SDE})$, $\underline{\Omega}(\underline{E}_\infty)$ is the *quantum supersitus* of $\underline{E}_\infty$. The proof can be conduced by following a line parallel to the ordinary case. The formal superquantization becomes effective if on the superquantum situs we recognize a (pre) spectral measure. In fact, in this way we can represent physical superobservables (represented as random variables on $\underline{\Omega}_s(\underline{E}_k)$) as linear operators. Then, in order to show this, we shall redefine the concept of super-Hilbert space. A *super-Hilbert space* is a $\mathbf{Z}_2$-graded commutative A-module, $\mathcal{H}$, such that: (i) A is endowed with a map $-$, *complex conjugation*, that satisfies the following properties: (a) $(\lambda + \lambda')^- = (\overline{\lambda} + \overline{\lambda}')$; (b) $(\lambda\lambda')^- = (\overline{\lambda}'\overline{\lambda})$; (c) If $\lambda \in \mathbf{C}$, $\overline{\lambda}$ is the ordinary c.c. ($\lambda \in A$ is called *realA-module* if $\overline{\lambda} = \lambda$ and *imaginaryA-module* if $\overline{\lambda} = -\lambda$). (ii) $\mathcal{H}$ is endowed with an *inner product*, i.e. by a one-to-one mapping $+ : \mathcal{H} \to \mathcal{H}^+ \equiv L_A(\mathcal{H}; A)$. [We shall denote the vectors of $\mathcal{H}$ by the Dirac-symbolism $|\phi > \in \mathcal{H}$, *(ket)* and the vectors of the dual $\mathcal{H}^+$ by $< \phi|$, *(bra)*. So we can write $|\phi >^+ = < \phi|$ and the inner product of $|\phi >, |\chi > \in \mathcal{H}$ will be denoted by $< \chi|\phi >$.] We recognize that the mapping satisfies the following axioms: (a) $||\phi>^+| = ||\phi>|$; (b) $(\lambda|\phi>)^+ =< \phi|\overline{\lambda}, \forall \lambda \in A$; (c) $(|\phi> +|\chi>)^+ =< \phi|+ < \chi|$; (d) $(< \chi|\phi>)^+ =< \phi|\chi>$. An element $|\phi>\in \mathcal{H}$ is called *physical observable* if it has non-vanishing body. Physical elements of $\mathcal{H}$ are also called *state vectors*. If $B \in L_A(\mathcal{H})$ we can assume, for abuse of notation (and by using the $+$-isomorphism) that $B \in L_A(\mathcal{H}^+)$. So we write indifferently $B(|\phi>) = B|\phi>, B(< \chi|) =< \chi|B$. One has the following properties: (a) $(\lambda < \chi|)B|\phi>= \lambda < \chi|B|\phi>$; (b) $(< \chi|+ < \phi|)B|\alpha>=< \chi|B|\alpha> + < \phi|B|\alpha>$; (c) $(\lambda < \chi|)B = \lambda < \chi|B$; (d) $(B + C)|\phi>= B|\phi> +C|\phi>$, $(B\lambda)|\phi>= B\lambda|\phi>$. As $L_A(\mathcal{H}) \cong L_A(\mathcal{H}^+)$ are $\mathbf{Z}_2$-graded commutative A-modules, one has $< \chi|\lambda B|\phi>= (-1)^{|\lambda||\chi|}\lambda < \chi|B|\phi>= (-1)^{|\lambda|(|B|+||\phi>|)} < \chi|B|\phi> \lambda = (-1)^{|\lambda||B|} < \chi|B\lambda|\phi>$, hence $\lambda B = (-1)^{|\lambda||B|}B\lambda$. The *adjoint* B^+ of $B \in L_A(\mathcal{H})$ is defined by $B^+|\phi>= (< \phi|B)^+$. The operator B is said to be *self-adjoint* iff $B^+ = B$. A linear operator $B \in L_A(\mathcal{H})$ is called *physical observable* iff : (i) B is self-adjoint; (ii) all its eigenvalues belong to A_0 (i.e. are *c-numbers*); (iii) for every eigenvalue there is at least one corresponding *physical eigenvector* (i.e. the corresponding body is non-vanishing); (iv) the set of physical eigenvectors that correspond to *soulless eigenvaules* (i.e. having no component in $A' = A_0' \bigoplus A_1) = A \setminus \mathbf{K}$, contains a complete basis. The soulless eigenvalues will be called *physical eigenvalues*. Then we can prove the following. All the eigenvalues of a physical observable are real. Eigenvectors corresponding to different physical eigenvalues of physical observable are orthogonal. A superquantum (pre-)spectral measure of $\underline{E}_\infty \subset J\mathcal{D}^\infty(W)$ identifies a representation of the algebra $\mathcal{F}_A(\underline{E}_\infty)$ of supernumerical functions $\underline{\Omega}(\underline{E}_\infty) \to A$: $\wedge : \mathcal{F}_A(\underline{E}_\infty) \to L(\mathcal{H}), f \mapsto \hat{f}$, such that $< \psi'|f|\psi>= \int_{\Omega(\underline{E}_\infty)} f dE_{\psi,\psi'}, \forall \psi \in \mathcal{H}, u, \psi' \in \mathcal{H}^+$. In practice the road that links superquantum spectral measures of SPDE's to quantized physical superobservables is reversed. This is obtained by means of the covariant quantization. (See below). We define *physical superobservable* for $\underline{E}_k \subset J\underline{\mathcal{D}}^k(W)$ any random variable $f[i] : \underline{\Omega}(\underline{E}_\infty)_c \to A$, such that $f[i](s) = \int_{M_B} i^*(\underline{f} \circ D^p s)d\mu$, where μ is a suitable measure on M_B, $\underline{f} : J\underline{\mathcal{D}}^p(W) \to A$ is a numerical function on $J\underline{\mathcal{D}}^p(W)$, $p \geq 0$, and i is a section of $b : M \to M_B$. Let us denote by $\mathcal{A}(\underline{E}_k)$ the set of physical superobservables of $\underline{E}_k$. Then covariant quantization of physical superobservables can be directly obtained by using a $\mathbf{Z}_2$-graded commutative Lie algebra just induced by $\underline{E}_k$ on the set $\mathcal{A}(\underline{E}_k)$ of physical superobservables on $\underline{E}_k$. The proceeding is similar to one that we developed for commutative PDE's. Let $\underline{E}_k = ker_\chi \kappa \subset J\underline{\mathcal{D}}^k(W)$ be a SPDE obtained as the kernel of a k-order superdifferential operator $\kappa : J\underline{\mathcal{D}}^k(W) \to K$, with respect to a G_w^∞ section $\chi : M \to K$, of the fiber bundle $\underline{\pi} : \kappa \to M$, defined in the category of supermanifolds. Then for any section $s \in G_w^\infty(W)$, solution of $\underline{E}_k$, and $j[\chi] = \partial\widetilde{\chi}$, being $\widetilde{\chi}$ a superdeformation of χ, we can associate to $\underline{E}_k$ an affine superdifferential equation $\underline{E}_k[s] = ker_{j[\chi]}J[s] \subset J\underline{\mathcal{D}}^k(s^*vTW)$, where $J[s] : C^\infty(s^*vTW) \to C^\infty(\chi^*vTK)$ is a A_0-linear map. We call $\underline{E}_k[s]$ *Jacobi superequation* of $\underline{E}_k$. For any section i of $b : M \to M_B$, we have the following natural extension of $J[s]$:

$$
\begin{array}{ccc}
(C_0^\infty(s^*vTW)^\phi)^+ & \xrightarrow{\;\;J[s|i]\;\;} & (C_0^\infty(\chi^*vTK)^\phi)^+ \\
\uparrow & & \uparrow \\
G^\infty(s^*vTW) & \xrightarrow[\;\;J[s]\;\;]{} & G^\infty(\chi^*vTK)
\end{array}
$$

So, the distribution solution of $\underline{E}_k[s]$ can be written as follows: $\nu = G[s|i]_{j[s]}$, where $G[s|i]$ is the Green's kernel of $J[s]$, (associated to the section i of b). Let us assume that f_i, $i = 1, 2$, are 0-order physical superobservables, i.e., defined by means of A-valued functions on W. We have the following structure of $\mathbf{Z}_2$-graded commutative Lie superalgebra, (for any section $i : M_B \to M$): $(f_1, f_2)[i](s) = G_0^+[s|i](jf_1(s) \otimes jf_2(s) \otimes \eta) - G_0^-[s|i](jf_1(s) \otimes jf_2(s) \otimes \eta) \equiv \widetilde{G}_0[s|i](jf_1(s) \otimes jf_2(s) \otimes \eta)$, where $jf : C^\infty(W) \to \bigcup_{s \in C^\infty(W)} C_0^\infty(s^*vT^+W)$ is the *current associated* to f as follows: $jf(s) = (vdf) \circ s$, where vdf is the vertical differential of the super differentiable function $f : W \to A$. $G^\pm[s|i]$ are advanced and retardet Green-kernels respectively, with respect to the space-time foliation on $i(M_B)$. We call $\widetilde{G}[s|i]$ the *superpropagator at the sections s and i*. The covariant quantization of physical superobservables of $\underline{E}_k$ is given by means of the following quantum bracket: $[\widehat{f}_1, \widehat{f}_2][i](s) = i\hbar\widetilde{G}_0[s|i](df_1(s) \otimes df_2(s) \otimes \eta) \otimes id_{\mathcal{H}}$, where $\mathcal{H}$ is a suitable super-Hilbert space, $(f_j)[i] \in \mathcal{A}(\underline{E}_k), j = 1, 2$. This process can be also extended to quantize p-order physical observables similarly to what just done for PDE's. Each superquantized random supervariable, identifies a quantum superspectral measure $E : (\Omega(\underline{E}_k)_c, \Sigma \hookrightarrow L_A(\mathcal{H}) \subset L_{\mathbf{K}}(\mathcal{H})$. In particular if we consider the p-order physical superobservables, $p \geq 0$, that are identified by Dirac measures on the base manifold M of $\pi : W \to M$, then we get a filtered Lie superalgebra $\widetilde{\mathcal{B}}_0 \subset \cdots \subset \widetilde{\mathcal{B}}_q \subset \cdots \subset \widetilde{\mathcal{B}}_\infty = \widetilde{\mathcal{B}}$, to which, by means of the covariant quantization we can associate a filtered quantum superalgebra $\hat{\mathcal{B}}_0 \subset \cdots \subset \hat{\mathcal{B}}_q \subset \cdots \subset \hat{\mathcal{B}}_\infty = \hat{\mathcal{B}}$ that identifies a quantum super PDE in the *general category of quantum superdifferential equations* $\hat{S\mathcal{D\mathcal{E}}}_g$. Therefore, one has a functor $\tilde{Q} : S\mathcal{D\mathcal{E}} \to \hat{S\mathcal{D\mathcal{E}}}_g$. Then the *canonical quantization* of super PDE's is obtained by introducing a *superframe*, $\tilde{\psi} \equiv (i, \psi)$, where $i : M_B \to M$ is an embedding of the body M_B, endowed with a frame $\psi \equiv (\tau : M_B \to \mathbf{R}, \phi : \mathbf{R} \times M_B \to M_B)$, and by considering also the set of physical pbservables $\tilde{\mathcal{P}}_{hys}^{can}(\underline{E}_k)$ given by $f \in \tilde{\mathcal{B}}$, restricted to M_B, beside their time derivative $\dot{f}$. Then, the covariant quantization of $\underline{E}_k$ identifies a filtered quantum superalgebra $\hat{\mathcal{B}}_t^{can}$, at any instant t, called *canonical quantization* of $\underline{E}_k$, at the time t and with respect to the superframe $\tilde{\psi}$. Finally, let us consider a relation between quantum cobordism and superpropagators. The superpropagator $\tilde{G}[s|i]$ identifies a quantum cobord $L \subset \Omega(\underline{E}_k)$, such that if $\tilde{G}[s|i]$ satisfies the boundary conditions $X_1 \bigcup X_2$, then $\partial L = X_1 \bigcup X_2$. In fact one has a canonical embedding $\underline{E}_k[s] \subset J\underline{\mathcal{D}}^k(W)$. So $E_k[s]$ can be considered also a SPDE of order k on the fiber bundle $\pi : W \to M$. In fact, taking into account the following commutative diagram:

$$
\begin{array}{ccccccc}
\underline{E}_k[s] & \subset & J\underline{\mathcal{D}}^k(s^*vTW) & \cong & (D^k s)^*vTJ\underline{\mathcal{D}}^k(W) & \to & vTJ\underline{\mathcal{D}}^k(W) \\
\downarrow & & \downarrow & & \downarrow & & \downarrow \\
M & = & M & = & M & \xrightarrow{D^k s} & J\underline{\mathcal{D}}^k(W)
\end{array}
$$

we can see that $J\underline{\mathcal{D}}^k(s^*vTW)$ can be identified with an open subset of $J\underline{\mathcal{D}}^k(W)$ (i.e., $J\underline{\mathcal{D}}^k(s^*vTW)$ is the vector fiber neighbourhood of $D^k s(M) \subset J\underline{\mathcal{D}}^k(s^*vTW)$). As a consequence, $\underline{E}_k[s]$ can be identified with a subbundle of $J\underline{\mathcal{D}}^k(W)$, hence with a k-order SPDE on the fiber bundle $\pi : W \to M$. Furthermore, we can also get a stronger result. Really, $\underline{E}_k[s]$ can be identified also with a subbundle of $\underline{E}_k$. In fact, $\underline{E}_k[s]$ is an affine equation, that is an affine bundle over M, with associated vector bundle the linearized equation of $\underline{E}_k$, $(D^k s)^*vT\underline{E}_k \subset (D^k s)^*vTJ\underline{\mathcal{D}}^k(W) \cong J\underline{\mathcal{D}}^k(s^*vTW)$; one has the following natural fiber bundle morphism over M:

$$
\begin{array}{ccc}
\underline{E}_k[s] \times_M (D^k s)^*vT\underline{E}_k & \to & \underline{E}_k[s] \\
\downarrow & & \downarrow \\
M & = & M
\end{array}
$$

As a consequence, with respect to a global regular section (solution) of $\underline{E}_k[s]$, one has the identification of $\underline{E}_k[s]$ with $(D^k s)^*vT\underline{E}_k$. Then, we can identify, even if not canonically, $\underline{E}_k[s]$ with a submanifold of $\underline{E}_k$. This means that any Cauchy hypersurface data for $\underline{E}_k[s]$ can be identified with a Cauchy hypersurface data for $\underline{E}_k$. Then, if $L \subset \underline{E}_{k+h}[s]$ is a singular solution of $\underline{E}_k[s]$ identified by means of the superpropagator $\widetilde{G}[s|i]$, one has that L can be identified with an integral supermanifold of $J\underline{D}^{k+h}(W)$. Moreover, if $X_1 \bigcup X_2 \subset J\underline{D}^{k+h}(s^*vTW)$ are the boundary conditions for $\widetilde{G}[s|i]$, then $\partial L = X_1 \bigcup X_2$, where X_1 and X_2 can be identifed with Cauchy hypersurface data for $\underline{E}_{k+h}[s]$. Hence, L is a quantum cobord cobording just X_1 and X_2. ∎

Example 3.65. (*Covariant and canonical quantization of quantum PDE's*). Covariant and canonical quantizations of quantum PDE's can be also considered by extending to quantum PDE's above formulations for PDE's and super PDE's. See refs.[64,67]. ∎

ADDENDUM I:
BORDISM GROUPS
AND THE (NS)-PROBLEM

Abstract - In refs.[65] are calculated, for the first time, the integral bordism groups of the $3D$ nonisothermal Navier-Stokes equation (NS). A direct consequence of these results is the proof of existence of global (smooth) solutions for (NS). Here we go in some further results emphasizing surgery techniques that allow us to better understand this geometric proof of existence of (smooth) global solutions for any (smooth) boundary condition. A theorem of non-uniqueness of such solutions for general boundary conditions is given on the ground of the symmetry properties of (NS) and just by using our results on the integral bordism groups of (NS) that allows us to recognize smooth solutions with singularities in their characteristic flows. Such solutions are not representable as sections of fibre bundles and allow us to solve the well known "non-analyticity paradox" and "turbulence paradox" [101(b)]. Sufficient conditions for (global) stability of regular (non-necessarily steady-state) solutions $V \subset J\mathcal{D}^2(W)$ are given that, in pseudoisothermal case, can be expressed in term of spectrum of a suitable differential operator associated to V, hence by means of a critical Reynolds number. Furthermore, we relate also the global (un)stability of regular solutions $V \subset (NS) \subset J\mathcal{D}^2(W)$ to a characteristic function, $\dot{f}$ (*fundamental global rate*), associated to V. Nullity of such characteristic function is a sufficient and necessary condition in order that Reynolds, Frank and Brindeman numbers, identified with characteristic adimensional numbers of compact 3-dimensional space-like Cauchy data of regular solutions $V \subset (NS) \subset J\mathcal{D}^2(W)$, should be characteristic numbers also of the same solutions. The isothermal, zero viscosity case, (Euler equation), is considered also in this framework.

4.0 - Introduction

The well known crucial problem in the theory of PDEs: the existence and (non)uniqueness of (smooth) global solutions for the Navier-Stokes equation for incompressible fluids, $((NS)\text{-}problem)^{63}$, is considered in the geometric framework of the integral bordism groups theory for PDE's. We aim to report on such results, in order to emphasize their importance just to solve

---

63 See, e.g., refs.[101(a),(b),(d),(e),(f),(i),(m),(n),(o)].

fundamental problems in the theory of PDEs [63–66]. Furthermore, we analize the uniqueness problem, and by using the symmetry properties of (NS) and their integral bordism groups of order $p \in \{1,2,3\}$, we prove the non-uniqueness of such solutions for general (smooth) boundary conditions. The addendum is divided in three sections. In section 1 we resume some our results on the space-time geometric structure of the (NS) equation and on theorems of existence of local and global solutions given in refs.[56,65]. Here (NS) is seen as a submanifold of the second order jet-derivative space $J\mathcal{D}^2(W)$ built on a suitable fiber bundle $\pi : W \to M$ over the 4-dimensional Galilean space-time M. In particular we recall that (NS) is not a completely integrable, and neither formally integrable, PDE as the canonical mapping $(NS)_{+1} \to (NS)$ is not surjective, where $(NS)_{+1}$ denotes first prolongation. This is enough to state that for any initial condition $q \in (NS)$, it is not possible to find a smooth solution of (NS).[64] However, we are able to identify in (NS) a sub-equation $\widehat{(NS)}$ that, instead, is formally integrable and also completely integrable. Thus for such a sub-equation we can find (local) smooth solutions for any initial condition $q \in \widehat{(NS)}$. Furthermore, by using symmetry properties of $\widehat{(NS)}$, we solve the Cauchy problem for (NS), admiting that the Cauchy data are integral manifolds contained into $\widehat{(NS)}$. In order to determine the structure of global solutions, we consider the integral bordism groups of (NS) as calculated in [65]. Note that at this level we see (NS) as a submanifold of $J^2_4(W)$, that is the second jet-derivative space for 4-dimensional submanifolds of W. This is possible as $J\mathcal{D}^2(W)$ is an open submanifold of $J^2_4(W)$. Furthermore, in this way we are able to recognize also nonregular solutions of (NS) with respect to the fibration $\pi : W \to M$.[65] Such non-regular solutions are very important in fluidodynamics. In fact, for example, waves and vortices can be, in gen-

---

[64] Note that in the language of the geometric theory of PDEs, an *initial condition* on a k-order PDE $E_k \subset J\mathcal{D}^k(W)$ is a point q belonging to the manifold representing E_k. This is not to confuse with a Cauchy data that, instead, is a $(n-1)$-dimensional integral manifold $N \subset E_k$, if $\dim M = n$, with $\pi : W \to M$ the fiber bundle on which is built the k-order jet-derivative space $J\mathcal{D}^k(W)$. (The non-complete integrability of the Navier-Stokes equation was first proved in a paper quoted in refs.[12].)

[65] Nonregular solutions of $(NS) \subset J\mathcal{D}^2(W)$ are 4-dimensional integral submanifolds V of (NS) that, except for a nowhere dense subset $\Sigma(V) \subset V$, can be locally represented as images of the second derivative $D^2 s$ of some (local) section s of $\pi : W \to M$. The set $\Sigma(V)$ is called

eral, described better by means of such solutions. Moreover, the bordism groups allow us to recognize also global (smooth) solutions with change of sectional topology (tunnel effect). We emphasize that the knowledge of the integral bordism groups of (NS) allows us to solve the problem of existence of global (smooth) solutions for (NS). This has been made in section 2 where we consider the problem of existence of global (smooth) solutions for a generic (smooth) boundary value problem. We prove that the existence of such global solutions can be assured for any initial condition into $(\widehat{NS})$, this sub-equation representing the "situs" where we can be sure to find solutions for (NS). Furthermore, we show that for any smooth boundary condition the corresponding global smooth solutions are not unique. In practical cases the seeming uniqueness is to ascribe to the choice at the beginning of some preferred flow. (For example for a fluid in a circular pipe, in general, one fixes the laminar flow.) However, a geometric study of the symmetry properties of (NS) allows us to solve the Cauchy problem by means of the method of characteristics and to recognize many flows at the same time for fixed boundary conditions.[66] Then the existence of many global solutions of a given boundary value problem has been proved by utilizing the property that the Navier-Stokes equation satisfies the (p)-homotopy principle, $0 \le p \le 3$ [65]. In this context we show, in particular, that Reynolds number, identified with a characteristic number associated to any space-like 3-dimensional, (resp. 1-dimensional closed), compact integral manifold, contained into a solution of (NS), does not identify a conservation law, (resp. 1-conservation laws), of (NS). The same considerations are made for the other adimensional numbers, depending on the velocity

---

the set of singular points of V. Note that nonregular solutions of $(NS) \subset J\mathcal{D}^2(W)$ can be regular solutions of $(NS) \subset J_4^2(W)$. In other words, if we consider (NS) a submanifold of $J_4^2(W)$, then a regular solution $V \subset (NS)$ is a 4-dimensional integral submanifold of (NS) that is diffeomorphic to a 4-dimensional submanifold of W by means of the canonical map $\pi_{2,0}: J_4^2(W) \to W$.

[66] Emphasize that to say that the method of characteristics solves the Cauchy problem does not necessarily mean that the characteristic distribution of the Navier-Stokes equation is non-trivial. In fact, it is true exactly the contrary: the characteristic distribution of (NS) and $(\widehat{NS})$ are zero! However, the existence of time-like vector fields in the infinitesimal symmetry algebra of (NS) allows us to generate solutions starting from space-like initial Cauchy data just by using the method of characteristics.

of the flows associated to solutions of (NS), i.e., the Frank and Brindeman numbers. This makes possible transitions of a flow from a stable situation to unstable one, and vice versa, into a same space-time solution of (NS). These last configurations produce singularities in flows with tunnel effects, at dimension 2, i.e., a space-like loop N can be transformed, by means of the flow, into n, $(n > 1)$, loops. (In fact, the first integral bordism groups $\Omega_1^{(NS)} = 0$.) Let us emphasize that the proved existence of smooth global solutions of (NS) does not necessarily means that to such solutions correspond characteristic flows without singularities. The smoothness is simply referred to the 4-dimensional integral manifolds that represent such solutions. There, however, the characteristic vector fields can have some points where the corresponding flows produce tunnel effects on some integral submanifolds. The integral bordism group approach, used by us, allows us to state that exist whether global smooth solutions with characteristic flows without singular points, or global solutions where the characteristic flows have some singular points. These last solutions present tunnel effects, i.e., change of sectional topology passing through singular points. Such solutions allow us to solve well known paradoxes related to the Navier-Stokes equations like "non-analyticity paradox" and "turbulunce paradox" [101(b)]. We give sufficient conditions for global stability of regular (non-necessarily steady-state) solutions $D^2 s(M) = V \subset (NS) \subset J\mathcal{D}^2(W)$ by means of the logaritmic derivative of a characteristic function defined on the set of regular solutions of the linearized Navier-Stokes equation $\widehat{(NS)_s} \subset (NS)$ at the solution $s : M \to W$. In particular for the pseudoisothermal case, i.e., when thermal and pressure perturbations are neglectable, we express such sufficient conditions in term of the spectrum of a suitable differential operator associated to V, hence by means of critical Reynolds number. The (un)stability of a solution V is related also to a characteristic function, $\mathfrak{f}$ (*fundamental global rate*), associated to V. The nullity of such a function guarantees that the Reynolds, Frank and Brindeman number are characteristic numbers of the solution. Therefore, if a solution has zero its fundamental global rate, and its initial Reynolds number is lower than the critical value, the solution is stable (assumed that the thermal and pressure perturbations are neglectable). In section 3 we consider also the previous results in the limit case of zero viscosity and thermal conducibility, i.e., in the case of the Euler equation for isothermal perfect incompressible fluids.

We show that all previous statements about existence and (non)uniqueness of global solutions continue to hold. However, in this case smooth global solutions, without singularities in the associated characteristic flows, preserve the vorticity. This fact is not guaranteed in the Navier-Stokes equation. Furthermore, for the Euler equation, in the $2D$ case, one has the conservation of the vorticity for any solution.

4.1 - GEOMETRY AND EXISTENCE THEOREMS FOR (NS)

In this section we shall recall and resume some our recent results on the geometric structure of the Navier-Stokes equation, (NS), that are important in order to solve the (NS)-problem. Let us start by recasting equation (NS) in its more natural geometric framework that allows us to give an intrinsic and fully covariant characterization of the Navier-Stokes equation, suitable to apply the geometric theory of PDEs.[67]

TAB.4.1 - Navier-Stokes equation: $(NS) \subset J\mathcal{D}^2(W)$

Name	Equation	Costitutive Equation
Continuity	div $(\rho v)=0$	
Motion	div $[\rho v \otimes v - P] - \rho B = 0$	$P = -pg + 2\chi \dot{e}$
Energy	$\rho \frac{\delta e}{\delta t} = <P, \dot{e}> - \text{div } (q)$	$q = \nu \text{grad } (\theta)$

The non-isothermal Navier-Stokes equation (NS), for incompressible fluids, on the Galilean space-time M results a 74-dimensional submanifold of the second order jet-derivative space $J\mathcal{D}^2(W)$ [12], where $W \equiv J\mathcal{D}(M) \times_M T_0^0 M \times_M T_0^0 M$, with $J\mathcal{D}(M) \equiv$ first order jet-derivative space for motions, i.e., first derivative of sections $m : T \to M$ of the Galilean affine fiber bundle $\tau : M \to T$, where T represents the time axis. $T_0^0 M \equiv M \times \mathbf{R}$. A section $s : M \to W$, of $\pi : W \to M$ is a triplet $s = (v, p, \theta)$, where

---

[67] In general the approaches adopted to study the Navier-Stokes equation are in a functional analysis framework [101(e),(h),(i),(m),(n)] or using tools of the differential geometry of infinite dimension manifolds [101(a),(f)]. In this paper we adopt, instead, a geometric setting that is natural in the geometric theory of PDEs. So the Navier-Stokes equation is seen as a submanifold of the second jet-derivative built on a suitable fiber bundle over the Galilean space-time.

$v \equiv$ velocity vector field, $p \equiv$ isotropic pressure field, $\theta \equiv$ temperature field. More precisely, with respect to an inertial frame ψ, (NS) is defined by the equations reported in Tab.4.1. There $\dot{e}$ denotes the infinitesimal strain, $g =$ the vertical metric field of M, $\chi =$ viscosity, $v =$ velocity field, $\nu =$ thermal conductivity, $e =$ internal energy, $B =$ body volume force. To the above equations we must add the thermodynamic constitutive equations for e, χ and ν. We can assume $e = e(\theta)$, $\nu = \nu(\theta)$ and $\chi = \chi(\theta)$. For sake of semplicity we shall assume that ν and χ are constant and $C_p \equiv (\partial\theta.e) =$ constant. Furthermore, we shall assume the body volume force conservative: $B = -\mathrm{grad}(f)$. Let us consider a coordinate system adapted to the frame ψ: $M : (x^\alpha)$, $0 \le \alpha \le 3$; $W : (x^\alpha, \dot{x}^i, p, \theta)$, $1 \le i \le 3$; $J\mathcal{D}(W) : (x^\alpha, \dot{x}^i, p, \theta, \dot{x}^i_\beta, p_\beta, \theta_\beta)$, $0 \le \beta \le 3$. Then, (NS) can be written in the following form:

$$
(NS)\left\{
\begin{array}{l}
(A)\left\{ F^0 \equiv \dot{x}^k G^j_{jk} + \dot{x}^i_s \delta^s_i = 0 \right\} \\[2ex]
(B)\left\{
\begin{array}{l}
F^j \equiv \dot{x}^s R^j_s + \dot{x}^s \dot{x}^i \rho G^j_{is} + \dot{x}^s \dot{x}^j_s \rho + \rho \dot{x}^j_0 \\[1.5ex]
\quad + \dot{x}^k_s S^{js}_k + \dot{x}^k_{is} T^{jis}_k + p_i g^{ij} + \rho(\partial x_i.f)g^{ij} = 0
\end{array}
\right\}, 1 \le j \le 3 \\[4ex]
(C)\left\{
\begin{array}{l}
F^4 \equiv \theta_0 \rho C_p + \dot{x}^k \theta_k \theta_{is} \bar{E}^{is} \rho C_p \\[1.5ex]
\quad + \dot{x}^k \dot{x}^P W_{kp} + \dot{x}^k \dot{x}^s_p \overline{W}^P_{ks} + \dot{x}^k_i \dot{x}^s_p Y^{ip}_{ks} = 0
\end{array}
\right\}
\end{array}
\right\}
$$

with

$$
\left\{
\begin{array}{l}
R^j_s \equiv -2\chi[G^i_{ik}(\partial x_s.g^{kj}) + G^j_{ki}(\partial x_s.g^{ki}) + (\partial x_i.\partial x_s.g^{ij})] \\[2ex]
S^{js}_k \equiv -2\chi[-G^i_{ik}g^{js} - G^i_{ip}g^{ps}\delta^j_k - 2G^j_{ik}g^{is} + \delta^j_k(\partial x_p.g^{ps})] \\[2ex]
T^{jis}_k \equiv 2\chi(g^{sj}\delta^i_k + g^{is}\delta^j_k) \\[2ex]
\bar{E}^{is} \equiv -\nu g^{is} \\[2ex]
W_{kp} \equiv -2\chi G^j_{ik}g_{js}(\partial x_p.g^{si}) \\[2ex]
\overline{W}^P_{ks} \equiv 2\chi[G^j_{ik}g_{js}g^{ip} + G^P_{sk} + (\partial x_k.g^{pe})g_{se}] \\[2ex]
Y^{ip}_{ks} \equiv 2\chi[g_{ks}g^{ip} + \delta^p_k\delta^i_s] \\[2ex]
1 \le i,j,k,p,s \le 3
\end{array}
\right\}.
$$

Here G^i_{jp} are the spatial components of the connection coefficients of the canonical connection on the Galilean space-time. Then, (NS) results an algebraic submanifold of $J\mathcal{D}^2(W)$. The following theorem gives an important structure property of the Navier-Stokes equation.[68]

---

[68] This and the following theorems in this section will be stated without proofs. These, and other informations, can be found in refs.[65].

Theorem 4.1. *Equation* (NS) *is an involutive but not formally integrable, and neither completely integrable, PDE of second order on the fiber bundle* $\pi : W \to M$. *Into* (NS) *we can distinguish an important sub-PDE* $\widehat{(NS)}$:

$$\widehat{(NS)} \;\subset\; (NS) \;\subset\; J\mathcal{D}^2(W) \;\subset\; J_4^2(W)$$

$$\boxed{70} \;<\; \boxed{74} \;<\; \boxed{79} \;=\; \boxed{79}$$

where the framed numbers denote the dimensions of the corresponding overstanding objects. $\widehat{(NS)}$ *is an involutive formally integrable PDE, but with zero characteristic distribution. (Also the characteristic distribution of* (NS) *is zero.) For the set of solutions* $Sol(-)$ *of these equations one has the following inclusions:* $Sol\widehat{(NS)} \subset Sol(NS)$. *Furthermore, one has the following exact commutative diagram:*

$$
\begin{array}{ccccc}
0 & \to & \widehat{(NS)}_{+1} & \to & J\mathcal{D}^3(W) \\
 & & \downarrow & & \downarrow \\
0 & \to & \widehat{(NS)} & \to & J\mathcal{D}^2(W) \\
 & & \downarrow & & \downarrow \\
 & & 0 & & 0
\end{array}
$$

where the index "+1" denotes "first prolongation". The local expression of $\widehat{(NS)}$ *is the following:*

$$
\widehat{(NS)}\left\{
\begin{array}{l}
(A)\left\{F^0 \equiv \dot{x}^k G^j_{jk} + \dot{x}^i_s \delta^s_i = 0\right\} \\[2mm]
(B)\left\{F^0_\alpha \equiv \dot{x}^k(\partial x_\alpha . G^j_{jk}) + \dot{x}^k_\alpha G^j_{jk} + \dot{x}^i_{s\alpha}\delta^s_i = 0\right\},\, 0\le\alpha\le 3 \\[2mm]
(C)\left\{
\begin{array}{l}
F^j \equiv \dot{x}^s R^j_s + \dot{x}^s \dot{x}^i \rho G^j_{is} + \dot{x}^s \dot{x}^j_s \rho + \rho \dot{x}^j_0 \\[1mm]
\quad + \dot{x}^k_s S^{js}_k + \dot{x}^k_{is} T^{jis}_k + p_i g^{ij} + \rho(\partial x_i . f)g^{ij} = 0
\end{array}
\right\},\, 1\le j\le 3 \\[2mm]
(D)\left\{
\begin{array}{l}
F^4 \equiv \theta_0 \rho C_p + \dot{x}^k \theta_k \theta_{is} \bar{E}^{is} \rho C_p \\[1mm]
\quad + \dot{x}^k \dot{x}^P W_{kp} + \dot{x}^k \dot{x}^s_p \overline{W}^p_{ks} + \dot{x}^k_i \dot{x}^s_p Y^{ip}_{ks} = 0
\end{array}
\right\}
\end{array}
\right\},
$$

where the symbols of the coefficients are like before. (NS) *is an affine fiber bundle over the affine submanifold* $(C) \subset J\mathcal{D}(W)$. *More precisely, we can write* $(NS) = \bigcup_{\bar{q}\in(C)}(NS)_{\bar{q}}$, *where* $(NS)_{\bar{q}}$ *is a 46-dimensional affine submanifold of* $\pi^{-1}_{2,1}(\bar{q})$ *with associated vector space the symbol* $(g_2)_q$ *of* (NS) *at any point* $q \in (NS)_{\bar{q}}$. *Furthermore,* (C) *is a fiber bundle over* W. *One has the following exact commutative diagram and short exact sequence:*

$$
\begin{array}{ccccc}
J\mathcal{D}^2(W) & \supset & (NS) & \to & W & \to & 0 \\
\downarrow \pi_{2,1} & & \downarrow & & \| \\
J\mathcal{D}(W) & \supset & (C) & \to & W & \to & 0 \qquad 0 \to g_2 \to vT(NS) \to \pi_{2,1}^* vT(C) \to 0 \\
\downarrow & & \downarrow \\
0 & & 0
\end{array}
$$

Remark 4.1. The maximal integral manifolds of $J_4^2(W)$ are of type $0 \le p \le 4$, with dimensions reported in the following table. (If V is a maximal integral manifold of $J_4^2(W)$ one has $type(V) = \dim(T(\pi_{2,0})(T_p V))$, $\pi_{2,0} : J_4^2(W) \to W$, $q \in V - \Sigma(V)$, $\Sigma(V) =$ set of Thom-Bordman singular points of V.)

TAB.4.2 - Dimensions of the maximal integral manifolds of $J_4^2(W)$

Type	Dimension	Remark
4	4	solutions of $J_4^2(W)$
3	8	
2	17	
1	31	
0	50	$\pi_{2,1}^{-1}(\bar{q})$, $\bar{q} \in J_4^1(W)$

The 4-dimensional maximal integral manifolds (of type 4) contained into (NS) are called solutions of (NS), and the integral manifolds $(NS) \cap \pi_{2,1}^{-1}(\bar{q})$, $\bar{q} \in \pi_{2,1}(NS)$, are integral manifolds of (NS) of dimension 46 transversal to the regular integral manifolds of dimension 4. Any integral manifold of $J_4^2(W)$ is transformed into integral manifolds by means of any Lie symmetry of $J_4^2(W)$, i.e., transformations of $J_4^2(W)$ that preserve the Cartan distribution of $J_4^2(W)$. Furthermore, any infinitesimal Lie symmetry that belongs to the Cartan distribution (characteristic vector field), transforms any maximal integral manifold into itself. The Cartan distribution $\mathbf{E}_2(\widehat{NS})$ of $(\widehat{NS})$ is the Cauchy characteristic distribution associated to the contact ideal $\mathcal{C}_2 \subset \Omega^\bullet(J_4^2(W))$ generated by the following differential forms:

$$
\left\{
\begin{aligned}
&\epsilon^0 \equiv dF^0 = (\partial x_\alpha . G_{jk}^j)\dot{x}^k dx^\alpha + G_{jk}^j d\dot{x}^k + d\dot{x}_s^k \delta_k^s \\[4pt]
&\epsilon_\alpha^0 \equiv dF_\alpha^0 = [(\partial x_\beta \partial x_\alpha . G_{jk}^j)\dot{x}^k + (\partial x_\beta . G_{jk}^j)\dot{x}_\alpha^k]dx^\beta + (\partial x_\alpha . G_{jk}^j)d\dot{x}^k + G_{jk}^j d\dot{x}_\alpha^k + d\dot{x}_{i\alpha}^i \\[4pt]
&\epsilon^j \equiv dF^j = A_\alpha^j dx^\alpha + B_i^j d\dot{x}^i + \rho d\dot{x}_0^j + S_k^{js} d\dot{x}_s^k + T_k^{jis} d\dot{x}_{is}^k + g^{kj} dp_k \\[4pt]
&\epsilon^4 \equiv dF^4 = \bar{A}_\beta dx^\beta + \bar{B}_k d\dot{x}^k + \bar{C}^\alpha d\theta_\alpha + \bar{D}_p^s d\dot{x}_s^p + \bar{E}^{is} d\theta_{is} \\[4pt]
&\qquad \omega^j \equiv d\dot{x}^j - \dot{x}_\alpha^j dx^\alpha,\ \omega_\alpha^j \equiv d\dot{x}_\alpha^j - \dot{x}_{\alpha\beta}^j dx^\beta,\ \omega^4 \equiv dp - p_\alpha dx^\alpha \\[4pt]
&\qquad \omega_\alpha^4 \equiv dp_\alpha - p_{\alpha\beta}dx^\beta,\ \omega^5 \equiv d\theta - \theta_\alpha dx^\alpha,\ \omega_\alpha^5 \equiv d\theta_\alpha - \theta_{\alpha\beta}dx^\beta \\[4pt]
&\qquad\quad 1\le j\le 3,\ 0\le\alpha\le 3
\end{aligned}
\right\} ,
$$

with

$$
\left\{
\begin{aligned}
A_\alpha^j \equiv\ & \dot{x}^s(\partial x_\alpha . R_s^j) + \dot{x}^i \dot{x}^s \rho(\partial x_\alpha . G_{is}^j) + \dot{x}_s^k(\partial x_\alpha . S_k^{js}) + \dot{x}_{is}^k(\partial x_\alpha . T_k^{jis}) \\[4pt]
& + \rho(\partial x_\alpha . g^{ij})(\partial x_i . f) + \rho g^{ij}(\partial x_\alpha \partial x_j . f) + (\partial x_\alpha . g^{ij})p_i \\[4pt]
B_i^j \equiv\ & \dot{x}^k 2\rho G_{ik}^j + \dot{x}_i^j \rho + R_i^j \\[4pt]
\bar{A}_\beta \equiv\ & \theta_{is}(\partial x_\beta . \bar{E}^{si}) + \dot{x}^k \dot{x}^p(\partial x_\beta . W_{kp}) + (\partial x_\beta . \bar{W}_{ks}^p) + \dot{x}^k \dot{x}_p^s(\partial x_\beta . Y_{ks}^{ip}) \\[4pt]
\bar{B}_k \equiv\ & \theta_k \rho C_p + \dot{x}^p(W_{kp} + W_{pk}) + \dot{x}_p^s \bar{W}_{ks}^p \\[4pt]
\bar{C}^\alpha \equiv\ & \dot{x}^\alpha \rho C_p\ (\dot{x}^0 = 1) \\[4pt]
\bar{D}_s^p \equiv\ & \dot{x}^k \bar{W}_{ks}^p + 2\dot{x}_i^k Y_{ks}^{ip} \\[4pt]
& 1\le i,j,k,p,s\le 3,\quad 0\le\alpha,\beta\le 3
\end{aligned}
\right\} .
$$

Therefore

$$
\mathcal{C}_2 = < \omega^I, \omega_\alpha^I, \epsilon^J, \epsilon_\alpha^0 >,\quad 1\le I\le 5,\quad 0\le J\le 4,\quad 0\le\alpha\le 3.
$$

The Cartan distribution $\mathbf{E}_2(\widehat{NS})$ is generated by vector fields of the following type:

$$
\begin{aligned}
(4.1)\qquad \zeta = X^\alpha[&\partial x_\alpha + \dot{x}_\alpha^j \partial\dot{x}_j + p_\alpha \partial p + \theta_\alpha \partial\theta + \dot{x}_{\alpha\beta}^j \partial\dot{x}_j^\beta + p_{\alpha\beta}\partial p^\beta + \theta_{\beta\alpha}\partial\theta^\beta] \\
& + \dot{X}_{\alpha\beta}^j \partial\dot{x}_j^{\alpha\beta} + Y_{\alpha\beta}\partial p^{\alpha\beta} + Z_{\alpha\beta}\partial\theta^{\alpha\beta}
\end{aligned}
$$

where the components X^α, $\dot{X}_{\alpha\beta}^j$, $Y_{\alpha\beta}$, $Z_{\alpha\beta}$ satisfy the following equations:

$$
(4.2)\qquad
\left\{
\begin{aligned}
& X^\alpha[A_\alpha^j + B_i^j \dot{x}_\alpha^i + \rho\dot{x}_{0\alpha}^j + S_k^{js}\dot{x}_{s\alpha}^k + g^{kj}p_{\alpha k}] + T_k^{jis}\dot{X}_{is}^k = 0 \\[4pt]
& X^\alpha[\bar{A}_\alpha + \bar{B}_k \dot{x}_\alpha^k + \bar{C}^s \theta_{\alpha s} + \rho C_p \theta_{0\alpha} + \bar{D}_k^s \dot{x}_{s\alpha}^k] + \bar{E}^{is} Z_{is} = 0 \\[4pt]
& X^\beta[(\partial x_\beta \partial x_\alpha . G_{jk}^j)\dot{x}^k + (\partial x_\beta . G_{jk}^j)\dot{x}_\alpha^k + \dot{x}_\beta^k(\partial x_\alpha . G_{jk}^j) + \dot{x}_{\beta\alpha}^s G_{jk}^j] + \dot{X}_{\alpha s}^s = 0 \\[4pt]
& 1\le j\le 3, 0\le\alpha\le 3
\end{aligned}
\right\} .
$$

Therefore, $\mathbf{E}_2(\widehat{NS})$ is a distribution of dimension 46: $\dim \mathbf{E}_2(\widehat{NS}) = 46$. Similarly one can see that the Cartan distribution $\mathbf{E}_2(NS)$ on (NS) is of dimension 50. Between the Cartan vector fields there are also the vertical vector fields, i.e., $(X^\alpha) = 0$, belonging to the symbol $\widehat{g}_2$ of $(\widehat{NS})$. In fact, for $(X^\alpha) = 0$ equations (4.2) reduce to the following ones:

$$\left\{\begin{array}{l} \bar{E}^{is}Z_{is}=0=g^{si}Z_{is} \\[2mm] T_k^{jis}\dot{X}_{is}^k=0=(g^{is}\delta_k^i+g^{is}\delta_k^j)\dot{X}_{is}^k \\[2mm] \dot{X}_{\alpha s}^s=0 \end{array}\right\}, \quad 1 \leq j \leq 3, \quad 0 \leq \beta \leq 3.$$

These are just the conditions that require for the vector fields on $(\widehat{NS})$ to belong to $\widehat{g}_2$. Furthermore, $\widehat{g}_2$ is an involutive distribution of dimension 42 on $(\widehat{NS})$. So through any point $q \in (\widehat{NS})$ passes a maximal integral manifold of $\widehat{g}_2 \to (\widehat{NS})$ that has just dimension 42 and its tangent space belongs to the kernel of the tangent mapping of $\pi_{2,1}|_{\widehat{(NS)}} : (\widehat{NS}) \to J\mathcal{D}(W)$.

The vector fields belonging to the involutive distribution $\widehat{g}_2 \to (\widehat{NS})$ cannot be characteristic vector fields of $(\widehat{NS})$. Hence, characteristic vector fields cannot have $(X^\alpha) = 0$ into the formula (4.1).

The Cartan distribution $\mathbf{E}_\infty(\widehat{NS})$, (that has dimension 4), of the infinity prolongation of $(\widehat{NS})$, coincides with the characteristic distribution of $(\widehat{NS})_{+\infty}$ and the 4-dimensional integral manifolds of $\mathbf{E}_\infty(\widehat{NS})$ represent regular integral solutions of $(\widehat{NS}) \subset (NS)$. (The same situation holds for (NS).) A *Cartan connection* on (NS) is a 4-dimensional sub-distribution $\mathbf{H} \subset \mathbf{E}_2(NS)$ such that $T(\pi_{2,1})(\mathbf{H}_q) = L_q, \forall\ q \in (NS)$, where $L_q \subset T_{\pi_{2,1}(q)}J_4^1(W)$ is the subspace uniquely identified by $q \equiv [N]_a^2 \in J_4^2(W)$. More precisely, $L_q = T_{\pi_{2,1}}N^{(1)}$. Note that $(NS)_{+1}$ is a submanifold of $J_4^3(W)$ but the mapping $(NS)_{+1} \to (NS)$ induced by $\pi_{3,2} : J_4^3(W) \to J_4^2(W)$ is not an epimorphism. A regular solution V of (NS) identifies a Cartan connection along the points of V. We call *curvature* of the Cartan connection $\mathbf{H}$ on (NS) the field of geometric objects

$$\Omega(\mathbf{H}) : (NS) \to \Lambda^2(\mathbf{H}^*)\bigotimes[TN \otimes \nu^*/Ann(g_2)]^*.$$

A Cartan connection $\mathbf{H}$ gives a splitting of the Cartan distribution of (NS): $\mathbf{E}_2 \cong g_2 \bigoplus \mathbf{H}$. Two Cartan connections $\mathbf{H}, \mathbf{H}'$ identify a field of geometric objects λ on (NS) called *soldering form*: $\lambda : (NS) \to \mathbf{H}^* \bigotimes g_2, \lambda(q) \in$

$T_a^* N \otimes g_{2,q}$. One has

$$\Omega(\mathbf{H}') = \Omega(\mathbf{H}) + \delta\lambda \quad \left\{ \begin{array}{ll} (\textit{Bianchi identity}) & \delta\Omega(\mathbf{H})=0 \\[2ex] \Omega(\mathbf{H})(q) \quad (\bmod & \delta(T_a^* N \otimes g_{2,q})) \in H^{1,2}(NS)_q \end{array} \right\}$$

where $H^{1,2}(NS)_q$ is the $(1,2)$-Spencer cohomology space of (NS) at $q \in (NS)$. Such a δ-cohomology class of $\Omega(\mathbf{H})(q)$ is called *Weyl tensor* of (NS) at $q \in (NS)$: $W_2(q) = [\Omega(\mathbf{H})(q)]$. Then, there exists a point $u \in (NS)_{+1}$ over $q \in (NS)$ iff $W_2(q) = 0$. Thus, for any point $q \in (\widehat{NS}) \subset (NS)$ one has $W_2(q) = 0$. $\blacksquare$

Theorem 4.2. (Existence of local regular solutions for (NS)). *For any point $q \in (\widehat{NS}) \subset (NS)$ we can construct a (smooth) regular solution of $(NS) \subset J_4^2(W)$.*

Theorem 4.3. (Cauchy problem for $(NS) \subset J_4^2(W)$). *Any (regular) integral submanifold $N \subset (\widehat{NS}) \subset (NS) \subset J_4^2(W)$ of dimension 3 that satisfies the initial conditions and the transversality conditions, with respect to a time-like vector field $\zeta \in \mathfrak{s}(C_2)$, (where C_2 is the ideal encoding $(\widehat{NS})$ given in Remark 4.1), generates a (regular, smooth) solution of (NS) if N has no frozen singularities. More precisely, the initial and transversal conditions are the following: $\zeta \notin \mathfrak{d}(TN)$, $\psi^* C_2 = 0$, $\psi^*(\zeta \rfloor C_2) = 0$, where $\psi : \Xi \to (\widehat{NS})$ is the mapping that represents the manifold N. Here $\mathfrak{d}(TN)$ denotes the $\Omega^0(N)$-module of all vector fields belonging to the tangent space of N, where $\Omega^0(N)$ is the space of smooth zero-forms on N. Moreover, $\mathfrak{s}(\mathcal{I})$ denotes the space of infinitesimal symmetries of an ideal $\mathcal{I}$ of differential forms. Furthermore, we say that N is space-like if it is contained into $(\widehat{NS})_t \equiv J\mathcal{D}^2(W)_t \bigcap (\widehat{NS})$, for some $t \in T$, where $J\mathcal{D}^2(W)_t$ is the fiber of $J\mathcal{D}^2(W)$ over $t \in T$, with respect to the following projection:*

$$J\mathcal{D}^2(W) \xrightarrow{\ \pi_{2,0}\ } W \xrightarrow{\ \pi\ } M \xrightarrow{\ \tau\ } T.$$

Remark 4.2. Emphasize that Theorem 4.3 is about existence of smooth solutions, and not simply analytic ones! In fact, even if from Theorem 4.2 we can state the existence of local analytic solutions as a consequence of the fact that $(\widehat{NS})$ is a formally integrable and analytic (or better algebraic) PDE, it is also true that even if the integral manifold N, considered in Theorem 4.3, should be analytic, the corresponding integral manifold generated by a vector field $\zeta \in \mathfrak{s}(C_2)$ will be smooth in general. $\blacksquare$

In order to pass from local existence theorems to global ones and characterize the topology of these global solutions, it is necessary to consider the integral bordism group in dimension 3 of (NS). In the following we will expound in some details the meaning of this concept.

Remark 4.3. We shall say that a p-dimensional, $p \leq 3$, compact closed smooth integral manifold $N \subset (NS)$ is *admissible* if the following conditions hold: (i) The set $\Sigma(N)$ of singular points of N has no open subsets and has no frozen singularities; (ii) For N passes at least one integral manifold V of dimension $p+1$ such that its set $\Sigma(V)$ of singular points has no open subset; (iii) $\Sigma(N)$ can be solved by means of integral deformations. From Theorem 4.3 it follows that the set of such admissible p-dimensional, $p \leq 3$, compact closed smooth integral manifolds is not empty. We define *integral bordism* in (NS) the usual bordism, but with the additional requirement that the manifolds contained in (NS) must be admissible integral ones. We denote by $\Omega_p^{(NS)}$ the set of integral bordism classes $[N]_{(NS)}$ of (NS) with $0 \leq p \leq 3$. The operation of taking disjoint union $\bigcup$ defines a sum $+$ on $\Omega_p^{(NS)}$ so that it becomes an abelian group (*integral p-bordism group* of (NS)). We call *bar singular chain complex* of (NS), with *coefficients into an abelian group* G, the chain complex $\{\bar{C}_p((NS); G), \bar{\partial}\}$ where $\bar{C}_p((NS); G)$ is the G-module of formal linear combinations with coefficients in G, $\sum_i \lambda_i c_i$, where c_i is a singular p-chain $f : \Delta^p \to (NS)$ that extends on a neighborhood $U \subset \mathbf{R}^{p+1}$, such that f on U is differentiable and $Tf(\Delta^p) \subset \mathbf{E}_2(NS)$. Denote by $\bar{H}_p((NS); G)$ the corresponding homology (*bar singular homology with coefficients in G of (NS)*). Let $\{\bar{C}^p((NS); G) \equiv Hom_{\mathbf{Z}}(\bar{C}_p((NS); \mathbf{Z}), G), \bar{\delta}\}$ be the corresponding dual complex and let $\bar{H}^p((NS); G)$ be the associated homology spaces (*bar singular cohomology with coefficients into G of (NS)*). Let us denote by $^G\Omega_{p,s}^{(NS)}$ the corresponding bordism groups in the singular case. Let us denote also by $^G[N]_{(NS)}$ the equivalence classes of integral singular bordisms. In the following, for semplicity, we will consider the case $G = \mathbf{R}$ only and we will omit the apex G in the symbols.   ■

Theorem 4.4. (*Integral bordism groups in (NS)*). *In the following table we summarize the calculated integral bordism groups of (NS).*

TAB.4.3 - Integral bordism groups in the Navier-Stokes equation

p	$\Omega_p^{(NS)}$	$\Omega_p^{(NS)+\infty}$	$\Omega_{p,s}^{(NS)}$	$\Omega_{p,s}^{(NS)+\infty}$
0	$\mathbf{Z}_2$	$\mathbf{Z}_2$	$\mathbf{R}$	$\mathbf{R}$
1	0	0	0	0
2	$\mathbf{Z}_2$	$\mathbf{Z}_2$	0	0
3	0	0	0	0

Proof. For a detailed proof of the results reported in Tab.4.3 see the last but one publication in refs.[65]. $\qquad\square$

Theorem 4.5. (Tunnel effects in the Navier-Stokes equation). *In the set of p-dimensional, $2 \le p \le 4$, integral manifolds of (NS) one has ones that change their sectional topology. More precisely one has the following: Let V be a p-dimensional compact admissible integral manifold of (NS) such that $\partial V = N_0 \bigcup N_1$, where N_0 and N_1 are closed compact smooth integral manifolds of (NS). Then, there exists an integral admissible Morse function $f : V \to [a,b]$ such that $f^{-1}(a) = N_0$, $f^{-1}(b) = N_1$, and such that:*

(A) (Simple tunnel effect). If f has a critical point q of index k, then there exists a k-cell $e^k \subset V - N_1$ and a $(p-k)$-cell $e_*^{\,p-k} \subset V - N_0$ such that:

(i) $e^k \cap N_0 = \partial e^k$;

(ii) $e_*^{\,p-k} \cap N_1 = \partial e_*^{\,p-k}$;

(iii) there is a deformation retraction of V onto $N_0 \cup e^k$;

(iv) there is a deformation retraction of V onto $N_1 \cup e_*^{\,p-k}$;

(v) $e_*^{\,p-k} \cap e^k = q$; $e_*^{\,p-k} \pitchfork e^k$. (Here $\pitchfork$ denotes transverse.)

(B) (Multi tunnel effect). If f is of type $(\nu_0, \dots, \nu_n)$, where ν_k denotes the number of critical points with index k, such that f has just one critical value c, $a < c < b$, then there are disjoint k-cells $e_i^k \subset V - N_1$, and disjoint $(p-k)$-cells $(e_*)_i^{\,p-k} \subset V - N_0$, $1 \le i \le \nu_k$, $k = 0, \dots, n$, such that:

(i) $e_i^k \cap N_0 = \partial e_i^k$;

(ii) $(e_*)_i^{\,p-k} \cap N_1 = \partial (e_*)_i^{\,p-k}$;

(iii) there is a deformation retraction of V onto $N_0 \cup \left\{ \bigcup_{i,k} e_i^k \right\}$;

(iv) there is a deformation retraction of V onto $N_1 \cup \left\{ \bigcup_{i,k} (e_*)_i^{\,p-k} \right\}$;

(v) $(e_*)_i^{\,p-k} \cap e_i^k = q_i \in V$; $(e_*)_i^{\,p-k} \pitchfork e_i^k$.

(C) (*No topology transition*). *If f has no critical point then $V \cong N_0 \times I$, where $I \equiv [0, 1]$.*

In particular, in the set $\mathcal{S}ol(NS)$, of all solutions of the Navier-Stokes equation, there are also solutions that present (multi)tunnel effects. These are not, in general, representable are image of second derivative of sections of the fiber bundle $\pi : W \equiv J\mathcal{D}(M) \times_M T_0^0 M \times_M T_0^0 M \to M$.[69]

4.2 - NAVIER-STOKES BOUNDARY VALUE PROBLEMS

In this section we consider the question if to each smooth boundary value problem on (NS) one can associate an unique global (smooth) solution. We shall prove that for the Navier-Stokes equation we have not, in general, the uniqueness of solutions for Cauchy problems, and neither the uniqueness of solutions for general boundary value problems. These facts are essentially related to the integral bordism groups of (NS) and its symmetry properties.

Definition 4.1. We define *global (smooth) solution* of (NS) any (smooth) solution of (NS) that is defined in any space-like region inside the boundary and for any $t \geq t_0$, where t_0 is an initial time.

Theorem 4.6. (Existence of global (smooth) solutions of (NS)). *For a generic (smooth) boundary value problem contained into $\widehat{(NS)} \subset (NS)$ exist global (smooth) solutions of (NS).*

Proof. A boundary value problem for (NS) can be directly described in the manifold $(NS) \subset J\mathcal{D}^2(W) \subset J_4^2(W)$ by requiring that a 3-dimensional compact space-like (for some $t = t_0$), admissible integral manifold $B \subset \widehat{(NS)} \subset (NS)$ propagates in (NS) in such a way that the boundary ∂B

---

[69] W is a trivial affine fiber bundle over the Galilean space-time M, (that is also an affine 4-dimensional manifold, with fiber the following 5-dimensional affine space: $\mathbf{I} \times \mathbf{R}^2$. Here $\mathbf{I} \subset \mathbf{M}$ is the 3-dimensional affine subspace of free vectors defined by $\mathbf{I} \equiv \{v \in \mathbf{M} | <v, \underline{\tau}> = 1\}$, where $\underline{\tau} \in \mathbf{M}^*$ is the 1-form on $\mathbf{M}$ identified by the differential of the affine universal time function $\tau : M \to T \equiv \mathbf{R}$. (on the time axis T one has fixed an affine frame, i.e., an origin of times and an unity of times.) We call $\mathbf{I}$ the space of *time-like* vectors. We call $\mathbf{S} \equiv \ker(\underline{\tau})$ the space of *space-like* vectors. Of course $\mathbf{I} \bigcap \mathbf{S} = \varnothing$. $\mathbf{I}$ is a 3-dimensional affine space having $\mathbf{S}$ as space of free vectors. On $\mathbf{S}$ is defined an Euclidean structure. The space-time M has the following structure $M = \bigcup_{t \in T} M_t$, where M_t are affine 3-dimensional Euclidean spaces with $\mathbf{S}$ as space of free vectors. Therefore for the vertical tangent space vTM one has the following identification $vTM \cong M \times \mathbf{S}$. One has also the following canonical isomorphism $u^*(vTJ\mathcal{D}(M)) \cong M \times \mathbf{S}$, for any section $u : M \to J\mathcal{D}(M)$ of the fiber bundle $J\mathcal{D}(M) \to M$.

describes a fixed 3-dimensional time-like integral manifold $Y \subset \widehat{(NS)} \subset (NS)$.[70] Y is not, in general, a closed (smooth) manifold. However, we can solder Y with two other compact 3-dimensional integral manifols X_i, $i = 1, 2$, in such a way that the result is a closed 3-dimensional (smooth) integral manifold $Z \subset \widehat{(NS)} \subset (NS)$. More precisely, we can take $X_1 \equiv B$, so that $\widetilde{Z} \equiv X_1 \bigcup_{\partial B} Y$ is a 3-dimensional compact integral manifold such that $\partial \widetilde{Z} \equiv C$ is a 2-dimensional space-like integral manifold. We can assume that C is an orientable manifold. Then from Tab.4.3 it follows that $\partial X_2 = C$, for some space-like compact 3-dimensional integral manifold $X_2 \subset \widehat{(NS)}$. Set $Z \equiv \widetilde{Z} \bigcup_C X_2$. Therefore, one has $Z = X_1 \bigcup_{\partial B} Y \bigcup_C X_2$. Then, from Theorem 4.4 (Tab.4.2) it follows that there exists a 4-dimensional integral (smooth) manifold $V \subset \widehat{(NS)} \subset (NS)$ such that $\partial V = Z$. It follows that the integral manifold V is a solution of our boundary value problem between the times t_0 and t_1, where t_0 and t_1 are the times corresponding to the boundaries where are soldered X_i, $i = 1, 2$ to Y. Now, this process can be extended for any $t_2 > t_1$. So we are able to find (smooth) solutions for any $t > t_0$. Therefore we are able to find global (smooth) solutions. Remark that such global solutions, obtained by gluing smooth manifolds together, are 4-dimensional topological manifolds that are, in general, piecewise smooth only. However, in order to assure the smoothness of the global solutions so built it is enough to develop such constructions in the infinity prolongation $\widehat{(NS)}_{+\infty}$ of $\widehat{(NS)}$. In fact, on $\widehat{(NS)}_{+\infty} \subset J_4^\infty(W)$ the Cartan distribution, at any point $q \equiv [N]_a^\infty \in \widehat{(NS)}_{+\infty}$, is a 4-dimensional vector space diffeomorphic to the 4-dimensional vector space $T_a N \subset T_a W$, where $N \subset W$ is a 4-dimensional submanifold of W.[71] $\qquad\qquad$ $\square$

---

[70] We shall require that the boundary ∂B of B is orientable.

[71] May be useful to emphasize that the use of integral bordism groups in the effective identification of solutions for initial values problems, as well as for boundary value problems, must be made in a correct way, otherwise it should appear that the Navier-Stokes equation can be solved in both directions of the time. In fact, when one fixes an "initial Cauchy data", one fixes also, for causality, the time orientation where we can find solutions. This is just explicitly considered in our approach, as solutions must satisfy also Reinhart bordism [65], i.e., bordism with vector fields. Vector fields inside for initial Cauchy data, and outside for final Cauchy data. Furthermore, a 3-dimensional space-time boundary, that is the propagation in the future, i.e. $t \geq t_0$, of some space-like initial $(t=t_0)$ 2-dimensial boundary, is necessarily time-like, hence the (4-dimensional) solution is necessarily a causal solution.

Theorem 4.7. *Boundary conditions do not assure, in general, the uniqueness of solutions.*

Remark 4.4. It is usefuel to emphasize that the non-uniqueness of solutions, in general, for the Navier-Stokes equation, stated in this theorem, does not exclude that adding some further constraints, (e.g. on the geometrical dimensions and on the class of solutions), one can obtain uniqueness of solutions. In fact, just from many decades one knows uniqueness theorems for some semplified situations. These are given in the framework of the functional analysis, (see, e.g., contributions by E.Hopf, O.A.Ladyzenskaya, J.Leray, J.L.Lions and G.Prodi [101(n)]), and also in the differential geometry of infinite dimensional manifolds, (see, e.g., contributions by V.I.Arnold & B.A.Khesin [101(a)], and D.G.Ebin & J.Marsden [101(f)]. See also the more recent work by J.Mattingly & Ya.Sinai [101(i)].)

Proof. We shall use the following lemmas.

Lemma 4.1. *For any admissible 3-dimensional space-like integral manifold there exist (global) solutions of (NS), but these are not unique.*

Remark 4.5. This lemma definitely solves a question on the non-uniqueness of solutions for Cauchy problems in (NS) that was been conjectured just many years ago starting from some partial results. See, e.g., the pioneering work by E.Hopf [101(h)] that has shown that the "initial value problem" is not well set for the Navier-Stokes equations.

Proof of Lemma 4.1. First of all note that even if the characteristic distribution of (NS) is zero we can use the method of characteristics to generate solutions. However, in general, we cannot use characteristic strips. In fact, recall that characteristic strips are vector fields ζ on (NS) belonging to the horizontal part $L_{\hat{q}}$ of the Cartan distribution $\mathbf{E}_2(NS) \equiv \bigcup_{q \in (NS)} \mathbf{E}_2(NS)_q$ in the splitting $(\star)$ $\mathbf{E}_2(NS)_q \cong (g_2)_q \oplus L_{\hat{q}}$, $q \in (NS)$, identified by means of a point $\hat{q} \in (NS)_{+1}$, $\pi_{3,2}(\hat{q}) = q$, such that $\zeta(q) \rfloor \delta(\xi) = 0$, for some vector $\xi \in (g_2)_q$. Here δ is the Spencer operator. Of course, in order to assure the existence of $\hat{q}$ over q, it is enough to consider $q \in \widehat{(NS)} \subset (NS)$, hence vector fields ζ on $\widehat{(NS)}$. Note that as $\dim \mathbf{E}_2(\widehat{NS})_q = 46$, and $\dim(\hat{g}_2)_q = 42$, for any point $q \in \widehat{(NS)}$, it follows that any 4-dimensional solution $V \subset \widehat{(NS)}$, passing for q, identifies a splitting like $(\star)$, in the Cartan distribution of $\widehat{(NS)}$ at q. Now, such a vector field ζ, with $\zeta(q) \in L_{\hat{q}}$, can be written in the following form:

$$(4.3) \qquad \zeta=\sum_{0\leq\alpha\leq 3} X^\alpha\Big(\partial x_\alpha+\dot{x}^i_\alpha\partial\dot{x}_i+p_\alpha\partial p+\theta_\alpha\partial\theta+\dot{x}^i_{\alpha\beta}\partial\dot{x}^\beta_i+p_{\alpha\beta}\partial p^\beta$$
$$+\theta_{\alpha\beta}\partial\theta^\beta+\dot{x}^i_{\alpha\beta\gamma}\partial\dot{x}^{\beta\gamma}_i+p_{\alpha\beta\gamma}\partial p^{\beta\gamma}+\theta_{\alpha\beta\gamma}\partial\theta^{\beta\gamma}\Big)$$

where $(\dot{x}^i_{\alpha\beta\gamma},p_{\alpha\beta\gamma},\theta_{\alpha\beta\gamma})$ satisfy equation $(\widehat{NS})_{+1}$, hence satisfy the following equations:

$$\left.\begin{aligned}
&F^0=0;\quad F^0_\alpha=0;\quad F^j=0;\quad F^4=0\\[4pt]
&F^0_{\alpha\beta}\equiv\dot{x}^k_\beta(\partial x_\alpha.G^j_{jk})+\dot{x}^k(\partial x_\beta(\partial x_\alpha.G^j_{jk}))+\dot{x}^k_{\alpha\beta}G^j_{jk}+\dot{x}^k_\alpha(\partial x_\beta.G^j_{jk})+\dot{x}^i_{s\alpha\beta}\delta^s_i=0\\[4pt]
&F^j_\beta\equiv\dot{x}^s_\beta R^j_s+\dot{x}^s(\partial x_\beta.R^j_s)+\dot{x}^s_\beta\dot{x}^i\rho G^j_{is}+\dot{x}^s\dot{x}^i_\beta\rho G^j_{is}+\dot{x}^s\dot{x}^i\rho(\partial x_\beta.G^j_{is})+\dot{x}^s_\beta\dot{x}^j_s\rho+\dot{x}^s\dot{x}^j_{s\beta}\rho+\rho\dot{x}^j_0\\[4pt]
&+\dot{x}^k_{s\beta}S^{js}_k+\dot{x}^k_s(\partial x_\beta.S^{js}_k)+\dot{x}^k_{is\beta}T^{jis}_k+\dot{x}^k_{is}(\partial x_\beta.T^{jis}_k)+p_{i\beta}g^{ij}+p_i(\partial x_\beta.g^{ij})\\[4pt]
&+\rho\Big(\partial x_\beta.((\partial x_i.f)g^{ij})\Big)=0\\[4pt]
&F^4_\beta\equiv\theta_{0\beta}\rho C_p+\dot{x}^k_\beta\theta_k\theta_{is}\overline{E}^{is}\rho C_p+\dot{x}^k\theta_{k\beta}\theta_{is}\overline{E}^{is}\rho C_p\\[4pt]
&+\dot{x}^k\theta_k\theta_{is\beta}\overline{E}^{is}\rho C_p+\dot{x}^k\theta_k\theta_{is}(\partial x_\beta.\overline{E}^{is})\rho C_p+\dot{x}^k_\beta\dot{x}^p W_{kp}+\dot{x}^k\dot{x}^p_\beta W_{kp}+\dot{x}^k\dot{x}^p(\partial x_\beta.W_{kp})\\[4pt]
&+\dot{x}^k_\beta\dot{x}^s_p\overline{W}^p_{ks}+\dot{x}^k\dot{x}^s_{p\beta}\overline{W}^p_{ks}+\dot{x}^k\dot{x}^s_p(\partial x_\beta.\overline{W}^p_{ks})+\dot{x}^k_{i\beta}\dot{x}^s_p Y^{ip}_{ks}+\dot{x}^k_i\dot{x}^s_{p\beta}Y^{ip}_{ks}+\dot{x}^k_i\dot{x}^s_p(\partial x_\beta.Y^{ip}_{ks})=0
\end{aligned}\right\}.$$

Then the system

$$\left\{\begin{aligned}
&\xi=\dot{X}^j_{\alpha\beta}\partial\dot{x}^{\alpha\beta}_j+Y_{\alpha\beta}\partial p^{\alpha\beta}+Z_{\alpha\beta}\partial\theta^{\alpha\beta}\\[4pt]
&v\rfloor\delta(\xi)=0\\[4pt]
&\left.\begin{aligned}
&\dot{X}^i_{s\alpha}\delta^s_i=0\\[4pt]
&\dot{X}^k_{is}T^{jis}_k=0\\[4pt]
&\rho C_p\dot{x}^k\theta_k\overline{E}^{is}Z_{is}=0
\end{aligned}\right\}(g_2(\widehat{NS}))
\end{aligned}\right\}
\Leftrightarrow
\left\{\begin{aligned}
&\dot{X}^j_{\alpha\beta}X^\alpha=0\\[4pt]
&Y_{\alpha\beta}X^\alpha=0\\[4pt]
&Z_{\alpha\beta}X^\alpha=0\\[4pt]
&\left.\begin{aligned}
&\dot{X}^i_{i\alpha}=0\\[4pt]
&\dot{X}^k_{is}T^{jis}_k=0\\[4pt]
&\rho C_p\dot{x}^k\theta_k\overline{E}^{is}Z_{is}=0
\end{aligned}\right\}(g_2(\widehat{NS}))
\end{aligned}\right\}$$

admits solutions for some $\dot{X}^j_{\alpha\beta},Y_{\alpha\beta},Z_{\alpha\beta}$ satisfying the subsystem $(g_2(\widehat{NS}))$. But not for any $\dot{X}^j_{\alpha\beta},Y_{\alpha\beta},Z_{\alpha\beta}$ satisfying the subsystem $(g_2(\widehat{NS}))$. In fact $(\widehat{NS})$, as well (NS), has zero characteristic distribution. In fact, for example, $(X^0\neq 0, X^i=0)$ is a possible solution for

$$(\dot{X}^j_{\alpha\beta})=0,$$

$$(Y_{\alpha\beta})=\begin{pmatrix}0 & 0 & 0 & 0\\ 0 & Y_{11} & Y_{12} & Y_{13}\\ 0 & Y_{21} & Y_{22} & Y_{23}\\ 0 & Y_{31} & Y_{32} & Y_{33}\end{pmatrix},$$

$$(Z_{\alpha\beta})=0.$$

Another solution is $(X^0 = 0, X^1 \neq 0, X^2 = X^3 = 0)$ for

$$(\dot{X}^j_{\alpha\beta}) = \begin{pmatrix} \dot{X}^j_{00} & 0 & \dot{X}^j_{02} & \dot{X}^j_{03} \\ 0 & 0 & 0 & 0 \\ \dot{X}^j_{20} & 0 & 0 & 0 \\ \dot{X}^j_{30} & 0 & 0 & 0 \end{pmatrix},$$

$$(Y_{\alpha\beta}) = \begin{pmatrix} Y_{00} & 0 & Y_{02} & Y_{03} \\ 0 & 0 & 0 & 0 \\ Y_{20} & 0 & Y_{22} & Y_{23} \\ Y_{30} & 0 & Y_{32} & Y_{33} \end{pmatrix},$$

$$(Z_{\alpha\beta}) = \begin{pmatrix} Z_{00} & 0 & Z_{02} & Z_{03} \\ 0 & 0 & 0 & 0 \\ Z_{20} & 0 & 0 & 0 \\ Z_{30} & 0 & 0 & 0 \end{pmatrix}$$

such that $\dot{X}^2_{20} + \dot{X}^3_{30} = 0$. Of course, for the construction of (causal) solutions of (NS) the most interesting ones are the characteristic strips with $X^0 \neq 0$. Now, in order to utilize such vector fields it should be necessary to recognize into $\widehat{(NS)}$ a sub-equation that admits such vector fields as generating its characteristic distribution and such that its Cartan distribution should be of dimension higher or equal to 4. (Otherwise it should not be possible to obtain 4-dimensional integral manifolds.) However, in the case before considered, $(X^0 \neq 0, X^j = 0)$, the sub-equation of $\widehat{(NS)}$ is obtained by adding the following equations: $(\star\star)$ $\{\dot{x}^j_{\alpha\beta} = 0, \theta_{\alpha\beta} = 0, p_{0\beta} = 0\}_{1 \leq j \leq 3, 0 \leq \alpha, \beta \leq 3}$. Such an augmented system $\widetilde{(NS)} \subset \widehat{(NS)}$ has dimension 26 and its Cartan distribution is of dimension $46 - (10 \times 3 + 10 + 4) = 46 - 44 = 2$. Too little to be utilized to generate 4-dimensional integral manifolds! (but enough to build 2-dimensional ones). Note that in the case $D = 2$ we have not more neither this possibility. In fact, in this case $\dim \mathbf{E}_2\widehat{(NS)} = 21$, $(\dim \hat{g}_2 = 18)$, and the augmented system $\widetilde{(NS)}$, is a 21-dimensional submanifold of $\widehat{(NS)}$ that has zero Cartan distribution. Therefore, $\widetilde{(NS)}$ cannot contain solutions of (NS). However, despite this situation, the method of characteristics continues to work in order to generate solutions of (NS), as we can use the infinitesimal symmetry algebra of $\widehat{(NS)}$. In fact, one has the following:

Lemma 4.2. (Infinitesimal symmetry algebra of $\widehat{(NS)}$). *The infinitesimal symmetry algebra of $\widehat{(NS)}$ is given by the following vector fields on W:* $\zeta = \bar{\zeta}^\alpha \partial x_\alpha + X^i \partial \dot{x}_i + Y \partial p + Z \partial \theta$, *where $\bar{\zeta}^\alpha \partial x_\alpha$ is the infinitesimal symmetry algebra of the oriented Galilean group with the unique restriction*

$(\partial x_0.\bar{\zeta}^s) = 0$, $s = 1, 2, 3$. Furthermore, $X^i = (\partial x_s.\bar{\zeta}^i)\dot{x}^s$, $Y = l(x^\alpha)$, $Z = \kappa \in \mathbf{R}$, where κ is an arbitrary constant and $l : M \to \mathbf{R}$ is a numerical function on M, such that $0 = \rho \mathcal{L}_{\bar{\zeta}}(\operatorname{grad} f)^j + (\operatorname{grad} l)^j$.

Remark 4.6. Note that in the isothermal case the infinitesimal symmetry algebra of $\widehat{(NS)}$ is considerably larger. In fact, it is identified with the following vector field $\zeta : W_0 \to TW_0$,

$$\zeta = \bar{\zeta}^\alpha \partial x_\alpha + X^i \partial \dot{x}_i + Y \partial p,$$

with $\bar{\zeta}^\alpha \partial x_\alpha$ the full infinitesimal symmetry algebra of the oriented Galilei group,

$$X^i = (\partial x_s.\bar{\zeta}^i)\dot{x}^s + h^i(x^\alpha),$$
$$Y = \rho g_{ks}[(\partial x_0.\bar{\zeta}^s) - h^s(x^\alpha)]\dot{x}^k + l(x^\alpha),$$

where l and h^i are numerical functions on M such that the following equations are satisfied:

$$0 = \operatorname{div} h^s,$$
$$0 = \rho \mathcal{L}_{\bar{\zeta}}(\operatorname{grad} f)^j + (\operatorname{grad} l)^j + (\partial x_q \partial x_p.h^i)T_i^{jqp}$$
$$+ h^i R_i^j + (\partial x_p.h^i)S_i^{jp} + \rho(\partial x_0.h^j),$$
$$0 = \rho[(\partial x_p.h^j) + 2(\partial x_0.\bar{\zeta}^s)G_{sp}^j - (\partial x_k.h^s)g_{ps}g^{ke}].$$

Here $W_0 \equiv J\mathcal{D}(M) \times_M T_0^0 M$.

Proof of Lemma 4.2. We must impose the following conditions: $\zeta^{(2)}.F^0|_{\widehat{(NS)}} = 0$, $\zeta^{(2)}.F_\omega^0|_{\widehat{(NS)}} = 0$, $\zeta^{(2)}.F^e|_{\widehat{(NS)}} = 0$, $\zeta^{(2)}.F^4|_{\widehat{(NS)}} = 0$, where the symbols are as defined in section 1. The conditions to stay on $\widehat{(NS)}$ are expressed by the following equations:

$$\dot{x}_s^p \delta_p^s = -\dot{x}^k G_{pk}^p,$$
$$\dot{x}_{k\omega}^q \delta_q^k = -\dot{x}^k(\partial x_\omega.G_{pk}^p) - \dot{x}_\omega^k G_{pk}^p,$$
$$\rho \dot{x}_0^e = -\dot{x}^p R_p^e - \dot{x}^p \dot{x}^q \rho G_{qp}^e - \dot{x}^p \dot{x}_p^e \rho - \dot{x}_p^k S_k^{ep} - \dot{x}_{qp}^k T_k^{eqp} - p_p g^{pe} - \rho(\partial x_p.f)g^{pe},$$
$$\rho C_p \theta_0 = -\dot{x}^k \theta_k \theta_{pq} \bar{E}^{pq} \rho C_p - \dot{x}^k \dot{x}^p W_{kp} - \dot{x}^k \dot{x}_p^q \bar{W}_{kq}^p - \dot{x}_q^k \dot{x}_p^m Y_{km}^{qp}.$$

The explicit espression of $\zeta^{(2)}$ is the following:

$$\zeta^{(2)} = \bar{\zeta}^\alpha \partial x_\alpha + X^i \partial \dot{x}_i + Y \partial p + Z \partial \theta$$

$$+ [-(\partial x_\alpha . \bar{\zeta}^\sigma) \dot{x}^i_\sigma + (\partial x_\alpha . X^i) + (\partial \dot{x}_s . X^i) \dot{x}^s_\alpha + (\partial p . X^i) p_\alpha + (\partial \theta . X^i) \theta_\alpha] \partial \dot{x}^\alpha_i$$

$$+ [-(\partial x_\alpha . \bar{\zeta}^\sigma) p_\sigma + (\partial x_\alpha . Y) + (\partial \dot{x}_s . Y) \dot{x}^s_\alpha + (\partial p . Y) p_\alpha + (\partial \theta . Y) \theta_\alpha] \partial p^\alpha$$

$$+ [-(\partial x_\alpha . \bar{\zeta}^\sigma) \theta_\sigma + (\partial x_\alpha . Z) + (\partial \dot{x}_s . Z) \dot{x}^s_\alpha + (\partial p . Z) p_\alpha + (\partial \theta . Z) \theta_\alpha] \partial \theta^\alpha$$

$$+ [-(\partial x_\alpha . \bar{\zeta}^\sigma) \dot{x}^i_{\beta\sigma} - (\partial x_\beta . \bar{\zeta}^\sigma) \dot{x}^i_{\alpha\sigma} - (\partial x_\alpha \partial x_\beta . \bar{\zeta}^\sigma) \dot{x}^i_\sigma$$

$$+ (\partial \dot{x}_j . X^i) \dot{x}^j_{\alpha\beta} + (\partial p . X^i) p_{\alpha\beta} + (\partial \theta . X^i) \theta_{\alpha\beta} + (\partial x_\alpha \partial \dot{x}_j . X^i) \dot{x}^j_\beta$$

$$+ (\partial x_\alpha . \partial p . X^i) p_\beta + (\partial x_\alpha \partial \theta . X^i) \theta_\beta + (\partial x_\alpha \partial x_\beta . X^i)$$

$$+ 2(\partial \dot{x}_j \partial \dot{x}_s . X^i) \dot{x}^j_\alpha \dot{x}^s_\beta + 2(\partial \dot{x}_j \partial p . X^i) \dot{x}^j_\alpha p_\beta + 2(\partial \dot{x}_j \partial \theta . X^i) \dot{x}^j_\alpha \theta_\beta + 2(\partial p \partial \theta . X^i) p_\alpha \theta_\beta] \partial \dot{x}^{\alpha\beta}_i$$

$$+ [-(\partial x_\alpha . \bar{\zeta}^\sigma) p_{\beta\sigma} - (\partial x_\beta . \bar{\zeta}^\sigma) p_{\alpha\sigma} - (\partial x_\alpha \partial x_\beta . \bar{\zeta}^\sigma) p_\sigma$$

$$+ (\partial \dot{x}_j . Y) \dot{x}^j_{\alpha\beta} + (\partial p . Y) p_{\alpha\beta} + (\partial \theta . Y) \theta_{\alpha\beta} + (\partial x_\alpha \partial \dot{x}_j . Y) \dot{x}^j_\beta$$

$$+ (\partial x_\alpha . \partial p . Y) p_\beta + (\partial x_\alpha \partial \theta . Y) \theta_\beta + (\partial x_\alpha \partial x_\beta . Y)$$

$$+ 2(\partial \dot{x}_j \partial \dot{x}_s . Y) \dot{x}^j_\alpha \dot{x}^s_\beta + 2(\partial \dot{x}_j \partial p . Y) \dot{x}^j_\alpha p_\beta + 2(\partial \dot{x}_j \partial \theta . Y) \dot{x}^j_\alpha \theta_\beta + 2(\partial p \partial \theta . Y) p_\alpha \theta_\beta] \partial p^{\alpha\beta}$$

$$+ [-(\partial x_\alpha . \bar{\zeta}^\sigma) \theta_{\beta\sigma} - (\partial x_\beta . \bar{\zeta}^\sigma) \theta_{\alpha\sigma} - (\partial x_\alpha \partial x_\beta . \bar{\zeta}^\sigma) \theta_\sigma$$

$$+ (\partial \dot{x}_j . Z) \dot{x}^j_{\alpha\beta} + (\partial p . Z) p_{\alpha\beta} + (\partial \theta . Z) \theta_{\alpha\beta} + (\partial x_\alpha \partial \dot{x}_j . Z) \dot{x}^j_\beta$$

$$+ (\partial x_\alpha . \partial p . Z) p_\beta + (\partial x_\alpha \partial \theta . Z) \theta_\beta + (\partial x_\alpha \partial x_\beta . Z)$$

$$+ 2(\partial \dot{x}_j \partial \dot{x}_s . Z) \dot{x}^j_\alpha \dot{x}^s_\beta + 2(\partial \dot{x}_j \partial p . Z) \dot{x}^j_\alpha p_\beta + 2(\partial \dot{x}_j \partial \theta . Z) \dot{x}^j_\alpha \theta_\beta + 2(\partial p \partial \theta . Z) p_\alpha \theta_\beta] \partial \theta^{\alpha\beta}.$$

So we have:

$$\zeta^{(2)} . F^0 \big|_{\widehat{(NS)}} = \bar{\zeta}^\alpha (\partial x_\alpha . G^p_{pk}) \dot{x}^k + X^i G^p_{pi}$$

$$+ [-(\partial x_i . \bar{\zeta}^s) \dot{x}^i_s + (\partial x_i . X^i) + (\partial \dot{x}_s . X^i) \dot{x}^s_i + (\partial p . X^i) p_i + (\partial \theta . X^i) \theta_i] = 0.$$

As the p_i, θ_i and $\dot{x}^i_s$ variables are arbitrary, it follows that their coefficients must be zero. From these conditions we get, after some simple calculations, that X^s, $1 \leq s \leq 3$, have the following expressions: $X^s = (\partial x_i . \bar{\zeta}^s) \dot{x}^i + h^s(x^\alpha)$, where h^s are arbitrary functions on M. Let us verify that above expression satisfies the equation $\zeta^{(2)} . F^0 \big|_{\widehat{(NS)}} = 0$. We get:

$$\bar{\zeta}^\alpha (\partial x_\alpha . G^p_{pk}) \dot{x}^k + G^p_{pi} [(\partial x_s . \bar{\zeta}^i) \dot{x}^s + h^i] - (\partial x_i . \bar{\zeta}^s) \dot{x}^i_s + (\partial x_i \partial x_s . \bar{\zeta}^i) \dot{x}^s + (\partial x_i . h^i) + (\partial x_s . \bar{\zeta}^i) \dot{x}^s_i = 0.$$

As $(\partial x_i \partial x_s . \bar{\zeta}^i) = 0$ for the Galilei group, we have: $\dot{x}^k \mathcal{L}_{\bar{\zeta}} \mathrm{tr}\,(G)_k + \mathrm{div}\, h^j = 0$. As $\mathcal{L}_{\bar{\zeta}} \mathrm{tr}\,(G)_k = 0$, we find the further condition: $\mathrm{div}\, h^j = 0$. One can easily verify, that with such expression for X^s is also verified the condition

$\zeta^{(2)}.F_\omega^0|_{\widehat{(NS)}} = 0$. Now the condition $\zeta^e.F^4|_{\widehat{(NS)}} = 0$ can be written in the following way:

$$0=\bar{\zeta}^\alpha(\partial x_\alpha.R_p^e)\dot{x}^P+\bar{\zeta}^\alpha(\partial x_\alpha.G_{qp}^e)\rho\dot{x}^P\dot{x}^q+\bar{\zeta}^\alpha(\partial x_\alpha.S_k^{ep})\dot{x}_p^k+\bar{\zeta}^\alpha(\partial x_\alpha.T_k^{eqp})\dot{x}_{qp}^k$$

$$+\bar{\zeta}^\alpha(\partial x_\alpha.g^{pe})p_p+\rho\bar{\zeta}^\alpha(\partial x_\alpha\partial x_p.f)g^{pe}+\rho\bar{\zeta}^\alpha(\partial x_p.f)(\partial x_\alpha.g^{pe})$$

$$+(\partial x_s.\bar{\zeta}^i)\dot{x}^s R_i^e+(\partial x_s.\bar{\zeta}^i)\dot{x}^s 2\rho G_{ip}^e\dot{x}^P+\rho(\partial x_s.\bar{\zeta}^i)\dot{x}^s\dot{x}_i^e+h^i R_i^e+2\rho h^i G_{ip}^e\dot{x}^P+\rho h^i\dot{x}_i^e$$

$$+[-(\partial x_p.\bar{\zeta}^s)\dot{x}_s^e+(\partial x_p.h^e)+(\partial x_s.\bar{\zeta}^e)\dot{x}_p^s]\rho\dot{x}^P$$

$$+[-(\partial x_0.\bar{\zeta}^s)\dot{x}_s^e+(\partial x_0.h^e)]\rho$$

$$+(\partial x_s.\bar{\zeta}^e)[-\dot{x}^P R_p^s-\dot{x}^P\dot{x}^q\rho G_{qp}^s-\dot{x}^P\dot{x}^s\rho-\dot{x}_p^k S_k^{sp}-\dot{x}_{qp}^k T_k^{sqp}-p_p g^{Ps}-\rho(\partial x_p.f)g^{Ps}]$$

$$+[-(\partial x_p.\bar{\zeta}^s)\dot{x}_s^i+(\partial x_p.h^i)\dot{x}_p^s]S_i^{ep}$$

$$+[-(\partial x_p.\bar{\zeta}^s)p_s+(\partial x_p.Y)+(\partial\dot{x}^s.Y)\dot{x}_p^s+(\partial p.Y)p_p+(\partial\theta.Y)\theta_p]g^{pe}$$

$$+[-(\partial x_q.\bar{\zeta}^s)\dot{x}_{ps}^i-(\partial x_p.\bar{\zeta}^s)\dot{x}_{qs}^i+(\partial x_j.\bar{\zeta}^i)\dot{x}_{qp}^j+(\partial x_q\partial x_p.h^i)]T_i^{epq}.$$

Then by annulling the θ_p-coefficient we get: $g^{pe}(\partial\theta.Y) = 0$, hence $Y = Y(x^\alpha, \dot{x}^i, p)$. Now by annulling the p_p-coefficient we get: $\mathcal{L}_{\bar{\zeta}}g^{pe}+g^{pe}(\partial p.Y) = 0$. As $\mathcal{L}_{\bar{\zeta}}g^{pe} = 0$, we have that $Y = Y(x^\alpha, \dot{x}^i)$. Another further information cames from the $\dot{x}_p^k$-coefficient that gives the following condition: $\mathcal{L}_{\bar{\zeta}}S_k^{ep} - \rho\delta_k^e[(\partial x_0.\bar{\zeta}^P) - h^P] + (\partial\dot{x}_k.Y)g^{pe} = 0$. As $\mathcal{L}_{\bar{\zeta}}S_k^{ep} = 0$, we get, after integration: $Yg^{pe} = \rho\delta_k^e[(\partial x_0.\bar{\zeta}^P) - h^P]\dot{x}^k + l^{pe}(x^\alpha)$. This gives also $Y = \rho g_{ks}[(\partial x_0.\bar{\zeta}^s) - h^s]\dot{x}^k + l(x^\alpha)$, with l an arbitrary numerical function on M. Now, by requering again that with such a determination of Y we should have $\zeta^{(2)}.F^e|_{\widehat{(NS)}} = 0$, we get the following new conditions:

$$\left\{ \begin{aligned} 0 &=\rho\mathcal{L}_{\bar{\zeta}}(\operatorname{grad} f)^e + (\operatorname{grad} l)^e + (\partial x_q\partial x_p.h^i)T_i^{eqp} + h^i R_i^e \\ &\quad +(\partial x_p.h^i)S_i^{ep} + \rho(\partial x_0.h^e) \\ 0 &=\rho[(\partial x_p.h^e) + 2(\partial x_0.\bar{\zeta}^s)G_{sp}^e - (\partial x_k.h^s)g_{ps}g^{ke}] \end{aligned} \right\}.$$

Let us now impose the condition $\zeta^{(2)}.F^4|_{\widehat{(NS)}} = 0$. We get the following equation:

$$0=\bar{\zeta}^\alpha(\partial x_\alpha.\bar{E}^{pq})\rho\dot{x}^k\theta_k\theta_{pq}C_p+\bar{\zeta}^\alpha(\partial x_\alpha.W_{kp})\dot{x}^k\dot{x}^p+\bar{\zeta}^\alpha(\partial x_\alpha.\bar{W}^p_{kq})\dot{x}^k\dot{x}^q_p+\bar{\zeta}^\alpha(\partial x_\alpha.Y^{qp}_{km})\dot{x}^k_q\dot{x}^m_p$$

$$+[(\partial x_s.\bar{\zeta}^i)\dot{x}^s+h^i]\theta_i\theta_{pq}\bar{E}^{pq}\rho C_p+[(\partial x_s.\bar{\zeta}^i)\dot{x}^s+h^i]\dot{x}^p W_{ip}+[(\partial x_s.\bar{\zeta}^i)\dot{x}^s+h^i]\dot{x}^k W_{ki}$$

$$+[(\partial x_s.\bar{\zeta}^i)\dot{x}^s+h^i]\dot{x}^q_p\bar{W}^p_{iq}$$

$$+[-(\partial x_p.\bar{\zeta}^s)\dot{x}^i_s+(\partial x_p.h^i)+(\partial x_s.\bar{\zeta}^i)\dot{x}^s_p]\bar{W}^p_{ki}\dot{x}^k+[-(\partial x_q.\bar{\zeta}^s)\dot{x}^i_s+(\partial x_q.h^i)+(\partial x_s.\bar{\zeta}^i)\dot{x}^s_q]Y^{qp}_{im}\dot{x}^m_p$$

$$+[-(\partial x_p.\bar{\zeta}^s)\dot{x}^i_s+(\partial x_p.h^i)+(\partial x_s.\bar{\zeta}^i)\dot{x}^s_p]Y^{qp}_{ki}\dot{x}^k_q+[-(\partial x_0.\bar{\zeta}^s)\theta_s+(\partial x_0.Z)+(\partial p.Z)p_0]\rho C_p$$

$$+(\partial\dot{x}_s.Z)C_p[-\dot{x}^P R^s_p-\dot{x}^P\dot{x}^q\rho G^s_{qp}-\dot{x}^P\dot{x}^s_p\rho-\dot{x}^k_p S^{sp}_k-\dot{x}^k_{qp}T^{sqp}_k-p_p g^{ps}-\rho(\partial x_p.f)g^{ps}]$$

$$+(\partial\theta.Z)[-\dot{x}^k\theta_k\theta_{pq}\bar{E}^{pq}\rho C_p-\dot{x}^k\dot{x}^P W_{kp}-\dot{x}^k\dot{x}^q_p\bar{W}^p_{kq}-\dot{x}^k_q\dot{x}^m_p Y^{qp}_{km}]$$

$$+[-(\partial x_k.\bar{\zeta}^s)\theta_s+(\partial x_k.Z)+(\partial\dot{x}_s.Z)\dot{x}^s_k+(\partial p.Z)p_k+(\partial\theta.Z)\theta_k]\dot{x}^k\theta_{pq}\bar{E}^{pq}\rho C_p$$

$$+[-(\partial x_p.\bar{\zeta}^s)\theta_{qs}-(\partial x_q.\bar{\zeta}^s)\theta_{ps}+(\partial\dot{x}_j.Z)\dot{x}^j_{pq}+(\partial p.Z)p_{pq}+(\partial\theta.Z)\theta_{pq}$$

$$+(\partial x_p\partial\dot{x}_j.Z)\dot{x}^j_q+(\partial x_p\partial p.Z)p_q+(\partial x_p\partial\theta.Z)\theta_q+(\partial x_p\partial x_q.Z)$$

$$+2(\partial\dot{x}_j\partial\dot{x}_s.Z)\dot{x}^j_p\dot{x}^s_q+2(\partial\dot{x}_j\partial p.Z)\dot{x}^j_p p_q+2(\partial\dot{x}_j\partial\theta.Z)\dot{x}^j_p\theta_q+2(\partial p\partial\theta.Z)p_p\theta_q]\dot{x}^k\theta_k\bar{E}^{pq}\rho C_p.$$

By annulling the p_{pq}-coefficient we get that $Z = Z(x^\alpha, \dot{x}^i, \theta)$. By annulling the $\theta_q\theta_k$ and $\theta_k\theta_{pq}$-coefficients we have that $h^i = 0$ and $Z = Z(x^\alpha, \dot{x}^i)$. By annulling the $\dot{x}^s_k\theta_{pq}$-coefficient we have $Z = Z(x^\alpha)$. Then by verifying that with these structure of Z equation $\zeta^{(2)}.F^4|_{\widehat{(NS)}} = 0$ is verified, we obtain the following new conditions: $(\partial x_0.\bar{\zeta}^s) = 0$ and $Z \equiv \kappa$, arbitrary constant.
□

Then, Theorem 4.3 guarantees the existence of solutions for any 3-dimensional integral space-like manifold contained into $\widehat{(NS)}$. More precisely, let

$$(\xi^A) \equiv (x^\alpha, y^I, y^I_\alpha, y^I_{\alpha\beta})$$

are fibered coordinates on $J\mathcal{D}^2(W)$ when

$$(x^\alpha, y^I) \equiv (x^\alpha, \dot{x}^i, p, \theta)$$

are fibered coordinates on W. Moreover, let $\zeta^{(2)}$ denote the second holonomic prolongation on $J\mathcal{D}^2(W)$ of a vector field $\zeta = \bar{\zeta}^\alpha\partial x_\alpha + \zeta^I\partial y_I$ on W that belongs to the infinitesimal symmetry algebra of $\widehat{(NS)}$. (See Lemma 4.2.) Hence we can write:

$$\zeta^{(2)}=\bar{\zeta}^\alpha\partial x_\alpha+\zeta^I\partial y_I$$

$$+[-(\partial x_\alpha.\bar{\zeta}^\sigma)y^I_\sigma+(\partial x_\alpha.\zeta^I)+(\partial y_K.\zeta^I)y^K_\alpha]\partial y^\alpha_I$$

$$+[-(\partial x_{\alpha_1}.\bar{\zeta}^\sigma)y^I_{\alpha_2\sigma}-(\partial x_{\alpha_2}.\bar{\zeta}^\sigma)y^I_{\alpha_1\sigma}-(\partial x_{\alpha_1}\partial x_{\alpha_2}.\bar{\zeta}^\sigma)y^I_\sigma$$

$$+(\partial y_K.\zeta^I)y^K_{\alpha_1\alpha_2}+(\partial x_{\alpha_1}\partial y_K.\zeta^I)y^K_{\alpha_2}+(\partial x_{\alpha_2}\partial y_K.\zeta^I)y^K_{\alpha_1}$$

$$+(\partial x_{\alpha_1}\partial x_{\alpha_2}.\zeta^I)+(\partial y_k\partial y_J.\zeta^I)y^K_{\alpha_1}y^J_{\alpha_2}]\partial y^{\alpha_1\alpha_2}_I$$

with the components $(\bar{\zeta}^\alpha, \zeta^I)$ given in Lemma 4.2. Such a vector field satisfies the conditions of transversality to any admissible 3-dimensional space-like integral manifold $N \subset \widehat{(NS)}$, required in Theorem 4.3. Hence, it guarantees a local smooth solution. However, as the infinitesimal symmetry algebra of $\widehat{(NS)}$ is infinite dimensional, we have that for any Cauchy problem, (i.e., any fixed N), we have infinite solutions. Therefore, Lemma 4.1 is proved.[72] $\qquad\qquad\square$

Let us consider, now, a boundary value problem for $(NS) \subset J\mathcal{D}^2(W) \subset J_4^2(W)$ defined by means of a 3-dimensional compact space-like (for some $t = t_0$), admissible integral manifold $B \subset (NS)$ propagating in (NS) in such a way that the boundary ∂B describes a fixed 3-dimensional time-like integral manifold $Y \subset \widehat{(NS)}$. Now, from Theorem 4.2 and Lemma 4.1, we can argue that there are many "characteristic vectors" that agree on the boundary Y, i.e., many global (smooth) solutions that satisfy fixed boundary conditions. In fact, let us consider Y diffeomorphic to a manifold $X \subset W$ by means of the projection $\pi_{2,0} : J_4^2(W) \to W$. Let us assume, for example, that X is topologically equivalent to S^3: $X \simeq S^3$. As W is a 9-dimensional affine fiber bundle over the affine manifold M, it follows that there are many 4-dimensional disks D^4 such that $\partial D^4 = S^3$. In other words, there are many compact smooth 4-dimensional manifolds $Z \subset W$ such that $\partial Z = X$. If we consider their second holonomic prolongation $Z^{(2)} \subset J_4^2(W)$ one has $\partial Z^{(2)} = Y$. In general $Z^{(2)}$ is not contained into (NS). However, as (NS) is a deformation retract of $J_4^2(W)$, we can deform $Z^{(2)}$ in such a way to become a regular integral manifold $V \subset (NS)$ with $\partial V = Y$. This is equivalent to say that (NS) is of (p)-homotopy type, $0 \le p \le 3$. (For a detailed proof see the third paper in refs.[65].) Of course, as $\dim(NS) = 74$ and $\dim \mathbf{E}_2(NS) = 50$, at different Z there correspond, in general, different integral manifolds V such that $\partial V = Y$.[73] Furthermore, taking into account the triviality of the first integral bordism group $\Omega_1^{(NS)}$

---

[72] Let us also emphasize that as (NS) and $\widehat{(NS)}$ are invariant for time-translations, generated by the vector field ∂t on (NS), it follows that for any space-like admissible integral 3-dimensional manifold $N \subset \widehat{(NS)}$, there exists a canonical solution that represents the dynamical propagation of N with respect to the inertial frame considered. By the way, the existence of this "canonical solution" does not justify to forget all the other ones.

[73] This conclusion on the non-uniqueness of the solutions for boundary value problems in the Navier-Stokes equation can be also reproduced in many boundary value problems, where

of (NS), we can have also solutions that produce some singular points in the corresponding flows. In fact, let us consider a loop in B. Its propagation could produce a tunnel effect forming, at some instant $t > t_0$, n loops, since the first integral bordism group, $\Omega_1^{(NS)}$, of (NS) is trivial: $\Omega_1^{(NS)} = 0$. This corresponds to the production of a tunnel effect in the integral flow lines of the fluid. These are, of course, compatible with the boundary conditions. Sources of such tunnel effects are eventual unstability of flows. In fact we have the following lemmas.

Lemma 4.3. *The Navier-Stokes equation (NS) can be rewritten in the following adimensional form:*

$$(4.4) \qquad (NS) \left\{ \begin{array}{l} \text{(sevenSl continuity equation) } \Phi^0 \equiv \bar{v}^k{}_{/k} \\[2mm] \text{(motion equation) } \Phi^j \equiv \frac{\delta \bar{v}^j}{\delta t} - \frac{2}{R_e}\dot{\bar{e}}^{kj}{}_{/k} + g^{kj}\bar{p}_{/k} - \frac{1}{F_r}\bar{B}^j = 0 \\[2mm] \text{(energy equation) } \Phi^4 \equiv \frac{\delta \bar{\theta}}{\delta t} - \frac{1}{R_e P_r}\Delta\bar{\theta} - 2\frac{B_r}{R_e P_r}<\dot{\bar{e}},\dot{\bar{e}}> = 0 \end{array} \right\},$$

where $\bar{x}^j = x^j/L$, $\bar{v}^j = v^j/L$, $\bar{p} = (p - p_0)/\rho U^2$, $\bar{t} = tU/L$, $\bar{B}^j = B^j/g$, $\bar{\theta} = (\theta - \theta_0)/(\theta_1 - \theta_0)$, $R_e = \frac{\rho U L}{\chi} =$ Reynolds number, $F_r = \frac{U^2}{gL} =$ Frank number, $P_r = \frac{C_p \chi}{\nu} =$ Prandtl number and $B_r = \frac{\chi U^2}{\nu(\theta_1 - \theta_0)} =$ Brindeman number. Here L is the fixed length scale in the problem, U is a mean value of the vector field and g is the gravity acceleration.

---

the classic theory assures the uniqueness of solutions. Note also that if V and V' are two smooth solutions of $(NS) \subset J_4^2(W)$ such that $\partial V = \partial V'$, this does not, in general, imply that $V \simeq V'$ (V homotopy equivalent to V'). However, from the short exact sequence in Theorem 4.1 it follows that each solution $V \subset (NS) \subset J_4^2(W)$ can be deformed by means of the flow associated to a vertical vector field $\zeta:(NS) \to g_2 \subset vT(NS)$, in such a way to generate another 4-dimensional submanifold $\widetilde{V}_\lambda \subset (NS)$, with $\partial \widetilde{V}_\lambda = \partial V$. This manifold is not, in general, a solution of (NS), but it is simply homotopy equivalent to V relatively to ∂V. On the other hand, if $V, V' \subset (NS)$ are two solutions of (NS), having the same boundary and homotopy type, they can differ from the set of singular points. For example, V can be regular, while V' is endowed with a nowhere dense subset $\Sigma(V') \neq \emptyset$ of singular points. In fact, (NS) is an affine fiber bundle over C, with associated vector bundle g_2. As $\dim(g_2)_q = 46$, $\forall q \in (NS)$, there is room for solutions with singular points. Of course this does not more happen on $(NS)_{+\infty}$, where the Cartan distribution has just dimension 4. However, also on $(NS)_{+\infty}$ we can distinguish between two types of solutions: one diffeomorphic to their projections on M and the other ones diffeomorphic to their projections on W. Of course the solutions of the first type are particular cases of the second ones.

Proof of Lemma 4.3. The proof is standard. $\qquad\qquad\qquad\qquad\qquad$ □

Lemma 4.4. *For the Navier-Stokes equation with viscosity* $\chi \neq 0$, *the Reynolds, Frank and Brindeman numbers do not identify conservation laws and neither 1-conservation laws. Furthermore, these adimensional numbers, identified with characteristic numbers of space-like domains of the fluids, are conserved into a solution* $V \subset (NS)$, *i.e., are characteristic numbers of* V, *iff the mean's shear strain is zero:* $\int_{\underline{B}_t} u^i u^j \dot{e}_{ij} \eta = 0$, *where* $\underline{B}_t \subset M$ *is any fixed 3-dimensional space-like domain corresponding to the 3-dimensional integral space-like manifold* $N_t \subset V_t$, $V_t \bigcap (NS)_t$, $t \in T$, u^i *are the space-like components of the distribution of velocity corresponding to the solution* V, $\dot{e}_{ij}$ *is the infinitesimal velocity of the shear strain,* $\dot{e}_{ij} = \frac{1}{2}(u_{i/j} + u_{j/i})$ *and* η *is the canonical vertical volume form on the Galilean space-time* M.

Proof of Lemma 4.4. Recall that a conservation law for (NS) can be identified with a differential 3-form β on $(NS)_{+\infty}$ such that for any solution $V \subset (NS)$ one has $d\beta|_V = 0$. This corresponds to the conservation of the characteristic number $\int_B \beta|_V$ on any space-like 3-dimensional compact domain $B \subset V$ associated to the solution V. Recall also that, more generally, a p-conservation law, $0 \leq p \leq 3$, for $(NS) \subset J_4^2(W)$ is any numerical function $f : \Omega_p^{(NS)+\infty} \to \mathbf{R}$ defined on the integral p-bordism group of the infinity prolongation $(NS)_{+\infty}$ of (NS). Now, it is enough to prove the theorem for regular solutions of $(NS) \subset J\mathcal{D}^2(W)$, i.e., 4-dimensional integral manifolds $V \subset (NS)$ that are diffeomorphic to their projection $\underline{V} \equiv \pi_2(V) \subset M$. Let $\zeta : V \to TV$ be the time-like characteristic vector field on V such that one can write $V = \bigcup_{\lambda \in \mathbf{T}} \phi_\lambda(B_t)$, where $B_t \subset (NS)$ is a space-like 3-dimensional integral manifold and $\partial\phi = \zeta$. Then, also $\underline{V}$ is generated by a characteristic vector field $v : \underline{V} \to T\underline{V}$ starting from a 3-dimensional space-like compact manifold $\underline{B}_t \subset M$: $\underline{V} = \bigcup_{\lambda \in \mathbf{T}} \underline{\phi}_\lambda(B)$, $\partial\underline{\phi} = v$. Let us define the following (adimensional) characteristic number (*Reynolds number*) $R_e(\zeta|B_t)$ associated to the couple (ζ, B):

$$(4.5) \qquad R_e(\zeta|B_t) = \frac{\rho}{\chi} L \sqrt{\frac{\int_{\underline{B}_t} u^2 \eta}{\int_{\underline{B}_t} \eta}} \equiv \frac{\rho}{\chi} < U_t > L$$

where $\int_{\underline{B}_t} \eta = \int_{\underline{B}_t} \sqrt{\det(g_{ij})} dx^1 \wedge dx^2 \wedge dx^3 = \text{vol}(\underline{B}_t)$, and L is the fixed length scale in the problem.[74] Then if $\underline{B}_{t'} = \phi_\lambda(\underline{B}_t)$, one has, (taking into account that the flow is necessarily isocoric),

$$(4.6) \qquad R_e(\zeta|B_{t'})^2 - R_e(\zeta|B_t)^2 = (\frac{\rho L}{\chi})^2 \frac{1}{\text{vol}(\underline{B}_t)} \int_{\underline{B}_t} u^i u^j (\mathfrak{U}_\lambda)_{ij} \eta$$

where $\mathfrak{U}_\lambda \equiv \underline{\phi}_\lambda^* g - g$ is the Ulmansi tensor of the flow. So $R_e(\zeta|B_t)^2 - R_e(\zeta|B_{t'})^2 = 0$ iff

$$(4.7) \qquad \int_{\underline{B}_t} u^i u^j (\mathfrak{U}_\lambda)_{ij} \eta = \int_{\underline{B}_t} u^i u^j (\mathfrak{U}_\lambda)_{ij} \sqrt{\det(g_{ij})} dx^1 \wedge dx^2 \wedge dx^3 = 0.$$

Now condition (4.7) is a very strong requirement that can be satisfied only in some particular cases. For examples: (i) $\mathfrak{U}_\lambda = 0$, i.e., the flow is isometric; (ii) $u^i u^j (\mathfrak{U}_\lambda)_{ij} = 0$, i.e., the vector field u is isotropic with respect to the metric $\mathfrak{U}_\lambda$. Thus, in general, the Reynolds number cannot identify a conservation law for (NS). Furthermore, let $N_t \subset B_t$ be an (integral) loop contained in B_t. The vector field ζ will propagate N_t describing a 2-dimensional integral manifold $Y \subset V$. Set $\gamma_t \equiv \pi_2(N_t) \subset \pi_2(Y) \equiv Z \subset \underline{V} \subset M$. We define 1-*Reynolds number* of the couple (ζ, N_t) the following (adimensional) characteristic number:

$$(4.8)$$

$$R_e^{(1)}(\zeta|N_t) = \frac{\rho}{\chi} \sqrt{\frac{\int_{\gamma_t} u^2 ds}{\int_{\gamma_t} ds}} \int_{\gamma_t} ds = \frac{\rho}{\chi} \sqrt{\text{vol}(\gamma_t) \int_{\gamma_t} u^2 ds} \equiv \frac{\rho}{\chi} <u_t> l_t$$

where $u \equiv v - \dot{\psi} = v^k \partial x_k = u^k \partial x_k$ is the space component of v with respect to an inertial frame ψ, l_t is the volume of the loop N_t, i.e. its length and $<u_t>$ is a mean value of the vector field u on the loop N_t. If we consider another admissible integral loop $N_{t'} \subset (NS)$, integral bording with N_t via the manifold Y, we get that its 1-Reynolds number $R_e^{(1)}(\zeta|N_{t'})$ is, in general, different from $R_e^{(1)}(\zeta|N_t)$. In fact, one has

$$R_e^{(1)}(\zeta|N_{t'})^2 - R_e^{(1)}(\zeta|N_t)^2 = (\tfrac{\rho}{\chi})^2 \left((\text{vol}(\gamma_{t'}) \int_{\gamma_{t'}} u^2 ds - \text{vol}(\gamma_t) \int_{\gamma_t} u^2 ds \right)$$

$$= (\tfrac{\rho}{\chi})^2 \left(\int_{\gamma_{t'}} \sqrt{(\underline{\phi}\lambda^* g)_{ij} \dot{x}^i \dot{x}^j} dt \int_{\gamma_{t'}} u^i u^j (\underline{\phi}_\lambda^* g)_{ij} ds - \int_{\gamma_t} \sqrt{g_{ij} \dot{x}^i \dot{x}^j} dt \int_{\gamma_t} u^i u^j g_{ij} ds \right)$$

---

[74] Let us define $U_{sup}(t) \equiv \sup_{p \in \underline{B}_t} |u(p)|$. Then, one has $R_e(\zeta|B_t) \leq \frac{\rho L}{\chi} \sqrt{\dfrac{\int_{\underline{B}_t} U_{sup}^2(t) \eta}{\int_{\underline{B}_t} \eta}} = \frac{\rho L U_{sup}(t)}{\chi} \equiv R_e^{sup}(t)$. Therefore, the characteristic Reynolds number defined into (4.5) admits as upper bound the Reynolds number $R_e^{sup}(t)$. Furthermore, setting $U_{sup} = \sup_{t \in T} U_{sup}(t)$, we get $R_e^{sup}(t) \leq R_e^{sup} \equiv \frac{\rho L U_{sup}}{\chi}$.

where $\lambda = t' - t$. As, in general, $\phi_\lambda^* g \neq g$, it results, in general, $R_e^{(1)}(\zeta|N_{t'}) - R_e^{(1)}(\zeta|N_t) \neq 0$. For example, if $l_{t'} \equiv vol(\gamma_{t'}) = vol(\gamma_t) \equiv l_t$, the conservation of the Reynolds number is not, in general, assured. In fact, in such cases we have:

$$R_e^{(1)}(\zeta|N_{t'})^2 - R_e^{(1)}(\zeta|N_t)^2 = (\frac{\rho}{\chi})^2 vol(\gamma_t) \int_{\gamma_t} u^i u^j (\mathfrak{U}_\lambda)_{ij} ds.$$

So $R_e^{(1)}(\zeta|N_{t'})^2 - R_e^{(1)}(\zeta|N_t)^2 = 0$ iff

$$(4.9) \qquad \int_{\gamma_t} u^i u^j (\mathfrak{U}_\lambda)_{ij} ds = 0.$$

Now equation (4.9) is verified iff one of the following conditions is satisfied: (i) $\mathfrak{U}_\lambda = 0$, i.e., the flow is isometric; (ii) $u^i u^j (\mathfrak{U}_\lambda)_{ij} = 0$, i.e., the vector field u is isotropic with respect to the metric $\mathfrak{U}_\lambda$; (iii) $u^i u^j (\mathfrak{U}_\lambda)_{ij} ds = d\varphi$, i.e., $u^i u^j (\mathfrak{U}_\lambda)_{ij} ds$ is an exact differential 1-form. Outside these particular cases (4.9) is not verified. Thus, in general, the 1-Reynolds number cannot identify a 1-conservation law for (NS). Similar considerations can be made for the Frank and Brindeman numbers also. See the following table.

TAB.4.4 - Adimensional characteristic numbers associated to the couple (ζ, B_t)

Reynolds	$R_e(\zeta\|B_t) \equiv \frac{\rho L}{\chi} \sqrt{\int_{\underline{B}_t} u^2 \eta / \int_{\underline{B}_t} \eta} \equiv \frac{\rho <U_t> L}{\chi}$
Frank	$F_r(\zeta\|B_t) \equiv \frac{1}{gL} (\int_{\underline{B}_t} u^2 \eta / \int_{\underline{B}_t} \eta) = \frac{<U_t>^2}{gL}$
Brindeman	$B_r(\zeta\|B_t) \equiv \frac{\chi}{\nu(\theta_1 - \theta_0)} (\int_{\underline{B}_t} u^2 \eta / \int_{\underline{B}_t} \eta) = \frac{\chi <U_t>^2}{\nu(\theta_1 - \theta_0)}$

Then it is easily seen that $F_r(\zeta|B_t)$ and $B_r(\zeta|B_t)$ are conserved iff $R_e(\zeta!B)$ is conserved, i.e., iff equation (4.7) holds.

Definition 4.2. For each solution $V \subset (NS)$ define the following characteristic function:

(4.10)(*fundamental global rate*)

$$\mathfrak{f} = \mathfrak{f}(t) \equiv \frac{d}{d\lambda} (R_e^2(\zeta|B_{t+\lambda}) - R_e^2(\zeta|B_t)) \,|_{\lambda=0}.$$

Lemma 4.5. *The fundamental global rate is proportional to the mean's shear rate:*

$$(4.11) \qquad \mathfrak{f}(t) = (\frac{\rho L}{\chi})^2 \frac{2}{vol(\underline{B}_t)} \int_{\underline{B}_t} u^i u^j \dot{e}_{ij} \eta.$$

Therefore, the Reynolds nunber $R_e(\zeta|B_t)$, (resp. the Frank number $F_r(\zeta|B_t)$, the Brindeman number $B_r(\zeta|B_t)$), is a characteristic number of the solution V, i.e., it is conserved, iff $\int_{\underline{B}_t} u^i u^j \dot{e}_{ij} \eta = 0$. (In such a case we write $R_e(V)$, (resp. $F_r(V)$, resp. $B_r(V)$), to denote the characteristic number of V, just associated to the Reynolds number, (resp. Frank number, resp. Brindeman number).)

Proof of Lemma 4.5. Equation (4.11) follows directly from equation (4.6). Then, from Lemma 2.3 it follows directly the remaining statement of the lemma. $\qquad\qquad\square$

Lemma 4.6. *In the set $Sol(NS)$ of solutions of the Navier-Stokes equation (NS), there are also ones where are present different domains, some corresponding to stable flows and other to unstable flows. Those solutions V such that $\dot{\mathfrak{f}} = 0$, and with the characteristic number $R_e(V)$ lower than the critical value, are stable solutions, (assumed that the temperature and pressure perturbations are neglectable).*

Proof of Lemma 4.6. Let us consider the linear Navier-Stokes equation $\overline{(NS)_s} \subset J\mathcal{D}^2(E)$, linearization of (NS) at the solution $s : M \to W$, where $E \equiv s^* vTW \cong M \times \mathbf{S} \times \mathbf{R}^2$. Here $\mathbf{S}$ denotes the 3-dimensional euclidean space of space-like vectors of the Galilean space-time. (The vertical tangent space vTW is calculated with respect to the fiber structure $\pi : W \to M$.) By considering (NS) written in the form (4.1) we can write:

$$\overline{(NS)_s} \begin{cases} (\partial \dot{x}_k.F^0)_s \nu^k + (\partial \dot{x}_k^p.F^0)_s \nu_p^k \equiv \Xi^0 = 0 \\[4pt] (\partial \dot{x}_k.F^j)_s \nu^k + (\partial \dot{x}_k^p.F^j)_s \nu_p^k + (\partial \dot{x}_k^0.F^j)_s \nu_0^k \\[2pt] \quad + (\partial \dot{x}_k^{ir}.F^j)_s \nu_{ir}^k + (\partial p^i.F^j)_s \nu_i^{(p)} \equiv \Xi^j = 0, \\[4pt] (\partial \dot{x}_k.F^4)_s \nu^k + (\partial \dot{x}_k^p.F^4)_s \nu_p^k + (\partial \theta^0.F^4)_s \nu_0^{(\theta)} \\[2pt] \quad + (\partial \theta^k.F^4)_s \nu_k^{(\theta)} + (\partial \theta^{ir}.F^4)_s \nu_{ir}^{(\theta)} \equiv \Xi^4 = 0 \\[4pt] \scriptstyle 1 \le j \le 3 \end{cases}$$

where the index s, below the round brackets, denotes the evaluation at the solution s. One can prove that $\overline{(NS)_s}$ is not a formally integrable and neither completely integrable PDE. However, into $\overline{(NS)_s}$ one can recognize a linear equation $\widehat{(NS)_s} \subset \overline{(NS)_s}$, that satisfies such integrability conditions. More precisely $\widehat{(NS)_s}$ is the linearization of $(\widehat{NS})$ at the solution s. This is enough to state that to any initial condition of $\widehat{(NS)_s}$, i.e., to any perturbation of the solution s, belonging to the equation $\widehat{(NS)}_s$, will correspond a local solution, i.e., a local perturbation of the solution s. Furthermore, the Cauchy problem for $\widehat{(NS)}_s$ admits regular solutions for any space-like regular 3-dimensional data. In fact, the pseudogroup G of symmetries of $\widehat{(NS)}$ induces a pseudogroup G_s of symmetries of $\widehat{(NS)}_s$. More precisely one has that $G_s \subset G$ is identified by means of all (local) fiber bundle transformations $(\phi, \bar{\phi})$ of $\pi : W \to M$, belonging to G, such that $s' \equiv \phi^* s \equiv \phi \circ s \circ \bar{\phi}^{-1} = s$. This can be immediately seen by means of the following commutative diagram:

$$
\begin{array}{ccccccccc}
M & = & M & \xrightarrow{\ \bar{\phi}\ } & & & M & = & M \\
\| & & \uparrow & & & & \uparrow & & \| \\
M & \xrightarrow{s} & W & \xrightarrow{\ \phi\ } & & & W & \xleftarrow{s'} & M \\
\nu\downarrow & & \uparrow & & & & \uparrow & & \downarrow\nu' \\
s^* vTW & \to & vTW & \xrightarrow{vT(\phi)} & & & vTW & \leftarrow & s'^* vTW \\
\uparrow & & \uparrow & & & & \uparrow & & \uparrow \\
(D^2 s)^* vT\widehat{(NS)} & \to & vT\widehat{(NS)} & \xrightarrow{vT(\phi^{(2)})} & & & vT\widehat{(NS)} & \leftarrow & (D^2 s')^* vT\widehat{(NS)} \\
\| & & \downarrow & & & & \downarrow & & \| \\
\widehat{(NS)}_s & \to & \widehat{(NS)} & \xrightarrow{\phi^{(2)}} & & & \widehat{(NS)} & \leftarrow & \widehat{(NS)}_{s'} \\
\end{array}
$$

$$
\begin{array}{ccccccc}
D^2\nu\uparrow & & \downarrow & & W \xrightarrow{\phi} W & & \uparrow D^2\nu' \\
M & = & M & \xrightarrow[\ \bar{\phi}\ ]{} & & M = & M
\end{array}
$$

Therefore

$$
\phi^{(2)}\left(\widehat{(NS)}_s\right) \subset \widehat{(NS)}_{s'}
$$

iff $s = s' = \phi^* s$. Now, a Cauchy problem for $\widehat{(NS)}_s$ cannot be solved with respect to a space-like Cauchy data manifold, by using time-translations, (as can be made, for example, for $\widehat{(NS)}$), as it should require that $\widehat{(NS)}_s$ is invariant for time translations. This happens only if s is time-translations invariant, (i.e., s is a steady-state solution). However, in general, the Cauchy problem for $\widehat{(NS)}_s$ admits solutions for any space-like regular 3-dimensional data, $C_t \subset \widehat{(NS)}_s$, as we can use the flow ϕ_λ induced by the characteristic flow that solves the corresponding Cauchy problem for s. In fact, for such a flow one has $\phi_\lambda^* s = s$. Furthermore, if $C_t = D^2\nu_0(B_t)$, where $B_t \subset M$

is a compact 3–dimensional space-like submanifold of $M_t \equiv \tau^{-1}(t) \subset M$, and $\nu_0 : B_t \to s^* vTW$ is a section over B_t, then

$$\phi_\lambda^* \nu_0 = vT(\phi_\lambda) \circ \nu_0 \circ \bar{\phi}_\lambda^{-1} : M \to s^* vTW$$

is a solution of $\widehat{\overline{(NS)}_s}$. In fact,

$$D^2(\phi_\lambda^* \nu_0)(p) = (J\mathcal{D}^2(vT(\phi_\lambda) \circ D^2\nu_0 \circ \phi_\lambda^{-1})(p) \in \widehat{\overline{(NS)}_s},$$

for any $p \in \bar{\phi}_\lambda(B_t)$, as

$$D^2\nu_0(\phi_\lambda^{-1}(p)) \in C_t \subset \widehat{\overline{(NS)}_s}$$

and $J\mathcal{D}^2(vT(\phi_\lambda))$ is a diffeomeorphism of $\widehat{\overline{(NS)}_s}$. Now, in order to discuss the uniqueness of such solutions, let us consider the following definitions and results. If $\pi : W \to X$ is a fibre bundle we denote the sheaf of sections of π by $\underline{W}$. For a fibre bundle morphism $A : J\mathcal{D}^k(W) \to K$, denote by

$$A^{(r)} : J\mathcal{D}^{k+r}(W) \to J\mathcal{D}^r(K)$$

its r-th prolongation. Let

$$(\diamond) \quad \cdots \to \underline{F_r} \xrightarrow{j_r} \underline{F_{r+1}} \xrightarrow{j_{r+1}} \underline{F_{r+2}} \to \cdots$$

be a sequence of linear differential operators of order $r, r+1, \cdots$ respectively. Let

$$\phi_r : J\mathcal{D}^r(F_r) \to F_{r+1}$$

be the corresponding fiber bundles homomorphisms. Then, we say that sequence $(\diamond)$ is *locally exact* if im $(j_r) = \ker(j_{r+1})$, $\forall r$. We say that $(\diamond)$ is *formally exact* if the corresponding sequences of prolongated fibered morphisms

$$\cdots \to J\mathcal{D}^{r+(r+1)+s}(F_r) \xrightarrow{\phi_r^{(r+1+s)}} J\mathcal{D}^{r+1+s}(F_{r+1}) \xrightarrow{\phi_{r+1}^{(s)}} J\mathcal{D}^s(F_{r+2}) \to \cdots$$

for all $s \geq 0$, are exact. We say that (4.11) is *exact* if it is locally and formally exact. The *Janet sequence of an involutive formally integrable linear PDE* $E_k \subset J\mathcal{D}^k(E)$ is the following sequence:

$$J^k(\Theta) : \quad 0 \to \Theta \to \underline{E} \xrightarrow{D} \underline{F_0} \xrightarrow{D} \underline{F_1} \xrightarrow{D} \underline{F_2} \xrightarrow{D} \cdots \xrightarrow{D} \underline{F_n} \to 0$$

where Θ is the sheaf of solutions of $E_k = \ker(\phi) \subset JD^k(E)$, $\phi : JD^k(E) \to JD^k(E)/E_k \equiv F_0$, $D_r \equiv \psi^r \circ D$, $\psi^r \equiv$ canonic epimorphism $\psi^r : JD(F_{r-1}) \to F_r \equiv \mathrm{coker}\,(\psi^{(r-1)})^{(1)}$, $\phi = \psi^0$, $D = \phi \circ D^k$. $J^k(\Theta)$ is formally exact. Furthermore, $D_1 \circ D = 0$. The operator D_1 is the compatibility condition for D and express the formal obstruction to solving the affine equation $D.s = f$, $s \in \underline{E}$, $f \in \underline{F_0}$. An example is the d-Poincaré sequence

$$0 \to \Theta \to \underline{\Lambda_0^0 M} \xrightarrow{d} \underline{\Lambda_1^0 M} \xrightarrow{d} \cdots \xrightarrow{d} \underline{\Lambda_n^0 M} \to 0$$

where Θ is the sheaf of functions locally constant on M. This is an exact Janet sequence associated to the exterior differential $d : JD(\Lambda_r^0 M) \to \Lambda_{r+1}^0 M$. The *Spencer operator* is the first order differential operator

$$\mathbf{D} : \underline{JD^k(W)} \to \underline{T^*M \bigotimes JD^{k-1}(vTW)}$$

defined by

$$\mathbf{D} = D \circ \pi_{k,k+1} - \beta_{1,k-1},$$

where $\beta_{1,k-1}$ is the canonical embedding $JD(W) \to JD(JD^{k-1}(W))$. One has

$$\ker(\mathbf{D}) = \left\{ u \in \underline{JD^k(W)} \,\middle|\, u = D^k s,\ s \in \underline{W} \right\}.$$

Associated to the Spencer operator one has, for any vector fiber bundle $\pi : E \to M$, the following locally exact complex:

$(S_1^q(E))$

$$0 \to \underline{E} \xrightarrow{D^q} \underline{JD^q(E)} \xrightarrow{\mathbf{D}} \underline{T^*M \bigotimes JD^{q-1}(E)} \xrightarrow{\mathbf{D}} \underline{\Lambda_2^0 M \bigotimes JD^{q-2}(E)} \xrightarrow{\mathbf{D}} \cdots \xrightarrow{\mathbf{D}} \underline{\Lambda_n^0 M \bigotimes JD^{q-n}(E)} \to 0$$

where $JD^{q-r}(E) = 0$ if $r > q$, and

$$\mathbf{D}(\omega \wedge u) = d\omega \wedge (\pi_{q,q-1} \circ u) + (-1)^j \omega \wedge \mathbf{D}, \quad \omega \in \Lambda_j^0 M,\ u \in \Lambda_j^0 M \bigotimes JD^{q-j}(E).$$

Then, the linear Spencer complex gives by restriction to $E_q \subset JD^q(E)$, $q = k + r$, the following complex (*first linear Spencer complex*) $S_1^q(\Theta)$ of E_k:

$(S_1^q(\Theta))$

$$0 \to \Theta \xrightarrow{D^{k+r}} \underline{E_{k+r}} \xrightarrow{\mathbf{D}} \underline{T^*M \bigotimes E_{k+r-1}} \xrightarrow{\mathbf{D}} \underline{\Lambda_2^0 M \bigotimes E_{k+r-2}} \xrightarrow{\mathbf{D}} \cdots \xrightarrow{\mathbf{D}} \underline{\Lambda_n^0 M \bigotimes E_{k+r-n}} \to 0$$

where Θ is the sheaf of solutions of E_k. We set $E_{k+r-s} = J\mathcal{D}^{k+r-s}(E)$ if $r < s \leq k + r$, $E_{k+r-s} = 0$ if $s > k + r$. $S_1^q(\Theta)$ is always locally exact at $\underline{E_{k+r}}$, but, in general, it is not formally exact. Hence $S_1^q(\Theta)$ is not, in general, exact. By restriction of $-\mathbf{D}$ on the symbol g_q of E_q, we get another complex (also called first Spencer complex of E_k.) The relation with the previous one is given by the following commutative diagram:

$$
\begin{array}{ccccccccc}
 & & 0 & & 0 & & & & 0 \\
 & & \downarrow & & \downarrow & & & & \downarrow \\
0 & \to & \underline{g_{k+r-1}} & \overset{-\delta}{\to} & \underline{T^*M \otimes g_{k+r}} & \overset{-\delta}{\to} \cdots \overset{-\delta}{\to} & \underline{\Lambda_n^0 M \otimes g_{k+r-n+1}} & \to 0 \\
\downarrow & & \downarrow & & \downarrow & & & & \downarrow \\
0 \to \Theta & \overset{D^{k+r+1}}{\to} & \underline{E_{k+r+1}} & \overset{D}{\to} & \underline{T^*M \otimes E_{k+r}} & \overset{D}{\to} \cdots \overset{D}{\to} & \underline{\Lambda_n^0 M \otimes E_{k+r-n-1}} & \to 0 \\
\| & & \downarrow & & \downarrow & & & & \downarrow \\
0 \to \Theta & \overset{D^{k+r}}{\to} & \underline{E_{k+r}} & \overset{D}{\to} & \underline{T^*M \otimes E_{k+r-1}} & \overset{D}{\to} \cdots \overset{D}{\to} & \underline{\Lambda_n^0 M \otimes E_{k+r-n}} & \to 0 \\
 & & \downarrow & & \downarrow & & & & \downarrow \\
 & & 0 & & 0 & & & & 0
\end{array}
$$

As g_{k+1} is involutive, the cohomology $H^{q-j,j}(E_k)$ of $S_1^q(\Theta)$ at $\Lambda_j^0 M \otimes E_{q-j}$ is independent of $s = q - j \geq k + 1$, and is called the $j - th$ *Spencer cohomology group* $H^j(E_k)$ of E_k. Note that for trivial linear PDEs, i.e., $E_k = J\mathcal{D}^k(E)$, the above commutative diagram has both rows and columns exact. The following is also called *first linear Spencer complex of E_k*:

$(S_1^q(E))$

$$
0 \to \underline{E_q} \overset{\mathbf{D}}{\to} \underline{T^*M \bigotimes J\mathcal{D}^{q-1}(E)} \overset{\mathbf{D}}{\to} \underline{\Lambda_2^0 M \bigotimes J\mathcal{D}^{q-2}(E)} \overset{\mathbf{D}}{\to} \cdots \overset{\mathbf{D}}{\to} \underline{\Lambda_n^0 M \bigotimes J\mathcal{D}^{q-n}(E)} \to 0
$$

with $q = k + r > k$, obtained from $S_1^q(\Theta)$ by dropping the term Θ. Of course $S_1^q(E_k)$ is not more locally exact at E_q. Let us consider, now, the interaction between a submanifold $X \subset M$ and above complexes. So let X be a s-dimensional submanifold of M and $\rho : X \to M$ the canonical embedding. Then we can associate to the first linear Spencer complex $S_1^q(\rho^* E)$ another complex $S_1^q(E)_\rho$ over X, such that the following sequence

$$
S_1^q(E)_\rho \to S_1^q(\rho^* E) \to 0
$$

is exact. More precisely, one has the following commutative diagram:

$$
\begin{array}{ccccccccc}
0\to & \underline{Z} & \overset{{}_\rho D^q}{\to} & \underline{\rho^*JD^q(E)} & \overset{D}{\to} & \underline{T^*X\otimes\rho^*JD^{q-1}(E)} & \overset{D}{\to}\cdots\overset{D}{\to} & \underline{\Lambda^0_sX\otimes\rho^*JD^{q-s}(E)} & \to 0 \\
& \downarrow & & \downarrow{\scriptstyle\rho_q} & & \downarrow{\scriptstyle\rho_{q-1}} & & \downarrow{\scriptstyle\rho_{q-s}} & \\
0\to & \underline{Z} & \overset{D^q}{\to} & \underline{JD^q(Z)} & \overset{D}{\to} & \underline{T^*X\otimes JD^{q-1}(Z)} & \overset{D}{\to}\cdots\overset{D}{\to} & \underline{\Lambda^0_sX\otimes JD^{q-s}(Z)} & \to 0 \\
& \downarrow & & \downarrow & & \downarrow & & \downarrow & \\
& 0 & & 0 & & 0 & & 0 &
\end{array}
$$

where $Z=\rho^*E$. The map ${}_\rho D^q$ is induced from $D^q:\underline{E}\to\underline{JD^q(E)}$ by ρ: ${}_\rho D^q(s\circ\rho)=(D^q s)\circ\rho$. $S^q_1(E)_\rho$ is exact at $\underline{\Lambda^0_i X\bigotimes\rho^*JD^{q-i}(E)}$, for $i>0$. Now, to $S^q_1(E)_\rho$ we can associate, by restriction on E_q and dropping the first term, the following sequence:

$$
0\to\underline{\rho^*E_q}\overset{D}{\to}\underline{T^*X\otimes\rho^*E_{q-1}}\overset{D}{\to}\cdots\overset{D}{\to}\underline{\Lambda^0_sX\otimes\rho^*E_{q-s}}\to 0.
$$

Let $H^{q,j}(E_k)_{\rho,x}$, (resp. $H^{q,j}(E_k)_{\rho(x)}$), be the cohomology of the complex

$$
S^q_1(E_k):\quad (\Lambda^0_jM\bigotimes E_{q-1})_{\rho(x)},\ x\in X.
$$

Then, one has a canonical map $\rho_*:H^{q,j}(E_k)_{\rho(x)}\to H^{q,j}(E_k)_{\rho,x}$, $\forall x\in X$, $q\ge k$, $j\ge 0$. The Cauchy problem for regular solutions of $E_k\subset JD^k(E)$ is the study of the above maps with $j=0$. Surjectivity of these maps corresponds to the existence theorems and injectivity to uniqueness. Then one can prove the following criterion of uniqueness:

Lemma 4.7. (H.Goldschmidt). *If g_{k+1} is involutive and X is non-characteristic for g_{k+1} [75] one has the existence and uniqueness for the Cauchy problem, i.e., there exists a unique section s of E defined on a neighbourhood U of X in M such that $D^k s(x)\in E_k$, $\forall x\in U$ and $s|_X=s_0\in C^\omega(E|_X)$, for some fixed s_0.*

---

[75] We say that X is *non-characteristic* for g_{k+1} if $((S^{k+1}_0 M)\otimes\rho^*E)\cap\rho^*g_{k+1}=0$, where $N\subset\rho^*T^*M$ is the conormal bundle of X in M and $S^r_0 N$ denotes its r-th symmetric product. Recall also that the normal bundle of X in M is identified with ρ^*TM/TX and the conormal bundle of X in M is just the dual bundle $N\equiv(\rho^*TM/TX)^*$ of the normal bundle. The embedding of N into ρ^*T^*M is obtained by means of the following exact sequence: $0\leftarrow T^*X\leftarrow\rho^*T^*M\leftarrow N\leftarrow 0$ obtained by duality from the following canonical short exact sequence: $0\to TX\to\rho^*TM\to\rho^*TM/TX\to 0$.

Proof of Lemma 4.7. See ref.[101(g)]. $\qquad\qquad\qquad\qquad\square$

Let us, now, apply above results to our involutive, formally integrable linear equation $\widehat{\overline{(NS)}}_s$, by specializing the submanifold $X \subset M$ with a 3-dimensional space-like submanifold. Then, in such a case we can see that the maps $\rho_* : H^{q,0}(\widehat{\overline{((NS)}}_s))_{\rho(x)} \to H^{q,0}(\widehat{\overline{((NS)}}_s))_{\rho,x}$, $\forall x \in X$, $q \geq 2$, is surjective. In fact, we have just seen that $\widehat{\overline{(NS)}}_s$ admits solutions for any space-like regular $3-$dimensional data. On the other hand the uniqueness fails. In fact, one can easily see that such 3-dimensional space-like submanifolds do not satisfy the condition to be non-characteristic for the symbol $\widehat{g}_{2+1}[s]$ of the first prolongation of $\widehat{\overline{(NS)}}_s$. (Note that in order to apply Lemma 4.7 it is necessary that the symbol $\widehat{g}_{2+1}[s]$ be involutive. This is just the case as $\widehat{g}_2[s]$ is involutive and hence also $\widehat{g}_{2+r}[s]$, $\forall r \geq 0$, are involutive for definition.) In fact, let us consider the submanifold identified by the equation $x^0 = const.$ in the coordinates (x^α) on M, adapted to an inertial frame. Then the induced coordinates on TM and TX are $(x^\alpha, \dot{x}^\alpha)$ and $(x^k, \dot{x}^k)$, respectively. Therefore, on the normal bundle $TM|_X/TX$ the induced coordinates are $(x^k, \dot{x}^0)$, hence on the conormal bundle N one has $(x^k, \dot{x}_0)$. This means that a vector $\zeta \in S^3(N) \bigotimes E$ iff its representation in coordinates is $\zeta = \zeta^I_{000}(x^k)dx^0 \otimes dx^0 \otimes dx^0 \otimes \partial y_I$, where $(y^I) \equiv (\dot{x}^k, p, \theta)$ are vertical coordinates on E. Now, the equation $\widehat{\overline{(NS)}}_s$ can be written in the following form:

$$\widehat{\overline{(NS)}}_s \left\{ \begin{aligned}
&(\partial\dot{x}_k.F^0)_s\nu^k + (\partial\dot{x}^p_k.F^0)_s\nu^k_p \equiv \Xi^0 = 0 \\[6pt]
&(\partial\dot{x}_k.F^0_\alpha)_s\nu^k + (\partial\dot{x}^\beta_k.F^0_\alpha)_s\nu^k_\beta + (\partial\dot{x}^{\beta s}_k.F^0_\alpha)_s\nu^k_{\beta s} \equiv \Xi^0_\alpha = 0 \\[6pt]
&(\partial\dot{x}_k.F^j)_s\nu^k + (\partial\dot{x}^p_k.F^j)_s\nu^k_p + (\partial\dot{x}^0_k.F^j)_s\nu^k_0 \\[6pt]
&+ (\partial\dot{x}^{ir}_k.F^j)_s\nu^k_{ir} + (\partial p^i.F^j)_s\nu^{(p)}_i \equiv \Xi^j = 0, \\[6pt]
&(\partial\dot{x}_k.F^4)_s\nu^k + (\partial\dot{x}^p_k.F^4)_s\nu^k_p + (\partial\theta^0.F^4)_s\nu^{(\theta)}_0 \\[6pt]
&+ (\partial\theta^k.F^4)_s\nu^{(\theta)}_k + (\partial\theta^{ir}.F^4)_s\nu^{(\theta)}_{ir} \equiv \Xi^4 = 0 \\[6pt]
&1 \leq j \leq 3
\end{aligned} \right\}$$

where the explicit expression of the coefficients are reported in the following note [76] . Therefore the first prolongation results:

$$
\widehat{((NS)_s)}_{+1}\left\{
\begin{aligned}
&(\partial\dot{x}_k.F^0)_s\nu^k+(\partial\dot{x}_k^p.F^0)_s\nu_p^k\equiv\Xi^0=0\\[4pt]
&(\partial\dot{x}_k.F_\alpha^0)_s\nu^k+(\partial\dot{x}_k^\beta.F_\alpha^0)_s\nu_\beta^k+(\partial\dot{x}_k^{\beta s}.F_\alpha^0)_s\nu_{\beta s}^k\equiv\Xi_\alpha^0=0\\[4pt]
&(\partial\dot{x}_k.F^j)_s\nu^k+(\partial\dot{x}_k^p.F^j)_s\nu_p^k\\[2pt]
&\quad+(\partial\dot{x}_k^0.F^j)_s\nu_0^k+(\partial\dot{x}_k^{ir}.F^j)_s\nu_{ir}^k+(\partial p^i.F^j)_s\nu_i^{(p)}\equiv\Xi^j=0,\\[4pt]
&(\partial\dot{x}_k.F^4)_s\nu^k+(\partial\dot{x}_k^p.F^4)_s\nu_p^k+(\partial\theta^0.F^4)_s\nu_0^{(\theta)}\\[2pt]
&\quad+(\partial\theta^k.F^4)_s\nu_k^{(\theta)}+(\partial\theta^{ir}.F^4)_s\nu_{ir}^{(\theta)}\equiv\Xi^4=0\\[4pt]
&(\partial x_\gamma.\Xi^0)_s+(\partial\dot{x}_k.F^0)_s\nu_\gamma^k+(\partial\dot{x}_i^r.F^0)_s\nu_{r\gamma}^i=0\\[4pt]
&(\partial x_\gamma.\Xi_\alpha^0)_s+(\partial\dot{x}_k.F_\alpha^0)_s\nu_\gamma^k+(\partial\dot{x}_k^\beta.F_\alpha^0)_s\nu_{\beta\gamma}^k+(\partial\dot{x}_k^{\beta r}.F_\alpha^0)_s\nu_{\beta r\gamma}^k=0\\[4pt]
&(\partial x_\gamma.\Xi^j)_s+(\partial\dot{x}_k.F^j)_s\nu_\gamma^k+(\partial\dot{x}_k^\beta.F^j)_s\nu_{\beta\gamma}^k+(\partial\dot{x}_k^{ir}.F^j)_s\nu_{ir\gamma}^k+(\partial p^i.F^j)_s\nu_{i\gamma}^{(p)}=0\\[4pt]
&(\partial x_\gamma.\Xi^4)_s+(\partial\dot{x}_k.F^4)_s\nu_\gamma^k+(\partial\dot{x}_k^p.F^4)_s\nu_{p\gamma}^k+(\partial\theta^\beta.F^4)_s\nu_{\beta\gamma}^{(\theta)}+(\partial\theta^{ir}.F^j)_s\nu_{ir\gamma}^{(\theta)}=0\\[4pt]
&1\le j\le 3
\end{aligned}
\right\}.
$$

As a consequence the corresponding symbol $\widehat{(g_2[s])}_{+1}\subset S^3(N)\bigotimes E$ is characterized by vector fields $\zeta_{\alpha\beta\gamma}^I dx^\alpha\odot dx^\beta\odot dx^\gamma\otimes\partial y_I=\zeta_{\alpha\beta\gamma}^I\partial y_I^{\alpha\beta\gamma}$ such that the following equations are satisfied:

$$
(4.12)\qquad\left\{
\begin{aligned}
&(\partial\dot{x}_k^{\beta p}.F_\alpha^0)_s\zeta_{\beta p\gamma}^k=0\\[4pt]
&(\partial\dot{x}_k^{ir}.F^j)_s\zeta_{ir\gamma}^k=0\\[4pt]
&(\partial\theta^{ir}.F^4)_s\zeta_{ir\gamma}^{(\theta)}=0
\end{aligned}
\right\}.
$$

Therefore, $(S^3(N)\bigotimes E)\bigcap\widehat{(g_2[s])}_{+1}\neq 0$ as from (4.12) we have not restrictions on the components ζ_{000}^I of $\zeta_{\alpha\beta\gamma}^I\partial y_I^{\alpha\beta\gamma}\in\widehat{(g_2[s])}_{+1}$. Finally, in order to recognize the existence of global smooth solutions of $\widehat{(NS)_s}$ it is enough to calculate their integral bordism groups. In particular, by applying our results on the integral bordism groups of PDEs, we can see that the 3-dimensional integral bordism group $\Omega_3^{\widehat{(NS)_s}}$ is zero. This is enough to state the existence of global (smooth) solutions of $\widehat{(NS)_s}$ for any admissible

---

[76] $(\partial\dot{x}_k.F^0)_s=G_{ik}^i$; $(\partial\dot{x}_k^p.F^0)_s=\delta_i^p$; $(\partial\dot{x}_k.F_\alpha^0)_s=\delta_\alpha^r(\partial x_r.G_{ik}^i)$; $(\partial\dot{x}_k^\beta.F_\alpha^0)_s=\delta_\alpha^\beta G_{ik}^i$; $(\partial\dot{x}_k^{\beta s}.F_\alpha^0)_s=\delta_k^s\delta_\alpha^\beta$; $(\dot{x}^0=1)$; $(\partial\dot{p}^i.F^j)_s=g^{ij}$; $(\partial\dot{x}_k^\beta.F^j)_s=\delta_\alpha^\beta\delta_k^j\rho\dot{x}^\alpha+S_k^{js}\delta_s^\beta$; $(\partial\dot{x}_k.F^j)_s=R_s^j+2\dot{x}^i\rho G_{is}^j+\dot{x}_s^j\rho$; $(\partial\dot{x}_k^{ip}.F^j)_s=T_k^{jip}$; $(\partial\dot{x}_k.F^4)_s=\theta_k\theta_{is}\bar{E}^{is}\rho C_p+\dot{x}^p W_{kp}+\dot{x}^i W_{ik}+\dot{x}_p^s\bar{W}_{ks}^p$; $(\partial\dot{x}_r^p.F^4)_s=\dot{x}^k\bar{W}_{kr}^p+\dot{x}_i^k Y_{kr}^{ip}+\dot{x}_i^k Y_{rk}^{pi}$; $(\partial\theta^\beta.F^4)_s=\delta_0^\beta\rho C_p+\delta_k^\beta\dot{x}^k\theta_{ir}\bar{E}^{ir}\rho C_p$; $(\partial\theta^{ir}.F^4)_s=\dot{x}^k\theta_k\bar{E}^{ir}\rho C_p$.

Cauchy data. If s is stationary the coefficients in the expression of $\widehat{\overline{(NS)}}_s$ are functions on M indipendent on x^0. Then the general regular solution of $\widehat{\overline{(NS)}}_s$ can be obtained as a linear combinations of solutions of the type $\nu = e^{-\lambda x^0}\mu(x^i)$. Then one find the following equations for λ and μ:

$$\left\{\begin{aligned}
&(1)\ (\partial \dot{x}_k.F^0)_s\mu^k+(\partial \dot{x}_k^p.F^0)_s\mu_p^k=0 \\
&(2)\ (\partial \dot{x}_k.F_\alpha^0)_s\mu^k-\lambda(\partial \dot{x}_k^0.F_\alpha^0)_s\mu^k \\
&\qquad +(\partial \dot{x}_k^p.F_\alpha^0)_s\mu_p^k-\lambda(\partial \dot{x}_k^{0r}.F_\alpha^0)_s\mu_r^k+(\partial \dot{x}_k^{pr}.F_\alpha^0)_s\mu_{pr}^k=0 \\
&(3)\ (\partial \dot{x}_k.F^j)_s\mu^k-\lambda(\partial \dot{x}_k^0.F^j)_s\mu^k \\
&\qquad +(\partial \dot{x}_k^p.F^j)_s\mu_p^k+(\partial \dot{x}_k^{ir}.F^j)_s\mu_{ir}^k+(\partial p^i.F^j)_s\mu_i^{(p)}=0,\ 1\leq j\leq 3 \\
&(4)\ (\partial \dot{x}_k.F^4)_s\mu^k+(\partial \dot{x}_k^p.F^4)_s\mu_p^k-\lambda(\partial \theta^0.F^4)_s\mu^{(\theta)}+(\partial \theta^k.F^4)_s\mu_k^{(\theta)}+(\partial \theta^{ir}.F^4)_s\mu_{ir}^{(\theta)}=0
\end{aligned}\right\}.$$

Then, taking into account that: $(\partial \dot{x}_k^0.F_\alpha^0)_s = \delta_\alpha^0 G_{ik}^i$, $(\partial \dot{x}_k^{0r}.F_\alpha^0)_s = 0$, $(\partial \dot{x}_k^0.F^j)_s = \rho\delta_k^j$, $(\partial \theta^0.F^4)_s = \rho C_p$, we get:

$$\left\{\begin{aligned}
&(1)\ (\partial \dot{x}_k.F^0)_s\mu^k+(\partial \dot{x}_k^p.F^0)_s\mu_p^k=0 \\
&(2)\ (\partial \dot{x}_k.F_\alpha^0)_s\mu^k+(\partial \dot{x}_k^p.F_\alpha^0)_s\mu_p^k+(\partial \dot{x}_k^{pr}.F_\alpha^0)_s\mu_{pr}^k=\lambda\delta_\alpha^0 G_{ik}^i\mu^k \\
&(3)\ \tfrac{1}{\rho}[(\partial \dot{x}_k.F^j)_s\mu^k+(\partial \dot{x}_k^p.F^j)_s\mu_p^k+(\partial \dot{x}_k^{ir}.F^j)_s\mu_{ir}^k+(\partial p^i.F^j)_s\mu_i^{(p)}]=\lambda\mu^j,\ 1\leq j\leq 3 \\
&(4)\ \tfrac{1}{\rho C_p}[(\partial \dot{x}_k.F^4)_s\mu^k+(\partial \dot{x}_k^p.F^4)_s\mu_p^k+(\partial \theta^k.F^4)_s\mu_k^{(\theta)}+(\partial \theta^{ir}.F^4)_s\mu_{ir}^{(\theta)}=\lambda\mu^{(\theta)}
\end{aligned}\right\}.$$

Thus the problem is reconduced to an eigenvalue problem (equations (3) and (4)) conditioned by the constraints (1) and (2). The linear differential operator involved is not, in general, self-adjoint. So the eigenvalues have both real and imaginary parts. If the real parts are all positive, then the amplitude of every parturbation decays exponentially with time and the underlying stationary solution s is then said to be *linearly stable*. In general, to a stationary solution will correspond eigenvalues with negative and positive real parts. Then the smallest Reynolds number coresponding to steady-state solutions with eigenvalues having positive real part is called *critical linear Reynolds number* $R_{e,c}^{linear}$. Of course solutions corresponding to eigenvalues with higher Reynolds numbers will be unstable. However, such analysis for unstability, works well only for stationary solutions. For non-stationary solutions we shall utilize the linearized equation $\widehat{\overline{(NS)}}_s$ with a different approach. In order to see this, let us emphasize that $\widehat{\overline{(NS)}}_s$ can be considered a vector neighbourhood of $D^2s(M)$ into $\widehat{(NS)}$ and that

$E \equiv s^* v T W$ is a vector neighbourhood of $s(M)$ into W. So a solution v of $\widehat{(NS)}_s$ can be identified with a vertical shift of the section s of $\pi : W \to M$ into a *perturbated section* $s' \equiv s + v$. Then, we can say that s is *stable* if any of such admissible perturbations v of s is such that the time-depending function $\mathfrak{p}(t) = \frac{1}{2} \int_{\underline{B}_t} v^2 \eta / \int_{\underline{B}_t} \eta$ is a decreasing function.[77] We call such a function the *square mean perturbation*.[78] One has:

(4.13)
$$\mathfrak{p}(t) = \frac{1}{2 vol(\underline{B}_t)} \left[\int_{\underline{B}_t} (v^{(v)})^2 \eta + \int_{\underline{B}_t} (v^{(p)})^2 \eta + \int_{\underline{B}_t} (v^{(\theta)})^2 \eta \right] \equiv \mathfrak{p}^{(v)} + \mathfrak{p}^{(p)} + \mathfrak{p}^{(\theta)}.$$

Let us, now, consider the rate $\dot{\mathfrak{p}}$ of variation of $\mathfrak{p}$ along the flow:[79]

(4.14)
$$\dot{\mathfrak{p}}(t) = \frac{1}{2} \lim_{\lambda \to 0} \left(\frac{\int_{\phi_\lambda(\underline{B}_t)} v^2 \eta}{\int_{\phi_\lambda(\underline{B}_t)} \eta} - \frac{\int_{\underline{B}_t} v^2 \eta}{\int_{\underline{B}_t} \eta} \right) / \lambda = \frac{1}{2} \frac{1}{vol(\underline{B}_t)} \int_{\underline{B}_t} \frac{\phi_\lambda^* v^2 - v^2}{\lambda} \eta = \frac{1}{2} \frac{1}{vol(\underline{B}_t)} \int_{\underline{B}_t} \frac{\delta v^2}{\delta t} \eta$$
$$= \dot{\mathfrak{p}}^{(v)} + \dot{\mathfrak{p}}^{(p)} + \dot{\mathfrak{p}}^{(\theta)}.$$

Then a sufficient condition for the stability is that

(4.15)
$$\dot{\mathfrak{p}}(t) \leq -c\, \mathfrak{p}(t)$$

where c is a positive constant. In fact, we shall use the following lemmas.

Lemma 4.8. (Gronwall's lemma). *Suppose $f(t)$ is a real function whose derivative is bounded according to:*

(4.16)
$$\frac{df}{dt} \leq g(t) f + h(t)$$

for some real functions $g(t)$ and $h(t)$. Then, $f(t)$ is bounded pointwise in time according to

(4.17)
$$f(t) \leq f(0) e^{G(t)} + \int_{[0,t]} e^{G(t-s)} h(s) ds$$

---

[77] Note that the vector fiber bundle $\pi : E \to M$ has a natural Euclidean structure on the fibers as these are identified with $\mathbf{S} \times \mathbf{R}^2$. In fact one has the canonical isomorphism: $s^* v T W \cong M \times \mathbf{S} \times \mathbf{R}^2$, for any section $s : M \to W$.

[78] We say that a solution $V \subset (NS)$ is *t-unstable* if the perturbations arised at some instant t, on the 3-dimensional compact space-like submanifold $B_t \subset V$, produce a square mean perturbation $\mathfrak{p}(t)$ that increases in the time. Otherwise we say that V is *t-stable*. If V is *t*-stable at any $t \in T$, we say that it is *globally stable*, (otherwise *globally unstable*).

[79] Note that formula (4.14) works well also if the flow ϕ_λ is not a diffeomorphism for all λ, i.e., if there are singular points in the flow.

where $G(t) = \int_{[0,t]} g(r)dr$.

Proof of Lemma 4.8. It is standard. (See e.g. ref.[101(e)].)   □

Lemma 4.9. *If equation* (4.15) *holds, one has:*

$$(4.18) \qquad\qquad \mathfrak{p}(t) = \mathfrak{p}(0)e^{-ct}.$$

So condition (4.15) *is sufficient for the stability.*

Proof of Lemma 4.9. It is enough to use Gronwall's lemma with $g(t) = -c$ and $h(t) = 0$.   □

Lemma 4.10. *If we consider neglectable the perturbations in temperature and pressure,*[80] *then for any admissible regular solution s of $(\widehat{NS})$ the ratio $\dot{\mathfrak{p}}/\mathfrak{p}$ is given by the following formula:*

$$(4.19) \qquad \frac{\dot{\mathfrak{p}}(t)}{\mathfrak{p}(t)} = 2\,\frac{-\left[\int_{\underline{B}_t}(\nabla_{\nu^{(v)}}v).\nu^{(v)}\eta + \frac{\chi}{\rho}(d\underline{\nu^{(v)}},d\underline{\nu^{(v)}})\right]}{\int_{\underline{B}_t}(\nu^{(v)})^2\eta}.$$

Proof of Lemma 4.10. Let us consider a perturbated solution $s' = s + \nu$ of $(\widehat{NS})$, where s is just a solution of $(\widehat{NS})$ and ν is a solution of the linearized equation $\overline{(\widehat{NS})}_s$. Let us find the intrinsic expression for the equation of $\nu = (\nu^{(v)},\nu^{(p)},\nu^{(\theta)}) = (\nu^k,\nu^{(p)},\nu^{(\theta)})$. Then by using $(\widehat{NS})$ written in the form:

$$(\widehat{NS})\quad\begin{cases}\delta\underline{v}=0\\[4pt](\partial x_\alpha.(\delta\underline{v}))=0\\[4pt]\rho\frac{\delta v}{\delta t}=\chi\overline{\Delta\underline{v}}-\mathrm{grad}\,(p+\rho f)\\[4pt]\rho C_p\frac{\delta\theta}{\delta t}=\nu\Delta\theta+2\chi<\dot{e},\dot{e}>\end{cases}$$

we get that the perturbation ν must satisfy the following equations:

$$\begin{cases}\delta\underline{\nu^{(v)}}=0\\[4pt](\partial x_\alpha.(\delta\underline{\nu^{(v)}}))=0\\[4pt]\rho\frac{\delta\nu^{(v)}}{\delta t}=-\rho(\nabla_{\nu^{(v)}}v+\nabla_{\nu^{(v)}}\nu^{(v)})+\chi\overline{\Delta\underline{\nu^{(v)}}}-\mathrm{grad}\,(\nu^{(p)})\\[4pt]\rho C_p\frac{\delta\nu^{(\theta)}}{\delta t}=\nu\Delta\nu^{(\theta)}+4\chi<\dot{e},\dot{e}^{\nu^{(v)}}>+2\chi<\dot{e}^{\nu^{(v)}},\dot{e}^{\nu^{(v)}}>+\nu^k\theta_{/k}+v^k\nu^{(\theta)}_{/k}\end{cases}.$$

So the corresponding linear equation $\overline{(\widehat{NS})}_s$ for the perturbation ν is the

---

[80] This is surely the case when we impose the isothermal constraint.

following:

$$(4.20) \qquad \widehat{\overline{(NS)}_s} \begin{cases} (1)\ \ \delta\underline{\nu^{(v)}}=0 \\[4pt] (2)\ \ (\partial x_\alpha.(\delta\underline{\nu^{(v)}}))=0 \\[4pt] (3)\ \ \rho\frac{\delta\nu^{(v)}}{\delta t}=-\rho(\nabla_{\nu^{(v)}}v)+\chi\overline{\Delta\underline{\nu^{(v)}}}-\mathrm{grad}\,(\nu^{(p)}) \\[4pt] (4)\ \ \rho C_p\frac{\delta\nu^{(\theta)}}{\delta t}=\nu\Delta\nu^{(\theta)}+4\chi<\dot e,\dot e^{\nu^{(v)}}>+\nu^k\theta_{/k}+v^k\nu^{(\theta)}_{/k} \end{cases}.$$

Recall that on a n-dimensional compact oriented Riemannian manifold, the Lapalce operator $\Delta=\delta d+d\delta$ for p-differential forms, locally written as $\alpha=\sum_{1\le i<i_1<\cdots<i_p\le n}\alpha_{i_1\cdots i_p}dx^{i_1}\wedge\cdots\wedge dx^{i_p}$ has the following expression:

$$\Delta\alpha=\sum_{1\le i<i_1<\cdots<i_p\le n}[-(\alpha_{i_1\cdots i_p/i})_{/j}$$
$$+\sum_{1\le s\le p}(-1)^s R^k_{i_s}\alpha_{ki_1\cdots\widehat{i_s}\cdots i_p}+2\sum_{r<p}(-1)^{r+s}R^k{}_{i_s}{}^t{}_{i_r}\alpha_{tki_1\cdots\widehat{i_r}\cdots\widehat{i_s}\cdots i_p}]dx^{i_1}\wedge\cdots\wedge dx^{i_p},$$

with $R^i{}_j{}^k{}_l$ the curvature tensor and $R^s_k=R^s{}_i{}^i{}_k$ the Ricci tensor. If we consider differential forms on a compact domain $X\subset E^3$ of an Euclidean affine 3-dimensional space, then $R^i{}_j{}^k{}_l=0$, $R^s_k=0$ and

$$\Delta\alpha=-(\alpha_{i_1\cdots i_p/i})_{/j}g^{ji}dx^{i_1}\wedge\cdots\wedge dx^{i_p}.$$

In particular, if α is a 1-form:

$$\Delta\alpha=-(\alpha_{k/i})_{/j}g^{ji}dx^k$$
$$=-[(\partial x_j\partial x_i.\alpha_k)-(\partial x_j.\Gamma^r_{ki})\alpha_r-\Gamma^s_{ij}(\partial x_s.\alpha_k)-\Gamma^s_{kj}(\partial x_i.\alpha_s)$$
$$+\Gamma^s_{ij}\Gamma^r_{ks}\alpha_r+\Gamma^s_{kj}\Gamma^r_{si}\alpha_r]g^{ji}dx^k.$$

Then from the third equation in (4.20), after scalar multiplication by $\nu^{(v)}$, we get

$$\frac{\rho}{2}\frac{\delta(\nu^{(v)})^2}{\delta t}=-\rho(\nabla_{\nu^{(v)}}v)\cdot\nu^{(v)}-\chi(\overline{\Delta\underline{\nu^{(v)}}})\cdot\nu^{(v)}-(\mathrm{grad}\,(\nu^{(p)}))\cdot\nu^{(v)}.$$

Let us untegrate above equation over $\underline{B}_t$, taking into account that on the vector space $\Omega^p(\underline{B}_t)$ of differential p-forms on $\underline{B}_t$ one has the following scalar product: $(\alpha,\beta)=\int_{\underline{B}_t}\alpha\wedge\star\beta$. Furthermore, for such scalar product one has $(\delta\alpha,\beta)=(\alpha,d\beta)$, where δ is the codifferential. Then, after considered the continuity equation $\delta\underline{\nu^{(v)}}=0$, we get:

$$(4.21)\qquad \int_{\underline{B}_t}\frac{\delta(\nu^{(v)})^2}{\delta t}\eta=-2\int_{\underline{B}_t}\left[(\nabla_{\nu^{(v)}}v)\cdot\nu^{(v)}-\frac{\chi}{\rho}(\overline{\Delta\underline{\nu^{(v)}}})\cdot\nu^{(v)}\right]\eta.$$

Thus, by considering equation (4.14) we get:

$$(4.22) \qquad \dot{\mathsf{p}}(t) = \frac{-1}{vol(\underline{B}_t)} \int_{\underline{B}_t} \left[(\nabla_{\nu^{(v)}} v) \cdot \nu^{(v)} - \frac{\chi}{\rho} (\overline{\triangle \underline{\nu^{(v)}}}) \cdot \nu^{(v)} \right] \eta.$$

So the lemma is proved. $\qquad\qquad\qquad\qquad\qquad\qquad\qquad\qquad\qquad\qquad\quad$ $\square$

Lemma 4.11. *If the solution* $V \subset \widehat{(NS)}$ *is such that there exists a constant* $c > 0$*, such that*

$$(4.23) \qquad c \int_{\underline{B}_t} (\nu^{(v)})^2 \eta \leq \int_{\underline{B}_t} \left[(\nabla_{\nu^{(v)}} v) \cdot \nu^{(v)} - \frac{\chi}{\rho} (\overline{\triangle \underline{\nu^{(v)}}}) \cdot \nu^{(v)} \right] \eta$$

for any admissible perturbation $\nu^{(v)}$ *of the velocity field of the space-time flow, then the solution* V *is stable, assumed that the perturbations of temperature and pressure are neglectable.*

Proof of Lemma 4.11. In fact under the conditions of this lemma, by using Lemma 4.10, one has the following inequality: $\dot{\mathsf{p}}/\mathsf{p} \leq -c$. So Lemma 4.9 assure that the solution is stable. $\qquad\qquad\qquad\qquad\qquad\qquad\qquad$ $\square$

Lemma 4.12. *Let* $V \subset \widehat{(NS)}$ *be a solution such that the perturbations of the temperature and pressure are neglectable. Then* V *is stable if the smallest eigenvalue* $\bar{\lambda}_1 = \bar{\lambda}_1(t)$ *of the self-adjoint differential operator* $P[s](\nu) = -\frac{\chi}{\rho} \triangle \underline{\nu^{(v)}} + \,< \dot{e}, \nu^{(v)} > +dp$*, for divergence free vector fields* $\nu^{(v)}$*, canonically associated to the solution* s*, is positive for any* $t \in T$ *and lower bounded:* $\bar{\lambda}_1(t) \geq \lambda_1 \geq 0$*.*

Proof of Lemma 4.12. From the above lemma it follows that to assure the stability of a solution it is enough that the infimum of the ratio

$$(4.24) \qquad \frac{\int_{\underline{B}_t} \left[(\nabla_{\nu^{(v)}} v) \cdot \nu^{(v)} - \frac{\chi}{\rho} (\overline{\triangle \underline{\nu^{(v)}}}) \cdot \nu^{(v)} \right] \eta}{\int_{\underline{B}_t} (\nu^{(v)})^2 \eta}$$

is a positive value $c \in \mathbf{R}$, $\forall t \in T$. As this ratio is invariant for transformations $\nu^{(v)} \mapsto \alpha(t)\nu^{(v)}$, $\forall \alpha(t) > 0$, we can also normalize $\nu^{(v)}$ to $\int_{\underline{B}_t} (\nu^{(v)})^2 \eta = 1$. So we can consider the infimum of the numerator only for such normalized perturbation $\nu^{(v)}$. This problem can be converted into a variational problem with the following action integral:

$$F[\nu^{(v)}] = \int_{\underline{B}_t} \left[-\frac{\chi}{\rho} (\overline{\triangle \underline{\nu^{(v)}}}) \cdot \nu^{(v)} + (\nabla_{\nu^{(v)}} v) \cdot \nu^{(v)} - p\delta \underline{\nu^{(v)}} - \lambda (\nu^{(v)})^2 \right] \eta, \ \forall t \in T$$

where p is the Lagrange multiplier (identified with the pressure) for the local constraint given in equation $(4.20)(1)$ and λ is the Lagrange multiplier for the global constraint $\int_{\underline{B}_t}(\nu^{(v)})^2\eta = 1$. The corresponding Euler-Lagrange equation is given by the following equations: Note that in equation $(4.25)(A)$ dp is the vertical differential with respect to the fiber bundle structure $\tau : M \to T$ (space-time on time) and $\lambda = \lambda(t)$ is a time-depending function.

$$(4.25) \qquad \left\{ \begin{array}{l} \text{(A)} \ -\frac{\chi}{\rho}\Delta\underline{\nu^{(v)}}+<\dot{e},\nu^{(v)}>+dp=2\lambda\underline{\nu^{(v)}} \\[2mm] \text{(B)} \ \delta\underline{\nu^{(v)}}=0 \end{array} \right\}.$$

Thus the problem is reconduced to an eigenvalue problem (equation (A)), constrained by equation (B). As the differential opertor involved is self-adjoint, it has a real spectrum and the stability of the solution is related to the sign of the smallest eigenvalue. If such eigenvalue is positive, $\forall t \in T$, then the ratio (4.24) is higher then a positive constant, hence the solution is stable. In fact, from (4.24) and (4.25) we get:

$$2\frac{\int_{\underline{B}_t}\left(-\frac{\chi}{\rho}(\overline{\Delta\nu^{(v)}})\cdot\nu^{(v)}+v_{i/j}\nu^i\nu^j\right)\eta}{\int_{\underline{B}_t}(\nu^{(v)})^2\eta} \geq \frac{\bar{\lambda}_1\int_{\underline{B}_t}(\nu^{(v)})^2\eta}{\int_{\underline{B}_t}(\nu^{(v)})^2\eta} = \bar{\lambda}_1(t) \geq \lambda_1.$$

So the lemma is proved. $\qquad\qquad\qquad\qquad\qquad\qquad\qquad\qquad\square$

Lemma 4.13. *Under the same hypotheses of above lemma and assumed that the domain $\underline{B}_t$ has the negative Laplacian $-\Delta$ for differential 1-forms, along with boundary conditions, strictly self-adjoint, with a discrete spectrum and smallest eigenvalue $\lambda_1(t) > 0$, that is lower bounded: $\lambda_1(t) \geq \widehat{\lambda_1} \geq 0$, then a sufficient condition for the stability of the solution $V \subset \widehat{(NS)}$ is that the Reynolds number R_e^{sup}, considered as the upper bound of the characteristic Reynolds number $R_e(t)$, satisfies the following condition:*[81]

$$(4.26) \qquad\qquad\qquad (R_e^{sup})^2 \leq \lambda_1.$$

Proof of Lemma 4.13. Let us define as Reynolds number the upper bound of the characteristic Reynolds number R_e, i.e. $R_e^{sup} = \frac{U_{sup}L\rho}{\chi}$, where $L = \frac{1}{\sqrt{\lambda_1}}$.

---

[81] Note that if condition (4.26) is satisfied, then as the characteristic Reynolds number $R_e < R_e^{sup}$, also $(R_e)^2 \leq \lambda_1$ holds. We call *nonlinear critical Reynolds number* the following adimensional number $R_{e,c}^{nonlinear}$ such that $(R_{e,c}^{nonlinear})^2 = \lambda_1$. Note that we denote also $R_e^{sup}(t) = U_{sup}(t)L\rho/\chi$, where $U_{sup}(t) = \sup_{p \in \underline{B}_t}|v(p)|$.

Let us consider also the adimensional variables: $\bar{x} = x/L$, $\bar{t} = \chi t/L^2$, $\bar{u} = u/U_{sup}$, $\bar{p} = pL/\rho\chi U_{sup}$. Then for the ratio (4.24) one has the following inequality:[82]

$$\frac{\int_{\underline{B}_t} \left(-(\overline{\triangle \underline{\nu}^{(v)}}) \cdot \nu^{(v)} + R_e^{sup}(\nabla_{\nu^{(v)}} v) \cdot \nu^{(v)} \right) \eta}{\int_{\underline{B}_t} (\nu^{(v)})^2 \eta} \geq \frac{1}{2} \left[\frac{(d\underline{\nu}^{(v)}, d\underline{\nu}^{(v)})}{\int_{\underline{B}_t} (\nu^{(v)})^2 \eta} - (R_e^{sup})^2 \right].$$

Therefore, the sufficient condition for stability can be given, by means of the Reynolds number, in the following form:

$$(4.27) \qquad (R_e^{sup})^2 \leq \inf \left(\frac{(d\underline{\nu}^{(v)}, d\underline{\nu}^{(v)})}{\int_{\underline{B}_t} (\nu^{(v)})^2 \eta} \right).$$

Let us, now, consider the following lemma.

Lemma 4.14. (Poincaré's inequality). *Let X be a 3-dimensional compact orientable Riemannian manifold. Let $\underline{\nu} \in \Omega^1(X)$ be a differential 1-form such that $\underline{\nu}^2$ and $(d\underline{\nu})^2$ are integrable on X. Then, one has:*

$$(4.28) \qquad \frac{-\int_X (\overline{\triangle \underline{\nu}}) \cdot \nu\eta}{\int_X \underline{\nu}^2 \eta} = \frac{(d\underline{\nu}, d\underline{\nu})}{\int_X \underline{\nu}^2 \eta} = \frac{\int_X (d\underline{\nu})^2 \eta}{\int_X \underline{\nu}^2 \eta} \geq \lambda_1,$$

where $\lambda_1 > 0$ is the smallest eigenvalue of $-\triangle$.

Proof of Lemma 4.14. It is standard. (See, e.g. ref.[101(m)].) $\qquad\square$

Then from (4.27) and (4.28) we get that the regular solution $V \subset \widehat{(NS)} \subset J\mathcal{D}^2(W)$ is stable if the upper bound R_e^{sup} of the characteristic Reynolds number R_e satisfies the inequality $(R_e)^2 \leq \lambda_1$. $\qquad\square$

As a by-product of above lemma we have that in a same solution of (NS) may be present 3-dimensional space-like domains, B_t where the Reynolds number $R_e^{sup}(t)$, upper bound of the Reynolds characteristic number $R_e(t)$, at the section B_t, exceds the critical value $R_{e,c}^{nonlinear}$ other the which one has unstability, and other ones where, instead, the $R_e^{sup}(t)$ is lower than the critical value. (In such cases $R_e^{sup} \geq R_e^{nonlinear}$ and so the solution is

---

[82] Note that if $E \to M$ is a Hermitian vector bundle over a Riemannian manifold M, then with respect to a metric connection, one has: $<\nabla_\zeta v, u> = -<v, \nabla_\zeta u> + <v, u> \mathrm{div}\, \zeta$, for any vector field $\zeta : M \to TM$ and section u, v of E over M.

globally unstable.[83] Furthermore, $\dot{\mathfrak{f}}$ is, in general, a function of the time, $\mathfrak{f} = \mathfrak{f}(t)$. So the Reynolds number is a characteristic number of the solution V iff $\dot{\mathfrak{f}} = 0$. In such a case if $R_e \leq R_{e,c}^{nonlinear}$, at the initial instant, then the solution is stable assumed that the thermal and pressure perturbations are neglectable. $\qquad\square$

Therefore, we can conclude the proof of Theorem 4.7 saying that, since (NS) is a PDE that satisfies the (p)-homotopy principle, $0 \leq p \leq 3$, and standing the triviality of the bordism groups $\Omega_k^{(NS)}$ and $\Omega_k^{(NS)+\infty}$, (as well as $\Omega_{k,s}^{(NS)}$ and $\Omega_{k,s}^{(NS)+\infty}$), for $k = 1, 3$, (and also for $k = 2$, for integral admissible oriented closed surfaces of (NS)), boundary conditions do not assure, in general, the uniqueness of solutions, even if smooth. $\qquad\square$

Remark 4.7. From above results it follows that (NS) can produce solutions which exihibit finite-time singularities. But this does not contradicts Theorem 4.6 that states that there are global smooth solutions for any *"initial condition"* in the sub-equation $\widehat{(NS)}$ of the Navier-Stokes equation. In fact a global smooth solution may be eventually unstable and, therefore, will not be possible to completely observe it in experiments. (Here "completely" means "for any time".) Furthermore, it is important to emphasize that smoothness for a solution $V \subset (NS)$, does not necessarily imply absence of singularities in the corresponding flow ϕ_λ generated by the characteristic vector field $\zeta = \partial\phi$ associated to the solution V, and by means of the which the initial Cauchy data $B_{t_0} \subset V$ is propagated to generate all the solution $V = \bigcup_\lambda \phi_\lambda(B_{t_0})$. In other words smoothness of the solution is not synonymous of absence of singularity in the flow. $\qquad\blacksquare$

More precisely one has the following.

Theorem 4.8. 1) *If $V \subset (NS)$ is a smooth (global) solution such that the corresponding characteristic flow is without singularities and such that the vorticity [84] is harmonic, $\Delta\underline{\omega} = 0$, (or equivalently rot $\omega = 0$, in particular*

---

[83] Note that if ν is a solution of $\widehat{(NS)}_s$ then $\nu' \equiv \phi^*\nu = vT(\phi)\circ\nu\circ\bar\phi^{-1}$ is a solution of $\widehat{(NS)}_{s'}$, $s' = \phi^* s$, $\forall\phi\in G =$ pseudogroup of symmetries of (NS). Then assuming that thermal and pressure perturbations are neglectable for s, and that s is stable, then also s' is stable. In fact, $\forall\phi\in G$ one has $\mathfrak{p}'(t) = \dfrac{\int_{\underline{B}_t}(\nu'(v))^2\eta}{vol(\underline{B}_t)} = \dfrac{\int_{\underline{B}_t}<\phi^*\nu(v),\phi^*\nu(v)>\eta}{vol(\underline{B}_t)} = \dfrac{\int_{\underline{B}_t}\phi^* g(\nu(v),\nu(v))\eta}{vol(\underline{B}_t)} = \dfrac{\int_{\underline{B}_t}(\nu(v))^2\eta}{vol(\underline{B}_t)} = \mathfrak{p}(t)$, $\forall t\in T$. So the symmetry pseudogroup G of $\widehat{(NS)}$ allows us to transform stable solutions into stable solutions.

[84] Recall that the spin, or vorticity, ω of a space-time flow with velocity v is defined, with

if $\underline{\omega} = 0$),[85] then the vorticity is conserved along the flow. The existence of nontrivial harmonic vorticity at any instant is related to the topology of the space-like 3-dimensional manifold B_t and its boundary ∂B_t.[86] If the Reynolds number R_e^{sup} is lower than the critical value $R_{e,c}^{nonlinear}$, then V is a globally stable solution (assumed that the thermal and pressure perturbations are neglectable). Furthermore, if $\dot{\mathfrak{f}}(V) = 0$, the characteristic Reynolds number R_e is a characteristic number of the solution.

2) If $V \subset (NS)$ is a smooth (global) solution such that the corresponding characteristic flow has some singularity, then the vorticity cannot be conserved, even if it is harmonic. If the initial Reynolds number R_e^{sup} is lower than the critical value $R_{e,c}^{nonlinear}$, then V is a globally stable solution (assumed that the thermal and pressure perturbations are neglectable). Furthermore, if $\dot{\mathfrak{f}}(V) = 0$, the characteristic Reynolds number R_e is a char-

---

respect to an inertial frame ψ, by $\omega \equiv \mathrm{rot}\ v_\wedge = \star\overline{dv_\wedge} : M \to vTM$, where $v_\wedge : M \to vT^*M$ is the vertical 1-form associated to $v_\wedge \equiv u \equiv v - \dot{\psi} : M \to vTM$, with $\dot{\psi}$ the velocity of the frame ψ and $\star : C^\infty(v\Lambda_p^0 M) \to C^\infty(v\Lambda_{3-p}^0 M)$ is the star Hodge operator induced by the vertical metric, and the overline denotes the vertical field associated to the dual one. However, it is usefuel to consider ω also as the exact vertical differential 2-form $\underline{\omega} \equiv dv_\wedge : M \to v\Lambda_2^0 M$, with potential the vertical differential 1-form $\underline{v_\wedge} : M \to vT^*M$. (Note that here the exterior differential is just the vertical differential with respect to the fibration $\tau : M \to T$, space-time on time.)

[85] $\underline{\omega}$ is harmonic iff $\delta dv_\wedge = 0$. In fact, a p-form α is harmonic, i.e., $\triangle\alpha=0$, where $\triangle=\delta d+d\delta$, iff $d\alpha=0$ and $\delta\alpha=0$. On the other hand, as $\underline{\omega}=dv_\wedge$, it follows that $\underline{\omega}$ is harmonic iff $\delta\underline{\omega}=0$. Finally note that if $\underline{\omega}$ is harmonic then must be harmonic also $\underline{v_\wedge}$. In fact, $\triangle v_\wedge = \delta dv_\wedge + d\delta v_\wedge = \delta\underline{\omega} - d(\mathrm{div}\ \underline{v_\wedge}) = \delta\underline{\omega} = 0$. This has as a consequence that $\mathrm{rot}\ \omega=0$. In fact, $0=\overline{\triangle v_\wedge}=\mathrm{grad}\ \mathrm{div}\ v_\wedge - \mathrm{rot}\ \mathrm{rot}\ v_\wedge = \mathrm{rot}\ \omega$. The vice versa it is also true.

[86] Note that $\ker(\triangle)\cong H^2(B_t;\mathbf{R})\cong\mathbf{R}^{\sharp\mathrm{comp}\partial B_t - \sharp\mathrm{comp}B_t}$, and $\dim H^2(B_t;\mathbf{R})=\sharp\mathrm{comp}\partial B_t - \sharp\mathrm{comp}B_t$. Then, the existence of nontrivial, (i.e., $\underline{\omega}\neq 0$), harmonic vorticity at the instant t, is related to the topology of the space-like 3-dimensional manifold B_t and its boundary ∂B_t. For example, if $B_t\simeq D^3$ (3-dimensional disk) then $\partial B_t\simeq S^2$ (2-dimensional space-like sphere). Since $H^2(B_t;\mathbf{R})=0$, it follows that all harmonic vorticities are trivial. So, if the boundary conditions impose that this topology of B_t should be conserved at any t, and there is not vorticity at some instant, we will have an irrotational flow. Furthermore, if B_t is the region between two concentric round sphere, then $H^2(B_t;\mathbf{R})=\mathbf{R}$, hence $\dim\ker(\triangle)=1$. In such a case we can have some nontrivial harmonic vortices. Therefore, if the fixed boundary conditions impose that such a topology is conserved at any instant, we will have a conserved non zero vorticity iff it is harmonic.

acteristic number of the solution.

3) *If $D = 2$ and the vorticity is harmonic, (or equivalently $\operatorname{rot}\omega = 0$, in particular if $\underline{\omega} = 0$), then it is conserved for any smooth (global) solution of the Navier-Stokes equation. Really, in this case we have $\omega = h\partial x_3$, $h \in \mathbf{R}$. (In other words, the fluid has the skew symmetric part, ω, of the full infinitesimal strain rate , $\dot{\varepsilon} = \dot{\varepsilon}_{(s)} + \dot{\varepsilon}_{(a)} = \dot{e} + \omega$, that is a simply rigid rotation.)*

Proof. 1) The *vorticity equation* for the Navier-Stokes equation can be written in the form ($\diamond$) $\rho\frac{\delta\omega}{\delta t} = \rho\nabla_\omega u + \chi\overline{\triangle\underline{\omega}}$ or equivalently ($\odot$) $\rho\mathcal{L}_v\omega = \chi\overline{\triangle\underline{\omega}}$.[87] Then if $\triangle\underline{\omega} = 0$ one has $\mathcal{L}_v\omega = 0$. As the flow is without singularities, above equation is equivalent to say $\phi_\lambda^*\omega = \omega$, $\forall\lambda$, where $\partial\phi = v$. In fact, in such cases ϕ_λ is a diffeomorphism between the related 3-dimensional integral space-like manifolds $B_t \subset V$. To conclude the proof of this theorem it is enough to consider the following remark and lemmas.

Remark 4.8. (*Hodge decomposition of vector fields on oriented compact domain $X \subset \mathbf{R}^3$*). In this remark we report the Hodge decomposition of vector fields on oriented compact domains of $\mathbf{R}^3$. (As these results are standard we omit any proof.) Such decompositions are usefuel in order to recognize the contributions to the spatial velocity distribution $u = u^k\partial x_k : M \to vTM$, corresponding to solutions of (NS), coming from the following 0-sequence generated by means of the differential operators grad , rot and div :

$$0 \to \eth_0 \xrightarrow{\ \text{grad}\ } \eth \xrightarrow{\ \text{rot}\ } \eth \xrightarrow{\ \text{div}\ } \eth_0 \to 0$$

---

[87] The term $\nabla_\omega u$ is called the *vortex streching*. Here $\frac{\delta}{\delta t}$ is the symbol of covariant derivative along the space-time flow. Furthermore, $\mathcal{L}_v$ is the symbol of Lie derivative with respect to the space-time velocity v of the fluid. Furthermore, here $\underline{\omega}:M{\to}vT^*M$ denotes the vertical 1-form associated to $\operatorname{rot} u\equiv\omega:M{\to}vTM$. Equation ($\diamond$) is obtained from the motion equation and continuity equation. In fact, these last can be rewritten in the following form ($\bullet\bullet$) $\rho\frac{\delta u}{\delta t} = \chi\overline{\triangle\underline{u}}-\operatorname{grad}(p+\rho f)$. In fact, we have $2\operatorname{div}\dot{e}=\gamma\circ\nabla^2 u+\operatorname{grad}\operatorname{div} u+R(u)$, where R is the the Ricci curvature vertical tensor of type $(0,2)$, associated to the vertical metric of the space-time, and $\gamma:vT^*M{\otimes}vT^*M{\otimes}vTM{\to}vTM$ is the canonical morphism induced by the vertical metric. On the other hand one has $\overline{\triangle\underline{u}}=\gamma\circ\nabla^2 u - R(u)$. Hence, $2\operatorname{div}\dot{e}=\overline{\triangle\underline{u}}+2R(u)$. So, taking into account that the canonical connection on the space-time is flat, and that on the shell equation $\operatorname{div} u=0$, we have $2\operatorname{div}\dot{e}=\overline{\triangle\underline{u}}$. Now, by taking the rot of both sides of the previous equation ($\bullet\bullet$), and taking into account that rot commutes with $\triangle$, and $\operatorname{rot}(\nabla_u u)=\nabla_{\operatorname{rot}\,u}u+\nabla_u\operatorname{rot} u$, we get formula ($\diamond$).

where $\mathfrak{d}_0 \equiv \mathfrak{d}_0(X)$ and $\mathfrak{d} \equiv \mathfrak{d}(X)$ represent respectively the space of C^∞ numerical functions and vector fields on X. In particular, it is usefuel the Hodge-structure of the kernel of div , where necessarily belongs the vertical vector fields u. (See Tab. 4.2.) In Tab.4.5 we report some definitions that will be used in Tab. 4.6, where are reported important slpittings. In particular, if X is closed we get the following splittings:

$$\mathfrak{d} \cong \triangle(\mathfrak{d}) \bigoplus H \cong \text{rot } (\mathfrak{d}) \bigoplus \text{grad } (\mathfrak{d}_0) \bigoplus H$$

where $\overline{\triangle\zeta} = (\text{grad div} - \text{rot}^2)\zeta$ is the Laplacian for vector fields, $H = \ker(\triangle)$. Here $\underline{\zeta} : X \to T^*X$ is the 1-form corresponding to the vector field $\zeta : X \to TX$. Furthermore, the bar over $\triangle\underline{\zeta}$ means the vector field corresponding to the differential 1-form $\triangle\underline{\zeta}$.

TAB.4.5 - Definitions of spaces $\mathfrak{d}(X)_{(-)}$

$\mathfrak{d}_{\text{fluxes knots}} \equiv \{\zeta \in \mathfrak{d} \mid \text{div } \zeta = 0, \zeta\mid_{\partial X} \cdot n = 0, \text{ all interior fluxes are zero}\}$
$\mathfrak{d}_{\text{harmonic knots}} \equiv \{\zeta \in \mathfrak{d} \mid \text{div } \zeta = 0, \text{rot } \zeta = 0, \zeta\mid_{\partial X} \cdot n = 0\} \cong H_1(X;\mathbf{R})$
$\mathfrak{d}_{\text{curly grads.}} \equiv \{\zeta \in \mathfrak{d} \mid \zeta = \text{grad } \varphi, \text{div } \zeta = \triangle\varphi = 0, \text{fluxes of } \zeta \text{ trough each comp.of } \partial X \text{ are zero}\}$
$\mathfrak{d}_{\text{harmonic grads.}} \equiv \{\zeta \in \mathfrak{d} \mid \zeta = \text{grad } \varphi, \text{div } \zeta = \triangle\varphi = 0, \varphi\mid_{\partial X} = \text{loc. const}\} \cong H_2(X;\mathbf{R})$
$\mathfrak{d}_{\text{grounded grads.}} \equiv \{\zeta \in \mathfrak{d} \mid \zeta = \text{grad } \varphi, \varphi\mid_{\partial X} = 0\}$
$\mathfrak{d}_{\text{knots}} \equiv \{\zeta \in \mathfrak{d} \mid \text{div } \zeta = 0, \zeta\mid_{\partial X} \cdot n = 0\} \cong \mathfrak{d}_{\text{fluxes knots}} \oplus \mathfrak{d}_{\text{harmonic knots}}$
$\mathfrak{d}_{\text{grads.}} \equiv \{\zeta \in \mathfrak{d} \mid \zeta = \text{grad } \varphi\} \cong \mathfrak{d}_{\text{divergence free grads.}} \oplus \mathfrak{d}_{\text{grounded grads.}}$
$\mathfrak{d}_{\text{divergence free grads.}} \equiv \{\zeta \in \mathfrak{d} \mid \zeta = \text{grad } \varphi, \text{div } \zeta = \triangle\varphi = 0\} \cong \mathfrak{d}_{\text{curly grads.}} \oplus \mathfrak{d}_{\text{harmonic grads.}}$

An interior flux of ζ is the flux trough every smooth surface $S \subset X$, with $\partial S \subset \partial X$.

TAB.4.6 - Hodge slpittings of $\mathfrak{d} \equiv \mathfrak{d}(X)$ and related splittings

$\mathfrak{d} \cong \mathfrak{d}_{\text{fluxes knots}} \oplus \mathfrak{d}_{\text{harmonic knots}} \oplus \mathfrak{d}_{\text{curly grads.}} \oplus \mathfrak{d}_{\text{harmonic}} \oplus \mathfrak{d}_{\text{grounded}}$ $\mathfrak{d} \cong \mathfrak{d}_{\text{knots}} \oplus \mathfrak{d}_{\text{grads.}}$
$\ker(\text{rot }) \cong \mathfrak{d}_{\text{harmonic knots}} \oplus \mathfrak{d}_{\text{curly grads.}} \oplus \mathfrak{d}_{\text{harmonic grads.}} \oplus \mathfrak{d}_{\text{grounded grads.}}$ $\ker(\text{div }) \cong \mathfrak{d}_{\text{fluxes knots}} \oplus \mathfrak{d}_{\text{harmonic knots}} \oplus \mathfrak{d}_{\text{curly grads.}} \oplus \mathfrak{d}_{\text{harmonic grads.}}$ $\text{im } (\text{grad }) \cong \mathfrak{d}_{\text{curly grads.}} \oplus \mathfrak{d}_{\text{harmonic grads.}} \oplus \mathfrak{d}_{\text{grounded grads.}}$ $\text{im } (\text{rot }) \cong \mathfrak{d}_{\text{fluxes knots}} \oplus \mathfrak{d}_{\text{harmonic knots}} \oplus \mathfrak{d}_{\text{curly grads.}}$

Lemma 4.15. *Let $X \subset \mathbf{R}^3$ be a compact domain in $\mathbf{R}^3$ with smooth boundary ∂X. Then, the structures of the homology groups $H_p(X; \mathbf{R})$, $H_p(X, \partial X; \mathbf{R})$, are reported in the following table.*

TAB.4.7 - Homology groups of oriented compact domain $X \subset \mathbf{R}^3$

p	$H_p(X;\mathbf{R})$	$H_p(X,\partial X;\mathbf{R})$
0	$\mathbf{R}^{(\sharp\text{comp. of } X)}$	0
1	$\mathbf{R}^{(\text{genus of } \partial X)}$	$\mathbf{R}^{(\sharp\text{comp. of } \partial X - \sharp\text{comp. of } X)}$
2	$\mathbf{R}^{(\sharp\text{comp. of } \partial X - \sharp\text{comp. of } X)}$	$\mathbf{R}^{(\text{genus of } \partial X)}$
3	0	$\mathbf{R}^{(\sharp\text{comp. of } X)}$

Proof of Lemma 4.15. These results are standard. (See e.g. refs.[101(c)]). Here emphasize only that the Mayer-Vietoris sequence is the following long exact sequence:

$$\cdots \to H_{k+1}(\mathbf{R}^3;\mathbf{R}) \to H_k(\partial X;\mathbf{R}) \to H_k(X;\mathbf{R}) \oplus H_k(X';\mathbf{R}) \to H_k(\mathbf{R}^3;\mathbf{R}) \to \cdots$$

where $X' \equiv$ closure of $\mathbf{R}^3 \setminus X$, and $H_p(\mathbf{R}^3;\mathbf{R}) = 0$, for $p \neq 0$. Furthermore, the long exact homology sequence for the pair $(X, \partial X)$ is the following:

$$\cdots \to H_{k+1}(X, \partial X;\mathbf{R}) \to H_k(\partial X;\mathbf{R}) \to H_k(X;\mathbf{R}) \to H_k(X, \partial X;\mathbf{R}) \to \cdots$$

From the Mayer-Vietoris sequence one has the following short exact sequences:

$$0 \to H_2(\partial X;\mathbf{R}) \to H_2(X;\mathbf{R}) \oplus H_2(X';\mathbf{R}) \to 0$$

$$0 \to H_1(\partial X;\mathbf{R}) \to H_1(X;\mathbf{R}) \oplus H_1(X';\mathbf{R}) \to 0$$

$$0 \to H_0(\partial X;\mathbf{R}) \to H_0(X;\mathbf{R}) \oplus H_0(X';\mathbf{R}) \to H_0(\mathbf{R}^3;\mathbf{R}) \to 0$$

Furthermore, from the long exact homology sequence one has the following short exact sequences.

$$0 \to H_3(X,\partial X;\mathbf{R}) \to H_2(\partial X;\mathbf{R}) \to H_2(X;\mathbf{R}) \to 0$$

$$0 \to H_2(X,\partial X;\mathbf{R}) \to H_1(\partial X;\mathbf{R}) \to H_1(X;\mathbf{R}) \to 0$$

$$0 \to H_1(X,\partial X;\mathbf{R}) \to H_0(\partial X;\mathbf{R}) \to H_0(X;\mathbf{R}) \to 0$$

Therefore, from the Poincarè duality, the Alexander duality, the Mayer-Vietoris duality and the long exact ohomology sequence, one has the splittings reported in the following table.

TAB.4.8 - Isomorphisms between homology groups of oriented compact domain $X \subset \mathbf{R}^3$

Origin Name	Isomorphisms
Poincarè Duality	$H_0(X;\mathbf{R}) \cong H_3(X,\partial X;\mathbf{R})$; $H_1(X;\mathbf{R}) \cong H_2(X,\partial X;\mathbf{R})$ $H_2(X;\mathbf{R}) \cong H_1(X,\partial X;\mathbf{R})$; $H_3(X;\mathbf{R}) \cong H_0(X,\partial X;\mathbf{R})$
Alexander Duality	$H_0(X;\mathbf{R}) \cong H_2(X;\mathbf{R})$; $H_1(X;\mathbf{R}) \cong H_1(X;\mathbf{R})$ $H_2(X;\mathbf{R}) \cong \widetilde{H}_0(X,\partial X;\mathbf{R})$
Mayer-Vietoris seq.	$H_2(\partial X;\mathbf{R}) \cong H_2(X,\partial X;\mathbf{R}) \oplus H_2(X';\mathbf{R})$ $H_1(\partial X;\mathbf{R}) \cong H_1(X;\mathbf{R}) \oplus H_1(X';\mathbf{R})$ $H_0(\partial X;\mathbf{R}) \cong H_0(X;\mathbf{R}) \oplus H_0(X';\mathbf{R})$ $H_0(X;\mathbf{R}) \oplus H_0(X';\mathbf{R}) \cong H_0(\partial X;\mathbf{R}) \oplus H_0(\mathbf{R}^3;\mathbf{R})$
Long exact hom. seq.	$H_2(\partial X;\mathbf{R}) \cong H_3(X,\partial X;\mathbf{R}) \oplus H_2(X;\mathbf{R})$ $H_1(\partial X;\mathbf{R}) \cong H_2(X,\partial X;\mathbf{R}) \oplus H_1(X;\mathbf{R})$ $H_0(\partial X;\mathbf{R}) \cong H_1(X,\partial X;\mathbf{R}) \oplus H_0(X;\mathbf{R})$

X' = closure of $\mathbf{R}^3 \backslash X$. $X' \cap X = \partial X$. The tilde over H_0 reduces the dimension of 1.

Poincarè duality can be also written: $H_p(X,\partial X;\mathbf{Z}) \cong H^{3-p}(X;\mathbf{Z})$, $H_p(X;\mathbf{Z}) \cong H^{3-p}(X,\partial X;\mathbf{Z})$.

Homology groups of X are torsion free, finitely generated free abelian groups.

These give the relations between homology spaces as it results from Tab.4.6.□
Lemma 4.16. (Hodge decomposition). *Let us X be a compact oriented manifolds of dimension n. Let us $\Omega^p(X)$ denote the vector space of smooth differential p-forms on M. Let us denote by $H^p \equiv \{\omega \in \Omega^p(X) | \triangle \omega = 0\}$ the space of harmonic p-forms on X. Then one has the following orthogonal direct sum decomposition:*

$$\Omega^p(X) \cong \triangle(\Omega^p(X)) \bigoplus H^p \cong d\delta(\Omega^p(X)) + \delta d(\Omega^p(X)) \bigoplus H^p$$
$$\cong d(\Omega^{p-1}(X)) \bigoplus \delta(\Omega^{p+1}(X)) \bigoplus H^p.$$

As a by-product we get that the equation $\triangle \omega = \alpha$ has a solution $\omega \in \Omega^p(X)$ iff the p-form α is orthogonal to the space of harmonic p-forms. Furthermore, H^p is finite dimensional and one has the following isomorphisms with the (co)homology spaces:

$$H^p \cong H^p(X) \cong H^p(X;\mathbf{R}) \cong H_p(X;\mathbf{R})^*$$

where $H^p(X)$ is the de Rham p-cohomology space of X. In particular, if $\dim X = 3$, and $X \subset \mathbf{R}^3$, the structure of H^p is reported in the following table.

TAB.4.9 - Spaces H^p of harmonic p-forms on oriented compact domain $X \subset \mathbf{R}^3$

p	H^p
0	$\mathbf{R}^{(\sharp \text{comp. of } X)}$
1	$\mathbf{R}^{(\text{genus of } \partial X)}$
2	$\mathbf{R}^{(\sharp \text{comp. of } \partial X - \sharp \text{comp. of } X)}$
3	0

Proof of Lemma 4.16. The proof is standard. (See, e.g., refs.[101(c)].) $\square$

2) In fact, even if one has $\triangle\underline{\omega} = 0$, hence $\mathcal{L}_v\omega = 0$, as the flow is not a diffeomorphism between the related 3-dimensional integral space-like manifolds $B_t \subset V$, from the equation $\mathcal{L}_v\omega = 0$ we cannot argue that it is also $\phi_\lambda^*\omega = \omega$, $\forall\lambda$. Emphasize, also, that even if equation ($\diamond$) can be rewritten in covariant form: $(\overline{\diamond})\ \rho\frac{\delta\underline{\omega}}{\delta t} = \rho\nabla_\omega\underline{u} + \chi\triangle\underline{\omega}$, equation $(\overline{\diamond})$ is not equivalent to the following equation: $\rho\mathcal{L}_v\underline{\omega} = \chi\nabla\underline{\omega}$.[88] Therefore, when on V there are singular points, in the flow of the characteristic vector field, the presence of the vortex streching does not allow that the vorticity should be conserved even if it is harmonic.

3) In fact, when the space-like domain B, where the fluid is defined at some instant, is 2-dimensional, orientable, and it is constrained to stay in a plane σ, then the vorticity can be written in the form $\underline{\omega}|_{\underline{B}_t} = h\eta_\sigma$, where h is, at any instant, a scalar function on σ and η_σ is the volume 2-form on σ. Then, the vorticity equation can be written as follows: $\rho\eta_\sigma\frac{\delta h}{\delta t} = \chi\triangle(h\eta_\sigma)$. Therefore the condition of conservation for the vorticity is equivalent to require the harmonicity of $\underline{\omega}$. This is equivalent to ask that $d\underline{\omega} = 0$ and $\delta\underline{\omega} = 0$. Now, we can consider a system of coordinates, adapted to the inertial frame, such that the spatial coordinates $\{x^k\}_{1\le k\le 3}$, are adapted also to the constraint for the fluid, i.e., such that $\partial x_3 \perp \sigma$. Then the vertical

---

[88] Note that in covariant form, the condition $\mathcal{L}_v\underline{\omega}=0$ is really equivalent to $\underline{\phi}_\lambda^*\underline{\omega}=\underline{\omega}$, $\forall\lambda$, even if $\underline{\phi}_\lambda$ is not a diffeomorphism for any λ.

metric g can be written: $g_{ij} = \begin{pmatrix} \gamma_{11} & \gamma_{12} & 0 \\ \gamma_{21} & \gamma_{22} & 0 \\ 0 & 0 & 1 \end{pmatrix}$, where $\gamma_{ab} = \gamma_{ab}(x^1, x^2)$, $a, b = 1, 2$, is the metric on σ. Then $\eta_\sigma = \sqrt{\det(\gamma_{ab})}dx^1 \wedge dx^2$ is the volume form on σ and the vorticity can be represented by $\underline{\omega} = h\eta_\sigma$. The first condition for the harmonicity of $\underline{\omega}$ is $0 = d\underline{\omega} = dh \wedge \eta_\sigma$, hence $dh = 0$. The second condition for the harmonicity of $\underline{\omega}$ gives: $0 = \delta\underline{\omega} = \delta(h\eta_\sigma) = \star^{-1}d\star(h\eta_\sigma) = \star^{-1}d(h.dx^3) = \star^{-1}(dh \wedge dx^3)$, hence $dh = 0$. Therefore, $\underline{\omega}$ is harmonic iff $\underline{\omega} = h\eta_\sigma$, with $h = h(t)$. But for such a function the vorticity equation requires $(\partial t.h) = 0$. This is equivalent to say that $\omega = h\partial x_3$, $h \in \mathbf{R}$. $\qquad\square$

Corollary 4.1. *By simply fixing the boundary conditions we cannot, in general, determine the conservation of the vorticity.*

Proof of Corollary 4.1. In fact, from Theorem 4.7 it follows that for any boundary condition we have not, in general, an unique solution. Hence, from Theorem 4.8 it follows that we can have solutions where the vorticity is conserved and other ones where, instead it is not conserved. $\qquad\square$

Remark 4.9. This solves the "non-analyticity paradox", i.e., in order to require vorticity, a region of fluid initially at rest (or in irrotational motion) must have a vorticity which is a non-analytic function of time [101(b)]. In fact, in solutions with singularities (solutions considered in Theorem 4.8(2)) the vorticity can start from zero and become different from zero, and again vanish and so on. Such solutions cannot, in fact, be realized as analytic sections of $\pi : W \to M$. $\qquad\blacksquare$

Remark 4.10. The exixtence of solutions with no singularities strongly depend on the boundary conditions. In fact some smooth boundary conditions generate by themselves singularities in the flow when the manifold describing the evolution of the boundary is with change of sectional topology. However, there are also boundaries value problems where such smooth solutions exist and for fixed type of flows are also unique. But this means to fix some further constraints in the Navier-Stokes equation. However there is a large class of boundary value problems where the method considered in Theorem 4.2 and Theorem 4.23 to build smooth global solutions, assures the existence of global smooth solutions without singularities in the corresponding characteristic flow, even if the uniqueness of such solutions it is not assured in the set of general smooth solutions. $\qquad\blacksquare$

In fact, from above results we can also state the following:

Theorem 4.9. *If at some instant t_0 the fluid is contained into a compact 3-dimensional smooth manifold B_t that is p-connected, $0 \le p \le 2$, and has the homotopy type of the 3-dimensional disk D^3,[89] and with smooth boundary, that evolves with a fixed law in such a way to describe a 3-dimensional smooth time-like manifold, that is q-connected, $0 \le q \le 1$, and such that the sectional topology is conserved, (i.e., it remains D^3), then exist global smooth solutions of (NS) that have the corresponding characteristic flows without singularities. In general, such solutions V are not unique and are not necessarily stable. However, if the Reynolds number $R_e^{sup} \le R_{e,c}^{nonlinear}$ and the thermal and pressure perturbations are neglectable, then V is globally stable. Furthermore, if $\dot{f}(V) = 0$, the characteristic Reynolds number R_e is a characteristic number of the solution V too.*

Proof. Note that, from the topological point of view, such a solution is D^4 and its boundary is S^3. It is well known that on S^3 there are nowhere zero vector fields. In fact, from the Hopf's theorem it follows that a smooth connected orientable closed n-dimensional manifold X has a nonzero tangent vector field iff $\chi(X) = 0$, where $\chi(X)$ is the Euler characteristic number of X. On the other hand as $\chi(X) = \sum_i (-1)^i \dim H_i(X)$, it follows that $\chi(S^3) = \dim H_0(S^3) - \dim H_3(S^3) = 1 - 1 = 0$. Furthermore, it is well known (see e.g. [26,101(c)]) that if $X \in [\emptyset] \in \Omega_n$, i.e., $X = \partial V$, then $\chi(X) = 2m$ and if $\dim X = 2s + 1$ one has $\chi(X) = 0$. This is a necessary condition that V has a nonzero vector field. In fact if V is a compact smooth manifold with $\chi(V) \ne 0$ then any tangent vector field ζ on V has a zero. (Index(ζ) = $\chi(V)$.) On the other hand any compact connected manifold with boundary has a nonvanishing vector field. (See, e.g., ref.[26].) Therefore, on such a solution V the characteristic vector field can be without zeros. Furthermore, as in such case one has that the space H^1 of harmonic 1-forms is zero, $(H^1 = H_1(B_t; \mathbf{R}) \cong H_1(S^2; \mathbf{R}) = 0)$, it follows that the unique harmonic vorticity is the zero vorticity. Then, from Theorem 4.8 we can conclude that there exists a laminar (irrotational) smooth global solution. In general, such solutions are not unique. In fact, if $s : X \subset M \to W$ is such a solution, we can consider the shifted smooth solution $s' = s + v$

---

[89] More precisely B_t has the following homotopy and homology groups respectively: $\pi_0(B_t) \cong \pi_p(B_t) \cong H_p(B_t; \mathbf{Z}) = 0$, $1 \le p \le 3$; $H_0(B_t; \mathbf{Z}) \cong \mathbf{Z}$.

such that $\nu : X \to s^* vTW$, $\nu|_{\partial X} = 0$ and $D^2\nu : X \to \widehat{(NS)}_s \subset \widehat{(NS)}$. So $D^2 s' : X \to (NS)$. Note that the existence of such solutions is assured as just proved in the proof of Lemma 4.6. Note also, that the existence of regular global smooth solutions of $(NS) \subset J_4^2(W)$, i.e., solutions diffeomorphic to 4-dimensional smooth submanifolds of W, but not necessarily diffeomorphic to their projections on M, guarantees the non-uniqueness of such solutions. Finally, by using Theorem 4.7 and Lemma 4.4 we can conclude the proof. $\qquad\square$

Example 4.1. (*Newtonian fluid: Isothermal nonsteady-state laminar flow in circular pipe*). Let us assume: (a) the mass density is constant: $\rho = $ constant; (b) the pipe is horizontal; (c) the pipe is very long (L); (d) there is not flow for $t < 0$; for $t = 0$ it is applied a gradient of pressure: $\frac{P_0 - P_L}{L}$; (e) the field of velocity has the following structure: $(v_j) = (v_r = 0, v_\theta = 0, v_z = v_z(r,t) \equiv u(r,t))$. Then the motion equation and continuity equations give the following equation:

(4.29)
$$\left\{ \rho(\partial t \cdot v_z) = \tfrac{P_0 - P_L}{L} + \chi \tfrac{1}{r}(\partial r \cdot (r(\partial r \cdot v_z))) \ +\text{boundary conditions:} \left\{ \begin{array}{l} v_z(r,0)=0, \quad 0 \leq r \leq R \\[4pt] v_z(0,t)=\text{finite} \\[4pt] v_z(R,t)=0. \end{array} \right\} \right\}.$$

As a consequence we get:

$$v_z(r,t) = \frac{(P_0 - P_L)R^2}{4\chi L}\left[(1 - (\tfrac{r}{R})^2) - 8\sum_{n=1}^{\infty} \frac{J_0(\alpha_n \tfrac{r}{R})}{\alpha_n^3 J_1(\alpha_n)} e^{-\alpha_n^2 \tau}\right]$$

where $\tau \equiv \chi t/\rho R^2$, $\alpha_n = $ zeros of Bessel function J_0, $J_i \equiv$ Bessel functions with $i \geq 0$. In fact, let us introduce the following adimensional variables: $\phi \equiv \frac{v_z}{(P_0 - P_L)R^2} 4\chi L$, $\xi \equiv \frac{r}{R}$, $\tau \equiv \frac{\chi t}{\rho R^2}$. Therefore, equation (4.29) can be written in the following adimensional form:

$$\left\{ (\partial \tau \cdot \phi) = 4 + \tfrac{1}{\xi}(\partial \xi \cdot (\xi(\partial \xi \cdot \phi))) \ +\text{boundary conditions:} \left\{ \begin{array}{l} \phi(\tau=0)=0 \\[4pt] \phi(\xi=1)=0 \\[4pt] \phi(\xi=0)=\text{finite} \\[4pt] \lim_{\tau \to \infty} \phi=\phi_\infty(\xi)\,(\text{steady-state}) \end{array} \right\} \right\}.$$

Note that $\phi_\infty(\xi)$ satisfies the following equation:

$$\left\{ 0 = 4 + \frac{1}{\xi}\frac{d}{d\xi}\left(\xi\frac{d\phi_\infty}{d\xi}\right) \ + \text{boundary conditions} : \phi_\infty(\xi = 1) = 0 \right\}.$$

As a consequence $\phi_\infty(\xi) = 1 - \xi^2$, (*Poiseuille velocity*). Therefore, let $\phi(\xi,\tau) = \phi_\infty(\xi) - \widehat{\phi}(\xi,\tau)$. Hence, $\widehat{\phi}$ must satisfy the following equation:

$$\left\{ (\partial\tau \cdot \widehat{\phi}) = \frac{1}{\xi}\partial\xi \cdot (\xi(\partial\xi \cdot \widehat{\phi})) \ + \text{boundary conditions}: \left\{ \begin{array}{c} \widehat{\phi}(\tau=\infty)=\phi_\infty \\[4pt] \widehat{\phi}(\xi=0)=\text{finite} \\[4pt] \widehat{\phi}(\xi=1)=0 \end{array} \right\} \right\}.$$

Let us find a solution of the form $\phi(\xi,\tau) = \Xi(\xi)T(\tau)$. Then above system splits in the following:

$$\frac{1}{T}\frac{dT}{d\tau} = \frac{1}{\Xi}\frac{1}{\xi}\frac{d}{d\xi}\left(\xi\frac{d\Xi}{d\xi}\right) \Rightarrow \left\{ \begin{array}{c} \frac{dT}{d\tau}=-\alpha^2 T \\[4pt] \frac{1}{\xi}\frac{d}{d\xi}(\xi\frac{d\Xi}{d\xi})+\alpha^2\Xi=0,\alpha^2=const. \end{array} \right\}.$$

Therefore we get the following solutions:

$$\left\{ T = c_0 e^{-\alpha^2\tau}, \ \Xi = c_1 J_0(\alpha\xi) + c_2 Y_0(\alpha\xi) \right\},$$

where c_0, c_1, c_2 are constant and J_0, Y_0 are Bessel functions of order 0. By considering the boundary conditions, we have:

$$\widehat{\phi}(\xi=0)=\text{finite}\Rightarrow\text{as} \quad \lim_{\xi\to 0} Y_0(\alpha\xi)=-\infty\Rightarrow c_2=0$$

$$\widehat{\phi}(\xi=1)=0\Rightarrow J_0(\alpha)=0.$$

As $J_0(\alpha)$ is an oscillating function, it has many zeros α_n. Hence, we have many solutions $\quad \Xi_n = C_{1n}J_0(\alpha_n)$, with $n = 1, 2, \cdots, n$. Then,

$$\widehat{\phi}(\xi,\tau) = \sum_{1\leq n\leq\infty} B_n e^{-\alpha_n\tau^2} J_0(\alpha_n\xi), \quad \text{with} \quad B_n \equiv c_0 c_{1n}.$$

Now, impose the condition:

$$\lim_{\tau\to\infty} \widehat{\phi}(\xi,\tau) = \phi_\infty(\xi) = 1 - \xi^2; \ 1 - \xi^2 = \sum_{1\leq n\leq\infty} B_n J_0(\alpha_n\xi).$$

Multiplying both sides for $\xi J_0(\alpha_m\xi)$ and integrate

$$\int_{[0,1]} J_0(\alpha_m\xi)(1 - \xi^2)\xi d\xi = \sum_{1\leq n\leq\infty} B_n \int_{[0,1]} J_0(\alpha_n\xi)J_0(\alpha_m\xi)\xi d\xi.$$

From the orthogonality of Bessel functions, $4\frac{J_1(\alpha_m)}{\alpha_m^3} = B_m\frac{1}{2}[J_1(\alpha_m)]^2$, it follows that $\frac{8}{J_1(\alpha_m)\alpha_m^3} = B_m$.

It should appear, from above calculations, that the solution is unique for the fixed boundary conditions. But all depends on the fact that we have just at the beginning fixed a flow type (laminar flow). Furthermore, note that in the flow considered, the vorticity is not zero. In fact, we have that the full infinitesimal strain rate $\dot{e} = \dot{e}_{(s)} + \dot{e}_{(a)} = \dot{e} + \omega$ is given by:

(4.30)

$$\dot{e} = \begin{pmatrix} 0 & 0 & 0 \\ 0 & 0 & 0 \\ (\partial r.u) & 0 & 0 \end{pmatrix} = \begin{pmatrix} 0 & 0 & \frac{(\partial r.u)}{2} \\ 0 & 0 & 0 \\ \frac{(\partial r.u)}{2} & 0 & 0 \end{pmatrix} + \begin{pmatrix} 0 & 0 & -\frac{(\partial r.u)}{2} \\ 0 & 0 & 0 \\ \frac{(\partial r.u)}{2} & 0 & 0 \end{pmatrix}.$$

The vertical vector field corresponding to ω is $\omega = \frac{(\partial r.u)}{2r}\partial\theta$ and the vertical 1-form is $\underline{\omega} = \frac{r(\partial r.u)}{2}d\theta$. Such a vorticity is not harmonic $\triangle\underline{\omega} \neq 0$. (In such a domain there are not harmonic forms other the null-forms.) In fact the vorticity is not conserved for nonsteady-state solutions: $\frac{\delta\omega}{\delta t} = (\partial t.(\frac{(\partial r.u)}{2r})) \neq 0$. Moreover, the solution $V \subset (NS)$ has not singularities in the characteristic flow and has not vortices. Of course many other flow types can occur with the same boundary conditions! In fact let us assume that a 3-dimensional time-like integral manifold X with two singular points, should be the boundary propagation of a 3-dimensional space-like integral manifold $N_{t_0} \subset B_{t_0} \subset \widehat{(NS)}$, considered as a compact submanifold of the compact integral manifold B_{t_0}, identified by the fluid at some instant t_0. Let us assume that the singular point on X produces a tunnel effect such that at some successive instant $t > t_0$, one has N_t divided into two parts $N_t = (N_t)_0 \bigcup (N_t)_1 \subset B_t$. Furthermore, let us assume that the second singular point on X recombines the separated sections into only one, so that the sectional topology of X returns that before the first singular point. Let Z be the evolution of the boundary ∂B_{t_0} such that, at any instant $t' \geq t_0$, $N_{t'} \subset B_{t'}$, and $\partial N_t \bigcap \partial B_{t'} = \emptyset$. Now, we shall prove that there exists a solution V such that $\partial V = Z$ and such that $X \subset V$. In fact, let us consider the boundary value problem $Z \bigcup X$. From Theorem 4.1 it follows that there exists a (smooth) integral manifold, $Y_{outside}$, solution of this boundary value problem. Now, let us consider the boundary value problem X. Again Theorem 4.1 guarantees that we have also a (smooth) solution Y_{inside} for this problem. Therefore the integral manifold $V = Y_{inside} \bigcup_X Y_{outside}$ is just a (smooth) solution of the boundary value problem $\partial V = Z$. (Note that Y_{inside} can be considered a sort of "wormhole solution" in V.) Of course as we can design many of such wormhole solutions, corresponding to different

X, the conclusion is that the boundary conditions do not assure the uniqueness of solutions. Note that the laminar flow, above considered, corresponds to an integral manifold X without singular points, i.e., a cylinder, and such that the soldered solution $Y_{inside} \bigcup_X Y_{outside}$ is a smooth manifold. On the other hand, it is well known that laminar flows are unstable for high Reynolds numbers. Of course these unstable and turbulent flows respect the same boundary conditions than laminar flows. Finally note that the fundamental global rate $\dot{f}$, (see Proof of Lemma 4.3), of a laminar solution is zero. In fact, for such a flow one has $(u^i) = (u^r, u^\theta, u^z) = (0, 0, v_z)$ and, by using equation (4.30), we get $\dot{e}_{ij} u^i u^j = 0$. Therefore, if at the beginning the characteristic Reynolds number $R_e(t)$ does not exced the critical value $R_{e,c}^{nonlinear}$, it will continue to remain lower than the critical value and the solution is globally stable. Furthermore, emphasize that the existence of singular solutions above considered solves the "turbulence paradox" [101(b)], i.e., the fact that the Poiseuille flow can become turbulent and eventually this turbulence can decay. $\blacksquare$

4.3 - THE LIMIT CASE: THE EULER EQUATION

The Euler equation for incompressible fluids can be considered the limit case of the Navier-Stokes equation (NS) for zero viscosity $(\chi = 0)$ and zero thermal conductivity $(\nu = 0)$, (or, equivalently, for isothermal case). In fact, the Euler equation can be written as a first order PDE on the fiber bundle $\bar{\pi} : W_0 \equiv J\mathcal{D}(M) \times_M T_0^0 M \to M$, coordinates $(x^\alpha, \dot{x}^i, p) \to (x^\alpha)$, by the equations reported in the following table:

TAB.4.10 - Euler equation: $(E) \subset J\mathcal{D}(W_0)$

$\dot{x}^k G^j_{jk} + \dot{x}^i_s \delta^s_i = 0$
$\dot{x}^s \dot{x}^i \rho G^j_{is} + \dot{x}^s \dot{x}^j_s \rho + \rho \dot{x}^j_0 + p_i g^{ij} + \rho(\partial x_i . f) g^{ij} = 0$

This equation is involutive formally integrable, as well as completely integrable [12]. So it is equivalent to its first prolongation $(E)_{+1} \subset J\mathcal{D}^2(W_0)$. One can easily see that equation $\widehat{(NS)}_{\chi=0}$, corresponding to the equation $\widehat{(NS)}$ for zero viscosity, contains the image of $(E)_{+1}$ into $J\mathcal{D}^2(W)$ by means of the second-holonomic prolongation of the canonical homomorphism over M: $j_{\bar{\theta}} : W_0 \to W$, given by $(v, p) \mapsto (v, p, \bar{\theta})$, where $\bar{\theta}$ is a fixed temperature. So we can consider that for zero viscosity and in the isothermal case, an incompressible fluid is represented by $(E)_{+1} \subset J\mathcal{D}^2(W)$ or equivalently by $(E) \subset J\mathcal{D}^2(W_0)$. Of course, for such an equation continue to hold Theorem 4.2, Theorem 4.3, Theorem 4.4, Theorem 4.5 and Theorem 4.6 by simply substituting (NS) and $\widehat{(NS)}$ with (E). Furthermore, an explicit calculation shows that the infinitesimal symmetry algebra of (E) is generated by vector fields $\zeta = \bar{\zeta}^\alpha \partial x_\alpha + X^i \partial \dot{x}_i + Y \partial p : W_0 \to TW_0$, with $\bar{\zeta}^\alpha \partial x_\alpha$ belonging to the infinitesimal symmetry algebra of the oriented Galilei group and X^i and Y numerical functions on W_0 given by: $X^i = \sum_{1 \leq r \leq 3}(\partial x_r.\bar{\zeta}^i)\dot{x}^r + h^i(x^\alpha)$, $Y = \rho \sum_{i,s} g_{is}[(\partial x_0.\bar{\zeta}^i) - h^i]\dot{x}^s + l(x^\alpha)$, with $l, h^i : M \to \mathbf{R}$ numerical functions on M such that the following equations are satisfied: $\text{div}\, h^s = 0$, $(\partial x_p.h^j) - (\partial x_i.h^s)g^{ij}g_{ps} = -2G^j_{sp}(\partial x_0.\bar{\zeta}^s)$, $\rho(\partial x_0.h^j) = -\rho\mathcal{L}_{\bar{\zeta}}(\text{grad}\, f)^j - (\text{grad}\, l)^j$, where $\mathcal{L}_{\bar{\zeta}}$ is the symbol of Lie derivative. For example if $\bar{\zeta}^s$ are independent of x^0, a solution for h^j and l is $h^s = 0$, $(\text{grad}\, l)^j = -\rho\mathcal{L}_{\bar{\zeta}}(\text{grad}\, f)^j$. (This is just the infinitesimal symmetry algebra that conserves $\widehat{(NS)}$.) Furthermore, if $\mathcal{L}_{\bar{\zeta}}(\text{grad}\, f)^j = 0$, (for example for gravitational potential with respect to time translation), $(\partial x_k.l) = 0$ and l can be choosen an arbitrary function $l = l(x^0)$. This shows that Theorem 4.8 continues to hold also in this limit case, even if there are not more informations coming from Reynolds number as it is always ∞-valued. In fact, in the case of the Euler equation we recognize that all the global smooth solutions, without singularities in the associated characteristic flows, guarantee the conservation of the vorticity. This property is, instead, absent in the Navier-Stokes equation, where also solutions with non singular characteristic flows may do not conserve the vorticity. (See Theorem 4.3.) In fact, we have the following theorem.

Theorem 4.10. 1) *Let* $V \subset (E)$ *be a smooth (global) solution of* (E) *such that the flow* ϕ_λ *generated by the corresponding characteristic vector field* ζ, $\partial\phi = \zeta$, *has not singular points. Then, the associated vorticity (or spin) identifies a conservation law (enstrophy) as well a 1-conservation law (strenght of the vortex)* $f : \Omega_1^{(E)+\infty} \to \mathbf{R}$ *on the first integral bordism group of the Euler equation. (Emphasize that such a conservation law does not exist for the Navier-Stokes equation.) This is a full obstruction to the change of the local vorticity.*

2) *Let* $V \subset (E)$ *be a smooth (global) solution of* (E) *such that the corresponding flow* ϕ_λ *has some singular points. Then, the associated vorticity does not identify more a conservation law, i.e. the enstrophy is not more*

conserved, and neither the strenght of the vortex identifies a 1-conservation law for the Euler equation.

3) In the case $D = 2$ one has the conservation of the enstrophy and strenght of the vortex for any smooth (global) solution of the Euler equation.

Proof. 1) The *vorticity equatuion* for the Euler equation is the following: ($\clubsuit$): $\frac{\delta\omega}{\delta t} = \nabla_\omega u$. This can be rewritten as $\mathcal{L}_v\omega = 0$. Then, as the solution is without singular points, it follows that $\underline{\phi}_\lambda^*\omega = \omega$, $\forall\lambda$, where $\underline{\phi}$ is the space-time flow associated to v, i.e., $\partial\underline{\phi} = v$. Now, let $N' \in [N]_{(E)} \in \Omega_1^{(E)}$, with $N' \subset B_{t'}$, $N \subset B_t$, $t' \neq t$, $B_t \equiv V \bigcap (E)_t$, where $(E)_t$ is the fiber of (E) over $t \in T$, and $V \subset (E)$ is a 4-dimensional integral manifold, regular solution of $(E) \subset JD(W_0)$ characterized by means of a charecteristic time-like vector field ζ. Let us denote by $\underline{B}_{tt}$, $\underline{B}_{tt'}$, γ, γ' and v the projections of B_t, $B_{t'}$, N, N' and ζ over M respectively. Let $u : M \to vTM$ be the space-like vector field corresponding to v by means of the inertial frame ψ, and let $\underline{\omega} : M \to vTM$ be the coresponding dual vector field. Then, one has $\int_{\gamma'} \underline{u} = \int_\gamma \underline{u} = \int_\gamma u \cdot \mathbf{T} ds$: (*Kelvin's circulation theorem*). Furthermore, let us, now, characterize the Kelvin's circulation theorem by means of a 1-conservation law on the causal integral 1-bordism group $\overline{\Omega}_1^{(E)}$. The equivalence classes of this group are represented by means of 1-dimensional closed compact admissible integral space-like manifolds contained into (E). Furthermore, $N' \in [N]_{(E)} \in \overline{\Omega}_1^{(E)}$ iff $N \subset (E)_t$, $N' \subset (E)_{t'}$, with $t \neq t'$, other than $N' \in [N]_{(E)} \in \Omega_1^{(E)}$. (One can see [14] that $\overline{\Omega}_1^{(E)} = \Omega_1^{(E)} = 0$.) Now, let $N'' \in [N]_{(E)} \in \Omega_1^{(E)}$ such that $N'' \subset B_t$, $N \subset B_t$. Moreover, let us assume that $\gamma'' \equiv \pi_2(N'')$ bords with $\gamma \equiv \pi_2(N)$ by means of the vorticity tube $\mathcal{T}$, that is the 3-dimensional integral submanifold of $\underline{B}_{tt} \equiv \pi_2(B_t) \subset M_t$, generated by the spin $\omega = \mathrm{rot}\, u$ starting from γ. Then, one can see that $\int_\gamma \underline{u} = \int_{\gamma''} \underline{u}$: (*Helmoltz' theorem*). This common values is called the *strength of the vortex tube* $\mathcal{T}$. On the other hand, as $\int_\gamma \underline{u} = \int_{\phi_\lambda(\gamma)} \underline{u}$, where $\partial\underline{\phi} = v$, it follows that we can characterize the strength of the vortex tube as a 1-conservation law on the causal integral group $\overline{\Omega}_1^{(E)}$: $f : \overline{\Omega}_1^{(E)} \to \mathbf{R}$.

2) Let us emphasize that, when the solution $V \subset (E)$ has the flow ϕ_λ associated to the characteristic vector field of V, with some singular points, ϕ_λ, is not necessarily a diffeomorphism between the related 3-dimensional space-like manifolds. (See Theorem 4.5.) Now, from the vorticity equation ($\clubsuit$) it follows that $\frac{\delta\omega}{\delta t} \neq 0$, i.e., the vorticity ω is not conserved along the flow lines by the presence of the vortex streching.

3) In the case $D = 2$ the vorticity equation ($\clubsuit$) becomes $[(\partial t.h) + \mathcal{L}_{v_\wedge} h].\eta_\sigma = 0$, hence $v.h = 0$, or equivalently $\frac{\delta\omega}{\delta t} = 0$. Therefore, in the $2D$-case the vorticity is conserved along any admissible flow. $\qquad\square$

ADDENDUM II: BORDISM GROUPS AND VARIATIONAL PDE's

Abstract - In the framework of the geometry of PDE's, we classify variational equations of any order with respect to their formal properties. A variational sequence is introduced for constrained variational PDEs that extends previous ones for variational calculus on fiber bundles. Such extended variational sequences allows us to locally and globally solve variational problems, constrained by PDEs of any order, $E_k \subset J_n^k(W)$, by means of some cohomological properties of E_k. Moreover, we relate constrained variational PDEs to the integral (co)bordism groups for PDEs. In this way we are able to characterize the structure properties of global solutions of constrained variational PDEs and to relate them to the structure of global solutions for the corresponding constraint equations.

5.0 - Introduction

In this addendum we consider some variational structures on PDE's and relate them with integral bordism groups and global solutions of such constrained variational problems. This addendum is divided in three sections. Section 1. Here we show how to recognize canonical connections on PDE's associated with their canonical formal structures. This is useful in order to obtain a process of "horizontal" and "vertical" splits for differential forms on PDE's that allows us to formulate constrained variational problems. Section 2. Here define the category of variational PDE's and characterize variational PDE's by means of a universal problem. Section 3. Here we introduce a variational sequence for constrained variational PDE's that extend previous ones considered for variational calculus on fiber bundles [102]. Such extended variational sequence allows us to locally and globally solve variational problems, constrained by PDE's of any order, $E_k \subset J_n^k(W)$, by means of some cohomological properties of E_k. Furthermore, the structures of global solutions of constrained variational PDE's are characterized by means of integral bordism groups. Some applications interesting the field theory (Yang-Mills equation) and the fluid dynamics (Navier-Stokes equation) are considered.

435

5.1 - HORIZONTAL AND VERTICAL FORMS ON PDE's

On the tangent space $TJ_n^k(W)$ of a finite order jet-space $J_n^k(W)$ on a $(m+n)$-dimensional manifold W, there is not, in general, a canonical connection. However, on the pull-back $\pi_{k+1,k}^* TJ_n^k(W)$, that is a vector bundle over $J_n^{k+1}(W)$, we recognize a canonical connection. In fact one has the following canonically split short exact sequence of vector bundles over $J_n^{k+1}(W)$:

$$0 \leftarrow \tau \xleftarrow[\underset{1}{\rightarrow}]{} \pi_{k+1,k}^* TJ_n^k(W) \xleftarrow[\underset{\Gamma}{\rightarrow}]{} vTJ_n^k(W) \leftarrow 0,$$

where

$$\tau = \bigcup_{\bar{q} \in J_n^{k+1}(W)} L_{\bar{q}}$$

and

$$vTJ_n^k(W) = \bigcup_{\bar{q} \in J_n^{k+1}(W)} T_{\pi_{k+1,k}(\bar{q})} J_n^k(W)/L_{\bar{q}}.$$

Hence

$$\pi_{k+1,k}^* TJ_n^k(W) \cong vTJ_n^k(W) \oplus \tau.$$

Dually one has, also, the canonical splitting:

$$\pi_{k+1,k}^* T^* J_n^k(W) \cong vT^* J_n^k(W) \oplus \tau^*,$$

and by functorial extension of this last splitting on the fiber bundle

$$\pi_{k+1,k}^* \Lambda_\bullet^0(J_n^k(W)) \to J_n^{k+1}(W),$$

where

$$\Lambda_\bullet^0(J_n^k(W)) \equiv \bigoplus_{q \geq 0} \Lambda^q(T^* J_n^k(W)),$$

we get a (*horizontal,vertical*)-bigradiation on the space of differential forms $\pi_{k+1,k}^* \Omega^\bullet(J_n^k(W))$. Furthermore, if $E_k \subset J_n^k(W)$ is a PDE of n-dimensional submanifolds of W, and if one has the following fiber bundle structure $\pi_{k+1,k} : E_{k+1} \to E_k$, where E_{k+1} is the first prolongation of E_k, we can reproduce above horizontalization-verticalization process, just by restriction on E_k. In fact, one has the following canonically split short exact sequences of vector bundles over E_{k+1}:

$$0 \leftarrow \tau \xleftarrow[\underset{1}{\rightarrow}]{} \pi_{k+1,k}^* TE_k \xleftarrow[\underset{\Gamma}{\rightarrow}]{} vTE_k \leftarrow 0,$$

where

$$\tau = \bigcup_{\bar{q} \in E_{k+1}} L_{\bar{q}}$$

and

$$vTE_k = \bigcup_{\bar{q} \in E_{k+1}} T_{\pi_{k+1,k}(\bar{q})} E_k / L_{\bar{q}}.$$

Furthermore, if on W there is also a fiber bundle structure $\pi : W \to M$, with $\dim(M) = n$, then we have canonically split sequences for PDEs of sections of π too, similarly to the previous ones, where now $\tau \cong \pi_{k+1}^* TM$.

5.2 - THE CATEGORY OF VARIATIONAL PDE's

Definition 5.1. The *category of differential equations*, $\underline{D\mathcal{E}}$ is defined by: (i) $X \in Ob(\underline{D\mathcal{E}})$ iff X is a manifold supplied with a finite-dimensional distribution $C(X) \subset TX$. (In particular, if $X \equiv E_k \subset J_n^k(W)$ is a PDE, we take $C(X) \equiv \mathbb{E}_n^k$, i.e., the Cartan distribution of E_k.) We set $C \dim X = \dim C(X) = $ *Cartan dimension* of X; (ii) $f \in Hom(\underline{D\mathcal{E}})$ iff it is a smooth map $f : X \to Y$, $X, Y \in Ob(\underline{D\mathcal{E}})$, which conserves the corresponding distributions: $f_* \equiv T(f) : C(X) \to C(Y)$, $f \in Hom_{\underline{D\mathcal{E}}}(X,Y)$, $C \dim X = n$, $C \dim Y = m$, $Crank\,(f) = r = \dim(T(f)_x)(C(X)_x)$, $x \in X$. (Thus the fibres $f^{-1}(y)$, $y \in im\,(f) \subset Y$, are $(n - r)$-Cartan dimensional objects of $\underline{D\mathcal{E}}$.) Isomorphisms of $\underline{D\mathcal{E}}$ are morphisms with fibres consisting of separate points. A *n-variational system* in the category $\underline{D\mathcal{E}}$ is a triplet $(X, \theta, \mathcal{C})$, where $X \in Ob(\underline{D\mathcal{E}})$ such that $C \dim C(X) \geq n$, $\theta \in \Omega^n(X)$ is a differential n-form on X, $\mathcal{C} \subset \Omega^\bullet(X)$ is the ideal of differential forms on X associated to the distribution $C(X) \subset TX$. In the particular case that $C(X) \equiv \mathbb{E}_n^k$ is the Cartan distribution of a PDE $X \equiv E_k \subset J_n^k(W)$, then $\mathcal{C} \equiv \mathcal{C}(E_k)$ is the contact ideal of E_k. If on W there is, also, a fiber bundle structure $\pi : W \to M$, where M is a n-dimensional manifold, then, we can talk also of the k-order jet-derivative space $J\mathcal{D}^k(W)$ for sections of π. $J\mathcal{D}^k(W)$ is a submanifold of $J_n^k(W)$ with the same dimension, (i.e., $J\mathcal{D}^k(W)$ is an open differential relation in $J_n^k(W)$). Let $\mathcal{S}ol(C(X))$ be the set of solutions of $C(X)$, i.e., the set of mappings $\alpha : N \to X$, where N is a (compact) n-dimensional oriented manifold such that $\alpha^* \mathcal{C} = 0$. We call *integral action* the numerical function $I : \mathcal{S}ol(C(X)) \to \mathbf{R}$ defined by $I(\alpha) \equiv \int_N \alpha^* \theta$.

Theorem 5.1. (First variational formula). *If $\zeta : X \to TX$ is a vector field on X and $\phi_t : X \to X$ is the corresponding (local) 1-parameter group of* diffeomorphisms of X we put:

$$I(\alpha_t) \equiv \int_N \alpha_t^* \theta, \ \alpha_t \equiv \phi_t \circ \alpha : N \to X.$$

Then, one has:

$$\frac{dI(\alpha_t)}{dt}\Big|_{t=0} = \int_N \alpha^*(\zeta \rfloor d\theta) + \int_N d\alpha^*(\zeta \rfloor \theta) = \int_N \alpha^*(\zeta \rfloor d\theta) + \int_{\partial N} \alpha^*(\zeta \rfloor \theta).$$

Definition 5.2. Let (X, θ, C) be a n-variational system in the category $\underline{D\mathcal{E}}$. We call (*Euler-Lagrange*) *extremal* of (X, θ, C) a n-dimensional submanifold map $\alpha : N \to X$ that belongs to $\mathcal{S}ol(C(X))$ and such that $\alpha^*(\zeta \rfloor d\theta) = 0$, for all vector fields $\zeta \in C^\infty(TX)$ that satisfy the following condition: $\alpha^* \mathcal{L}_\zeta \gamma = 0, \forall \gamma \in C.$[90]

Proposition 5.1. *If* $\alpha : N \to X$ *is an extremal of the n-variational system* (X, θ, C), *it is also an extremal of the n-variational system* (X, θ', C), *where* $\theta' = \theta + \gamma, \forall \gamma \in C \bigcap \Omega^n(X).$

Definition 5.3. A *Cartan form* of the n-variational problem (X, θ, C) is any differential form $\theta' \in [\theta] \equiv \theta + C$ such that $d\theta' \in C$. We shall denote such a Cartan form with $\sigma[\theta]$. Let (X, θ, C) be a n-variational problem such that θ is a Cartan form. Let $\{\theta^a\}_{1 \leq a \leq m}$ be a basis of homogeneous degree for the ideal C (which we assume to be graded). Then, we can write $d\theta = \theta^a \wedge \omega_a$, where $(\omega_1, \ldots, \omega_m)$ is a set of differential forms on X of homogeneous degree. Let $\mathcal{E} =< C; \omega_1, \ldots, \omega_m >$ be the exterior differential system generated by the forms $\{C; \omega_1, \ldots, \omega_m\}$. We call *Cartan extremal* associated to such a n-variational system, any submanifold $\alpha : N \to X$, $\dim N = n$, such that $\alpha^* \mathcal{E} = 0$.

Theorem 5.2. *A Cartan extremal is also an Euler-Lagrange extremal.*

Proof. Let ζ be a vector field on X. Then, one has: $\zeta \rfloor d\theta = (\zeta \rfloor \theta^a) \wedge \omega_a \pm \theta^a \wedge (\zeta \rfloor \omega_a)$. Let $\alpha : N \to X$ be a solution of (X, θ, C), i.e., $\alpha^* C = 0$, that is a Cartan extremal. Then one has also: $\alpha^*(\zeta \rfloor d\theta) = \alpha^*(\zeta \rfloor \theta^a) \wedge \alpha^*(\omega^a) = 0$. But this equation is the condition that α be extremal. $\qquad\square$

---

[90] This condition means that the vector fields ζ must preserve the constraint that α remains a solution after the induced flow by ζ. For example, if $X \equiv E_k \subset J_n^k(W)$, then ζ is any infinitesimal Lie symmetry of E_k. Note also that, when $E_k = J_n^k(W)$, in order to obtain the usual expression for the Euler-Lagrange equation we shall also add "suitable boundary conditions".

As the converse it is not, in general, assured, i.e., an Euler-Lagrange extremal does not necessitate to be a Cartan extremal, we are conduced to the following.

Definition 5.4. We say that a n-variational system $(X,\theta,\mathcal{C})$ in $\underline{D\mathcal{E}}$ admits an *universal Cartan form* $\sigma[\theta]$ if the Euler-Lagrange extremals are also Cartan extremals.

Example 5.1. Let $X \equiv J\mathcal{D}^k(W)$, $\dim W = n+m$, $\pi : W \to M$, $\dim M = n$, $\dim W = n+m$. Let us consider the following fiber coordinates (x^α, y^j) on W and $(x^\alpha, y^j, y^j_\alpha, \ldots, y^j_{\alpha_1 \cdots \alpha_k})$ on $J\mathcal{D}^k(W)$. Let $C(X) \equiv$ Cartan distribution on $J\mathcal{D}^k(W)$ so that $\mathcal{C}$ is generated by the contact differential 1-forms $\omega^j_\sigma \equiv dy^j_\sigma - y^j_{\sigma\alpha}dx^\alpha$, $|\sigma| \leq k-1$. Then a Cartan form for the n-variational system $(X, \theta \equiv Ldx^1 \wedge \ldots \wedge dx^n, \mathcal{C})$, where $L : J\mathcal{D}^k(W) \to \mathbf{R}$ is a numerical function on $J\mathcal{D}^k(W)$ is given by:

$$\left\{\begin{array}{l} \sigma[\theta] = Adx^1 \wedge \ldots \wedge dx^n + \sum_{0 \leq \alpha \leq k-1} A_j^{\beta i_1 \cdots i_\alpha} dy^j_{i_1 \cdots i_\alpha} dx^1 \wedge \ldots \wedge \widehat{dx^\beta} \wedge \ldots \wedge dx^\alpha \\[2mm] A \equiv L - \sum_{0 \leq \alpha \leq k-1} (-1)^{\beta+1} A_j^{\beta i_1 \cdots i_\alpha} y^j_{i_1 \cdots i_\alpha \beta} \\[2mm] A_j^{\beta i_1 \cdots i_\alpha} = (-1)^{\beta+1}[(\partial y_j^{\beta i_1 \cdots i_\alpha}.L) \\[2mm] \quad -(-1)^{\gamma+1}(\partial x_\gamma . A_j^{\gamma\beta i_1 \cdots i_\alpha}) - (-1)^{\gamma+1} \sum_{\omega > 0, 0 \leq \alpha \leq k-2} (\partial y_e^{e_1 \cdots e_\omega} . A_j^{\gamma\beta i_1 \cdots i_\alpha}) y^e_{e_1 \cdots e_\omega \gamma}] \\[2mm] A_j^{\beta i_1 \cdots i_{k-1}} = (-1)^{\beta+1}(\partial y_j^{\beta i_1 \cdots i_{k-1}}.L) \end{array}\right\}.$$

The corresponding Euler-Lagrange form $\mathcal{E}[\theta] : J\mathcal{D}^{2k}(W) \to \Lambda_0^{n+1}(J\mathcal{D}^{2k}(W))$ identifies the Euler-Lagrange equation $E[\theta] \subset J\mathcal{D}^{2k}(W)$ that is a PDE of order $2k$. (See refs [58,102] and references quoted there.) In particular, we get: 1) If $k = 0$, we can write $\sigma[\theta] = Ldx^1 \wedge \ldots \wedge dx^n = \theta$. $k = 1$, $\sigma[\theta] = (-1)^{\beta+1}(\partial y^\beta_\alpha.L)dy^\alpha \wedge dx^1 \wedge \ldots \wedge \widehat{dx^\beta} \wedge \ldots \wedge dx^n - Hdx^1 \wedge \ldots \wedge dx^n$, where $H = -L + (\partial y^\beta_\alpha.L)y^\alpha_\beta$ is the *Hamiltonian*. This is the Cartan form for first order unconstrained Lagrangian theories. Note that $\sigma[\theta] \equiv \theta + (\partial y^\alpha_j.L)\theta^j \wedge (\partial x_\alpha \rfloor \eta)$, we get $d\sigma[\theta] = \theta^j \wedge \omega_j$, with $\omega_j \equiv d((\partial y^\alpha_j.L)(\partial x_\alpha \rfloor \eta)) - (\partial x_\alpha.L)\eta$. Therefore $\sigma[\theta]$ is a Cartan form and the differential ideal $\mathcal{E}$ is generated by $\{\theta^j, \omega_j\}$. Then the Cartan extremals are n-dimensional submanifolds $V \subset J\mathcal{D}(W)$ image of holonomic sections $u = Ds$ such that $V^{(1)} = D^2s(M)$ satisfy the following classical Euler-Lagrange equation $E[\theta] \subset J\mathcal{D}^2(W)$, ($\bullet$) $(\partial x_\alpha.(\partial y^\alpha_j.L)) - (\partial y_j.L) = 0$. Vice versa, as the Euler-Lagrange extremals of the integral action $I = \int L\eta$ satisfy the same equation ($\bullet$), it follows that such extremals are also Cartan extremals.[91] ∎

 Let $(X,\theta,\mathcal{C})$ be a n-variational system of $\underline{D\mathcal{E}}$. Let $t:X \to \mathbf{R}$ be a numerical function

Definition 5.5. We say that two objects $X, Y \in Ob(\underline{D\mathcal{E}})$ are *equivalent* if there exists an isomorphism $f \in Hom_{\underline{D\mathcal{E}}}(X, Y)$. This definition identifies an equivalence relation R in $Ob(\underline{D\mathcal{E}})$. Let us denote $\mathcal{D} \equiv \underline{D\mathcal{E}}/R$ the category obtained by means of equivalence classes.

Remark 5.1. Let $E_k \subset J_n^k(W)$ be a formally integrable PDE of order k on the $(m + n)$-dimensional manifold W. Then, in general, $E_\infty \notin [E_k] \in \mathcal{D}$. $\blacksquare$ So we are conduced to the following:

Definition 5.6. Let $E_k \subset J_n^k(W)$, $E'_{k'} \subset J_{n'}^{k'}(W')$ be two objects of $\underline{D\mathcal{E}}$. We say that E_k is *formally equivalent* to $E'_{k'}$ if the following conditions are satisfied: (i) $n = n'$; (ii) If $k' \geq k$, $(E_k)_{+h} \cong E'_{k'}$, for some $h \geq 0$, such that $k + h = k'$; (iii) one has the following exact sequences: $(E_k)_{+h'} \to (E_k)_{+(h'-1)} \to 0$, $1 \leq h' \leq h$; (iv) $(E_{k+h})_{+r} \cong (E'_{k'})_{+r}$, for $r > 0$.[The isomorphisms considered in this definition are meant in the category $\underline{D\mathcal{E}}$.]

Proposition 5.2. *Let us denote by* $\underline{Sol}(E_k)$ *the set of regular solutions of* $E_k \subset J_n^k(W)$. *Then, if* E_k *is formally equivalent to* $E'_{k'}$ *one has:* $\underline{Sol}(E_k) = \underline{Sol}(E'_{k'})$. *In particular, if* E_k *is formally integrable, then* E_k *is formally equivalent to* E_∞, *and* $\underline{Sol}(E_k) = \underline{Sol}(E_\infty)$. *Furthermore, two formally equivalent PDEs* E_k, $E'_{k'}$, *do not necessitate to be equivalent in* $\underline{D\mathcal{E}}$.

So we are conduced to the following:

Definition 5.7. Let $\mathcal{D}_\bullet$ be the subcategory of $\mathcal{D}$ made by objects that are PDEs. Then the formal equivalence is an equivalence relation R_f in $\mathcal{D}_\bullet$. So we put: $\mathcal{D}_f \equiv \mathcal{D}_\bullet/R_f$. We call *universal PDE* in the class $[E_k] \in \mathcal{D}_f$ the one $E'_{k'}$ such that k' =maximum degree, i.e., any other $E''_h \in [E_k]$ has $h \leq k'$. All the completely integrable PDEs, as well as the formally integrable PDEs, are represented by an universal PDE with $k' = \infty$.[We can make each equivalent class $[E_k]_f \in \mathcal{D}_f$ into a category. Then the problem to find the universal PDE in this category is an universal problem, as an universal PDE is an initial object in such a category.]

Remark 5.2. Note that two formally equivalent PDEs $E_k, E'_{k'} \in Ob(\mathcal{D})$

---

on X such that $dt \neq 0$. Then, there are, locally, uniquelly determined forms $H \in \Omega^{n-1}(X)$, $\theta_1 \in \Omega^n(X)$, such that the following relations are satisfied: $\theta = \theta_1 + H \wedge dt$, where θ_1 and H do not contain the form dt. H is called the *Hamiltonian form* associated to the *time function t*. Furtermore, let $\alpha: N \to X$ be a compact submanifold mapping such that $\alpha^* dt \neq 0$. Then we call $E_{t_0} \equiv \int_{N_{t_0}} \alpha^* H = \int_{\alpha(N_{t_0})} H$ the *energy* of the n-variational system, with respect to the time function $t: X \to \mathbf{R}$, at the time t_0.

have the same set of regular solutions: $(\bullet\bullet)$ $\underline{Sol}(E_k) \cong \underline{Sol}(E'_{k'})$. But the converse is not necessarily true. So two PDEs $E_k, E'_{k'} \in Ob(\mathcal{D})$ that satisfy condition $(\bullet\bullet)$ are not necessarily formally equivalent. In fact, in order condition $(\bullet\bullet)$ should be satisfied it is enough that the infinity prolongations E_∞ and E'_∞ of E_k and $E'_{k'}$ respectively should be equivalent in $\underline{\mathcal{DE}}$. On the other hand E_k is not formally equivalent to its infinity prolongation E_∞ if E_k is not formally integrable. So we are conduced to give the following definition. $\blacksquare$

Definition 5.8. We say that two objects $E_k, E'_{k'} \in Ob(\mathcal{D}_\bullet)$ are *universally equivalent* if $E_\infty \cong E'_\infty$. This is an equivalence relation R_u. Set $\mathcal{D}_u \equiv \mathcal{D}_\bullet/R_u$. Let $[E_\infty]_u$ be an equivalence class in $\mathcal{D}_u$. We call E_∞ the *universal PDE* in the class $[E_\infty]_u$.[E_∞ results an initial object in the category $[E_\infty]_u$, hence the problem of the determination of E_∞ is an universal problem.]

Proposition 5.3. *If $E_k, E'_{k'} \in Ob(\mathcal{D}_\bullet)$ are formally equivalent they are also universally equivalent. Then one has the following short exact sequence of functors:* $\mathcal{D}_f \overset{u}{\to} \mathcal{D}_u \to 0$, $[E_k]_f \mapsto [E_\infty]_u$. *The functor u is surjective and the fiber over $[E_\infty]_u$ is given by all the classes $[E_k]_f$ that have the same E_∞.*

Definition 5.9. We define *universal n-variational system* one defined on some universal PDE in the category $\mathcal{D}_f$ or in the category $\mathcal{D}_u$.

Remark 5.3. As in the fiber bundle structure $u : \mathcal{D}_f \to \mathcal{D}_u$, $\mathcal{D}_f$ is the total space, it happears more suitable to refer to the category $\mathcal{D}_f$ when one talk of universal problems for PDEs. $\blacksquare$

Example 5.2. Let $E_\infty \subset J_n^\infty(W) \in Ob(\mathcal{D}_u)$ and let us denote by $\Omega^i(E_\infty)$ the $\Omega^0(E_\infty)$-module of smooth differential i-forms on the manifold E_∞. Let us denote by $C\Omega^i(E_\infty) \equiv \{\omega \in \Omega^i(E_\infty)|\omega|_{N^{(\infty)}} = 0, \quad \forall N \subset W, \quad \dim N = n, N^{(\infty)} \subset E_\infty\}$. Set $\bar{\Omega}^i \equiv \bar{\Omega}^i(E_\infty) \equiv \Omega^i(E_\infty)/C\Omega^i(E_\infty)$. One has $C\Omega^i(E_\infty) = \Omega^i(E_\infty)$ if $i > n$, hence $\bar{\Omega}^i = 0$ for $i > n$. In particular, if $E_\infty = J_n^\infty(W)$, we call $\theta \in \bar{\Omega}^n(J_n^\infty(W))$ a *Lagrangian density*. Its local expression looks like $\theta = L(x^\alpha, y^j, y^j_\alpha, \ldots)dx^1 \wedge \ldots \wedge dx^n$ in the neighborhood of any point $q \equiv [N]_a^\infty \in J_n^\infty(W)$, where $x^1, \ldots, x^n$ are local coordinates on N, in a neighborood of the point $a \equiv \pi_{k,0}(q) \in N \subset W$. Furthermore, if $\theta \in \bar{\Omega}^n(E_\infty)$, with $E_\infty \subset J_n^\infty(W)$, we say that θ is a *Lagrangian density with constraint* the PDE $E_\infty \subset J_n^\infty(W)$. One has the following short exact sequence of complexes:

$$0 \to C\Omega^\bullet \to \Omega^\bullet \to \bar{\Omega}^\bullet \to 0.$$

We call *bar de Rham cohomology* of E_∞ the corresponding homology $\bar{H}^q$ of the first complex on the left. Furthermore, we have the following filtration compatible with the exterior differential:

$$C^0\Omega^0 \equiv \Omega^q \supset C^1\Omega^q \supset C^2\Omega^q \supset \ldots \supset C^q\Omega^q \supset 0,$$

where

$$C^P\Omega^q(E_\infty) \equiv \left\{ \beta \in \Omega^q(E_\infty) \,\middle|\, \begin{array}{l} \beta(\zeta_1,\ldots,\zeta_q)=0 \\[4pt] \text{if at least } q-p+1 \text{ of the fields } \zeta_s \text{ belong to} \\[4pt] \mathbf{E}_n^\infty \subset TE_\infty \end{array} \right\}.$$

$[C^P\Omega^q(E_\infty) = 0$ for $p > q$; $C^P\Omega^q(E_\infty) = \Omega^q(E_\infty)$ if $p \le 0$; $\Omega^q(E_\infty) = 0$ if $q < 0$; $C\Omega^0(E_\infty) = 0$; $C\Omega^q = \Omega^q$ if $q > n$; $C^{q-n}\Omega^q = \Omega^q$ if $q > n$; $C^P\Omega^q \supset c^{p+1}\Omega^q$. Note also that $d^{-1}C^{p+1}\Omega^{q+p+1} \equiv \{\omega \in C^P\Omega^{p+q} | d\omega \in C^{p+1}\Omega^{p+q+1}\}.]$ As a consequence, we have associated a spectral sequence $\{E_r^{p,q}(E_\infty), d_r^{p,q}\}$ of E_∞. Of particular importance are the first two terms of above spectral sequence:

$$E_0^{p,q} \cong \frac{C^P\Omega^{p+q}}{C^{p+1}\Omega^{p+q}} \Rightarrow E_0^{0,q} \cong \bar{\Omega}^q, d_0^{0,q} = \bar{d}$$

$$E_1^{p,q} \cong \frac{C^P\Omega^{p+q} \cap d^{-1}C^{p+1}\Omega^{p+q+1}}{C^{p+1}\Omega^{p+q} \oplus dC^P\Omega^{p+q-1}} \Rightarrow E_1^{0,q} \cong \frac{\Omega^q \cap d^{-1}C\Omega^{q+1}}{d\Omega^{q-1} \oplus C\Omega^q} \cong \bar{H}^q$$

$(E_1^{p,q} = 0$ if $p > 0$, $q \ne n - 1, n$.) Then $E_1^{0,n} \cong \bar{H}^n$ can be identified with the space of all constrained Lagrangian densities on E_∞. Furthermore, the following term in the above spectral sequence $d_1 : E_1^{0,n} \to E_1^{1,n}$ represents the Euler-Lagrange operator that associates to a Lagrangian density θ, the corresponding Euler-Lagrange $(n + 1)$-form $d_1^{0,n}(\theta) \equiv \mathcal{E}[\theta] \in E_1^{1,n}$. An extremal for θ is a n-dimensional submanifold $N \subset W$ such that $N^{(\infty)} \subset E_\infty$ and $d_1(\theta)|_{N^{(\infty)}} = 0$. It is well known that $E_1^{0,n-1}$ is the space of conservation laws of E_∞ that can be represented by differential $(n-1)$-forms. In the following we shall denote by $E[\theta] \subset E_\infty \subset J_n^\infty(W)$ the PDE (*Euler-Lagrange equation*) corresponding to the Lagrangian density θ: $E[\theta] \equiv \{[N]_a^\infty \in E_\infty \,|\, \mathcal{E}[\theta]|_{N^{(\infty)}} = 0\} \subset E_\infty \subset J_n^\infty(W)$. The local expression of $\mathcal{E}[\theta]$ is the following: $\mathcal{E}[\theta] = \sum_{1 \le j \le m}[\sum_{i \in I}(-1)^{[i]}(\partial_i(\partial y_j^i.L))]\delta y^j \wedge dx^1 \wedge \ldots \wedge dx^n$, with $[i] = i_1 + \ldots + i_n$, $\partial_i = (\partial_i)^{i_1}\cdots(\partial_n)^{i_n}$, where $\partial_\mu = \partial x_\mu + \sum_{1 \le j \le m} A_\mu^j \partial y_j$, $\mu = 1,\ldots,n$, are basis Cartan vector fields for E_∞ and $\delta y^j \equiv dy^j - \sum_{1 \le \mu \le n} A_\mu^j(x,y)dx^\mu$ are the basis Cartan 1-forms on E_∞. In particular, if $E_\infty = J_n^\infty(W)$ and L is a function on $J_n^r(W)$ the Euler-Lagrange equation is the usual Euler-Lagrange equation for unconstrained higher order variational calculus [1,2]. ∎

Definition 5.10. We call *contact form* on a PDE $E_k \subset J_n^k(W)$, $\alpha \in \Omega^\bullet(E_k)$ such that $\alpha|_{N^{(k)}} = 0$ for all the k-prolongations $N^{(k)}$ of n-dimensional submanifolds $N \subset W$, such that $N^{(k)} \subset E_k$. Let us denote such a subspace of $\Omega^\bullet(E_k)$ by $\Omega_c^\bullet(E_k)$.

Proposition 5.4. *Let $C(E_k)$ be the Pfaffian ideal generated by the differential 1-forms $\{\omega_\beta^j \equiv dy_\beta^j - y_{\alpha\beta}^j dx^\alpha, \omega^I \equiv dF^I\}_{|\beta|\leq k-1}$, where $\{x^\alpha, y_\gamma^j\}_{|\gamma|\leq k}$ are local coordinates on $J_n^k(W)$ in the neighbourhood of $q = [N]_a^k \in E_k \subset J_n^k(W)$, with $\{x^\alpha\}$ local coordinates on N in the corresponding neighbourhood of $a \equiv \pi_{k,0}(q) \in N$. Furthermore, $F^I = 0$ are the local equations for E_k. We call $C(E_k)$ the contact ideal of E_k.[Note that $C(E_k) = j_k^* C(J_n^k(W))$, where $j_k : E_k \to J_n^k(W)$ is the canonical embedding.] Then $\Omega_c^\bullet(E_k)$ can be identified with the closed ideal $\overline{C}(E_k)$ associated to $C(E_k)$ and generated by $\{\omega_\beta^j, \omega^I, d\omega_{\beta_1 \cdots \beta_{k-1}}^j\}_{|\beta|\leq k-1}$. We call $\overline{C}(E_k)$ the contact differential system of E_k. One has $\Omega_c^0(E_k) = 0$ and $\Omega_c^q(E_k) = \Omega^q(E_k)$ if $q > n$. Furthermore, the distribution on E_k associated to the contact differential system $\overline{C}(E_k)$ is just the characteristic distribution $\mathbb{C}\mathrm{har}(E_k) \subset TE_k$.*

Proof. In fact all the differential forms α on E_k that are zero on all the possible n-dimensional integral manifolds $N^{(k)}$, are that of $\overline{C}(E_k)$ that can be represented as $\alpha = \omega_\sigma^j \wedge \alpha_j^\sigma + \omega^I \wedge \bar{\alpha}_I + d\omega_\sigma^j \wedge \beta_j^\sigma$, with $|\sigma| \leq k - 1$ and $\alpha_j^\sigma, \bar{\alpha}_I, \beta_j^\sigma$ arbitrary differential forms on E_k. Furthermore, recall that to an ideal $\mathcal{I} \subset \Omega^\bullet(X)$ of differential forms on a manifold X, it is associated a distribution, *Cauchy characteristic distribution*, $\mathbb{D}_\mathcal{I} \equiv \{\zeta \in \mathfrak{d}(TX)|\zeta\rfloor\mathcal{I} \subset \mathcal{I}\} \subset TX$. Here $\mathfrak{d}(TX)$ is the $\Omega^0(X)$-module of vector fields on X. The corresponding characteristic vector fields are defined by $Char(\mathcal{I}) \equiv \{\zeta \in \mathfrak{d}(TX)|\zeta \in \mathfrak{d}(\mathbb{D}_\mathcal{I})|\zeta\rfloor d\mathcal{I} \subset \mathcal{I}\}$. If $\mathcal{I}$ is Pfaffian $\mathcal{I} \equiv < \omega^j >$, $\mathbb{D}_\mathcal{I} \equiv \{\zeta\rfloor\omega^j = 0\}$ and $Char(\mathcal{I}) = \mathfrak{s}(\mathcal{I}) \cap \mathfrak{d}(\mathbb{D}_\mathcal{I}) \cong \mathfrak{d}(\mathbb{D}_{\overline{\mathcal{I}}})$, where $\mathfrak{s}(\mathcal{I})$ is the infinitesimal symmetry algebra of $\mathcal{I}$. In our case $\mathcal{I} = C(E_k)$, hence we get $Char(C(E_k)) \cong \mathfrak{d}(\mathbb{D}_{\overline{C}(E_k)})$. So $\zeta\rfloor\overline{C}(E_k) \subset \overline{C}(E_k)$ for any characteristic vector field ζ of E_k. □

Remark 5.4. In the particular case that $k = \infty$, i.e., $E_\infty \subset J_n^\infty(W)$, then $\mathbb{C}\mathrm{har}(E_\infty) = \mathbb{C}\mathrm{har}(C(E_\infty)) = \mathfrak{d}E_\infty$, hence $\overline{C}(E_\infty) = C(E_\infty)$. ∎

Proposition 5.5. *Let $E_{k+1} \in [E_k] \in \mathcal{D}_f$. Then, to any form $\alpha \in \Omega^q(E_k)$ we can associate the q-form $\pi_{k+1,k}^* \alpha \in \Omega^q(E_{k+1})$. Then we can decompose*

$\pi^*_{k+1,k}\alpha = h(\alpha) + p_1(\alpha) + \ldots + p_q(\alpha) \equiv \sum_{0\leq i\leq q} p_i(\alpha)$, where $p_0(\alpha) = h(\alpha)$ is the *horizontal component* of α and $p_i(\alpha)$ is the *i-contact component* of α. More precisely

$$p_i(\alpha)(\zeta_1,\cdots,\zeta_q)(\bar{q})=\tfrac{1}{i!(q-i)!}\,\epsilon^{j_1\cdots j_i j_{i+1}\cdots j_q}\,\alpha(p(\zeta_{j_1}),\cdots,p(\zeta_{j_i}),h(\zeta_{j_{i+1}}),\cdots,h(\zeta_{j_q}))$$

for any $\bar{q} \in E_{k+1}$, $\forall\zeta_s \in T_{\bar{q}}E_{k+1}$, where $(\pi_{k+1,k})_*(\zeta_s) = h(\zeta_s) + p(\zeta_s) \in T_{\pi_{k+1,k}(\bar{q})}E_k \cong L_{\bar{q}} \bigoplus vT_{\pi_{k+1,k}(\bar{q})}E_k$ is the canonical splitting in the horizontal and vertical components respectively of $(\pi_{k+1,k})_*(\zeta_s)$. Here

$$vT_{\pi_{k+1,k}(\bar{q})}E_k \equiv T_{\pi_{k+1,k}(\bar{q})}E_k/L_{\bar{q}},$$

with $L_{\bar{q}} \subset (\mathbb{E}^k_n)_q$ the integral n-plane identified with $L_{\bar{q}} \cong T_qN^{(k)}$, where $\bar{q} = [N]^{k+1}_a$, $a \equiv \pi_{k+1,0}(\bar{q}) \in N \subset W$, $q \equiv \pi_{k+1,k}(\bar{q}) \in E_k$.[Note that this splitting, restricted to the Cartan distribution, gives: $(\mathbb{E}^k_n)_q \cong L_{\bar{q}} \oplus g_k$, where g_k is the symbol of E_k.] If $k + 1 \leq q \leq \dim(E_k)$ and $\alpha \in \Omega^q(E_k)$, one has $p_i(\alpha) = 0$, $0 \leq i \leq q - k - 1$. In such cases $\pi^*_{k+1,k}\alpha = p_{q-n}(\alpha)+\cdots+p_q(\alpha)$. Then, we say that α is strongly contact if $p_{q-n}(\alpha) = 0$. The differential of a strongly contact form does not necessitate to be a strongly contact form. One has the following properties: (i) $p_i(\alpha \wedge \beta) = \sum_{r+s=i} p_r(\alpha) \wedge p_s(\beta)$; (ii) $\zeta\rfloor p_i(\alpha) = p_{i-1}(p(\zeta)\rfloor\alpha) + p_i(h(\zeta)\rfloor\alpha)$. If $\alpha = j^*_k\beta \in \Omega^q(E_k)$, where

$$\beta = \sum_{0\leq s\leq q} \frac{1}{s!(q-s)!}\beta^{\sigma_1\cdots\sigma_s}_{j_1\cdots j_s i_{s+1}\cdots i_q}\,dy^{j_1}_{\sigma_1} \wedge \cdots \wedge dy^{j_s}_{\sigma_s} \wedge dx^{i_{s+1}} \wedge \cdots \wedge dx^{i_q}$$
$$\in\Omega^q(J^k_n(W)),$$

with $|\sigma_i| \leq k$ and $j_k : E_k \left\{ F^I(x^\alpha, y^j_\sigma) = 0 \right\} \to J^k_n(W)$ the canonical embedding, then one has the following local representation of the forms $p_i(\alpha)$:

$$\left\{ \begin{array}{l} p_i(\alpha)=j^*_{k+1}p_i(\beta)\in\Omega^q(E_{k+1}) \\[2mm] p_i(\beta)=\tfrac{1}{i!(q-i)!} B^{\sigma_1\cdots\sigma_i}_{j_1\cdots j_i i_{i+1}\cdots i_q}\,\omega^{j_1}_{\sigma_1} \wedge\cdots\wedge\omega^{j_i}_{\sigma_i} \wedge dx^{i_{i+1}} \wedge\cdots\wedge dx^{i_q} \in\Omega^q(J^{k+1}_n(W)) \\[2mm] B^{\sigma_1\cdots\sigma_i}_{j_1\cdots j_i i_{i+1}\cdots i_q}=\sum_{i\leq s\leq q}\left(\tfrac{q-i}{q-s}\right)\beta^{\sigma_1\cdots\sigma_i\sigma_{i+1}\cdots\sigma_s}_{j_1\cdots j_i j_{i+1}\cdots j_s i_{s+1}\cdots i_q}\,y^{j_{i+1}}_{\sigma_{i+1}}\cdots y^{j_s}_{\sigma_s}\,alt(i_{i+1}\cdots i_s i_{s+1}\cdots i_q) \\[2mm] j_{k+1}:E_{k+1}\left\{ \begin{array}{l} F^I(x^\alpha,y^j_\sigma)=0, \quad |\sigma|\leq k \\[2mm] (\partial x_\alpha.F^I)+\sum_{|\sigma|\leq k} y^j_\sigma(\partial y^\sigma_j.F^I)=0 \end{array} \right\}\to J^{k+1}_n(W) \end{array} \right\}.$$

Therefore $p_i(\alpha)(\zeta_1,\cdots,\zeta_q)(\bar{q}) = 0$ if at least i of the vectors ζ_s belong to the Cartan distribution $(\mathbb{E}_n^{k+1})_{\bar{q}}$. So if we put:

$$C^p\Omega^q(E_{k+1}) \equiv \left\{ \beta \in \Omega^q(E_{k+1}) \,\middle|\, \begin{array}{l} \beta(\zeta_1,\cdots,\zeta_q)=0 \\[4pt] \text{if at least } q-p+1 \text{ of the vectors } \zeta_s \text{ belong to} \\[4pt] (\mathbb{E}_n^{k+1})_{\bar{q}} \end{array} \right\},$$

and call it the *space of p-Cartan q-forms on $E_{k+1} \subset J_n^{k+1}(W)$,* we get that $p_i(\alpha) \in C^{i+1}\Omega^q(E_{k+1})$. In particular $\alpha \in \Omega^q(E_k)$, with $q > n$, is a *strongly contact form* if $\pi_{k+1,k}^*\alpha$ has not forms belonging to $C^{n+1}\Omega^q(E_{k+1})$. So α is strongly contact iff $\alpha = j_k^*\beta$, with

$$\beta = \sum_{q-k+1 \le p+s,\, p+2s \le q,\, |\sigma_i| \le k-1,\, |\gamma_i| \le n-1} \omega_{\sigma_1}^{j_1} \wedge \cdots \wedge \omega_{\sigma_p}^{j_p} \wedge d\omega_{\gamma_1}^{i_1} \wedge \cdots \wedge d\omega_{\gamma_s}^{i_s} \, \Phi_{j_1 \cdots j_p i_1 \cdots i_s}^{\sigma_1 \cdots \sigma_p \gamma_1 \cdots \gamma_s},$$

where $j_k : E_k\{F^I(x^\alpha, y_\sigma^j)=0, \ |\sigma| \le k\} \to J_n^k(W)$ is the canonical embedding, and $\Phi_{\cdots}^{\cdots}$ are some $(q-p-2s)$-forms on $J_n^k(W)$.

Proof. It follows directly from above considerations and results reported in refs.[58,102]. $\qquad\square$

Definition 5.11. For any PDE $E_k \subset J_n^k(W)$ we denote by

$$\Xi^n(E_k) \equiv \frac{\Omega^n(E_k)}{\Omega_c^n(E_k) \oplus d\Omega_c^{n-1}(E_k)}$$

the *space of constrained generalized Lagrangian densities* on E_k. Here $\Omega_c^n(E_k) \equiv \Omega^n(E_k) \bigcap \overline{C}(E_k)$.

Theorem 5.3. Let $E_k \subset J_n^k(W)$ be a PDE such that $\pi_{k+1,k} : (E_k)_{+1} \to E_k$ is a surjective fiber bundle. Then, a differential n-form $\alpha \in \Omega^n(E_k)$ identifies a constrained Lagrangian density on $(E_k)_{+1}$. Furthermore, the following constrained n-variational systems

$$(E_k, \alpha, \mathcal{C}(E_k))$$

and

$$(E_{k+1}, h(\alpha), \mathcal{C}(E_{k+1}))$$

are universally equivalent, i.e., the corresponding Euler-Lagrange equations are universally equivalent. Furthermore, they are formally equivalent, i.e., the corresponding Euler-Lagrange equations are formally equivalent, iff one has the following exact sequences $(E_k)_{+s} \to (E_k)_{+(s-1)} \to 0$, with $1 \le s \le k+1$.

Proof. Let us first note that, by means of the embedding $j_k : E_k \to J_n^k(W)$, we can represent any differential p-form on E_k by means of some p-form on $J_n^k(W)$. So, if α is an n-form on E_k we can consider any n-form β on $J_n^k(W)$ such that $\beta|_{E_k} = \alpha$ and associate to such a β its horizontal correspondent $h(\beta)$ on $J_n^{k+1}(W)$. Put $\theta = h(\beta)|_{E_{k+1}}$. θ does not depend on the particular β used, in the sense that if $\beta' \in \Omega^n(J_n^{k+1}(W))$ such that $\beta'|_{E_{k+1}} = \beta|_{E_{k+1}}$ then $h(\beta')|_{E_{k+1}} = h(\beta)|_{E_{k+1}}$. Now, the unconstrained n-variational systems $(J_n^k(W), \beta, C(J_n^k(W)))$ and $(J_n^{k+1}(W), h(\beta), C(J_n^{k+1}(W)))$ have the same Euler-Lagrange $(n+1)$-form

$$\mathcal{E}[\beta] = \mathcal{E}[h(\beta)] : J_n^{2k+1}(W) \to \Lambda_0^{n+1}(J_n^{2k+1}(W)).$$

Therefore, the constrained n-variational systems

$$(E_k, \alpha, C(E_k))$$

and

$$(E_{k+1}, \theta, C(E_{k+1}))$$

will be formally equivalent iff conditions in theorem are verified. In fact, in such a case one has that E_k and E_{k+1} are both formally equivalent to $(E_k)_{+s}$. Hence the corresponding constrained Euler-Lagrange equations one has:

$$E[\alpha] = E[\theta] \subset (E_k)_{k+1} \subset J_n^{2k+1}(W).$$

Of course as $E[\alpha]_{+\infty} = E[\theta]_{+\infty} \subset E_\infty \subset J_n^\infty(W)$, above constrained variational systems are universally equivalent. $\square$

Theorem 5.4. *In particular, if $[E_k] \in \mathcal{D}_f$ is the equivalence class of a formally integrable PDE, and*

$$(E_k, \theta, C(E_k))$$

is a constrained n-variational system, then this system is universally and formally equivalent to

$$(E_{k+1}, h(\theta), C(E_{k+1})).$$

Proof. It is a direct consequence of the fact that E_k is formally integrable and above results. $\qquad\qquad\square$

Remark 5.5. The formal equivalence allows us to unlarge the set of extremals (solutions) of constrained variational systems, including also singular extremals. $\qquad\qquad\blacksquare$

Definition 5.12. Let $\mathcal{L} \subset \mathcal{D}_f$ be the subcategory of $\mathcal{D}_f$ defined by variational PDEs, i.e., $[E_k] \in Ob(\mathcal{L})$ iff there exists a Lagrangian density $\theta \in \bar{\Omega}^n(\bar{E}_\infty)$, $\bar{E}_\infty \subset J_n^\infty(W)$, such that for some $E_{k'} \in [E_k]$ one has $E_{k'} = E[\theta]$.

Definition 5.13. We say that two Lagrangian densities $\theta \in \bar{\Omega}^n(E_\infty)$, $\theta' \in \bar{\Omega}^n(E'_\infty)$, are *equivalent*, $\theta \sim \theta'$, if the corresponding Euler-Lagrange equations $E[\theta]$, $E[\theta']$, belong to the same equivalence class in $\mathcal{L}$: $E[\theta] \in [E[\theta']] \in \mathcal{L}$.

Proposition 5.6. *Let $\sigma[\theta] \in [\theta] \in \bar{\Omega}^n(E_\infty)$ and $\sigma[\theta'] \in [\theta'] \in \bar{\Omega}^n(E'_\infty)$ be two Cartan forms associated to two Lagrangian densities θ and θ'. Let us assume that $\theta \sim \theta'$. Then the Cartan extremals of $\sigma[\theta]$ are Euler-Lagrange extremals of θ' and the Cartan extremals of $\sigma[\theta']$ are Euler-Lagrange extremals of θ. Whether, in advance, $\sigma[\theta]$ and $\sigma[\theta']$ are universal Cartan forms, their extremals are in correspondence one-to-one.*

Proof. In fact, as the Cartan extremals are Euler-Lagrange extremals, the proposition follows directly from the assumption that θ and θ' are equivalent. In the case that $\sigma[\theta]$ and $\sigma[\theta']$ are universal Cartan forms, their extremals are in correspondence one-to-one as any Euler-Lagrange extremal is also a Cartan extremal. $\qquad\qquad\square$

Definition 5.14. We say *equivalent* two Cartan forms $\sigma[\theta]$ and $\sigma[\theta']$ if the following conditions are verified: (i) $\sigma[\theta] \in [\theta]$ and $\sigma[\theta'] \in [\theta']$, where θ and θ' are equivalent Lagrange densities. (ii) $\sigma[\theta]$ and $\sigma[\theta']$ are universal Cartan forms.

Theorem 5.5. *For two equivalent Cartan forms $\sigma[\theta]$ and $\sigma[\theta']$ their sets of extremals are in correspondence one-to-one.*

Proof. It is a direct consequence of Proposition 25 and Definition 26. $\qquad\square$

5.3 - CONSTRAINED VARIATIONAL SEQUENCES

Theorem 5.6. *Let $[E_k] \in \mathcal{D}_f$. Then to each PDE E_k in this equivalence* *class we can associate a sequence, (constrained variational sequence):*

$$0 \to \Xi^0(E_k) \xrightarrow{\epsilon_k^0} \Xi^1(E_k) \xrightarrow{\epsilon_k^1} \Xi^2(E_k) \xrightarrow{\epsilon_k^2} \cdots$$

Furthermore, if $E'_{k'} \in [E_k]$, with $k' > k$, the constrained variational sequence, $(\widetilde{\Xi}^\bullet(E_k), \tilde{\epsilon}_k^\bullet)$, associated to $E'_{k'}$ is the extension of a sequence, reduced variational sequence, associated to the couple $(E_k, E'_{k'})$. In other words, one has the following short exact sequence of complexes:

$$0 \to \widetilde{\Xi}^\bullet(E_k) \to \Xi^\bullet(E'_{k'}) \to \widehat{\Xi}^\bullet(E'_{k'}) \to 0.$$

If $E'_{k'}$ has trivial q-cohomology, $q \neq 0, \dim(E'_{k'})$, one has $\widetilde{\Xi}^\bullet(E_k) \cong \Xi^\bullet(E_k)$. Furthermore, if $E'_{k'}$ is an universal PDE in $[E_k] \in \mathcal{D}_f$, then all the constrained variational sequences of this class are represented in the constrained variational sequence of $E'_{k'}$, universal constrained variational sequence (of the class $[E_k]$)), that we will denote as follows:

$$0 \to \Xi^0[E_k] \xrightarrow{\epsilon_k^0} \Xi^1[E_k] \xrightarrow{\epsilon_k^1} \Xi^2[E_k] \xrightarrow{\epsilon_k^3} \cdots.$$

This is a final object in the category of constrained variational sequences of PDEs in the class $[E_k]$. As a consequence it is also an extension of all reduced variational sequences of PDEs belonging to $[E_k]$, and with respect to $E'_{k'}$. We will call the mapping $\epsilon_k^n : \Xi^n(E_k) \to \Xi^{n+1}(E_k)$ the generalized Euler-Lagrange operator for generalized Lagrangian densities in $\Xi^n(E_k)$, and the mapping $\epsilon_k^{n+1} : \Xi^{n+1}(E_k) \to \Xi^{n+2}(E_k)$, the generalized Helmoltz-Sonin mapping. We denote by $H^\bullet_{var}(E_k)$ the homology of the constrained variational sequence of E_k and we call it the variational cohomology of E_k. The condition that a $(n+1)$-form α, belonging to $\Xi^{n+1}(E_k)$, ϵ_k^{n+1}-closed, *globally represents* the generalized Euler-Lagrange form *of some generalized* Lagrangian θ on E_k, i.e., $\alpha = \epsilon_k^n(\theta)$, is given by $H^{n+1}_{var} = 0$. We call universal variational cohomology of $[E_k]$ the homology of $\Xi^\bullet[E_k]$.

Proof. In fact, to E_k one can associate its constrained variational sequence defined by means of the following short exact sequence of complexes:

$$0 \to \Theta^\bullet(E_k) \to \Omega^\bullet(E_k) \to \Xi^\bullet(E_k) \to 0,$$

where $\Theta^p(E_k) \equiv \Omega_c^p(E_k) \bigoplus d\Omega_c^{p-1}(E_k)$, with $\Omega_c^p(E_k) \equiv \Omega^p(E_k) \bigcap \overline{C}(E_k)$, for $p \leq n$, and $\Omega_c^p(E_k)$ the space of strongly contact p-forms on E_k for

$p > n$. On the other hand as $k' > k$, we get the following short exact sequence $E'_{k'} \to E_k \to 0$ that on the related differential forms induces the following exact commutative diagram of complexes:

$$
\begin{array}{ccccccc}
& & & & 0 & & 0 \\
& & & & \uparrow & & \uparrow \\
0 & \to & K^\bullet(E_k) & \to & \Omega^\bullet(E_k)/\Theta^\bullet(E_k) & \to & \Omega^\bullet(E'_{k'})/\Theta^\bullet(E'_{k'}) \\
& & & & \uparrow & & \uparrow \\
& & 0 & \to & \Omega^\bullet(E_k) & \to & \Omega^\bullet(E'_{k'}) \\
& & & & \uparrow & & \uparrow \\
& & 0 & \to & \Theta^\bullet(E_k) & \to & \Theta^\bullet(E'_{k'}) \\
& & & & \uparrow & & \uparrow \\
& & & & 0 & & 0
\end{array}
$$

Now, let us see under which conditions $K^\bullet(E_k) = 0$. Note that if $q \leq n$, then if $\pi^*_{k',k}\alpha \in \Theta^p(E'_{k'}) \cong \Omega^p_c(E'_{k'})$, then must necessarily be $\alpha \in \Theta^p(E_k)$. In fact, if $\pi^*_{k',k}\alpha$ is a contact form, i.e., $i^*_{k'}(\pi^*_{k',k}\alpha) = 0$, where $i_{k'} : N \to E'_{k'}$ denotes the embedding of the k'-prolongation $N^{(k')}$ of any n-dimensional submanifold $N \subset W$, solution of $E'_{k'}$, one has also $i^*_k\alpha = 0$, where $i_k : N \to E_k$ is the k-prolongation of $N \subset W$. In fact, one has: $i^*_k\alpha = i^*_{k'}(\pi^*_{k',k}\alpha)$. Hence, if $i^*_{k'}(\pi^*_{k',k}\alpha) = 0$ one has also $i^*_k\alpha = 0$. Therefore, α is a contact form. Furthermore, if $d(\pi^*_{k',k}\alpha) \in \overline{C}_{k'}(E'_{k'})$ must necessarily be also that $d\alpha \in \overline{C}_k(E_k)$. In conclusion, $\alpha \in \Theta^p(E_k)$ for $p \leq n$. Let us, now, prove if this property holds also for $p > n$. So, let us assume that $\pi^*_{k',k}\alpha \in \Theta^p(E'_{k'})$ for $p > n$. We have three possibilities: (i) $\pi^*_{k',k}\alpha \equiv \alpha_0 \in \Omega^p_c(E'_{k'})$; (ii) $\pi^*_{k',k}\alpha \equiv \alpha_1 \in d\Omega^{p-1}_c(E'_{k'})$; (iii) $\pi^*_{k',k}\alpha \equiv \alpha_0 + \alpha_1$. We shall require that $E_{k+1}, E'_{k'+1} \in [E_k] \in \mathcal{D}_f$. Let us first consider the case (i). Then, one has the following commutative diagram:

$$
\begin{array}{ccccccccc}
0 & \to & \Omega^q_c(E'_{k'}) & \to & \Omega^q(E'_{k'}) & \xrightarrow{p^{k'}_{q-n}} & \pi^*_{k'+1,k'}\Omega^q(E'_{k'}) & \subset & \Omega^q(E'_{k'+1}) \\
& & \uparrow & & a\uparrow & @ & \uparrow b & & \uparrow \\
0 & \to & \Omega^q_c(E_k) & \to & \Omega^q(E_k) & \xrightarrow{p^k_{q-n}} & \pi^*_{k+1,k}\Omega^q(E_k) & \subset & \Omega^q(E_{k+1}) \\
& & \uparrow & & \uparrow & & \uparrow & & \uparrow \\
& & 0 & & 0 & & 0 & & 0
\end{array}
$$

Let us assume that $\pi^*_{k',k}\alpha \in \Omega^q_c(E'_{k'})$ for $\alpha \in \Omega^q(E_k)$. Then, for the commutativity of the @-diagram $(p^{k'}_{q-n} \circ a)(\alpha) = 0$, hence $(b \circ p^k_{q-n})(\alpha) = 0$. This means that $p^k_{q-n}(\alpha) \in \ker(b)$. On the other hand as b is injective, must necessarily be $p^k_{q-n}(\alpha) = 0$. Hence $\alpha \in \Omega^q_c(E_k)$. Let us consider, now, the case (ii). Let us assume that $\pi^*_{k',k}\alpha = d\beta \in \Omega^{q-1}_c(E'_{k'})$. By exterior differentiation of both sides of above equation we get $d\pi^*_{k',k}\alpha = 0$. Hence, at least locally (or globally if $H^q(E'_{k'}) = 0$) we can write $\alpha = d\gamma$. This means that $d\pi^*_{k',k}\gamma = d\beta$, hence $\pi^*_{k',k}\gamma \in \Omega^{q-1}_c(E'_{k'})$. But from the case (i) we can conclude that must be $\gamma \in \Omega^{q-1}_c(E_k)$. Therefore, $\alpha = d\gamma \in \Omega^{q-1}_c(E_k)$. Finally, the third case, being the sum of cases (i) and (ii) allows us to conclude that if $\pi^*_{k',k}\alpha \in \Theta^q(E'_{k'})$ then $\alpha \in \Theta^q(E_k)$. Therefore, in general, i.e., without assuming that $H^q(E'_{k'}) = 0$ for $q \neq 0, \dim(E'_{k'})$, we get the following short sequence of complexes:

$$0 \to K^\bullet(E_k) \to \Xi^\bullet(E_k) \to \Xi^\bullet(E'_{k'}).$$

As a consequence we get the extension property claimed in the theorem. In the particular case that $E'_{k'}$ is the universal PDE in $[E_k] \in \mathcal{D}_f$, it follows that it is an initial object in the category of PDEs identified with the equivalence class $[E_k]$. Therefore the theorem is proved. $\qquad\square$

Theorem 5.7. *For the variational cohomology of E_k one has the following canonical isomorphism: $H^\bullet_{var}(E_k) \cong H^\bullet(E_k; \mathbf{R})$.*

Proof. Let us consider the sheaves sequence associated to the variational sequence $\Xi(E_k)$, i.e. the following sequence:

$$0 \to \mathcal{R} \to \underline{\Xi^1(E_k)} \to \underline{\Xi^2(E_k)} \to \cdots \to \underline{\Xi^p(E_k)} \to \cdots,$$

where $\mathcal{R} \equiv E_k \times \mathbf{R}$ denotes the constant sheaf of real numbers over E_k and $\underline{\Xi^p(E_k)}$, $p > 0$, denotes the sheaf of sections of $\Xi^p(E_k)$ over E_k. Such a sequence is an acyclic resolution of $\mathcal{R}$ that, for well known results of sheaves theory, has the cohomology of the sequence of global sections, i.e. the variational cohomology $H^\bullet_{var}(E_k)$, coinciding with the de Rham cohomology $H^\bullet(E_k, \mathbf{R})$ of E_k. $\qquad\square$

Theorem 5.8. *The constrained variational sequence of E_∞, infinity prolongation of some PDE $E_k \subset J^k_n(W)$, is isomorphic to the C-sequence of E_k. [The C-sequence of a PDE has been introduced by the Moscow mathematical school of geometry of PDEs. See, e.g., refs.[39] and references quoted there.] In the particular case that E_k in $[E_k] \in \mathcal{D}_f$ is formally integrable, one has a representation of the constrained variational sequence of E_k into the universal one, hence in the C-sequence of E_k. One has $H^q_{var}(J^\infty_n(W)) \cong H^q(W; \mathbf{R})$, and the following morphisms: $H^q(W; \mathbf{R}) \to$*

$H^q_{var}(E_k) \cong H^q_{var}(E_{k+s})$, $s \geq 1$. We say that E_k is *wholly variational* if $H^q_{var}(E_k) \cong H^q_{var}(E_\infty) \cong H^q(W; \mathbf{R})$, at least for $q = n + 1$. [If E_k is formally integrable and it is also an affine subbundle of $J^k_n(W) \to J^{k-1}_n(W)$, over $J^{k-1}_n(W)$, then E_k is wholly variational and $H^q_{var}(E_k) \cong H^q(W; \mathbf{R})$.] If E_k is wholly variational the condition of global universal variationality $H^{n+1}_{var}(E_\infty) = 0$ is equivalent to the condition of global variationality on E_k, i.e., $H^{n+1}_{var}(E_k) = 0$, and to the condition $H^{n+1}(W; \mathbf{R}) = 0$.

Proof. Note that the space of generalized Lagrangian densities on E_k is $\Xi^n(E_k) \cong \frac{\Omega^n(E_k)}{\Omega^n_c(E_k)}$. Furthermore, in the particular case that $k = \infty$ one has $\Xi^n(E_\infty) \cong \bar{\Omega}^n(E_\infty)$. (In this case E_∞ is necessarily the universal PDE in $[E_k]$.) Now, let us assume that E_k is formally integrable, (hence E_∞ is just the universal PDE in $[E_k]$). Then the sequences $\{\bar{\Omega}^\bullet(E_\infty), \bar{d}\}$, $\{\Xi^\bullet(E_\infty), \epsilon^\bullet_\infty\}$ and $\{E_1^{\bullet,n}, d_1^{\bullet,n}\}$ can be combined toghether to obtain the following commutative diagram:

$$
\begin{array}{ccccccccccc}
0 \to & \Xi^0(E_\infty) & \xrightarrow{\epsilon^0_\infty} & \Xi^1(E_\infty) & \xrightarrow{\epsilon^1_\infty} \cdots \xrightarrow{\epsilon^{n-1}_\infty} & \Xi^n(E_\infty) & \xrightarrow{\epsilon^n_\infty} & \Xi^{n+1}(E_\infty) & \xrightarrow{\epsilon^{n+1}_\infty} & \Xi^{n+2}\cdots \\
& \| \wr & & \| \wr & & \| \wr & & \| \wr & & \| \wr \\
0 \to & \bar{\Omega}^0(E_\infty) & \xrightarrow{\bar{d}^0} & \bar{\Omega}^1(E_\infty) & \xrightarrow{\bar{d}^1} \cdots \xrightarrow{\bar{d}^{n-1}} & \bar{\Omega}^n(E_\infty) & \xrightarrow{\mathcal{E}} & E_1^{1,n} & \xrightarrow{\bar{d}_1^{1,n}} & E_1^{2,n}\cdots \\
& & & & & \| \downarrow & & \uparrow d_1^{0,n} & & \\
& & & & & \bar{H}^n(E_\infty) & \cong & E_1^{0,n} & &
\end{array}
$$

The sequence on the middle is called *C-sequence* of E_k. The sequence on the top is the universal variational sequence of the class $[E_k] \in \mathcal{D}_f$ and the corresponding homology is the universal variational cohomology of $[E_k]$. (If $E_\infty = J^\infty_n(W)$, then $H^\bullet_{var}(E_\infty) = H^\bullet(J^\infty_n(W); \mathbf{R}) \cong H^\bullet(W; \mathbf{R})$.) Above sequences can be related to the variational sequence $\Xi^\bullet(E_k)$ to get the following commutative diagram:

$$
\begin{array}{ccccccccccc}
0 \to & \Xi^0(E_k) & \xrightarrow{\epsilon^0_k} & \Xi^1(E_k) & \xrightarrow{\epsilon^1_k} \cdots \xrightarrow{\epsilon^{n-1}_k} & \Xi^n(E_k) & \xrightarrow{\epsilon^n_k} & \Xi^{n+1}(E_k) & \xrightarrow{\epsilon^{n+1}_k} & \Xi^{n+2}(E_k)\cdots \\
& \downarrow & & \downarrow & & \downarrow & & \downarrow & & \downarrow \\
0 \to & \Xi^0(E_\infty) & \xrightarrow{\epsilon^0_\infty} & \Xi^1(E_\infty) & \xrightarrow{\epsilon^1_\infty} \cdots \xrightarrow{\epsilon^{n-1}_\infty} & \Xi^n(E_\infty) & \xrightarrow{\epsilon^n_\infty} & \Xi^{n+1}(E_\infty) & \xrightarrow{\epsilon^{n+1}_\infty} & \Xi^{n+2}(E_\infty)\cdots \\
& \| \wr & & \| \wr & & \| \wr & & \| \wr & & \| \wr \\
0 \to & \bar{\Omega}^0(E_\infty) & \xrightarrow{\bar{d}^0} & \bar{\Omega}^1(E_\infty) & \xrightarrow{\bar{d}^1} \cdots \xrightarrow{\bar{d}^{n-1}} & \bar{\Omega}^n(E_\infty) & \xrightarrow{\mathcal{E}} & E_1^{1,n} & \xrightarrow{\bar{d}_1^{1,n}} & E_1^{2,n}\cdots
\end{array}
$$

that gives the representation, *universal representation*, of the variational sequence of E_k into the universal one of the class $[E_k] \in \mathcal{D}_f$ and in the C-sequence. As $E_{k+s} \to E_{k+(s-1)}$, $s \geq 1$, is an affine subbundle of the affine bundle $J_n^{k+s}(W) \to J_n^{k+(s-1)}(W)$, it follows that $H_{var}^q(E_{k+s}) \cong H^q(J_n^{k+s}(W)|_{E_{k+(s-1)}}; \mathbf{R}) \cong H^q(E_k; \mathbf{R})$, $s \geq 1$. As a consequence we get $H_{var}^q(E_\infty) \cong H^q(E_k; \mathbf{R})$. In general $H^q(W; \mathbf{R})$ has a representation in $H_{var}^q(E_{k+s})$, i.e., one has a morphism $H^q(W; \mathbf{R}) \to H_{var}^q(E_k) \cong H_{var}^q(E_{k+s})$. If E_k is wholly variational the condition of global universal variationality $H_{var}^{n+1}(E_\infty) = 0$ is equivalent to the condition of global variationality on E_k, i.e., $H_{var}^{n+1}(E_k) = 0$, and to the condition $H^{n+1}(W; \mathbf{R}) = 0$. $\qquad\square$

Remark 5.6. If E_k is not formally integrable we cannot more reproduce the above conclusions on the relations between E_k and E_∞, as the lost of surjectivity in the mapping $E_\infty \to E_k$ does not allow us to build the representative mapping $\Xi^\bullet(E_k) \to \Xi^\bullet(E_\infty)$. However, we can give the following definition. $\qquad\blacksquare$

Definition 5.15. We say that a PDE $E_k \subset J_n^k(W)$ is *universally regular* if there exists a subequation $E_k^{(\infty)} \subset E_k \subset J_n^k(W)$ that is formally integrable and such that $(E_k^{(\infty)})_{+\infty} \cong E_\infty \subset J_n^\infty(W)$.

Example 5.3. The Navier-Stokes equation is not formally integrable but it is universally regular. $\qquad\blacksquare$

Proposition 5.7. *If E_k is universally regular one has:*

$$\underline{Sol}(E_k) \cong \underline{Sol}(E_k^{(\infty)}) \cong \underline{Sol}(E_\infty),$$

i.e., the regular solutions of E_k are all and only the regular solutions of $E_k^{(\infty)}$. Furthermore, if E_k is universally regular, then $E_k^{(\infty)} \in [E_\infty] \in \mathcal{D}_f$ and E_∞ is the universal PDE of the class $[E_k]$.

Remark 5.7. Therefore, in order to represent the variational sequence of E_k to one of the infinity prolongation E_∞, and in the C-sequence of E_k, we need to have that E_k is universally regular. In this a case we must substitute E_k with $E_k^{(\infty)}$. Furthermore, note that two formally integrable equations $E_k \subset J_n^k(W)$, $E_{k'} \subset J_n^{k'}(W')$, are formally equivalent iff their infinity prolongations E_∞ and E_∞' respectively are isomorphic in $\underline{\mathcal{DE}}$. In this case we can consider the constrained variational sequences on the universal PDE E_∞, $\Xi^\bullet(E_\infty)$ and represent any finite order variational sequence of the equivalence class $[E_k] \in \mathcal{D}_f$ into $\Xi^\bullet(E_\infty)$. Furthermore, if two formally integrable PDEs E_k and $E_{k'}'$ are related by a surjective morphism $E_\infty' \to E_\infty \to 0$, then the relation between the corresponding constrained variational sequences is given by the following sequence

$$0 \to \Xi^\bullet(E_\infty) \to \Xi^\bullet(E_\infty')$$

that is globally exact. In particular, this is the case when $E'_\infty \to E_\infty$ is a covering [39]. ∎

Remark 5.8. If $E_k = J\mathcal{D}^k(W)$, i.e., E_k is a trivial PDE on the fiber bundle $\pi : W \to M$, the constrained variational sequence coincides with the variational sequence previously considered in the literature [102]. ∎

Theorem 5.9. *Associated to any class* $[E_k] \in \mathcal{D}_f$ *one has the following exact sequence, (cohomological constrained variational sequence of* $[E_k]$*):*

$$\cdots \to H^m_{var}(\widetilde{\Xi}^\bullet(E_k)) \to H^m_{var}(\Xi^\bullet[E_k]) \to H^m_{var}(\Xi^\bullet[E_k]/\widetilde{\Xi}^\bullet(E_k))$$
$$\overset{\delta}{\to} H^{m-1}_{var}(\widetilde{\Xi}^\bullet(E_k)) \to \cdots \to H^0_{var}(\Xi^\bullet[E_k]/\widetilde{\Xi}^\bullet(E_k)) \to 0,$$

where $H^\bullet_{var}(-)$ *is the homology group of the variational sequence* $(-)$.
Proof. From above theorem we get the following short exact sequence of complexes:

$$0 \to \widetilde{\Xi}^\bullet(E_k) \to \Xi^\bullet[E_k] \to \Xi^\bullet[E_k]/\widetilde{\Xi}^\bullet(E_k) \to 0.$$

This is enough to assure that the connecting linear mappings:

$$H^p_{var}(\Xi^\bullet[E_k]/\widetilde{\Xi}^\bullet(E_k)) \overset{\delta}{\to} H^{p-1}_{var}(\widetilde{\Xi}^\bullet(E_k))$$

are defined. Hence one has the exact homological sequence given in the theorem as a consequence of standard results of homological algebra. □

Theorem 5.10. *Let* $(\Xi^\bullet, \epsilon^\bullet)$ *be the constrained variational sequence associated to* $[E_k] \in \mathcal{D}_f$. *Then, one has the following exact commutative diagram:*

$$
\begin{array}{ccccccccc}
 & & & & 0 & & 0 & & \\
 & & & & \downarrow & & \downarrow & & \\
 & & 0 & \to & B^\bullet_{var} & \to & Z^\bullet_{var} & \to & H^\bullet_{var} & \to & 0 \\
 & & & & \downarrow & & \downarrow & & \\
0 & \to & \Omega^\bullet_{var} & \to & Bor^\bullet_{var} & \to & Cyc^\bullet_{var} & \to & 0 \\
 & & & & \downarrow & & \downarrow & & \\
 & & & & 0 & & 0 & &
\end{array}
$$

where $B^\bullet_{var} \equiv \ker(\epsilon^\bullet)$, $Z^\bullet_{var} \equiv \mathrm{im}\,(\epsilon^\bullet)$. *Furthermore,* $b \in [a] \in Bor^\bullet_{var}$ *iff exists* $c \in \Xi^\bullet$ *with* $\epsilon(c) = a - b$; $b \in [a] \in Cyc^\bullet_{var}$ *iff* $\epsilon(a - b) = 0$; $b \in [a] \in \Omega^\bullet_{var}$ *iff* $\epsilon(a) = \epsilon(b) = 0$ *and exists* $c \in \Xi^\bullet$ *such that* $\epsilon(c) = a - b$. $H^\bullet_{var}$ *is the variational cohomology group of* E_k, *i.e.,* $H^\bullet_{var} = H^\bullet_{var}(E_k)$. *We call* $Bor^\bullet_{var}$ *the variational cobordism group of* E_k, $Cyc^\bullet_{var}$ *the variational cyclism group* of E_k and $\Omega^\bullet_{var}$ the closed variational cobordism group of E_k. Furthermore, we call the bottom horizontal line the variational cobordism (short exact) sequence of E_k. One has the following isomorphism: $\Omega^\bullet_{var} \cong H^\bullet_{var}$. [This allows us to interpret the constrained variational cohomology as singular bordism groups on the constraint PDE.] Furthermore, if $E'_{k'} \in [E_k] \in \mathcal{D}_f$, with $k' > k$, one has the commutative exact diagram below reported, where the sequence with widetilde are associated to the reduced variational sequences. In particular, if $E'_{k'}$ is the universal PDE in the class $[E_k]$, we put $\Omega^\bullet_{var}[E_k] \equiv \Omega^\bullet_{var}(E'_{k'})$, $Bor^\bullet_{var}[E_k] \equiv Bor^\bullet_{var}(E'_{k'})$, $Cyc^\bullet_{var}[E_k] \equiv Cyc^\bullet_{var}(E'_{k'})$ and $H^\bullet_{var}[E_k] \equiv H^\bullet_{var}(E'_{k'})$. We call

$$0 \to \widetilde{\Omega}^\bullet_{var}(E_k) \to \widetilde{B}^\bullet_{var}(E_k) \to \widetilde{Cyc}^\bullet_{var}(E_k) \to 0$$

the universal variational cobordism (short exact) sequence of $[E_k]$. Therefore, the universal variational cobordism (short exact) sequence of $[E_k]$ is an extension of the reduced variational cobordism (short exact) sequence of PDEs in $[E_k]$. One has the following exact sequence, (universal variational closed cobordism sequence of $[E_k]$):

$$\cdots \to \widetilde{\Omega}^{n-1}_{var}(E_{k'}) \to \Omega^{n-1}_{var}[E_k] \to \Omega^{n-1}_{var}[E_k|E_{k'}] \xrightarrow{\delta} \widetilde{\Omega}^{n-2}_{var}(E_k) \to$$

$$\cdots \xrightarrow{\delta} \widetilde{\Omega}^{0}_{var}(E_{k'}) \to \Omega^{0}_{var}[E_k] \to \Omega^{0}_{var}[E_k|E_{k'}] \to 0.$$

$$
\begin{array}{ccccccc}
& 0 & & 0 & & 0 & \\
& \downarrow & & \downarrow & & \downarrow & \\
0 \to & \widetilde{\Omega}^\bullet_{var}(E_k) & \to & \widetilde{B}^\bullet_{var}(E_k) & \to & \widetilde{Cyc}^\bullet_{var}(E_k) & \to 0 \\
& \downarrow & & \downarrow & & \downarrow & \\
0 \to & \Omega^\bullet_{var}(E'_{k'}) & \to & B^\bullet_{var}(E'_{k'}) & \to & Cyc^\bullet_{var}(E'_{k'}) & \to 0 \\
& \downarrow & & \downarrow & & \downarrow & \\
0 \to & \Omega^\bullet_{var}(E'_{k'})/\widetilde{\Omega}^\bullet_{var}(E_k) & \to & B^\bullet_{var}(E'_{k'})/\widetilde{B}^\bullet_{var}(E_k) & \to & Cyc^\bullet_{var}(E'_{k'})/\widetilde{Cyc}^\bullet_{var}(E_k) & \to 0 \\
& \downarrow & & \downarrow & & \downarrow & \\
& 0 & & 0 & & 0 &
\end{array}
$$

Proof. The proof follows directly from definitions and above results, taking into account that $(\Xi^\bullet, \epsilon^\bullet)$ is a cochain complex and considering also the extended homological sequence of $[E_k]$ and the fact that $\Omega^p_{var}(E_k) \cong$

$H^p_{var}(E_k)$. Note that $\Omega^p_{var}[E_k|E_k]$ denotes the closed p-cobordism group corresponding to the sequence $\Xi^\bullet[E_k]/\Xi^\bullet(E_{k'})$. $\qquad\square$

Classification of global solutions of constrained variational equations can be made by means of integral bordism goups as considered by A.Prástaro. (See in this book and quoted references.) Relations between global solutions of variational equations and their constraints are given by the following theorem.

Theorem 5.11. *The relation between the integral bordism group* $\Omega_p^{(E_\infty)var}$ *of a constrained variational PDE* $(E_\infty)_{var} \subset (E_\infty)_{const.} \subset J_n^\infty(W)$, *and the ones* $\Omega_p^{(E_\infty)const.}$ *of the constraint equation* $(E_\infty)_{const.}$ *is described by means of the following exact commutative diagram:*

$$
\begin{array}{ccccccccc}
 & & & & 0 & & & & \\
 & & & & \downarrow & & & & \\
 & & & & \bar{K}_p(E_\infty)_{var} & & & & \\
 & & & & \downarrow & & & & \\
0 & \to & \bar{K}_p^{(E_\infty)var} & \to & \Omega_p^{(E_\infty)var} & \to & \Omega_p((E_\infty)_{const.},(E_\infty)_{var}) & \to & 0 \\
 & & & & & & \downarrow & & \\
 & & & & & & \Omega_p^{(E_\infty)const.} & & \\
 & & & & & & \downarrow & & \\
 & & & & & & 0 & &
\end{array}
$$

From which it results that $\Omega_p^{(E_\infty)var}$ *is an extension of the integral bordism group* $\Omega_p((E_\infty)_{const.},(E_\infty)_{var})$ *of* $(E_\infty)_{const.}$ *relatively to* $(E_\infty)_{var}$.

Proof. The proof is directly obtained considering results in refs.[2]. $\qquad\square$

Example 5.4. (*Yang-Mills equation*). Let M be a globally hyperbolic oriented 4-dimensional smooth space-time with volume form

$$
\eta = \sqrt{|\det(g_{\alpha\beta})|}dx^0 \wedge dx^1 \wedge dx^2 \wedge dx^3,
$$

where $g = g_{\alpha\beta}dx^\alpha \otimes dx^\beta$ is the hyperbolic metric. Let us denote $\Omega^\bullet(M;\mathfrak{g}) \cong \mathfrak{g} \otimes \Omega^\bullet(M)$ the **Z**-graded vector space of $\mathfrak{g}$-valued differential forms on M, where $\mathfrak{g}$ is a semisimple Lie algebra. One has the following: (i) A sequence $\{\Omega^\bullet(M;\mathfrak{g});d\}$ where d is the exterior differential defined by $d\omega = d(a\otimes\alpha) = a\otimes d\alpha$; (ii) A symmetric bilinear map $<,>: \Omega^\bullet(M;\mathfrak{g})\times\Omega^\bullet(M;\mathfrak{g}) \to \Omega^0(M)$, $< a \otimes \alpha, b \otimes \beta >=< \alpha,\beta >< a,b >$, where $< a,b >$ is the Cartan metric of $\mathfrak{g}$; (iii) A bilinear map $[,] : \Omega^\bullet(M;\mathfrak{g})\times\Omega^\bullet(M;\mathfrak{g}) \to \Omega^\bullet(M;\mathfrak{g})$, $[a\otimes\alpha,b\otimes\beta] = [a,b]\otimes\alpha\wedge\beta$; (iv) Any $\mu \in \Omega^1(M;\mathfrak{g})$ defines the following *covariant exterior* *differential* [This concept is related to nonholonomic connections on principal fiber bundles. (See refs.[2].)]: $_\mu d\alpha = d\alpha - [\mu, \alpha] \in \Omega^{p+1}(M; \mathfrak{g})$, if $\alpha \in \Omega^p(M; \mathfrak{g})$. One has $_\mu d^2\alpha = [\mu, [\mu, \alpha]] - [d\mu, \alpha]$. (v) For any $\mu \in \Omega^\bullet(M; \mathfrak{g})$, one has a linear map $_\mu Ad : \Omega^p(M; \mathfrak{g}) \to \Omega^{p+1}(M; \mathfrak{g})$, $_\mu Ad(\alpha) = [\mu, \alpha]$. So we can write $_\mu d^2 = {}_\mu Ad^2 - Ad(d\mu)$. (v) A non linear operator, *Cartan-Maurer curvature*, $F : \Omega^1(M; \mathfrak{g}) \to \Omega^2(M; \mathfrak{g})$, $F(\alpha) = d\alpha - \frac{1}{2}[\alpha, \alpha]$. $F(\alpha)$ is called the *Yang-Mills field* of α. One has $_\mu d^2 = 0$ iff $F(\mu) \equiv d\mu - \frac{1}{2}[\mu, \mu] = 0$. Then the *Yang-Mills Lagrangian density* $\lambda : J\mathcal{D}(\mathfrak{g} \otimes \Lambda_1^0 M) \to \Lambda_4^0(M)$ is defined for any $\alpha \in \Omega^1(M; \mathfrak{g})$ by $\lambda \circ D\alpha \equiv \lambda(\alpha) = \frac{1}{2} < F(\alpha), F(\alpha) >= \frac{1}{2} < d\alpha - \frac{1}{2}[\alpha, \alpha], d\alpha - \frac{1}{2}[\alpha, \alpha] > \eta$. The corresponding Euler-Lagrange operator is $\Diamond : \Omega^1(M; \mathfrak{g}) \to \Omega^1(M; \mathfrak{g})$, $\Diamond(\alpha) = \delta F(\alpha) - Ad^d(\alpha)(F(\alpha))$, where $\delta : \Omega^p(M; \mathfrak{g}) \to \Omega^{p-1}(M; \mathfrak{g})$ is the Hodge codifferential induced by the codifferential on $\Omega^\bullet(M)$, i.e., $\delta\omega = \delta(a \otimes \alpha) = a \otimes \delta\alpha$. Here Ad^d is the dual of Ad with respect to the inner product $<,>$, i.e., for any $\mu \in \Omega^1(M; \mathfrak{g})$, $Ad^d : \Omega^p(M; \mathfrak{g}) \to \Omega^{p-1}(M; \mathfrak{g})$ is the zero order, linear differential operator such that $< \beta', [\mu, \beta] >=< Ad^d(\mu)(\beta'), \beta >$, for $\beta \in \Omega^1(M; \mathfrak{g})$, $\beta' \in \Omega^2(M; \mathfrak{g})$. With respect to coordinates (x^α) on M and a basis $\{Z_a\}_{1 \leq a \leq m}$ on $\mathfrak{g}$, one has the following local representation of $\alpha \in \Omega^1(M; \mathfrak{g})$: $\alpha = Z_a \otimes \alpha_\alpha^a dx^\alpha$ and $F(\alpha) \in \Omega^2(M; \mathfrak{g})$:

$$F(\alpha) = \sum_{0 \leq \alpha, \beta \leq 3} Z_a \otimes [\alpha_{[\alpha\beta]}^a - \frac{1}{2} C_{bc}^a \alpha_\alpha^b \alpha_\beta^c] dx^\alpha \wedge dx^\beta$$

$$\equiv \sum_{0 \leq \alpha, \beta \leq 3} Z_a \otimes F_{\alpha\beta}^a dx^\alpha \wedge dx^\beta,$$

with

$$F_{\alpha\beta}^a \equiv \alpha_{\alpha\beta}^a - \alpha_{\beta\alpha}^a - \frac{1}{2} C_{bc}^a \alpha_\alpha^b \alpha_\beta^c.$$

Then the Yang-Mills Lagrangian can be written as follows: $L = \frac{1}{2} F_{\alpha\beta}^a F_a^{\alpha\beta}$, where $F_a^{\alpha\beta} = F_{\gamma\delta}^b g^{\gamma\alpha} g^{\delta\beta} g_{ba}$. (($g_{ba}$) is the matrix Cartan metric). We call *Yang-Mills variational system* the following 4-variational system: $(X, \theta, \mathcal{C})$, where $X \equiv J\mathcal{D}(E)$, $\pi : E \equiv \mathfrak{g} \otimes \Lambda_1^0 M \to M$, with coordinates $(x^\alpha, A_\alpha^a, A_{\alpha\beta}^a)$; $\mathcal{C}$ is the Pfaffian ideal in $\Omega^\bullet(X)$ generated by the following differential forms: $\theta_\alpha^a \equiv dA_\alpha^a - A_{\alpha\beta}^a dx^\beta$. Furthermore $\theta = \frac{1}{2} F_{\alpha\beta}^a F_a^{\alpha\beta} \eta$. The corresponding Euler-Lagrange equation is locally given by the following:

$$(YM) \subset J\mathcal{D}^2(E) \quad \left\{ \begin{array}{ll} A_{\alpha\beta}^a = (\partial x_\beta . A_\alpha^a) & \text{(Yang-Mills potential)} \\[2mm] F_{\alpha\beta}^a = A_{\alpha\beta}^a - A_{\beta\alpha}^a - \frac{1}{2} C_{bc}^a A_\alpha^b A_\beta^c & \text{(Yang-Mills fields)} \\[2mm] 2(\partial x_\beta . F_b^{\alpha\beta}) = C_{bc}^a F_a^{\alpha\beta} A_\beta^c & \text{(variational equations)} \end{array} \right\}.$$

Furthermore, one has the following Cartan form: $\sigma[\theta] = \theta - 2F^\alpha_{\beta a} dx^\beta \wedge \theta^a_\alpha$. In fact one has $d\sigma[\theta] = \omega^\alpha_a \wedge \theta^a_\alpha \in \mathcal{C}$, where $\omega^\alpha_b \equiv 2dF^\alpha_{\beta b} \wedge dx^\beta - C^a_{bc} F^{\alpha\beta}_a A^c_\beta \eta$. Thus the Cartan extremals are the integral submanifolds $\alpha : \mathbb{R}^4 \to X$ of the following differential ideal $\mathcal{E}$ generated by $\{\mathcal{C}, \omega^\alpha_a\}$. Hence they are solutions of the following PDEs:

$$\begin{cases} A^a_{\alpha\beta} = (\partial x_\beta . A^a_\alpha), \\ F^a_{\alpha\beta} = A^a_{\alpha\beta} - A^a_{\beta\alpha} - \dfrac{1}{2} C^a_{bc} A^b_\alpha A^c_\beta, \\ 2(\partial x_\beta . F^{\alpha\beta}_b) = C^a_{bc} F^{\alpha\beta}_a A^c_\beta. \end{cases}$$

These equations are just the equations that identify the Yang-Mills equation (YM), as above written, as a constrained variational system. Therefore, the Cartan extremals coincide with the Euler-Lagrange extremals. One can prove that $(YM) \subset J\mathcal{D}^2(E)$ is an involutive formally integrable PDE of second order on $\pi : E \to M$, hence it is also completely integrable (as it is analytic). Hence, for any initial condition $q \in (YM)$, passes an analytic solution of (YM). Furthermore, for any 3-dimensional space-like (with respect to a relativistic frame on M) integral manifold $N \subset (YM)$, (Cauchy data), we can identify a smooth solution of (YM). In fact, it is enough to identify a vector field ζ of symmetry for (YM) that is time-like. Such a vector field surely exists as in (YM) does not explicitly happear the time-coordinate x^0. Hence $\zeta = \partial x_0 : (YM) \to T(YM)$ is such a vector field transversal to N. Therefore its flow ϕ_λ, $\partial\phi = \zeta$ locally generates a solution V, starting from N: $V = \bigcup_\lambda \phi_\lambda(N) \subset (YM)$. Let us, now, consider the relation between the variational sequence and the constraint equation in this problem, that results just $J\mathcal{D}(E)$ or $J^1_4(E)$. So, by referring to the above results, we have now to consider the following sequence:

$$\cdots \to \Xi^4(J^\infty_4(E)) \to \Xi^{4+1}(J^\infty_4(E)) \to \Xi^{4+2}(J^\infty_4(E)) \to \cdots .$$

We get $H^{q+1}_{var} \cong H^{4+1}(E) \cong H^5(M) = 0$. Therefore, any 5-form α such that $[\alpha] \in E^{1,4}_1(J^\infty_4(E))$ that is $d^{1,4}_1$-closed is globally the Euler-Lagrange form of some Lagrangian density β on $J^\infty_4(E)$ such that $[\beta] \in E^{0,4}_1(J^\infty_4(E))$. Let us calculate, now, the integral bordism group of (YM) for regular smooth solutions of the equation $(YM) \subset J^2_4(E)$. For this let us emphasize also that the total space E is p-connected, $p \in \{0, 1, 2, 3\}$. Furthermore, any compact closed smooth p-dimensional manifold can be embedded into E

up to bordisms. In fact, $\dim(E) = 4 + m.4 = 4(1 + m) \geq 2p + 1$. In fact, for $m = 1$ and $p = 3$ we get $8 > 7$. Therefore for $m \geq 1$ and $p \leq 3$ above condition is surely verified. Thus (YM) satisfies the p-homotopy principle for $p \in \{0, 1, 2, 3\}$. Under these conditions we can say, that $\Omega_p^{(YM)} \cong \Omega_p^{(YM)+s} \cong \mathbf{Z}_2$ for $p = 0, 2$, and $\Omega_p^{(YM)} \cong \Omega_p^{(YM)+s} \cong 0$ for $p = 1, 3$, and with $s \geq 1$. Then, we have the following exact commutative diagrams:

$$
\begin{array}{ccccccccc}
 & & & & & & 0 & & \\
 & & & & & & \downarrow & & \\
 & & & & 0 & = & \overline{K}_p(YM) & & \\
 & & & & \downarrow & & \downarrow & & \\
0 & \to & \overline{K}_p^{(YM)+s} & \to & \Omega_p^{(YM)+s} & \to & \Omega_p(J_4^{2+s}(W), (YM)_{+s}) & \to & 0 \\
 & & \| & & \|\wr & & \downarrow & & \\
0 & \to & A & \cong & \Omega_p^{J_4^{2+s}(E)} & & & \to & 0 \\
 & & \downarrow & & \downarrow & & & & \\
 & & 0 & = & 0 & & & &
\end{array}
$$

where $A = \mathbf{Z}_2$ for $p = 0, 2$, and $A = 0$ for $p = 1, 3$. This proves that the integral bordism groups of (YM) do not come from a homology theory in $\underline{\mathcal{DE}}$. In fact, $\Omega_p(J_4^{2+s}(W), (YM)_{+s}) \neq 0$ for $p = 0, 2$, even if the inclusion $(YM)_{+s} \to J_4^{2+s}(E)$ is a homotopy equivalence for $s \geq 1$. These results allow us to completely characterize the structure of global smooth regular solutions of (YM). In particular, we have that for any time-like 3-dimensional smooth integral boundary condition there exist global smooth solutions of (YM).   ■

Example 5.5. (*Variational problems with Navier-Stokes PDE*). Let us consider the following 4-variational system $(X = (NS), \mathcal{C}(X), \theta)$, where $(NS) \subset J\mathcal{D}^2(W)$ is the non-isothermal Navier-Stokes equation for incompressible fluids on the Galilean space-time M. $W \equiv J\mathcal{D}(M) \times_M T_0^0 M \times_M T_0^0 M \cong M \times \mathbf{I} \times \mathbf{R}^2$ is the fiber bundle of space-time velocity, isotropic pressure and temperature. So we can identify (NS) with a submanifold of $J\mathcal{D}^2(W)$. Then $\mathcal{C}(X)$ is the contact ideal of (NS). Furthermore, θ is a Lagrangian density $\theta = L\sqrt{\det(g_{ij})}dx^0 \wedge dx^1 \wedge dx^2 \wedge dx^3$ with $L : (NS) \to \mathbf{R}$ and (g_{ij}) the vertical metric on M. (NS) is not formally integrable, but it is universally regular. In fact, there exists a subequation $\widehat{(NS)} \subset (NS) \subset J\mathcal{D}^2(W)$ that is involutive, formally integrable (and completely integrable). $\widehat{(NS)}$ is obtained adding to (NS) the first prolongation of the continuity equation (that is an equation of first order). Therefore, the equivalence class

$[(\widehat{NS})] \in \mathcal{D}_f$ has $(\widehat{NS})_{+\infty} \subset J\mathcal{D}^\infty(W)$ as universal PDE and the Euler-Lagrange equation of the variational system can be identified with a sub-equation $E_\infty \subset (\widehat{NS})_{+\infty} \subset J\mathcal{D}^\infty(W)$. Furthermore $H^{4+1}_{var}((NS)_{+\infty}) = H^{4+1}_{var}((\widehat{NS})_{+\infty}) = H^5_{var}((\widehat{NS})) = H^5_{var}(NS) = 0$ as $(\widehat{NS})$ and (NS) are affine subbundles of $J\mathcal{D}^2(W) \to J\mathcal{D}(W)$, and (NS) is universally regular. Hence (NS) is also wholly variational. Therefore the inverse variational problems admit global solutions. (An important example of variational problem constrained by the Navier-Stokes equation is to find solutions with the smallest possible energy dissipation rate for fixed boundary conditions.) For such equations, by calculating their integral bordism groups, following our general methods, we can prove the existence of global (smooth) solutions for any (smooth) boundary condition. ∎

REFERENCES

[1]   E. Abe, *Hopf algebras*, Cambridge Univ.Press, Cambridge 1980.

[2]   M. Atiyah, *Topological quantum field theory*, Pub. Math. I.H.E.S. **68**(1988), 175–186; *The Geometry and Physics of Knots*, Cambridge University Press, Gambridge 1990.

[3]   D. Barnes and J. M. Mack, *An Algebraic Introduction to Mathematic Logic*, Springer-Verlag, Berlin 1975.

[4]   N. Bourbaki, *Algèbre*, Hermann, Paris 1970.

[5]   N. Bourbaki, *Theorie Spectrales*, Hermann, Paris 1969.

[6]   N. Bourbaki, *Topologie Générale, I.II.*, Hermann, Paris 1971, 1974.

[7]   N. Bourbaki, *Topological Vector Spaces*, Springer-Verlag, Berlin 2002.

[8]   R. L. Bryant, S. Chern, R. B. Gardner, H. L. Goldschmidt and P. A. Griffiths, *Exterior differential systems*, Springer-Verlag, New York 1991.

[9]   R . L. Bryant and P. A. Griffiths, *Characteristic chomology of differential systems, I.*, J. Amer. Math. Soc. **83**(1995), 507–596; II. Duke Math. J. **78(3)**(1995), 531–676.

[10]   S. Buoncristiano, P. Rourke and R. J., Sanderson, *A geometric approach to homology theory*, London Math. Soc. Lect. Notes **18**, Cambridge Univ. Press 1976.

[11]   A. Connes, *Noncommutative differential geometry*, Publ. Math. IHES **39**(1985), 257–360; **62**(1986), 44–144.

[12]   A. Connes, *Noncommutative Geometry*, Academic Press, San Diego, 1994.

[13]   D. Deturk, H. Goldschmidt and J. Talvacchia, *Connections with prescribed curvature and Yang-Mills currents: The semi-simple case*, Ann. Sci. Ec. Norm. Sup. **24**(4)(1991), 57–112.

[14]   B. de Witt, *Supermanifolds*, Cambridge Univers. Press, Cambridge 1986.

[15] J. Dieudonné, *Fondements de la géométrie algébrique moderne*, Advances Math. **3**(1969), 322-413.

[16] V. G. Drinfield, *Quantum groups*, Proceedings Int. Congr. Math., vol. 1,2(Berkeley Calif., 1986), 798–820, Amer. Math. Soc., Providence, R.I., 1987.

[17] B. A. Dubrovin, A. T. Fomenko and S. P. Novikov, *Modern Geometry - Methods and Applications*, vols. I–III, Springer-Verlag, Berlin 1984, 1985, 1990.

[18] J.Eliashberg, *Cobordism des solutions de relations differentielles*, Traveaux en cours, Paris 1986.

[19] H. Goldschmidt, *Integrability criteria for systems of non-linear partial differential equations*, J. Diff. Geom. **1**(1967), 269–307.

[20] H. Goldschmidt, *Points singuliers d'un opérateur differentiel analytique*, Inventiones Math. **9**(1970), 165–174.

[21] M. Gromov, *Partial Differential Relations*, Springer-Verlag, Berlin 1986.

[22] A. Grothendieck, *Eléments de géométrie algébrique I: Le lengage des schémas*, Publ. Math. IHES **4**(1960), 1–228; *II: Étude globale élémentaire de quelques classes de morphismes*, **8**(1961), 1–222; *III: Étude cohomologique des faisceaux coérents I*, **11**(1961), 1–167; *III: Étude cohomologique des faisceaux coérents II*, **17**(1963), 1–91; *IV: Étude locale des schémes et des morphismes des schémes I*, **20**(1964), 1–259; *IV: Étude locale des schémes et des morphismes des schémes II*, **24**(1965), 1–231; *IV: Étude locale des schémes et des morphismes des schémes III*, **28**(1966), 1–255; *IV: Étude locale des schémes et des morphismes des schémes IV*, **32**(1967), 1–361.

[23] A. Grothendieck and J. A. Dieudonné, *Eléments de Géométrie Algébrique*, Springer-Verlag, Berlin, 1971.

[24] S. W. Hawking, *Particle freation by black holes* Commun. Math. Phys. **43(3)**(1975), 199–220; *Erratum* **46(2)**(1976), 206. (See also all the references by S.W.Hawking quoted in refs.[97].)

[25] J. Hilton and V. Stanmbach, *A course in homological algebra*, Springer-Verlag, Berlin 1971.

[26] M. W. Hirsh, *Differential Topology*, Springer-Verlag, Berlin 1976.

[27] G. Hochschild, B. Kostant and A. Rosenberg, *Differential forms on regular affine algebras*, Trans. Amer. Math. Soc. **102**(1962), 383–408.

[28] J. Johnson, *A notion of Krull dimension for differential rings*, Comm. Math. Helv. **44**(2)(1969), 207–216.

[29] J. Johnson, *Kähler differentials and differential algebra*, Ann. Math. **89**(1969), 92–98.

[30] J. Johnson, *Differential dimension polynomials and fundamental theorem on differential modules*, Amer. J. Math. **96**(1)(1969), 239–248.

[31] J. Johnson, *Extensions of differential modules over formal power series rings*, Amer. J. Math. **93**(3)(1971), 731–741.

[32] J. Johnson, *On Spencer's cohomology theory for linear partial differential operators*, Trans. Amer. Math. Soc. **154**(1971), 137–149.

[33] J. Johnson, *Kähler differentials and differential algebra in arbitrary characteristic*, Trans. Amer. Math. Soc. **192**(2)(1974), 201–208.

[34] J. Johnson, *A notion of regularity for differential local algebras*, Contributions to Algebra, A Collection of Ppaers Dedicated to Ellis Kolckin, Academic Press, 19747, 211–232.

[35] J. Johnson, *Systems of n-partial differential equations in n unknown functions: The conjecture of M. Janet*, Trans. Amer. Math. Soc. **242**(1978), 329–334.

[36] J. Johnson, *Prolongations of integral domains*, J. Algebra **94(2)** (1983), 173–210.

[37] M. Kashiwara, *Algebraic Study of Systems of Partial Differential Equations* (Master Thesis, Tokyo University, December 1970), Mémoires de la Société Mathématiques de France **63**, 1996, 86 pp.

[38] H. H. Keller, Lect. Notes Math. **417**(1974), Springer-Verlag, Berlin.

[39] I. S. Krasil'shchik, V. Lychagin and A. M. Vinogradov, *Geometry of jet spaces and nonlinear partial differential equations*, Gordon and Breach, New York 1986.

[40]   V. Lychagin, *Calculus and quantizations over Hopf algebras*, Acta Appl. Math. **51**(1998), 303–352.

[41]   V. Lychagin, *Quantizations of differential equations*, Nonlinear Analysis **47(4)**(2001), 2621–2632.

[42]   V. Lychagin and A. Prástaro, *Singularities of Cauchy data, characteristics, cocharacteristics and integral cobordism*, Diff. Geom. Appl. 4(1994), 287–300.

[43]   J.-L. Loday, *Cyclic Homology*, Springer-Verlag, Berlin, 1992.

[44]   S. Mac Lane, *Categories for the Working Mathematician*, Springer-Verlag, New York, Heidelberg, Berlin, 1971.

[45]   I. B. Madsen and R. J. Milgram, *The Classifying Spaces for Surgery and Bordism of Manifolds*, Ann. Math. Studies, Princeton Univ. Press, 1979.

[46]   Yu. Manin, *Topics in Noncommutative Geometry*, Princeton University Press, Princeton, NJ., 1991.

[47]   J. McCleary, *User's Guide to Spectral Sequences*, Publish or Perish in., USA 1985.

[48]   J. C. McConnell and J. C. Robson, *Noncommutative Noetherian rings*, J.Wiley & Sons, G.B., 1987.

[49]   J. Milnor, *On manifolds homeomorphic to 7-sphere*, Ann. Math. **69(1)**(1956), 399–405.

[50]   J. Milnor and J. Stasheff, *Characteristic Classes*, Ann. Math. Studies **76**, Princeton Univ. Press, 1974.

[51]   J. Morgan and D. Sullivan, *Transversality characteristic class and linking cycles in surgery theory*, Ann. Math. **99**(1974), 463–544.

[52]   A. Nijenhuis, *Theory of geometric objects*, Doctoral Thesis, Amsterdam, 1952.

[53]   V. P. Palamodov, *Differential operators on the class of convergent power series and the Weirstrass auxiliary lemma*, Funktsional'nyi Analizi ego Prilozhenija, **2(3)**(1968), 58–69.

[54]   V. P. Palamodov, *Linear Differential Operators with Constant Coefficients*, Grundleheren der Mathematishen Wissenshaften, **168** (1970), Springer-Verlag.

[55] L. S. Pontrjagin, *Smooth manifolds and their applications homotopy theory*, Amer. Math. Soc. Transl. **11**(1959), 1–114.

[56] A. Prástaro, *On the general structure of continuum physics I, II, III*, Boll. Un. Mat. Ital. **5(7-B)**(1980), 704–726; **5(S-FM)**(1981), 69–106; **5(S-FM)**(1981), 107–129. *Geometrodynamics of non-relativistic continuous media I,II*, Rend. 89–116.

[57] A. Prástaro, *Spinor super-bundles of geometric objects on spin^G space-times*, Boll. Un. Mat. Ital. **6(1-B)**(1982), 1015–1028. *Gauge geometrodynamics*, Riv. Nuovo Cimento **5(4)**(1982), 1–122.

[58] A. Prástaro, *Dynamic conservation laws*, Geometrodynamics Proceedings 1985, (ed. A. Prástaro), World Scientific, Singapore (1985), 283–420.

[59] A. Prástaro, *Geometry of quantized PDE's*, Differential Geometry and Applications, (eds. J. Janyska and D. Krupka), World Scientific, Singapore (1990), 392–404.

[60] A. Prástaro, *Cobordism of PDE's*, Boll. Un. Mat. Ital. **30(5-B)**(1991), 977–1001.

[61] A. Prástaro, *Quantum geometry of PDE's*, Rep. Math. Phys. **30(3)**(1991), 273–354.

[62] A. Prástaro, *Geometry of super PDE's.*, Geometry in Partial Differential Equations, (eds. A. Prástaro and Th. Rassias), World Scientific, Singapore (1994), 259–315; *Geometry of quantized super PDE's*, Amer. Math. Soc. Transl. **167(2)**(1995), 165–192; *Quantum geometry of super PDE's*, Rep. Math. Phys **37(1)**(1996), 23–140.

[63] A. Prástaro, *Geometry of PDEs and Mechanics*, World Scientific, Singapore 1996.

[64] A. Prástaro, *(Co)bordisms in PDE's and quantum PDE's*, Rep. Math. Phys. **38(3)**(1996), 443–455.

[65] A. Prástaro, *Quantum and integral (co)bordism groups in partial differential equations*, Acta Appl. Math. **51(3)**(1998), 243–302; *Quantum and integral bordism groups in the Navier-Stokes equation*, New Developments in Differential Geometry, Budapest

1996, (ed. J. Szenthe), Kluwer Academic Publishers, Dordrecht 1998, 343–360; *(Co)bordism groups in PDE's*, Acta Appl. Math. **59(2)**(1999), 111–202.

[66] A. Prástaro, *Local and global solutions of the Navier-Stokes equation*, Steps in Differential Geometry, Proceedings of the Colloquium on Differential Geometry, 25–30 July, 2000, Debrecen, Hungary, (eds. L. Kozma, P. T. Nagy and L. Tomassy), Univ. Debrecen (2001), 263–271, (http://pc121.math.klte.hu/diffgeo/); *Navier-Stokes equation. Global existence and uniqueness*, (to appear).

[67] A. Prástaro, *(Co)bordism groups in quantum PDE's*, Acta Appl. Math. **64(2/3)**(2000), 111–217.

[68] A. Prástaro, *Theorems of existence of local and global solutions of PDEs in the category of noncommutative quaternionic manifolds*, Quaternionic Structures in Mathematics and Physics, (eds. S. Marchiafava, P. Piccinni and M. Pontecorvo), World Scientific, Singapore (2001), 329–337.

[69] A. Prástaro, *Quantum manifolds and integral (co)bordism groups in quantum partial differential equations*, Nonlinear Analysis **47(4)** (2001), 2609–2620.

[70] A. Prástaro, *Integral bordisms and Green kernels in PDE's*, Cubo Matemat. Educ. **4(2)**(2002), 315–370.

[71] A. Prástaro, *(Co)bordism groups in quantum super PDE's*, Int. J. Math. Games Theory Algebra,(to appear).

[72] A. Prástaro and Th. M. Rassias, *A geometric approach to a noncommutative generalized d'Alembert equation*, C. R. Acad. Sc. Paris **330(I-7)**(2000), 545–550; *Results on the J.D'Alembert equation*, Ann Acad. Paed. Cracoviensis Studia Math. **1**(2001), 117–128; *On the Ulam stability in geometry of PDE's*, Functional Equations, Integrability and Applications, Th. M. Rassias (ed.), Kluwer Academic Publishers, Dordrecht, (2003), 139–147; *Ulam stability in geometry of PDE's*, Nonlinear Funct. Anal. Appl. **8(2)**(2003), 259–278.

[73] A. Prástaro and T. Regge, *The group structure of supergravity*, Ann. Inst. H. Poincaré Phys. Théor. **44(1)**(1986), 39–89.

[74] D. Quillen, *Elementary proofs of some results of cobordism theory using Steenrod operations*, Advances Math. **7**(1971), 29–56.

[75] M. Rothshtein, *The axioms of supermanifolds and a new structure arising from them*, Trans. Amer. Math. Soc. **297(1)**(1986), 159–180.

[76] J. Rotman, *An Introduction to Homological Algebra*, Academic Press, San Diego, 1979.

[77] Yu. B. Rudyak, *On Thom Spectra, Orientability and Cobordism*, Springer-Verlag, Berlin, 1998.

[78] H. H. Schaefer, *Topological Vector Spaces*, Springer-Verlag, New York, 1971.

[79] D. C. Spencer, *Determination of structures on manifolds defined by transitive continuous pseudogroups I, II, III*, Ann. Math. **75**(1965), 1–114; *Overdetermined systems of linear partial differential equations*, Bull. Amer. Math. Soc. **75**(1969), 179–239.

[80] D. C. Spencer, *Overdetermined systems of linear partial differential equations*, Bull. Amer. Math. Soc. **75**(1969), 179–239.

[81] R. E. Stong, *Notes on Bordism Theories*, Ann. Math. Studies. Princeton Univ. Press 1968.

[82] M. E. Sweedler, *Hopf Algebras*, Benjamin, New York, 1969.

[83] R. Switzer, *Algebraic Topology-Homotopy and Homology*, Springer-Verlag, Berlin 1975.

[84] R. Thom, *Quelques propriétés globales des variétés differentiables*, Comm. Math. Helv. **28**(1954), 17–88; *Remarques sur les problèmes comportant des inéqualities différentiables globales*, Bull. Soc. Math. France, **89**(1959), 455–461.

[85] F. Treves, *Topological Vector Spaces, Distributions and Kernels*, Academic Press, New York, 1967.

[86] P. van Nieuvwenhuizen, *Supergravity*, Phys. Rep. **68(4)**(1981), 189–398.

[87] C. T. Wall, *Determination of the cobordism ring*, Ann. Math. **7**(1960), 292–311.

[88] J. Wess and J. Bagger, *Supersymmetry and Supergravity*, Princeton Series in Physics, Princeton University Press, Princeton, New Jersey, 1983.

[89] J. Wess and B. Zumino, *Covariant differential calculus on the quantum hyperplane* , Nucl. Phys. **18B**(1990), 302–312.

[90] P. West, *Introduction to Supersymmetry and Supergravity*, World Scientific, Singapore, 1986.

[91] J. A. Wheeler, *On the nature of quantum geometrodynamics*, Ann. Phys. **2**(1957), 604–614.

[92] M. Wodzicki, *Cyclic homology of differential operators*, Duke Math. J. **54**(1987), 641–647; *Excision in cyclic homology and in rational algebraic K-theory*, Ann. Math. **129**(1989), 591–639.

[93] S. L, Woronowicz, *Twisted SU(2) group. An example of non-commutative differential calculus*, Publ. Res. Inst. Math. Sci. **23(1)**(1987), 117–181; *Compact matrix pseudogroups*, Commun. Math. Phys. **111(4)**(1987), 613–665; *Differential calculus on compact matrix pseudogroups (quantum groups)* **122(1)**(1989), 125–170.

[94] K. Yosida, *Functional Analysis*, Springer-Verlag, Berlin 1965.

[95] V. V. Zharinov, *Geometric Aspects of Partial Differential Equations*, World Scientific, Singapore 1992.

[96] For informations on the quantum field theory see, e.g., the following references:

(a) J. Baez, I. E. Segal and Z. Zhou, *Introduction to Algebraic and Constructive Quantum Field Theory*, Princeton University Press, Princeton 1992.

(b) P. M. A. Dirac, *The Principles of Quantum Mechanics*, Oxford University Press, London 1958.

(c) S. Doplicher, R. Haag and J. E. Roberts, *Fields, observables and gauge transformations I, II*, Commun. Math. Phys. **13**(1969), 1; **15**(1989), 173; *Local observables and particle statistics I, II*, **23**(1971), 199; **35**(1974), 49.

(d) J. Glimm and A. Jaffe, *Quantum Physics. A Functional Integral Point of View*, Springer-Verlag, Berlin 1981.

(e) R. Haag, *Local Quantum Physics, Fields, Particles, Algebras*, Springer-Verlag, Berlin 1981.

(f) S. S. Horzhy, *Introduction to Algebraic Quantum Field Theory*, Kluwer Academic Publishers, Dordrecht 1990.

(g) A. Lichnerowicz, *Champs spinoriels et prpagateurs en relativité générale*, Bull. Soc. Math. France **92**(1964), 11-100.

(h) N. Nakaniskhi and I. Ojima, *Covariant Operator Formalism of Gauge Theories and Quantum Gravity*, World Scientific, Singapore 1990.

(i) Y. Ne'eman and E. Eizenberg, *Membranes & Other Extendons (p-Branes)*, World Scientific, Singapore 1995.

(l) E. Witten, *Supersymmetric Yang-Mills theory on a four-manifold*, J. Math. Phys. **35**(1994), 5101–5135.

[97] For informations on the physics of quantum black holes see e.g. the following book and references quoted there:
M. D. Birrel and P. C. Davies, *Quantum Fields in Curved Spaces*, Cambridge University Press, London, 1982.

[98] Existence theorems of black hole solutions in Einstein-Yang-Mills theories, can be found, e.g., in the following references:
(a) R. Arnowitt, S. Deser and C. W. Misner, *The dynamics of general relativity*, Gravitation, Witten, L. (ed.), Wiley, N.Y. (1962), 227–263.

(b) R. Bartnik and J. McKinnon, *Particle-like solutions of the Einstein-Yang-Mills equations*, Phys. Rev. Lett. **61**(1988), 141–144.

(c) H. Kunzle, *Analysis of static spherically symmetric $SU(n)$-Einstein-Yang-Mills equations*, Commun. Math. Phys. **162**(1994), 371–397.

(d) N. E. Mavromatos and E. Winstenley, *Existence theorems for hairy black holes in $SU(N)$ Einstein-Yang-Mills theories*, J. Math. Phys. **39**(1998), 4849–4873.

(e) W. H. Ruan, *Existence of infinitely many black holes in $SU(3)$ Einstein-Yang-Mills theory*, Nonlinear Analysis **47**(2001), 6109–6119.

(f) J. Smoller, A. Wasserman, S.-T. Yau and J. B. McLeod, *Smooth static solutions of the Einstein/Yang-Mills equations*, Commun.

Math. Phys. **143**(1991), 115–147.

(g) J. Smoller and A. Wasserman, *Existence of infinitely many smooth, static, global solutions of the Einstein-Yang-Mills equations*, Commun. Math. Phys. **151**(1993), 303–325.

(h) J. Smoller and A. Wasserman and S.-T. Yau, *Existence of black-hole solutions for the Einstein-Yang-Mills equations*, Commun. Math. Phys. **151**(1993), 377–401.

[99] From our geometric theory of covariant and canonical quantization of PDE's it follows that quantizations of PDE's can be interpreted as phenomena governed by deformations of the same equations around their solutions by means of physical observables. This fact agrees with the usually accepted point of view that quantization is synonymous of deformation. See, e.g., the following references:

(a) F. Bayer, M. Flato, C. Fronsdal and A. Lichnerowicz, *Quantum mechanics as a deformation of classical mechanics*, Lett. Math. Phys. **1**(1975/77), 521–570.

(b) V. Kac and P. Cheung, *Quantum Calculus*, Springer-Verlag, Berlin 2001.

(c) M. Ja. Vilenkin and A. V. Klimyk, *Representations of Lie Groups and Special Functions*, Vols.I–III, Kluwer Acad. Publ. Dordrecht, 1991/93.

[100] For a panorama on the actual status of art of quantum field theory in condensed matter see the following book and references quoted there:

N. Nagosa, *Quantum Field Theory in Condensed Matter Physics*, Springer-Verlag, Berlin 1999.

[101] Further references for Addendum I.

(a) V. I., Arnold and B. A., Khesin, B. A., *Topological Methods in Hydrodynamics*, Springer-Verlag, New-York 1998.

(b) G., Birkoff, *Hydrodynamics*, Princeton University Press, Princeton 1960.

(c) G. E., Bredon, *Topology and Geometry*, Springer-Verlag, Berlin 1993.

(d) A., Chorin, *Vorticity and Turbulence*, Springer Verlag, Berlin 1994.

(e) G., Doering and J., Gibbon, *Applied Analysis of the Navier-Stokes Equations*, Cambridge University Press, 1995.

(f) D. G., Ebin and J., Marsden, *Groups of diffeomorphisms and the motion of an incompressible fluid*, Ann. Math. **92**(1970), 102–163; *Groups of diffeomerphisms and the solution of the classical Euler equation for a perfect fluid*, Bull. Amer. Math. Soc. **75**(1969), 962–967.

(g) H., Goldschmidt and D., Spencer, *Submanifolds and overdetermined differential operators*, Complex Analysis & Algebraic Geometry, Cambridge (1977), 319–356.

(h) E., Hopf, *Statistical hydromechanics and functional calculus*, J. Rat. Mech Anal. **1**(1952), 87–141.

(i) J., Mattingly and Ya. G., Sinai, *An elementary proof of the existence and uniqueness theorem for 2D-Navier-Stokes systems*, (submitted to Ann.Math.)

(l) F. W., Warner, *Foundations of Differentiable Manifolds and Lie Groups*, Scott, Foresman and C., Glenview, Illinois, USA, 1971.

(m) M. E., Taylor, *Partial Differential Equations*, vols. I–III, Springer-Verlag, New York 1996.

(n) R., Temam, *Navier-Stokes Equations*, North-Holland Publishing Company, Amsterdam, New York, Oxford, 1979.

(o) Notices Amer. Math. Soc. **47(8)**(2000).

[102] Further references for Addendum II.

(a) I. Anderson and G. Thompson, *The Inverse problem of the Calculus of Variations for Ordinary Differential Equations*, Memoirs Amer. Math. Soc. **473(98)**(1992).

(b) H. Goldschmidt and S. Sternberg, *The Hamilton-Cartan formalism in the calculus of variations*, Ann. Inst. Fourier, Grenoble, **27(1)**(1973), 203–267.

(c) D. Krupka, *Lectures on variational sequences*, Advanced Texts in Mathematics, Open Education & Sciences, Opava, Czech Republic (1995), 1–86.

(d) B. A. Kupershmidt, *Geometry of jet bundles and the structure of Lagrangian and Hamiltonian formalism*, Lecture Notes Math., Springer-Verlag, **775**(1980), 162–218.

INDEX

H

I